ADVANCES IN CHEMICAL ENGINEERING

Edited by **Zeeshan Nawaz** and **Shahid Naveed**

INTECHOPEN.COM

Advances in Chemical Engineering
Edited by Zeeshan Nawaz and Shahid Naveed

CBS, Edition **2017**

Published by InTech
Janeza Trdine 9, 51000 Rijeka, Croatia

Publishing Process Manager Bojan Rafaj

Technical Editor InTech DTP team

Cover Designer InTech Design Team

Additional hard copies can be obtained from orders@intechopen.com

Advances in Chemical Engineering, Edited by Zeeshan Nawaz and Shahid Naveed
p. cm.
ISBN 978-953-51-0392-9

INTECH

open science | open minds

Contents

Contents

Contents

Preface

This book addresses the evolutionary stage of Chemical Engineering and provides an overview to the state of the art and technological advancements. Chemical Engineering applications have always been challenging. Optimization of plausible solutions to problems in economic manner through technology is worth. The script has been designed to enable the reader to access the desired knowledge on fundamentals and advancements in Chemical Engineering in a single text. The molecular perspective of Chemical Engineering is increasingly becoming important in the refinement of kinetic and thermodynamic molding. As a result many of the theories, subject matters and approaches are being revisited and improved. The approach to industrial problems has been reviewed with modern trend in technology.

The subject of primary interest in this text is to highlight recent advances in chemical engineering knowledge. Therefore, the book is divided into four section fundamentals, catalysis & reaction engineering, process engineering and separation technology. Fundamentals covers, application of chebyshev polynomials to analyze soft equation of state, UV-VIS Spectra analysis, proton NMR technique applications, and modeling approach for process redesign. Catalysis and reaction engineering discusses; asymmetric catalyst design, intensification, modeling, prepration, characterization and recycling, carbohydrate-based surfactants, bio-fuel production form micro-algae, biodiesel using triglycerides, and pollutant formation during combustion. Third section of process engineering focused on systematic multiobjective process optimization, CFD modeling of fluidized bed, optimization of evaporation system, spouted bed scale-up, desalination, and economic evaluation of macro scale production processes. Last section of the book has emphasized on separation technology includes phase diagram analysis, membranes developments and applications, and phase transitions study. Molecular chemistry, reaction engineering and modeling have been demonstrated to be interrelated and of value for practical applications. A rational and robust industrial design can be conceived with this understanding. The book can be of interest for undergraduate, graduate and professionals for a number of reasons besides the incorporation of innovation in the text.

Dr. Zeeshan Nawaz
SABIC Technology and Innovation,
Saudi Basic Industries Corporation,
Riyadh
Kingdom of Saudi Arabia

Prof Dr. Shahid Naveed
Department of Chemical Engineering,
University of Engineering and Technology,
Lahore
Pakistan

Part 1

Fundamentals

1

Application of Chebyshev Polynomials to Calculate Density and Fugacity Using SAFT Equation of State to Predict Asphaltene Precipitation Conditions

Seyyed Alireza Tabatabaei-Nejad and Elnaz Khodapanah
Chemical Engineering Department, Sahand University of Technology, Tabriz
Iran

1. Introduction

Equations of state are the essential tools to model physical and chemical processes in which fluids are involved. The majority of PVT calculations carried out for oil and gas mixtures are based on a cubic equation of state (EoS). This type of equations dates back more than a century to the famous Van der Waals equation (Van der Waals, 1873). The cubic equations of state most commonly used in the petroleum industry today are very similar to the Van der Waals equation, but it took almost a century for the petroleum industry to accept this type of equation as a valuable engineering tool. The Redlich and Kwong EoS (Redlich & Kwong, 1949) was modified from the VdW with a different attractive term, the repulsive term being the same. Since 1949 when Redlich and Kwong (RK) formulated their two-parameter cubic EoS, many investigators have introduced various modifications to improve ability of RK-EoS. Two other well-known cubic equations are Soave-Redlich-Kwong (SRK), (Soave, 1972) and Peng-Robinson (PR) (Peng & Robinson, 1976) equations which have different formulation of the attractive term and are popular in the oil industry in the thermodynamic modeling of hydrocarbon fluids.

There are thousands of cubic equations of states, and many noncubic equations. The noncubic equations such as the Benedict-Webb-Rubin equation (Benedict et al., 1942), and its modification by Starling (Starling, 1973) have a large number of constants; they describe accurately the volumetric behavior of pure substances. But for hydrocarbon mixtures and crude oils, because of mixing rule complexities, they may not be suitable (Katz & Firoozabadi, 1978). Cubic equations with more than two constants also may not improve the volumetric behavior prediction of complex reservoir fluids. In fact, most of the cubic equations have the same accuracy for phase behavior prediction of complex hydrocarbon systems; the simpler often do better (Firoozabadi, 1999).

Hydrocarbons and other non-polar fluid vapor–liquid equilibrium properties can be satisfactorily modeled using a symmetric approach to model both, the vapor and the liquid phase fugacity with the use of a Van der Waals type equation model (Segura et al., 2008), the Soave–Redlich–Kwong or Peng–Robinson equations being the most popular ones. When polar fluids are involved at moderate pressures, activity coefficient models are more suitable for modeling the liquid phase. When a higher pressure range is also a concern, a symmetric EoS approach with complex mixing rules including an excess Gibbs energy term from an activity coefficient model can provide good results. Unfortunately, even those approaches show limitations for complex fluids and can drastically fail near the critical region, unless a specific treatment is included (Llovell et al., 2004, 2008).

Since the early 1980's, there has been increased interest in developing an EoS for pure fluids and mixtures of large polyatomic molecules that does not rely on a lattice description of molecular configurations. A rigorous statistical-mechanical theory for large polyatomic molecules in continuous space is difficult because of their asymmetric structure, large number of internal degrees of freedom, and strong coupling between intra- and intermolecular interactions. Nevertheless, a relatively simple model represents chain-like as freely joined tangent hard spheres (Chapman et al., 1984; Song et al., 1994; Wertheim, 1984). A hard-sphere-chain (HSC) EoS can be used as the reference system in place of the hard-sphere reference used in most existing equations of state for simple fluids. Despite their simplicity, hard-sphere-chain models take into account some significant features of real fluids containing chain-like molecules including excluded volume effects and chain connectivity. To describe the properties of fluids consisting of large polyatomic molecules, it is necessary to introduce attractive forces by adding a perturbation to a HSC EoS. Assuming that the influence of attractive forces on fluid structure is week, a Van der Waals type or other mean-field term (e.g. square-well fluids) is usually used to add attractive forces to the reference hard-sphere-chain EoS (Prausnitz & Tavares, 2004).

Molecular-based equations of state, also routed in statistical mechanics, retain their interest in chemical engineering calculations as they apply to a wide spectrum of thermodynamic conditions and compounds, being computationally much less demanding than molecular simulations. Among them, the Statistical Associating Fluid Theory (SAFT) EoS has become very popular because of its capability of predicting thermodynamics properties of several complex fluids, including chain, aromatic and chlorinated hydrocarbons, esters alkanols, carboxylic acids, etc. (Huang & Radosz, 1990). SAFT was envisioned as an application of Wertheim's theory of association (Wertheim, 1984, 1986) through the use of a first-order thermodynamic perturbation theory (TPT) to formulate a physically based EoS (Chapman et al., 1990; Huang & Radosz, 1991). The ambition of making SAFT an accurate equation for engineering purposes has promoted the development of different versions that tried to overcome the limitations of the original one (Economou, 2002; Muller & Gubbins, 1995).

SAFT has a similar form to group contribution theories in that the fluid of interest is initially considered to be a mixture of unconnected groups or segments. SAFT includes a chain connectivity term to account for the bonding of various groups to form polymers and an explicit intermolecular hydrogen bonding term. A theory based in statistical mechanics offers several advantages. The first advantage is that each of the approximations made in the development of SAFT has been tested versus molecular simulation results. In this way, the range of applicability of each term in the EoS has been determined. The second advantage is that the EoS can be systematically refined. Since any weak approximations in SAFT can be identified, improvement is made upon the EoS by making better approximations or by extending the theory. Like most thermodynamic models, SAFT approaches require the evaluation of several parameters relating the model to the experimental system. A third advantage of SAFT-type equations versus other approaches is that, as they are based on statistical mechanics, parameters have a clear physical meaning; when carefully fitted they can be used with predictive power to explore other regions of the phase diagram far from the data and operating conditions used in the parameter regression, performing better than other models for interacting compounds like activity coefficient models (Prausnitz et al., 1999). In SAFT a chain molecule is characterized by the diameter or volume of a segment, the number of segments in the chain, and the segment–segment dispersion energy. For an associating or hydrogen bonding molecule, two more physical parameters are necessary: the association energy related to the change in enthalpy of association and the bond volume related to the change in entropy on association. The SAFT equation has found some impressive engineering applications on those fluids with chain bonding and hydrogen bonding (Chapman et al., 2004).

Asphaltenes are operationally defined as the portion of crude oil insoluble in light normal alkanes (e.g., n-heptane or n-pentane), but soluble in aromatic solvents (e.g., benzene or toluene). This solubility class definition of asphaltenes suggests a broad distribution of asphaltene molecular structures that vary greatly among crude sources. In general, asphaltenes possess fused ring aromaticity, small aliphatic side chains, and polar heteroatom-containing functional groups capable of donating or accepting protons inter- and intra-molecularly. Although asphaltene fractions can be complex molecular species mixtures, they convey, as a whole, an obvious chemical similarity, irrespective of crude geographic origin (Ting, 2003). Asphaltene stability depends on a number of factors including pressure, temperature, and compositions of the fluid; the latter incorporates the addition of light gases, solvents and other oils commingled operation or charges due to contamination. During pressure depletion at constant temperature, asphaltene aggregate formation is observed within a range above and below the bubble point. As pressure drops during production from the reservoir pressure, asphaltene precipitatin can appear due to changes in the solubility of asphaltene in crude oil. The maximum asphaltene precipitation occurs at or around the bubble point pressure. Below the bubble point light gases come out of the solution increasing the asphaltene solubility again (Ting, 2003). Temperature changes also affect asphaltene precipitation, For hydrocarbons deposited in shallow structure, the wellhead flowing temperatures are typically not excessive, 110-140 °F. However, sea bottom temperature in deep water is cold, often near or below 40 °F, even in equatorial waters. Cooling of flow streams during transportation can lead to asphaltene precipitation (Huang & Radosz, 1991). Increases in temperature at constant pressure normally stabilize the asphaltene in crude oil. Depending on the composition of the oil, it is possible to find cases where precipitation first decreases and then increases with increasing temperature (Verdier et al., 2006). Also, depending on the temperature level, significant temperature effects can be observed (Buenrostro-Gonzales & Lira-Galeana, 2004). Changes in composition occur during gas injection processes employed in Enhanced Oil Recovery (EOR). Gas injection includes processes such as miscible flooding with CO_2 , N_2 or natural gas or artificial gas lifting. The dissolved gas decreases asphaltene solubility and the asphaltene becomes more unstable (Verdier et al., 2006).

The tendency of petroleum asphaltenes to associate in solution and adsorb at interfaces can cause significant problems during the production, recovery, pipeline transportation, and refining of crude oil. Therefore, it is necessary to predict the conditions where precipitation occurs and the amount of precipitate. The approach we have taken here to model is to use the SAFT EoS, as it explicitly builds on the association interaction and the chain connectivity term to account for the bonding of various groups. Therefore, the equation is able to provide insights on the asphaltene precipitation behavior. By some algebraic manipulations on this equation, we derive a simplified form of the compressibility factor or pressure as a function of density. Due to pressure explicit form of the SAFT EoS, an approximation technique based on Chebyshev polynomials to calculate density and hence fugacity requisite to perform phase equilibrium calculations is applied. To demonstrate the ability of SAFT EoS a binary system composed of ethanol and toluene is tested. Applying Chebysheve polynomial approximation, density is calculated for the above system at different temperatures in a range of 283.15 K to 353.15 K and for pressures up to 45 MPa. Evaluating fugacity is a necessary step in phase equilibrium calculations. Hence, fugacity is derived using SAFT EoS. Then the model is used to predict phase behavior of oil-asphaltene systems.

2. Formulation of the problem

2.1 SAFT equation of state

The statistical association fluid theory (SAFT) (Chapman et al., 1990) is based on the first order perturbation theory of Wertheim (Wertheim, 1987). The essence of this theory is that the residual Helmholtz energy is given by a sum of expressions to account not only for the effects of short-range repulsions and long-range dispersion forces but also for two other effects : chemically bonded aggregation (e.g. formation of chemically stable chains) and association and/or solvation (e.g. hydrogen bonding) between different molecules (or chains). For a pure component a three step process for formation of stable aggregates (e.g. chains) and subsequent association of these aggregates is shown in figure 1. Initially, a fluid consists of equal-sized, single hard spheres. Intermolecular attractive forces are added

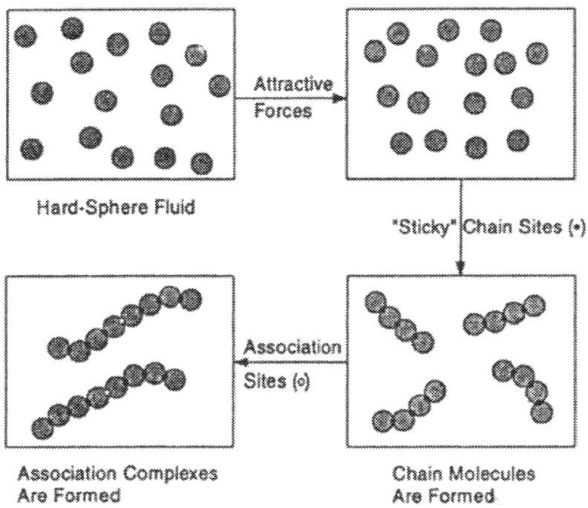

Fig. 1. Three steps to form chain molecules and association complexes from hard spheres in the SAFT model (Prausnitz et al., 1999).

which are described by an appropriate potential function, such as the square-well potential. Next, each sphere is given one, two or more "sticky spots", such that the spheres can stick together (covalent bonding) to form dimmers, trimers and higher stable aggregates as chains. Finally, specific interaction sites are introduced at some position in the chain to form association complex through some attractive interaction (e.g. hydrogen bonding). Each step provides a contribution to the Helmholtz energy.

Using SAFT EoS, the residual molar Helmholtz energy A^R contributes from formation of hard spheres, chains, dispersion (attraction), and association which would be in the form of:

$$A^R = A_{hs} + A_{ch} + A_{disp} + A_{assoc} \tag{1}$$

Here the sum of the first two terms is the hard-sphere-chain reference system accounting for molecular repulsion and chain connectivity (chemical bonding); the sum of the last two terms is the perturbation accounting for molecular attraction and for association due to specific interactions like hydrogen bonding. Application of the relation between molar Helmholtz energy, A, and the equation of state, gives the SAFT EoS for pure fluids (Prausnitz et al., 1999). We can write for compressibility factor of a real fluid:

$$Z = \frac{P}{\rho RT} = Z^{id} + Z_{hs} + Z_{ch} + Z_{disp} + Z_{Assoc} \tag{2}$$

with $Z^{id} = 1.0$ and for mixtures,

$$Z_{hs} = \frac{6}{\pi N_A \rho}\left[\frac{\xi_0 \xi_3}{1-\xi_3} + \frac{3\xi_1 \xi_2}{(1-\xi_3)^2} + \frac{(3-\xi_3)\xi_2^3}{(1-\xi_3)^2}\right] \tag{3}$$

With

$$\xi_k = \frac{\pi N_A \rho}{6}\sum_{i=1}^{N_c} z_i r_i (d_i)^k \qquad k = 0, 1, 2, 3 \tag{4}$$

$$d_i = \sigma_i[1 - 0.12exp(-3\varepsilon_i/kT)] \tag{5}$$

$$Z_{hs} = \frac{hs_1\rho + hs_2\rho^2 + hs_3\rho^3}{1 + hs_4\rho + hs_5\rho^2 + hs_6\rho^3} \tag{6}$$

here ρ is the total molar density, z_i, is the mole fraction of component i, r_i is the number of segments per molecule i, and d_i is the temperature dependent segment diameter. The parameters T, N_A, k, ε_i and σ_i are temperature, Avogadro's and Boltzmann's constants, segment energy and diameter, respectively. By simple algebraic manipulation on Eq. (3) , we arrive at the following simplified form of the hard sphere term:

where,

$$hs_1 = (\bar{r}S_3 + 3S_1S_2)P_{na} \tag{7}$$

$$hs_2 = (-2\bar{r}S_3^2 - 3S_1S_2S_3 + 3S_2^3)P_{na}^2 \tag{8}$$

$$hs_3 = (\bar{r}S_3^3 - S_2^3S_3)P_{na}^3 \tag{9}$$

$$hs_4 = -3S_3P_{na} \tag{10}$$

$$hs_5 = 3S_3^2P_{na}^2 \tag{11}$$

$$hs_6 = -S_3^3P_{na}^3 \tag{12}$$

The parameters used in Eqs. (7) - (12) are defined as the following:

$$\bar{r} = \sum_{i=1}^{N_c} z_i r_i \tag{13}$$

$$S_1 = \sum_{i=1}^{N_c} z_i r_i d_i \tag{14}$$

$$S_2 = \sum_{i=1}^{N_c} z_i r_i d_i^2 \tag{15}$$

$$S_3 = \sum_{i=1}^{N_c} z_i r_i\, d_i^3 \tag{16}$$

$$P_{na} = \frac{\pi N_A}{6} \tag{17}$$

The contribution accounting for the formation of chain molecules of the various components in the mixture is

$$Z_{ch} = \sum_{i=1}^{N_c} z_i(1 - r_i)L(d_i) \tag{18}$$

$$L(d_i) = \frac{2\xi_3 + 3d_i\xi_2 - 4\xi_3^2 + 2d_i^2\xi_2^2 + 2\xi_3^3 + d_i^2\xi_2^2\xi_3 - 3d_i\xi_2\xi_3^2}{(1-\xi_3)\left(2 - 4\xi_3 + 3d_i\xi_2 + 2\xi_3^2 + d_i^2\xi_2^2 - 3d_i\xi_2\xi_3\right)} \tag{19}$$

where $\xi_k (k = 2,3)$ is given by Eq. (4). It is remarkable that no mixing rules are necessary in Eq. (3) and (18). After some arithmetic operations on Eq. (18), the following simplified density dependent equation for the chain term of SAFT EoS is presented as:

$$Z_{ch} = \sum_{i=1}^{N_c} z_i(1-r_i)\frac{ch_1(d_i)\rho + ch_2(d_i)\rho^2 + ch_3(d_i)\rho^3}{2 + ch_4(d_i)\rho + ch_5(d_i)\rho^2 + ch_6(d_i)\rho^3 + ch_7(d_i)\rho^4} \tag{20}$$

where,

$$ch_1(d_i) = [2S_3 + 3d_iS_2]P_{na} \tag{21}$$

$$ch_2(d_i) = \left[-4S_3^2 + 2d_i^2S_2^2\right]P_{na}^2 \tag{22}$$

$$ch_3(d_i) = \left[d_i^2S_2^2S_3 + 2S_3^3 - 3d_iS_2S_3^2\right]P_{na}^3 \tag{23}$$

$$ch_4(d_i) = [-6S_3 + 3d_iS_2]P_{na} \tag{24}$$

$$ch_5(d_i) = \left[6S_3^2 + d_i^2S_2^2 - 9d_iS_2S_3\right]P_{na}^2 \tag{25}$$

$$ch_6(d_i) = \left[9d_iS_2S_3^2 - 2S_3^3 - d_i^2S_2^2S_3\right]P_{na}^3 \tag{26}$$

$$ch_7(d_i) = [-3d_iS_2S_3^3]P_{na}^3 \tag{27}$$

SAFT uses the following expression for the dispersion contribution to the compressibility factor (Pedersen & Christensen, 2007):

$$Z_{disp} = -2\pi\rho\frac{\partial(\xi_3 I_1)}{\partial(\xi_3)}\overline{r^2\varepsilon\sigma^3} - \pi\rho\bar{r}\left[C_1\frac{\partial(\xi_3 I_2)}{\partial(\xi_3)} + C_2\xi_3 I_2\right]\overline{r^2\varepsilon^2\sigma^3} \tag{28}$$

where

$$C_1 = 1 + \bar{r}\frac{8\xi_3 - 2\xi_3^2}{(1-\xi_3)^4} + (1-\bar{r})\frac{20\xi_3 - 27\xi_3^2 + 12\xi_3^3 - 2\xi_3^4}{[(1-\xi_3)(2-\xi_3)]^2} \tag{29}$$

$$C_2 = -C_1^2\left[\bar{r}\frac{-4\xi_3^2 + 20\xi_3 + 8}{(1-\xi_3)^5} + (1-\bar{r})\frac{2\xi_3^3 + 12\xi_3^2 - 48\xi_3 + 40}{[(1-\xi_3)(2-\xi_3)]^3}\right] \tag{30}$$

$$\overline{r^2\varepsilon\sigma^3} = \sum_{i=1}^{N_c}\sum_{j=1}^{N_c} z_i z_j r_i r_j \left(\frac{\varepsilon_{ij}}{kT}\right)\sigma_{ij}^3 \tag{31}$$

$$\overline{r^2\varepsilon^2\sigma^3} = \sum_{i=1}^{N_c}\sum_{j=1}^{N_c} z_i z_j r_i r_j \left(\frac{\varepsilon_{ij}}{kT}\right)^2\sigma_{ij}^3 \tag{32}$$

$$I_1 = \sum_{j=0}^{6} a_j(\bar{r})\,\xi_3^j \tag{33}$$

$$I_2 = \sum_{j=0}^{6} b_j(\bar{r})\,\xi_3^j \tag{34}$$

In equations (31) and (32):

$$\varepsilon_{ij} = \sqrt{\varepsilon_i \varepsilon_j}\left(1 - k_{ij}\right) \tag{35}$$

$$\sigma_{ij} = \frac{1}{2}\left(\sigma_i + \sigma_j\right) \tag{36}$$

where k_{ij} is a binary interaction parameter similar to that in the mixing rule for the a-parameter of a cubic EoS (Pedersen & Christensen, 2007). In equationa (33) and (34):

$$a_j(\bar{r}) = a_{0j} + \frac{\bar{r}-1}{\bar{r}} a_{1j} + \frac{\bar{r}-1}{\bar{r}} . \frac{\bar{r}-2}{\bar{r}} a_{2j}\ , \quad j = 0, 1, \ldots, 6 \tag{37}$$

$$b_j(\bar{r}) = b_{0j} + \frac{\bar{r}-1}{\bar{r}} b_{1j} + \frac{\bar{r}-1}{\bar{r}} . \frac{\bar{r}-2}{\bar{r}} b_{2j}\ , \quad j = 0, 1, \ldots, 6 \tag{38}$$

The universal constants for a_{0j}, a_{1j}, a_{2j}, b_{0j}, b_{1j} and b_{2j} are given in Table 1.

J	a_{0j}	a_{1j}	a_{2j}	b_{0j}	b_{1j}	b_{2j}
0	0.9105631	- 0.3084017	-0.0906148	0.7240947	0.5755498	0.0976883
1	0.6361281	0.1860531	0.4527843	2.2382792	0.6995095	-0.2557575
2	2.6861348	-2.5030047	0.5962701	-4.0025849	3.8925674	-9.1558561
3	-26.547362	21.419793	-1.7241829	-21.003577	-17.215472	20.642076
4	97.759209	-65.255885	-4.1302112	26.855641	192.67226	-38.804430
5	-159.59154	83.318680	13.776632	206.55134	-161.82646	93.626774
6	91.297774	-33.746923	-8.6728470	-355.60235	-165.20769	-29.666905

Table 1. The universal constants for a_{0j} , a_{1j}, a_{2j}, b_{0j}, b_{1j} and b_{2j} parameters used in SAFT EoS (Pedersen & Christensen, 2007).

Again, simplification of Eq. (28), would yield the following density dependent form of the dispersion term in SAFT EoS:

$$Z_{disp} = \left[dI_1.\rho + dI_2.\rho^2 + dI_3.\rho^3 + dI_4.\rho^4 + dI_5.\rho^5 + dI_6.\rho^6 + dI_7.\rho^7 \right] - \left(6P_{na}\bar{r}.Adisp_2\right)$$
$$\cdot\left[\left(\frac{\rho + disp_1.\rho^2 + disp_2.\rho^3 + disp_3.\rho^4 + disp_4.\rho^5 + disp_5.\rho^6 + disp_6.\rho^7}{4 + disp_7.\rho + disp_8.\rho^2 + disp_9.\rho^3 + disp_{10}.\rho^4 + disp_{11}.\rho^5 + disp_{12}.\rho^6}\right)\right.$$
$$\times\left(b_0 + dI_8.\rho + dI_9.\rho^2 + dI_{10}.\rho^3 + dI_{11}.\rho^4 + dI_{12}.\rho^5 + dI_{13}.\rho^6\right) - \left(P_{na}S_3\rho^2\right)$$
$$\times\left(\frac{1 + disp_1.\rho + disp_2.\rho^2 + disp_3.\rho^3 + disp_4.\rho^4 + disp_5.\rho^5 + disp_6.\rho^6}{4 + disp_7.\rho + disp_8.\rho^2 + disp_9.\rho^3 + disp_{10}.\rho^4 + disp_{11}.\rho^5 + disp_{12}.\rho^6}\right)^2 \tag{39}$$
$$\times\left(\frac{disp_{13} + disp_{14}.\rho + disp_{15}.\rho^2 + disp_{16}.\rho^3 + disp_{17}.\rho^4 + disp_{18}.\rho^5}{8 + disp_{19}.\rho + disp_{20}.\rho^2 + disp_{21}.\rho^3 + disp_{22}.\rho^4 + disp_{23}.\rho^5 + disp_{24}.\rho^6 + disp_{25}.\rho^7 + disp_{26}.\rho^8}\right)$$
$$\left.\times\left(b_0 + Idisp_1.\rho + Idisp_2.\rho^2 + Idisp_3.\rho^3 + Idisp_4.\rho^4 + Idisp_5.\rho^5 + Idisp_6.\rho^6\right)\right]$$

where the parameters used in the equation are given below,

$$disp_1 = 4(3\bar{r} - 4)S_3P_{na} \tag{40}$$

$$disp_2 = (27\bar{r} - 6)S_3^2P_{na}^2 \tag{41}$$

$$disp_3 = (82 - 70\bar{r})S_3^3P_{na}^3 \tag{42}$$

$$disp_4 = (51\bar{r} - 52)S_3^4P_{na}^4 \tag{43}$$

$$disp_5 = 16(1 - \bar{r})S_3^5P_{na}^5 \tag{44}$$

$$disp_6 = 2(\bar{r} - 1)S_3^6P_{na}^6 \tag{45}$$

$$disp_7 = -20_3S_3P_{na} \tag{46}$$

$$disp_8 = 41S_3^2P_{na}^2 \tag{47}$$

$$disp_9 = -44S_3^3P_{na}^3 \tag{48}$$

$$disp_{10} = 26S_3^4P_{na}^4 \tag{49}$$

$$disp_{11} = -8S_3^5P_{na}^5 \tag{50}$$

$$disp_{12} = S_3^6P_{na}^6 \tag{51}$$

$$disp_{13} = 40 + 24\bar{r} \tag{52}$$

$$disp_{14} = (192\bar{r} - 128)S_3P_{na} \tag{53}$$

$$disp_{15} = (148 - 372\bar{r})S_3^2P_{na}^2 \tag{54}$$

$$disp_{16} = (230\bar{r} - 70)S_3^3P_{na}^3 \tag{55}$$

$$disp_{17} = (8 - 52\bar{r})S_3^4P_{na}^4 \tag{56}$$

$$disp_{18} = (2 + 2\bar{r})S_3^5 P_{na}^5 \quad (57)$$

$$disp_{19} = -52 S_3 P_{na} \quad (58)$$

$$disp_{20} = 146 S_3^2 P_{na}^2 \quad (59)$$

$$disp_{21} = -231 S_3^3 P_{na}^3 \quad (60)$$

$$disp_{22} = 225 S_3^4 P_{na}^4 \quad (61)$$

$$disp_{23} = -138 S_3^5 P_{na}^5 \quad (62)$$

$$disp_{24} = 52 S_3^6 P_{na}^6 \quad (63)$$

$$disp_{25} = -11 S_3^7 P_{na}^7 \quad (64)$$

$$disp_{26} = S_3^8 P_{na}^8 \quad (65)$$

$$Idisp_1 = b_1 S_3 P_{na} \quad (66)$$

$$Idisp_2 = b_2 S_3^2 P_{na}^2 \quad (67)$$

$$Idisp_3 = b_3 S_3^3 P_{na}^3 \quad (68)$$

$$Idisp_4 = b_4 S_3^4 P_{na}^4 \quad (69)$$

$$Idisp_5 = b_5 S_3^5 P_{na}^5 \quad (70)$$

$$Idisp_6 = b_6 S_3^6 P_{na}^6 \quad (71)$$

$$dIdisp_1 = 2a_1 S_3 P_{na} \quad (72)$$

$$dIdisp_2 = 3a_2 S_3^2 P_{na}^2 \quad (73)$$

$$dIdisp_3 = 4a_3 S_3^3 P_{na}^3 \quad (74)$$

$$dIdisp_4 = 5a_4 S_3^4 P_{na}^4 \quad (75)$$

$$dIdisp_5 = 6a_5 S_3^5 P_{na}^5 \quad (76)$$

$$dIdisp_6 = 7a_6 S_3^6 P_{na}^6 \quad (77)$$

$$dI_1 = -12P_{na}.Adisp_1.a_0 \quad (78)$$

$$dI_2 = -12P_{na}.Adisp_1.dIdisp_1 \quad (79)$$

$$dI_3 = -12P_{na}.Adisp_1.dIdisp_2 \quad (80)$$

$$dI_4 = -12P_{na}.Adisp_1.dIdisp_3 \quad (81)$$

$$dI_5 = -12P_{na}.Adisp_1.dIdisp_4 \quad (82)$$

$$dI_6 = -12P_{na}.Adisp_1.dIdisp_5 \tag{83}$$

$$dI_7 = -12P_{na}.Adisp_1.dIdisp_6 \tag{84}$$

$$dI_8 = 2b_1S_3P_{na} \tag{85}$$

$$dI_9 = 3b_2S_3^2P_{na}^2 \tag{86}$$

$$dI_{10} = 4b_3S_3^3P_{na}^3 \tag{87}$$

$$dI_{11} = 5b_4S_3^4P_{na}^4 \tag{88}$$

$$dI_{12} = 6b_5S_3^5P_{na}^5 \tag{89}$$

$$dI_{13} = 7b_6S_3^6P_{na}^6 \tag{90}$$

$$Adisp_1 = \overline{r^2\varepsilon\sigma^3} = \sum_{i=1}^{N_c}\sum_{j=1}^{N_c} z_iz_jr_ir_j(\varepsilon_{ij}/kT)\sigma_{ij}^3 \tag{91}$$

$$Adisp_2 = \overline{r^2\varepsilon^2\sigma^3} = \sum_{i=1}^{N_c}\sum_{j=1}^{N_c} z_iz_jr_ir_j(\varepsilon_{ij}/kT)^2\sigma_{ij}^3 \tag{92}$$

Similarly, Z_{assoc} can be derived rigorously from statistical mechanics (Chapman et al., 1990). The relation is a mole fraction average of the corresponding pure-component equations:

$$Z_{assoc} = \rho\sum_{i=1}^{N_c} z_i\left[\sum_{S_i}\left(\frac{1}{X^{S_i}}-\frac{1}{2}\right)\frac{\partial X^{S_i}}{\partial\rho}\right] \tag{93}$$

where X^{S_i}, the mole fraction of component i in the mixture not bonded with other components at site S, is given by:

$$X^{S_i} = \left(1 + N_A\sum_{j=1}^{N_c}\sum_{Y_j} z_i\rho X^{Y_j}W_{ij}\right)^{-1} \tag{94}$$

with

$$W_{ij} = \left[\frac{1}{1-\xi_3} + \frac{3d_id_j}{d_i+d_j}\frac{\xi_2}{(1-\xi_3)^2} + 2\left(\frac{d_id_j}{d_i+d_j}\right)^2\frac{\xi_2^2}{(1-\xi_3)^3}\right](\sigma_{ij}k^{S_iY_j})\left[exp\left(\frac{\varepsilon^{S_iY_j}}{kT}\right)-1\right] \tag{95}$$

In Eq. (94), summation $\sum_{Y_j}$ is over all specific interaction sites on molecule j and summation $\sum_j$ is over all N_c components. The association/salvation $\varepsilon^{S_iY_j}$ and the dimensionless parameter $k^{S_iY_j}$ characterize, respectively, the association $(i = j)$ and solvation $(i \neq j)$ energy and volume for the specific interaction between sites S and Y. These parameters are adjustable. Equation (93) requires no mixing rules. As it can be seen in Eq. (94), X^{S_i}'s satisfy a non-linear system of equations which can be solved using any iterative technique such as Gauss-Seidel, Successive-Over-Relaxation (SOR) or Jacobi iterative method. The derivative of the function X^{S_i} with respect to ρ yields the following equation:

$$\left(\frac{\partial X^{S_i}}{\partial \rho}\right) = -\left(X^{S_i}\right)^2 N_A \left[\begin{array}{c} \sum_{j=1}^{N_c}\sum_{Y_j} z_j X^{Y_j} W_{ij} + \sum_{j=1}^{N_c}\sum_{Y_j} z_j \rho X^{Y_j}\left(\frac{\partial W_{ij}}{\partial \rho}\right) \\ + \sum_{j=1}^{N_c}\sum_{Y_j} z_j \rho W_{ij} X^{Y_j}\left(\frac{\partial X^{Y_j}}{\partial \rho}\right) \end{array}\right] \tag{96}$$

where,

$$\left(\frac{\partial W_{ij}}{\partial \rho}\right) = \left(\sigma_{ij} k^{S_iY_j}\right)\left[exp\left(\frac{\varepsilon^{S_iY_j}}{kT}\right) - 1\right]$$
$$\times \left[\begin{array}{c} (1-\xi_3)^{-2}\frac{\partial \xi_3}{\partial \rho} + \frac{3d_id_j}{d_i+d_j}\left[(1-\xi_3)^{-2}\left(\frac{\partial \xi_2}{\partial \rho}\right) + 2(1-\xi_3)^{-3}\xi_2\left(\frac{\partial \xi_3}{\partial \rho}\right)\right] \\ +2\left(\frac{d_id_j}{d_i+d_j}\right)^2\left[2\xi_2\left(\frac{\partial \xi_2}{\partial \rho}\right)(1-\xi_3)^{-3} + 3(1-\xi_3)^{-4}\left(\frac{\partial \xi_3}{\partial \rho}\right)\xi_2^2\right] \end{array}\right] \tag{97}$$

As it can be seen from Eq. (96), $(\partial X^{S_i}/\partial \rho)$'s are solutions of a linear system of equations which can be estimated using a known technique such as Gaussian Elimination, Gauss-Jordan or Least Square method (Burden et al., 1981).

2.2 Derivation of fugacity using SAFT EoS

The fugacity of component i in terms of independent variables V and T is given by the following equation for a given phase (Danesh , 1998; Prausnitz et al., 1999; Tabatabaei-Nejad, & Khodapanah, 2009):

$$RTln\varphi_k^\alpha = $$
$$= RTln\frac{f_k^\alpha}{z_k^\alpha P} = \int_{V^\alpha}^{\infty}\left[\left(\frac{\partial P}{\partial n_k}\right)_{T,V,n_{j\neq k}} - \frac{RT}{V}\right]dV - RTlnZ^\alpha \tag{98}$$
$$k = 1,2,\ldots,N_c \quad \alpha = L,V$$

where f_k , φ_k , n_k, V , Z and P are fugacity, fugacity coefficient and the number of moles of component k , volume, compressibility factor, and pressure, respectively. The superscript α denotes liquid (L) and vapor phases (V).

The compressibility factor is related to the volume by the following equations:

$$Z^{\alpha} = \frac{P}{\rho^{\alpha} RT} \tag{99}$$

$$\rho^{\alpha} = \frac{n^{\alpha}}{V^{\alpha}} \tag{100}$$

where n refers to the total number of moles of the known phase.

To use equation (98), we require a suitable EoS that holds for the entire range of possible mole fractions z at the system temperature and for the density range between 0 and $\left(\frac{n}{V}\right)$. Application of the SAFT EoS in Eq. (98) yields the following equation for calculating the fugacity of the components:

$$\begin{aligned} RTln\left(Z^{\alpha}\varphi_k^{\alpha}\right) = & \int_V^{\infty} \frac{RT}{V}\left[\frac{\partial(nZ_{hs})}{\partial n_k}\right]_{T,V,n_{j\neq k}} dV + \int_V^{\infty} \frac{RT}{V}\left[\frac{\partial(nZ_{ch})}{\partial n_k}\right]_{T,V,n_{j\neq k}} dV \\ & + \int_V^{\infty} \frac{RT}{V}\left[\frac{\partial(nZ_{disp})}{\partial n_k}\right]_{T,V,n_{j\neq k}} dV + \int_V^{\infty} \frac{RT}{V}\left[\frac{\partial(nZ_{assoc})}{\partial n_k}\right]_{T,V,n_{j\neq k}} dV \\ & k = 1,2,\ldots,N_c \end{aligned} \tag{101}$$

The following equations are derived for the first term in Eq. (101) accounting for the hard sphere contribution of SAFT:

$$\begin{aligned} IR\_HS = \int_V^{\infty} \frac{RT}{V}\left[\frac{\partial(nZ_{hs})}{\partial n_k}\right]_{T,V,n_{j\neq k}} dV = \int_0^{\rho} \frac{RT}{\rho}\left[\frac{\partial(nZ_{hs})}{\partial n_k}\right]_{T,V,n_{j\neq k}} d\rho \\ k = 1,2,\ldots,N_c \end{aligned} \tag{102}$$

$$\begin{aligned} \left[\frac{\partial(nZ_{hs})}{\partial n_k}\right]_{T,V,n_{j\neq k}} = & \left[(hs_{1k} + 2hs_1).\rho + (hs_{2k} + 3hs_2).\rho^2 + (hs_{3k} + 4hs_3).\rho^3\right] \\ & \times\left[1 + hs_4.\rho + hs_5.\rho^2 + hs_6.\rho^3\right]^{-1} - \left[1 + hs_4.\rho + hs_5.\rho^2 + hs_6.\rho^3\right]^{-2} \\ & \times\left[(hs_{4k} + hs_4).\rho + (hs_{5k} + 2hs_5).\rho^2 + (hs_{6k} + 3hs_6).\rho^3\right] \times \left[hs_1.\rho + hs_2.\rho^2 + hs_3.\rho^3\right] \end{aligned} \tag{103}$$

The parameters used in Eq. (103) are given by equations (7) - (17) and the following equations:

$$hs_{1k} = (\bar{r}_k S_3 + \bar{r} S_{3k} + 3S_{1k}S_2 + 3S_1 S_{2k}) P_{na} \tag{104}$$

$$hs_{2k} = (-2\bar{r}_k S_3^2 - 4\bar{r} S_{3k} S_3 - 3S_{1k}S_2S_3 - 3S_{2k}S_1S_3 - 3S_1S_2S_{3k} + 9S_{2k}S_2^2) P_{na}^2 \tag{105}$$

$$hs_{3k} = (\bar{r}_k S_3^3 + 3\bar{r} S_3^2 S_{3k} - 3S_2^2 S_{2k} S_3 - S_2^3 S_{3k}) P_{na}^3 \tag{106}$$

$$hs_{4k} = -3S_{3k} P_{na} \tag{107}$$

$$hs_{5k} = 6S_3 S_{3k} P_{na}^2 \tag{108}$$

$$hs_{6k} = -3S_3^2 S_{3k} P_{na}^3 \tag{109}$$

where,

$$\bar{r}_k = (-\bar{r} + r_k) \tag{110}$$

$$S_{1k} = -S_1 + r_k d_k \tag{111}$$

$$S_{2k} = -S_2 + r_k d_k^2 \tag{112}$$

$$\begin{gathered} S_{3k} = -S_3 + r_k d_k^3 \\ k = 1,2,\ldots,N_c \end{gathered} \tag{113}$$

The non-ideality of the mixture due to formation of chain molecules of the various components which was described using the second term in Eq. (101) is derived as the following form:

$$IR\_Chain = \int_V^\infty \frac{RT}{V}\left[\frac{\partial(nZ_{ch})}{\partial n_k}\right]_{T,V,n_{j\neq k}} dV = \int_0^\rho \frac{RT}{\rho}\left[\frac{\partial(nZ_{ch})}{\partial n_k}\right]_{T,V,n_{j\neq k}} d\rho \tag{114}$$

$$k = 1,2,\ldots,N_c$$

$$\left[\frac{\partial(nZ_{ch})}{\partial n_k}\right]_{T,V,n_{j\neq k}}$$

$$\times\sum_{i=1}^{N_c}\left[\begin{gathered} \delta_{ik}.(1-r_i)\times\left[ch_{1k}(d_i).\rho + ch_{2k}(d_i).\rho^2 + ch_{3k}(d_i).\rho^3\right] \\ \times\left[2 + ch_{4k}(d_i).\rho + ch_{5k}(d_i).\rho^2 + ch_{6k}(d_i).\rho^3 + ch_{7k}(d_i).\rho^4\right]^{-1} \\ +z_i(1-r_i)\times\left[2 + ch_{4k}(d_i).\rho + ch_{5k}(d_i).\rho^2 + ch_{6k}(d_i).\rho^3 + ch_{7k}(d_i).\rho^4\right]^{-1} \\ \left[\begin{gathered} (ch_{1k}(d_i) + ch_1(d_i)).\rho \\ +(ch_{2k}(d_i) + 2ch_2(d_i)).\rho^2 + (ch_{3k}(d_i) + 3ch_3(d_i)).\rho^3 \end{gathered}\right] \\ -z_i(1-r_i)\times\left[ch_1(d_i).\rho + ch_2(d_i).\rho^2 + ch_3(d_i).\rho^3\right] \\ \times\left[2 + ch_4(d_i).\rho + ch_5(d_i).\rho^2 + ch_6(d_i).\rho^3 + ch_7(d_i).\rho^4\right]^{-2} \\ \left[\begin{gathered} (ch_{4k}(d_i) + ch_4(d_i)).\rho + (ch_{5k}(d_i) + 2ch_5(d_i)).\rho^2 \\ +(ch_{6k}(d_i) + 3ch_6(d_i)).\rho^3 + (ch_{7k}(d_i) + 4ch_7(d_i)).\rho^4 \end{gathered}\right] \end{gathered}\right] \tag{115}$$

$$k = 1,2,\ldots,N_c$$

where the parameters are described using equations (21) - (27) and the equations given below:

$$ch_{1k}(d_i) = (2S_{3k} + 3d_i S_{2k})P_{na} \tag{116}$$

$$ch_{2k}(d_i) = \left(-8S_3 S_{3k} + 4d_i^2 S_2 S_{2k}\right)P_{na}^2 \tag{117}$$

$$ch_{3k}(d_i) = \left(2d_i^2 S_2 S_{2k} S_3 + d_i^2 S_2^2 S_{3k} + 6S_3^2 S_{3k} - 3d_i S_{2k} S_3^2 - 6d_i S_2 S_3 S_{3k}\right)P_{na}^3 \tag{118}$$

$$ch_{4k}(d_i) = (-6S_{3k} + 3d_i S_{2k})P_{na} \tag{119}$$

$$ch_{5k}(d_i) = \left(12S_3 S_{3k} + 2d_i^2 S_2 S_{2k} - 9d_i S_{2k} S_3 - 9d_i S_2 S_{3k}\right)P_{na}^2 \tag{120}$$

$$ch_{6k}(d_i) = \left(9d_i S_{2k} S_3^2 + 18d_i S_2 S_3 S_{3k} - 6S_3^2 S_{3k} - 2d_i^2 S_2 S_{2k} S_3 - d_i^2 S_2^2 S_{3k}\right)P_{na}^3 \tag{121}$$

$$ch_{7k}(d_i) = \left(-3d_i S_{2k} S_3^3 - 9d_i S_2 S_3^2 S_{3k}\right)P_{na}^4 \qquad k = 1,2,\ldots,N_c \tag{122}$$

The dispersion contribution to the non-ideal behavior of the mixture (the third term in the right hand side of Eq. (101) is derived as the following forms:

$$IR\_Disp = \int_V^\infty \frac{RT}{V}\left[\frac{\partial(nZ_{disp})}{\partial n_k}\right]_{T,V,n_{j\neq k}} dV = \int_0^\rho \frac{RT}{\rho}\left[\frac{\partial(nZ_{disp})}{\partial n_k}\right]_{T,V,n_{j\neq k}} d\rho \qquad k = 1,2,\ldots,N_c \tag{123}$$

$$\left[\frac{\partial(nZ_{disp})}{\partial n_k}\right]_{T,V,n_{j\neq k}} = Z_{disp} + (Z_{disp})_k \tag{124}$$

$$(Z_{disp})_k = A_{1k} + A_{2k}A_3 + A_2 A_{3k}, \quad k = 1,2,\ldots,N_c$$

where,

$$A_1 = dI_1.\rho + dI_2.\rho^2 + dI_3.\rho^3 + dI_4.\rho^4 + dI_5.\rho^5 + dI_6.\rho^6 + dI_7.\rho^7 \tag{125}$$

$$A_{1k} = (dI_{1k} + dI_1).\rho + (dI_{2k} + 2dI_2).\rho^2 + (dI_{3k} + 3dI_3).\rho^3 + (dI_{4k} + 4dI_4).\rho^4 + (dI_{5k} + 5dI_5).\rho^5 + (dI_{6k} + 6dI_6).\rho^6 + (dI_{7k} + 7dI_7).\rho^7 \qquad k = 1,2,\ldots,N_c \tag{126}$$

$$A_2 = -6P_{na}\bar{r}Adisp_2 \tag{127}$$

$$A_{2k} = -6P_{na}\left[\bar{r}_k.Adisp_2 + \bar{r}.Adisp_{2k}\right] \qquad k = 1,2,\ldots,N_c \tag{128}$$

$$A_3 = \left(\frac{B_1}{B_2}\right) \cdot B_3 - \left(\frac{B_1}{\rho B_2}\right)^2 \cdot \left(\frac{B_4}{B_5}\right) \cdot B_6 \cdot B_7 \tag{129}$$

$$B_1 = 4\rho + disp_1 \cdot \rho^2 + disp_2 \cdot \rho^3 + disp_3 \cdot \rho^4 + disp_4 \cdot \rho^5 + disp_5 \cdot \rho^6 + disp_6 \cdot \rho^7 \tag{130}$$

$$\begin{gathered} B_{1k} = 4\rho + \left(d_{1k} + 2disp_1\right) \cdot \rho^2 + \left(d_{2k} + 3disp_2\right) \cdot \rho^3 + \left(d_{3k} + 4disp_3\right) \cdot \rho^4 \\ + \left(d_{4k} + 5disp_4\right) \cdot \rho^5 + \left(d_{5k} + 6disp_5\right) \cdot \rho^6 + \left(d_{6k} + 7disp_6\right) \cdot \rho^7 \\ k = 1,2,\ldots,N_c \end{gathered} \tag{131}$$

$$B_2 = 4 + disp_7 \cdot \rho + disp_8 \cdot \rho^2 + disp_9 \cdot \rho^3 + disp_{10} \cdot \rho^4 + disp_{11} \cdot \rho^5 + disp_{12} \cdot \rho^6 \tag{132}$$

$$\begin{gathered} B_{2k} = \left(d_{7k} + disp_7\right) \cdot \rho + \left(d_{8k} + 2disp_8\right) \cdot \rho^2 + \left(d_{9k} + 3disp_9\right) \cdot \rho^3 \\ + \left(d_{10k} + 4disp_{10}\right) \cdot \rho^4 + \left(d_{11k} + 5disp_{11}\right) \cdot \rho^5 + \left(d_{12k} + 6disp_{12}\right) \cdot \rho^6 \end{gathered} \tag{133}$$

$$B_3 = b_0 + dI_8 \cdot \rho + dI_9 \cdot \rho^2 + dI_{10} \cdot \rho^3 + dI_{11} \cdot \rho^4 + dI_{12} \cdot \rho^5 + dI_{13} \cdot \rho^6 \tag{134}$$

$$\begin{gathered} B_{3k} = b_{0k} + \left(dI_{8k} + dI_8\right) \cdot \rho + \left(dI_{9k} + 2dI_9\right) \cdot \rho^2 + \left(dI_{10k} + 3dI_{10}\right) \cdot \rho^3 \\ + \left(dI_{11k} + 4dI_{11}\right) \cdot \rho^4 + \left(dI_{12k} + 5dI_{12}\right) \cdot \rho^5 + \left(dI_{13k} + 6dI_{13}\right) \cdot \rho^6 \end{gathered} \tag{135}$$

$$B_4 = disp_{13} + disp_{14} \cdot \rho + disp_{15} \cdot \rho^2 + disp_{16} \cdot \rho^3 + disp_{17} \cdot \rho^4 + disp_{18} \cdot \rho^5 \tag{136}$$

$$\begin{gathered} B_{4k} = d_{13k} + \left(d_{14k} + disp_{14}\right) \cdot \rho + \left(d_{15k} + 2disp_{15}\right) \cdot \rho^2 \\ + \left(d_{16k} + 3disp_{16}\right) \cdot \rho^3 + \left(d_{17k} + 4disp_{17}\right) \cdot \rho^4 + \left(d_{18k} + 5disp_{18}\right) \cdot \rho^5 \end{gathered} \tag{137}$$

$$\begin{gathered} B_5 = 8 + disp_{19} \cdot \rho + disp_{20} \cdot \rho^2 + disp_{21} \cdot \rho^3 + disp_{22} \cdot \rho^4 + disp_{23} \cdot \rho^5 \\ + disp_{24} \cdot \rho^6 + disp_{25} \cdot \rho^7 + disp_{26} \cdot \rho^8 \end{gathered} \tag{138}$$

$$\begin{gathered} B_{5k} = \left(d_{19k} + disp_{19}\right) \cdot \rho + \left(d_{20k} + 2disp_{20}\right) \cdot \rho^2 \\ + \left(d_{21k} + 3disp_{21}\right) \cdot \rho^3 + \left(d_{22k} + 4disp_{22}\right) \cdot \rho^4 + \left(d_{23k} + 5disp_{23}\right) \cdot \rho^5 \\ + \left(d_{24k} + 6disp_{24}\right) \cdot \rho^6 + \left(d_{25k} + 7disp_{25}\right) \cdot \rho^7 + \left(d_{26k} + 8disp_{26}\right) \cdot \rho^8 \end{gathered} \tag{139}$$

$$B_6 = b_0 \cdot \rho^2 + Idisp_1 \cdot \rho^3 + Idisp_2 \cdot \rho^4 + Idisp_3 \cdot \rho^5 + Idisp_4 \cdot \rho^6 + Idisp_5 \cdot \rho^7 + Idisp_6 \cdot \rho^8 \tag{140}$$

$$\begin{gathered} B_{6k} = \left(b_{0k} + 2b_0\right) \cdot \rho^2 + \left(Idisp_{1k} + 3Idisp_1\right) \cdot \rho^3 \\ + \left(Idisp_{2k} + 4Idisp_2\right) \cdot \rho^4 + \left(Idisp_{3k} + 5Idisp_3\right) \cdot \rho^5 \\ + \left(Idisp_{4k} + 6Idisp_4\right) \cdot \rho^6 + \left(Idisp_{5k} + 7Idisp_5\right) \cdot \rho^7 + \left(Idisp_{6k} + 8Idisp_6\right) \cdot \rho^8 \end{gathered} \tag{141}$$

$$B_7 = P_{na} S_3 \tag{142}$$

$$B_{7k} = P_{na} S_{3k} \quad k = 1,2,\ldots,N_c \tag{143}$$

The parameters used in Eqs. (125) - (143) are given through Eqs. (40) - (92) and the following equations:

$$d_{1k} = 12[\bar{r}_k S_3 + \bar{r} S_{3k}]P_{na} \tag{144}$$

$$d_{2k} = [27\bar{r}_k S_3^2 + 2(27\bar{r} - 26)S_3 S_{3k}]P_{na}^2 \tag{145}$$

$$d_{3k} = [70\bar{r}_k S_3^3 + 3S_3^2 S_{3k}(70\bar{r} + 42)]P_{na}^3 \tag{146}$$

$$d_{4k} = [51\bar{r}_k S_3^4 + 3S_3^3 S_{3k}(51\bar{r} - 27)]P_{na}^4 \tag{147}$$

$$d_{5k} = \left[-16\bar{r}_k S_3^5 + 5S_3^4 S_{3k}(8 - 16\bar{r})\right]P_{na}^5 \tag{148}$$

$$d_{6k} = \left[2\bar{r}_k S_3^6 + 6S_3^5 S_{3k}(2\bar{r} - 1)\right]P_{na}^6 \tag{149}$$

$$d_{7k} = -20 S_{3k} P_{na} \tag{150}$$

$$d_{8k} = 82 S_3 S_{3k} P_{na}^2 \tag{151}$$

$$d_{9k} = -132 S_3^2 S_{3k} P_{na}^3 \tag{152}$$

$$d_{10k} = 104 S_3^3 S_{3k} P_{na}^4 \tag{153}$$

$$d_{11k} = -40 S_3^4 S_{3k} P_{na}^5 \tag{154}$$

$$d_{12k} = 6 S_3^5 S_{3k} P_{na}^6 \tag{155}$$

$$d_{13k} = 24\bar{r}_k \tag{156}$$

$$d_{14k} = [192\bar{r}_k S_3 + (192\bar{r} - 128)S_{3k}]P_{na} \tag{157}$$

$$d_{15k} = [-372\bar{r}_k S_3^2 + 2S_3 S_{3k}(148 - 372\bar{r})]P_{na}^2 \tag{158}$$

$$d_{16k} = [230\bar{r}_k S_3^3 + 3S_3^2 S_{3k}(230\bar{r} - 70)]P_{na}^3 \tag{159}$$

$$d_{17k} = [-52\bar{r}_k S_3^4 + 4S_3^3 S_{3k}(8 - 52\bar{r})]P_{na}^4 \tag{160}$$

$$d_{18k} = \left[2\bar{r}_k S_3^5 + 10 S_3^4 S_{3k}(1 + \bar{r})\right]P_{na}^5 \tag{161}$$

$$d_{19k} = -52 S_{3k} P_{na} \tag{162}$$

$$d_{20k} = 292 S_3 S_{3k} P_{na}^2 \tag{163}$$

$$d_{21k} = -693 S_3^2 S_{3k} P_{na}^3 \tag{164}$$

$$d_{22k} = 900 S_3^3 S_{3k} P_{na}^4 \tag{165}$$

$$d_{23k} = -690 S_3^4 S_{3k} P_{na}^5 \tag{166}$$

$$d_{24k} = 312 S_3^5 S_{3k} P_{na}^6 \tag{167}$$

$$d_{25k} = -77 S_3^6 S_{3k} P_{na}^7 \tag{168}$$

$$d_{26k} = 8S_3^7 S_{3k} P_{na}^8 \qquad k = 1,2,\ldots,N_c \tag{169}$$

$$Adisp_{1k} = 2 \times \left[-\sum_{i=1}^{N_c}\sum_{j=1}^{N_c} z_i z_j r_i r_j \left(\varepsilon_{ij}/kT\right)\sigma_{ij}^3 + \sum_{i=1}^{N_c} z_i r_i r_k (\varepsilon_{ik}/kT)\sigma_{ik}^3 \right] \tag{170}$$

$$Adisp_{2k} = 2 \times \left[-\sum_{i=1}^{N_c}\sum_{j=1}^{N_c} z_i z_j r_i r_j \left(\varepsilon_{ij}/kT\right)^2\sigma_{ij}^3 + \sum_{i=1}^{N_c} z_i r_i r_k (\varepsilon_{ik}/kT)^2\sigma_{ik}^3 \right] \tag{171}$$

$$a_{jk}(\bar{r}) = \begin{bmatrix} \left[\dfrac{\bar{r}_k}{\bar{r}} - \dfrac{\bar{r}_k \cdot (\bar{r}-1)}{\bar{r}^2}\right] a_{1j} \\ + \left[\dfrac{\bar{r}_k \cdot (\bar{r}-2)}{\bar{r}^2} + \dfrac{\bar{r}_k \cdot (\bar{r}-1)}{\bar{r}^2} - 2\dfrac{\bar{r}_k \cdot (\bar{r}-1)\cdot(\bar{r}-2)}{\bar{r}^3}\right] a_{2j} \end{bmatrix} \tag{172}$$

$$b_{jk}(\bar{r}) = \begin{bmatrix} \left[\dfrac{\bar{r}_k}{\bar{r}} - \dfrac{\bar{r}_k \cdot (\bar{r}-1)}{\bar{r}^2}\right] b_{1j} \\ + \left[\dfrac{\bar{r}_k \cdot (\bar{r}-2)}{\bar{r}^2} + \dfrac{\bar{r}_k \cdot (\bar{r}-1)}{\bar{r}^2} - 2\dfrac{\bar{r}_k \cdot (\bar{r}-1)\cdot(\bar{r}-2)}{\bar{r}^3}\right] b_{2j} \end{bmatrix} \tag{173}$$

$$j = 0, 1, \ldots, 6 \qquad k = 1, 2, \ldots, N_c$$

$$Idisp_{1k} = [b_{1k}S_3 + b_1 S_{3k}]P_{na} \tag{174}$$

$$Idisp_{2k} = [b_{2k}S_3^2 + 2b_2 S_3 S_{3k}]P_{na}^2 \tag{175}$$

$$Idisp_{3k} = [b_{3k}S_3^3 + 3b_3 S_3^2 S_{3k}]P_{na}^3 \tag{176}$$

$$Idisp_{4k} = [b_{4k}S_3^4 + 4b_4 S_3^3 S_{3k}]P_{na}^4 \tag{177}$$

$$Idisp_{5k} = [b_{5k}S_3^5 + 5b_5 S_3^4 S_{3k}]P_{na}^5 \tag{178}$$

$$Idisp_{6k} = [b_{6k}S_3^6 + 6b_6 S_3^5 S_{3k}]P_{na}^6 \tag{179}$$

$$dIdisp_{7k} = 2[a_{1k}S_3 + a_1 S_{3k}]P_{na} \tag{180}$$

$$dIdisp_{2k} = 3[a_{2k}S_3^2 + 2a_2 S_3 S_{3k}]P_{na}^2 \tag{181}$$

$$dIdisp_{3k} = 4[a_{3k}S_3^3 + 3a_3 S_3^2 S_{3k}]P_{na}^3 \tag{182}$$

$$dIdisp_{4k} = 5[a_{4k}S_3^4 + 4a_4 S_3^3 S_{3k}]P_{na}^4 \tag{183}$$

$$dIdisp_{5k} = 6[a_{5k}S_3^5 + 5a_5 S_3^4 S_{3k}]P_{na}^5 \tag{184}$$

$$dIdisp_{6k} = 7[a_{6k}S_3^6 + 6a_6 S_3^5 S_{3k}]P_{na}^6 \tag{185}$$

$$dI_{1k} = -12[Adisp_{1k} \cdot a_0 + Adisp_1 \cdot a_{0k}]P_{na} \tag{186}$$

$$dI_{2k} = -12[Adisp_{1k} \cdot dIdisp_1 + Adisp_1 \cdot dIdisp_{1k}]P_{na} \tag{187}$$

$$dI_{3k} = -12[Adisp_{1k} \cdot dIdisp_2 + Adisp_1 \cdot dIdisp_{2k}]P_{na} \tag{188}$$

$$dI_{4k} = -12[Adisp_{1k} \cdot dIdisp_3 + Adisp_1 \cdot dIdisp_{3k}]P_{na} \tag{189}$$

$$dI_{5k} = -12[Adisp_{1k} \cdot dIdisp_4 + Adisp_1 \cdot dIdisp_{4k}]P_{na} \tag{190}$$

$$dI_{6k} = -12[Adisp_{1k} \cdot dIdisp_5 + Adisp_1 \cdot dIdisp_{5k}]P_{na} \tag{191}$$

$$dI_{7k} = -12[Adisp_{1k} \cdot dIdisp_6 + Adisp_1 \cdot dIdisp_{6k}]P_{na} \tag{192}$$

$$dI_{8k} = 2[b_{1k}S_3 + b_1S_{3k}]P_{na} \tag{193}$$

$$dI_{9k} = 3[b_{2k}S_3^2 + 2b_2S_3S_{3k}]P_{na}^2 \tag{194}$$

$$dI_{10k} = 4[b_{3k}S_3^3 + 3b_3S_3^2S_{3k}]P_{nc}^3 \tag{195}$$

$$dI_{11k} = 5[b_{4k}S_3^4 + 4b_4S_3^3S_{3k}]P_{nc}^4 \tag{196}$$

$$dI_{12k} = 6\left[b_{5k}S_3^5 + 5b_5S_3^4S_{3k}\right]P_{na}^5 \tag{197}$$

$$dI_{13k} = 7\left[b_{6k}S_3^6 + 6b_6S_3^5S_{3k}\right]P_{na}^6 \tag{198}$$

The association contribution to the non-ideal behavior of mixtures containing associating compounds (the last term given by the right hand side of Eq. (101) is derived as the following forms:

$$IR\_Assoc = \int_V^{\infty} \frac{RT}{V}\left[\frac{\partial(nZ_{assoc})}{\partial n_k}\right]_{T,V,n_{j\neq k}} dV = \int_0^{\rho} \frac{RT}{\rho}\left[\frac{\partial(nZ_{assoc})}{\partial n_k}\right]_{T,V,n_{j\neq k}} d\rho \tag{199}$$

$$k = 1, 2, \dots, N_c$$

$$\left[\frac{\partial(nZ_{assoc})}{\partial n_k}\right]_{T,V,n_{j\neq k}} = Z_{assoc} + \rho(A_k) + \rho^2 \sum_{i=1}^{N_c} z_i \left(\frac{\partial A_i}{\partial n_k}\right)_{T,V,n_{j\neq k}} \tag{200}$$

$$k = 1, 2, \dots, N_c$$

where,

$$\left(\frac{\partial A_i}{\partial n_k}\right)_{T,V,n_{j\neq k}} = \sum_{S_i} -\frac{1}{(X^{S_i})^2}\left(\frac{\partial X^{S_i}}{\partial n_k}\right)_{T,V,n_{j\neq k}} \cdot \left(\frac{\partial X^{S_i}}{\partial \rho}\right) + \left(\frac{1}{X^{S_i}} - \frac{1}{2}\right)\left[\frac{\partial}{\partial n_k}\left(\frac{\partial X^{S_i}}{\partial \rho}\right)\right]_{T,V,n_{j\neq k}} \tag{201}$$

$$k = 1, 2, \dots, N_c$$

$$A_i = \sum_{S_i}\left(\frac{1}{X^{S_i}} - \frac{1}{2}\right)\left(\frac{\partial X^{S_i}}{\partial \rho}\right) \tag{202}$$

where $(\partial X^{S_i}/\partial \rho)$ is given by Eq. (96), other derivatives of X^{S_i} used in Eq. (202) are given below:

$$\left(\frac{\partial X^{S_i}}{\partial n_k}\right)_{T,V,n_j\neq k} = -(X^{S_i})^2 N_A \left[\begin{array}{c} \sum_{j=1}^{N_c}\sum_{Y_j}\frac{1}{V}\delta_{jk}X^{Y_j}W_{ij} + \sum_{j=1}^{N_c}\sum_{Y_j} z_j\rho X^{Y_j}\left(\frac{\partial W_{ij}}{\partial n_k}\right)_{T,V,n_j\neq k} + \\ \sum_{j=1}^{N_c}\sum_{Y_j} z_j\rho W_{ij}\left(\frac{\partial X^{Y_j}}{\partial n_k}\right)_{T,V,n_j\neq k} \end{array}\right] \tag{203}$$

$$k = 1, 2, \dots, N_c$$

$$\left[\frac{\partial}{\partial n_k}\left(\frac{\partial X^{S_i}}{\partial \rho}\right)\right]_{T,V,n_j\neq k} = 2\left(X^{S_i}\right)^{-1}\left(\frac{\partial X^{S_i}}{\partial n_k}\right)_{T,V,n_{j\neq k}} \cdot \left(\frac{\partial X^{S_i}}{\partial \rho}\right)$$

$$-\left(X^{S_i}\right)^2 N_A \left[\begin{array}{c} \sum_{j=1}^{N_c}\sum_{Y_j}\frac{\delta_{jk}-z_j}{n}X^{Y_j}W_{ij} + \sum_{j=1}^{N_c}\sum_{Y_j} z_jW_{ij}\left(\frac{\partial X^{Y_j}}{\partial n_k}\right)_{T,V,n_{j\neq k}} + \sum_{j=1}^{N_c}\sum_{Y_j} z_jX^{Y_j}\left(\frac{\partial W_{ij}}{\partial n_k}\right)_{T,V,n_{j\neq k}} \\ + \sum_{j=1}^{N_c}\sum_{Y_j}\frac{\delta_{jk}}{V}X^{Y_j}\left(\frac{\partial W_{ij}}{\partial \rho}\right) + \sum_{j=1}^{N_c}\sum_{Y_j} z_j\rho\left(\frac{\partial X^{Y_j}}{\partial n_k}\right)_{T,V,n_{j\neq k}} \cdot \left(\frac{\partial W_{ij}}{\partial \rho}\right) \\ + \sum_{j=1}^{N_c}\sum_{Y_j} z_j\rho X^{Y_j}\left(\frac{\partial}{\partial n_k}\left(\frac{\partial W_{ij}}{\partial \rho}\right)\right)_{T,V,n_{j\neq k}} + \sum_{j=1}^{N_c}\sum_{Y_j}\frac{\delta_{jk}}{V}W_{ij}\left(\frac{\partial X^{Y_j}}{\partial \rho}\right) \\ + \sum_{j=1}^{N_c}\sum_{Y_j} z_j\rho\left(\frac{\partial W_{ij}}{\partial n_k}\right)_{T,V,n_{j\neq k}} \cdot \left(\frac{\partial X^{Y_j}}{\partial \rho}\right) + \sum_{j=1}^{N_c}\sum_{Y_j} z_j\rho W_{ij}\left(\frac{\partial}{\partial n_k}\left(\frac{\partial X^{Y_j}}{\partial \rho}\right)\right)_{T,V,n_{j\neq k}} \end{array}\right] \tag{204}$$

where $\left(\partial W_{ij}/\partial \rho\right)$ is given by Eq. (97), other derivatives of W_{ij} are given as the following:

$$\left(\frac{\partial W_{ij}}{\partial n_k}\right)_{T,V,n_{j\neq k}} = \left(\sigma_{ij}k^{S_iY_j}\right)\left[exp\left(\frac{\varepsilon^{S_iY_j}}{kT}\right) - 1\right]$$

$$\times\left[\begin{array}{c} (1-\xi_3)^{-2}\frac{\partial \xi_3}{\partial n_k} + 3\frac{d_id_j}{d_i+d_j}\left[(1-\xi_3)^{-2}\left(\frac{\partial \xi_2}{\partial n_k}\right) + 2(1-\xi_3)^{-3}\xi_2\left(\frac{\partial \xi_3}{\partial n_k}\right)\right] \\ +2\left(\frac{d_id_j}{d_i+d_j}\right)^2\left[2(1-\xi_3)^{-3}\xi_2\left(\frac{\partial \xi_2}{\partial n_k}\right) + 3(1-\xi_3)^{-4}\left(\frac{\partial \xi_3}{\partial n_k}\right)\xi_2^2\right] \end{array}\right] \tag{205}$$

$$k = 1, 2, \dots, N_c$$

$$\left(\frac{\partial}{\partial n_k}\left(\frac{\partial W_{ij}}{\partial \rho}\right)\right)_{T,V,n_{j\neq k}} = \left(\sigma_{ij}k^{S_iY_j}\right)\left[exp\left(\frac{\varepsilon^{S_iY_j}}{kT}\right)-1\right]$$
$$\times\left[\begin{array}{c} 2\left(1-\xi_3\right)^{-3}\left(\frac{\partial \xi_3}{\partial n_k}\right)\left(\frac{\partial \xi_3}{\partial \rho}\right)+\left(1-\xi_3\right)^{-2}\frac{\partial}{\partial n_k}\left(\frac{\partial \xi_3}{\partial \rho}\right) \\ +\frac{3d_id_j}{d_i+d_j}\left[\begin{array}{c} 2\left(1-\xi_3\right)^{-3}\left(\frac{\partial \xi_3}{\partial n_k}\right)\left(\frac{\partial \xi_2}{\partial \rho}\right)+\left(1-\xi_3\right)^{-2}\frac{\partial}{\partial n_k}\left(\frac{\partial \xi_2}{\partial \rho}\right) \\ +6\left(1-\xi_3\right)^{-4}\left(\frac{\partial \xi_3}{\partial n_k}\right)\xi_2\left(\frac{\partial \xi_3}{\partial \rho}\right)+2\left(1-\xi_3\right)^{-3}\left(\frac{\partial \xi_2}{\partial n_k}\right)\left(\frac{\partial \xi_3}{\partial \rho}\right) \\ +2\left(1-\xi_3\right)^{-3}\xi_2\frac{\partial}{\partial n_k}\left(\frac{\partial \xi_3}{\partial \rho}\right) \end{array}\right] \\ +2\left(\frac{d_id_j}{d_i+d_j}\right)^2\left[\begin{array}{c} 2\left(1-\xi_3\right)^{-3}\left(\frac{\partial \xi_2}{\partial n_k}\right)\left(\frac{\partial \xi_2}{\partial \rho}\right)+2\xi_2\left(1-\xi_3\right)^{-3}\frac{\partial}{\partial n_k}\left(\frac{\partial \xi_2}{\partial \rho}\right) \\ +6\left(1-\xi_3\right)^{-4}\left(\frac{\partial \xi_3}{\partial n_k}\right)\xi_2\left(\frac{\partial \xi_2}{\partial \rho}\right)+12\left(1-\xi_3\right)^{-5}\left(\frac{\partial \xi_3}{\partial n_k}\right)\left(\frac{\partial \xi_3}{\partial \rho}\right)\xi_2^2 \\ +3\left(1-\xi_3\right)^{-4}\frac{\partial}{\partial n_k}\left(\frac{\partial \xi_3}{\partial \rho}\right)\xi_2^2+6\left(1-\xi_3\right)^{-4}\left(\frac{\partial \xi_2}{\partial n_k}\right)\xi_2\left(\frac{\partial \xi_3}{\partial \rho}\right) \end{array}\right] \end{array}\right] \tag{206}$$

$$k = 1, 2, \ldots, N_c$$

here,

$$\frac{\partial \xi_2}{\partial \rho} = P_{na}S_2 \tag{207}$$

$$\frac{\partial \xi_3}{\partial \rho} = P_{na}S_3 \tag{208}$$

$$\left(\frac{\partial \xi_2}{\partial n_k}\right)_{T,V,n_{j\neq k}} = \frac{1}{V}P_{na}r_kd_k^2 \tag{209}$$

$$\left(\frac{\partial \xi_3}{\partial n_k}\right)_{T,V,n_{j\neq k}} = \frac{1}{V}P_{na}r_kd_k^3 \tag{210}$$

$$\left(\frac{\partial}{\partial n_k}\left(\frac{\partial \xi_2}{\partial \rho}\right)\right)_{T,V,n_{j\neq k}} = P_{na}.\sum_{i=1}^{N_c}\left(\frac{\delta_{ik}-z_i}{n}\right)r_id_i^2 = \frac{1}{V}.P_{na}.\frac{1}{\rho}\left[r_kd_k^2-S_2\right] \tag{211}$$

$$\left(\frac{\partial}{\partial n_k}\left(\frac{\partial \xi_3}{\partial \rho}\right)\right)_{T,V,n_{j\neq k}} = P_{na}.\sum_{i=1}^{N_c}\left(\frac{\delta_{ik}-z_i}{n}\right) r_i d_i^3 = \frac{1}{V}.P_{na}.\frac{1}{\rho}\left[r_k d_k^3 - S_3\right] \tag{212}$$

$$kk = 1, 2, \dots, N_{c_c}$$

In the above equations , δ_{jk} refers to dirac delta function which is defined as following:

$$\delta_{jk}\delta_{ik} = \begin{cases} 0 & j \neq k \\ 1 & j = k \end{cases} \tag{213}$$

3. Application of Chebyshev polynomials to calculate density

The integration of the terms used in equations (102), (114), (123) and (119) for calculating the fugacity coefficients are performed numerically using Gaussian quadrature method. We found that five point quadrature method leads to a result with acceptable accuracy. As it can be seen from Eq. (101) the fugacity coefficient is a function of temperature, pressure, composition and the properties of the components. In order to calculate the fugacity coefficient of each component, we should first calculate the density of mixture at a given pressure, temperature and composition using Eq. (2). As it can be seen, from the mentioned equation, the density as function of the pressure is not known explicitly. Therefore, the estimation of the density at a given pressure should be performed using an iterative procedure, starting from initial guesses because of the multiplicity of the solution. A solution which is obtained by an iterative technique depends on the choice of the initial guess. Therefore, iterative procedures can not cover all acceptable roots unless the number of roots and the approximate values of the solutions (i.e. initial guesses) had already been known. Hence, an alternative, robust, fast and accurate technique that can predict all acceptable solutions is proposed. The proposed method is based on a numerical interpolation using Chebyshev polynomials in a finite interval (Burden et al., 1981).

It should be pointed out that Chebyshev series provide high accuracy and can be transformed to power series which are suitable for root finding procedure. More general accounts of root finding through Chebyshev approximations are given in (Boyd, 2006). The aforementioned method enables us to calculate all possible solutions and select among them those which are physically interpretable.

It should be considered that using Chebyshev polynomials to approximate a given function will become more efficient when it has non-zero values at both end points of the interval. It can be shown that the pressure vs. density function in SAFT EoS linearly goes to zero for negligible values of the density. In order to avoid this problem, P/ρ vs. density using Chebyshev polynomials was interpolated.

Another advantage of using Chebyshev polynomials for approximating a function is that for a specific number of basic functions, it always leads to a well-conditioned matrix during the calculation of the unknown coefficients of the basis functions, which is more accurate than the other interpolation techniques.

Figure 2 shows the interpolation error using Chebyshev polynomials of degree 15 for approximating pressure vs. density of a binary mixture of ethanol and toluene containing 37.5 mole% ethanol. Figure 3 shows the error in interpolation for another system (oil sample) for which the composition is given in Table 2 (Jamaluddin et al., 2000).

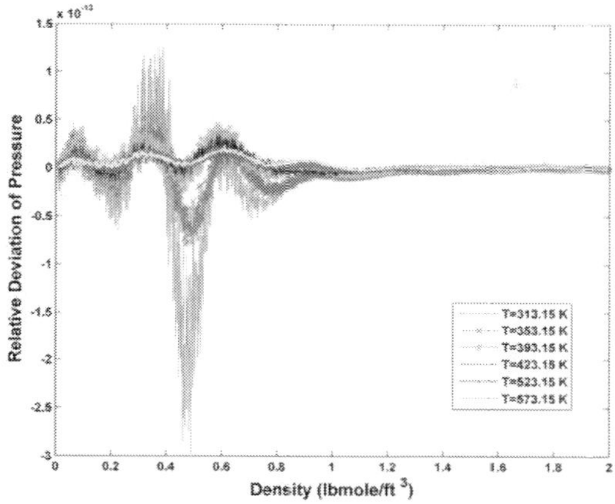

Fig. 2. Interpolation error using Chebyshev polynomials for approximating preesure vs. density of a binary mixture of ethanol and toluene containing 37.5 mole% of ethanol at different temperatures.

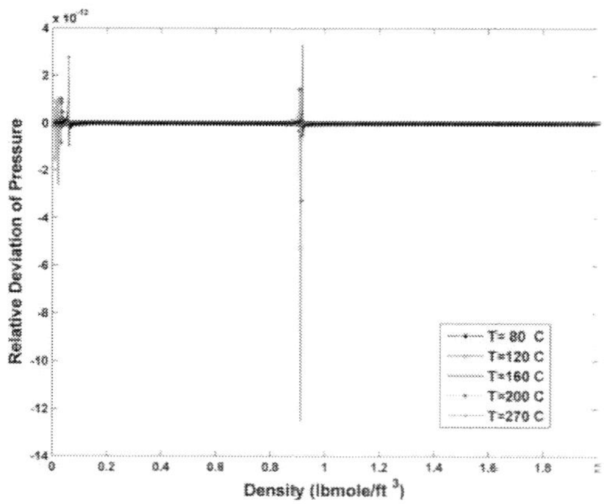

Fig. 3. Interpolation error using Chebyshev polynomials for approximating preesure vs. density at different temperatures for an oil sample of the composition given in Table 2.

After approximating the $P(\rho)$ function using Chebyshev polynomials, it is necessary to find solutions for density values at the given pressure(s) and select those which are physically interpretable. In doing so, the complex and negative solutions and those which make $\partial P/\partial \rho$

Component and Properties	Oil
N_2	0.49
CO_2	11.37
H_2S	3.22
C_1	27.36
C_2	9.41
C_3	6.70
iC_4	0.81
nC_4	3.17
iC_5	1.22
nC_5	1.98
C_6	2.49
C_{7+}	31.79
C_{7+} molecular weight	248.3
C_{7+} density (g/cm^3)	0.877

Table 2. Composition (mole%) and properties of the oil sample used to investigate the effect of temperature and pressure on asphaltene precipitation (Jamaluddin et al., 2000).

negative, are discarded because they have no physical meaning. Figure 4 shows a typical plot of pressure versus density for SAFT EoS in the positive region of density. As it can be seen in Figure 4, the derivative of pressure with respect to density ($\partial P/\partial \rho$) has two zeros in this region for different values of the shown temperatures. For pressures between the maximum and minimum of the $P(\rho)$ function (e.g. the pressure region between two parallel lines passing through the maximum and minimum of the middle curve), the system has three zeros one of which is not acceptable. The smaller root corresponds to the vapor phase density and the larger root corresponds to the liquid phase density. At pressures below the minimum of $P(\rho)$, the function has only a single root which is identified as the vapor phase density. At pressures above the maximum of $P(\rho)$, only a single zero is detected for the function which is identified as the liquid phase density. By increasing the temperature (Figure 5), the roots of $\partial P/\partial \rho$ approaches to each other. At some temperature they coincide above which $\partial P/\partial \rho$ has not any zero. At these temperatures the system has only a single root for any value of the pressure which is identified as the vapor phase density. Therefore, the procedure for finding roots of the SAFT EoS at the given pressure can be summarized as the following:

1. The pressure versus density of SAFT EoS is approximated using Chebyshev polynomials.
2. The derivative of pressure with respect to density is calculated to find zeros of $\partial P/\partial \rho$. The complex and negative zeros are eliminated.
3. The roots of the fitting polynomial are estimated at the given pressure using a proper root finding algorithm for polynomials. The negative and complex roots and those which make $\partial P/\partial \rho$ negative are eliminated.

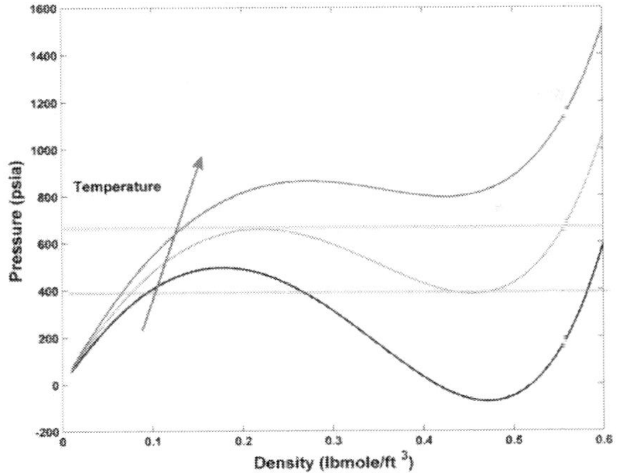

Fig. 4. A typical plot of pressure versus density of SAFT EoS at different temperatures.

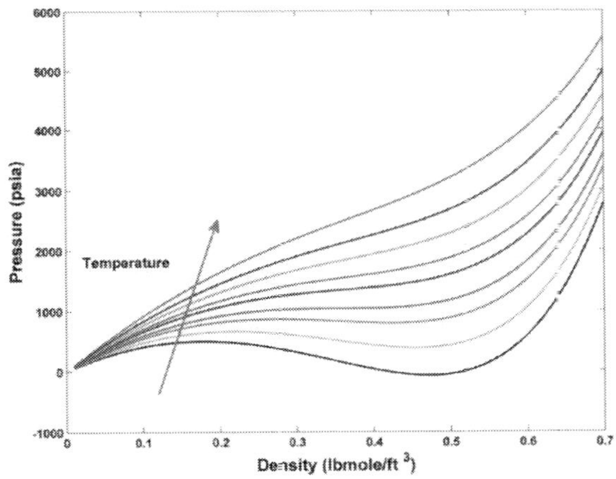

Fig. 5. A typical plot of pressure versus density of SAFT EoS at different temperatures.

4. If two physically meaning roots are obtained at the given pressure, the smaller root corresponds to vapor phase density and the larger one corresponds to the liquid phase density.
5. If the system has only a single root at the given pressure and $\partial P/\partial \rho$ has two zeros, if the obtained root is larger than the larger root of $\partial P/\partial \rho$, it is identified as the liquid phase density, otherwise, if the estimated root is smaller than the smaller root of $\partial P/\partial \rho$, it corresponds to the vapor phase density.
6. If the system has only a single root at the given pressure and $\partial P/\partial \rho$ has not any zero, the calculated root is identified as the vapor phase density.

4. Results and discussion

4.1 Density calculation for binary systems of ethanol and toluene

The SAFT EoS was first applied to calculate densities of the asymmetrical binary systems composed of ethanol and toluene. Experimental liquid densities for ethanol (1) and toluene (2) and seven of their binary mixtures in the temperature range 283.15-353.15 K at each 10 K and for pressures up to 45 MPa in steps of 5 MPa are given in (Zeberg-Mikkelsen et al., 2005). No density measurements were performed at 353.15 K and 0.1 MPa for ethanol as well as for mixtures containing more than 25 mole% ethanol, since ethanol and all mixtures with a composition higher than 25 mole% ethanol is either located in the two phase region or the gaseous phase (Zeberg-Mikkelsen et al., 2005). A comparison of the experimental density values of the aforementioned binary mixtures and pure compounds with the values calculated using SAFT EoS has been performed in this work. Figure 6 shows plots of the compressibility factor (Z-factor) of ethanol for different pressures of 0.1, 25 and 45 MPa using the SAFT EoS. As can be seen in this figure the contribution from the hard chain term ($Z_{hc} = Z_{hs} + Z_{ch}$), the dispersion term (Z_{disp}), and the association term (Z_{assoc}) are shown at different pressures versus density. Each point on a constant pressure curve corresponds to a certain temperature. Increasing the temperature, the liquid density decreases. A comparison between experimental and calculated densities using SAFT equation are presented in figures 7-10 versus pressure for different temperatures. The average absolute values of the relative deviations (*AAD*) found between experimental and calculated densities for different compositions of the binary mixtures of ethanol and toluene at different pressures and temperatures is 0.143%. Figure 11 represents relative deviations for different mixtures of ethanol and toluene on a 3D diagram.

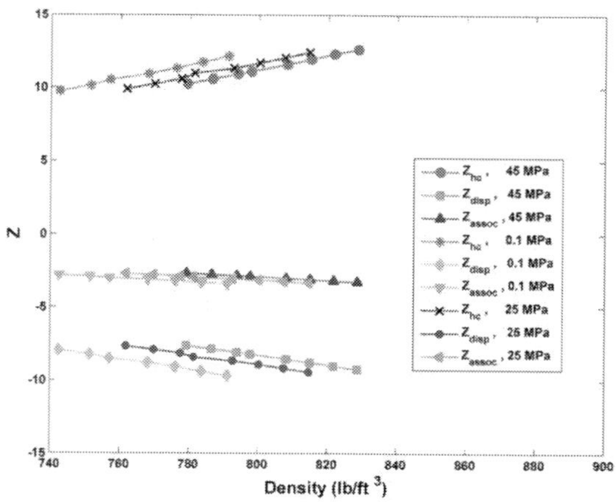

Fig. 6. Contributions to Z-factor of ethanol at different pressures and temperatures according to SAFT EoS.

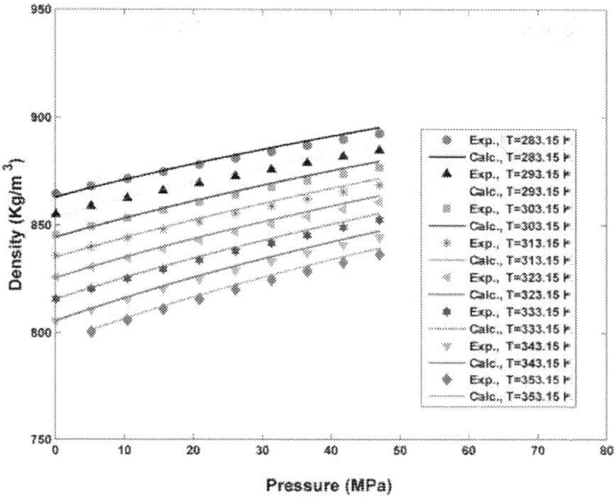

Fig. 7. Experimental and calculated densities versus pressure at different temperatures using SAFT EoS for binary system composed of ethanol and toluene at 25.0 mole% of ethanol.

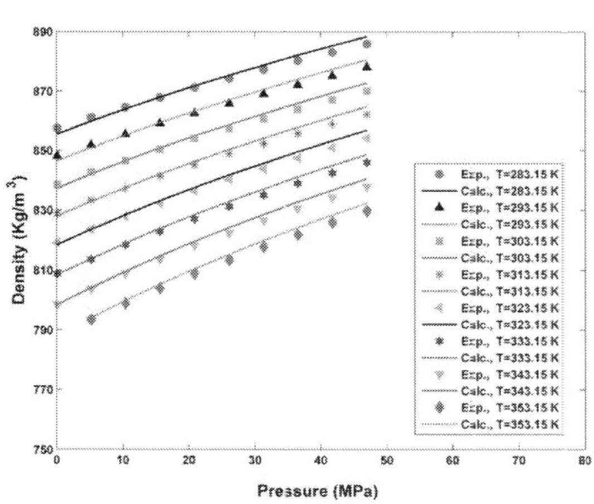

Fig. 8. Experimental and calculated densities versus pressure at different temperatures using SAFT EoS for binary system composed of ethanol and toluene at 37.5 mole% of ethanol.

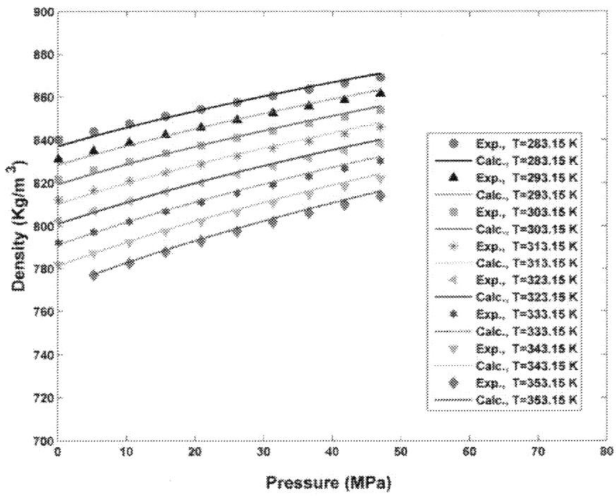

Fig. 9. Experimental and calculated densities versus pressure at different temperatures using SAFT EoS for binary system composed of ethanol and toluene at 62.5 mole% of ethanol.

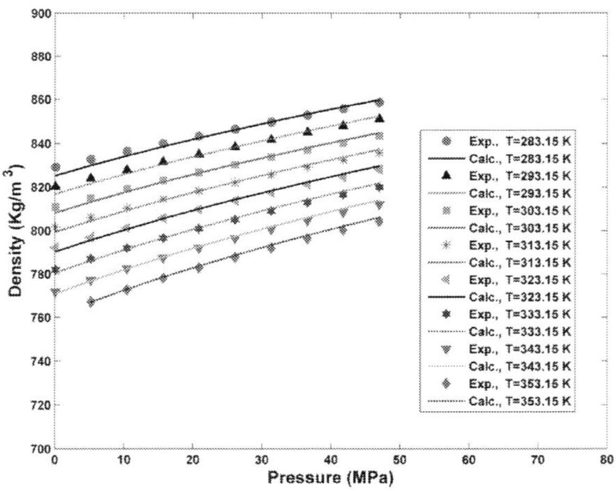

Fig. 10. Experimental and calculated densities versus pressure at different temperatures using SAFT EoS for binary system composed of ethanol and toluene at 75.0 mole% of ethanol.

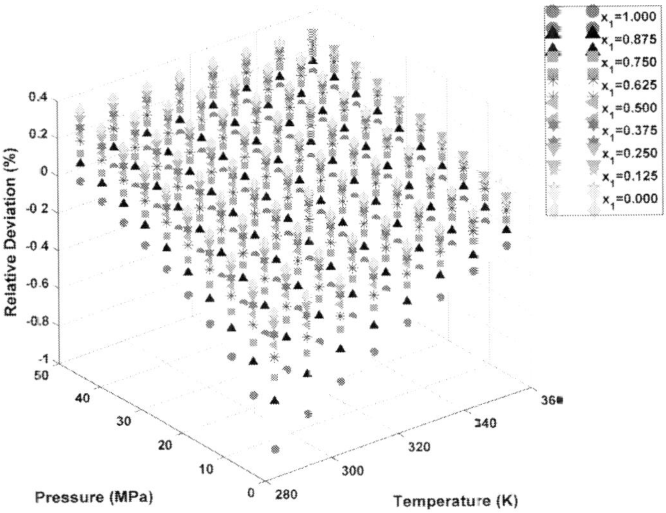

Fig. 11. Percent relative deviation of the calculated densities using SAFT EoS for different compositions of ethanol.

4.2 Phase equilibrium calculations for asphaltene-oil systems

Given a pressure, P, a temperature, T, and a mixture with global composition $z_i(i = 1, \dots, n_c)$, flash calculations determine the phase molar fraction and composition by solving equations (214) and (215),

$$f_i^V = f_i^L \qquad i = 1, \dots, N_c \tag{214}$$

$$f_{N_c}^L = f_{N_c}^S \tag{215}$$

f_i^V, f_i^L are the fugacity of component i in the vapor and liquid phase, respectively. $f_{N_c}^L$ and $f_{N_c}^S$ represent the fugacity of the asphaltene component in the liquid and solid phases respectively. The fugacity of the components in the vapor and liquid phases are calculated using the SAFT model as mentioned previously. In order to calculate the fugacity of the asphaltene component in the solid phase an experimental value of the asphaltene precipitation amount at a given temperature and pressure is required. Then, the thermodynamic equality of the fugacities at equilibrium conditions is used to calculate the reference value of the fugacity of the asphaltene component. The Quasi Newton Successive Substitution (QNSS) method, is used to solve the equilibrium equations. In conjunction with the above equations, the following material balance equations can be derived:

$$\sum_{i=1}^{N_c} \frac{(K_i^{VL} - 1) z_i}{1 + V(K_i^{VL} - 1) + S(K_i^{SL} - 1)} = 0 \qquad i = 1, \dots, N_c \tag{216}$$

$$\sum_{i=1}^{N_c} \frac{(K_i^{SL} - 1)z_i}{1 + V(K_i^{VL} - 1) + S(K_i^{SL} - 1)} = 0 \qquad i = 1, \dots, N_c \tag{217}$$

where, V and S are the mole fraction of the gas and solid phases, respectively. K_i^{VL} and K_i^{SL} are the equilibrium ratios of the vapor-liquid and liquid-solid phase equilibrium, respectively. In order to find the values of V and S using the above system of equations, we used the Newton-Raphson method. The complexity of multiphase flash calculations is due to the fact that the number of phases in equilibrium is not known a priori. The stability test for performing multiphase flash calculations has been performed using the stage-wise method developed by Michelsen (Michelsen, 1982a, 1982b). The parameters of the SAFT EoS including r, σ and ε have been given by Gross and Sadowski (2001) for N_2, CO_2 and hydrocarbons from C_1 to C_{20}. The values of these parameters for heavier and lumped component, also the volume and energy parameters used in association term have been estimated using a tuning approach. The thermodynamic model was applied to predict the precipitation behavior of petroleum fluids. The SAFT model has been used to describe the oil and gas phases. Table 3 presents the composition of an oil sample and a solvent from Burke et al. (1990). The oil sample was mixed with various amounts of solvent. Table 4 shows the precipitation and saturation pressure data for different concentrations of solvent in the oil sample for which the compositions are given in Table 3. The weight percent corresponds to the percentages with respect to the original mass of the oil. The last column in Table 4 shows the total amount of precipitates. To estimate the fugacity of pure asphaltene phase at a reference state, data at 0 mole% of solvent and $P^* =$ 3014.7 psia and T = 218 °F was used. The amount of precipitate, 0.14 weight%, was removed from the feed. The fugacity of the asphaltene component in the remaining mixture was then calculated using SAFT equation of state and equated to f_s^* (reference state fugacity of the asphaltene). The amounts of asphaltene precipitation and saturation pressures are also calculated using WinProp (CMG software) in which the fluid phases are described with a cubic equation of state and the fugacities of components in the solid phase are predicted using the solid model desribed in (Nghiem & Heidemann, 1982). In this work the Peng-Robinson equation of state has been used to describe the fluid phases non-ideality. Figures 12-13 represent a comparison between experimental and calculated asphaltene precipitation amounts and saturation pressures at different concentrations of solvent in the oil sample using SAFT EoS and WinProp software. As it can be seen in Figure 12 and Table 4, for solvent concentration above 78 mole%, the measured values show a substantial decrease in the precipitate. For this case, WinProp does not show a drop in the amount of precipitate. Howevere, it shows that the amount of precipitate levels off at high concentration of the solvent. Burke et al. (1990), and Chaback (1991) attributed the decrease in the precipitate at high solvent concentration to the switching of the mixture from bubble point fluid to a dew point fluid. WinProp does not show a decrease in the precipitate, while SAFT EoS based on the developed method for the calculation of densities and fugacities shows a very good agreement with measured data. The saturation pressures calculated using the proposed model at solvent concentrations above 78 mole % correspond to the upper dew point pressures while those calculated using WinProp are the buuble point pressues (Figure 13). The average relative deviations of saturation pressure and asphaltene precipitation amount using SAFT EoS are 4.6% and 3.8%, respectively. The values of AAD obtained using WinProp are higher than 17% for both parameters.

Component and Properties	Oil	Solvent
N_2	0.51	3.17
CO_2	1.42	17.76
C_1	6.04	30.33
C_2	7.00	26.92
C_3	6.86	13.09
iC_4	0.83	1.26
nC_4	3.35	4.66
iC_5	0.70	0.77
nC_5	3.46	1.26
C_6	3.16	0.78
C_{7+}	66.68	
C_{7+} molecular weight	281	
C_{7+} specific gravity	0.9020	
Oil molecular weight	202.4	
API gravity of stock tank oil	24.0	
Reservoir temperature, °F	218	
Saturation pressure, psia	600	

Table 3. Composition (mole%) and properties of oil ad solvent from Burke et al. (1990).

Solvent (mol%)	Mixture Saturation Pressure (psia)		Test Pressure (psia)	Precipitates from oil (wt%)		Total Precipitaes (wt%)
	Exp.	Calc.		Exp.	Calc.	
0	600	596.14	3014.7	0.14	0.1578	8.97
20	1050	1053.3	3014.7	0.27	0.2667	7.83
50	2310	2142	3014.7	1.46	1.4700	6.96
70	3750	3467	4214.7	1.65	1.6503	4.34
78	4510	4565	5014.7	3.21	3.4025	7.84
85	5000	5395	5014.7	1.29	1.2714	8.02
90	4250	4578	5014.7	1.10	1.0428	7.17

Table 4. Experimental and calculated values of the amount of asphaltene precipitation and saturation pressure for different mixtures of the oil sample and solvent given in Table 3.

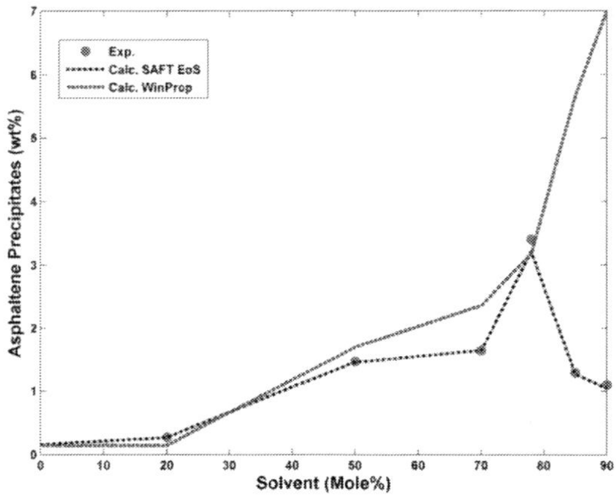

Fig. 12. Comparison between experimental and calculated values of the amount of asphaltene precipitation for different mixtures of the oil sample and solvent given in Table 3 using SAFT EoS and WinProp.

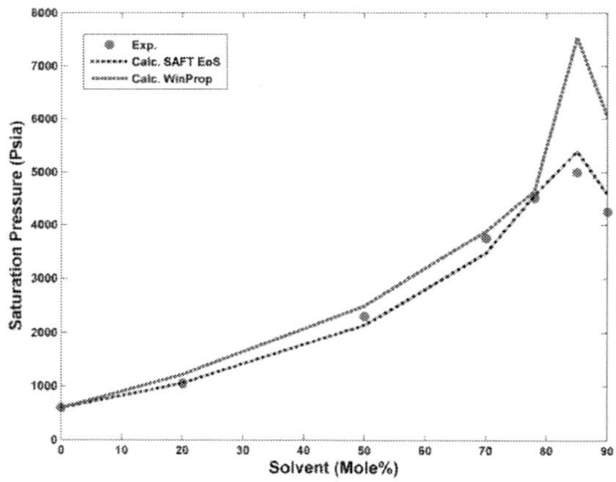

Fig. 13. Comparison between experimental and calculated values of the saturation pressure for different mixtures of the oil sample and solvent given Table 3 using SAFT EoS and WinProp.

The effect of temperature and pressure on the solid model prediction results has also been investigated. To do so, different hydrocarbon mixtures have been used. Figure 14 shows a comparison between the calculated and experimental values of the bubble point pressure and upper asphaltene onset pressure (AOP) for an oil sample (Jamaluddin et al., 2002). As it can be seen in this figure, excellent agreement is observed between experimental and predicted values of the upper AOP using SAFT EoS. The average values of the relative

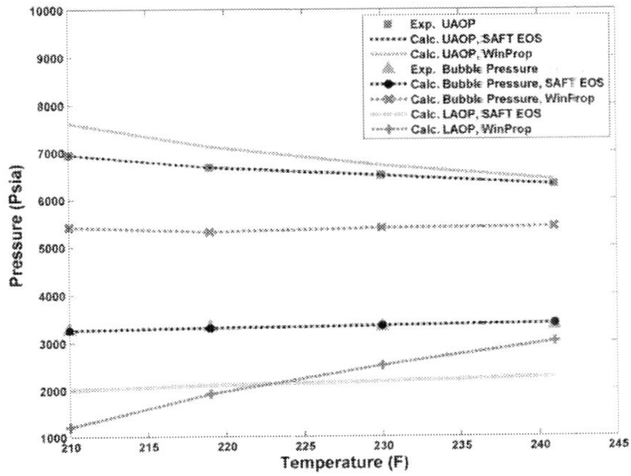

Fig. 14. Comparison between experimental and calculated values of the saturation pressure, upper and lower AOP's for the oil sample given in (Jamaluddin et al., 2002) using SAFT EoS and WinProp.

deviation in calculating the upper AOP using SAFT EoS and WinProp are 0.002% and 5.26%, respectively. Also, shown in Figure 14 are the calculated values of the lower AOP at different temperatures. The amounts of asphaltene precipitation vs. pressure at different temperatures is presented in Figure 15 for SAFT EoS. As it shown (Fig. 15) the maximum values of the asphaltene precipitation occures at the bubble point pressure of the mixture above which the amount of precipitaes decreases by increasing pressure up to upper AOP. In the pressure range below the bubble point pressure, decreasing pressure leads to a decrease in the amount of asphaltene precipitation and becomes infinitesimal at lower AOP. The effect of the injection gas on the asphaltene precipitation conditions has also been investigated for an oil sample given in (Rydahl et al., 1997) (Figures 16-17). Figure 17 shows that by increasing the amount of the injection gas added to the initial oil, the pressure interval of the asphaltene precipitation increases and shifts to the right side of the figure. Also, the amount of asphaltene precipitation increases by increasing the amount of the injection gas. A similar scenario happens when using WinProp.

The amount of asphalltene precipitation at different dilution ratios of normal heptane (nC_7) and stock tank conditions, are also calculated using both models for two Iranian oil samples, Sarvak oil A and Fahliyan oil (Bagheri et al., 2009). Figures 18-19 shows a comparison between experimental and calculated amount of asphaltene precipitation vs. dilution ration of nC_7 using SAFT EoS and WinProp. Again an excellent agreement has been observed using SAFT EoS with the experimental data. As it can be seen in these figues using WinProp, the amount of precipitaes increases rapidly at low dilution ratios after which the slope of the curve does not change considerably. The results show that SAFT EoS with average relative deviations of 2.32% and 1.73% for Sarvak oil A and Fahliyan oil, agrees well with the experimental data in comparison with the results obtained using WinProp (10.26% and 13.19%).

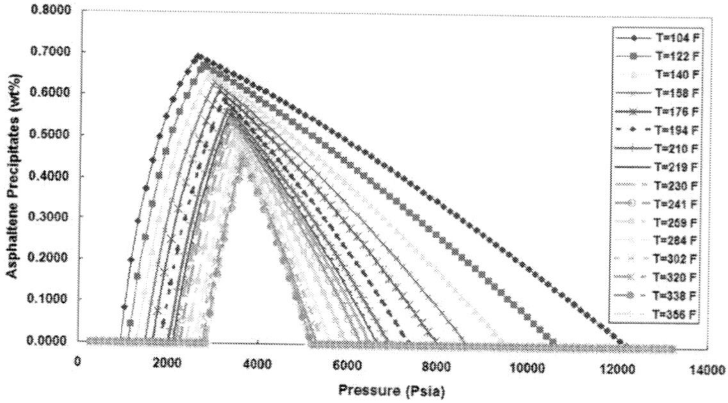

Fig. 15. The amount of asphaltene precipitation vs. pressure at different temperatures for the oil sample given in (Jamaluddin et al., 2002) using SAFT EoS.

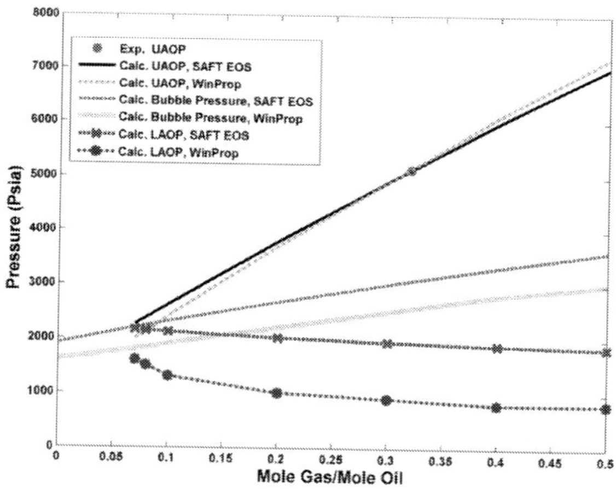

Fig. 16. Calculated values of the saturation pressure, upper and lower asphaltene AOP's vs. the amount of injection gas for the oil sample given in (Rydahl et al., 1997) using SAFT EoS and WinProp.

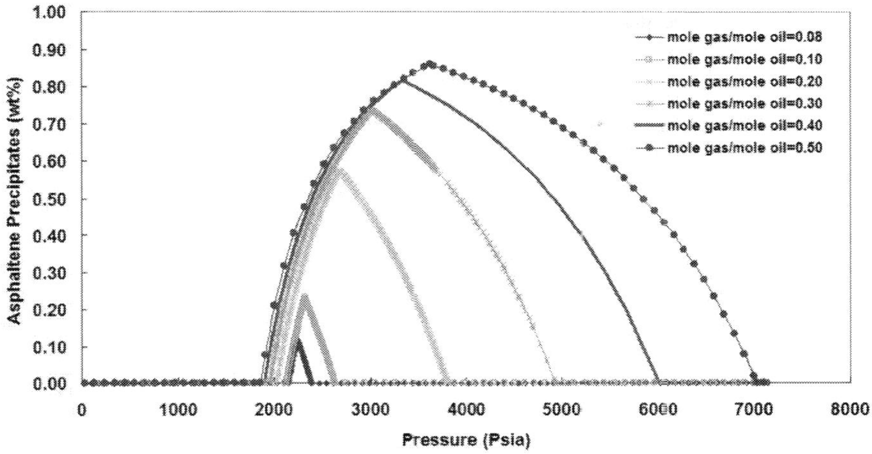

(a)

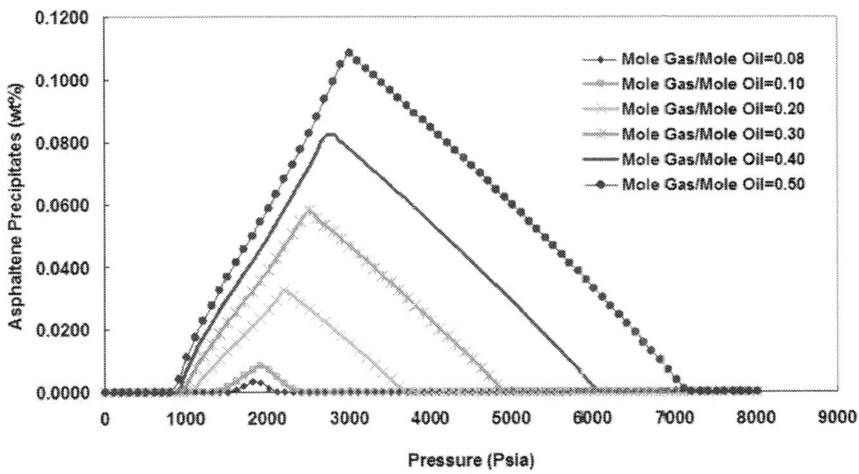

(b)

Fig. 17. The amount of asphaltene precipitation vs. pressure at different values of the injection gas for the oil and gas samples given in (Rydahl et al., 1997) using (a) SAFT EoS and (b) WinProp.

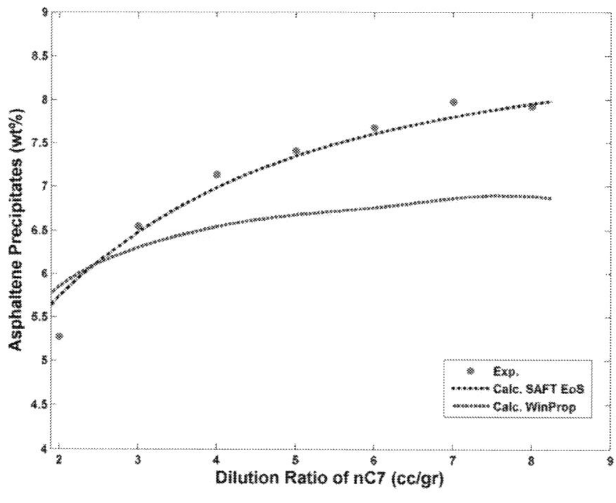

Fig. 18. Comparison between experimental and calculated values of the amount of asphaltene precipitation vs. dilution ration of nC_7 for Sarvak Oil A (Bagheri et al., 2009) using SAFT EoS and WinProp.

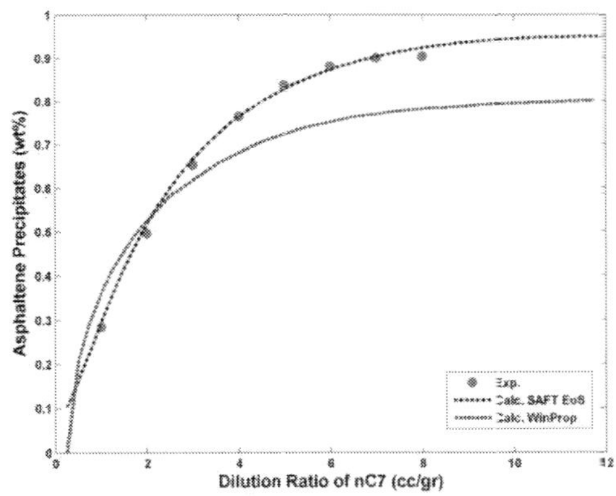

Fig. 19. Comparison between experimental and calculated values of the amount of asphaltene precipitation vs. dilution ration of nC_7 for Fahliyan Oil (Bagheri et al., 2009) using SAFT EoS and WinProp.

5. Conclusion

In this study a model based on statistical association fluid theory (SAFT) has been developed to predict phase behavior of hydrocarbon systems containing asphaltene and associating components. A robust, fast and accurate method based on Chebyshev polynomial approximation was proposed to find density using SAFT EoS which plays an important role in the calculation of the fugacity coefficients. The model was first evaluated using binary systems of ethanol and toluene. A good agreement between experimental and calculated liquid densities at different pressures, temperatures and compositions was obtained. The proposed model was then used to investigate the effect of solvent addition on the amount of asphaltene precipitate. The results showed a good agreement between experimental and calculated values of the amount of precipitate for different solvent–oil mixtures. In addition, the effect of temperature on the onset pressure of asphaltene precipitation and bubble point pressure was investigated. An excellent agreement was observed between experimental and predicted values of the asphaltene onset pressure at different temperatures.

6. References

Bagheri, MB; Kharrat, R. & Mirzabozorg, A. (2009). A Novel Method to Develop a New Scaling Equation for Modeling of Asphaltene Precipitation, *Proceedings of the 2009 SPE/EAGE Reservoir Characterization and Simulation*, Abu Dhabi, UAE, October 19-21, 2009.

Benedict, M.; Webb, G.B. & Rubin, L.C. (1942). An Empirical Equation for Thermodynamic Properties of Light Hydrocarbons and Their Mixtures II: Mixtures of Methane, Ethane, Propane, and n-Butane. *J. Chem. Physics*, Vol. 10, pp. 747.

Boyd, J.P. (2006). Computing Real Roots of a Polynomial in Chebyshev Series form Through Subdivision. *Appl. Numer. Math.*, Vol. 56, No. 8, pp. 1077–1091.

Buenrostro-Gonzales, E. & Lira-Galeana, C. (2004). Asphaltene Precipitation in Crude Oils, Ttheory and Experiments. *AIChE J.*, Vol. 50, pp. 2552-2570.

Burden, R.L.; Faures, J.D. & Reynolds, A.C. (1981). *Numerical analysis* (Second Edition), Prindle, Weber & Schmidt, Boston.

Burke, N.E.; Hobbs, R.E. & Kashou, S.F.R. (1990). Measurement and Modeling of Asphaltene Precipitation. *J. Petroleum Tech.*, Vol. 42, pp.1440.

Chaback JJ. (1991). Discussion on Measurement and Modeling of Asphaltene Precipitation. *Journal of Petroleum Technology*. Vol. 43, pp. 1519-1520.

Chapman, W.G.; Jackson, G. & Gubbins, K.E. (1988). Phase Equilibria for Associating Fluids Chain Molecules with Multiple Bonding Sites. *Molec. Phys.*, Vol. 65, pp. 1057.

Chapman, W.G.; Gubbins, K.E., Jackson, G. & Radosz, M. (1990). New Reference Equation of State for Associating Liquids. *Ind. Eng. Chem. Res.*, Vol. 29, pp. 1709-1721.

Chapman, W.G.; Sauer, Sh.G.; Ting, D. & Ghosh, A. (2004). Phase Behavior Applications of SAFT Based Equations of State from Associating Fluids to Polydisperse, Polar Copolymers. *Fluid Phase Equilibria*, Vol. 217, pp. 137–143.

Danesh, A. (1998). *PVT and Phase Behavior of Petroleum Reservoir Fluids*, Elsevier Science, ISBN 0444 821961, Netherlands.

Economou, I.G. (2002).A Successful Model for the Calculation of Thermodynamics and Phase Equilibria Propeties of Complex Fluid Mixtures. *Ind. Eng. Chem. Res.*, Vol. 41, pp. 953–962.

Firoozabadi. (1999). *Thermodynamics of Hydrocarbon Reservoirs*, McGraw-Hill, NewYork.

Gross, J. & Sadowski, G. (2001). Perturbed-Chain SAFT : An Equation of State Based on Perturbation Theory for Chain Molecules. *Ind. Eng. Chem. Res.*, Vol. 40, pp. 1244-1260.

Huang, S.H. & Radosz, M. (1990). Equation of State for Small, Large, Polydisperse and Associating molecules, Ind. Eng. Chem. *Res.*, Vol. 29, No. 11, pp. 2284–2294.

Huang, S.H. & Radosz, M. (1991). Equation of State for Small, Large, Polydisperse and Associating molecules, Ind. Eng. Chem : Extension to Fluid Mixtures. *Ind. Eng. Chem. Res.*, Vol. 30, pp. 1994–2005.

Jamaluddin, A.K.M.; Joshi, N.; Joseph, M.T.; D'Cruz, D.; Ross, B.; Greek, J.; Kabir. C.S. & Fadden, J.D. (2000). Laboratory Techniques to Defines the Asphaltene Deposition Envelop, *Proceedings of the Petroleum Society's Canadian International Petroleum Conference*, Calgary, Canada, Jane 4-8, 2000.

Jamaluddin, K.M.; Joshi, N.; Iwere, F. & Gurpinar, F. (2002). An Investigation of Asphaltene Instability under Nitrogen Injection, *Proceedings of SPE International Petroleum Conference and Exhibition*, SPE 74393, Villahermosa, Mexico, February 10-12, 2002.

Katz, D.L. & Firoozabadi, A. (1978). Predicting Phase Behavior of Condensate/Crude Oil Systems Using Methane Interaction Coefficients. *J. Pet. Tech.*, Vol. 27, pp. 1649.

Llovell, F.; Pamies, J.C. & Vega, L.F. (2004). Thermodynamic Properties of Lennard-Jones Chain Molecules: Renormalization Group Corrections to a Modified Statistical Associated Fluid Theory. *J. Chem. Phys.*, Vol. 121, No 21, pp. 10715–10724.

Llovell, F.; Vega, L.F.; Seiltgens, D.; Mejia, A. & Segura, H. (2008). An Accurate Direct Technique for Parametrizing Cubic Equations of State: Part III. Application of a Crossover Treatment. *Fluid Phase Equilibria*, Vol. 264, No. 1–2, pp. 201–210.

Michelsen, M.L. (1982a). The Isothermal Flash Problem, I. Stability. *Fluid Phase Equilibria*, Vol. 9, pp. 1-19.

Michelsen, M.L. (1982b). The Isothermal Flash Problem, II. Phase Split Calculation. *Fluid Phase Equilibria*, Vol. 9, pp. 21-40.

Muller, E.A. & Gubbins, K.E. (1995). An Equation of State for Water from a Simplified Intermolecular Potential. *Ind. End. Chem. Res.*, Vol. 34, pp. 3662–3673.

Nghiem, L.X., Heidemann, R.A. (1982). General Accelaration Procedure for Multiphase Flash Calculation with Application to Oil-Gas-Water Systems, *Proceedings of the 2nd European Symposium on Enhanced Oil Recovery*, Paris, France, November 8-10, 1982.

Pedersen, K.S. & Christensen, P.L. (2007). *Phase Behavior of Petroleum Reservoir Fluids*, CRC Press, Talor & Francis Group LLC.

Peng, D.Y. & Robinson, D.B. (1976). A New Two Constant Equation of State. *Ind. Eng. Chem. Fundam.*, Vol. 15, pp. 59-64.

Prausnitz, J.M.; Lichtenthaler, R.N. & de Azevedo, E.G. (1999). *Molecular Thermodynamics of Fluid Phase Equilibria* (Third Edition), Prentice Hall PTR, pp. 390.

Prausnitz, J.M. & Tavares, F.W. (2004). Thermodynamics of Fluid-Phase Equilibria for Standard Chemical Engineering Operations. *AIChE J*, Vol. 50, No.4, pp. 739–761.

Redlich, O. & Kwong, J.N.S. (1949). On the Thermodynamics of Solutions. V. An Equation of Sate. Fugacities of Gaseous Solutions. *Chem. Rev.*, Vol. 44, pp. 233–244.

Rydahl, A.K.; Pedersen, K.S. & Hjermstad, H.P. (1997). Modeling of Live Oil Asphaltene, *Precipitation Proceedings of AIChE Spring Meeting*, Houston, March 9-13, 1997.

Segura, H.; Seiltgens, D.; Mejia, A.; Llovell, F. & Vega, L.F. (2008). An Accurate Direct Technique for Parameterizing Cubic Equations of State. Part II. Specializing Models for Predicting Vapor Pressures and Phase Densities. *Fluid Phase Equilibria*, Vol. 265, pp. 155–172.

Soave, G. (1972). Equilibrium Constants from a Modified Redlich-Kwong Equation of State. *Chem. Eng. Sci.*, Vol. 27, pp. 1197-1203.

Song, Y.; Lambert, S.M. & Prausnitz, J.M. (1994). Equation of State for Mixtures of Hard Sphere Chains Including Copolymers. *Macromolecules*, Vol. 27, pp. 441.

Starling, K.E. (1973). *Fluid Thermodynamic Properties for Light Petroleum Systems*, Gulf Publishing Company, Houston.

Tabatabaei-Nejad, S.A.R. & Khodapanah, E. (2009). An Investigation on the Sensitivity Analysis of the Parameters of Proposed Wax Precipitation Model. *Journal of Petroleum Science & Engineering*, Vol. 68, pp. 89–98.

Ting, D.L., (2003). Thermodynamic Stability and Phase Behavior of Asphaltenes in Oil and of Other Highly Asymmetric Mixtures, PhD Thesis, Rice University, Houston, USA.

Van der Waals, J.D. (1873). Over de Continuiteit van der Gas en Vloeistoftoestand Leiden, Doctoral Dissertation, The Netherlands.

Verdier, S.; Carrier, H.; Andersen, S.I. & Daridon, J.L. (2006). Study of Pressure and Temperature Effects on Asphaltene Stability in Presence of CO_2. *Energy and Fuels*, Vol. 20, pp. 1584-1590.

Wertheim, M.S. (1984). Fluids with Highly Directional Attractive Forces. I. Statistical Thermodynamics. *J. Stat. Phys.*, Vol. 35, No.1, pp. 19–47.

Wertheim, M.S. (1986). Fluids with Highly Directional Attractive Forces. III. Multiple Attraction Sites. *J. Stat. Phys.*, Vol. 42, No.3, pp. 459–476.

Wertheim, M.S. (1987). Thermodynamic Perturbation Theory of Polymerization. *J. Chem. Phys.*, Vol. 87 , pp. 7323.

Zeberg-Mikkelsen, C.K.; Lugo, L.; Garcia, J. & Fernandez, J. (2005). Volumetric Properties under Pressure for the Binary System Ethanol + Toluene, *Fluid Phase Equilibria*, Vol. 235, pp. 139–151.

2

Based on Common Inverted Microscope to Measure UV-VIS Spectra of Single Oil-Gas Inclusions and Colour Analysis

Ailing Yang
Ocean University of China
China

1. Introduction

Hydrocarbon fluid inclusions occur in a variety of geological environments, most commonly within carbonate rocks from petroliferous sedimentary (Stasiuk & Snowdon,1997). Oil-gas inclusions (OGIs) belong to hydrocarbon inclusions. The sizes of OGIs are usually in the range of 5-20μm. A micro-mass (ng-fg) palaeo-oils were trapped in a single OGI. The interaction between OGIs and outside is relative weak although a long geological age elapsing. The OGI can be seen as a closed system. In this way, an intact (not cracked) OGI can be seen as a micro-oil-gas reservoir. The OGIs take rich information about the palaeo-oils. Generally, in the evolution of the oil-gas reservoir, with the deposition depth increasing, the stratum temperature increases. The organic macromolecules were decomposed into micromolecules. The oil maturity evolutes from low to high. Similarly, in the different digenetic stages, the types, colours and composites of the OGIs are also different. With the maturity of organism from low to high, the types of OGIs are mainly from liquid phase, liquid and gas phase to gas phase. The colours change from colourless, yellow, and brown to black (Liu, Y. R. *et al.*, 2003, as cited in Burruss R.C.,1991). Micro-fluorescence properties of OGIs, largely controlled by the aromatic characteristics of the hydrocarbons, are a signature of the organic chemical composition. This phenomenon was often used to distinguish OGIs from saltwater inclusions. By different fluorescence colours of OGIs, the oil and gas charge history and oil-gas maturity are qualitatively determined. And by the abundance of grains containing OGIs (GOI)(Liu, K. Y. & Eadington, 2005, as cited in Eadington, 1996), the petroleum characteristics of the corresponding strata can be qualitatively estimated. The micro-spectroscopy is very important to OGIs and mainly includes VIS and UV-VIS spectra.

1.1 The VIS spectra technique

The VIS spectra technique includes fluorescence micro photometry (FMP), fluorescence alteration of multiple macerals (FAMM) and laser scanning confocal microscope (LSCM).

1.1.1 Fluorescence micro photometry (FMP)

FMP is a technique combined microscope and photometer (see Fig.1). Generally, the exciting source is UV light (365 nm, called internal light source in this chapter) from a mercury arc

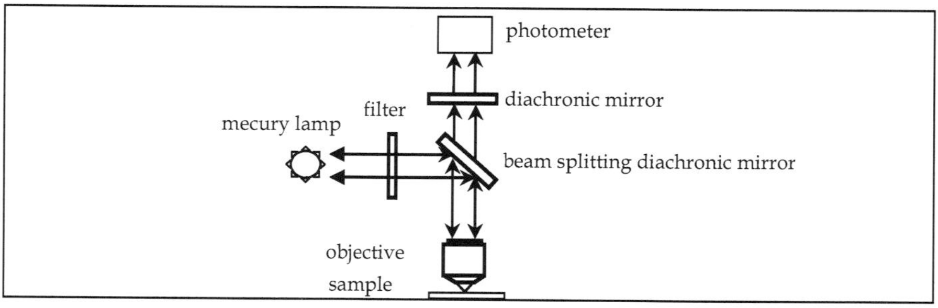

Fig. 1. A schematic diagram of a FMP. The upright microscope includes in a beam splitting diachronic mirror (with higher reflectivity to the incident light and also a higher transmittance to the fluorescence) and a diachronic mirror (absorbing the scattered light from the sample, with a higher transmission to the fluorescence).

lamp in a microscope. The UV light is focused by an objective and incidents onto a sample. The sample absorbs the exciting photons then emits fluorescence. The wavelengths of the fluorescence are longer than that of the incident light.

Some fluorescence backs (upward for the upright microscope and downward for the inverted microscope) and is collected by the same microscope objective. That light enters into the aperture of the photometer and the intensity is measured. For the inverted microscope, except for the downward fluorescence, one can measure the upward (transmitted) fluorescence because there is a big space above the sample stage to connect a photometer with the body of the microscope.

According to different study contents, one can choose different filter, beam splitting diachronic mirror, diachronic mirror and detector to realize fluorescence, transmission, reflective and absorption measurement. For example, in Fig. 1, if the photometer is an energy meter, one can measure the intensity of the fluorescence. And if the diachronic mirror has a higher absorption to the fluorescence and a higher transmission to the scattered light from the sample, the reflectivity of the sample can be measured. MPV3-microphotometer is an example. It was widely used to measure the vitrinite reflectivity of coal and source rock (Xiao *et al.*, 1998, 2000).

The superiors of the above FMP are simple and low cost. But there are still some shortcomings. The first is one can only gain visible spectra (400-780 nm) for the exciting light is at 365nm. But some aromatic hydrocarbons emit UV fluorescence which can not be measured by above method. The second is the pin hole (about 0.15 mm) to limit the incident light spot often larger than the under studied OGI. Not only the under studied OGI emits fluorescence, but also other OGIs, the cements around the mineral grain trapping the OGI and even the mineral grain itself. The other OGIs may not be the same generation as the under studied OGI. What' more, the intensities of the cements are usually stronger than that of the under studied OGI. That means the fluorescence of the under studied OGI often mixes with other fluorescence. This makes the OGI looks like involving into a colourful cloud. So one can see if the size of the incident light spot is larger than that of the under studied OGI, the measured fluorescence spectrum may include the fluorescence of other OGIs, the cements around the mineral grain trapping the OGI and even the mineral grain itself. This will lead to a decreasing reliability. So one should be cautious to judge the oil and gas charge history and oil-gas maturity by the fluorescence colours of the OGIs.

In 1997, Stasiuk *et al.* (Stasiuk & Snowdon,1997) used Zeiss MPV II to study the fluorescence of OGIs. The wavelength of the exciting light is 365 nm. A 2.5-5 μm light spot was obtained by a pin hole. The size of the light spot is less than that of the OGIs. The authors measured the micro-fluorescence spectra of artificially synthesized and natural hydrocarbon inclusions. The fluorescence spectra were related with the chemical constitutes, API degree and other geochemistry parameters of crude oils. The migration of oil and gas was revealed. The crude oils used to synthesize the artificial hydrocarbon inclusions composite a wide range of saturated hydrocarbons, aromatic hydrocarbons and NSO compounds. And also span a long geologic time (Ordovician, Devonian, Carboniferous, Jurassic and Cretaceous). For the artificial hydrocarbon inclusions, the spectra are the real spectra for single hydrocarbon inclusions. And for the natural inclusions, there are some organic compounds in the mineral grains difficult to be observed even under the microscope, so strictly speaking, the fluorescence spectra of natural inclusions are the addition of the inclusions and the mineral matrix. The results showed that most artificially synthesized hydrocarbon inclusions have main colours. A small amount have other colours. The authors gave a detailed analysis between the peak position L_{max}, Q index (Q=Intensity @ 650 nm / Intensity @ 500 nm) of the fluorescence spectra and saturated hydrocarbons, aromatic hydrocarbons, NSO compounds, nC_{18}/Ph 、nC_{18}/Pr , API degree of the crude oils. The results showed the correlative factors between the peak position L_{max} and the saturated hydrocarbons, Q index and saturated hydrocarbons are all up to 0.84. Q index can be used to indicate the maturities of the crude oils.

Caja *et al.* (Caja, 2007, 2009) measured the fluorescence spectra of hydrocarbon inclusions. The results showed that Q_{580} index (580-700nm vs 400-580nm areas) is very sensitive to the total composites of the petroleum. The higher the Q_{580} index, the heavier the crude oil and the tighter relation with the API degree. For the hydrocarbon inclusions emitting yellow fluorescence, the API degree is in the range of 15-22. And for the hydrocarbon inclusions radiating blue fluorescence, the API degree is in the range of 23-45.

1.1.2 Fluorescence alteration of multiple macerals (FAMM)

In 1990s, CSITRO of Australia established FAMM. This technique combined laser with microscope. For laser has good monochromaticity and high intensity, a micrometer light spot can be easily obtained, which enhances the space resolution of the setup. By means of micro-Raman setup, assuming Ar+ laser as exciting light source (488 nm), a focal spot as less as several micrometer was obtained. The emitting intensities of source rocks at 625 nm were detected with time changing. Wilkins (Wilkins *et al.*, 1992, 1995) probed the relation between vitrinite inhibiting and the fluorescence intensity changing with time. The result showed that FAMM can effectively correct inhibiting effect of vitrinite reflectivity before the stage of high maturity. Lo (Lo, *et al.*, 1997) and Veld (Veld *et al.*,1997) used this technique to measure the maturity of source rock. With the help of FAMM, Xiao (Xiao *et al.*, 2002) used a laser induced fluorescence microscopy (LIFM) to study the maturities of carbonate rocks with higher maturities.

FAMM and LIFM techniques are not very fitful to measure the fluorescence spectra of single OGIs. After assuming laser, one can measure the fluorescence spectra, but can not directively observe the sample by eyes because the laser safety threshold for human eyes is less than 5 μW (Sliney D. H. ,1995). For most matters, the absorption bands are in the range of UV. But the wavelengths of Ar^+ ion laser are 488 nm and 515 nm, they don't mach well with the absorption bands of materials. So it is not very well to use Ar^+ ion laser to excite the fluorescence of single OGI. The focused laser has a higher radiation flux density at the sample surface and easily damages the sample. The changing fluorescence intensities at 625 nm with time may be relevant to the Photolysis (Sanches S., 2011) and photo polymerization of vitrinite in the coal and source rock.

1.1.3 Laser scanning confocal microscope (LSCM)

LSCM was realized in 1980s. Fig. 2 is a schematic diagram of a LSCM. The illuminating pinhole and the detector pinhole are conjugate to the focal plane of the objective. A point on the focal plane of the objective is focused at the illuminating and detecting pinholes at the same time. The other points out of the focal plane of the objective have not images at the detecting pinhole. This means confocal. A point light source illuminating and a point image are realized by an illuminating pinhole before the laser and a detecting pinhole before the detector. The light passes through the detecting pinhole will be received by a PMT or a cCCD and quickly imaged in a computer. The images have higher space resolution and good quality. By controlling the movement of the objective, LSCM can realize continues optical section by tomography similar to CT. After computer 3D imaging, a 3D profile of a micro sample can be recovered.

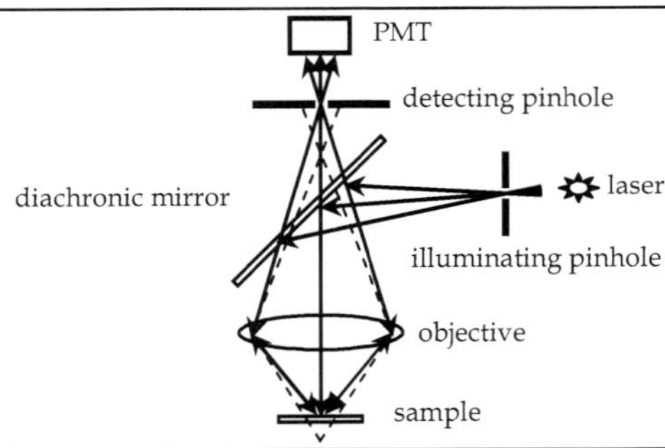

Fig. 2. The schematic diagram of a LSCM.

LSCM has been widely used in cytobiology, cytophysiology, neurophysiology and other modern medicine and biology relative to cells (Damaskinos *et al.*, 1995; Kevin, 2003). It has been a forceful tool in the field of bio-science. Recently, LSCM was used to obtain precise volume ratios between liquid and gas phases in OGIs. Combining homogenous temperature, GC-MS analysis and PVTsim software, the pressure of palaeo-liquid can be obtained (Aplin *et al.*, 1999; Thiéry *et al.* 2002, Liu, D. H. *et al.*, 2003). Further more, the depths of the strata existing palaeo-liquids can also be obtained by the pressure. This is meaningful to study the migration of oil, gas and reservoir formation.

LSCM has special superiors as resolving the fluorescent macerals with micrometer sizes (source rock and oil shale). Stasiuk (Stasiuk, 1999) used LSCM to observe algae in oil shale.

LSCM can effectively resolve the cellular tissue in the algae. The author observed that algae have stable outermost cell wall predominance. The result assured kerogen has a selected preservation during the formation. LSCM was also used to obtain images about the space distribution of organism in mineral slides. Stasiuk (Stasiuk, 2001) measured fluorescence spectra of lipoids and chlorophyll originated from diatom. The mineral slides in the Saanich Inlet region are in rich spore oil and reproductive spore, which is relative to the periodic spore blooming in the spring. The results under the LSCM may be the preservation of spore oil at suitable depth.

Although LSCM has a higher space resolution and good image quality, till now the shortest laser wavelength of semiconductor laser is 370nm, the fluorescence of single OGI is still in the range of VIS.

1.2 The UV-VIS spectra technique for single OGIs

As above mentioned, some organic composites in OGIs emit fluorescence at UV range. So it is necessary to develop UV-VIS spectra technique.

For overcoming above drawbacks, Kihle (Kihle, 1995) established an UV-VIS micro-spectroscopy for measuring excitation-emission spectra of single OGIs. The setup includes an upright tri-ocular deep ultraviolet (DUV) microscope, a condensed lens, a collimated objective, two fiber adaptors, two fiber cables, and an UV-VIS spectrometer. The DUV microscope is very expensive comparing with the common fluorescence microscope. But for measuring the UV-VIS spectra of single OGIs, it had to be used. The spectrometer was connected with the microscope by the two fiber cables, the condensed lens, the collimated objective and the two fiber adaptors. The spectrometer supplies the exciting light in the range of 220-900 nm and record the UV-VIS spectra.

The setup could successfully measure the UV-VIS spectra of single OGIs. But there are four points to be noted. The first is for obtaining a small focal point, one has to replace the original focusing mirror of the microscope itself with a condensed mirror which was settled down on a 3D adjustable adaptor. The second is the C-mount was connected with the fiber cable not a videograph head as origin. So one can only observe the sample by ocular and can't take photos of the samples or get the sizes of the single OGIs and focusing point precisely. The third is the expensive DUV microscope, its price is 9-20 times of a common fluorescence microscope. This will greatly limit it using. The fourth is that for effective collecting fluorescence an oil immersed objective had to be used, which is trouble to clean.

The above experimental system is so expensive and one has to rebuild the microscope. Is it possible to replace the DUV microscope with a common microscope to lower cost and still detect UV-VIS fluorescence spectra of single OGIs? Such kind of micro-spectroscopy system to meet this aim was established (Yang, 2009a, 2009b, 2011) This will be given in the third section of this chapter.

2. The geology background of the samples

The sand rocks came from the core drilling samples of five oil wells (Bai 95#, Hua5#, Jian 22#, Nong 29# and Fu 4#), Jinlin Oil Filed, Songliao Basin, northeast China. Bai 95# oil well localizes in the west slope of the basin, the other are in the south to the basin. The earth strata belong to middle to shallow, upper Cretaceous Period. The thickness of inclusion thin slide is about 150 nm.

Table 1 shows the information about the strata, depths, GOIs, average homogeneous temperatures ($\overline{T}_h s$) and salinities of the sand rocks.

There are two episodes OGIs in these samples. The first episode is in the earlier stage of quartz overgrowth, the second in the later stage of the quartz overgrowth.

Oil well	Stratum	Depth	The first episode			The second episode		
			GOI (%)	$\overline{T}_h$ (°C)	Salinity (%)	GOI (%)	$\overline{T}_h$ (°C)	Salinity (%)
Bai 95#	K_2y_{2+3}	419.3-422.1	2-3±	75.3	9.93	5-6±	113.3	4.03
	K_2y_1	426.8-436.5	3±	74.6	10.31	2±	95.4	2.98
	K_2qn_2	494.6	4±	72.3	10.95	2±	98.7	5.09
Hua 5#	K_2n_1	1449.8 1524.0	2±	\	\	0.5±	110.3	3.79
Jian 22#	K_2q_4	377.8 420.0	<1	81.0	6.77	20±	104.2	4.89
Nong 29#	K_2q_4	571.0-595.9	2±	77.0	11.34	20±	116.8	7.86
Fu 4 #	K_2q_3	419.1-426.6	2-3±	75.0	7.17	70±	135.2	4.35

Table 1. Information about the strata, depths, the measurement results of GOIs, $\overline{T}_h s$ and salinities of the sand rocks.

The GOIs, $\overline{T}_h$ s and salinities are different in the two stages. In the first episode and for all strata, except for Jian 22# oil well, the GOIs belong to middle value (Liu, K. Y. & Eadington, P., 2005, as cited in Eadington, P, 1996). In the second episode, the GOIs of Bai 95# K_2y_1, K_2qn_2 and Hua 5# K_2n_1 decrease. But for Bai 95# K_2y_{2+3} stratum, the GOI doubles. The GOIs in the second episode increase from 10 to 23 times of Jian 22#, Nong 29# and Fu 4 # oil wells, which shows there are lots of oils charging these strata in the second episode.

In the first episode, $\overline{T}_h$ s are in the range of 72.3-81.0 °C. And in the second episode, $\overline{T}_h s$ increase. The increasing amplitude is in the range of 23- 60 °C

The salinities of all strata decrease in the second episode. The $\overline{T}_h$ s and salinities show that the deposition environment has a great change in the second episode. The second episode may be in the water expanding time.

Under the fluorescence microscope, the OGIs of the first episode are liquid phase and very small. They are dense and often distribute along a line or a stripe. Their colors are often deep brown or grey brown. The fluorescence intensities are very weak. It is very difficult to measure the fluorescence of such kind of oil inclusions. The second episode OGIs are liquid or liquid-gas phases. In a mineral grain, they distribute in scattering or in group. The liquids in the OGIs are light yellow, light brown yellow, brown and grey black. Under excited at

UV, blue or green light, they often emit fluorescence. Fig.3 shows the typical OGIs in the Bai 95# oil well. Fig.3 a is a micro-photo of the first episode oil inclusions, and Fig. 3b is the second episode OGIs. Fig.3 c is the fluorescence image of Fig.3 b as excited at 365 nm (internal light source). The OGI emits light yellow fluorescence.

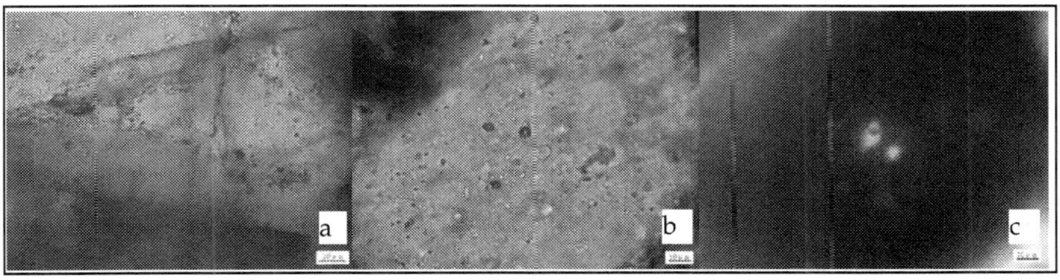

Fig. 3. The two episodes OGIs in Bai 95# oil well, in which, a is the first episode, the depth is 426.8 m; b the second episode, the depth 420.1 m; and c the fluorescence image of b under excited at 365 nm internal light source in the microscope. The scale in a is 20 μm, and in b and c 10 μm.

3. Based on common inverted microscope to measure UV-VIS spectra of single OGIs and colour analysis

3.1 The experimental setup

Fig.4 shows the schematic diagram of the experimental setup (Yang, 2009a, 2009b), which includes seven parts: (1) an inverted fluorescence microscope (IMF); (2) a reflecting microscope objective (RMO); (3) a 3D adaptor; (4) a micro lens; (5) a fiber cable; (6) a spectrometer and (7) a computer.

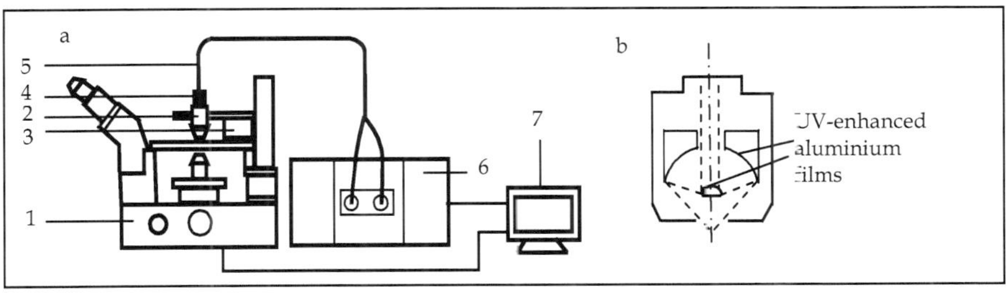

Fig. 4. Schematic diagram of the micro-spectroscopy system (a) based on common inverted microscope and the reflecting microscope objective (b). In a, 1—IMF; 2—RMO; 3—3D adaptor; 4—micro-lens; 5 —fiber cable; 6—UV-VIS spectrometer and 7—computer.

The IFM includes an internal light source (mercury arc lamp), three (violet, blue and green) standard fluorescence "cubes" bandpass filters, a dichronic mirrors and beam splitters in the optical light path. The mercury arc lamp can be used as an internal excitation source. Using bandpass filters and fluorescence "cubes", a beam with tens of nm band width can be obtained. There are five differences between this system and Kihle's setup (Kihle, 1995). (1)

A common and cheap IFM was used but not an expensive upright DUV microscope. The 3D adjustable adaptor can conveniently connected to the main body of the IFM and not need any rebuilding to the microscope. (2) It is easy to adjust the RMO and the IMF coaxial. When the switch of the IMF is on and kept the least intensity, one can adjust the 3D adjustable adaptor while observe the reflected light intensity from the RMO. When the beam intensity is maximal, the two are coaxial. (3) The RMO has many advantages over refracting objective. It is an all-reflecting construction and free from chromatic aberration. The primary spherical aberration, primary coma and primary astigmatism have been corrected. The specific mirror coating is UV-enhanced aluminium film, which has as high as 89% average reflectivity in the range of 190 nm-10 μm and is highly recommended for most UV use. Comparison with the refracting objective, the RMO has stronger focusing ability. (4) The RMO has a relative big numerical aperture (0.65), so it can be used as an excellent focusing element and also a good component to effectively collect fluorescence; (5) One can use the videohead to take photos for the samples and the focusing point in time.

In the setup, the spectrometer is FluroMax-4 (Horiba Jobin Yvon) with a 150 W xeon lamp and a single photon PMT. The RMO is made of Ealing (52X). The micro-lens and the fiber cable are from Avertise. The diameter of the fiber is 400 μm and has a 90% transmission in the range of 240-800 nm. There are seven fibers in the fiber cable. The fibres are arranged like a club. The middle one is as exciting fiber to guide the exciting light from the xenon lamp in the spectrometer to the micro-lens. The peripheral six are as emitting fibers to guide the fluorescence into the spectrometer.

The system integrates the functions of micro-area location, DUV light excitation, weak fluorescence detection and real-time taking photos together. A computer program to calculate the chromaticity coordinates of the OGIs by the spectra excited at 365 nm was also established.

3.2 The main measuring steps

For observing, taking photos and measuring the fluorescence spectra of single OGIs, there are several key steps to be noted in the experiment. To decreasing influence from environment, the experiment was done in the dark. The main steps are as following: (1) Find and localize an OGI in the middle of the visual field. Take photos when white light illuminates the OGI. (2)Take photos as exciting by UV, blue and green light of the internal light source. Be careful not make the CCD satiate. (3) Adjust the 3D adaptor and make the RMO and the objective of the IMF coaxial. When the two are coaxial, one can observe a brightest and roundest spot over the RMO. Note the power of the green light from the mercury arc lamp least and ware a goggle glasses to protect eyes. (4) Obtain a least focal spot. Connect the micro-lens with the 3D adaptor, and also the fibre cable with the micro-lens. Switch on the spectrometer, let the exciting fibre guide the 550 nm green light out of the spectrometer and get to the micro-lens and then to the RMO. The RMO focuses the exciting light into a small spot and incident on the inclusion thin slide. If the adjustment is good enough, one can see a green focal spot on the slide. According to the position of the focal spot, one should repeatedly adjust x, y, and z axes of the adjustable adaptor and decrease the size of the focal spot. For obtaining a least focal spot, one should also adjust the relative distance between the micro-lens and the RMO. Once a least focal spot appearance, this distance should be fixed and not change it in the whole experiment. (5) Take photos of single OGI excited by external light source (Xeon lamp) and record the fluorescence spectra.

The exciting and emitting slits are 10 nm and 8 nm respectively. A proper filter should be placed before the emitting window of the spectrometer. (6) Measure the micro fluorescence spectra of OGI excited by the internal light source (UV, blue and green). Note that the exciting slit of the spectrometer should be off. The emitting slit is 5 nm. (7) Measue the background fluorescence spectra when excited by external light source. The exciting wavelengths and the two slits widths are as same as step (5).

3.3 The lest focal spot of the RMO

Fig. 5 is the least focal spot of the RMO. The background is a quartz grain in an inclusion thin slide. The size of focal spot is about 12 μm. The halo light around the focal spot is the result of the diffraction. It is difficult to be eliminated for a RMO. The focal spot includes most energy.

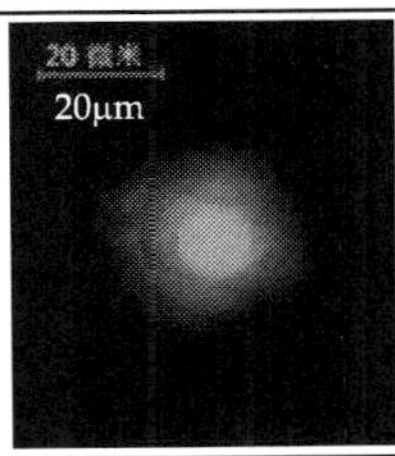

Fig. 5. The least light spot

3.4 The background fluorescence and the subtract factor

The focal spot of the RMO is not small enough. When the size of the focal spot is larger than that of the OGI, not only the OGI emits fluorescence, but also the background. The background fluorescence is complex. It may origins as following: (1) The mineral grain may include in a few even lots of micro oil inclusions (Stasiuk & Snowdon, 1997; Meng, 2009), but they can't be resolved under the IMF. (2) The surface of the mineral grain was contaminated by oils existing in the sand rock as preparing the inclusion thin slide. (3) The mineral grain may contain some thulium, which will emit fluorescence when they are excited by light. (4) The inclusion thin slide may be contaminated by organism from air or by people's hands.

Fig.6 shows the transmission micro-images of a single OGI (a, c, and e) and background (quartz grain)(b, d and f). In Fig.6 a and Fig.6 b, the exciting wavelengths are 250 nm and a violet fluorescence filter cube was used. In Fig.6 c and Fig.6 d, the exciting wavelength is 470 nm and a blue fluorescence filter cube was used. And in Fig.6 e and Fig.6 f, the 550 nm light was as exciting beam, the fluorescence filter cube was violet (The green fluorescence filter cube blocks the green light to get to the videohead, but the violet fluorescence filter cube permits the green light arrive to the videohead).

At the 550 nm light exciting, the borderline of the OGI and the focal spot are clear. This will advantage to determine the sizes of the OGI and the focal spot.

From Fig.6, one can see that the fluorescence of single OGI is the addition of the background and the OGI. For decreasing the influence of the background, a subtract factor was determined.

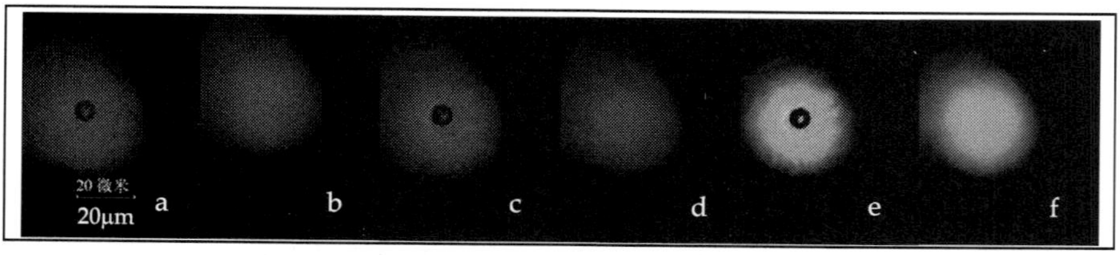

Fig. 6. Transmission micro-images of single OGI (a, c and e) and the background (b, d and f). In which, a and b, c and d, and e and f are excited at 250 nm, 470 nm and 550 nm respectively. In a, b, e, and f, the violet fluorescence filter cube was used, which has a higher transmittance to the green light. In c and d, the blue fluorescence filter cube was used.

The focal spot includes most exciting energy. The energy distribution is approximated to be homogeneous on the spot area. The OGI and the background are seen as surface light sources. Assuming the areas of the OGI and the light spot are S_1 and S_2. The two areas are easily obtained from Fig.6 e and Fig.6 f. The luminous intensities from unit surfaces of the OGI and the background are I_i and I_b. If the total fluorescence intensity of OGI and part background (S_2-S_1) is $I_1(\lambda)$, the intensity of the background is $I_2(\lambda)$. Then

$$I_1 = S_1 I_i + (S_2 - S_1) I_b \quad (1)$$

$$I_2 = S_2 I_b. \quad (2)$$

From Eq. (1) and (2), the intensity of the OGI is

$$I_i S_1 = I_1 - (1 - \frac{S_1}{S_2}) I_2 = I_1 - F I_2 \quad (3)$$

$$F = 1 - \frac{S_1}{S_2}. \quad (4)$$

Here F is the subtract factor. It can be obtained from the micro-photos of the OGI and the background. By the experiment, one can obtain I_1, I_2, S_1, and S_2. According to Eq. (3) and (4), one can further obtain the fluorescence spectra of single OGI. The following results based on above analysis.

3.5 The UV-VIS spectra of single OGIs

The first episode oil inclusions are very small and dense. The fluorescence is very weak. The background fluorescence greatly interferes the signal. The OGIs in the second episode are relative big and in scattering or in group in the mineral grains. It is possible to obtain the fluorescence spectra of single OGIs of the second episode, so these inclusions were chosen as samples.

In the experiment, we first observed dozens of OGIs in an inclusion thin slide then chose some typical OGIs to take photos and measure fluorescence spectra. The typical OGIs in an inclusion thin slide often have similar fluorescence colors. A relative clear area in the quartz grain was chosen as background. There are 38 typical OGIs to be measured. Table 2 shows the depths, strata, sizes, phases, F factors and corresponding oil wells of 38 OGIs.

Oil well	Stratum	Depth (m)	No.	Size (μm)	Phase	F (%)
Bai 95#	K_2y_{2+3}	419.3	1	7.0, 12.3	L+G	86.0
		420.1	2	7.7, 10.1	L+G	79.0
	K_2y_1	426.8	3	4.9, 10.7	L+G	94.2
		431.0	4	8.3, 9.7	L+G	86.0
		434.5	5	7.0, 10.7	L+G	88.7
			6	7.7, 10.8	L+G	35.3
		435.0	7	12.5, 37.1	L	86.4
	K_2qn_2	494.6	8	14.1, 14.6	L+G	57.2
			9	6.5, 10.1	L	75.0
Hua 5#	K_2n_1	1449.8	10	4.5, 4.7	L+G	90.0
		1522.4	11	17.3, 53.7	L+G	42.7
		1523.0-1524.0	12*	11.1, 12.5	L+G	0.0
		1523.3	13	4.0, 4.0	L+G	96.0
Jian 22#	K_2q_4	377.8	14	16.6, 19.6	L+G	59.8
		396.6	15	9.42, 9.38	L+G	49.0
			16	15.0, 15.4	L+G	61.9
		401.9	17	9.6, 15.3	L+G	69.9
		404.7	18	5.7, 6.6	L+G	91.5
			19	21.6, 22.9	L+G	49.1
			20	10.8, 16.7	L+G	64.7
		420.0	21*	26.4, 27.1	L+G	38.3
			22	17.1, 17.1	L+G	36.6
Nong 29#	K_2q_4	571.0-572.0	23	11.0, 17.3	L+G	72.0
			24	27.2, 30.0	L+G	0.0
		573.0-574.0	25	7.4, 10.6	L+G	82.0
		595.9	26	8.7, 9.7	L+G	86.0
			27	12.7, 25.4	L+G	65.6
			28*	37.2, 37.4	L+G	0.0
			29	12.2, 25.3	L+G	50.0
Fu 4 #	K_2q_3	419.1	30	9.8, 11.7	L	81.6
			31	20.8, 24.2	L+G	39.0
		422.8	32	51.1, 71.3	L+G	0.0
			33	10.3, 15.7	L+G	62.2
			34	7.8, 13.6	L	78.0
		424.3	35	6.8, 7.0	L+G	89.2
			36*	22.6, 23.0	L+G	0.0
		425.2	37	15.9, 18.3	L+G	65.7
			38	15.5, 25.3	L+G	55.6

Table 2. The depths, strata, sizes, phases (L is liquid, and G gas), F factor and corresponding oil wells of 38 OGIs. * means the OGI is in the cement.

Fig.7-Fig.22 are the transmission, reflective micro photos and fluorescence spectra of OGIs of No.6, No.7, No.12, No.15, No.20, No.25, No.26 and No.34. No.6 and No.7 (Fig.7-Fig.10) are in Bai 95# oil well, the depths are 434.5 m and 435.0 m. No.12 (Fig.11-Fig.12) is in Hua 5# oil well, the depth is 1523-1524 m. No.15 (Fig. 13-Fig.14) and No.20 (Fig.15-Fig.16) are in Jian 22# oil well, the depths are 396.6 m and 404.7 m. No. 25 (Fig.17-Fig.18) and No. 26 (Fig.19-Fig.20) are in Nong 29# oil well, the depths are 573-574 m and 595.6 m. No.34 (Fig.21-Fig.22) is in Fu 4# oil well, the depth is 422.8 m.

No.6, No.15, No.20, No.25 and No.26 are two phases (liquid and gas) inclusions. No. 7 and No.34 are single phase (liquid) oil OGIs. No.12, No.21 and No.36 are in cements. The others are all in the quartz grains.

From the micro photos, one can see when the internal light source excites the aimed OGI, not only the OGI emits fluorescence, but also other OGIs and the cements around the mineral grains. The intensities of the cements are often stronger than that of the OGIs. Even the mineral grains also emit weak fluorescence. These are the reasons why the spectra look like each other even the OGIs with different phases and colours. For example, No. 6 and No.7 OGIs are in the same oil well, the former is liquid-gas inclusion, the later is liquid phase. The spectra of the two excited by internal light source (see Fig.8 a and Fi9.10 a) are similar. Obviously, the spectra were influenced by the background. According to the spectra excited by violet light of the microscope (see Fig. 8 a and Fig.10 a, solid line), the peak positions are at 400 nm. The main aromatic hydrocarbons in the OGIs may be three or four cyclic hydrocarbons. But the other aromatic hydrocarbons in the OGIs are difficult to determine because except for the main peak, the other parts of the spectra are almost flat. The peak positions excited by the internal blue and green light are at 500-510 nm and 575 nm-600 nm respectively.

When the external light source (250 nm, 365 nm, 440 nm, 470 nm and 546 nm) excited the aimed OGI, the focal spot size is small. The cements were not illuminated, so the fluorescence spectra of the single OGIs weren't interfered by the spectra of the cements. If the OGIs are not dense in the grain, only one OGI is excited, one can obtain the spectra of single OGI after subtracting the background.

From Fig. 8 b, Fig.10 b, Fig.12 b, Fig.14 b, Fig.16 b, Fig.18 b, Fig.20 b and Fig.22 b, one can see that the spectra are in the range of UV-VIS at 250nm exciting. These results show that the experimental setup can measure the UV-VIS spectra of single OGIs. Combined the spectra excited by different external lights with different wavelengths, the characteristics of the OGIs are as following:

1. For most OGIs, when excited at 250 nm, there are three main peaks in the range of 400-470 nm, which are corresponding to three and four condensed aromatic rings, the palaeo oils are the medium Oils (Liu, K. Y., & Eadington P., 2005; Abbas *et al.*, 2006). There are also two secondary peaks near to 535 nm and 610 nm, the former is the fluorescence of four or more condensed aromatic rings, the later comes from resin and asphaltene, the palaeo oils are medium and heavy Oils (Liu, K. Y., & Eadington P., 2005; Abbas *et al.*, 2006). The spectra in the range of 280-400 nm belong to two or three condensed aromatic rings, the palaeo oils are light Oils (Liu, K. Y., & Eadington P., 2005; Abbas *et al.*, 2006). So the OGIs of No.6, No.7, No.12, No.15, No.20, No.25 and No.34 are all filled with light, medium and heavy oils. And medium oils are the major part. The aromatic hydrocarbons in these OGIs are mainly three, four and five cyclic hydrocarbons. There

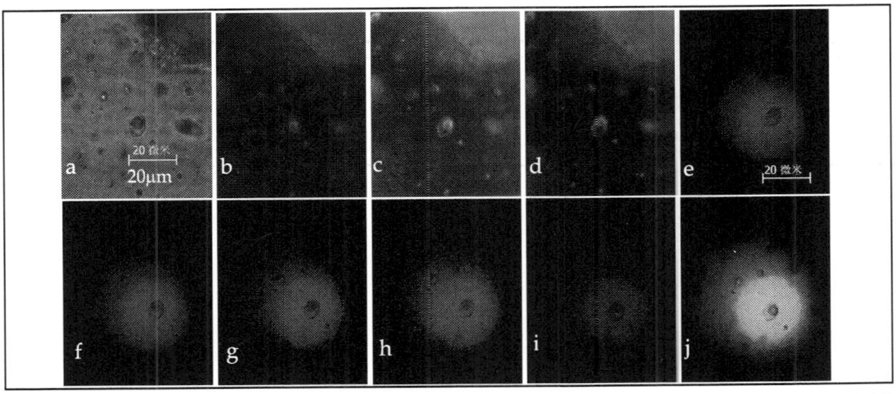

Fig. 7. The transmission (a, e, f, g, h, i and j) and reflective (b, c and d) micro-photos of No.6 OGI (Bai 95#, 434.5 m) when excited by white light (a), internal light source (c violet; d:blue and e:green) and external light source (e:250 nm; f:365 nm; g:440 nm; h:470 nm; and i:546 nm), j is a transmission image when the 550 nm light exciting and violet fluorescence filter cube was used.

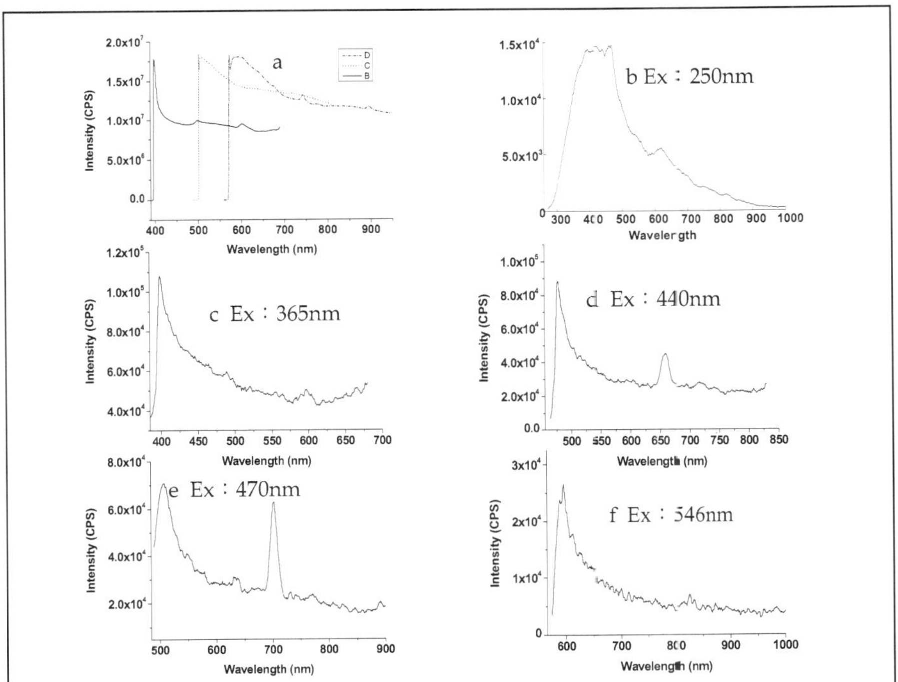

Fig. 8. The fluorescence spectra of No.6 OGI excited by internal light source (in a, solid: violet, short dot: blue, and dash dot: green) and external light source (b-f).

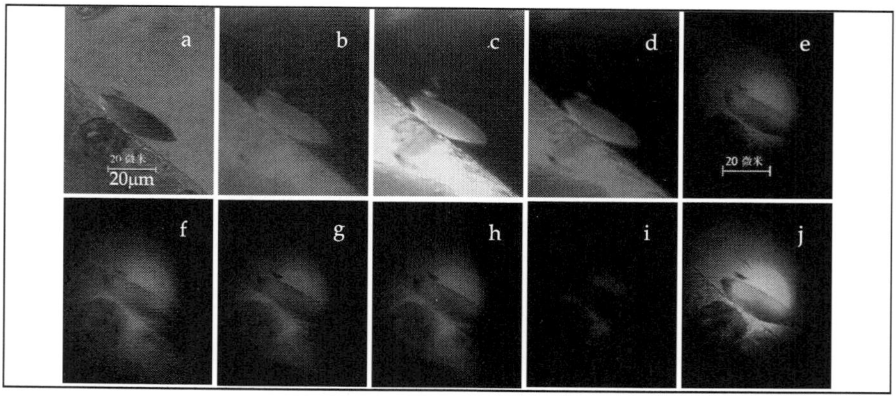

Fig. 9. The transmission (a, e, f, g, h, i and j) and reflective (b, c and d) micro-photos of No.7 OGI (Bai 95#, 435.0 m) when excited by white light (a and b), internal light source (b:violet; c:blue and d:green) and external light source (e:250 nm; f:365 nm; g:440 nm; h:470 nm and i:546 nm), j is a transmission image when the 550 nm light exciting and violet fluorescence filter cube was used.

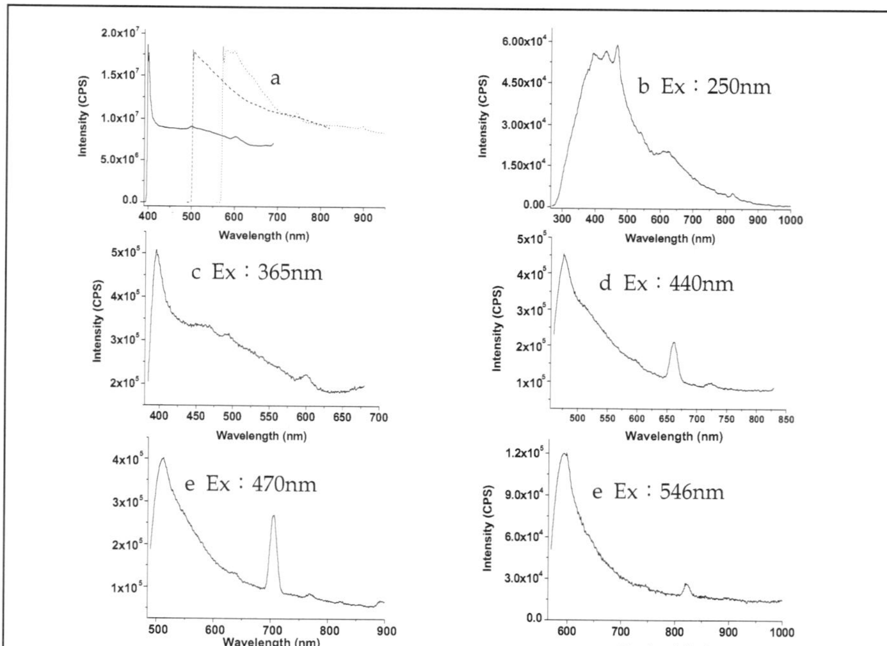

Fig. 10. The fluorescence spectra of No.7 OGI excited by internal light source (in a, solid: violet, long dot: blue, and short dot: green) and external light source (b-f).

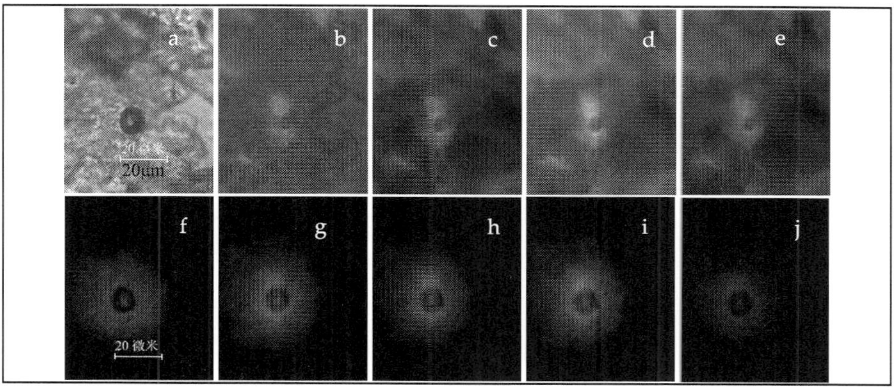

Fig. 11. The transmission (a, b, f, g, h, i and j) and reflective (b, c, d and e) micro-photos of No.12 OGI (Hua 5#, 1523.4 m) in cement when excited by white light (a and b), internal light source (b and c:violet; d:blue and e:green) and external light source (f:250 nm; g:365 nm; h:440 nm; i:470 nm and j:546 nm), b is a transmission and reflection image when white and violet light illuminating at the same time.

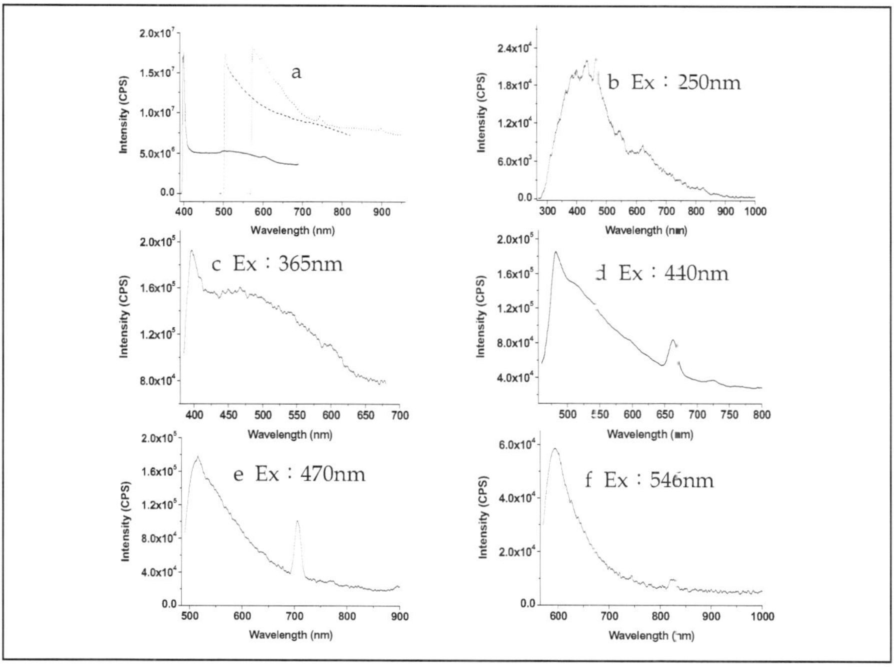

Fig. 12. The fluorescence spectra of No.12 OGI excited by internal light source (in a, solid: violet, long dot: blue, and short dot: green) and external light source (b-f).

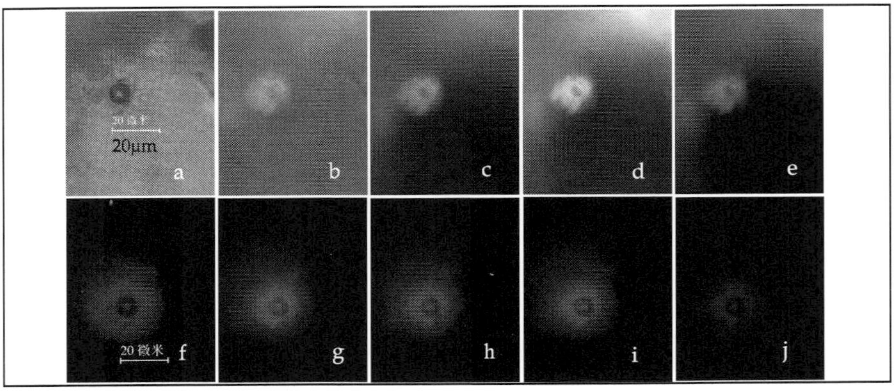

Fig. 13. The transmission (a, b, f, g, h, i and j) and reflective (b, c and d) micro-photos of No.15 OGI (Jian 22#, 396.6 m) when excited by white light (a and b), internal light source (b and c:violet, d:blue and e:green) and external light source (e:250 nm; f:365 nm; g:440 nm; h:470 nm and i:546 nm), b is a transmission and reflection image when white and violet light illuminating.

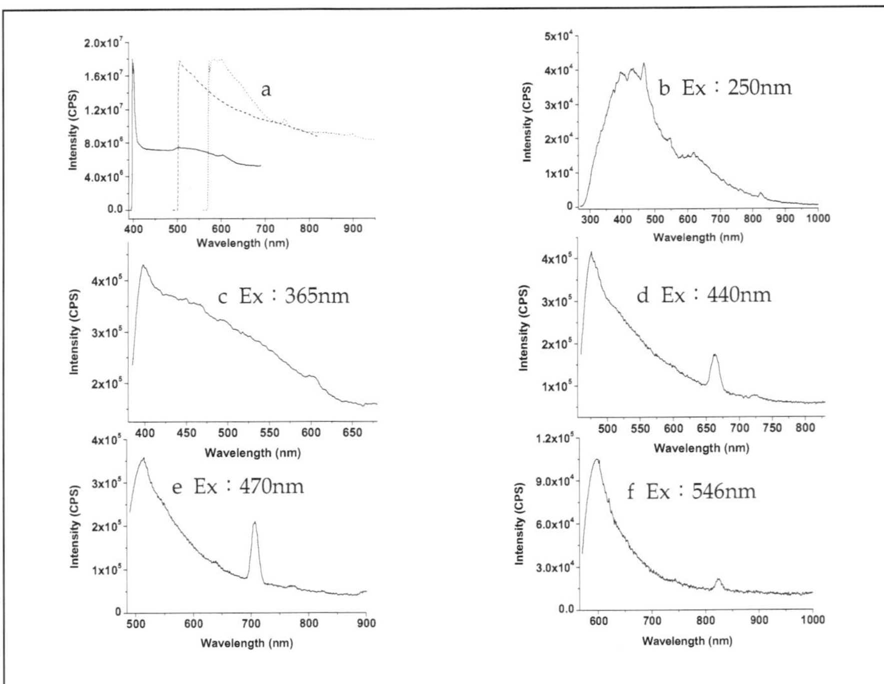

Fig. 14. The fluorescence spectra of No.15 OGI excited by internal light source (in a, solid: violet, long dot: blue, and short dot: green) and external light source (b-f).

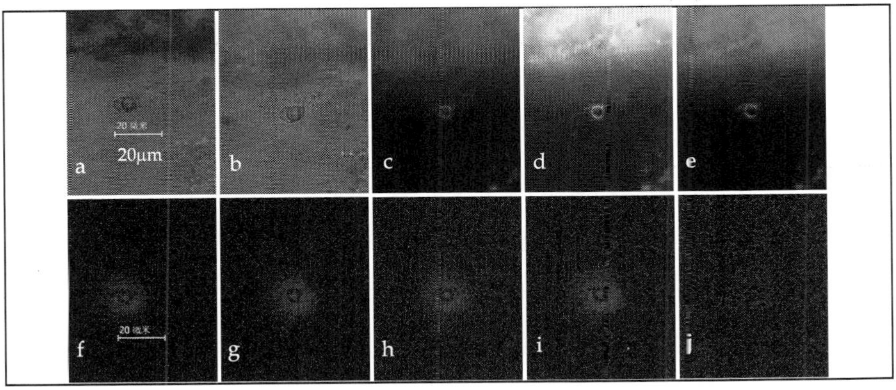

Fig. 15. The transmission (a, b, f, g, h, i and j) and reflective (b, c, d and e) micro-photos of No.20 OGI (Jian 22#, 404.7 m) when excited by white light (a and b), internal light source (b and c:violet; d:blue and e:green) and external light source (f:250 nm; g:365 nm; h:440 nm; i:470 nm and j:546 nm), b is a transmission and reflection image when white and violet light illuminating.

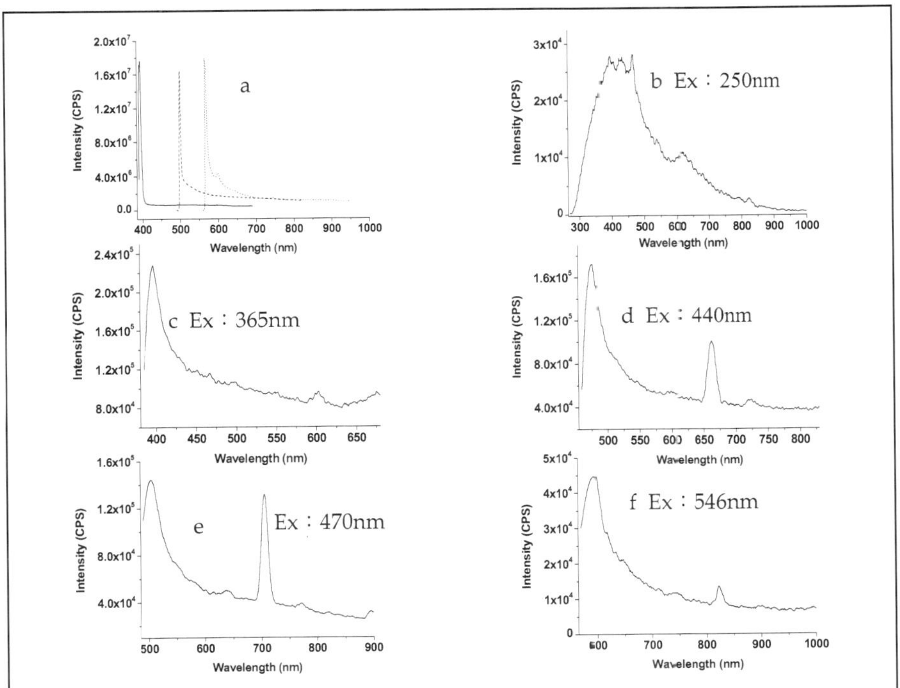

Fig. 16. The fluorescence spectra of No.20 OGI excited by internal light source (in a, solid: violet, long dot: blue, and short dot: green) and external light source (b-f).

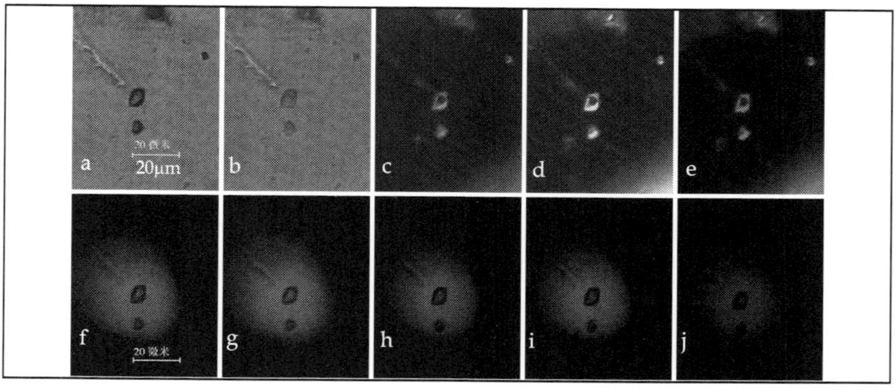

Fig. 17. The transmission (a, b, f, g, h, i and j) and reflective (b, c, d and e) micro-photos of No.25 OGI (Nong 29#, 573-574 m) when excited by white light (a and b), internal light source (b and c:violet; d:blue and e:green) and external light source (f:250 nm; g:365 nm; h:440 nm; i:470 nm and j:546 nm). b is a transmission and reflection image when white and violet light illuminating.

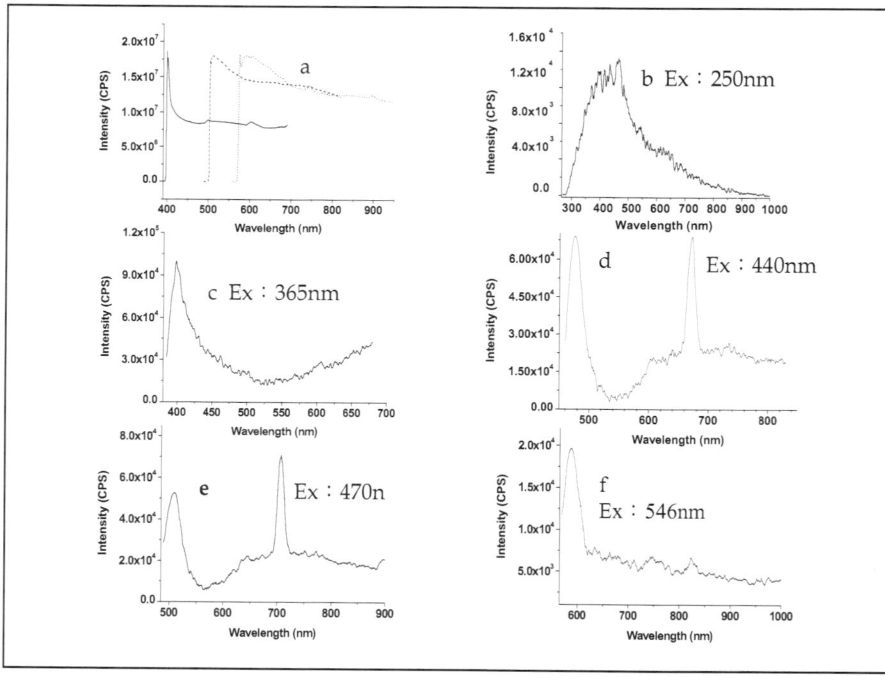

Fig. 18. The fluorescence spectra of No.25 OGI excited by internal light source (in a, solid: violet, long dot: blue, and short dot: green) and external light source (b-f).

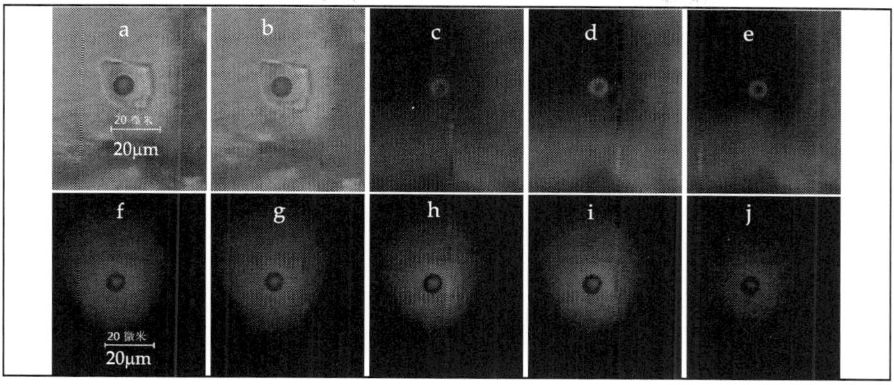

Fig. 19. The transmission (a, b, f, g, h, i and j) and reflective (b, c, d and e) micro photos of No.26 OGI (Nong 29#, 595.9 m) when excited by white light (a and b), internal light source(b and c: violet; d: blue and e: green) and external light source (f:250 nm; g:365 nm; h:440 nm; i:470 nm and j:546 nm), b is a transmission and reflection image when white and violet light illuminating.

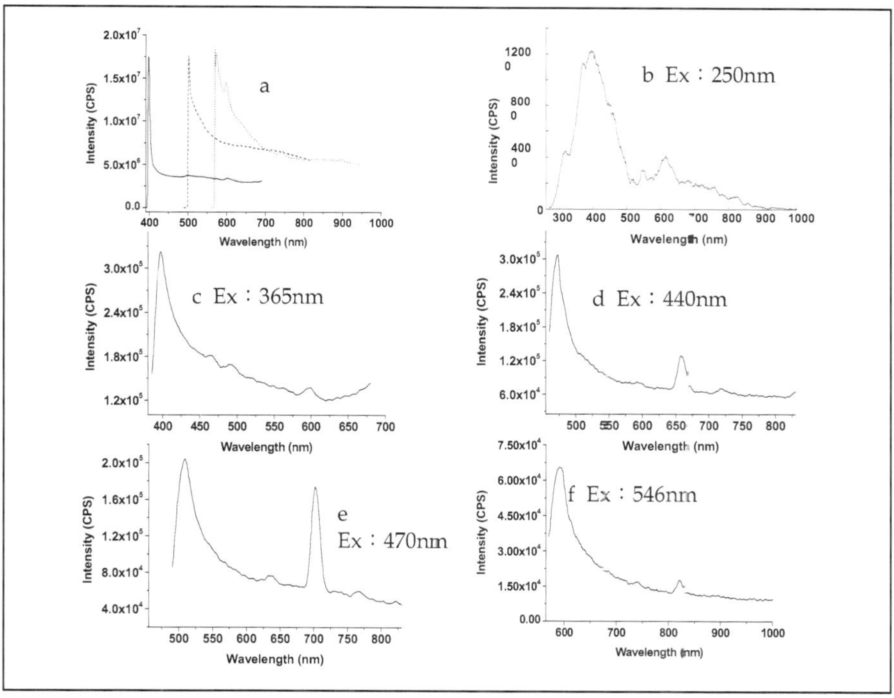

Fig. 20. The fluorescence spectra of No.25 OGI excited by internal light source (in a, solid: violet, long dot: blue, and short dot: green) and external light source (b-f).

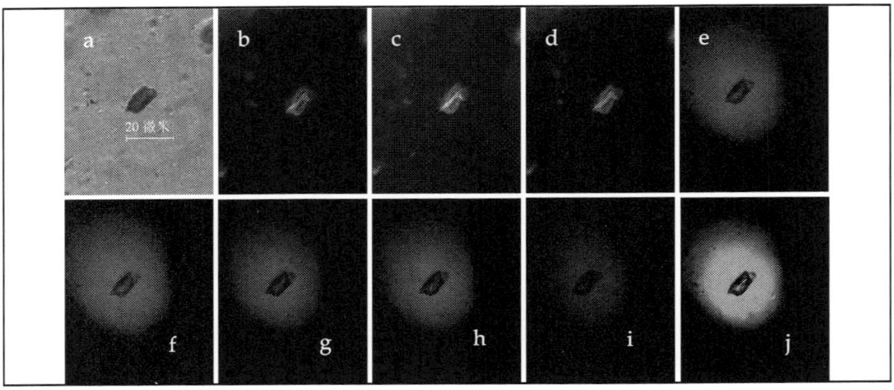

Fig. 21. The transmission (a, e, f, g, h, i and j) and reflective (b, c and d) micro-photos of No.34 OGI (Fu4# 422.8 m) when excited by white light (a), internal light source (b: iolet; c:blue and d:green) and external light source (e:250 nm; f:365 nm; g:440 nm; h:470 nm and i:546nm), j is a transmission image when the 550 nm light exciting and violet fluorescence filter cube was used.

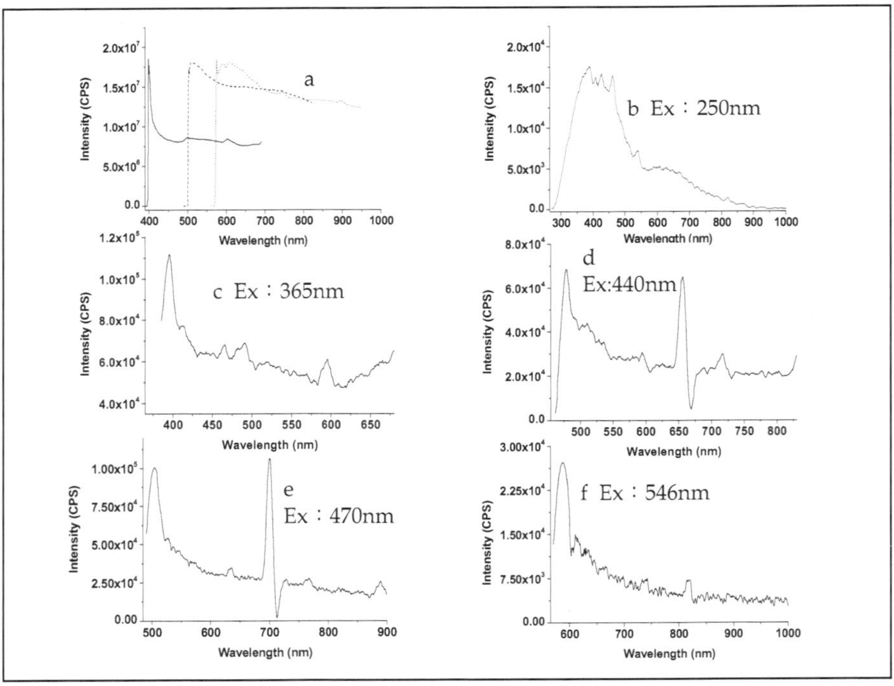

Fig. 22. The fluorescence spectra of No.34 OGI excited by internal light source (in a, solid: violet, long dot: blue, and short dot: green) and external light source (b-f).

are still some one and two cyclic aromatic hydrocarbons and heavy hydrocarbons. The spectrum (Fig.20 b) of No. 26 OGI is different from others. There is one main peak at 400 nm. The full width of half maximum (FWHM) is obviously less than the others, which shows the main hydrocarbons in this OGI are light-medium oils. The secondary peaks in 575-605 nm show there are also some heavy oils in this OGI. The aromatic hydrocarbons in No. 26 OGI are mainly three and four cyclic hydrocarbons. There are still some two cyclic aromatic hydrocarbons and heavy hydrocarbons. For all 38 testing OGIs, the spectra of 90% are similar to No.6 when exciting at 250 nm.This result shows that the palaeo-oils in the OGIs may come from two maternal sources with different maturities, one is high, another is low. So in the later time of quartz overgrowth, there may be two oil sources charging these strata at the same time.

2. In the same oil well, the FWHMs of liquid oil inclusions are wider than that of the liquid-gas OGIs at exciting 250 nm. For example, the FWHMs of No. 6 (liquid and gas) and No.7 (liquid) inclusions are 175 nm and 185 nm respectively. But in different oil wells, this result is not always right. For example, the FWHMs of No.12 and No. 15 (liquids and gas) are 189 nm and 190 nm, which are wider than that of No.7. So one should be cautious to say that the palaeo-oils in the liquid oil inclusions are more heavier than that of in the liquid and gas OGIs.
3. For all the 38 testing OGIs, when exciting at 365 nm (external light source), the spectra can be classified into two types. The first type has an obvious "bump" following the main peak. The wavelength range is in 425-650 nm. This range is relative to the medium and heavy hydrocarbons, which indicate that these inclusions including in more medium and heavy hydrocarbons (see Fig.10 c, Fig.12 c, Fig.14 c). The second type has not an obvious "bump", which shows that these OGIs (see Fig. 8 c, Fig. 16 c, Fig.18 c, Fig. 20 c and Fig.22 c) don't include in enough medium and heavy hydrocarbons as in the first type. The palaeo-oil maturities of the second type are higher than the first one.
4. When exciting at 440 nm, except for the main peak near to 478 nm, there is another peak at 657 nm for all 38 OGIs . No.25 OGI has another peak at 672 nm.
5. When exciting at 470 nm, except for the main peak near to 502 nm, there is another peak at 706 nm for all 38 OGIs.
6. The results of (4) and (5) show that a kind of characteristic matter exist in the OGIs. 440 nm and 470 nm are the effective exciting wavelengths for this kind of matter. This needs to make further analysis to determine what it is.
7. The characteristics of the spectra excited at 546 nm are not obvious. The reason is the absorption bands for the most hydrocarbons are in the range of UV and violet.

3.6 The colours of single OGIs

For simplicity and intuition, the colours of the OGIs under excited at 365 nm are often to be used to judge the maturities of palaeo-oils in the inclusions. But the judgment by eyes are subjective. The results will be influenced by psychology, physiology or environment. So it is need to calculate the chromaticity coordinates of the OGIs.

The calculation theory (Xu & Su,2004) is as following. Assume $P(\lambda)$ is the power spectrum distribution of an OGI, the tristimulus values are given by equation (5),

$$\begin{aligned} X &= \int_{380}^{780} P(\lambda)\overline{x}(\lambda)d\lambda \\ Y &= \int_{380}^{780} P(\lambda)\overline{y}(\lambda)d\lambda \\ X &= \int_{380}^{780} P(\lambda)\overline{z}(\lambda)d\lambda. \end{aligned} \tag{5}$$

In which, $\overline{x}(\lambda)$, $\overline{y}(\lambda)$ and $\overline{z}(\lambda)$ are the tristimulus values of standard light source. One can get them by looking up the CIE 1931 system. The chromaticity coordinates of an OGI are

$$\begin{aligned} x &= \frac{X}{X+Y+Z} \\ y &= \frac{Y}{X+Y+Z} \\ z &= \frac{Z}{X+Y+Z}. \end{aligned} \tag{6}$$

In which, $x+y+z=1$ (*x*:red; *y*:green and *z*:blue). According to above theory, a computing program based on Matlab was established to calculate the chromaticity coordinates of OGIs.

The chromaticity coordinates show that for all OGIs, the chromaticity coordinates excited by internal light source (365 nm) are larger than that of external light source (365 nm). The main reason is that not only the OGI emitting fluorescence, but also the cements and the mineral grain itself when excited by internal light source, which influenced the spectra of the under studied OGI. The chromaticity coordinates deflect to light blue white or yellow white, and relative centralize for different OGIs. When the external light source excites the OGIs, the focal spot is small. The cements don't emit fluorescence and not influence the spectra of single OGIs. The fluorescence of the mineral grain around the OGI has been subtracted as the background. Such spectra are near to the real spectra of single OGIs.

For the OGIs in the cements in Jian 22# and Fu 4# oil wells, the chromaticity coordinates of these OGIs are less than the OGIs in the mineral grains whatever external or internal light source exciting (Fig.23 c and Fig.23 e), which indicates the hydrocarbons in the cements are lighter than those in mineral grains. The formation time of the OGIs in the cements may be later than that in the mineral grains. But for OGI in cements of Nong 29# oil well, the chromaticity coordinates basic overlap with the OGIs in the mineral grains. The OGIs in the cements and in the mineral grains may be formed at the same geology age.

The chromaticity coordinates are dispersive for single OGIs under external light source excited. This result shows that it is easier to distinguish OGIs with external light source exciting. For OGIs in Bai 95# oil well, the chromaticity coordinates are in the range of light blue white and light yellow white. For OGIs in Hua 5# oil well, the chromaticity coordinates are in the range of blue green and light yellow white. For OGIs in Jian 22# oil well, the chromaticity coordinates are in the range of green blue and light yellow white. For OGIs in Nong 29# oil well, the chromaticity coordinates are in the range of light blue white and light green blue. For OGIs in Fu 4# oil well, the chromaticity coordinates are in the range of light green blue, light blue white and white. Above results indicate that in the later time of quartz overgrowth, there may be two maternal source to charge these strata, one with high maturity, another with low maturity.

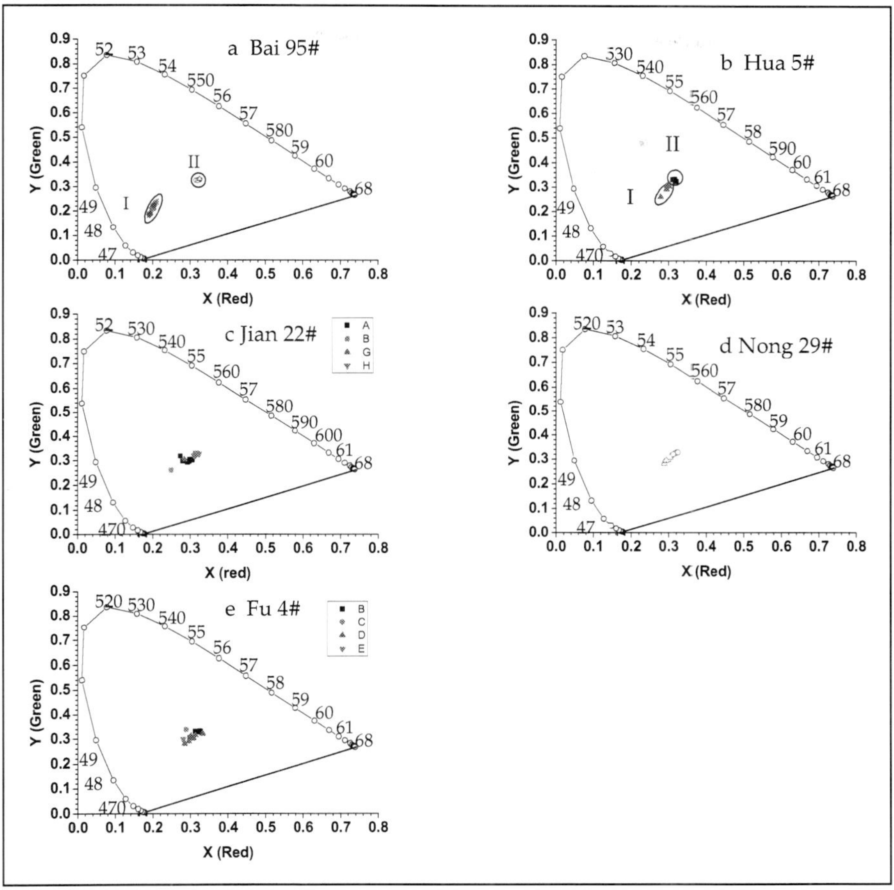

Fig. 23. The chromaticity coordinates of OGIs in the five oil wells excited by external and internal light sources, in which, a and b are Bai 95# , Hua 5# respectively, external (I) and internal (II); c—Jian 22#, external (A, B) and internal (G, H), B(G) is in the cement; d—Nong 29#, external (triangle) and internal (circle); e—Fu 4#, external (D, E) and internal (B, C), D(C) is in the cement.

4. Conclusion

The inclusion samples came from the core drilling samples of five oil wells, Jinlin Oil Filed, Songliao Basin, northeast China(upper Cretaceous Period). The microscope testing shows:

1. The experimental setup based on common inverted microscope, no need any rebuilding to the microscope and can measure the UV-VIS spectra of single OGIs. The cost was decreased greatly.

2. By the spectra excited at 250 nm, the main aromatic hydrocarbons in an OGI can be qualitative determined. Almost all OGIs contain light, medium and heavy petroleum. The main aromatic hydrocarbons are three, four and five cyclic hydrocarbons. There are also some heavy hydrocarbons and non-hydrocarbons in these OGIs. In the same oil well, the FWHMs of liquid oil inclusions are wider than that of the liquid-gas OGIs at exciting 250 nm. But in different oil wells, this result is not always right.
3. The peak positions of the spectra excited at 365 nm are in the range of 395-400 nm, the main aromatic hydrocarbons are three and four cyclic aromatics. The "bump" (425-650nm) following the main peak shows there are more medium and heavy hydrocarbons in the OGIs. The OGIs without the "bump" don't have many medium and heavy hydrocarbons. The palaeo-oil maturities of the second type are higher than the first one.
4. The special peaks near to 657 nm (440 nm exciting) and 706 nm (470 nm exciting) for all OGIs indicate there may be same special matters in these OGIs. This need further analysis combined with GC-MS and palaeo-biology.
5. The chromaticity coordinates are more dispersive for single OGIs under external light source excited than internal light source. The calculating colours excited at external light source are more objective than judging by eyes under exciting at internal light source. But one should note the different spectra may have same colour. So the spectra are important.
6. The colours, the spectra excited at 250 nm and chromaticity coordinates show that there may be two maternal sources to charge these strata in the later time of quartz overgrowth. One is high maturity, another is low maturity.

This optical system can measure UV-VIS spectra of single OGIs. For further decreasing the focal spot, one can try a 74x RMO or decrease the fiber diameter and increase the coupling efficient between the fibre and the spectrometer.

The UV-VIS spectra should be combined with GC-MS analysis to find the characteristic matter in these OGIs.

This setup is promising in measuring the fluorescence spectra of micro areas, such as bio-inclusions, micro fractures in mineral slides, special mineral grains and so on.

5. Acknowledgment

This work was supported by Chinese National Program for High Technology Research and Development (national 863 plan, granted No.2006AA09Z336) and National Natural Science Foundation of China (granted No. 41172110). The author is appreciated for Jilin Oil Field to supply sand rocks. The author greatly thanks professor Zhang, J. L., Dr. Tang., M.M., Ren, W. W. and Yang, Y. for help in the experiment.

6. References

Abbas, O., Rébufa, C., Dupuy, N., Permanyer, A., Kister, J. & Azevedo, D.A. (2006). Application of chemometric methods to synchronous UV fluorescence spectra of petroleum oils. *Fuel*, Vol. 85, No.17-18, (Dec. 2006), pp. 2653-2661, ISSN 0016-2361

Aplin, A.c., Macleod, G., Larter, S.R., Pedersen, K.S., Sorensen H. & Booth, T. (1999). Combined use of Confocal Laser Scanning Microscopyand PVT simulation for estimating the composition and physical properties of petroleum in fluid inclusions. *Marine and Petroleum Geology*, Vol.16, No.2, (Mar., 1999), pp. 97-110. ISSN 9153-7061

Caja, M. A., Permanyer, A., Munz, I. A. & Johansen, H. (2007). Preliminary data on oil and aqueous fluid inclusions of the fracture-fill in the Corones and Armancies Fms, Eocene, SEPyrenees. *GEOGACETA*, Vol.101, (n. d. 2007), pp. 127-130, ISSN 0213683X

Caja, M. A., Permanyer, A., Kihle, J., Munz, I. A. & Johansen, H. (2009). Fluorescence quantification of oil fluid inclusions and oil shows:Implications for oil migration (Armancies Fm, South-eastern Pyrenees, Spain). *Journal of Geochemica Exploration*, Vol.101, No.1, (April, 2009), pp.16, ISSN 0375-6742

Damaskinos, S., Dixon, A. E., Ellis, K. A., & Diehl-Jones, W. L. (1995). Imaging biological specimens with the confocal scanning laser microscope/macroscope. *Micron*, Vol.26, No.6, (n.d.,1995), pp. 493-502. ISSN 0968-4328

Kevin, B., Liesbeth, P., Niek, N. S., Stefaan, C. D. S. & Joseph, D. (2003). Three-Dimensional Fluorescence Recovery after Photobleaching with the Confocal Scanning Laser Microscope. *Biophysical Journal*, Vol.85, No.4, (Oct., 2003), pp. 2240-2252, ISSN 0006-3495

Kihle, J. (1995). Adaptation of fluorescence excitation-emission micro-spectroscopy for characterization of single hydrocarbon fluid inclusions. *Organic Geochemistry*, Vol. 23, No.11-12, (Nov.-Dec. 1995), pp. 1029-1042,ISSN 0146-6380

Liu, D. H., Xiao, X. M., Mi, J. K., Li, X. Q., Shen, J. K., Song, Z. G. & Peng, P. A. (2003). Determination of trapping pressure and temperature of petroleum inclusions using PVT simulation software—a case study of Lower Ordovician carbonates from the Lunnan Low Uplift, Tarim Basin. *Marine and Petroleum Geology*, Vol.20, No.1, (Jan., 2003), pp. 29-43, ISSN 9153-7061

Liu, K. Y. & Eadington, P. (2005). Quantitative fluorescence techniques for detecting residual oils and reconstructing hydrocarbon charge history. *Organic Geochemistry*, Vol. 36, No. 7, (July, 2005), pp. 1023-1036, ISSN 0146-6380

Liu, Y. R., Lu, X. B. & He, M. C. (2003). Progress on Application of Fluid Inclusion in Petroleum Exploration. *Geology Bullet in of Mineralogy, Petrology and Geoch emistry*, Vol.22, No.3, (July, 2003), pp. 245-250, ISSN 1007-2802

Lo, H. B., Wilkins, R. W. T., Ellacott, M. V. & Buckingham, C. P. (1997). Assessing the maturity of coals and other rocks from North America using the fluorescence alteration of multiple macerals (FAMM) technique. *International Journal of Coal Geology*, Vol.33, No.1, (Jan. , 1997), pp. 61-71, ISSN 0166-5162

Meng, D. W., Wu, X.L., Fan, X.Y., Meng, X. & Zheng, J.P. & Mason, R. (2009). Submicron-sized fluid inclusions and distribution of hydrous components in jadeite, quartz and symplectite-forming minerals from UHP jadeite–quartzite in the Dabie Mountains, China: TEM and FTIR investigation. *Applied Geochemistry*, Vol.24, No.4, (April, 2009), pp. 517-526, ISSN 0883-2927

Sanches S. , Leitão C., Penetra A., Cardoso V.V. , Ferreira E., Benoliel M.J., Barreto Crespo M.T., Pereira V.J. (2011). Direct photolysis of polycyclic aromatic hydrocarbons in drinking water sources. *Journal of Hazardous Materials*, Vol. 192, (Jun. 2011), pp. 1458-1465, ISSN 0304-3894

Siegwart, R. (2001). Indirect Manipulation of a Sphere on a Flat Disk Using Force Information. *International Journal of Advanced Robotic Systems*, Vol.6, No.4, (Dec. 2009), pp. 12-16, ISSN 1729-8806

Sliney D. H. (1995). Risk assessment and laser safety. *Optics & Laser Technology*, Vol. 27, No. 5, (n. d. 1995), pp. 279-284, ISSN 0030-3092

Stasiuk, L. D. & Snowdon, L. R. (1997). Fluorescence micro-spectrometry of synthetic and natural hydrocarbon fluid inclusions:crude oil chemistry, density and application to petroleum. *Applied Geochemistry*, Vol.12, No.3, (May, 1997), pp. 229-241, ISSN 0883-2927

Stasiuk, L.D. (1999). Confocal laser scanning fluorescence microscopy of Botryococcus alginite from boghead oil shale, Boltysk, Ukraine: selective preservation of various micro-algal components. *Organic Geochemistry*, Vol.30, No.8, (Aug.,1999), pp. 1021-1026, ISSN 0146-6380

Stasiuk, L.D. & Sanei, H. (2001). Characterization of diatom-derived lipids and chlorophyll within Holocene laminites, Saanich Inlet, British Columbia, using conventional and laser scanning fluorescence microscopy. *Organic Geochemistry*, Vol.32, No. 12, (Dec., 2001), pp. 1417-1428, ISSN 0146-6380

Thiéry, R., Pironon, J., Walgenwitz, F. & Montel F. (2002). Individual characterization of petroleum fluid inclusions (composition and P–T trapping conditions) by microthermometry and confocal laser scanning microscopy: inferences from applied thermodynamics of oils. *Marine and Petroleum Geology*, Vol.19, No.7, (Aug., 2002), pp. 847-859, ISSN 9153-7061

Veld, H., Wilkins, R. W. T., Xiao, X. M. & Buckingham, C. P. (1997). A fluorescence alteration of multiple macerals (FAMM) study of Netherlands coals with "normal" and "deviating" vitrinite reflectance. *Organic Geochemistry*, Vol.26, No.3-4, (Feb., 1997), pp. 247-255, ISSN 0146-6380

Wilkins, R. W. T., Wilmshurst, J. R., Russell, N. J., Hladky, G., Ellacott, M.V. & Buckingham C. (1992). Fluorescence alteration and the suppression of vitrinite reflectance. *Organic Geochemistry*, Vol.18, No.5, (Sept., 1992), pp. 629-640, ISSN 0146-6380

Wilkins, R. W. T., Wilmshurst J.R., Hladky, G., Ellacott, M. V. & Buckingham, C. P. (1995). Should fluorescence alteration replace vitrinite reflectance as a major tool for thermal maturity determination in oil exploration? *Organic Geochemistry*, Vol.22, No.1, (Jan. , 1995), pp. 191-209, ISSN 0146-6380

Xiao, X.M., Wilkins R. W. T., Liu Z.F. & Fu, J.M. (1998). A preliminary investigation of the optical properties of asphaltene and their application to source rock evaluation. *Organic Geochemistry*, Vol.28, No. 11, (July,1998), pp. 669-676, ISSN 0146-6380

Xiao, X.M., Wilkins, R. W. T., Liu D.H. Liu, Z. F. & Fu, J. M. (2000). Investigation of thermal maturity of lower Palaeozoic hydrocarbon source rocks by means of vitrinite-like maceral reflectance — a Tarim Basin case study. *Organic Geochemistry*, Vol.31, No.10, (Oct., 2000), pp.1041-1052, ISSN 0146-6380

Xiao, X. M., Wilkins R. W. T., Liu, D. H. & Shen, J.Q. (2002). Laser-induced fluorescence microscopy—application to possible high rank and carbonate source rocks. *International Journal of Coal Geology*, Vol.5, No.2, (July, 2002), pp. 129-141, ISSN 0166-5162

Xu, X.R. & Su, Z. M. (Oct., 2004). Luminescence Theory & Luminescence Materials, Chemical Industry, ISBN: 7502554106, 9787502554101, Beijing

Yang, A.l., Zhang, J.L., Ren. W.W. & Tang. M.M. (2009a). Micro-spectroscopy system based on common inverted microscope to measure UV-VIS spectra of a micro-area. *Proceedings of SPIE 7283*, ISBN 0277-786X, Chengdu, China, Nov., 2008

Yang, A.l., Ren. W.W., Zhang, J.L., & Tang. M.M. (2009b). A micro-spectroscopy system to measure UV-VIS spectra of single hydrocarbon inclusions. *Proceedings of SPIE 7384*, ISBN 0277-786X, Beijing, China, April, 2009

Yang A.L.,Tang, M.M.,Ren. W.W.,Yang, Y. & Zhang, J.L. (2011). Investigation of the ultraviolet-visible micro-fluorescence-spectra and chromaticity of single oil inclusion. *ACTA OPTICA SINICA*, Vol.31, No.3, (Mar., 2011), pp. 0318002-1-6, ISSN 0253-2239

3

Challenging Evaluation of the Hybrid Technique of Chemical Engineering – Proton NMR Technique for Food Engineering

Yasuyuki Konishi[1,*] and Masayoshi Kobayashi[2]
[1]*Hokkaido Industrial Technology Center, 379 Kikyo-cho, Hakodate, Hokkaido*
[2]*Advanced Technology Institute of Northern Resources, 8-6, Tonden 2-Jo, 2-Chome, Sapporo, Hokkaido*
Japan

1. Introduction

In the field of food engineering, chemical engineering and bioengineering have commonly used the data obtained from physicochemical techniques. In particular, water species are strongly related to food quality, as has been demonstrated in the International Symposium on Properties of Water (Eds. Rockland and Stewart, 1981). The water species retained in foods, as is well known, expose their multifunctional nature by dynamically responding to environmental conditions such as temperature, pressure, dehydration processes, water content, etc. For scientific analysis of the multifunctional water species retained in foods, of particular difficulty is the influence of nonlinear dynamic change on food quality. To quantitatively evaluate the nature of the water species, a large number of researchers have demonstrated the diversity of their biological and physicochemical nature using various parameters. The water activity (a_w) has, for example, commonly been used as a parameter to evaluate the deterioration of foods (Fennema, 1976). The effective water diffusivity parameter (*De*) in foods has also frequently been employed to evaluate the dehydration rate (Jason, 1958). From the chemical engineering point of view, this is a typical procedure in evaluating water mobility in the food tissue matrix. Conversely, for direct identification of the water species at the molecular level, magnetic resonance techniques have been applied to evaluate food quality, as demonstrated in the International Conferences on Applications of Magnetic Resonance in Food Science (Eds. Belton *et al.*, 2003).

In our previous papers (Konishi *et al.*, 2001, 2003), using separately chemical engineering and proton NMR techniques, water species retained in fish paste sausage and squid were roughly distinguished into two different species as a function of water content. The two species were water species A_1, to be recognized as the higher water content at W_0 >120%-d.b. accompanied by higher water diffusivity, *De*, and water species A_2, W_0 <120%-d.b., with lower *De*. This delineation, however, was unclear because of a vagueness of the boundary

* Corresponding Author

between the two water species. To clearly discriminate between the two water species, a new technique is needed for the design of food products. In the present study, responding to this need, a hybrid technique derived from joint research using chemical engineering techniques and proton NMR techniques was proposed. The aims of this study are: (1) to demonstrate the dynamism of water species influenced by the change of environmental conditions and the dynamic transformation between the two water species, A_1 and A_2; (2) to quantitatively visualize the limitations of both the chemical engineering technique and the proton NMR technique; (3) to apply a hybrid technique in choosing specified design parameters in order to clearly identify differences between the two water species and among a variety of food products; and (4) to classify twelve commercial jerky products, as a typical example, into a few groups based on the characterization of design parameters using the hybrid technique.

2. Experimental methods of the hybrid technique

2.1 Methods

Six commercially distributed beef and pork jerkies, (BJ-1~BJ-6) and (PJ-1~PJ-6), respectively, a pork meat (designated P_H, used as a reference meat) produced in Hokkaido, two beef meats (B_A: produced in Australia, and B_H: produced in Hokkaido), salmon (SM), and squid (SQ) were chosen as the food samples. All samples used in this study were tabulated in Table 1. The commercial jerkies used were characterized as having a water content range

	Features	Group No.
PJ-1	soy-sauce base taste, about 1.0 mm thick	2
PJ-2	soy-sauce base taste, 1.1~1.4 mm thick	1
PJ-3	light taste, oily, smoked flavor, 0.8~5.9 mm thick	2
PJ-4	salt base taste, smoked flavor, strong taste, hard, 2.2~4.8 mm thick	2
PJ-5	salt base taste, smoked flavor, strong taste, soft, 1.3~2.7 mm thick	3
PJ-6	soy-sauce base taste, light taste, hard, 1.9~3.5 mm thick	3
BJ-1	salt base taste, light taste, hard, 1.4~2.2 mm thick	2
BJ-2	salt base taste, strong taste, soft, about 2.2 mm thick	3
BJ-3	miso-sauce base taste, strong taste, soft, 4.2~5.9 mm thick	3
BJ-4	salt base taste, light taste, soft, 1.6~3.1 mm thick	2
BJ-5	soy-sauce base taste, strong taste, hard, 3.1 mm thick	1
BJ-6	salt base taste, light taste, hard, 0.9~2.8 mm thick	2
P_H	pork meat produced in Hokkaido, soy-sauce base taste, hard, about 6±1.0 mm thick	
B_H	beef meat produced in Hokkaido, about 8±1.0 mm thick	
B_A	beef meat produced in Australia, about 8±1.0 mm thick	
SQ	squid mantle, 50×50×7±1.5 mm	
SM	salmon, 50×15×12~26 mm	

Table 1. Features of the pork meat (P_H), beef meat (B_A and B_H), squid (SQ), salmon (SM), and the commercial jerkies used for this study. PJ-1~-6: pork jerkies; BJ-1~-6: beef jerkies.

of 30~80%-d.b. (dry base, W_D), the pork and beef meats had the initial water contents of 230~320%-d.b. and 230~280%-d.b., respectively, and the initial water content of the salmon and the squid commonly ranged from 300~360%-d.b. To evaluate the effective diffusivity (De) of water species, each of the samples was placed in a stainless steel net tray (4 meshes) that was mechanically hung from a strain gage transducer in the dryer. The sample weight was, to evaluate the dehydration rate, continuously recorded by the output of strain-gage transducer using a data-logger. With molecular-level analysis, to evaluate the correlation time (τ_C) of the water species in the salmon, squid, 12 jerkies, B_H, B_A, and P_H, a nuclear magnetic resonance (NMR) technique was applied to measure the ^{1}H-NMR spectra and the spin-spin relaxation time (T_2) of water protons. All samples, cut into 2×2×10 mm pieces, were inserted into an NMR sample tube (4mm in inner diameter, 180mm in length). ^{1}H-NMR spectra were obtained using a JEOL A-500 FT-NMR spectrometer operating at 500MHz for protons. The observed frequency width was 20 kHz. The 90° pulse width was 12.5 μs, and the number of pulse repetitions was 8. The proton chemical shifts were measured by using a slight amount of water containing deuterium oxide as an external reference. All the NMR measurements were performed at 23.5±0.5°C. The spin-spin relaxation times, T_2, were obtained by the spin locking method. The hardness (N/m^2) of the samples was measured using a creep tester equipped with a V-shaped plunger (30 mm wide, 1 mm thick) to press a 60% of the sample size of 2~8×10×50 mm.

2.2 Evaluation of the parameters

2.2.1 Effective diffusion coefficient (*De*) of water species

As has been reported by a large number of researchers (Andreu and Stamatopoulos, 1986; Waananen and Okos, 1996; Litchfield and Okos, 1992; Kannan and Bandyopadhyay, 1995), diverse mathematical equations for evaluating the effective diffusion coefficient (De) of water in foods have been demonstrated in various forms. In the present study, De's of all the samples were evaluated by applying Equation (1) to the dehydration response curves obtained. In the present experimental drying conditions, it was reconfirmed that the drying operations were within a falling-rate period. Through all the evaluation of De, the effect of the shrinkage of the samples (derived from the dehydration) on the values obtained was previously evaluated, and the sizes of the samples were corrected.

$$\frac{W - We}{W_D - We} = \left(\frac{8}{\pi^2}\right)^3 exp\left(\frac{-\pi^2 \cdot De \cdot t}{4} \cdot \left(L_a^{-2} + L_b^{-2} + L_c^{-2}\right)\right) \tag{1}$$

Where W is the water content (%-d.b.) of the sample at $t = t$ (h), We is the equilibrium water content (%-d.b.), W_D is the water content (%-d.b.) at the given initial condition, De is the effective diffusion coefficient (m^2/h) of water species retained in the foods, L_a, L_b, and L_c (m) are the half distances of the sample width, and t is dehydration time (h).

De is represented as a function of the structure parameters (porosity and labyrinth factor) of food and the activation energy, as shown in Equation (2).

$$De = \left(\frac{\varepsilon}{\chi}\right) \cdot D = \delta \cdot D_0 \cdot exp\left[\frac{-E_D}{R \cdot (T_D + 273)}\right] \tag{2}$$

Where ε is porosity, χ is labyrinth factor, D is diffusivity (m²/h), D_0 is the frequency factor of D, δ is diffusibility (= ε/χ), E_D is activation energy of De, R is gas constant, and T_D is drying temperature.

The pre-exponential factor, $\delta \cdot D_0$, can be a useful tool to discriminate between the water species retained in different kinds of foods, as is described in upcoming sections.

2.2.2 Correlation time (τ_C) of water protons

The spin-locking pulse technique used was effective in detecting a very fast relaxation signal at low water contents. For evaluation of the relaxation time, T_2, the equation $M_t = M_0\, exp(-ts/T_2)$ was used, where M_t is the magnitude of the magnetization vector after the spin locking pulse, M_0 is magnitude of the macroscopic equilibrium magnetization vector, and ts is the spin locking pulse length. In the present study, the plot of $ln\,[M_t / M_0]$ vs. ts indicated good linearity (which demonstrates a correlation coefficient higher than 0.99) through all water content of the foods, suggesting that the evaluated T_2 value was reasonable. From T_2, the correlation time of a water proton, τ_C, was evaluated using Equation (3) as described by Abragam (1963):

$$\frac{1}{T_2} = \frac{\gamma^4 \cdot \hbar^2 \cdot I(I+1)}{5r^6}\left(3\tau_c + \frac{5\tau_c}{1+\omega_0^2 \cdot \tau_c^2} + \frac{2\tau_c}{1+4\omega_0^2 \cdot \tau_c^2}\right), \quad (3)$$

where γ is the gyromagnetic ratio of a proton (= 2.675×10^8 rad·T⁻¹·s⁻¹), $\hbar$ is the modified Plank's constant (6.63×10^{-34} J · s), I is the nuclear-spin quantum number of a water proton (= 0.5), r is the proton-proton distance of a water molecule (0.16 nm), ω_0 is the resonance frequency of NMR (= 3.14×10^9 s⁻¹), and τ_C is the correlation time of a water proton (s).

3. Dynamic transformation of water species retained in foods

3.1 Variety of water species in foods

3.1.1 Visualization from the proton NMR technique

The water species retained in foods are dynamically changed into multifunctional states according to environmental conditions such as temperature, atmospheric pressure, atmospheric relative humidity, seasoning components such as salt and sugar, concentration of seasoning components, etc. Figure 1 is a schematic explanation of the water species adsorbed on macromolecules such as proteins in foods. The water species are classified into three groups as: (1) monolayer adsorbed water localized in region-I, which is referred to as bond water, immobilized water, or non-freezing water; (2) multi-adsorbed water localized in region-II, which is referred to as strongly restricted water, glass-state water, etc.; and (3) free water localized in region-III, which is referred to as normal isotropic bulk water, capillary condensation water, etc. The proton NMR technique produced correlation times of the three regions, I, II, and III, as $\tau_C = 10^{-6\sim-7}$s, $10^{-6\sim-11}$s, and 10^{-12}s, respectively. This classification is interestingly correlated to the three regions of the adsorption isotherm evaluated by the chemical engineering technique, as shown in Figure 2 by curve 1. As has been reported in previous papers (Konishi *et al.*, 2003, 2010), water species in the three regions are dynamically transformed by each other depending on environmental conditions in

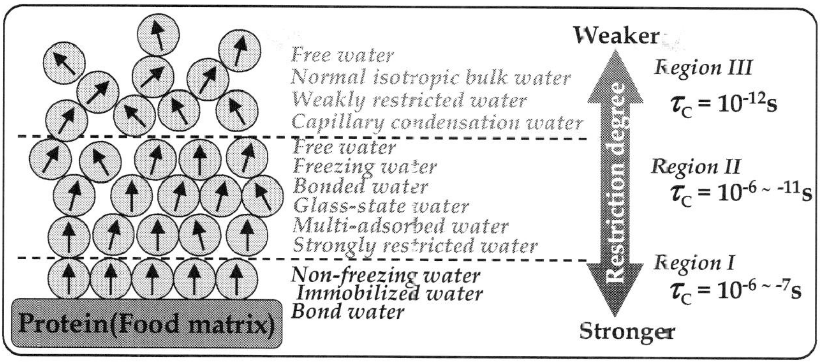

Fig. 1. Schematic explanation of the water species adsorbed on macromolecules in food.

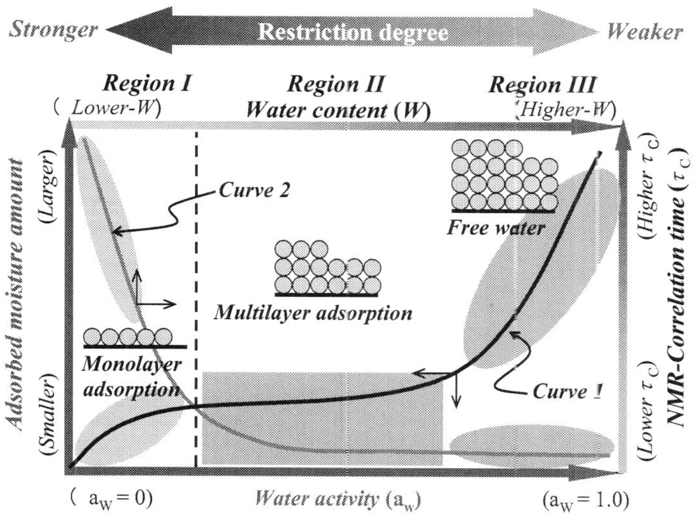

Fig. 2. Classification of the adsorbed water species in the food tissue matrix.

conjunction with the progress of dehydration. In atmospheric conditions, according to the progress of dehydration, the free water species in region III is preferentially removed rather than the two water species in regions I and II.

3.1.2 Visualization from the chemical engineering technique

Taking into account the water activity map presented by Labuza *et al.* (1970) and Schmidt (2004), one may visualize three regions on a water activity space. Fig. 2 shows a schematic explanation of the specified two curves: curve 1 for the water adsorption isotherm of food systems as a function of water activity (the isotherm was divided again into three regions, I, II, and III) and curve 2 for the NMR correlation time (τ_C) of water species retained in food as a function of water content (W_0). As can be seen from the comparison of the two curves, the water species adsorbed in region I with lower water activity contributes to the higher correlation time (τ_C). The higher τ_C means higher restriction strength derived from the food tissues. The water species adsorbed in the region III with higher water activity, contributing to lower τ_C, which means lower restriction strength. These results bring about an inverse relation between the two parameters, τ_C (curve 2), and the adsorbed water amount (curve 1) as shown in Fig. 2. This relation can be usefully employed in the hybrid technique as is described in upcoming sections.

The three water species distributed in regions I, II, and III presented in Figs. 1 and 2 can be further classified as species A_1 and species A_2. The water species in regions I and III belong to species A_2 and species A_1, respectively. The water species in region II belongs to an intermediate species between species A_1 and species A_2. Figure 3 schematically illustrates the water species A_1 and A_2 retained in foods as a model description. From the schematic visualization, one may recognize water species A_1 to be weakly restricted by macromolecules in foods, such as proteins, at greater water content and water species A_2 to be strongly restricted due to the progress of dehydration. The space formed due to the dehydration in region III in Fig. 1 would create micropores resulting in higher porosity (ε) and higher labyrinth factor (χ) in the food tissue matrix. These micropores would be filled again with water species when re-hydration occurred, indicating a reproducibility of the adsorbed water species as is discussed in upcoming sections (see Fig. 5).

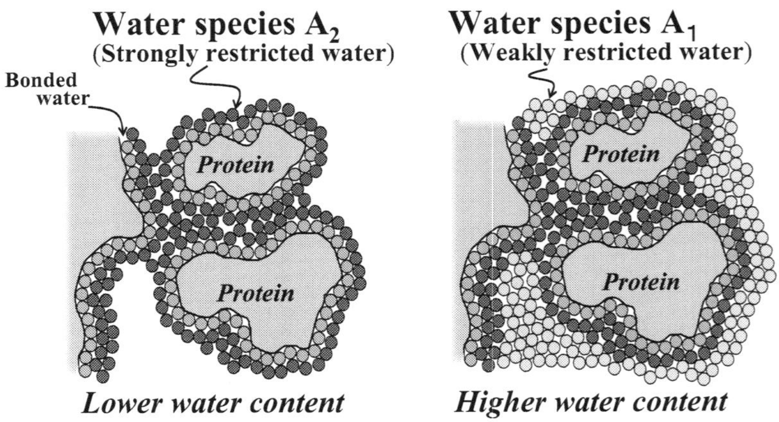

Fig. 3. Schematic explanations of the adsorbed states for water species in foods.

3.2 Dynamism and reversibility of the water species in foods identified by the proton NMR technique

The proton NMR spectrum clearly demonstrates the dynamism of water species derived from the dehydration of the sample (Konishi *et al.*, 2010). Figure 4 demonstrates the dynamism of the reduced amount of water species (the dehydration of squid) as a steep reduction of the proton NMR spectra of peak-A. The three peeks, A, B, and C, were identified in our previous work (Konishi *et al.*, 2010) as weakly restricted water, strongly

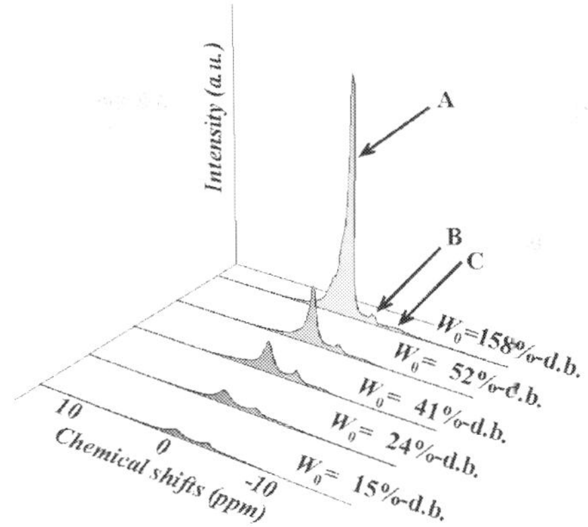

Fig. 4. Dynamic behavior of ^{1}H-NMR spectra for species A, B, and C in the squid muscle in the course of a drying operation.

restricted water, and lipids, respectively. One may recognize the reduction of only peak-A due to dehydration, whereas peaks-B and -C gave an identical value indicating that the two species are not influenced by dehydration.

Focusing on the reversibility of peak-A, one can easily recognize the dynamism of proton NMR spectra of water species in the squid, as shown in Figure 5. The strength of peak-A was steeply reduced by dehydration (see curve (1)), and the peak height was restored by re-hydration (see curve (2)), whereas the peaks-B and -C kept the same peak height regardless of dehydration and re-hydration. This reproducibility of peak-A strongly demonstrates the ability

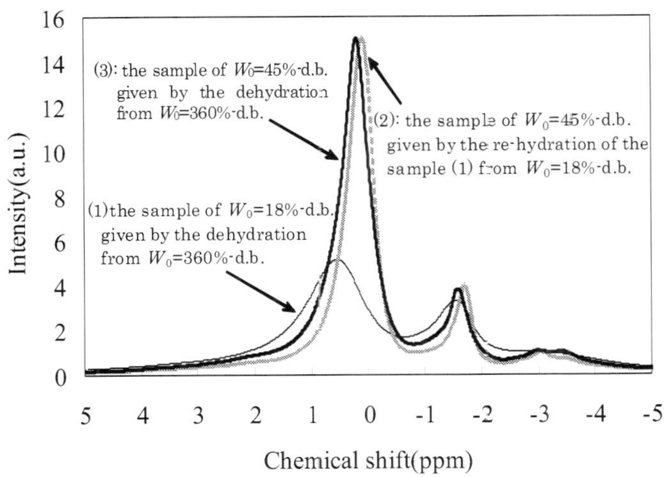

Fig. 5. Reproduced dynamic behavior of ^{1}H-NMR spectra for species-A, -B and -C in the squid muscle.

of water species A_1 and A_2 to be reformed in the food tissue matrix through re- hydration. This reversibility is a very important factor for dried food products because, generally speaking, dried foods are frequently re-hydrated before being consumed (dried kelp, dried green onions, and dried noodles, etc.). Exploitation of reversible water species in dried foods could produce the same taste as when the food is freshly prepared. To evaluate the reproducibility of dried food taste, one should focus on τ_C because the τ_C value is closely related to the concentration of seasoning and the kind of food. The reproducibility of τ_C is, therefore, an important factor for taste reproducibility. Figure 6 demonstrates the reproducibility of τ_C as a function of W_0. The results show positive reproducibility of τ_C, indicating favorable taste reproducibility in the case of dried squid.

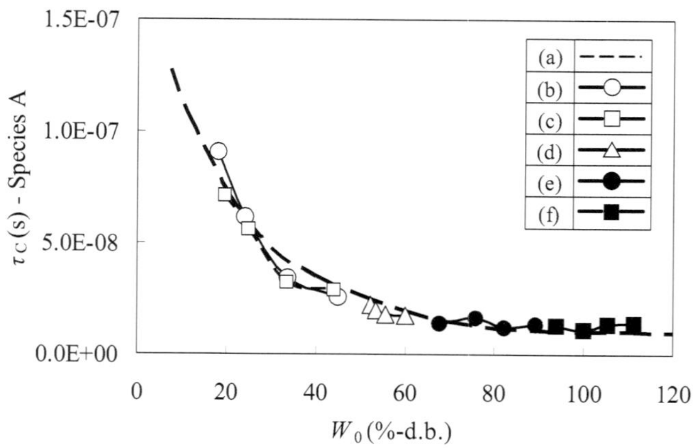

Fig. 6. Reproducibility of τ_C(species A) observed between the dehydration and the water-readsorption operations for the squid muscle: (a) under the continuous drying operation from W_0 = 120%-d.b. to 10%-d.b., (b) after the gradual readsorption of water from W_0 = 18%-d.b. to W_0 = 45%-d.b., (c) from W_0 = 20%-d.b. to W_0 = 44%-d.b., (d) from W_0 = 50%-d.b. to W_0 = 60%-d.b., (e) from W_0 = 68%-d.b. to W_0 = 89%-d.b., and (f) from W_0 = 94%-d.b. to W_0 = 111%-d.b.

4. Limitations of the proton NMR technique

To design the food products requested by the markets, the design parameters chosen should be able to clearly discriminate between various kinds of foods depending on requirements. In particular, the water species retained in the foods should be characterized by the kinds of foods and environmental influences such as temperature, atmospheric pressure, water content, etc. Although the proton NMR water species evaluation technique is a useful tool, one should recognize that the method has some difficulty discriminating among water species retained in foods. Figure 7 illustrates the τ_C values as a function of W_0 for B_A, B_H, and P_H. The τ_C~W_0 curves obtained could not discriminate among the three meats, B_A, B_H, and P_H (Konishi and Kobayashi, 2009), even though the water species A_1 and A_2 are roughly distinguished at the critical point of τ_C = $C\tau_C$ = 10^{-8} s. This unclear discrimination between the three curves can be understood as a limitation of the proton NMR technique.

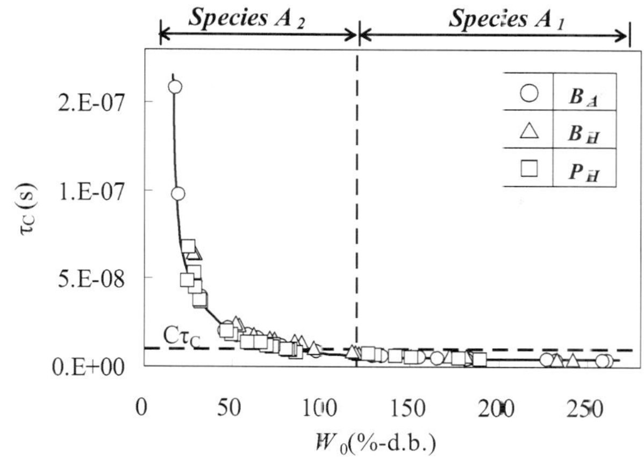

Fig. 7. τ_C as a function of W_0 for B_A, B_H, and P_H.

Figure 8 illustrates τ_C as a function of water content (W_0) of the 12 jerkies and the P_H. The solid line for P_H indicated a steep increase with decreasing W_0, between the values of τ_C = 1.0×10^{-8}s and τ_C = 1.1×10^{-7}s. Since τ_C indicates rotation time of the water species, the steep increase of τ_C demonstrates that the restriction strength of the water species in this region increases markedly upon dehydration from W_0 = 120 to 20%-d.b. Similar results were also obtained for the beef (Konishi and Kobayashi, 2009). The τ_C for the jerkies, on the other hand, fell around the solid line of P_H. From these results, one can recognize that it is difficult to distinguish among the three τ_C -W_0 curves for the pork jerkies, beef jerkies, and P_H, indicating again a limitation of the proton NMR technique.

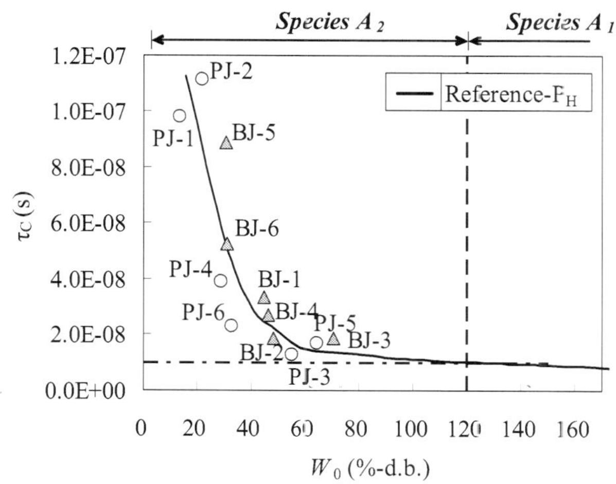

Fig. 8. τ_C as a function of W_0 for the pork and beef jerkies (open circle and closed triangle) and the pork (solid line).

5. Limitation of the chemical engineering technique

5.1 Unclear discrimination of water activity in foods in the chemical engineering technique

Water activity (a_w) has conveniently been used to design food products because of the ease of evaluating the variable using the chemical engineering technique (Fennema, 1976). Figure 9 illustrates a_w as a function of W_0 for the squid and the salmon. As can be seen from the comparison between the two curves, an identical curve was obtained without detecting any difference between the squid and the salmon. This result strongly indicates a limitation of the chemical engineering technique. To develop a new food product, one should choose another parameter for discriminating among products. To address this issue, a computer simulation method for developing new food design parameters is proposed in the next section.

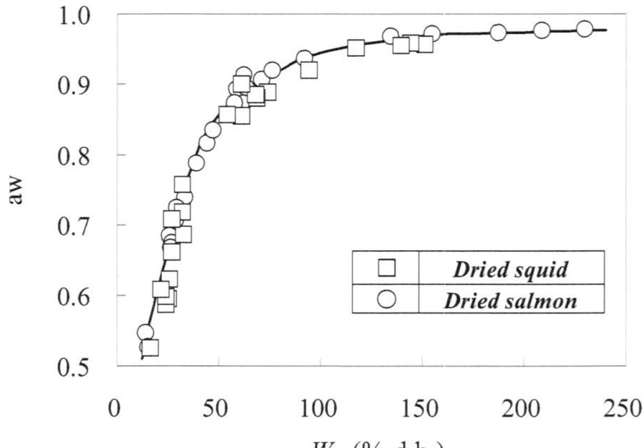

Fig. 9. a_w as a function of W_0 for the salmon and squid.

5.2 Ambiguous physical meaning of the chemical engineering parameters derived from a computer simulation for the dehydration dynamism of salmon

5.2.1 Dehydration dynamism of the water species in the salmon

To characterize the water species retained in the salmon, dehydration response curves could effectively be employed. Figure 10 illustrates the dehydration response curves as a function of the water content of the salmon. The five samples of W_0 = 54, 99, 144, 200, and 252%-d.b. were prepared from a sample of W_0 = 342%-d.b., and each of the samples was treated by a PUP-operation (poultice up: the samples were stored in a dark room at T = 2°C for 24 h before use; this operation was conducted to provide homogeneity of water distribution in the sample). The PUP operation contributes to the stepwise increase of the dehydration rate, as shown in Figure 11, by the curves of $b' \sim f'$. Focusing on the five points of $b \sim f$ on the dehydration curve of W_0 = 342%-d.b. in Fig. 11(A), the dehydration rates of the five samples clearly demonstrate a stepwise increase, as shown in Fig. 11(B), by the curves of $b' \sim f'$ due to PUP operation. One can recognize that curves $b' \sim f'$ demonstrate a monotonous decay

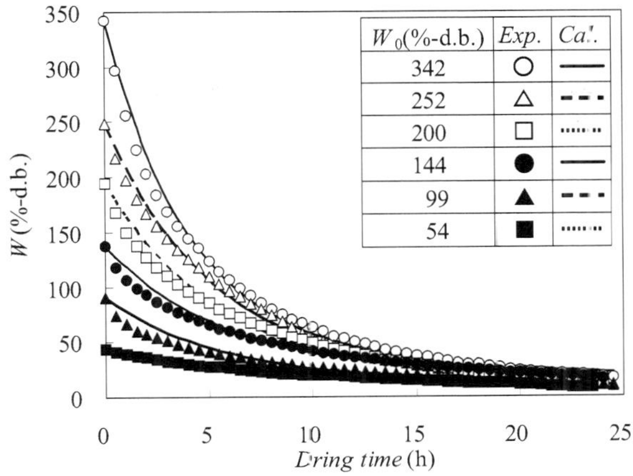

Fig. 10. Dehydration response curves of the salmon (T_D = 45°C).

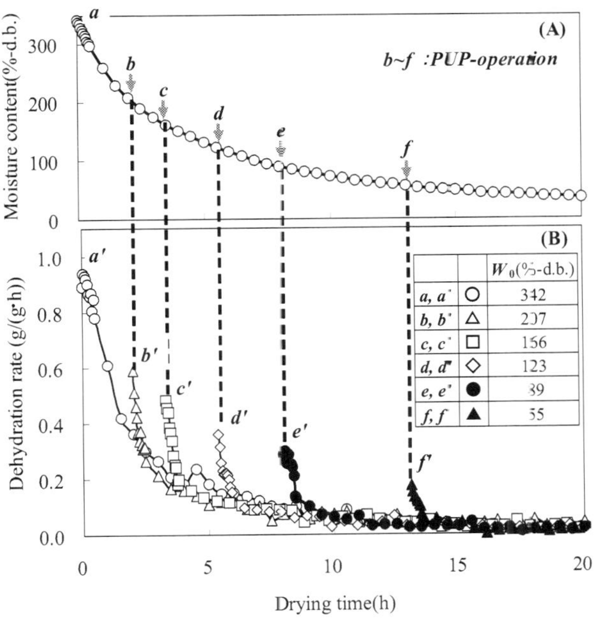

Fig. 11. Acceleration behavior of dehydration rate derived from the PUP operation for the salmon (T_D = 45°C).

curve reaching finally to the curve of *a'* with an increase in the elapsed time. Using the dehydration rates obtained, the effective diffusivity of the water species, *De*, can be evaluated as a function of W_0 by applying Equation (1).

5.2.2 Physicochemical mobility (*De*) of the water species

One can recognize the *De* value as a physicochemical mobility of the water species retained in the foods. Figure 12 illustrates *De* for the salmon as a function of W_0 and temperature. Although the data are widely scattered, one can roughly recognize two different water species (species A_1 and A_2) formed in the salmon, depending on the values of W_0. The two water species were divided at about W_0 = 120%-d.b., indicating an identical *De* for the water species A_1 region between W_0 = 120 and 360%-d.b. and a gradual decrease of *De* for the water species A_2 region with decreasing W_0 at 35~55°C. Using a rough evaluation of *De* obtained at the three temperatures, one would roughly construct the Arrhenius plots as shown in Figure 13. The slopes of the straight lines seem to give different activation energies depending on the water content of the samples.

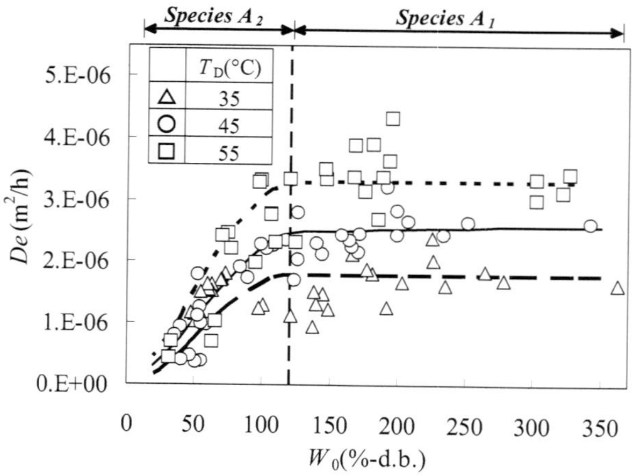

Fig. 12. *De* as a function of W_0 in the salmon.

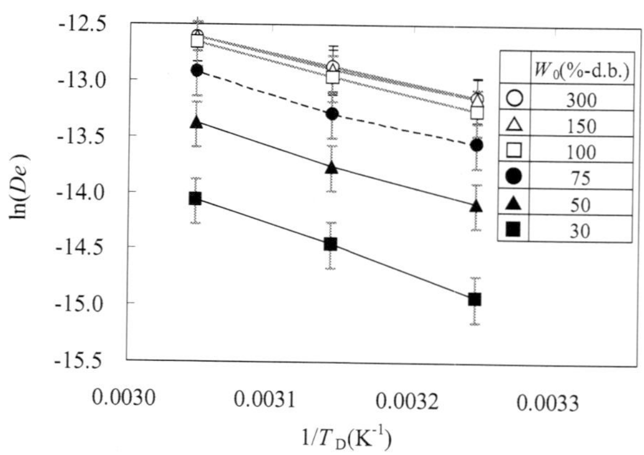

Fig. 13. Arrhenius plots of *De* for the salmon.

Figure 14 demonstrates E_D as a function of W_0. The activation energy obtained, the E_D values, demonstrates again two water species, A_1 and A_2, divided at W_0 = 120%-d.b. E_D in the water species A_1 region gave an identical value of 22 (±4) kJ/mol and a gradual increase from 20 to 38 (±8)kJ/mol with decreasing W_0. Taking into account the restriction of water species growth that accompanies increased dehydration, one can observe evidence of the activation energy growth (indicating a difficulty of the molecular diffusion of water species in the food tissue matrix due to the growth of the restriction strength) as a result of the decrease of water species mobility (*De*).

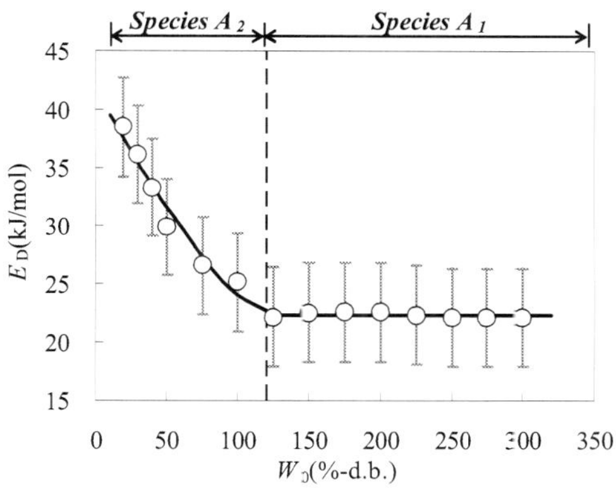

Fig. 14. E_D as a function of W_0 for the salmon.

As for a reason that the activation energy was drastically changed at W_0 = 120 %-d.b., it can be attributed to a change in the physical structure of the food tissue matrix due to progressive dehydration. The value of W_0 = 120%-d.b. corresponds exactly to the critical value of τ_C ($C\tau_C = 10^{-8}$s), at which water species A_1 and A_2 were divided according to their restriction strength, and the pre-exponential factor (PF = $\delta \cdot D_C$) of *De* showed steep growth with decreasing W_0 (as will be discussed in the upcoming section, see Fig. 16). The steep growth of PF resulted from the drastic change in the porosity and labyrinth factor of the salmon at W_0 = 120 %-d.b. accompanied by the gradual reduction of *De* decreasing W_0.

5.2.3 Water tank model for the computer simulation of the dehydration response curves

The transient response curves of dehydration in the drying operation of foods have been simulated using various mathematical models (Chhinnan, 1984; Madamba *et al.*, 1996; Thompson *et al.*, 1968; Noomhorm and Verma, 1986). In the present study, the dehydration response curves of the salmon presented in Fig. 10 are simulated using a water tank model as a first approximation. The water species retained in the sample are portioned to two different water tanks and the two tanks are respectively filled by the two different water species, A_1 (weakly restricted water species) and A_2 (strongly restricted water species), as shown in Figure 15 (Konishi *et al.*, 2001). During the drying operation, the two water species in the tanks are simultaneously drained by regulating the two valves as mass transfer coefficient, k_W (h^{-1}) and k_S (h^{-1}).

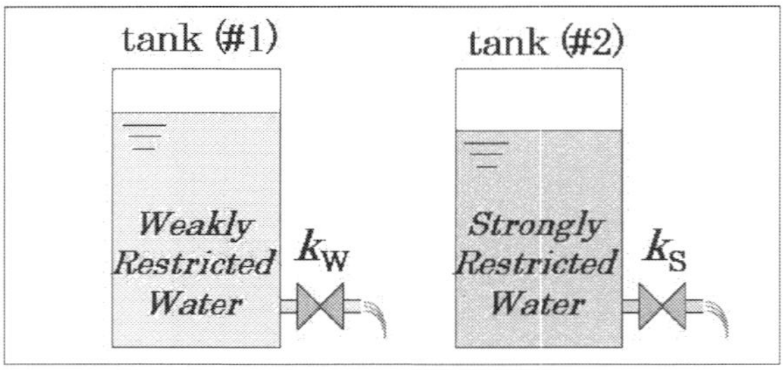

Fig. 15. Schematic explanation of the water tank model proposed as a dehydration of food.

On the material balance equation in the course of the dehydration operation presented in Fig. 10, the water content (W), presented as a perpendicular axis of Fig. 10, can be replaced by the water ratio, $W_R = (W\text{-}We)/(W_D\text{-}We)$, where We is the equilibrium water content (%-d.b.) and W_D is the initial water content of the sample. The dynamism of W_R for the two tanks is expressed by Equation (4).

$$\left(-\frac{dW_{RI}}{dt}\right)=\left(-\frac{dW_{Rw}}{dt}\right)+\left(-\frac{dW_{Rs}}{dt}\right) \tag{4}$$

$$-\frac{dW_{Rw}}{dt}=k_w\cdot W_{Rw} \tag{5}$$

$$-\frac{dW_{Rs}}{dt}=k_s\cdot W_{Rs}\ , \tag{6}$$

where W_{RW} is the ratio of the weakly restricted water species (=(W_W-We) / (W- W_D)), W_{RS} is the ratio of the strongly restricted water species (=(W_S-We) / (W- W_D)), W_W is the amount of weakly restricted water species (%-d.b.), and W_S is the amount of strongly restricted water species (%-d.b.).

The differential equation (4) is easily solved and expressed by Equation (7).

$$\begin{aligned}W_R=\frac{W-We}{W_D-We}&=W_{R1}\exp\left[k_s\exp\left(\frac{-E_D}{R\cdot(T_D+273)}\right)\cdot t\right]\\&+W_{R2}\exp\left[k_w\exp\left(\frac{-E_D}{R\cdot(T_D+273)}\right)\cdot t\right]\end{aligned} \tag{7}$$

$$k_s=\beta_1\cdot De^0 \tag{8}$$

$$k_w=\beta_2\cdot De^0\ , \tag{9}$$

where W_{R1} and W_{R2} are a proportion of the strongly restricted water amount and the weakly restricted water amount, respectively, and k_S (h^{-1}) and k_W (h^{-1}) are the mass transfer coefficient of the strongly restricted water species and of the weakly restricted water species, respectively. De^0 (= $\delta \cdot D_0$) is the pre-exponential factor of De, and β_1 and β_2 are the constants evaluated by a curve fitting simulation between the experimental and the calculated curves as shown in Fig. 10. The curve fitting obtained showed favorable agreement with the experimental dehydration curves as shown in Fig. 10 by solid curves.

5.2.4 Discrimination of water species A_1 and A_2 due to the pre-exponential factor (De^0) of De

The pre-exponential factor, De^0, can easily be evaluated by the extrapolation of the Arrhenius plot lines against the perpendicular axes in Fig. 13. Figure 15(A) illustrates De^0 as a function of W. The De^0 values obtained were of identical value regardless of W in the water species A_1 region, whereas in the water species A_2 region, there was a steep increase with decreasing W_0.

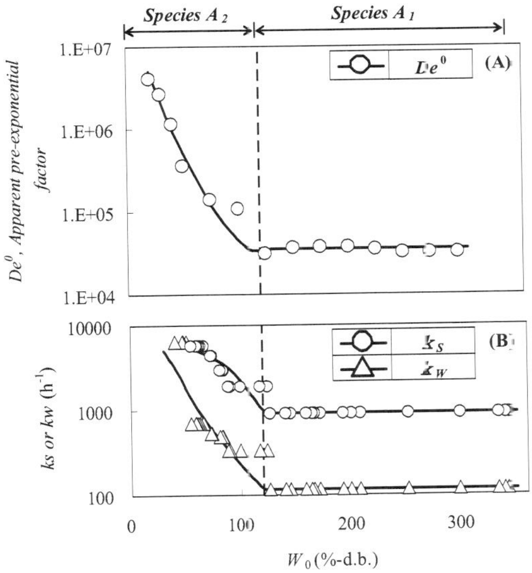

Fig. 16. Pre-exponential factor (De^0) and the dehydration parameters for the computer simulation (k_S, k_W) as a function of W_0 (T_D = 45°C) for the salmon.

Figure 16(B) illustrates k_S and k_W as a function of W_0. The two values show tendencies similar to De^0 even though they can be evaluated by the computer simulation. Comparing the two values between k_S (or k_W) and De^0, one can generate a linear relation between the two values as shown in Figure 17. The parameters of k_S and k_W roughly demonstrate the equations of k_S = 0.025(±0.008) De^0 and k_W = 0.007(±0.004) De^0. Using Equations (8) and (9), one may evaluate β_1 = 0.025(±0.008) and β_2 = 0.007(±0.004). Concerning the physical meaning of β_1 and β_2, since the two parameters were evaluated by the computer fitting simulation, they seem to be meaningless. This evidence indicates a disadvantage of the chemical engineering technique, even though the linear relation between k_S (or k_W) and De^0 would cause the presumption of a relationship between the two parameters.

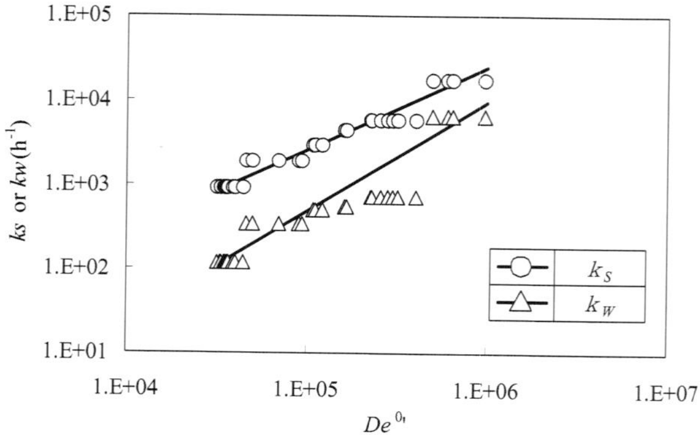

Fig. 17. Dehydration parameters (*k*s and *k*w) as a function of pre-exponential factor (De^0)(T_D=45°C) for the salmon.

6. Advantages derived from the hybrid technique

6.1 Visualized correlation between molecular mobility (τ_C) and physicochemical mobility (*De*)

Physicochemical mobility, *De*, which is evaluated by the chemical engineering technique (Fig. 12), and molecular mobility, τ_C, which is evaluated by the proton NMR technique (Fig. 8), are important parameters for characterizing the water species retained in foods as the diffusion rate in the three dimensional space of the food tissue matrix and as the rotation time of water molecules, respectively. The two parameters (*De* and τ_C) for the pork meat (P_H) and 12 jerkies are interestingly related, as shown in Figure 18. This clearly shows an advantage of the hybrid technique. *De* for P_H demonstrates a drastic change at $\tau_C = 10^{-8}$ s,

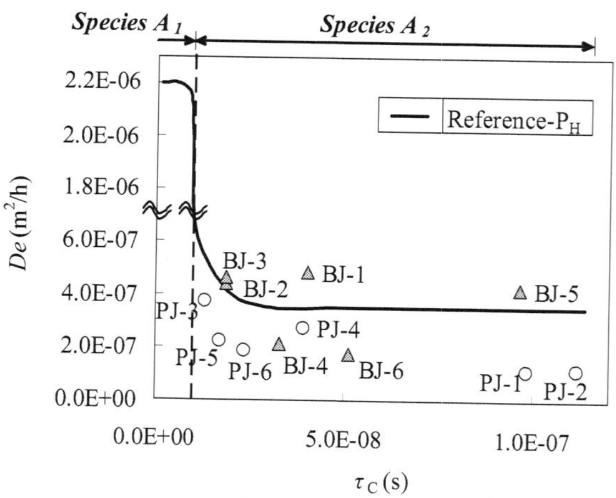

Fig. 18. *De* as a function of τ_C for jerkies and P_H.

indicating an identical De = 2.2(±0.2)×10^{-6} m^2/h in the species A_1 region and De = 3.5(±1.0) ×10^{-7} m^2/h in the water species A_2 region. All the De values for the commercial jerkies are distributed in the water species A_2 region, indicating an identical De = 3.0(±1.8)×10^{-7} m^2/h, which, even though the data are widely scattered, is the same as P_H. Regarding the jerky data, although all the samples distributed in the market commonly use the water species A_2 not A_1 and the effective diffusivity, De, falling in the range of 1.0~5.0×10^{-7} m^2/h, τ_C is widely distributed in the range of 10^{-8} ~ 1.1×10^{-7} s. Since each of the jerkies used in this study has own taste different from other, the wide distribution of τ_C has an important meaning in developing different tastes in jerky. This idea is also supported in an upcoming section.

6.2 Clear discrimination of the two water species due to diffusivity (*De*)

As has been demonstrated in Fig. 12, the De~W_0 plots could not clearly discriminate the water species A_1 and A_2. The De~τ_C plots derived from the hybrid technique, on the other hand, are conveniently able to visualize the discrimination between beef and pork and the two water species, A_1 and A_2, as shown in Figure 19. In particular, the boundary of the two water species is clearly recognized as $C\tau_C$ =10^{-8} s, indicating an advantage of the hybrid technique. Focusing on the behavior of B_A, B_H, and P_H, the De values gave a constant value in the two water species regions: in the A_1 region, as 3.1(±0.5) ×10^{-6}, 4.7(±0.1) ×10^{-6}, and 2.2(±0.3) ×10^{-6} m^2/h, respectively; in the A_2 region, as 1.8(±0.5) ×10^{-6}, 2.0(±0.4) ×10^{-6}, and 7.0(±1.0) ×10^{-7} m^2/h, respectively, even though τ_C is significantly changed from 10^{-8} to 4.5×10^{-8} s. These results suggest the water species A_2 has the interesting characteristic of being able to make a wide shift of the molecular mobility, τ_C, without changing in the De values. This result indicates that, by changing the value of τ_C as a design parameter for food products, the character of various meats can be significantly altered. The hardness of various meats and salmon, for example, can also be controlled by a change in τ_C, as shown in the next section.

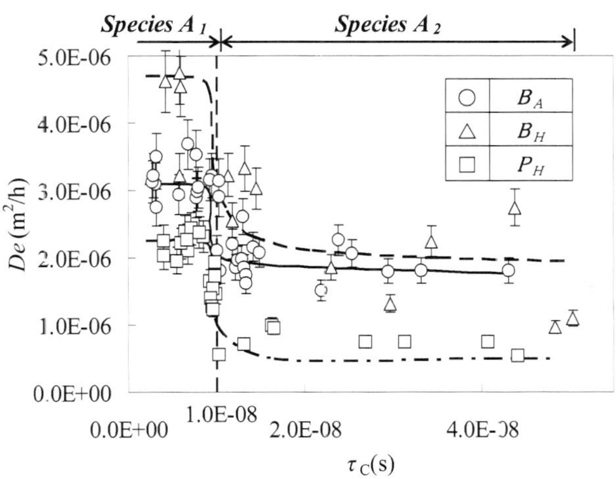

Fig. 19. Comparing behavior of De among B_A, B_H, and P_H as a function of τ_C at 70°C.

6.3 Clear discrimination of the two water species by hardness

Figure 20 demonstrates the hardness of B_A, B_H, P_H, and salmon as a function of τ_C. The hardness of all the samples shows a drastic increase at $C\tau_C = 10^{-8}$ s with increasing τ_C in the water species A_2 region. This clear discrimination between the water species A_1 and A_2 is an advantage of the hybrid technique. One may recognize that the N_P value in the water species A_1 region commonly showed 1.0(±0.5) ×10^6 N/m^2 regardless of τ_C in all the samples, whereas in the water species A_2 region and at the range of τ_C = 3×10^{-8} ~ 6×10^{-8} s, it varied widely depending on the samples, as with N_P = 1.5×10^7 for B_A, 1.3×10^7 for B_H, and 1×10^7 N/m^2 for P_H at τ_C = 5.0×10^{-8} s. From these results, one can recognize that it is possible to design food products with different N_P values without changing the τ_C values, and it's possible to design products with different τ_C values without changing the N_P values. Keeping N_P = 1.0×10^7 N/m^2, for example, one can have different τ_C values such as τ_C = 2.9×10^{-8} s for B_A, τ_C = 3.5×10^{-8} s for B_H, and τ_C = 4.8×10^{-8} s for P_H. Although the physical meaning of this τ_C value shift is still unclear, it can be attributed to the taste difference between B_A, B_H, and P_H because concrete evidence that the τ_C value changes depending on the kind of food, the concentration of seasonings, and the kind of seasonings has been obtained. Details of these results will be reported in an upcoming work (Konishi and Kobayashi, 2011).

The visualization of N_P ~τ_C exhibited in Fig. 20 clearly demonstrates the hybrid technique as a useful tool for discriminating between the two water species and among the kinds of food.

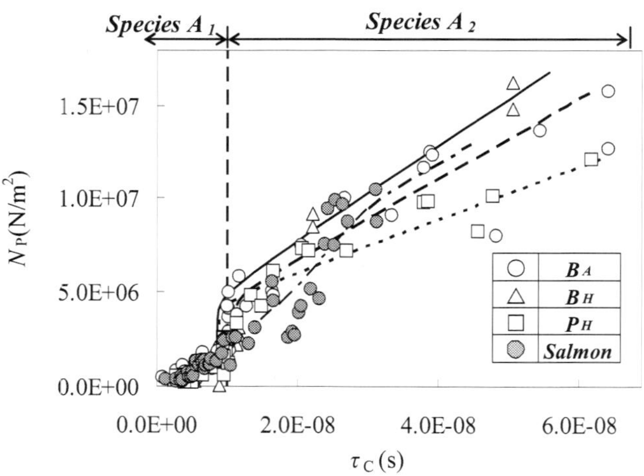

Fig. 20. Comparing behavior of N_P as a function of τ_C between B_A, B_H, and P_H at 50°C.

6.4 Physical meaning of $C\tau_C = 10^{-8}$ s derived from the pre-exponential factor (De^0) of De

Focusing on the boundary ($C\tau_C$ = 10^{-8} s) between water species A_1 and A_2, one may be interested in what is happening in the food tissue matrix at the $C\tau_C$ = 10^{-8} s. The information for the food tissue matrix can be obtained from Equation (2), and the equation is rewritten by Equation (10). The pre-exponential factor (δD_0) of Equation (10) consists of a function of ε,

χ, and ΔS. Since the δD_0 value can be evaluated from the extrapolation of the Arrhenius plots against the perpendicular axes in Fig. 13, these evaluated for B_A, B_H, and P_H can be plotted as a function of τ_C, as shown in Figure 21. Three δD_0 curves obtained demonstrate a steep decay at the $C\tau_C = 10^{-8}$ s. Based on the rough evaluation of the steep decay for the three curves to be 1/30 ~1/150, the reduction of the δD_0 values should be attributed to a simultaneous change in the ε value, which becomes smaller while the χ value becomes larger and the ΔS value becomes smaller because of the steep change in the physical structure of the B_A, B_H, and P_H's tissue matrix at the $C\tau_C = 10^{-8}$ s derived from dehydration.

$$De = \delta D_0 \cdot exp\left(\frac{-E_D}{R \cdot (T_D + 273)}\right) = \left(\frac{\varepsilon}{\chi}\right) \cdot \left(\frac{k \cdot (T_D + 273)}{h}\right) \left[exp\left(\frac{\Delta S}{R}\right)\right] \cdot exp\left(\frac{-E_D}{R \cdot (T_D + 273)}\right) \quad (10)$$

where k is Boltzmann constant, h is Planck constant, $\delta(=\varepsilon/\chi)$ is diffusibility, ε is porosity, χ is labyrinth factor, and ΔS is activation entropy.

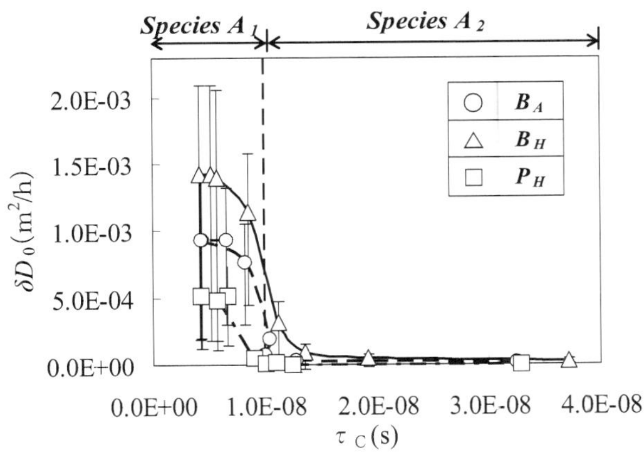

Fig. 21. Comparing behavior of the pre-exponential factor between B_A, B_H, and P_H at 50°C as a function of τ_C.

Figure 22 illustrates the optical microscope photographs of the cross-section for B_H. Comparing Figs. 22(A) and (B), one can clearly recognize that a large number of pores (complicated structures) appeared in the sample of W_0 = 54%-d.b. rather than that of W_0 = 127%-d.b., indicating a growth of the labyrinth factor (χ). This growth of χ contributes to the reduction of δ and is supported by the experimental evidences of the steep reductions of both δD_0 in Fig. 21 and De in Fig. 19 in the water species A_2 region.

The δD_0 ~ τ_C plots again demonstrate the hybrid technique to be a useful tool for discriminating between the two water species and among the kind of foods. The drastic change of δD_0 should derive the change of water species from A_1 to A_2, suggesting the diffusion mechanism change between the two regions. This idea is strongly supported by experimental evidence as shown in Figure 23. All E_D's of water species A_2 for B_A, B_H, and P_H demonstrated a drastic reduction at $C\tau_C$ in the water species A_2 region. This reduction of E_D strongly suggests a change in the diffusion mechanism from the water species A_1 region.

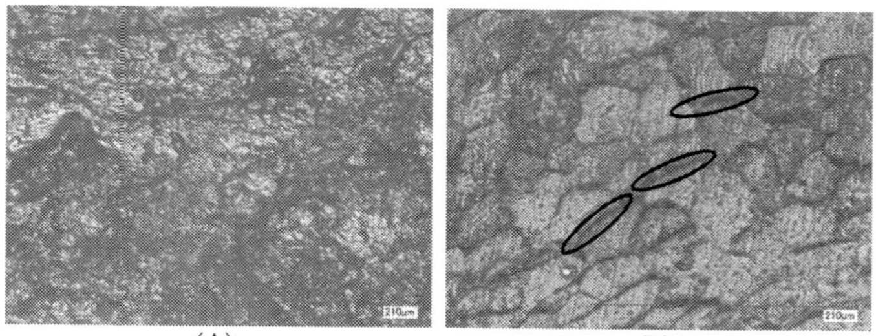

(A) (B)

Fig. 22. Optical microscope photographs (× 1000) of the cross-section for P_H, (A): 127%-d.b., (b): 54%-d.b.

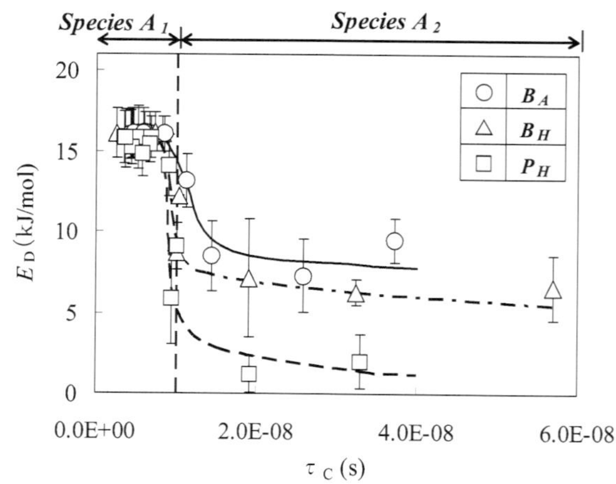

Fig. 23. Comparing behavior of E_D among B_A, B_H, and P_H as a function of τ_C at 50~70°C.

7. A new method for jerky product design using the parameters induced by the hybrid technique

Our interest is focused on a classification of the 12 jerky products influenced by multifunctional water species. Hills *et al.* (1999) empirically demonstrated a linear relation between the NMR relaxation rates ($1/T_2$) and a_w. In addition, Hills (1999) derived a theoretical equation to explain the linear relation. Since the relaxation time, T_2, can be replaced by τ_C using Equation (3), it is proposed that a_w should be related to τ_C. Figure 24 demonstrates a_w as a function of τ_C for the 12 jerky samples. All the samples fell roughly around three straight lines, indicating the existing of three groups. The three lines started from a_w = 0.92, which is an identical value independent of the samples and empirically evaluated, meaning a specified value for all kinds of foods (Konishi and Kobayashi, 2010). The a_w = 0.92 value always gave $\tau_C = 10^{-8}$ s independent of the kinds of foods and seasoning species. As has been discussed in the previous section, one can recognize both a_w = 0.92 and $\tau_C = 10^{-8}$ s to be critical values for foods in which the water species drastically changes from

A_1 to A_2 and the pore structure of the food tissue matrix provides regulation of ε and χ values accompanied by drastic increases or decreases.

Focusing on the three straight lines in Fig. 24, line-1 generates Group-1 (BJ-5 and PJ-2), line-2, Group-2 (BJ-4, BJ-1, PJ-4, BJ-6, and PJ-1), and line-3, Group-3 (PJ-5, BJ-3, BJ-2, and PJ-6). Based on the experimental data presented here, although one cannot demonstrate an exact physical meaning of the three groups' difference, it might be presumed that each group should have the specified seasonings different from others. Concerning the seasonings in the three groups, Group-1 is characterized by soy sauce, Group-2, by a mixed seasoning such as soy sauce, smoke flavoring, and salt, and Group-3, by a mixed seasoning such as salt, soy-sauce, and miso sauce. Although each of the three groups could not be distinguished by the chemical engineering (the aw ~ W_0 curves and the De ~ W_0 curves) and proton NMR techniques (the τ_C ~ W_0 curves), the hybrid technique (the aw ~ τ_C linear relation) clearly distinguished the characteristics of the seasonings for the 12 different food products. This evidence strongly demonstrates the hybrid technique to be a useful tool to design various food products requested by the commercial fields.

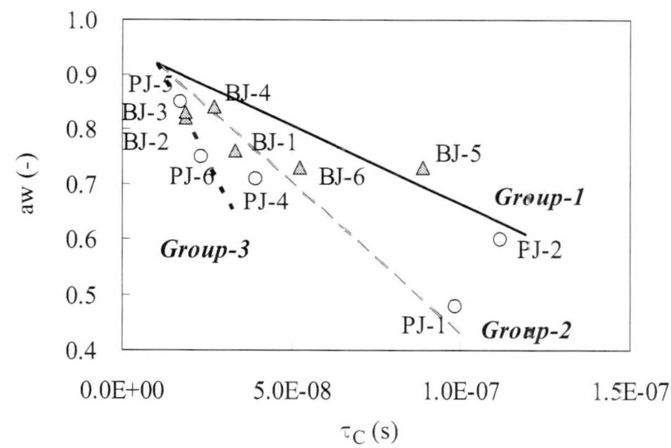

Fig. 24. a_w as a function of τ_C for the beef and pork jerkies.

8. Conclusions

The hybrid of the proton NMR technique and the chemical engineering technique was applied to discriminate the 12 commercially distributed jerkies, salmon, B_A, B_H, and P_H and to distinguish between the two water species, A_1 and A_2, by using four parameters: De, τ_C, N_P, and W_0.

1. The dehydration and adsorption of the water species, A_1 and A_2, retained in the foods were reversibly repeated and the transformation between the two water species was reproducible.
2. The proton NMR technique and the chemical engineering technique could respectively not discriminate among the τ_C ~ W_0 curves and among the a_w ~ W_0 curves obtained for the foods. Although the computer simulation model proposed by the chemical engineering technique generated a good fitting against the dehydration response

curves, the unknown parameters evaluated by the curve fitting, β_1 and β_2, were recognized as meaningless.

3. The hybrid technique proposed demonstrates a clear discrimination between water species A_1 and A_2, divided at the critical value of $C\tau_C = 10^{-8}$ s, where the drastic change in the values of *De*, E_D, and N_P appeared. The physical meaning of $C\tau_C$ was understood as a drastic change in the pre-exponential factor of *De* derived from the change in the food tissue matrix due to dehydration.
4. Using the hybrid technique, a variety of food products could easily be designed by changing the value of τ_C even though the values of a_w are identical. The meat jerkies commercially distributed in Japan were reasonably classified into three groups characterized by the three $a_w \sim \tau_C$ straight lines. The advantage of the hybrid technique brought a conclusion that the $a_w \sim \tau_C$ straight lines give characteristic slopes depending on the seasonings of the 12 jerkies. This could be useful in designing a variety of food products.

9. Nomenclature

B_A	*beef meat produced in Australia* (-)
B_H	*beef meat produced in Hokkaido, Japan* (-)
D	*moisture diffusion coefficient* (m^2/h)
D_0	*frequency factor of D* (m^2/h)
De	*effective water diffusion coefficient* (m^2/h)
De^0	*pre-exponential factor of De* (PF=$\delta \cdot D_0$, m^2/h)
E_D	*activation energy of water diffusivity* (kJ/mol)
H	*Planck constant* (erg. s)
I	*nuclear spin quantum number of water proton* (= 0.5) (--)
K	*Boltzmann constant* (erg/deg)
k_S	*mass transfer coefficient of strongly restricted water* (h^{-1})
k_w	*mass transfer coefficient of weakly restricted water* (h^{-1})
L_a	*half distance of a-axis of the rectangular sample* (m)
L_b	*half distance of b-axis of the rectangular sample* (m)
L_c	*half distance of c-axis of the rectangular sample* (m)
M_t	*magnitude of magnetization vector* (-)
M_0	*magnitude of macroscopic equilibrium magnetisation vector* (-)
N_P	*hardness of meat products* (Newton/m^2)
P_H	*pork meat produced in Hokkaido, Japan* (-)
R	*gas constant* (=8.314J/K mol)
r	*proton-proton distance of water molecule* (= 0.16 nm)
ΔS	*activation entropy* (kJ/mol)
T_2	*spin-spin relaxation time of water proton* (s)
T_D	*drying temperature* (°C)
t	*drying time* (h)
ts	*spin locking pulse length* (s)
W	*water content at the drying time t* (%-d.b.)
W_0	*initial water content at the time of PUP operated* (%-d.b.)
W_D	*initial water content of drying flesh sample* (%-d.b.)
We	*equilibrium water content* (%-d.b.)

W_R	*the water ratio* (= (W-We)/(W$_D$-We))
W_{R1}	*the proportion of the strongly restricted water amount* (-)
W_{R2}	*the proportion of the weakly restricted water amount* (-)
W_{RS}	*the ratio of the strongly restricted water species* (-)
W_{RW}	*the ratio of the weakly restricted water species* (-)
W_S	*the amount of strongly restricted water species* (%-d.b.)
W_W	*the amount of weakly restricted water species* (%-d.b.)

Greek letters

β_1	*constant evaluated by the curve fitting simulation* (m^{-2})
β_2	*constant evaluated by the curve fitting simulation* (m^{-2})
ε	*porosity of the food tissue* (-)
π	*the ratio of the circumference of a circle to its diameter* (=3.14)
γ	*gyromagnetic ratio of proton* (=2.675×10^8 rad·T^{-1} s^{-1})
$\hbar$	*modified Plank's constant* (=6.63×10^{-34} J·s)
ω_0	*resonance frequency* (=3.14×10^9 s^{-1})
τ_C	*correlation time of water proton* (s)
$C\tau_C$	*critical correlation time of water proton* (s)
χ	*labyrinth factor of the meat tissue* (-)
δ	*diffusibility* (=ε / χ)(-)

10. Acknowledgment

This work was financially supported by the Cooperation of Innovation Technology and Advanced Research in Evolution Area (City Area) from the Japanese Ministry of Education, Culture, Sports, Science, and Technology. The authors wish to thank Associate Professor Koichi Miura, Kitami Institute of Technology, for his assistance with the proton NMR analysis.

11. References

Abragam, A., (1963). The Principles of Nuclear Magnetism. *Oxford at the Clarend Press*, p347-349.

Andreu, J. & Stamatopoulos, A.A. (1986). Durum wheat pasta drying kinetics. *Lebensm. Wiss. u.-Technol.*, 19 , 448-456.

Belton, P.S.; Gil, A.M.; Webb, G.A., & Rutledge, D., Eds.(2003). Magnetic resonance in food science-latest developments, *The Royal Society of Chemistry, Cambridge.*

Chhinnan, M.S. (1984). Evaluation of selected mathematical models for describing thin-layer drying of in-shell pecans. *Trans. ASAE*, 27(2), 610-615.

Fennema, O.(1976). In principles of food science, Part 1, Marcel Dekker, New York.

Hills, B.P. (1999). NMR studies of water mobility in foods. In"Water Management in the Design and Distribution of Quality Foods" ISOPOW 7(Y.H. Roos, R.B. Leslie, and P.J.Lillford, eds) Technomic Publishing, Lancaster, PA.

Hills, B. P., Manning, C. E. & Godward, J. (1999). A multistate theory of water relations in biopolymer systems. In: *Advances in Magnetic Resonance in Food Science* (edited by P. S. Belton, B. P. Hill and G. A. Webb). Pp.45-62. Cambridge, UK: Royal Society of Chemistry.

Jason, A.C.(1958). A Study of evaporation and diffusion processes in the drying of fish muscle in fundamental aspects of dehydration of food stuffs, ed. Society of Chemical Industry. *McMillan, London*, pp.103-134.

Kannan, D. and Bandyopadhyay, S. (1995). Drying characteristics of a tropical marine fish slab. *J.Food Sci. Technol.*, 32(1), 13-16.

Konishi, Y.; Horiuchi, J., & Kobayashi, M.(2001). Dynamic evaluation of the dehydration response curves of foods characterized by a poultice-up process using a fish-paste sausage-I. Determination of the mechanisms for moisture transfer, *Drying Technology*, 19(7), 1253-1269

Konishi, Y.; Horiuchi, J., & Kobayashi, M.(2001). Dynamic evaluation of the dehydration response curves of foods characterized by a poultice-up process using a fish-paste sausage-II, A new tank model for a computer simulation, *Drying Technology*, 19(7), 1271-1285.

Konishi, Y.;Miura, K. & Kobayashi, M.(2003). Drying efficiency design using multifunctional dynamics of water molecules in foods-H-NMR analysis of a fish paste sausage and squid, *AIDIC Conference series*, vol.6, 183-190.

Konishi, Y. & Kobayashi, M.(2009). Quantitative evaluation of the design parameters requested in beef and pork drying operation, *AIDIC Conference Series*, 9, 177-186.

Konishi, Y., & Kobayashi, M., (2010). Food product design using the water species as a probe molecule. Annual meeting of Japanese Society of Food Engineering.

Konishi, Y., Kobayashi, M., & Miura, K. (2010). Characeterization of water species revealed in the drying operation of Todarodes pacificus Steenstrup using water proton NMR analysis, *International Journal of Food Science and Technology.*, 45.1889-1894.

Konishi, Y. & Kobayashi, M.(2011). Dynamism of the water species as a probe molecule in food, *Chemical Engineering Transactions*, Ed. Sauro Pierucci, 24, 475-480.

Labuza, T.P., Tannenbaum, S.R., & Karel,M. (1970), Water Content and Stability of Low moisture and intermediate moisture foods. Food Technol.24, 543-550.

Litchfield, J.B. & Okos, M.R. ,1992. Moisture diffusivity in pasta during drying. *J. Food Engng.*, 17, 117-142.

Madamba, P. S., Driscoll, R. H. & Buckle, K. A.(1996). The thin-layer drying characteristics of garlic slices. *J. Food Engng.*, 29, 75-97.

Noomhorm, A. & Verma, L.R.(1986). Generalized single layer rice drying models. *Trans. ASAE.*, 29(2), 587-591.

Rockland, L.B. and Stewart, G.F., Eds.(1981). International symposium on properties of water, water activity: Influences on food quality, *Academic Press Inc., London.*

Schmidt, S.J., Water and Solids Mobility in Foods, (2004). Advances in Food and Nutrition research, volume 48, 2004, Edited by S.L. Taylor, Elsevier

Thompson, T.L., Peart, R.M. & Foster, G.H. (1968). Mathematical simulation of corn drying (A new model). *Trans.ASAE.*,24(3),582.

Waananen, K.M. & Okos, M.R. (1996). Effect of porosity on moisture diffusion during drying of pasta. *J. Food Engng.*, 28, 121-137.

4

Modelling Approach for Redesign of Technical Processes

Ivan Lopez-Arevalo[1], Victor Sosa-Sosa[1] and Saul Lopez-Arevalo[2]
[1]Information Technology Laboratory, Cinvestav-Tamaulipas
[2]School of Electrical and Electronic Engineering
University of Nottingham, University Park, Nottingham
[1]Mexico
[2]UK

1. Introduction

Nowadays the design and development of new products or modification of existent ones (redesign) is a key and fundamental element to enhance innovation and competitiveness of industrial companies. Design has an increasing importance to differentiate one product from another.

In general, design is the process of specifying a description of a product that satisfies a set of requirements (Umeda 90). Redesign is the process of changing the description of an existent product to satisfy a new set of requirements (Brown 98). Design engineering includes both design and redesign. In the literature we can find diverse terms to narrow design and redesign, such as preliminary, conceptual, functional, creative, routinary, non-routinary, personified, parametric, innovative, etc., but the characteristic activities of the global design engineering can be divided as follows [Subba-Rao 99], see Figure 1:

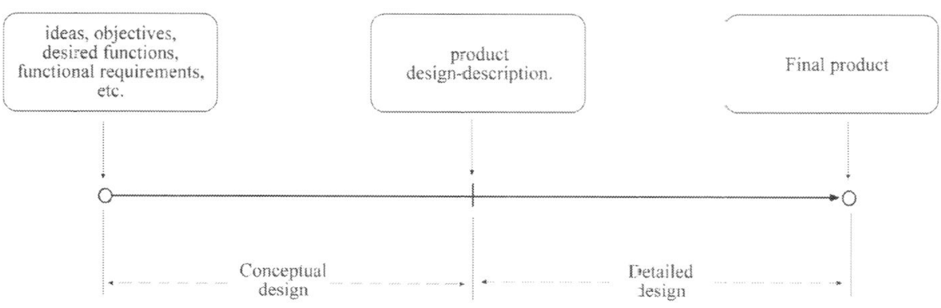

Fig. 1. Product design path.

- Conceptual (re)design, the phase where the global goals, requirements and operation of the product are established based on abstract concepts. The research presented in this chapter deals with this aspect.

- Detailed (re)design, the phase where the results of the conceptual design are used to physically implement a product.

Design engineering involves a wide range of activities. Design engineering can appear in a broad variety of domains, from the assembly of brakes to complex industrial plants and from simple chips to the most advanced super computers. Both design and redesign consist of two main components: the (re)design process and the (re)design object. The (re)design process involves all the (re)design activities performed over the (re)design object, which is the subject entity to be (re)designed. In engineering domains is common to refer to the (re)design object as the artefact. An artefact is a type of product to denote physical and technical devices. The (re)design process over an artefact is characterised by the map between functional requirements to structural requirements. Design and redesign, both can be considered as a dialectic process between goals (what it is desired) and possibilities (real constraints), directed to the satisfaction of functional specifications and performance (Stephanopoulos 90b).

The industry deals with complex technical processes where their behaviour is only predicted by means of complex numerical simulators. The redesign of a process is sometimes necessary when certain time has passed from its implantation or when they must adapt to economical, technological, or environmental requirements. The redesign is not part of the maintenance stage but must be considered into the process life cycle.

From a general point of view, the redesign is done typically in three steps: design-description acquisition (modelling), problem analysis (diagnosis) and proposal of modifications (generation of alternatives). In real redesign situations, human designers intuitively create mental abstract models by removing superfluous information about the process. Such models are based on functions of the components of the process and its context.

From the early 60's, Artificial Intelligence techniques have been used for design, such as constraint-based systems, case-based reasoning, model-based reasoning, planning, neural networks, and genetic algorithms. Although in these approaches the modelling and simulation of the processes has been solved in acceptable way, another problem has been generated, the used knowledge representations require so detailed information that sometimes it is difficult to understand.

The main objective of this work is to obtain a support framework to assist the human designer in the redesign of complex technical processes. The structure of this framework must be based on the common redesign activities performed by human designers on real redesign situations. Therefore, the framework must able to reduce the complexity of the processes to be redesigned, and therefore facilitate the redesign activities.

The framework was obtained integrating model-based reasoning and case-based reasoning techniques. Using model-based reasoning the original process can be modelled hierarchically. Using case-based reasoning alternative process parts can be obtained from other processes, which have to be adapted into the original process. The framework was implemented in the Chemical Engineering domain due to the complexity of the processes involved and the interaction with experts in the area.

The proposed redesign framework has to be able to deal with complex technical processes. In this sense, the type of processes we are referring to follow some assumptions:

1. The complexity of the process must be high.
2. Complex numerical simulators can be used to model the behaviour of the process.
3. The process is already implanted, which means there is a design solution that satisfies the original requirements of such process.
4. Human designers can understand the process intuitively identifying its functional sections.
5. The process can be represented by functional abstract concepts.

The chapter is organised as follows. In section 2 the contextualisation background about redesign and modelling is given. Section 3 describes the proposed modelling approach for redesign. Section 4 shows the experimentation carried out and the obtained results of a first prototype for the proposed approach. Finally, the section 5 presents some conclusions and remarks.

2. Background

2.1 The process of redesign

The research work related to the process of redesign is huge. Following the different research approaches are presented from a general point of view to more specific one. Some of the most interesting definitions of design that we have found in the literature are summarised as:

- Design is formally a search problem in a large space of objects that satisfy multiple constraints (Chandrasekaran 90).
- Design is the task of devising courses of action to change or create better ones (Simon 96).
- Design starts with an intended activity or use (Maher 97b) and uses available knowledge to arrive at a description of an artefact which will produce those results (Gero 90b).

Defining design is difficult because the term refers both to a product (the object to be designed) and a process (the process of design). The reasoning process involved in design allows moving from a functional concept as a starting point to a product solution. Therefore, the design activity can be seen as an activity of synthesis, which is strongly influenced by the skills and mental models of the designer.

Regarding the process of redesign we have identified the following definitions in the literature:

- Redesign is considered as design in which there is a priori knowledge on the general and specialised functions to be performed and on the working principles to be selected [Salomons 95].
- Redesign is an inherent part of most design processes; in which new requirements or new domain knowledge influence the original design process (Brazier 96); but can also be seen as a family of design methods in itself (Pos 97).
- Redesign is part of design, which proposes suitable modifications free for inconvenience of existent artefacts (Kitamura 99).

Independently of the point of view of redesign, three types of redesign can be identified (Dixon 89): parametric redesign, component redesign, and structural redesign. In order to perform any of these redesign types, it is essential that some form of knowledge is available that allows the adaptation of existing designs. According to Pos (Pos 97) and based on the previous mentioned definitions of design/redesign, is possible distinguish two general points of view about the relationship between design and redesign, these are:

1. Viewing the design as total set which contains redesign as a subset. In order to satisfy this relationship, all the components of the design reasoning process should be satisfied for redesign. However redesign as a specialised subset would not be applicable in the same contexts as the more general notion of design. Here, design is viewed as an iterative process that uses intermediate results to get a final design description that fulfills the requirements. The task of redesign on the basis of a design created earlier produces a new temporary design description that is closer to the specification than the former design description.
2. Viewing both design and redesign as independent sets joined by a small common subset. For this relationship to be satisfied there is an expectation that some crossover or overlap will occur, thus only some of the components of design reasoning will be applicable in the redesign context and viceversa. Here, redesign starts with a previously constructed design description and a new set of requirements. The previously constructed design description must now be modified to fulfill the new set of requirements.

Adopting any of the above points of view, basically minimal differences can be distinguished. In both cases the important issue is to bridge the gap between a set of requirements and an existing design description. We can see that design starts from scratch, however, redesign starts with an existing design description, which is modified until it fulfills the current requirements. Both points of view can be captured by a single spectrum of problem-solving methods for redesign.

Several authors (Akin 82, Chandrasekaran 93, Eldonk 96, Brazier 96, Bridge 97, Pos 97) state that the required knowledge for redesign is based on the following two principles:

- Minimise changes in the current design, and
- Maximise existing properties and benefits of the current design.

Many systems that solve redesign problems have been described in literature (Mitchell 83, Howe 86, Fischer 87, Daube 89, Goel 91, Smyth 96, Eldonk 96, Kitamura 99). However when one takes a closer look at the different variants of the redesign task, subtle differences exist that have an impact on how the task can be performed and what kinds of knowledge are involved.

There are a variety of research works referring to design or redesign; from (re)design of abstract (for example, components in software engineering) to physical entities (for example, a reactor in chemical engineering), for a general review see (Brown 97), for some details see (Akin 82, Mitchell 83, Howe 86, Fischer 87, Mostow 89, Goel 91, Bras 92, Stroulia 92a, Chandrasekaran 93, French 93, Brazier 96, Eldonk 96, Pos 97, Price 97, Umeda 97, Gero 98, Culley 99, Culley 99, Kitamura 99, Kraslawski 00, Grossmann 00, Arana 01, Maher 01).

Models in design and redesign are particularly important to guarantee they represent the intentions by which they were created. In general, the models are abstractions of the reality that guarantees communication of ideas by joining concepts, aggregations and relations

(Bridge 97). Akin (Akin 82) outlines that the representational aspects to determine the utility of a model in design are:

- The represented information must be in a level of abstraction suitable for its intention.
- The contents must be in such a way that they are compatible with the expected results according to the mental representations of the designer.
- The model must be consistent with the reality that tries to reflect.

A substantial amount of research has focused on defining models of design (French 85, Tomiyama 87, Treur 89, Brown 89, Chandrasekaran 90, Gero 90a, Takeda 90a, Alberts 92, Vescovi 93, Ohsuga 97, Brown 97). Most of this research highlights that the modelling of the functionality (or properties) of the design object description is an important aspect of the overall design process.

It is possible to represent explicit knowledge in (re)design by means of modelling functions of artefacts. This facilitates the systematisation of the reasoning and some tasks of (re)design. The reasoning based on functions allows abstracting information of the design on the same way as it is made in the reasoning of the initial stages of the design. The process of design of an artefact starts with the conceptual or functional design followed by the basic design and the detailed design (Stephanopoulos 90a). Within these, the functional design plays the central role since it guarantees the quality of the design and the innovation of the product (Umeda 97, Culley 99). The idea of function is fundamental in design since the work of the designers is to design artefacts that must achieve explicit functions (Chandrasekaran 00). Functional modelling is useful to model the object of (re)design, this modelling of objects enhances the formulation of (re)design strategies and the overall (re)design process. Functional modelling "hides" sections of the artefact structure at a lower abstraction level facilitating the manipulation of the artefact description.

Most of the research work on (re)design considers redesign as a knowledge-intensive field; wherein the processes (e.g., tasks) performed, descriptions of sequencing of processes, descriptions of the information within the system, and knowledge employed to perform a task are explicitly modelled most of the times by means of knowledge-based systems. These modelling frameworks try to model the (re)design so the (re)design object as well as the (re)design process are understandable by humans. To do this, human designers use the object specifications to propose a reasonable (re)design approach need to be understood (Leveson 00). Reasoning strategies employed in (re)design are derived or extensions of the commonly named *problem-solving strategies*. Examples of strategies are hypothesis and test (Hempel 66, White 05), pattern recognition (Doyle 62, Kirsch 64, Mitchell 97), skeletal plan refinement (Friedland 85, Tu 89), heuristic classification (Clancey 85), propose and revise (Goel 89), propose critique modify (Chandrasekaran 90), decision tree search (Raiffa 68, Qi 92), means-ends analysis (Newell 63, Rasmussen 86), and reasoning by analogy (Gick 80, Gentner 83).

In the above strategies, the human designer needs to formulate an explicit model of expertise as an integration of two types of models: a domain model and problem solving method. The domain model corresponds to the (re)design object and the problem solving method model corresponds to the (re)design process. Work on domain modelling has only recently attracted the attention of knowledge based system researchers (Stephanopoulos 90a, Schoen 91, Gruber 93, Skuce 93, Sowa 95, Kitamura 98, Fensel 01b, Gomez-Perez 04).

The problem solving method determines how those entities in the model will be used in the actual problem solving process. Domain specific concepts, relationships, and knowledge pertaining to them are captured in the domain model through ontologies (Chittaro 93, Kitamura 99, Fensel 01b, Kuraoka 03).

Independently of models and strategies employed in the (re)design, it is important that such data and knowledge can be recorded in a consistent manner for the future understanding of the (re)design; this constitutes what is called *(re)design rationale.*

2.2 The role of function in the design process

Functions in design play the central role since it guarantees the quality of the design and the innovation of the product (Umeda 97, Culley 99). Function is regarded as what a design object is supposed to do; it is a manageable representation of the overall behaviour of the object (Price 98). Some authors define function as an abstraction of its intended behaviour strongly related to its context (Gero 90a, Goel 92, Stroulia 92a, Chittaro 93, Brown 97, Chandrasekaran 00). Initially the human designers think in functions before they are concerned with specific properties. Functions can exist at different levels of abstraction, depending on the design phase that one is in and the current focus of the design interest. In preliminary design phases, functions usually are independent of working principle, whereas in later design phases, when the functions have been detailed, they become more and more dependent on the working principle that has been selected. In the following, a distinction between three levels or categories of functions is made:

- *General functions.* (Keuneke 91, Lind 94, Kitamura 98, Bo 99) proposed a restricted list of general functions dealing with the transformation of matter, energy and/or information, which are independent of the working principle.
- *Specialised functions or subfunctions.* Act on flows, forces, moments etc., independent of the working principle.
- *Working principle dependent function.* Salomons (Salomons 95) defines it as the realisation of a specialised function (by means of physical phenomena). Several alternative solutions for fulfilling working principle dependent functions can exist without changing the working principle itself.

2.3 The design process

The design process is a complex and not yet well understood cognitive process conducted by humans (Salomons 95). The design process is related to the process of actions and decisions that are taken during design in order to arrive at completed product design. Models of design processes provide a structured description of a process of design. The models differ in their underlying formalisations and have been represented in structures such as:

- blackboard architectures (Ball 92),
- algorithms (Alberts 93b),
- SOAR (cognitive architecture for developing systems with intelligent behaviour) (Steier 91),
- task models or problem solving methods (Brown 89, Brazier 94, Wielinga 97), or

- agent architectures (Dunskus 95, Berker 96, Lander 97).

The following models of the design process can be distinguished:

- Prescriptive models (Salomons 95)
- Descriptive models (Stephanopoulos 90b, Ohsuga 97, Sumi 97)
- Opportunistic design (French 93)
- Decision support problem (Bras 92, Thornton 93, Ullman 91)
- Theorem problem solving process (Takeda 94a)
- Human learning process (Gero 04)
- Multiagent design (Wood 01, DeLoach 04).

2.4 The design object

The design object is the central "actor" that receives the attention during the overall design process. This can be a model of a component, artefact, process or system. Traditionally, the design object was created by technical drafts; but with the advent of computers, the design object has become a computer model that can be shown, modified and deleted easily. Thus, several models of artefacts have been used in design.

Some authors (Rasmussen 86, Douglas 88, Hoover 91, Lind 94, Turton 98, Leveson 00) have observed that abstractions of the design object are important during the design process to manipulate design objects. In this sense, Hoover (Hoover 91) has observed that:

- the design object evolves through abstractions and refinements.
- abstractions and refinements are selected opportunistically and are characterised by the designer focusing on a few aspects of the design object at a time.
- refinements are made within the framework of abstractions.
- conceptual, layout, and detailed stages are not distinct steps in the design process.

Several research works have been developed about (re)design object manipulation. The most relevant approaches in this issue are model-based design and case-based design.

2.4.1 Model-based design

One of the most used approaches in the manipulation of the (re)design object is model-based design which really is a branch of model-based reasoning (MBR) applied to (re)design. Model-based reasoning constitutes a set of techniques applied in several domains and it is used to create models and reasoning about the domain. Mainly the most used technique from MBR has been *compositional modelling* (Falkenhainer 91, Falkenhainer 92, Nayak and Joskowicz 96), which is an approach to construct a model of an artefact (components, devices, processes, systems, etc.) on the basis of a description of the artefact and a query about the composition of the artefact. The modelling of functions (*functional representation*) (Sembugamoorthy 86, Chittaro 98, Chandrasekaran 00) is crucial in compositional modelling. Functional Representation is a top-down approach to describe functions on devices (function), its structure (structure) and its causal processes (behaviour) of the device that culminate with the achievement of the function. Functional modelling reduces drastically the amount of information if simulation is required (Price 98). The approaches of functional modelling can be classified in two groups:

- *State-based representations.* It uses units of function representation, which are abstractions of behaviour states. Behaviour states and hence functions may be associated even with static objects which do not cause any state change.
- *Flow-based representations.* Flow-based representations are based on the concepts of flow and effort. In this approach exists a predefined set of functions, and functions of all existent components are expressed in terms of these primitive functions. This approach is based on the System Theory (Bertalanffy 50) and its derivatives (Abstraction Hierarchy (Rasmussen 86), Qualitative Process Theory (Forbus 84), and Multilevel Flow Modelling (Lind 90, Lind 94)).

2.4.2 Case-based design

Case-based design is a branch of case-based reasoning (CBR). CBR is a general paradigm to solve problems based on the recovery, reuse, revision and retention of specific experiences (cases) (Aamodt 94). CBR has been applied to component-based systems (Maher 97a) which is, however, mostly concerned with the manipulation of design object descriptions. This paradigm is particularly attractive in domains where explicit models do not exist or its understanding is difficult (Kolodner 93). In CBR, similarities between formal methods implemented in computer programs and informal observations from designers are taken into account (Maher 97a). The applications of CBR can be for classification or synthesis tasks. (Re)design problems are within the synthesis tasks.

The direct or analogical use of previous designs or plans of design can reduce and improve the quality of design because take advantages of previous experiences (Maher 95). CBR is viewed as a redesign process for the adaptation of a case where a new artefact (named goal) is designed to achieve certain function, its physical structure can be inferred in analogical way from some physical, chemical or biological object (named source) whose function is similar to the required function.

CBR has been applied to solve problems of real world; there are several works about case-based design for example (Goel 89, Qian 92, Sycara 92, Bhatta 94, Borner 96, de Silva Garza 96, Maher 01, Price 97).

2.5 Redesign approaches

Here the division between research on design and redesign is remarked, only research from the redesign perspective is presented.

2.5.1 Generic approaches in engineering

Goel et. al. [Stroulia 92a, Stroulia 92b, Goel 94b, Goel 97a] presented a control architecture for model-based redesign in the context of case-based redesign. They state that the redesign task is characterised by small differences in the functions desired of and delivered by an existent known design. The redesign is divided in three subtasks: a) generation of modifications to the structure of the old design, b) realisation of the modifications on the structure, and c) evaluation of the new design. Eldonk et. al. [Alberts 93a, Bakker 94, Eldonk 96] presented a redesign approach based on techniques developed in model-based diagnosis. Eldonk et. al. state that redesign activities are diagnosis and respecification. The objective of this approach is to find the part of the system, which causes the discrepancy between a formal specification of the system to be redesigned and the description of the existing technical system. Kitamura and Mizoguchi [Sasajima 95, Kitamura 99] proposed a redesign approach based on ontologies of functional concepts. They focus on capturing the rationales of design of an artefact and in organising general strategies of redesign. For the first point, they use an ontology of functional concepts that allows to identify functional structures and to represent automatically part of the design rationale. For the second point, they use an ontology of redesign strategies. This approach consists of the following stages: functional understanding, analysis of requirements, proposal of alternative and evaluation.

2.5.2 Mechanical engineering

Aranna et. al. [Fothergill 95, Forster 96, Forster 97b, Arana 00] proposed a redesign environment called DEKLARE, which supports acquisition, representation and reuse of redesign knowledge. It allows the designer to use design techniques to suggest alternative designs that fulfill specific requirements. Gupta et. al. [Das 94] proposed a methodology that automatically provides suggestions of redesign for reducing setup costs for mechanical parts. This approach is based on the interpretation of the design as a collection of mechanical features. The objective is to generate alternative mechanical features by means of geometric changes of the original parts and adding them to the feature set of the original part. Kim [Kim 93] proposed an approach for redesign of assemblies by means of planning techniques. Kim deals with the absence of required design information using the replay and modify principle. He employs a reverse engineering model to infer information about the process executed when creating a given design, and using the inferred information for design recreation or redesign. The propose model consists of the three stages: knowledge acquisition, construction of the default design plan, and redesign based on cases.

2.5.3 Electrical and electronic engineering

Steinberg and Mitchell developed a system to redesign VLSI circuits [Steinberg 85]. This redesign approach is based on planning techniques and causal and teleological reasoning [de Kleer 79]. The subtasks of this approach are: a) focus on an appropriate section of the circuit, b) generate redesign options to the level of proposed specifications for individual modules, c) rank the generated redesign options, d) implement the selected redesign option, and e) detect and repair of side effects resulting from the redesign. Maulik et. al. [Maulik 92] proposed the use of optimisation techniques to redesign CMOS analog circuits. The optimisation approach is guided by three principles. First, equations that describe device characteristics are encapsulated and separated from equations that describe the performance of the circuit topologies. Secondly, constrained optimisation techniques are employed to synthesise the redesigned-scaled CMOS circuit. Finally, constrained optimisation allows the solution of some final constraints over specific variables. Based on the approach of Umeda et. al. [Umeda 92], Tomiyama et. al. [Umeda 94] describe an extension of their approach taking into account the potential functions of the components of an artefact to redesign it. The architecture consists of sensors, which monitor the machine, and a model-based reasoner diagnoses faults and plans repairs. The system generates a FBS model (Function-Behaviour-State) based on the design object and then searches the model for candidate redundant function. The FBS model consists of a function hierarchy that represents the designer's intentions, and a behaviour network that describes how the function hierarchy is carried out. The system first tries a control type strategy that adjusts various machine parameters. If the strategy fails the system applies a strategy based on functional redundancy, it uses the potential functions of existing parts in a slightly different way from the original design. Heo et. al. [Heo 98] presented a redesign approach of digital electronic systems by means of evolutive programming. They use directed acyclic graphs known as task flow graphs (TFG) to represent the redesign object. Each node of the graph represents computational tasks; an edge represents a transfer of data. The design process consists of five tiers: a) system-level design, b) architectural design, c) logic design, d) circuit design, and e) physical design.

2.5.4 Chemical engineering

The (re)design of chemical processes is made with the purpose of adapting existing processes to changes in economic, technological or environmental requirements. In the eighties, there were mainly significant advances on saving energy by means of two constraint-based approaches: a) pinch methodology (Tjoe 86, Smith 87, Linnho. 88) and b) mathematical programming on synthesis and design of processes [Papoulias 83, Pistikopoulos 87, Vaselenak 87]. In the nineties, Gundersen [Gundersen 90] made a revision of systematic methods of redesign of processes, which were broadly tackled. In such revision, he emphasised two important observations:

- Most of the projects in the industry of processes were redesign projects.
- The systematic methods of redesign of processes are based on methods of design of processes.

Doherty et. al. [Fischer 87] developed a systematic procedure of redesign by means of opportunistic searches; the procedure considers modifications in the structure of the flowsheet and in the dimension of equipments. Kirkwood et. al. [Kirkwood 88] implemented a methodology of redesign by means of an expert system by using heuristic rules to construct hierarchical structures. Nelson and Douglas [Nelson 90] developed a systematic procedure considering alternative reaction routes; the procedure is hierarchical and provides guides to identify viable processes. Rapoport et. al. [Rapoport 94] proposed an algorithm to design units of process by means of the redesign of already existing ones. The algorithm consists on hierarchical levels and heuristic rules; this approach is similar to synthesis of processes. Stephanopoulos et. al. [Han 95] developed an approach based on agents to synthesis of processes; they modeled the process of design like a set of tasks that can be executed by agents. Systems have also been developed to satisfy economic, environmental and safety constraints. Kraslawski et. al. [Kraslawski 00] developed a methodology centred on the identification and elimination of bottlenecks in reaction and separation sections. Sylvester et. al. [Sylvester 00] optimised processes within the concept of Greener Process. Hertwig et. al. [Hertwig 01] applied techniques of MINLP (Mixed-Integer Non-Linear Programming) to optimise configuration of processes. Pasanen [Pasanen 01] developed a tool for conceptual design of processes, this is called Phenomenon Driven Process Design (PDPD). This methodology focuses on the systematisation of design conceptual of chemical processes. Uerdingen et. al. [Uerdingen 01] presented a screening method based on an analysis of the flow path pattern. They use performance indicators to rate the economic impact of each component of the flowsheet in the flow path.

2.6 Modelling

2.6.1 Modelling the redesign process

The overall redesign process depends on the problem-solving strategy used. In order to start the redesign process, the problem must be specified in terms of objectives that the original artefact must satisfy and the criteria that can be used to rank the alternative designs. Then a synthesis process takes place and the results are a set of alternative designs. Each of these alternatives is analysed and evaluated in terms of the predefined objectives and design criteria. Finally, one alternative is selected to be implemented. The process is highly iterative; the results from later stages are fedback to early stages to modify objectives, criteria, design alternatives, and so on.

Design alternatives are generated through a process of analysis of system composition. The designer breaks down the system (artefact) into a set of subsystems (components), together with the functions and constraints imposed upon the individual subsystem designs. These aspects are analysed with respect to desired system performance features and constraints. The process is iterative until an acceptable design alternative is achieved. At the end of this process all components must be described in such detail that an implementation of the whole object can be performed.

2.6.2 Modelling the redesign object

The understanding of the redesign object depends strongly on the mental models of the human designer. Usually designers communicate their ideas more easily in terms of abstract, high-level descriptions to describe complex concepts (Price 03). The description of the redesign object can be done in many different ways, depending on the context and purpose for which the description is to be used. In the early phases of redesign, highly abstract descriptions (e.g. qualitative or causal) might be helpful, whereas in later phases, more detailed and quantitative descriptions provide more suitable information.

Considering the notion of function, some researchers (Sembugamoorthy 86, Goel 89, Franke 92, Keuneke 91, Chittaro 93, Iwasaki 93) organise the knowledge in a domain by means of functional concepts. The main claim of these approaches is that functions and intentions can provide important additional information for understanding and reasoning about the structure and behaviour of physical systems. In addition, other researchers have directed their extentions to hierarchical modelling by means of different aggregation levels (Liu 91, Rajamoney 91) or different approximations (Weld 86, Kuipers 87, Struss 91, Falkenhainer 91) to organise the knowledge. Independently of the tools and representations employed, several authors (Fischo. 78, Checkland 81, Jaffe 91, Vicente 92) suggest that two important aspects must be addressed if computer tools are used to tackle activities of complex systems:

- *Content*, the semantic information that should be contained in the representation given the goals and tasks of the users. The content gives the basic issues to understand the information about the redesign object. Independently of the amount and complexity of the information, the designer can conceive, in general terms, the objectives of the redesign object.
- *Structure*, how to design the representation to facilitate that the user can extract the required information. The structure concerns to the organisation of the process components.

Rasmussen (Rasmussen 85) observed that the complexity of a system depends on the level of resolution in which the system is considered. The complexity can only be measured comparing with other systems observed at the same level of abstraction. The complexity can be manageable with more or less detail in the representations: then, hierarchical modelling can be seen as a way to handle complex systems. Models of complex artefacts can be expressed in terms of a hierarchy of levels of organisation, each one more complex than the previous. This modelling approach is named *Hierarchy Theory* (Rasmussen 86). Rasmussen studied the protocols developed by people working on complex systems and found that human users structure the system along two dimensions: a part-whole abstraction and a means-ends abstraction.

Some authors (Umeda 90, Franke 92, Lind 94) propose representation approaches for physical systems which maintain a clear separation between knowledge of structure and behaviour on one side and knowledge of function or purposes on the other side. This feature makes them useful in redesign of technical complex systems. The hierarchical functional modelling approaches employed in the present work are Multilevel Flow Modelling (MFM) (Lind 90, Lind 94, Lind 96, Lind 99) and Multimodelling (Ziegler 79, Praehofer 91, Chittaro 93), which are following described. Both approaches provide a more intuitive vision of reasoning on each task to be performed, and thus the redesign activities are enhanced. These approaches have been applied successfully in diagnosis and control domains.

2.6.2.1 Multilevel flow modelling

MFM provides a graphical and systematic basis for using means-end and whole-part hierarchical decompositions in the modelling of complex systems such as industrial plants. By the distinction between means and ends, a system is described in terms of goals, functions and the physical components that involves. At the same time, each of these descriptions can be given on different levels of whole-part decompositions. The main types of decomposition are illustrated in Figure 2. These are functional models with a very high level of abstraction, combined with a teleological representation of goals, or purposes, of the modelled system. Lind has suggested a syntax for a formal language and given the general ideas on how to use the MFM representation.

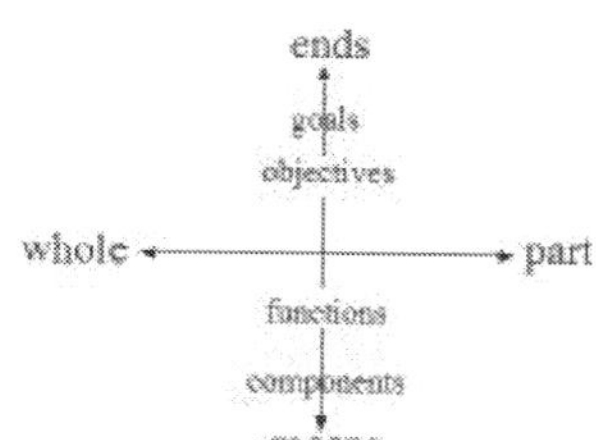

Fig. 2. Means-ends and part-whole dimensions in MFM.

An MFM model is a prescriptive description of a system, a representation of what it has been designed to do, how it should do it, and with which information it should do it. Thus, the three basic concept types of MFM are:

- *goals*, which are the objectives or purposes of the system, i.e., the ends that the designers and operators want that the system reaches.
- *functions*, which are the means by which the goals are obtained, i.e., the powers or capabilities of the system.
- *physical components*, which are the different elements of the system, the equipment of which it consists.

The concept of goal is central to MFM, as it is the "descriptor object" for teleological information. It is important to be able to recognise and describe goals, as they play an important role in every activity using means-end information. Without knowing the goals, it is virtually impossible to know the available functions. Three different types of goals can be recognised:

- *production goals*, which are used to express what enables production. For example, a specific process variable should be kept within a given interval.
- *safety goals, which are used to express reasons of safe operation. For instance, a particular process variable should be kept above or below some value, or inside or outside an interval.*
- *economy goals, which are used to express considerations of overall process optimisation.*

The function is the second important concept on MFM. A function is always associated with a goal, and correspondingly, goals are always associated with functions MFM describes the functional structure of a system as a set of interrelated flow structures on different abstraction levels. The levels are connected via achievement and condition relations; the flow structures consist of connected flow functions. Thus, the following types of flow structures can be:

- mass flows
- energy flows
- information flows

These flows are of completely different types, but they have many properties in common. Most flow functions can appear in each type of flow structure, thus, there are three flow types of flow functions. In MFM plant functions are represented by a set of mass, energy, activity and information flow structures on several levels of abstraction. The levels are interdependent and form means-end structures. Mass and energy flow structures are used to model the functions of the plant and activity and information flow structures are used to model the functions of the operator and the control systems. The mass and energy flow functions are:

- *source*, the capability of a physical system to act as an infinite reservoir of mass, energy, or information.
- *transport*, the capability of a system to transfer mass, energy, or information from one part of the system to another (from one medium to another).
- *barrier*, the capability of a system to prevent the transfer of mass, energy or information from one part of the system to another (from one medium to another).
- *storage*, the capability of a system to accumulate mass, energy, or information.
- *balance*, the capability of a system to provide a balance between the total rates of incoming and outgoing flows.

- *sink*, represents the capability of a system to act as an infinite drain of mass, energy or information.

These functions can be used to describe information flows. There are also some specific information flow functions:

- *observer*, the capability of a system to translate physical observations to information.
- *decision maker*, represents the decision-making capabilities of a system.
- *actor*, represents the capability of a system to turn information into physical consequences.

In addition to the flow functions, some organisational functions are used. They are concerned with expressing support and control:

- *network*, which is used to group a flow structure and connect it to a goal.
- *manager*, which describes control and supervisory systems, including human operators.

2.6.2.2 Multimodelling

The Multimodelling approach represents many diverse and explicit models of a system, which are used in a cooperative way in specific problem solving tasks. The fundamental assumptions about knowledge modelling and reasoning mechanisms do not identify a unique way of representing a physical system and reasoning about it. On the contrary, the Multimodelling approach is an abstract and general framework that allows for a variety of specific implementations. The fundamental concepts in Multimodelling are:

1. Ontologies. An ontology contains the descriptions of entities in the real system. Two types of ontologies can be distinguished:
    - Object-centred ontology. The real world is made up of individual objects whose properties can be stated in an objective, context independent and general way.
    - System-centred ontology. The real world is made up of systems, intended as organised units, whose elements cannot be defined in isolation.
2. Representational assumptions. This issue concerns about what to represent of the real system in the model. This involves two basic aspects:
    - The scope of the model, i.e., the aspects of the real system which are considered relevant to the purpose of the model.
    - The precision of the model, i.e., the degree of accuracy of the representation
3. Epistemological types. The type of knowledge represented in the model. These types can be:
    - *Structural*. The knowledge about system topology, i.e., the components that constitute the system and how they are linked.
    - *Behavioural*. The knowledge that describes how components work and interact in terms of the physical quantities (variables and parameters).
    - *Functional*. The knowledge about the role components play in the physical processes in which they take part. This knowledge relates the behaviour of the system to its goals, and deals with functional roles, processes, and phenomena.
    - *Teleological*. The knowledge about the goals assigned to the system by its designer and about the operational conditions that allow their achievement through correct operation.

- *Empirical*. The knowledge concerning the explicit representation of the system properties through empirical associations (such as observation, experimentation, and experience). This knowledge may include subjective competence that usually human experts acquire through direct interaction with the system.

4. Aggregation levels. The degree of granularity of the represented knowledge.

3. The redesign framework

The Multilevel Flow Modelling (MFM) and the Multimodelling approaches are able to represent how the human designers behave during the redesign process. MFM is used for high-levels of abstractions and Multimodelling is more suitable for the intermediate and lower levels. Thus, the structure and behaviour of the components (equipments) are abstracted using the Multimodelling approach and then this abstraction is mapped to the MFM approach. The bridge between both approaches is the functions for the equipments in the domain. This functional modelling is the basis to manipulate the process during all the redesign process. The main idea is to model hierarchically the process and reason by using functional abstract concepts. In this way the designer can "navigate" in top-down and bottom-up directions in the representation in similar way as when the designer creates its mental models about the process. From an abstract point of view, there are three actors that play an independent role in this framework:

1. *The simulator*. The commercial software used to obtain the design description of the process and to implement and evaluate the generated alternative process designs.
2. *The reasoner*. The software modules required to model the process, identify the suitable equipment/section to be modified and obtain similar equipments/sections based on the selected equipment/section.
3. *The human designer*. The human user interpreting the results.

The stages of the proposed redesign framework (Figure 3) are:

1. Identification of objectives and criteria. This stage covers the design-description acquisition and the identification of candidates stages of the framework.
2. Generation of design alternatives. This stage is similar to the obtaining alternatives and adaptation stages in the framework.
3. Evaluation of alternatives. This stage is carried out manually by the human designer.
4. Implementation of alternatives. This stage is also carried out manually by the human designer.

This redesign framework can deal with complex technical processes (the redesign object). The modelling approach was chosen to mimic the behaviour of human designers in real redesign situation of such processes. The final intention is to support human designers, not to carry out the redesign automatically without human intervention.

3.1 Design-description acquisition

The first stage of the framework is to obtain the design description of the process to be redesigned. This description is enough to carry out the redesign activities, and just few adaptations are necessaries to fulfill the redesign objective. This stage is carried out in two substages: data acquisition and functional identification.

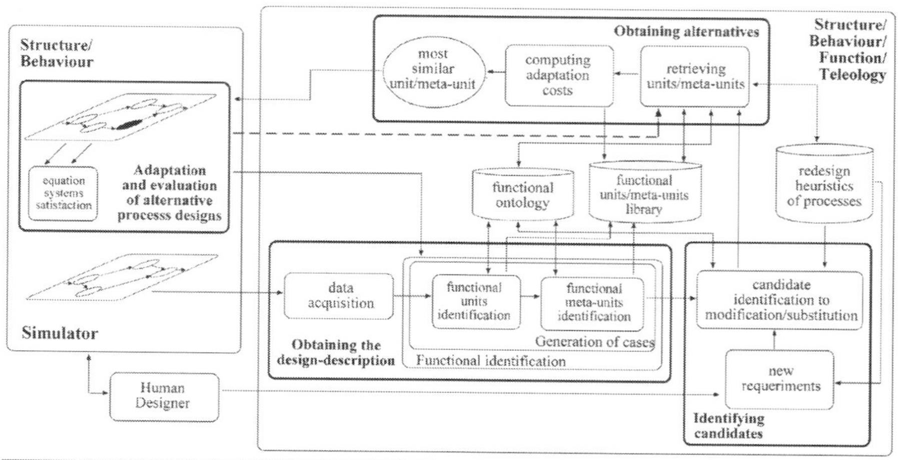

Fig. 3. The proposed redesign framework.

3.1.1 Data acquisition

It deals with data extracted from the specialised simulator used to implement the process to be redesigned. It was conceived as an appropriate step to reduce human intervention on the introduction of data to the reasoner module. The aim of the data acquisition is to obtain only the most useful data to generate the appropriate knowledge useful for the redesign of the process; thus irrelevant or superfluous data is ignored. Based on this data, the following types of knowledge are generated:

- *Structural.* Knowledge related to the topology of the process, i.e., the equipments conforming the process and the connection between them.
- *Behavioural.* Knowledge related to the values of variables and parameters that characterise the behaviour of each equipment.

3.1.2 Functional identification

The data obtained from the simulator is used to model hierarchically the process. To do this, the functions of each equipment in the process must be identified. Based on the identified functions, the functional sections of the process can incrementally be identified. In the rest of the chapter any equipment will be named *unit* and a functional section will be named *meta-unit*. This stage is divided into functional unit identification and functional meta-units identification. Here it is necessary to specify an ontology about the functional issues of the existing equipments in the process. By using this ontology, it is also required to specify a priority order of functions and the process variables related to each one. The grouping of functions depends strongly on such priority order.

Functional unit identification

The function of each unit is inferred by analysing their inputs (preconditions) and their outputs (postconditions), the variables involved, and the neighbour units. This process involves the analysis of the behaviour of the unit and its consequences in the surrounding units (the units connected to it). The next classification of functions was obtained based on MFM and Multimodelling:

- *Broad function*. Denotes a process-independent function that can be achieved considering only flows of mass, energy, or information.
- *General function*. Denotes a function that can be achieved by several equipments in a domain. These functions deal with the transformation of mass and energy and are independent of the physical phenomena.
- *Specific function*. Denotes the abstract function as it is known in the domain of the process. These functions relate flow variables with a specific physical process. According to the domain, these are functions to denote functional sections into the process.
- *Working function*. Denotes a function that can be achieved by a specific single unit. These functions relate specific flow variables with a specific physical phenomenon.

A unit can have several functions but only one goal (the objective, intention or purpose of the artefact). Several units can have a common goal. Therefore, based on the data extracted from the simulator it is possible to infer the following knowledge:

- *Functional*. Knowledge about the roles of each unit. The functional knowledge connects the behaviour (physical phenomena and processes) of the unit to its goal.
- *Teleological*. Knowledge related to the goals of each unit by considering the required input operational conditions and the output operational conditions that were meant to be produced.

The functional knowledge is independent of the process (the same functional knowledge can be found in others processes), while the teleological knowledge are the goals assigned to the units by the designer).

As result of this stage, each unit is represented by structural, behavioural, functional, and teleological models. The aim of this stage is to model the process in a higher level of abstraction (respect to the simulator).

Functional meta-unit identification

Based on the functions inferred in the functional unit identification stage, it is possible to identify the functional sections of the process (named *meta-units*). The incremental identification of these functional sections denotes the most important sections of the process. This incremental identification is carried out by generating different representations of the process at different levels of abstraction. The function of a unit is a working function because the unit (representing real equipment) was designed to perform only such function. The functional sections of the process denote specific and general functions by means of meta-units. Meta-units representing general functions are composed by meta-units with specific functions, not necessarily of the same type. A meta-unit represents a functional section at an abstract level. Thus a meta-unit at a higher abstract level can contain several units and other meta-units. Two or more meta-units can generate a more abstract meta-unit. Units/meta-units with lower priority functions are "absorbed" by units/meta-units with higher priority functions, as shows Figure 4.

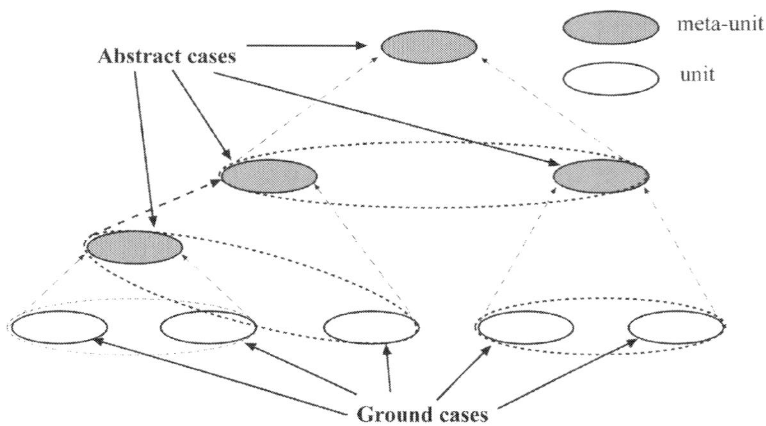

Fig. 4. Grouping of units/meta-units.

Every functional section forms a hierarchy of meta-units and units where meta-models are connected in a same level (intralevel), as shows Figure 5, and at different levels (interlevel), as shown in Figure 6.

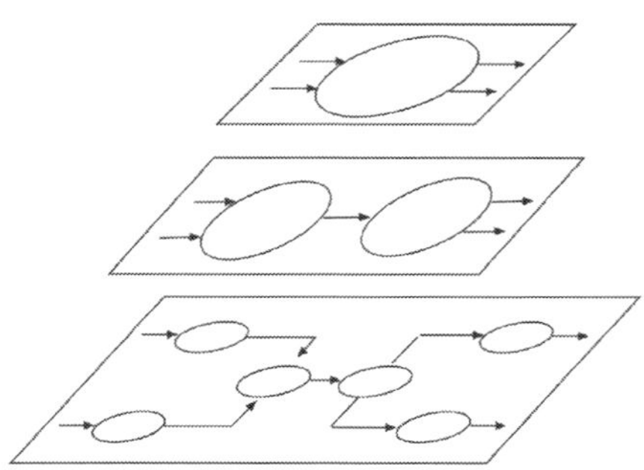

Fig. 5. Intralevel meta-models.

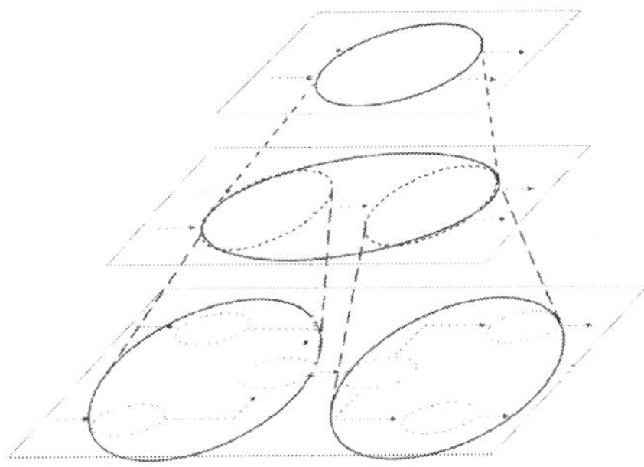

Fig. 6. Interlevel meta-models.

Then, the overall process is represented by several functional sections denoting flow structures. Incrementally the identification of such functional sections denotes the most important sections of the process. This process finishes until the most important functional sections are identified. This corresponds to the "blackbox" from which the original design could begin (Figure 7).

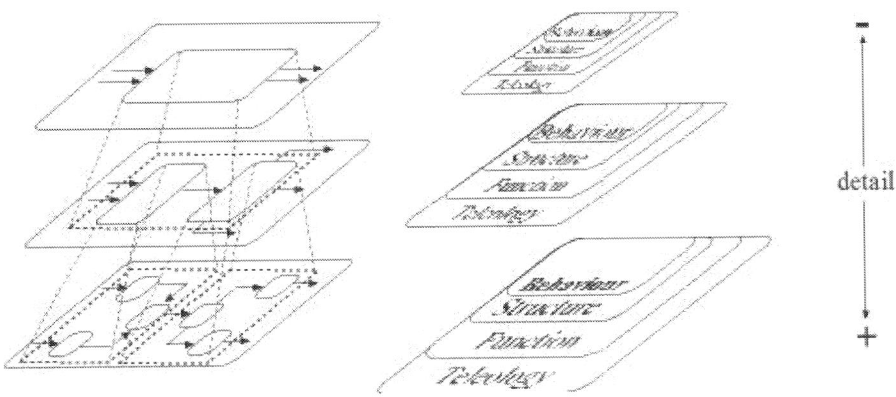

Fig. 7. Abstraction of a process.

3.2 Candidate identification

The aim of this stage is to get the suitable unit or meta-unit to be modified to fulfill the redesign objectives. In a first instance, the redesign must be focused on a process variable. Once the variable is identified, a diagnostic algorithm is used to identify the units/meta-units affecting such process variable. This reasoning process is based on the functions identified at the functional analysis stage. This stage is composed of two substages: specification of redesign requirements and identification of the suitable unit/meta-unit for modification or substitution.

3.2.1 Specification of redesign requirements

Here the human designer must specify the new requirements that the process must satisfy. Two categories of redesign requirements can be identified: functional requirements and physical requirements. A design specification always contains a single functional requirement; it may also contain a set of physical requirements. A functional requirement represents an abstraction of the intended behaviour of the artefact. It can be a general, specific, or working function. There is no direct association between the function that has to be provided and the physical mechanism that provides it. A physical requirement represents an abstraction of the physical process variables, which satisfy the functional requirement specified in the design specification. It denotes preferences about the designer intentions regarding some aspect of the process. The redesign specification can be represented by the functional and physical requirements or only by physical requirements. A redesign specification is a mean (goal) which is defined in terms of functions that must be embodied in a process in order to provide some higher level functionality.

3.2.2 Identification of candidates

Here In this stage, a diagnostic algorithm is used, that works on the functional concepts identified in the functional analysis. Diagnosis helps to detect "faulty" components (those that do not satisfy the global performance of the process). The design description and the new specifications are used to identify the possible candidates for modification or substitution. The diagnostic algorithm returns an ordered list of units or meta-units. Because the diagnostic algorithm operates over abstract functional concepts, no simulation is required. The diagnostic algorithm does not return the exact unit or meta-unit responsible for the "faulty" behaviour, it returns a list of units/meta-units that do not fulfill the global performance of the process represented by the redesign specification. The human designer is responsible to choose, from the resulting list, the appropriate unit/meta-unit that has to be modified or substituted into the process. Since this unit/meta-unit is connected to others by a flow path, the "cause" and "consequence" units/meta-units also must be identified. A cause unit/meta-unit is(are) the unit(s)/meta-unit(s) situated before the current unit/meta-unit in the flow path. They are responsible to provide the appropriate operational conditions to the involved process variables in the function of the unit/meta-unit of interest. A consequence unit/meta-unit is(are) the unit(s)/meta-unit(s) situated after of the current unit/meta-unit in the flow path. They are the unit(s)/meta-unit(s) affected by the operational conditions given by the unit/meta-unit of interest. Both, the cause and the consequence units/meta-units, are not necessarily the closer neighbours.

3.3 Generation of alternatives

The aim of this stage is to obtain similar units (equipments) or meta-units (sections) to adapt them into the current process based on the suitable unit/meta-unit identified by the human designer at the last stage. The best way to obtain similar units/meta-units is from similar processes. With the adaptation of any retrieved unit/meta-unit into the process of interest, then the alternative process design is obtained, which is the final goal of the redesign framework. An appropriate approach to perform this stage is case-based reasoning (CBR) for reusing past experiences on new situations.

Starting with the selected unit/meta-unit at the candidate identification stage, similar units/meta-units can be retrieved from other processes. The retrieved unit/meta-unit, which functional and teleological models are the most approximate to the functional and teleological models of the unit/meta-unit of interest, is adapted. This process requires that the performance and operational conditions of the cause and consequence units/meta-units associated with the retrieved unit must be similar to the original case.

According to the later stage, the overall target process is modelled as a graph denoting a hierarchy of functions. Therefore hierarchical case-based reasoning (Smyth 01) is required. Thus, based on the levels of abstraction, two kinds of cases are distinguished:

- *ground cases*. Cases located at the lowest level of abstraction, units (real equipments)
- *abstract cases*. Cases represented at higher levels of abstraction, meta-units (non-existent "meta-equipments")

The organisation of cases into the library is performed according to the type of functions of the unit/meta-unit. In this way, several groups can be distinguished according to the general function type: *source, transport, barrier, storage, balance, and sink*. Within each functional group, units and meta-units are grouped based on their specific functions. Again, within specific functions groups, the units/meta-units are grouped based on the working function achieved. There are not distinctions between units and meta-units with the same specific function. The case library is organised by an abstract hierarchy based on function groups. This structure denotes the organisation of the functional ontology used in the framework.

To retrieve cases from the case library, a similarity engine is used. Only units/meta-units of the same specific functional group are considered. The similarity engine uses functional and teleological targets to search into the library of cases. Functional and teleological models denote strongly the relationship between the units/meta-units and its neighbours. Two types of similarity are computed, local and global, which are defined as follows:

- *Local similarity*. Similarity between two cases is based on the local similarity between each feature of such cases. The computation depends on the type of the feature and the value that it could take.
- *Global similarity*. Once a set of local similarities has been computed for each known feature-value pair, the CBR system computes the global similarity of the candidate cases based on such set.

As final result, a set of cases is obtained which contains meta-units (with its corresponding units/meta-units) or units. The set is ranked according to the global similarity between the target case (the unit/meta-unit of interest) and the source cases (the retrieved units/meta-units).

3.4 Adaptation and evaluation

For this stage the human designer intervention is required since the use of the specialised simulator employed in the data acquisition is mandatory to test the proposed candidates. The adaptation is highly domain-dependent and requires online simulation of the process to verify its correct performance. Since information of abstract cases can not be used directly, the adaptation and revision of equipment must use information of ground cases (real equipments on the simulator). The human designer must fit the ground cases with the process variables involved.

To facilitate the adaptation, an adaptation cost is computed to suggest the human designer the adaptability of the chosen unit/meta-unit. The adaptation cost is based on the differences of the selected unit (source case) and the cause and consequence units/meta-units identified with the diagnostic algorithm. Thus, the cost is a normalised numerical value denoting the difference on the values of the process variables involved in the performance of the unit and the values of the process variables involved in the performance of the neighbour units/meta-units.

The adaptation cost has a value between 0 and 1. Values close to zero mean the adaptation is difficult. The designer experience is determinant because modifications on equipments may affect the overall performance of the process. The modification of the original process, based on the adaptation of a retrieved case, generates an alternative of the process for every unit/meta-unit adapted.

4. Evaluation and results

The framework has been applied to the Chemical Engineering because chemical processes are suitable for the proposed approach. A chemical plant can be constituted of one or more chemical processes.

The software modules of the redesign framework have been implemented in Java (Sun 05). Additional libraries have been used such as JESS (JESS 08), Ozone (Ozone 08), and The Selection Engine (Wetzel 00). The interaction with the user is done through a graphical interface.

The implementation of the concepts of the proposed redesign framework is based on the following ontological commitments.

- The chemical processes typically operate at steady-state. That means that values of variables do not change with respect to time.
- A chemical process is constituted of real and abstract units. The abstract units are the sections of the process that appear as atomic elements in conceptual models. All real equipments can be viewed as descendants of the generic real equipment.
- A generic real equipment can be modelled as an object having four attributes: structure, behaviour, function, and teleology to describe all the properties of any real equipment.

Since the framework requires functional concepts, a crucial point is to define the type of functions by means of an ontology. The functional ontology obtained is formed by high-level and low-level concepts in a similar way to the SUMO (Suggested Upper Merged Ontology) ontology structure (Niles 01). SUMO structures the concepts using meta-concepts, where terminology of general purpose is situated at higher levels, while terminology to specific domains is situated at lower levels. The ontology developed has extended generic concepts of SUMO such as process, objects and mereological and topological concepts. Additional specific concepts have been defined: physico-chemical processes, thermodynamic processes, substances (mass and energy), substance roles (of chemical compounds), functional roles, devices (equipments and connections), measure units, tasks, operations, and relations.

Most of the concepts in the ontology correspond to physical entities. The high-level concepts denote very abstract concepts, which can be found in several domains. The middle-level includes the functional concepts proposed in the Multilevel Flow Modelling and Multimodelling approaches, which are: *source, transport, barrier, storage, balance, and sink*. The low-level functional concepts come from the well-known chemical process design methodologies developed by Douglas (Douglas 88) and Turton (Turton 98). The low-level functional concepts can be grouped as: *reaction, separation, temperature change, pressure change, and flow change*. These concepts are called *general functions*. Each specific function is divided into more specific ones named *specific functions*, which denote the function of the equipment into the process. Also each specific function is divided in more specific ones, called *working functions*. A working function can be associated with one or more units and a unit can be related to more than one function. But from the several working functions, only one is the main function in the process.

Over the identified functions an importance functional order and the variables involved in such functions (carried out by the equipments of the process) have been defined. This order was defined with the aim of forming groups of functions where more important functions "absorb" functions with minor importance, see Figure 8.

The framework was tested over 50 chemical processes (Lopez-Arevalo 05):

1. Acetaldehide from ethanol
2. Acetaldehide from ethylene and oxygen
3. Ethyl acetate
4. Vinyl acetate
5. Acetone
6. Acetic acid
7. Acrylic acid
8. Cyanhydric acid
9. Nitric acid
10. Acrolein
11. Ammonia from natural gas and pure N2
12. Ammonia from pure N2 and H2
13. Phthalic anhydride from naphtalene
14. Phthalic anhydride from o-Xylene
15. Maleic anhydride
16. Bencene and methane
17. Bencene and o-Xylene
18. Bencene, Toluene and Styrene
19. Separation of Chlorine-Bencene and Bencene
20. Ethyl-Bencene
21. Cumene
22. 1,3-Butadiene
23. Cyclohexane
24. Allyl Chloride

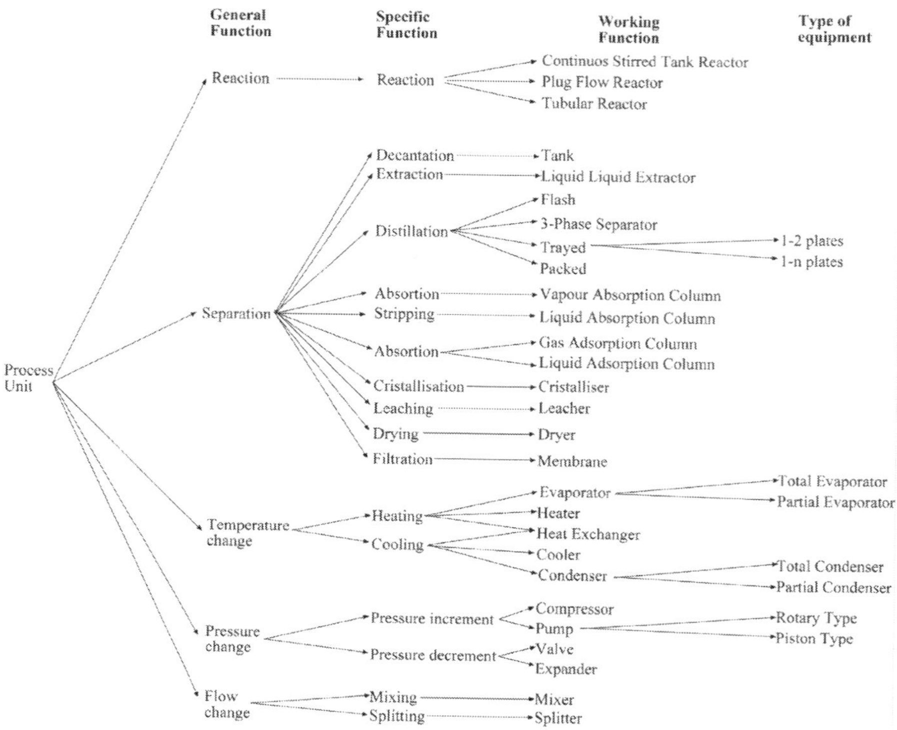

Fig. 8. The hierarchy of functions

25. Separation of Ciclohexane
26. Chloroform
27. Ethanol
28. Purification of Ethanol
29. Dimethyl ether
30. Ethyl tert-butylic ether (ETBE)
31. Methyl tert-butylic ether (MTBE)
32. Tert-amyl Methyl ether (TAME)
33. Styrene
34. Separation of ethane, n-heptane y n-octane
35. Ethylene
36. Ethylene oxide
37. Formaldehyde
38. Formaline
39. Methyl formate
40. HP gas
41. Heptane

42. Hydrogen
43. Separation of metane
44. Separation of metane and ethane
45. Methanol from natural gas
46. Methanol from carbon monoxide
47. Oxygen and nitrogen
48. Purification of parafins
49. Propyleneglycol and dipropylene glycol
50. Vinyl chloride

4.1 Modelling the ammonia process

The ammonia production process (Figure 9) has been selected as case study because it is one of the most relevant chemical processes in the industry. Ammonia is one of the most important chemicals commodities because of its role in the production of fertiliser and hence of food. It is produced in over 80 countries worldwide with a volume of 130 million tonnes annually (GIA 04).

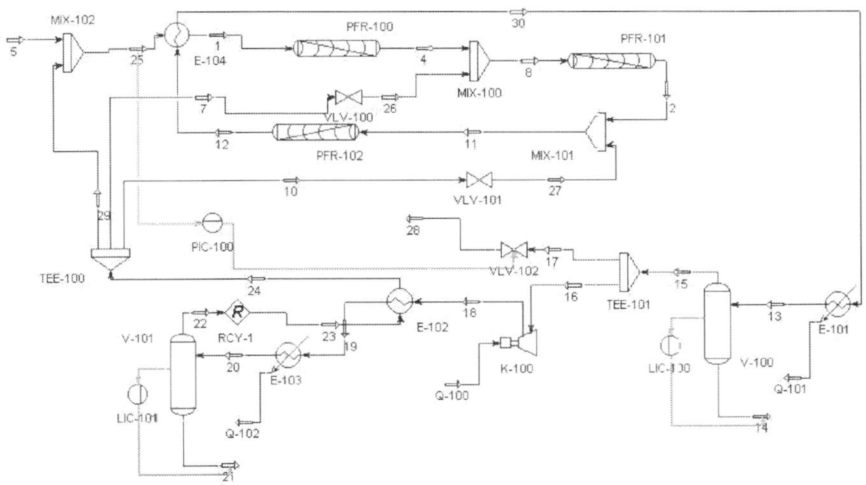

Fig. 9. The ammonia production process

Technical changes on process equipments were taken into account and some other issues such as economical costs, changes in pipes, environment impact, etc. were not considered. To illustrate the performance only the ammonia production process is used as case study.

First, the data is extracted from the Hysys simulator (Hysys 04), and then the "roles" of chemical substances are asked to the human designer. The first level of the process (abstraction level 0) is shown in Figure 10. This GUI allows the user to interact in two ways,

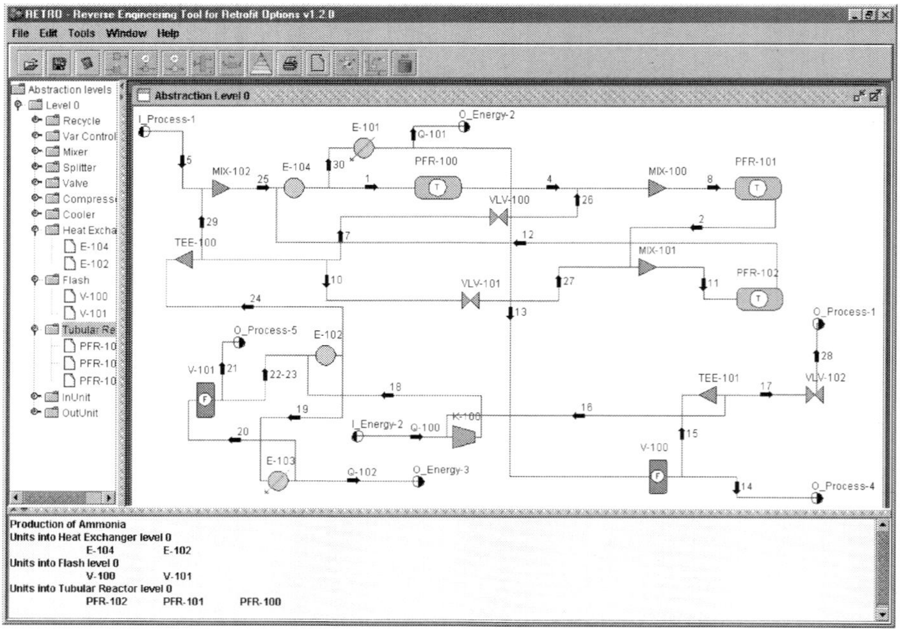

Fig. 10. First representation of the ammonia production process.

through menus and panels. The diagram panel (upper-right panel) allows the user to manipulate the process layout; the user can organise the components of the process according to its needs. The navigation panel (upper-left panel) is used to navigate into the levels of the process; abstraction levels and its corresponding components. The information panel (bottom panel) displays information about operations carried out in the prototypes. Table 1 summarises the type of equipments and its corresponding functions.

For the generation of meta-units, the flow change equipments (mixers and splitters) are grouped to other ones in the abstraction level 1, as it is shown in Figure 11.

The different abstract levels were automatically generated until the final abstract models were created at level 6 (see Figure 12). The abstraction process continues until the whole process is represented by just one meta-unit. Only new units and meta-units are represented to illustrate how the functional groups are created. In this sense, connections between units and meta-units in the same level have not been represented. The scheme represents the process by means of groups of the general class of type of equipment; its general function can be deduced from them. Figure 12 represents all the unit and meta-units in each level, in similar way that they are presented to the designer.

General Function	Specific Function	Working Function	Label
Flow change	Flow increment	Mixer	MIX-100
			MIX-101
			MIX-102
	Flow decrement	Splitter	TEE-100
			TEE-101
Pressure change	Pressure increment	Compressor	K-100
	Pressure decrement	Valve	VLV-100
			VLV-101
			VLV-10
Temperature change	Temperature increment	Cooler	E-101
			E-103
	Temperature exchange	Heat Exchanger	E-102
			E-104
Separation	Distillation	Flash Separator	V-100
			V-101
Reaction	Reaction	Tubular Reactor	PFR-100
			PFR-101
			PFR-102

Table 1. Equipment and functions in the ammonia production process.

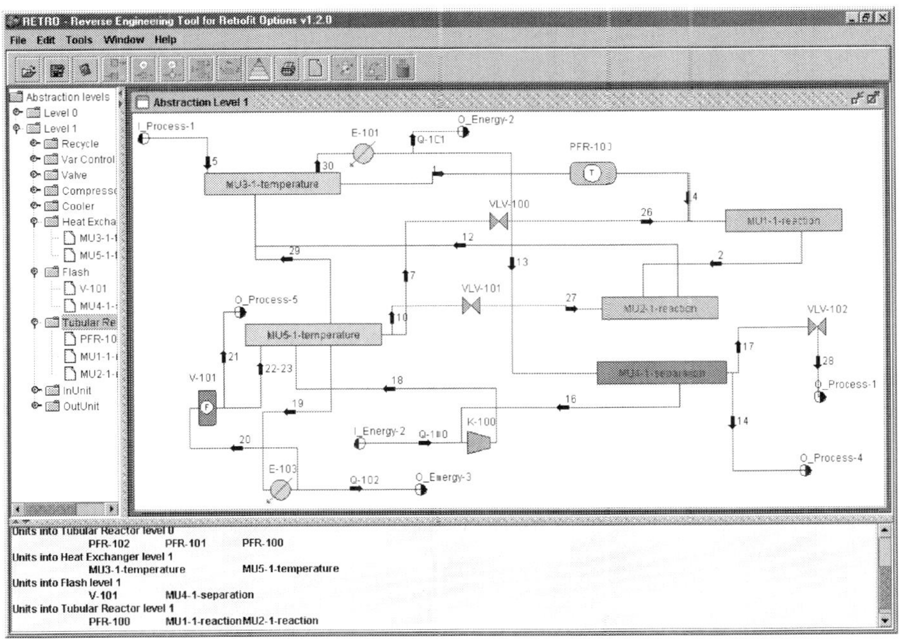

Fig. 11. Grouping of flow change units.

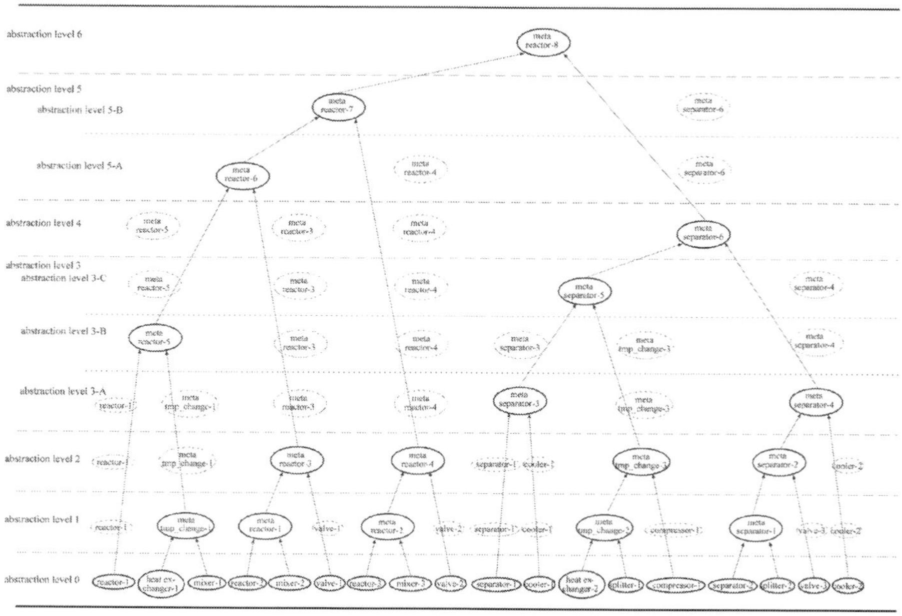

Fig. 12. Hierarchical representation of the ammonia production process in bottom-up direction.

4.2 Identification of candidates

According to the new design objective(s) the process must fulfill, modifications on reactors are used to illustrate the results of the framework:

Requirement

> *The redesign problem is the increase of the production of ammonia by 15% in the plant represented by the scheme of Figure 10.*

The human designer first needs to identify the variables that may affect directly the production of ammonia. Then the designer identifies that the increase of production can be achieved by modifying any of the next conditions:

- pressure
- temperature
- concentration

This gives an idea on the types of equipment the diagnosis must focus on. Assuming that the concentration variable is selected, this is affected by reactors and separators. Initially, reactors affect the concentration of product because they produce the main product, and separators affect it in secondary manner by incrementing the purity of the product.

Therefore, the focus will be on reactors, where the ammonia is originated. Since all the roles of the chemical substances are known, the diagnostic module focuses on the concentration of ammonia (which has the main product role). Thus all values related to this substance are analysed.

We are interested in finding where the main product is produced. The search starts at the highest level in the hierarchy -abstraction level 6- following the flow direction, from left to right. As result of the search, all the units and meta-units affecting the concentration variable have been identified, as shows Table 2.

Abstraction level	Identified components
6	meta-reactor-8
5	meta-reactor-7
4	meta-reactor-6, meta-reactor-4
3	meta-reactor-5, meta-reactor-3, meta-reactor-2
2	reactor-1, meta-reactor-1, reactor-3
1	reactor-2

Table 2. Identified candidates

Since modifications to the process can be performed only at ground level, the cause and consequence units are searched in this level. To illustrate the cause and consequence identification, assume that we focus on the meta-reactor-3, which at ground level includes the units PFR-101, MIX-100, and VLV-100 (Figure 13). In the state analysis, the state

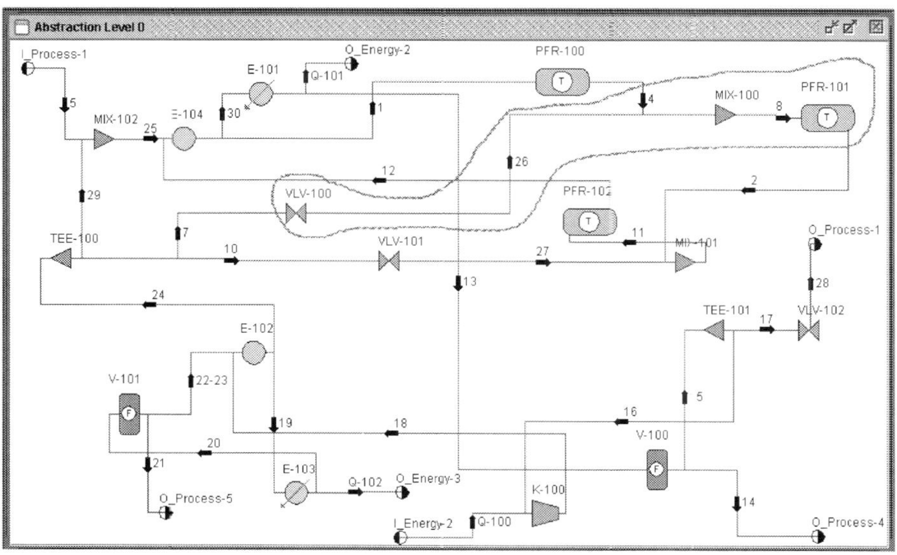

Fig. 13. Units composing the meta-reactor-3.

conditions are propagated to the units connected to the meta-reactor-3 in the flow path. The analysis in backward/forward stream directions finishes when a closer primary function is reached.

The state of the meta-reactor-3 is set to low capacity because the production of the main product is not enough. Since this meta-unit is not an initial unit in the path, its state may be originated by the effect of the performance of other units. Then back units are analysed. Considering the stream-4, the function to analyse is a source (PFR-100), which directly affects the concentration variable. Perhaps other back units in the same direction may affect the variable, but this is the closer primary function affecting the concentration variable. It may have low volume state, which originates the low capacity of meta-reactor-3. Therefore, the unit associated to this function is identified as cause unit and this branch of backward analysis in this direction finishes. Considering the stream-7, the function to analyse is a balance (TEE-100), which does not affect the variable. Then, the next function is analysed, which is a balance of temperature (E-102) that again does not affect directly the variable. The next function is storage (V-101), which affects the variable. This is another primary function affecting the concentration variable; it may have low volume state. Then its associated unit is a cause unit and this branch of backward analysis finishes.

Now, forward analysis is carried out following the output stream of the functional group. The low capacity state of meta-reactor-3 originates a low flow state and a low volume state, and consequently affects the following source functions producing a low capacity state in such function. Thus, considering the stream-2, the function to analyse is a balance (MIX-101), which does not affect the concentration variable. The next function is source (PFR-102), which affects the variable; it may have low capacity state originated for the low capacity state of the meta-reactor-3. Then, the unit associated with this function is a consequence unit. Since is the closer primary function affected in the forward stream direction, the forward analysis finishes.

Therefore, the identified cause and consequence units for meta-reactor-3 are shown in Table 3, which also represents the cause and consequence units for all the meta-units.

Candidate	**Cause units**	**Consequence units**
reactor-1 (PFR-100)	separator-2 (V-101)	reactor-2 (PFR-101)
reactor-2 (PFR-101)	reactor-1 (PFR-100)	reactor-3 (PFR-102)
reactor-3 (PFR-102)	reactor-2 (PFR-101)	separator-1 (V-101)
meta-reactor-1	reactor-1 (PFR-100) separator-2 (V-101)	reactor-3 (PFR-102)
meta-reactor-2	reactor-2 (PFR-101)	separator-1 (V-101)
meta-reactor-3	reactor-1 (PFR-100) separator-2 (V-101)	reactor-3 (PFR-102)
meta-reactor-4	reactor-2 (PFR-101)	separator-1 (V-101)
meta-reactor-5	separator-2 (V-101)	reactor-2 (PFR-101)
meta-reactor-6	separator-2 (V-101)	reactor-3 (PFR-102)
meta-reactor-7	separator-2 (V-101)	separator-1 (V-101)

Table 3. Cause and consequence units.

4.3 Generation of alternatives

A CBR module is used to obtain alternatives units/meta-units that may be adapted into the ammonia process. Again, the process representation showed in Figure 12 is used to denote the composition of meta-units; and the process representation showed in Figure 13 to denote the presence of units/meta-units in each abstraction level.

Assuming a threshold of 14 items, the most similar source cases from computing the global similarities are summarised in Table 4. It shows the percentage of similarity, the specific function, the inlet and outlet functions of the source case.

Rank	Similarity	Function	Inlet Function	Outlet Function
1	56%	meta-reactor	reaction	separation
2	43%	meta-reactor	reaction	flow change
3	37%	tubular reactor	tmp change	separation
4	31%	plug flow reactor	pres change	tmp change
5	30%	meta-reactor	separation	pres change
6	29%	tubular reactor	reaction	separation
7	29%	meta-reactor	tmp change	tmp change
8	27%	meta-reactor	pres change	separation
9	25%	tubular reactor	tmp change	tmp change
10	23%	tubular reactor	tmp change	separation
11	20%	plug flow reactor	flow change	pres change
12	20%	meta-reactor	pres change	tmp change
13	16%	meta-reactor	reaction	tmp change
14	15%	meta-reactor	flow change	tmp change

Table 4. Result of the global similarity computation for meta-reactor-3.

The modelling of the 50 processes generated 1590 cases in the case library. Therefore, the software prototypes were continually enhanced according to the needs of these processes.

The aspects considered in the evaluation of the framework were:

1. Modelling of the process
   - use of simplified models
   - suitable grouping of equipment/sections
   - intuitive goal-driven approach
   - comprehensive and clear representations of equipment/sections
   - easy and intuitive graphical interface

- transparent integration with the numerical simulator

2. Identification of candidates
   - clear and easy search over simple but consistent concepts
   - module easy to use
   - intuitive interpretation of results
3. Suggestion of equipment/sections
   - suggestions according to purpose-driven strategy
   - appropriate guidelines for modification/substitution
   - reuse of past design solutions
   - easy access to abstract and detailed data of proposed solutions
   - rapid response making agile the creation of alternative prototypes

5. Conclusions

This chapter describes a redesign support framework for technical processes based on hierarchical modelling. This hierarchical modelling is based on means-end and whole-parts aspects. The hierarchical representation enhances the reasoning mechanism to identify the elements to be modified and the possible alternatives. The framework focused on conceptual redesign issues where abstract models are employed. The processes are modelled hierarchically based on their functions and goals. The framework consists of four stages: design-description acquisition, identification of candidates to redesigned, generation of alternatives, and adaptation and evaluation of alternatives. The implementation of the framework was on the Chemical Engineering domain.

The redesign framework combines model-based reasoning and case-base reasoning techniques. This framework enables the designer to work directly with the conceptual design of an existing process (i.e. a process already in operation) to automatically generate abstract multiple-models, which can be modified to develop alternative process designs. The procedure can be seen as the reverse engineering approach of "replay and modify". This model-based approach provides an appropriate way of combining hierarchical and functional modelling to represent and reason about complex technical processes. The hierarchical case-based approach provides a systematic way of reusing the sections of previous processes.

The framework extends the use of Multimodelling and Multilevel Flow Modelling approaches to integrate mental abstract models about the behaviour of processes in the redesign activities. These models provide a more intuitive vision of reasoning on each task to be performed, and thus the redesign activities are enhanced.

The research has some limitations, some of the major ones are:

- The framework was implemented only in one domain. The ideas can be applied to another domain, but a new implementation will be necessary.
- The framework was tested with simulated plants. We did not have access to real plant information, but the results obtained were validated by a team of chemical engineers specialised in design of processes.
- The implementation of the framework is not manageable by novice users because important human designer decisions must be taken.

6. Acknowledgment

This research was partially funded by project number 175537 from "Fondo Mixto Conacyt-Gobierno del Estado de Tamaulipas".

7. References

(Aamodt 94) A. Aamodt & E. Plaza. Case-based reasoning: Foundational issues, methodological variations and system approaches. AI Communications, vol. 7, pages 39–59, 1994.

(Arana 00) I. Arana H. Ahriz & P. Fothergill. Improving re-design support. In Proceedings of the Fifth World Conference on Integrated Design and Process Technology (IDPT2000), 2000.

(Arana 01) I. Arana H. Ahriz & P. Fothergill. Redesign Knowledge Analysis, Representation and Reuse. Industrial Knowledge Management A Micro-level Approach, 2001. Springer-Verlag.

(Akin 82) O. Akin. Representation and architecture. Silver Spring, O. Akin and E.F. Weinel eds., Maryland, Information Dynamics Inc., 1982.

(Alberts 92) L.K. Alberts P.M. Wognum & N.J. Mars. Structuring design knowledge on the basis of generic components. Artificial Intelligence in Design (AID'92), vol. In: Gero, J.S. (Ed.), pages 639–656, 1992. Kluwer, Dordrecht.

(Alberts 93a) L.K. Alberts. YMIR: An ontology for engineering design. PhD thesis, University of Twente, 1993. Netherlands.

(Alberts 93b) L.K. Alberts R.R. Bakker D. Beekman & P.M. Wognum. Model-based redesign of technical systems. Proceedings of the 4th international workshop on principles of diagnosis, 1993.

(Bakker 94) R.R. Bakker S.J. van Eldonk P.M. Wognum & N.I. Mars. The use of model-based diagnosis in redesign. In Proc. Reasoning about physical systems. 11th European Conference on Artificial Intelligence, pages 647–651, A. Cohn Ed., 1994.

(Ball 92) N.R. Ball & F. Bauert. The Integrated Design Framework: supporting the design process using a blackboard system. Artificial Intelligence in Design (AID'92), vol. Gero, J.S. (Ed.), pages 327–348, 1992. Kluwer academic publishers.

(Berker 96) I. Berker & D.C. Brown. Conflicts and Negotiation in Single Function Agent Based Design Systems. Concurrent Engineering: Research and Applications, Journal, Special Issue: Multi Agent Systems in Concurrent Engineering, vol. 4, no. 1, pages 17–33, 1996. Brown, D.C., Landes, S.E. and Petrie, C.J. (Eds.), Technomic Publishing Inc.

(Bertalanffy 50) L.von Bertalan.y. An Outline of General Systems Theory. British Journal for the Philosophy of Science, vol. 1, no. 2, 1950.

(Bhatta 94) S. Bhatta A. Goel & S. Prabhakar. Innovation in Analogical Design: A Model-Based Approach. Proceedings of Artificial Intelligence in Design. pages 57–74, 1994. Kluwer Academic Publishers, Dordrecht, The Netherlands.

(Bo 99) Y. Bo & F. Salustri. Function Modeling Based on Interactions of Mass, Energy and Information. Florida Artificial Intelligence Research Society, 1999.

(Borner 96) K. Borner E. Pipping E. Tammer & C. Coulon. Structural Similarity and Adaptation. Proceedings of the Third European Workshop on Case-Based Reasoning, pages 58–75, 1996. Springer-Verlag.

(Bras 92) B.A. Bras. Foundation for designing decision-based design processes. PhD thesis, University of Houston, 1992.

(Brazier 94) F.M. Brazier P.H. Van Langen Zs. Ruttkay & J. Treur. On formal speci.-cation of design tasks. Arti.cial Intelligence in Design (AID'94), vol. Gero, J.S. and Sudweeks, F. (Eds.), pages 535–552, 1994. Kluwer Academic Publishers, Dordrecht.

(Brazier 96) F.M. Brazier P.H. Van Langen J. Treur & N.J. Wijngaards. Redesign and reuse in compositional knowledge-based systems. Knowledge Based Systems, Special Issue on Models and Techniques for Reuse of Designs, vol. 9, no. 2, pages 105–119, 1996.

(Bridge 97) C. Bridge. From design to redesign: Revisiting design models to highlight similarities and differences. Seminar Presentation, 1997. Department of Architectural and Design Science, University of Sydney, Australia.

(Brown 89) D.C. Brown & B. Chandrasekaran. Design problem solving: knowledge structures and control strategies. Morgan Kaufmann, 1989.

(Brown 97) D.C. Brown & W.P. Birmingham. Understanding the Nature of Design. IEEE Expert, vol. 12, no. 2, pages 14–16, 1997.

(Brown 98) D.C. Brown. Intelligent Computer-Aided Design. AI in Design Group, Com-puter Science Department, WPI, September 1998. Worcester, MA, USA.

(Buckingham 91) S. Buckingham Shum. Cognitive Dimensions of Design Rationale. In People and Computers VI: Proceedings of HCI'91, pages 331–344, Cambridge Univer-sity Press. Cambridge, 1991.

(Chang 2003) Chang WC, Van YT. Researching design trends for the redesign of product form. Design Studies 24(2):173-180.

(Chandrasekaran 90) B. Chandrasekaran. Design problem solving: a task analysis. AI Magazine, vol. 11, no. 4, pages 59–71, 1990.

(Chandrasekaran 93) B. Chandrasekaran A. Goel & Y. Iwasaki. Functional Representation as Design Rationale. IEEE Computer, no. Special Issue on Concurrent Engineering, pages 48–56, 1993.

(Chandrasekaran 00) B. Chandrasekaran & J.R. Josephson. Function in Device Representation. Engineering with Computers, Special Issue on Computer Aided Engineering, vol. 16, pages 162–177, 2000.

(Checkland 81) P. Checkland. Systems thinking, systems practice. John Wiley and Sons, 1981.

(Chen 2009) Li, S. and Chen, L., 2009, Pattern-based Reasoning for Rapid Redesign: A Proactive Approach. Research in Engineering Design, Vol. 21, pp. 25-42.

(Chittaro 93) L. Chittaro G. Guida C. Tasso & E. Toppano. Functional and Teleological Knowledge in the Multimodeling Approach for Reasoning about Physical Systems: A case study in diagnosis. IEEE Transactions on Systems Man. and Cybernetics., vol. 23, no. 6, pages 1718–1751, 1993.

(Chittaro 98) L. Chittaro & A.N. Kumar. Reasoning about function and its applications to engineering. Artificial Intelligence in Engineering, vol. 12, pages 331–336, 1998.

(Culley 99) S.J. Culley. Final Report -Future Issues For Design Research Workshop. Technical Report, Faculty of Engineering and Design, University of Bath, 1999.

(Das 94) D. Das S. Gupta & D. Nau. Reducing setup cost by automated generation of redesign suggestions. Proc. ASME Computers in Engineering Conference, pages 159- 170, 1994.

(Daube 89) F. Daube & B. Hayes-Roth. A Case-Based Mechanical Redesign System. In Proceedings of the 11th International Joint Conference on Artificial Intelligence (IJCAI-89) Vol. 2, pages 1402–1407, Morgan Kaufmann Publishers, 1989.

(de Kleer 79) J. de Kleer. Causal and Teleological Reasoning in Circuit Recongnition. PhD thesis, Massachusetts Institute of Technology, 1979.

(de Silva Garza 96) A. Gomez de Silva Garza & M.L. Maher. Design by interactive exploration using memory based techniques. Knowledge based systems, vol. 9, no. 1, pages 151–161, 1996.

(DeLoach 04) S.A. DeLoach. The MaSE Methodology. Methodologies and Software Engineering for Agent Systems. The Agent-Oriented Software Engineering Handbook Series : Multiagent Systems, Artificial Societies and Simulated Organizations, vol. 11, 2004. Bergenti, F., Gleizes, M.P., Zambonelli, F. (Eds.) Kluwer Academic Publishing.

(Dixon 89) J.R. Dixon M.J. Guenette R.K. Irani E.H. Nielsen M.F. Orelup & R.V. Welch. Computer-based models of design processes: the evaluation of design for redesign. NSF Engineering Design Research Conference, pages 491–506, 1989. University of Massachusetts, USA.

(Douglas 88) J.M. Douglas. Conceptual design of chemical processes. Mc Graw Hill, New York, 1988.

(Dove 2001) Dove R. Response ability: the language, structure, and culture of the agile enterprise. Wiley, New York.

(Dunskus 95) B.V. Dunskus D.L. Grecu D.C. Brown & I. Berker. Using Single Function Agents to Investigate Con.ict. Artificial Intelligence in Engineering Design and Manufacturing (AIEDAM), Special Issue: Conflict Management in Design, vol. 9, no. 4, pages 299–312, 1995.

(Eldonk 96) S.J. van Eldonk L.K. Alberts R.R. Bakker F. Diker & P.M. Wognum. Redesign of technical systems. Knowledge-based Systems, Special Issue on Models and Techniques for Reuse of Designs, vol. 9, pages 93–104, 1996.

(Falkenhainer 91) B. Falkenhainer & K.D. Forbus. Compositional Modeling: Finding the Right Model for the Job. Artificial Intelligence, vol. 51, pages 95–143, 1991.

(Falkenhainer 92) B. Falkenhainer & K.D. Forbus. Composing task-specific models. Automated Modelling, ASME, vol. 41, pages 1–9, 1992.

(Fensel 01b) D. Fensel. Ontologies: A silver bullet for knowledge management and electronic commerce. Heidelberg, Germany, 2001.

(Fischer 87) G. Fischer A.C. Lemke & C. Rathke. From design to redesign. In Proceedings of the 9th International Conference on Software Engineering, pages 369–376, IEEE Computer Society Press, 1987.

(Fischo. 78) B. Fischo. P. Slovic & S. Lichtenstein. Fault trees: Sensitivity of estimated failure probabilities to problem representation. Journal of Experimental Psychology: Human Perception and Performance, vol. 4, 1978.

(Forbus 84) K.D. Forbus. Qualitative Process Theory. Artificial Intelligence, vol. 24, pages 85–168, 1984.

(Forster 96) J. Forster I. Arana & P. Fothergill. Redesign knowledge representation with DEKLARE. In Proceedings of KEML'96: 6th Workshop on Knowledge Engineering: Methods and Languages, Paris, France, 1996.

(Forster 97b) J. Forster P. Fothergill & I. Arana. Enabling intelligent variant design using constraints. IEE Intelligent Design Systems Colloquium, vol. Digest No: 97/016, 1997. London.

(Fothergill 95) P. Fothergill J. Forster J. A. Lacunza F. Plaza & I. Arana. DEKLARE: A methodological approach to re-design. Proceedings of Conference on Integration in Manufacturing, pages 109–122, 1995. K. R. von Barisani, P. A. MacConaill and K. Tierney (Eds.), Vienna. IOS Press.

(Franke 92) D.W. Franke. A theory of teleology. PhD thesis, University of Texas at Austin, 1992.

(French 85) R. French & J. Mostow. Toward Better Models of the Design Process. AI Magazine, vol. 6, no. 1, pages 44–57, 1985.

(French 93) M.J. French R.V. Chaplin & P.M. Langdon. A creativity aid for designers. International Conference on Engineering Design, ICED'93, pages 53–59, 1993. The Hague, Netherlands.

(Fricke 2005) Fricke E, Schulz A P. Design for changeability (DfC): principles to enable changes in systems throughout their entire lifecycle. Systems Engineering 8(4):342-359. Wiley Periodicals.

(Gero 90a) J.S. Gero. Design prototypes: a knowledge representation schema for design. AI Magazine, vol. 11, no. 4, pages 26–36, 1990.

(Gero 90b) J.S. Gero & A. Rosenman. A conceptual framework for knowledge-based de-sign research at Sydney University's design computing unit. Artificial Intelligence in Engineering, vol. 5, no. 2, pages 65–77, 1990.

(Gero 04) J.S. Gero & U. Kannengiesser. The situated Function-Behaviour-Structure framework. Design Studies, vol. 25, no. 4, pages 373–391, 2004.

(GIA 04) GIA. Greener Industry Association: Ammonia Process. http://www.uyseg.org/greener industry/index.htm, 2004.

(Goel 89) A. Goel & B. Chandrasekaran. Functional representation of designs and re-design problem solving. Proc. Eleventh International Joint Conference on Arti.cial Intelligence, pages 1388–1394, 1989. Los Altos, California: Morgan Kaufmann Publishers.

(Goel 91) A. Goel. A model-based approach to case adaptation. In In Proceedings of the 13th Annual Conference of the Cognitive Science Society (CogSci'91), pages 143–148, Chicago, Illinois, 1991.

(Goel 92) A. Goel & B. Chandrasekaran. Case-Based Design: A Task Analysis. Artificial Intelligence Approaches to Engineering Design, Innovative Design, vol. II, no. 6, pages 165–184, 1992. Academic Press, Tong y D. Sriram (editors).

(Goel 94b) A. Goel & S. Prabhakar. A control architecture for redesign and design veri.-cation. Proc. Second Australian and New Zealand Conference on Intelligent Information Systems Conference, pages 377–381, 1994. Brisbane, Qld., Australia.

(Goel 97a) A. Goel A. Gomez de Silva Garza N. Grue J.W. Murdock & M.M. Recker. Functional Explanations in Design. IJCAI-97 Workshop on Modeling and Reasoning, 1997.

(Gomez-Perez 04) A. Gomez-Perez R. Gonzalez-Cabero & M. Lama. Development of Semantic Web Services at the Knowledge Level. In The European Conference on Web Service (ECOWS'04), pages 72–86, 2004.

(Grossmann 00) I.E. Grossmann & A.W. Westerberg. Research Challenges in Process Systems Engineering. AICHE Journal, vol. 46, no. 9, pages 1700–1703, 2000.

(Gruber 93) T.R. Gruber. A translation approach to portable ontology specifications. Knowledge Acquisition, vol. 5, no. 2, pages 199–220, 1993.

(Gundersen 90) T. Gundersen. Retrofit Process Design -Research and Applications of Systematic Methods. Foundations of Computer-Aided Process Design, pages 213–240, 1990. J.J. Siirola, I. E. Grossmann and G. Stephanopoulos (Eds.).

(Han 95) Ch. Han J.M. Douglas G. Stephanopoulos. Agent-Based Approach to a Design Support System For the Synthesis of Continuous Chemical Processes. Computers in Chemical Engineering, vol. 19 (Supplement), pages S63–S69, 1995.

(Heo 98) D.H. Heo A.C. Parker & C.P. Ravikumar. An Evolutionary Approach to System Redesign. Proc. Eleventh International Conference on VLSI Design, pages 359–362, 1998. Chennai, India.

(Hertwig 01) T.A. Hertwig A. Xu A.B. Nagy R.W. Pike J.R. Hopper & C.L. Yaws. A prototype system for economic, environmental and sustainable optimization of a chemical complex. Proc. 11th European Symposium on Computer Aided Process Engineering, pages 1017–1022, 2001 R. Gani and S.B. Jorgensen eds, Elsevier.

(Hoover 91) S.F. Hoover J.R. Rinderle S. Finger. Models and abstractions in design. In Proceedings of the International Conf. on Engineering Design, ICED'91, pages 46–57, Zurich, 1991.

(Howe 86) A. Howe P. Cohen J. Dixon & M.D. Simmons. A domain-independent program for mechanical engineering design. In Proceedings of the 1st International Conference on Applications of AI in Engineering Problems, pages 289–299, Southampton University, U.K., 1986.

(Hysys 04) Hysys. Introduction to hysys.plant ver. 3.1. AEA Technology Engineering Software, Hyprotech, Calgary, Canada, 2004.

(JESS 08) JESS. Jess 7.1 manual. http://herzberg.ca.sandia.gov/jess/docs/index.shtml, 2008.

(Keuneke 91) A.M. Keuneke. Device Representation: The Significance of Functional Knowledge. IEEE Expert, vol. 6, no. 2, pages 22–25, 1991.

(Kim 93) G.Y. Kim & G. Bekey. Construting Design Plans for DFA Redesign. Proc. IEEE International Conference on Robotics and Automation, pages 312–318, 1993.

(Kirkwood 88) R.L. Kirkwood M.H. Locke & J.M. Douglas. A prototype expert system for synthesizing chemical process flowsheets. Comp. Chem. Eng., vol. 12, no. 4, pages 329–343, 1988.

(Kitamura 98) Y. Kitamura & R. Mizoguchi. Functional Ontology for Functional Understanding. Proceedings of The Twelfth International Workshop on Qualitative Reasoning, AAAI Press, pages 77–87, 1998. Cape Cod. USA.

(Kitamura 99) Y. Kitamura & R. Mizoguchi. Towards Redesign based on Ontologies of Functional Concepts and Redesign Strategies. In Second International Workshop On Strategic Knowledge And Concept Formation, page JSPS8, Iwate, Japan, 1999.

(Kolodner 93) J. Kolodner. Case-based reasoning. Morgan Kaufman Publishers, Inc., 1993.

(Kraslawski 00) R. Ben-Guang H. Fan-Yu A. Kraslawski & L. Nystrom. Review: study on the methodology for retrofitting chemical processes. Chemical Engineering Technology, vol. 23, no. 6, pages 479–484, 2000.

(Kuraoka 03) K. Kuraoka & R. Batres. An Ontological Approach to Represent HAZOP Information. Technical Report TR-2003-01, Process Systems En-gineering Laboratory, Tokyo Institute of Technology, 2003.

(Lander 97) S.E. Lander. Issues in Multi-agent Design Systems. IEEE Expert, vol. 12, no. 2, pages 18–26, 1997.

(Leveson 00) N.G. Leveson. Intent specifications: an approach to building human-centered specifications. IEEE Transactions on Software Engineering, vol. 26, no. 1, pages 15–35, 2000.

(Lind 90) M. Lind. Representing Goals and Functions of Complex Systems. Technical Report 90-D-38, Institute of Automatic Control Systems, Technical University of Denmark, 1990.

(Lind 94) M. Lind. Modelling goals and functions of complex industrial plants. Applied Artificial Intelligence, vol. 8, pages 259–284, 1994.

(Lind 96) M. Lind. Status and challenges of intelligent plant control. Annual Reviews in Control, vol. 20, pages 23–41, 1996.

(Lind 99) M. Lind. Plant Modeling for Human Supervisory Control. Transactions of the Institute of Measurement and Control, vol. 21, no. 4/5, pages 171–180, 1999.

(Linnho. 88) B. Linnho. G.T. Polley & V. Sahdev. General Process Improvements Through Pinch Technology. Chem. Eng. Progr., vol. 84, pages 51–58, 1988.

(Maher 95) M.L. Maher B. Balachandran & D.M. Zhang. Case-based reasoning in design. Lawrence Erlbaum Associates, Sydney, 1995.

(Maher 97a) M.L. Maher & A. Gomez de Silva Garza. Case-Based Reasoning in Design. IEEE Expert, vol. 12, no. 2, pages 34–41, 1997.

(Maher 97b) M.L. Maher S. Simno. & J. Mitchell. Formalising building requirements using an activity/space model. Automation in construction, vol. 6, pages 77–95, 1997.

(Maher 01) M.L. Maher & A. Gomez de Silva Garza. GENCAD: A Hybrid Analogi-cal/Evolutionary Model of Creative Design. Proc. of the 4th International Conference on Computational Models of Creative Design, 2001. J.S. Gero and M.L. Maher (eds).

(Maulik 92) P.C. Maulik M.J. Flynn D.J. Allstot & L.R. Carley. Rapid Redesign of Analog Standard Cells Using Constrained Optimization Techniques. Proc. IEEE Custom Integrated Circuits Conference, pages 8.1.1–81.3, 1992.

(Mitchell 83) T.M. Mitchell L.I. Steinberg S. Kedar-Cabelli V.E. Kelly J. Shul-Man & T. Weinrich. An intelligent aid for circuit redesign. In Proceedings of the 3th National Conference on Artificial Intelligence (AAAI-83), Washington, USA, 1983.

(Mostow 89) J. Mostow. Design by derivational analogy: Issues in the automated replay of design plans. Artificial Intelligence, vol. 40, no. 1-3, pages 119–184, 1989.

(Nelson 90) D.A. Nelson & J.M. Douglas. A systematic procedure for retro.tting chemical plants to operate utilising different reaction paths. Ind. Eng. Chem. Res., vol. 29, pages 819–829, 1990.

(Niles 01) I. Niles & A. Pease. Towards a Standard Upper Ontology. In 2nd International Conference on Formal Ontology in Information Systems (FOIS-2001), pages 2–9, 2001.

(Ohsuga 97) S. Ohsuga. Strategic Knowledge Makes Knowledge Based Systems Truly In-telligent. In Proceedings of the First International Workshop on Strategic Knowledge and Concept Formation, pages 1–24, Lutchi Research Centre, 1997.

(Ozone 08) Ozone. Ozone Developers' Guide. http://www.ozone-db.org/frames/documentation/overview.html, 2008.

(Papoulias 83) S.A. Papoulias & I.E. Grossman. A Structural Optimization Approach in Process Synthesis II: Heat Recovery Networks. Computers and Chemical Engineering, vol. 7, pages 707–721, 1983.

(Pasanen 01) A. Pasanen. Phenomenon-Driven Process Design Methodology. PhD the-sis, Department of Process and Environmental Engineering of the University of Oulu, Finland, 2001.

(Pistikopoulos 87) E.N. Pistikopoulos & I.E. Grossman. Optimal Retrofit Design for Improving Process Flexibility in Linear Systems. Technical Report EDRC-06-24-87, Engineering Design Research Center. Carnegie Mellon University, 1987.

(Pos 97) A. Pos. Problem Solving for Redesign. 10th European Knowledge Workshop on Knowledge Acquisition. Modeling and Management, pages 205–220, 1997. Sant Feliu de Guíxols, Catalonia.

(Price 97) C.J. Price I.S. Pegler M.B. Ratcli.e & A. McManus. From troubleshooting to process design: closing the manufacturing loop. In 2nd International Conference on Case-based Reasoning Research and Development, pages 114–121, London, UK, 1997. LNCS Vol. 1266, Springer-Verlag.

(Price 98) C.J. Price. Function-directed electrical design analysis. Artificial Intelligence in Engineering, vol. 12, pages 445–456, 1998. Elsevier.

(Price 03) C.J. Price N.A. Snooke & S.D. Lewis. Adaptable Modeling of Electrical Systems. In 17th International Workshop on Qualitative Reasoning, pages 147–153, 2003. Brasilia, Brazil.

(Qian 92) L. Qian & J.S. Gero. A design support system using analogy. Proceedings of the Second International Conference on AI in Design, pages 795–813, 1992. Kluwer Academic Publishers.

(Rapoport 94) H. Rapoport R. Lavie & E. Kehat. Retrofit design of new units into an existing plant: Case study: Adding new units to an aromatics plant. Computers and Chemical Engineering, vol. 18, no. 8, pages 743–753, 1994.

(Rasmussen 85) J. Rasmussen. The Role of hierarchical knowledge representation in decision making and system management. IEEE Transactions on Systems, Man and Cybernetics, vol. SMC-15, no. 2, 1985.

(Rasmussen 86) J. Rasmussen. Information processing and human-machine interaction: An approach to cognitive engineering. North Holland, 1986.

(Salomons 95) O.W. Salomons. Computer Support in the Design of Mechanical Products. PhD thesis, University of Twente, 1995.

(Sasajima 95) M. Sasajima Y. Kitamura M. Ikeda & R. Mizoguchi. FBRL: A Function and Behavior Representation Language. Proceedings of IJCAI, pages 1830–1836, 1995.

(Schoen 91) E. Schoen. Intelligent Assistance for the Design of Knowledge based Systems. PhD thesis, Stanford University, 1991.

(Sembugamoorthy 86) V. Sembugamoorthy & B. Chandrasekaran. Functional Represen-tation of Devices and Compilation of Diagnostic Problem-Solving Systems. Experience, Memory and Reasoning, J.L. Kolodner and C.K. Riesbeck, eds, Lawrence Erlbaum, Hillsdale. N.J., pages 47–73, 1986.

(Sylvester 00) R.W. Sylvester W.D. Smith & J. Carberry. Information and modelling for greener process design. AIChE Symp. Series, vol. 96, no. 323, pages 26–30, 2000.

(Simon 96) H.A. Simon. The sciences of the artificial. MIT Press, Cambridge, Massachusetts, 1996.

(Skuce 93) D. Skuce. A multi functional knowledge management system. Knowledge Acquisition, vol. 5, pages 305–346, 1993.

(Smith 87) R. Smith & B. Linnho. Process integration using pinch technology. Proceedings of ATEE Symp. Energy Management in Industry, 1987. Paris.

(Smyth 96) B. Smyth & M.T. Keane. Using adaptation knowledge to retrieve and adapt design cases. Knowledge-based Systems, Special Issue on Models and Techniques for Reuse of Designs, vol. 9, no. 2, pages 127–136, 1996.

(Smyth 01) B. Smyth M.T. Keane & P. Cunningham. Hierarchical Case-Based Reasoning Integrating Case-Based and Decompositional Problem-Solving Techniques for Plant-Control Software Design. IEEE Trans. on Knowledge and data engineering, vol. 13, no. 5, pages 793–812, 2001.

(Sowa 95) J. Sowa. Distinctions, Combinations and Constraints. In Proceedings of IJCAI 95 Workshop on Basic Ontological Issues in Knowledge Sharing, Montreal, Canada, 1995.

(Steier 91) D. Steier. Automating Algorithm Design within a General Architecture for Intelligence. Automating Software Design, pages 577–602, 1991. Lowry, M.R. and Mccartney, R.D. (Eds.).

(Steinberg 85) L. Steinberg & T. Mitchell. The Redesign System: A Knowledge-base approach to VLSI CAD. IEEE Design and Test of Computers, vol. 2, no. 45-54, 1985.

(Stephanopoulos 90a) G. Henning G. Stephanopoulos & H. Leone. MODEL.LA A modelling framework language for process engineering II. Multifaced modelling of processing systems. Computers and Chemical Engineering, vol. 14, no. 8, pages 847–869, 1990.

(Stephanopoulos 90b) G. Stephanopoulos G. Henning & H. Leone. MODEL.LA A mod-elling framework language for process engineering I. The formal framework. Computers and Chemical Engineering, vol. 14, no. 8, pages 813–846, 1990.

(Stroulia 92a) E. Stroulia M. Shankar A. Goel & L. Penberthy. A Model-Based Approach to Blame Assignment in Design. Proceedings of the Second International Conference on AI in Design, pages 519–537, 1992. J.S. Gero (editor).

(Stroulia 92b) E. Stroulia & A. Goel. Generic Teleological Mechanisms and their use in Case Adaptation. Proceedings of the Fourteenth Anual Conference of the Cognitive Science Society, pages 319–324, 1992. Indiana, USA.

(Struss 91) P. Struss. A theory of model simplification and abstraction for diagnosis. In Proceedings of the Fifth International Workshop on Qualitative Reasoning, pages 25- 57, 1991.

(Subba-Rao 99) S. Subba-Rao A. Nahm Z. Shi X. Deng & A. Syamil. Artificial intelligence and expert systems applications in new product development-a survey. Journal of Intelligent Manufacturing, vol. 10, no. 3-4, pages 231–244, 199. Kluwer Academic Publishers.

(Sumi 97) Y. Sumi. Supporting the Acquisition and Modelling of Requirements in Software Design. In Proceedings of the First International Workshop on Strategic Knowledge and Concept Formation, pages 205–216, Lutchi Research Centre, 1997.

(Sun 08) Sun Microsystems. Java Language Specification. http://java.sun.com/docs/index.html, 2008.

(Sycara 92) K. Sycara. CADET: A case-based synthesis tool for engineering design. In-ternational Journal for Expert Systems, no. 2, pages 157–188, 1992.

(Takeda 90a) H. Takeda P. Verkaamp T. Tomiyama & H. Yoshikawa. Modelling Design Processes. AI Magazine, vol. 11, no. 4, pages 37–48, 1990.

(Takeda 94a) H. Takeda. Abduction for design. In Proceddings of Formal Design Methods for CAD, pages 221-243, Tallinn, Estonia, 1994.

(Tay 2003) Tay FEH, Gu J. A methodology for evolutionary product design. Engineering with Computers 19:160-173.

(Thornton 93) A.C. Thornton & A. Johnson. Constraint specification and satisfaction in embodiment design. In Procedings of the International Conference on Engineering Design. ICED 93, pages 1319–1326, The Hague, 1993.

(Tjoe 86) T.N. Tjoe & B. Linnho. Using pinch technology for processes retrofit. Chem. Eng., vol. 93, pages 47–60, 1986.

(Tomiyama 87) T. Tomiyama & H. Yoshikawa. Extended General Design Theory. In Proceedings of the IFIP WG 5.2 Working Conference on Design Theory for CAD, pages 95–125, 1987.

(Treur 89) J. Treur. A logical analysis of design tasks for expert systems. International Journal of Expert Systems, vol. 2, pages 233–253, 1989.

(Turton 98) R. Turton R.C. Bailie W.B. Whiting & J.A. Shaeiwitz. Analysis, synthesis and design of chemical processes. Prentice-Hall, New Jersey, 1998.

(Uerdingen 01) E. Uerdingen U. Fischer & K. Hunderbuhler. A screening method for identifying economic improvement potentials in retrofit design. In ESCAPE-11 (European Symposium on Computer Aided Process Engineering), pages 573–578, 2001.

(Ullman 91) D.G. Ullman. Design histories: archiving the evolution of products. In Proceedings of the DARPA Workshop on Manufacturing, Salt Lake City, USA, 1991.

(Umeda 90) Y. Umeda T. Tomiyama & H. Yoshikawa. Function, Behaviour and Structure. Applications of Artificial Intelligence in Engineering V, pages 177–193, 1990. J.S. Gero (Ed).

(Umeda 92) Y. Umeda T. Tomiyama & H. Yoshikawa. A design methodology for a self-maintenance machine based on funtional redundancy. Design Theory and Methodology -DTM'92, pages 317–324, 1992.

(Umeda 94) Y. Umeda T. Tomiyama H. Yoshikawa & Y. Shomimura. Using Functional maintenance to improve fault tolereance. IEEE Expert, vol. 9, no. 3, pages 25–31, 1994.

(Umeda 97) Y. Umeda & T. Tomiyama. Functional Reasoning in Design. IEEE Expert, vol. 12, no. 2, pages 42–48, 1997.

(Vaselenak 87) J.A. Vaselenak I.E. Grossman & A.W. Westerberg. Optimal Retro.t De-sign of Multiproduct Batch Plant. Industrial and Engineering Chemistry Research, vol. 26, no. 4, pages 718–726, 1987.

(Vescovi 93) M. Vescovi & Y. Iwasaki. Device design as functional and structural re.ne-ment. In Working Notes of the IJCAI'93 Workshop on AI in Design, pages 55–60, 1993.

(Vicente 92) K.J. Vicente & J. Rasmussen. Ecological interface design: Theoretical foundations. IEEE Trans. on Systems, Man and Cybernetics, vol. 22, no. 4, 1992.

(Wetzel 00) B. Wetzel. Selection Engine. http://selectionengine.sourceforge.net/, 2000.

(Wielinga 97) B.J. Wielinga & A.Th. Schreiber. Con.guration-design problem solving. IEEE Expert, vol. 12, no. 2, pages 49–56, 1997.

(Wood 01) M. Wood & S.A. DeLoach. An Overview of the Multiagent Systems Engi-neering Methodology. Agent-Oriented Software Engineering, LNAI, vol. 1957, 2001. P. Ciancarini, M. Wooldridge, (Eds.).

(Xin 01) W. Xin & X. Guangleng. Supporting design reuse based on integrated design rationale. IEEE International Conference on Systems, Man and Cybernetics, vol. 3, pages 1909-1912, 2001.

(Yoshikawa 91) H. Yoshikawa F. Arbab & T. Tomiyama. Intelligent CAD III: Selected and Reviewed Papers and Reports. In Third International Workshop on Computer Aided Design, 1991.

5

Application Potential of Food Protein Modification

Harmen H.J. de Jongh[1] and Kerensa Broersen[2,*]
[1]TI Food and Nutrition, AN Wageningen
[2]Faculty of Science and Technology, Nanobiophysics,
MIRA Institute for Biomedical Technology and Technical Medicine,
University of Twente, Enschede
The Netherlands

1. Introduction

Proteins are essential in foods, not only for their nutritional value, but also as modulator of structure and perception of a food product. The functional behavior of a protein is inherently susceptible to physico-chemical conditions as pH, ionic strength, temperature, or pressure, making them also an unpredictable, and at the same time, opportune component in food production. Proteins are generally also industrially costly, and with increasing world population and welfare the pressure on protein-availability for food purposes gives rise to some concerns. In view of a more sustainable use of protein-sources a number of routes have been followed in the past decades that provided big steps forward in protein availability: (i) more efficient production or protein refinery methods, (ii) use of alternative protein sources, and (iii) optimized usage of protein functionality. Especially in wheat production correlations between genetic expression and functional product behavior allowed breeders to optimize cultivars for geographic location (e.g. Payne et al., 1984). Alternatively, one has the ability to express specific proteins in non-original sources, for example human milk proteins in plants, such as rice (e.g. Lönnerdal, 2002). Directed alterations in the genome of food-producing organisms can lead to changes in the primary sequences of relevant proteins and thereby introduce potentially new functionality. If sufficient quantities of the novel protein are synthesized and become admixed with the basal levels of protein in the food, the functional properties of the food system (texture-formation) may become improved. Alternatively, the modified protein can be isolated for use as food ingredient. More recently, a number of proteins from less-conventional origin have been identified as human food ingredients that one has started to exploit, e.g. algae, leafs, insects, and various seeds. Successful utilization of these new proteinaceous materials has thus far been rather limited, requiring breakthroughs in extractability, their digestibility, nutritive value, and overall functional and organoleptic properties. More downstream in the process is the modulation of protein functional behavior at an ingredient level. This can be physical-chemically, enzymatically, or via chemical engineering.

* Corresponding Author

This review will focus on the use of chemical engineering to study or better exploit protein functionality in food products. Reasons to employ chemical modification of proteins will be discussed in the context of its relevance in understanding the fundamental principles of proteins as structuring agents in food. These include improving shelf life and sensory properties as well as the development of new functionalities of food proteins, such as the application of plant proteins as meat-texturizers. Further we discuss how these insights have contributed thus far to a more sustainable utilization of protein, including aspects as consumer acceptance and existing/changing legislations for the use as ingredients.

2. Functional role of proteins as food ingredient

This paragraph will briefly summarize the functional properties of proteins as food ingredients. The molecular mechanisms of their roles in food products will be highlighted. In the next paragraph these molecular mechanisms will be discussed as target for chemical engineering. On each of these topics extensive reviews have been published which will be pointed out.

2.1 Surface properties: Emulsions and foams

Surfaces or interfaces in food products are abundantly present in terms of emulsions (oil-water interface) or foams (air-water interface). Examples of foams and emulsions in food products include ice cream, dressings or margarine. In foams and emulsions respectively air cells or oil droplets are dispersed or in an aqueous phase; the latter case also water in oil emulsions exist. As a result of their amphiphilic character, proteins, being composed of polar and non-polar amino acids, can contribute to the formation and stability of such dispersions by adsorbing to the interface and developing stabilizing films by coating the oil or air droplets and hence act as emulsifiers. The ability of proteins to induce film formation depends on a number of parameters of molecular nature which have been studied into detail in the past and is collectively governed by a net energy gain from absorbing at an interface. Milk proteins have been identified as good foaming agents as a result of their aggregation state, molecular stability and flexibility, electrostatics, and (surface) exposed hydrophobicity (Hunter et al., 1991; Luey et al., 1991; Shirahama et al., 1990; Suttiprasit et al., 1992; Waniska & Kinsella, 1985). Chemical modifications have been employed in the past to improve surface activity of less performing proteins. These studies and consequences of the used modifications will be discussed in more detail in the following paragraph. A wide range of methods has been employed in literature to study the chemical and molecular properties of proteins adsorbed at interfaces which lead to a detailed understanding of the principle forces of importance to surface activity. These methods include ellipsometry and infrared reflection absorption spectroscopy (IRRAS) which provide information on denaturation and concentration of adsorbed proteins (de Jongh & Wierenga, 2006; Grigoriev et al., 2007; Martin et al., 2003; McClellan et al., 2003). Stabilization of an air-water or oil-water interface is governed by a multiple step process. First, absorption at the interface requires proteins to diffuse to the interface and their retention at the interface is governed by the kinetic barrier of absorbance which, in turn, is influenced by factors such as exposed hydrophobicity (Wierenga et al., 2003), and net charge (Wierenga et al., 2005). Effective absorption onto the interface is followed by rearrangement of the protein molecules to form a thermodynamically stable but dynamic monolayer of molecules coating the droplets (de

Jongh & Wierenga, 2006; Graham & Phillips, 1979a, 1979b). Some publications report that the rearrangement process coincides with structural extension of the protein molecules (reviewed in MacRitchie, 1978) while other results suggest that secondary structure does not undergo variation subsequent to adsorption onto the interface (de Jongh & Wierenga, 2006; Graham & Phillips, 1979c). Despite the finding that local concentrations of 150-300 mg/ml can be reached at the interface (Meinders et al., 2001), proteins are still able to laterally diffuse as was shown by thiolated ovalbumin (de Jongh & Wierenga, 2006), illustrating that chemical modification has been used to underpin the molecular mechanisms of the surface activity of proteins.

2.1.1 Emulsions

Emulsions consist of two immiscible liquids, oil and water, in which the droplets are termed dispersed phase and the liquid surrounding the droplets is called continuous phase. Depending on the concentrations of each liquid and the environmental conditions, oil-in-water emulsions or water-in-oil emulsions can be formed. These consist of oil droplets in a continuous water phase and water droplets in a continuous oil phase, respectively. Examples of food related emulsions are milk, vinaigrette, and mayonnaise. Emulsions are often unstable by nature and will phase separate or coalesce over extended time, or storage. To stabilize emulsions, so-called emulsifiers can be used which form a small layer on the surface of the dispersed phase, thereby physically separating the continuous phase from the dispersed phase. Such treatment will allow the incorporation of emulsified solutions in food products which can be stored over an extended time frame. Amphiphilic proteins, containing both hydrophilic and hydrophobic regions, are effective emulsifiers. These proteins adsorb onto the interface between the oil and water phase, and stabilize the oil and water phase by selective interaction with both surfaces, thereby preventing the individual droplets from coalescing (reviewed in Rodríguez Patino et al., 2008). Other types of emulsifiers used in food products include lipids, phospholipids, surfactants or polysaccharides (Dickinson, 1992; McClements, 2005). Apart from a texture perspective, the absorption of proteins at the oil-water interface is of interest to the delivery of nutrients (reviewed in Malaki Nik et al., 2010). Upon digestion, changes occur at the oil-water interface of emulsified food products as a function of emulsifier stability, affecting the digestibility and subsequently the availability of encapsulated nutrients. The amphiphilic character of proteins can be modulated for example by the covalent attachment of lipid chains (see paragraph 3.2). Attachment of lipid chains of various lengths to a protein renders it more hydrophobic resulting in an improved ability to stabilize emulsions and/or foams (Aewsiri et al., 2011a, 2011b). Apart from exposed hydrophobicity, other factors have also been identified to determine the affinity of a protein for an oil-water interface. Aggregation or molecular weight of proteins for example adversely affects the emulsifying activity of proteins (Baldursdottir et al., 2010; Corzo-Martínez, 2011). That the aggregation effect is more complex than originally postulated was shown by single molecule total internal reflectance fluorescence microscopy of bovine serum albumin showing that the orientation at which the aggregates absorb to the interface largely determine the rate of absorbance to the interface (Walder & Schwartz, 2010). The identification of factors contributing to the surface activity of proteins allows the improvement of emulsifying activity by means of targeted chemical modification.

2.1.2 Foams

Foams consist of gas bubbles dispersed into a liquid. The stability of the air bubbles in a foam is determined by the foaming agent which forms a layer of adsorbed molecules separating the air bubbles from the continuous liquid phase, similar to the emulsifying activity described in paragraph 2.1.1 (reviewed in Halling, 1981; Wilde, 2000). Adsorption of a protein to the air-water interface induces partial dehydration of the molecule promoting protein-protein interactions. This effect is further amplified by the finding that local protein concentrations at the air-water interface can reach up to 150 to 300 mg/ml (Meinders et al., 2001). The rate of absorption to the air-water interface has been reported to depend largely on the hydrophobic nature of the protein under investigation (Kudryashova et al., 2003; Wierenga et al., 2003). Increasing the exposed hydrophobicity of proteins by means of conjugation with lipid chains was shown to increase the adsorption rate to the air-water interface (Wierenga et al., 2003). Net charge is a second parameter of interest determining adsorption kinetics, in which higher net charge slows down the adsorption process due to the electric repulsive forces involved (de Jongh et al., 2004; Kudryashova et al., 2005; Le Floch-Fouéré et al., 2011). Highly aggregated heat-treated ovalbumin was further shown to induce a ten-fold decrease the diffusion rate of proteins to the interface compared to the native protein (Kudryashova et al., 2005). However, the tendency of aggregated ovalbumin to remain adhered to the interface upon first interaction is significantly larger than for non-aggregated protein, which was found to rapidly desorb from the surface after absorption (Kudryashova et al., 2005). Collectively, it has been shown that surface activity of proteins is not determined by a single molecular characteristic but rather depends on a combination of factors. Hence, several types of chemical modification can be employed to improve the air-water interface activity of proteins.

2.2 Aggregation and gelation

Protein aggregation is a major topic in the field of food science, the regulation of which is believed to markedly affect the texture of food products (Zhou et al., 2008; reviewed in Doi, 1993). Aggregated protein can act as a nucleation prerequisite to induce gelation, albeit at high protein concentration (Alting et al., 2003; Barbut & Foegeding, 1993; Ju & Kilara, 1998). Processing conditions and storage can induce protein aggregation, even at ambient temperatures (Promeyrat et al., 2010; Santé-Lhoutellier et al., 2008). The resulting protein aggregates can vary widely in size and morphology as a result of the environmental conditions under which they were formed, among other factors. For example, upon inspection using electron microscopy, amylin (Patil et al., 2011), hen egg white lysozyme (Arnaudov & de Vries, 2005), and β-lactoglobulin (Arnaudov et al., 2003; Veerman et al. 2002) were found to form negatively stained and long fibrillar aggregates at pH values far away from the isoelectric point of the respective proteins. Near the isoelectric pH (Arnaudov & de Vries, 2005), or at high salt concentration (Arnaudov & de Vries, 2006; Veerman et al. 2002), spherical or amorphous aggregates are formed. That aggregate morphologies cannot always be categorized as either fibrillar or amorphous in a clear-cut manner was shown by various groups observing substantial heterogeneity in aggregate morphology, also called polymorphism, within the same preparation (Bauer et al., 1995; Jiménez et al., 2002, 1999). As a consequence, gels formed under these conditions of β-lactoglobulin are particulate and the particle size depends on heating temperature and heating rate (Bromley et al., 2006).

These observations lead to the understanding that net charge plays a major role in determining aggregate and gel morphology (Krebs et al., 2009; Langton & Hermansson, 1992). Cryo-EM investigation of aggregates formed from ovalbumin which had been succinylated to various degrees provided further evidence that net charge dominantly determines aggregate morphology (Weijers et al., 2008). The propensity of proteins to aggregate, or rate of aggregation, has been shown to vary as a function of protein conformational stability (Chiti et al., 2000; Hurle et al., 1994; Kelly, 1998; Quintas et al., 1997; Ramirez-Alvarado et al., 2000; Siepen & Westhead, 2002), rate of unfolding (Broersen et al., 2007a), net charge (Calamai et al., 2003; DuBay et al., 2004), and secondary structure propensity (Fernandez-Escamilla et al., 2004). Exposed hydrophobicity (Calamai et al., 2003) and the possibility to form disulfide bonds naturally affect the aggregation and gelation propensity of proteins as these two forces are primarily driving the assembly process. This has been exemplified in a study which showed that β-lactoglobulin A modified with *N*-ethylmaleimide largely resisted aggregation induced by heating (Kitabatake et al., 2001). Extensive knowledge of the molecular factors driving the aggregation process of proteins has lead to the development of a number of algorithms able to predict protein aggregation with high fidelity (Chiti et al., 2003; DuBay et al., 2004; Fernandez-Escamilla et al., 2004; Maurer-Stroh et al., 2010).

The energetics and kinetics of protein aggregation have been subject of many publications to date and has been reviewed on numerous occasions (e.g. Luheshi & Dobson, 2009; Straub & Thirumalai, 2011). It has been recognized that, regardless of primary sequence or physicochemical properties, all proteins have an inherent tendency to form aggregates *in vitro* under certain conditions (reviewed in Chiti & Dobson, 2006; Dobson, 1999). Protein aggregation is a multiple step complex process which can be viewed as a cascade of steps of assembly which may vary in molecular detail as a function of the protein studied or the environmental conditions. Nevertheless, the aggregate growth mechanisms of many different proteins share essential characteristics which have been elucidated both by experimental and computational methods (Teplow et al., 2006). The onset of aggregation often requires the (partial) unfolding or conformational rearrangement of proteins (Calamai et al., 2003; Dobson, 1999; Kelly, 1996; Rochet & Lansbury, 2000). Using human lysozyme (Canet et al., 1999) and HypF-N (Marcon et al., 2005) it has been shown that a population of less than 1% of partially folded protein can be sufficient to trigger the onset of the aggregation process. The resulting exposure of hydrophobic regions which normally reside in the core of a folded protein drives the self-assembly process to form small oligomers. These oligomers, or nuclei, are metastable and their transient and short-lived nature dictates that they can dissociate into monomeric protein, which has been shown for various proteins. Many efforts in the field of protein aggregation suggest that the nucleus has to reach a critical size which then allows further assembly by monomer addition to ultimately form mature fibrils or aggregated networks (Jarrett & Lansbury, 1993; Lomakin et al., 1996, 1997; Sorci et al., 2011). An alternative scenario is the formation of intermediate protofibrils along the pathway which subsequently assemble into fibrils (Harper et al., 1997; Walsh et al., 1997). Mature fibrils have classically been viewed as the stable end-stage of the aggregation process which are not susceptible for dissociating conditions. Recently, using calorimetric methods (Morel et al., 2010), molecular dynamics simulations (Zidar & Merzel, 2011), and mechanical deformation studies (Paparcone & Buehler, 2011; Paparcone et al., 2010; Xu et al., 2010) it has been observed that fibrils can be dissociated albeit at high temperature.

Detailed knowledge of the molecular parameters determining aggregation propensity, rate and morphology resulted in the ability to tune protein aggregation through chemical engineering. The attachment of sugar chains to proteins has been shown to inhibit self-association (Marquardt & Helenius, 1992; reviewed in Helenius et al., 1997; Land & Braakman, 2001; Song et al., 2001). This effect was largely attributed to covalently linked sugar moieties affecting kinetic partitioning between folding and aggregation from an (partially) unfolded state. For example, glycosylation was found to increase the folding rate of the protein rapidly shielding exposed hydrophobic regions which could potentially act as a driving force for aggregation at ambient temperature (Broersen et al., 2007b; Shental-Bechor & Levy, 2008; Wang et al., 1996). Interestingly, high temperatures induced more rapid aggregation of glycosylated proteins compared to their non-glycosylated counterparts (Broersen et al., 2007b; Chobert et al., 2006). However, another study involving glycated bovine serum albumin concluded that glycation of the protein inhibited its aggregation upon incubation at moderate temperatures (Rondeau et al., 2010, 2007). Glycosylation of proteins has also been shown to affect gel properties: attachment of a ketohexose to ovalbumin by means of the Maillard reaction resulted in the formation of gels with enhanced breaking strength (Sun et al., 2004). The effects of various types of chemical protein engineering on the physico-chemical functionality of proteins will be discussed in more detail in paragraph 3 of this review.

2.3 Protein structural integrity

Proteins in food products can lose their native structure as a result of processing conditions including storage, heat treatment, acidification, dehydration, mechanical processing or shear, and microbial hydrolysis. For example, long-term storage of milk powder has been found to induce lactosylation of the proteins present in the preparation which, in turn, results in affected powder solubility and emulsifying and foaming properties (reviewed in Thomas et al., 2004). Unfolded or hydrolyzed protein molecules can exert very different functionality to food products compared to folded proteins, a classical example being the boiling of an egg which converts the liquid-like transparent egg white into an opaque semi-solid structure with very different textural properties. This paragraph will shortly discuss the principles of protein folding and structure and the forces that are implied. It was first recognized by Anfinsen (1973) that the primary sequence of a protein dictates the specific folded, or native, conformation a protein will assume to allow functional activity. Following urea-induced denaturation of ribonuclease A, the protein was allowed to refold by removal of urea. The protein was found to regain its native structure and functionality after this treatment suggesting that proteins can adopt their native conformation spontaneously (Anfinsen, 1973). This finding was awarded with the Nobel Prize in Chemistry in 1972 and opened up an avenue of experimental and theoretical work in the field of protein folding and unfolding. The structural insights into the folding and unfolding processes of many proteins have since then been explored using a vast range of biophysical instrumentation, both at the ensemble (reviewed in Buchner et al., 2011; Sanchez-Ruiz, 2011) as well as at single molecule (reviewed in Borgia et al., 2008; Ferreon & Deniz, 2011) level. Protein conformational stability can be defined as the ability of the natively folded structure of a protein to resist unfolding. Two types of stability can be distinguished: the difference in energy content between the folded state and unfolded state of a protein is termed thermodynamic stability. Boltzmann's distribution law defines the distribution of folded and unfolded protein molecules at a certain time. The rate at which the unfolded protein collapses to a folded state is reflected by the folding rate and represents the kinetic stability of a protein. Energy landscape theories and the folding funnel hypothesis have both been used as models to understand the energetic barriers of a protein separating the folded from the unfolded state (reviewed in Onuchic et al., 1997; Plotkin & Onuchic, 2002; Wolynes, 2005). Both models start from a similar principle in which there is an energy difference between the folded state of a protein and its unfolded state and that the folded state is defined as the favored entropic state of a protein. Local energy minima in the folding process can result in the accumulation of transient intermediate structures, which are neither folded, nor unfolded, to a larger or lesser extent, depending on the environmental conditions of folding or the primary sequence of the protein (reviewed in Baldwin, 2008; Englander et al., 2007). These intermediate structures, which often lack biological activity, are sometimes sufficiently stable to allow substantial accumulation (reviewed in Englander et al., 2007). Stable folding intermediates are related to a high propensity of aggregation as the hydrophobic core is not sufficiently shielded while opposing assembly forces are absent (reviewed in Ferreira et al., 2006). Kinetic protein stability is defined by a variety of molecular parameters including the proximity of native contacts in the primary sequence of a protein (Cieplak et al., 2004; Plaxco et al., 1998, 2000), internal friction or the energetic of intrachain interactions, energy barriers to backbone rotations and long-range residue interactions (Pabit et al., 2004; Qiu & Hagen, 2004), rate of diffusional motion of an unfolded peptide chain through the solvent (Pabit et al., 2004), and the presence or absence of intermediate state(s) (Baumketner, 2003; Onuchic et al., 1997). Forces that contribute to thermodynamic stability are the strength of intramolecular hydrogen bonds and solvent-protein interactions, both enthalpic in nature. The entropic contribution is mainly defined by the hydrophobic effect of folding through an increase in disorder of water molecules upon folding. Many of the forces retaining a protein structure intact can be disrupted, removed or introduced by chemical modification. For example, reaction of proteins with sugars by means of the Maillard reaction can lead to distinct changes in protein stability. The stability of proteins in food products can be affected by using proteins which have been modified by means of glycosylation or charge modification. Succinylation with the aim to increase net charge of soy protein hydrolysate lead to improved digestibility of the protein as investigated by a multienzyme method involving trypsin, chymotrypsin, and peptidase (Achouri & Zhang, 2001). This finding suggests that the protein had undergone structural rearrangement as a result of the succinylation process. Succinylation also lead to destabilization of Faba bean legumin (Schwenke et al., 1998). This group used differential scanning calorimetry (DSC) to study protein stability of legumin and found a decreased specific enthalpy for unfolding upon succinylation. Succinylation also resulted in an increased surface hydrophobicity of the protein suggesting at least partial unfolding of the molecule. Kosters and colleagues (2003) compared the effects of many different types of chemical modification on protein stability, including lipophilization using capric acid, glycosylation and succinylation of ovalbumin. Ovalbumin stability was probed by DSC, tryptophan fluorescence and circular dichroism (CD). Lipophilization resulted in a decreased denaturation temperature of ovalbumin reflected in an enthalpy decrease and a lower stability upon guanidine titration. Glycosylation was found to stabilize structural integrity of ovalbumin A similar finding using CD and DSC to study temperature-induced unfolding has been reported upon glucosylation (Broersen et al., 2004; van Teeffelen et al., 2005) and fructosylation (Broersen et al., 2004) of β-lactoglobulin or glucosylation of codfish parvalbumin (de Jongh et al., 2011). Interestingly, assays performed at ambient temperature but involving denaturant-induced unfolding reported a decrease in protein stability upon conjugation of glucose to β-lactoglobulin. This phenomenon, which appeared unique for a glycosylation reaction, has been studied in further detail by van Teeffelen and colleagues (2005). The observations could be explained in terms of a decreased change in heat capacity upon unfolding as a result of glucosylation indicating that the hydration pattern of proteins upon glycosylation is significantly affected.

2.4 Shelf life

Proteins can affect shelf life and stability of food products by enhancing antioxidant activity, affecting gas exchange, antimicrobial activity or by stabilization of emulsion or foam-based food products (del Rosario Moreira et al., 2011; Emmambux et al., 2004; Mendis et al., 2005). Protein films can be used as as packaging biomaterials as a result of their ability to form networks with rheologically advantageous characteristics (Arvanitoyannis, 1999; Audic & Chaufer, 2005, Longares et al., 2005). However, mixtures of for example proteins and polysaccharides have been found to exert superior functional properties compared to proteins or polysaccharides in isolation (reviewed by Pogaku et al., 2007). For example, the application of an edible coating of storage proteins obtained from sorghum, called kafirins, has been shown to extend the shelf life of freshly harvested pears (Buchner et al., 2011). The shelf life of meat was shown to be extended upon application of a collagen and gelatin coating and led to reduced decoloration, antioxidant activity, and reduction of microbial spoilage (Havard & Harmony, 1869). Such coatings extend quality and shelf-life by acting as a slow-release gas barrier (Baldwin, 1994; Buchner et al., 2011; Park, 1999). Nanobiocomposites of maize prolamin protein zein have also been employed as a gas barrier by coating tomatoes (Park et al., 1994), and apples (Bai et al., 2003). Even though collagen and gelatin coatings were reported to both effectively retain water in meat products (Antoniewski et al., 2007; Farouk et al., 1990), extensive moisture loss of kafirin coated pears compared to the uncoated product left them unacceptable toward consumers (Buchner et al., 2011). The authors (Buchner et al., 2011) therefore suggested to prepare wax or triglycerides/kafirin mixtures instead of pure kafirin coatings to prevent moisture loss as kafirin films themselves do not function effectively as water barriers (Emmambux et al., 2004; Gillgren & Stading, 2008). Because lipids form a very suitable moisture barrier as a result of their hydrophobic character, lipophilized proteins possibly form more effective coatings for fruit. However, to date, no work has been published to demonstrate the effect of lipid-incorporation into proteins to prepare stable coatings for fruit.

Some proteins and peptides are known to have antimicrobial activity (Nizet, 2006; reviewed by Wimley, 2010). Some of these are also applied as food preservatives such as nisin, which is a potent antibacterial 34 amino acid peptide containing a number of uncommon amino acids. Nisin has been employed as an approved food preservative in cheese (Martins et al., 2010), fish, meat, and beverages (reviewed in Lubelski et al., 2008). Another known antimicrobial peptide is ε-poly-L-lysine which exhibits antimicrobial activity against bacteria and fungi and is used as a food preservative (reviewed in Hamano, 2011). This asset has also been explored within a food based environment, for example by applying mixed formulations of chitosan, a linear polysaccharide, and casein polymers to a number of food products including carrot, cheese, and salami (del Rosario Moreira et al., 2011). Pure caseinate films applied to squash slices showed limited antimicrobial activity (Ponce et al., 2008). While chitosan alone exerts significant anti-microbial activity, the inclusion of casein polymers into the formulation improved bactericidal properties even more (del Rosario Moreira et al., 2011) as a result of the ionic interaction between the two biopolymers (Pereda et al., 2008, 2009). Cao-Hoang and colleagues (2010) produced a nisin-containing sodium caseinate film to investigate the antimicrobial activity of both surface- and in-depth *Listeria innocua* inoculated soft cheese. The presence of the film reduced surface contamination with *L. Innocua* significantly, while antimicrobial activity within the cheese matrix depended on the distance from the film-coated surface. Antimicrobial films prepared from a mixture of oregano oil and whey protein isolate showed inhibition of growth of lactic acid bacteria, reduction of pseudomonads, total flora and growth rates when applied to fresh beef (Zinoviadou et al., 2009). Even though in many cases complex formulae have been employed, containing both a protein component as well as a carbohydrate or oil component, no publications are known that show the effects of covalently linked lipid or sugar to protein films as a potential edible film.

2.5 Sensory: Color, flavor, odor, texture

Sensory aspects of food products include sensation of flavor, odor, color, and texture. These factors play a large role in consumer acceptation of food products and the effects of various types of protein chemical engineering and their applications will be discussed in this paragraph. Many types of modification target the amino groups of lysine residues, including succinylation, lipidation and glycosylation through the Maillard reaction. Textural properties, including emulsifying, foam and gelling capacities, have been discussed in detail in paragraphs 2.1 and 2.2.

2.5.1 Flavor

Even though most proteins are tasteless, ingestion of a small number of proteins is perceived as sweet. These include thaumatin (Ohta et al., 2008; van der Wel & Loeve, 1972), monellin (Morris & Cagan, 1972), brazein (Ming & Hellekant, 1994), and lysozyme (Masuda et al., 2001). The sweetness of lysozyme results from the abundance of lysine residues which was shown by alanine substitution in lysozyme (Masuda et al., 2005a). It is therefore perceivable that modification of lysine residues by conjugation of a chemical group has consequences for the sweetness of the protein. Extensive acetylation and phosphopyridoxylation of lysine residues of lysozyme decreased the perceived sweetness of the protein further demonstrating that lysine residues play a major role in sensory aspects of this protein (Masuda et al., 2005, Kaneko & Kitabatake, 2001). No other reports on flavor modulating aspects as a result of chemical protein engineering have been reported.

2.5.2 Odor

One publication which studied the effect of acylation by acetic and succinic anhydride of flaxseed protein isolates reported no off-odors upon modification, although no results were presented (Wanasundara & Shahidi, 1997).

2.5.3 Color

Three types of modifications have been reported to affect the color of the protein preparation. First, succinylation was reported to convert the color of soy isolate from tan to chalk-white upon visual inspection (Franzen & Kinsella, 1976a). Upon measurement of surface reflectance using a colorimeter, increasing degrees of succinylation of flax seed protein lead to brighter protein preparations (Wanasundara & Shahidi, 1997). A similar observation has been reported for succinylation of fish muscle (Groninger, 1973), alfa-alfa leaf protein (Franzen & Kinsella, 1976b), and soy bean protein preparations in a U.S. patent for coffee whitener (Melnychym & Stapley, 1973). Acylation by treatment of protein with acetic anhydride also lead to *brighter* flaxseed protein isolate, although the effect of succinylation of the same protein was stronger (Wanasundara & Shahidi, 1997). Franzen and Kinsella (1976a) showed no effect on color upon acetylation of soy isolate. A third type of modification with a strong effect on color is glycosylation through the Maillard reaction. This reaction is a complex cascade of reactions initiated by the interaction of a reducing sugar with an amino group. Colored products are formed only at later stages of the reaction pathway and include aldols and melanoidins which are high molecular weight compounds (reviewed in Zamora & Hidalgo, 2006). Paragraph 3.1 further extends on the formation of browning products and Amadori compounds related to the Maillard reaction.

2.5.4 Texture

Texture is a complex consumer perception of mouthfeel, tastants and afterfeel. A number of most relevant texture attributes, like 'spreadability' (essential for attributes like 'creamy'), 'crumbliness' or 'separating/wateriness' have been shown to be directly related to the energy household in protein-based products when energy is exerted onto the system. This applied energy may arise from oral processing, exposure to heat, gravity or applied pressure during for example industrial processing. The energy balance in protein-based food structures has been proposed by van Vliet and Walstra (1995): $W_{applied} = W_{stored} + W_{dissipated} + W_{fracture}$. This model implies that all energy applied to the gel can be used either for fracture, can dissipate or can be stored in the network (and regained after release of exerted forces). At a microstructural level this translates into fracture nucleation points and fracture propagation modes, whereas energy dissipation is often assumed to be controlled by serum flow properties. This latter factor is directly determined by the porosity of the gel as set-up by the microstructure and the pore deformation propensity when the system is put under strain. Van den Berg and colleagues (2008) showed that an attribute like 'spreadability' is directly related to directing as much energy as possible to fracture. When energy can be efficiently stored in the protein matrix, this directs the attribute 'crumbliness'. The effective interaction between protein-building blocks that make up the product matrix can be engineered. Strengthening this interaction, using for example transglutaminase (e.g. Dondero et al., 2006), will direct the energy flow from fracture to storage and gels will become less spreadable and more crumbly. Sala and co-workers (2008) showed that by modulating the interaction between a filler (like a fat-particle) and the protein matrix the texture of these protein gels could be strongly affected. This balance between active-inactive property of the filler could be delicately controlled by means of chemical modification.

3. Types of chemical protein engineering – Exploring potential functionality

Proteins can be chemically modified *in vitro* by covalently cross-linking the protein with a molecule of interest. Many of these reactions will also take place post-translationally in the strictly regulated environment of the cell, such as phosphorylation of cAMP-dependent protein kinases which plays a role in the enhancement of glycogen degradation (Soderling,

1975). However, this paragraph will entirely focus on the kind of chemical engineering intentionally brought about to link specific molecules to proteins which act as functional ingredients in food. These molecules change the behavior of the protein and are largely hypothesized to infer characteristics to the protein which are little present in the unmodified protein, such as improved foaming properties, inhibition of aggregation or enhanced surface activity. Rationale for chemical modification of proteins is multiple but can be categorized into four main reasons.

i. Waste control: For example the re-use of fish gelatin from waste requires less natural resources for their production. Other examples include the production of a peptide with anti-oxidant activity from algae (Sheih et al., 2009). Protein rich by-products are also recovered upon electrocoagulation of wastewater resulting from egg processing (Xu et al., 2002). Chemical modification can be used to increase the functional properties of these proteins.
ii. Health considerations: An example of this is the replacement of meat or soy proteins by (other) vegetable proteins (O'Kane et al., 2004; Pedroche et al., 2004; Vioque et al., 1999; reviewed in Moure et al., 2006). Similar to proteins obtained as by-product from waste material, unmodified plant proteins often have limited functional applicability. Hence, chemical modification of plant proteins can be used to improve molecular functionality.
iii. Cost effectivity by extending molecular functionality of a protein. Chemical modification can be employed to enhance the functional properties of a protein, such that less material is required to obtain a product with similar structural characteristics.
iv. Structure-function relationships: Chemical modification is often used to investigate the contribution of specific molecular parameters, such as surface hydrophobicity, to functionality of the protein at ingredient level. Ample examples of protein functional extension are discussed in the rest of this paragraph.

The types of modification described in literature are extensive and include phosphorylation (attachment of a phosphate to serine, tyrosine or threonine), methylation (attachment attachment of a methyl group to arginine or the N-terminus of the protein), glycosylation (attachment of carbohydrates to lysine or the N-terminus), acetylation (attachment of acetyl to an amino group such as lysine or the N-terminus), and many more. The various types of modification and their impact on molecular behavior of proteins have been reviewed in a vast number of publications and book chapters (Feeney et al., 1982; Means & Feeney, 1971; Tawfik, 2002). This paragraph aims to shortly discuss the current state-of-the-art of the various reactive groups in proteins which can be targeted by chemical engineering. The chemical basis of these modifications will be discussed and applications from literature will be reviewed merely to illustrate the wide variety of applications of the chemical modification procedure.

3.1 Glycosylation and deglycosylation

Food storage and preparation processes such as heating by pasteurization or sterilization often provide for conditions which induce spontaneous and uncontrolled reaction of a reducing carbohydrate present in the food matrix with proteins. This reaction, termed the Maillard reaction, which is actually a complex cascade of reactions, is responsible for the formation of browning products and can have substantial impact on the flavor and color of food products. The Maillard reaction is initiated by a condensation reaction between the ε-amino group of lysine and the reducing group of a sugar to form Amadori or Heyn's rearrangement products via *N*-substituted glycosylamine. During the advanced stages of this reaction, the Amadori and Heyn's rearrangement products are degraded via a number of pathways (Mossine et al., 1994; Röper et al., 1983). The last stages of the Maillard reaction involve extensive protein cross-linking reactions and the formation of so-called melanoidins (Pellegrino et al., 1999). As a result of the wide range of intermediate chemical structures formed, several of these intermediates can be employed as indicators of the Maillard reaction to monitor the extent of the reaction in food products as a measure of quality control. For example, ε-*N*-(furoylmethyl)-L-lysine (furosine) formation, an intermediate in the Maillard reaction, was shown to be the result of lactosylation upon storage of milk powder (Le et al., 2011). Also prolonged storage of high-protein nutrition bars showed non-enzymatic Maillard browning as a result of interaction between whey protein isolate and high-fructose corn syrup or sorbitol syrup (McMahon et al., 2009). The baking process of bread was found to affect color formation determined by furosine and hydroxymethylfurfural concentrations (Ramírez-Jiménez et al., 2000). Variation of baking temperature and dough composition determines the extent of furosine formation and loss, acid-released lysine, and carboxymethyllysine formation of cookies (Charissou et al., 2007). Glycosylation via the Maillard reaction has also been brought about intentionally to study the effects of covalent sugar linkage to proteins in terms of e.g. protein stability and aggregation (Feeney et al., 1975). Glycosylation of proteins by means of the Maillard reaction has been observed to both induce and protect against aggregation. Incubation of hazelnut proteins with glucose resulted in the formation of high molecular weight protein aggregates detected by SDS-PAGE (Cucu et al., 2011). At the same time, others have shown that Maillardation with glucose inhibited the aggregation of cod fish parvalbumin (de Jongh et al., 2011). As the Maillard reaction is a reaction involving many steps, one possibility which can be raised to explain the discrepancy between these observations is that the various intermediates may display differences in resistance against aggregation, some of which may be protective, others which may be inducing aggregation. Increased thermal stability upon glycosylation was found for many proteins including the apple allergen Mal d3 upon reaction with glucose (Sancho et al., 2005), and β-lactoglobulin reaction with glucose and fructose (Broersen et al., 2005). Alternatively, deglycosylation was shown to induce denaturation and aggregation of ovalbumin (de Groot et al., 2007). Apart from the Maillard reaction, protein glycosylation can be achieved by several other routes. *N*-glycosylation takes place by modification of the side chains of asparagine or arginine (Kornfeld & Kornfeld, 1985). *O*-glycosylation is brought about by modification of serine, threonine, or tyrosine (Hart, 1992). Many different glycan structures have been identified to be involved in these two types of modification and they are often necessarily involved in biological function of the protein (Rudd et al., 2001). These last two types of glycosylation take place mainly as a result of post-translational processing of proteins *in vivo* and are not used as means to induce glycosylation of proteins applied in the food industry. Hence, *N*- and *O*-glycosylation will not be discussed in this paragraph.

3.2 Lipophilization

Covalent linkage of lipids to proteins results in increased hydrophobic exposure of a protein with interesting applications related to altered surface properties. Increased exposed hydrophobicity of proteins has for example been related to an improved capacity to form and stabilize emulsions and foams which is the result of improved potential to interact with hydrophobic surfaces, both the air-water and oil-water interface, and including (model)membranes (Nakai, 1983; Wierenga et al., 2003; reviewed in Wilde, 2000; Wilde et al., 2004). Various saturated and unsaturated fatty acids have been employed to induce lipophilization of proteins including caproic acid (Liu et al., 2000), capric acid (Aewsiri et al., 2010; Kosters et al., 2003; Liu et al., 2000), lauric acid (Aewsiri et al., 2010), myristic acid (Aewsiri et al., 2010; Ibrahim et al., 1993; Liu et al., 2000), palmitic acid (Haque et al., 1982; Haque & Kito, 1983a, 1983b; Ibrahim et al., 1991), stearic acid (Djagny et al., 2001; Ibrahim et al., 1993), and oxidized forms of linoleic acid (Aewsiri et al., 2011a, 2011b), and the efficiency of the lipophilization reaction was found to be inversely proportional to the length of the lipid chains used (Liu et al., 2000). Reaction of 28% of the available free amino groups of ovalbumin with activated capric acid was shown to result in retained secondary structure while inducing oligomerization and destabilization of the protein structure as a result of lowering the enthalpy for unfolding (Kosters et al., 2003). The presence of acyl chains was thought to cause significant dehydration of the protein. In another study, hen egg white lysozyme was lipophilized with short and middle chain saturated fatty acids including caproic (C6:0), capric (C10:0), and myristic (C14:0) acid (Liu et al., 2000). Lipophilization of lysozyme was reported to decrease the thermal stability of lysozyme as a result of partial loss of α-helical content of the protein, and this molecular destabilization appeared to be proportionally related to the chain length and the number of bound fatty acids. The lysine residues involved in the modification were thought to be located in the helical region and to subsequently induce partial unfolding of the α-helical region surrounding these residues (Liu et al., 2000). Lysozyme has also been chemically modified using palmitic acid (Ibrahim et al., 1991, 1993) with the primary aim to study the effect of lipophilization on the antimicrobial effect of the protein. Even though increasing extents of covalent linkage with palmitoyl residues lead to insoluble protein, as spectrophotometrically determined by solution turbidity at 500 nm, foaming stability and emulsifying activity were progressively improved by linkage of palmitic acid to the protein molecule. More groups showed that the foaming or emulsifying activities of a wide range of proteins could be improved upon lipophilization, including soybean glycinin (Haque et al., 1982), α_{s1}-casein (Haque & Kito, 1983b), and cuttlefish skin gelatin (Awesiri et al., 2011a, 2011b). A further effect resulting from the incorporation of myristic and stearic acids into lysozyme was related to antimicrobial activity and stearic and palmitic acid conjugation resulted in more effective antimicrobial agents against *E. coli*, than the attachment of myristic acid or the unmodified protein (Ibrahim, 1993). Myristoylation was found to induce lysozyme aggregation resulting in concurrent loss of antimicrobial function. The effects of palmitoylation on the structural and functional properties of α_{s1}-casein have also been explored (Haque & Kito, 1983a, 1983b). The conjugation of the ε-amino groups of α_{s1}-casein with palmitic acid lead to micelle formation as a result of increased hydrophobicity while negative net charge was increased (Haque & Kito 1983a). Further work by this group showed that palmitic acid linkage to α_{s1}-casein did not lead to large scale structural rearrangement of the molecule, both at a secondary and a tertiary structure level using circular dichroism. Interestingly, Aewsiri and colleagues have also investigated the antioxidative activity of cuttlefish skin gelatin modified with a combination of oxidized linoleic acid and oxidized tannic acid, a potent antioxidant (Aewsiri et al., 2011a). Oxidation of lipids and proteins in foams primarily takes place at the air-water interface and the addition of a hydrophilic anti-oxidant alone reduces surface activity (Aewsiri et al., 2011a). Co-conjugation of tannic acid and linoleic acid to gelatin both improves migration of the protein to the air-water or oil-water interface improving foaming and emulsifying activity, respectively, while retaining anti-oxidant activity (Aewsiri et al., 2011a).

3.3 Chemical-reactive groups

Sulfhydryl groups play an important role in regulating the self-assembly of proteins as well as their stability driven by disulfide interchange reactions (Sawyer, 1968). Hence, the presence of these groups has substantial impact on the aggregation and gelation behavior of a wide variety of proteins which has been confirmed by many researchers (Arntfield et al., 1991; Broersen et al., 2006; Graña-Montes et al., 2011; Hayakawa & Nakai, 1985; Hoffmann & van Mil, 1997; Margoshes, 1990; Mine, 1992; Sawyer, 1968; Shimada & Cheftel, 1989). A variety of modifications can be performed targeting sulfhydryl groups, which are part of the cysteine residues. Sulfhydryl groups are highly reactive against various reactants and are thus suitable targets for modification. Sulfhydryls can be blocked to prevent cross-linking by S-methyl methanethiosulfonate (MMTS), *N*-ethylmaleimide (NEM) (Kitabatake et al., 2001), or iodoacetamide (Anson, 1940; Huggins & Jensen, 1949; Smythe, 1936), or additional sulfhydryl-groups can be attached to primary amines (SATA) of proteins. Further, *N*-hydroxysuccinimide esters can react irreversibly with primary amines releasing *N*-hydroxysuccinimide.

Disulfide bonds are thought to play a crucial role determining the stability of proteins (Betz, 1993; Zavodszky et al., 2001; reviewed by Creighton, 1988) as well as to impact on the aggregation process and gel formation of various proteins including ovalbumin (Broersen et al., 2006; Kato et al., 1983), vicilin (Arntfield et al., 1991), and β-lactoglobulin (Sawyer, 1968). Aggregates and gel networks are often the result of combined action of hydrophobic and electrostatic interactions and covalent interactions, in the form of disulfide bonds, are sometimes present (Kato et al., 1983; Koseki et al., 1989; Sun & Hayakawa, 2002). Thiolation of ovalbumin mediated through the reaction of S-acetylmercaptosuccinic anhydride (S-AMSA) with primary amines results in the formation of acetylthio groups and the acetyl group can be cleaved off to yield reactive sulfhydryl groups by the addition of hydroxylamine (Klotz & Heiney, 1962). A range of modification degrees can be obtained by varying the S-AMSA:lysine ratio (Broersen et al., 2006). Next to the linkage of a sulfhydryl group, additional carboxyl groups are conjugated through this reaction introducing additional charge variation which can lead to an additional parameter which can induce variation in aggregation, gelation or stability of a protein. To circumvent this variation, proteins with activated sulfhydryl groups are best compared with similarly modified proteins with blocked (i.e. not reacted with hydroxylamine) acetylthio groups, rather than directly with the unmodified protein (Broersen et al., 2006). Thiolation of ovalbumin in this way lead to limited changes at a secondary and tertiary structure level at high degrees of modification suggesting that the original molecular fold was largely retained upon modification. High degrees of thiolation resulted in a decrease of thermal stability of ovalbumin while fibril morphology was affected. Interestingly, the rate of aggregate formation was not modified by the presence of additional sulfhydryl groups available for disulfide formation. It was concluded that disulfide formation does not represent the prime driving force for aggregation of ovalbumin which was further illustrated by the finding that at room temperature, where significant sulfhydryl groups are already exposed upon modification, no disulfide bond aggregation was observed. Final aggregate morphology, gel formation and stability are affected as a result of rapid covalent network formation which does not allow rearrangement into more stable networks, as illustrated by lower gel Young's moduli obtained upon thiolation (Broersen et al., 2006).

3.4 Charge modification by methylation and succinylation

Net charge and local charge densities of proteins have been implicated in the regulation of protein stability, aggregation, and aggregate morphology affecting the visual appearance of food products. These hypotheses have been substantiated by a range of observations which involved charge introduction, removal or reversal through succinylation and methylation reactions (Broersen et al., 2007a; Weijers et al., 2008). The reactions of succinylation and acetylation both lead to blockage of the reactive amino groups of proteins with an acyl residue and are hence collectively termed acylation reactions. The rate of acylation reaction depends on the rate of nucleophilic attack. Succinylation leads to increased net negative charge by the covalent linkage of succinate anions to the cationic amino groups of a protein thereby converting a cationic group into an anionic residue having implications for the distribution of net charge of a protein. Upon acetylation, ammonium cations are replaced by neutral acetyl groups resulting in electrostatically neutral groups. Large extents of succinylation have been reported to affect the integrity of secondary and tertiary structure of soy protein hydrolysate as shown by intrinsic tryptophan fluorescence and circular dichroism (Achouri & Zhang, 2001). Similar conformational rearrangements have been reported upon succinylation of whey protein isolate (Gruener & Ismond, 1997), bovine serum albumin (Jonas & Weber, 1970), canola protein (Lakkis & Villota, 1992), Faba bean legumin (Schwenke et al., 1998), rapeseed 12S globulin (Gueguen et al., 1990), and winged bean protein (Narayana & Rao, 1991). As a result of co-incubation of soy protein hydrolysate with succinic anhydride, which is a common compound used to succinylate proteins, heterogeneous reaction mixtures were obtained. Next to the aimed amine groups, this method of succinylation also commonly results in *O*-succinylation, involving threonine or serine hydroxyl groups or tyrosine succinylation (Achouri & Zhang, 2001; Chang & Sun, 1978; Schwenke et al., 1998). This latter reaction was found to be reversible upon treatment with hydroxylamine (Habeeb & Atassi, 1969), but, when present, to induce substantial expansion of Faba bean legumin as observed by viscometric studies (Schwenke et al., 1998). It was postulated that the high accumulation of negative charge upon extensive succinylation leads to dissociation and expansion of the individual subunits legumin is composed of (Schwenke et al., 1998). Other functional properties are equally affected by succinylation. For example, protein solubility has been reported to increase upon succinylation as has been demonstrated for rapeseed preparations (Dua et al., 1996), flax protein isolate (Wanasundara & Shahidi, 1997), oat protein isolate (Mirmoghatadaie et al., 2009), and soy protein isolate (Franzen & Kinsella, 1976a). Improved solubility has been related to the ability of proteins to perform more efficiently as stabilizers in emulsions and foams (Nakai & Li-Chan, 1988; Waniska & Kinsella, 1979), which, in turn, is greatly affected by their ability to absorb at the air-water interface (Wierenga et al., 2005). It has indeed been shown that treatment of a variety of proteins with succinic anhydride leads to increased foam capacity (Dua et al., 1996; Franzen & Kinsella, 1976a; Mirmoghatadaie et al., 2009), although others suggest that succinylation leads to decreased foam expansion capacity (Wanasundara & Shahidi, 1997). These seemingly contradictive findings may be explained by the demonstration by Wierenga and colleagues (2005) that the likelihood of a protein molecule to adsorb at an interface is the result of a balance between hydrophobic and steric effects: highly charged molecules may be adsorbing to the interface as a result of hydrophobic interaction, but the density at which protein molecules continue adsorbing to the interface is mainly determined by the repulsive nature of the charged proteins. Other effects observed upon charge modification of proteins are related to emulsification properties: methylation increased while succinylation decreased the emulsifying activity of rapeseed preparations (Dua et al., 1996). Other studies show an increased emulsifying activity and stability upon succinylation of soy protein (Franzen & Kinsella, 1976a), and oat protein isolate (Mirmoghatadaie et al., 2009). In terms of gelation, an increase in net charge lead to more transparent gels upon gelation of ovalbumin which is related to the morphology of the aggregated network making up the gel structure (Weijers et al., 2008). Overall, many and detailed efforts have been made employing net charge modification of proteins in the field of food science. These studies have lead to in-depth knowledge of the role of electrostatics to common protein functionalities such as emulsification, foaming, and aggregation propensity.

3.5 PEGylation

The covalent attachment of a polyethylene glycol (PEG) polymer chain to a protein, also termed 'PEGylation' is mostly applied in the field of pharmaceutics as the conjugation of non-toxic PEG imparts substantial advantages to support drug delivery (reviewed in Damodaran & Fee, 2010; DeSantis & Jones, 1999; Francis et al., 1998). The protein targets for PEG modification are regarded as non-specific and include the ε-amino groups of lysine and other nucleophilic groups such as glutamic acid, aspartic acid, threonine, serine or tyrosine on the surface of the protein resulting in highly heterogeneous protein-PEG conjugates upon modification (Losso & Nakai, 2002). Commercially available PEG is available as mixtures of different oligomer sizes in various molecular weight ranges enabling the variation of exposed hydrophobicity of proteins. Conjugation of hydrophilic PEG to a hydrophobic protein generally results in an increase in hydrodynamic size and water solubility (Damodaran & Fee, 2010). From a pharmaceutical viewpoint, PEGylation has been reported to enhance circulation life of bovine liver catalase in the blood of mice while the presence of PEG does not induce an immune response upon injection (Abuchowski et al., 1977). Modification of peroxidase from turnip was shown to enhance catalytic activity of the enzyme with increased stability in organic solvents as well as increased temperature resistance (Quintanilla-Guerrero et al., 2008). Similar results were found upon PEGylation of trichosanthin, which showed prolonged plasma half-life and reduced immunogenicity (He et al., 1999). PEGylation of lysozyme similarly lead to stabilization of the protein against pH and temperature variation as well as resistance against proteolysis (Silva Freitas & Abrahão-Neto, 2010). Most of these studies have been carried out in the context of pharmaceutical application. The study of PEGylation as a potential route to bring about modification of physicochemical parameters of proteins applied in food products has been less well explored. The only known effort in the field of food science combined oxidative sulfitolysis with conjugation of 5000 dalton activated PEG to investigate the impact on the emulsifying properties of β-lactoglobulin A (Losso & Nakai, 2002). PEG molecules were found to cover the entire surface of unmodified and sulfitolyzed β-lactoglobulin and PEGylation alone did not improve emulsifying activity or emulsion stability. The combination of sulfitolysis and

PEGylation however lead to more stable emulsions and improved emulsion activity index as a result of better stabilization of individual droplets against coalescence by the absorption of PEG onto the surface of the sulfitolyzed β-lactoglobulin (Losso & Nakai, 2002). From the limited number of studies available it is not possible to derive direct conclusions on the applicability of PEGylation on the advancement of functionality of proteins in food products.

3.6 Deamidation

Deamidation involves the hydrolysis of the amino acids glutamine and asparagine into glutamic and aspartic acid and is achieved by acid, alkaline, or enzymatic treatment (Liao et al., 2009; Shih, 1990; reviewed in Wright & Urry, 1991). Consequently, deamidated protein is often obtained as a by-product of food processing. For example, the extrusion of wheat flour induces deamidation of wheat proteins (Izzo et al., 1993). Deamidation has been shown to affect protein functionality. Functional properties, such as solubility, emulsifying and foaming properties, of gluten have been reported to improve upon low levels of deamidation brought about by mild acid hydrolysis (Hamada & Marshall, 1989; Matsudomi et al., 1982, 1985). Deamidation also was reported to increase exposed hydrophobicity of gluten induced by a conformational change and, subsequently, to increase surface activity (Matsudomi et al., 1982). A number of studies investigated the molecular mechanism for protein structural destabilization upon deamidation. For example, the deamidation treatment has direct implications for charge density and, in turn, affects electrostatic interactions the protein may undergo by interacting with water or upon self-assembly (Finley, 1975; reviewed in Riha et al., 1996). This role of electrostatics to deamidation-induced disruption of protein structure was supported by further observations on wheat gluten: both acetic acid and HCl induced deamidation had substantial consequences for the secondary structure of wheat gluten. It was thus postulated that strong deamidation induced protein unfolding as a result of electrostatic repulsion (Liao et al., 2010). Acetic acid induced deamidation of wheat gluten was further found to inhibit SDS-stable aggregate formation whilst largely retaining its ability to form disulfide bonds (Liao et al., 2010). The rate of the deamidation reaction has been found to depend on primary sequence and pH under which the reaction takes place, but was independent of ionic strength for model peptides (Patel & Borchardt, 1990a, 1990b; Robinson & Rudd, 1974; Tyler-Cross & Schirch, 1991), soy protein and egg white lysozyme (Zhang et al., 1993).

3.7 Cross-linking

A variety of aldehydes, including gluteraldehyde, formaldehyde, and α-hydroxyadipaldehyde, have been used to induce chemical cross-linking of proteins. Between these reagents, gluteraldehyde was found to cross-link bovine serum albumin most efficiently forming large insoluble networks (Hopwood, 1969). Also for other proteins gluteraldehyde has been reported as the most efficient cross-linking agent yielding thermally and chemically stable cross-links (Bowes & Cater, 1968; Nimni et al., 1987). Primary amino groups have been reported to act as prime target to initiate the aldehyde-induced cross-linking reaction (Quiocho & Richards, 1966), next to aromatic amino acids (Hopwood et al., 1970). Depending on environmental conditions, gluteraldehyde can bring about cross-linking through a wide variety of reaction mechanisms (reviewed in Migneault et al., 2004). This is caused by the large number of different molecular structures gluteraldehyde can assume in solution (Hardy et al., 1969; Korn et al., 1972; Richard & Knowles, 1968; reviewed in Migneault et al., 2004) although the mechanistic details for this is unknown. As a result of its reported toxicity, gluteraldehyde-induced cross-linking has not been employed in the field of food technology, other than as a tool to enable the investigation of intermediates in the aggregation pathway or to immobilize proteins onto a surface to allow further investigation.

3.8 Measuring the degree of modification

To evaluate the effect of a specific type of modification on the functional aspects under study, the success of the chemical engineering process on proteins is evaluated by most researchers. To this end, targeted chemical and biophysical assays have been developed which are now widely used. These quantitative assays are mostly based on the formation of a chromogenic or fluorogenic product upon specific interaction with reactive groups of a protein. Some assays will provide information on the average degree of modification in the entire ensemble of protein molecules in a solution. Examples of such assays are Ellman's reagent (Ellman, 1959) or the sulfhydryl-disulfide exchange (SEI) index (Owusu Apenten et al., 2003). Others are also useful to obtain information on the distribution of the degree of modification obtained, such as mass spectrometry. The type of assays developed can be categorized by the type of aimed conjugated chemical group of the protein they probe, such as amine groups, thiol groups or carboxyl groups. Some other researchers use methods which rather probe for the attached molecule, such as the use of gas liquid chromatography (GLC) to determine the degree of lipid incorporation (Haque et al., 1982).

3.8.1 Amine groups

Amine groups in proteins originate either from free amino groups of proteins or from N-terminal residues of proteins (Skraup & Kaas, 1906; Chibnall, 1942). From 1906 it was recognized that lysines in proteins were largely responsible for the free amino groups present in proteins (Skraup & Kaas, 1906). This finding triggered the development of various assays to study accessible ε-amino groups of lysines in proteins (Gurin & Clarke, 1934; Sanger, 1945). Three of the most commonly used assays in the field of protein engineering to determine the number of available amino groups of a chemically modified protein involve the use of chemicals 2,4,6-trinitrobenzenesulphonic acid (TNBS) (Fields, 1971), ninhydrin (Yemm & Cocking, 1955; Schilling et al.; 1963; Samejima et al., 1971), and ortho-phtaldialdehyde (OPA) (Roth, 1971). The latter assay is based on the reaction of the OPA compound with free amino groups in proteins in the presence of β-mercaptoethanol under alkaline conditions. This reaction results in the formation of highly fluorescent alkyl-iso-indole derivatives which emit at a wavelength of 455 nm upon excitation at 340 nm. High concentrations (*i.e.* 10%) of sodium dodecyl sulfate (SDS) are often added to the protein solution to aid the exposure of all amino groups which are sometimes buried in the folded protein. The extinction coefficient for the formed adducts of both α- and ε-amino groups are similar with an absorptivity of 6000 M^{-1} cm^{-1}. The OPA assay was further developed to evaluate the degree of proteolysis of dairy proteins by determining the number of α-amino groups released upon hydrolysis (Church et al., 1983). The assay can also be used to quantify all types of reactions involving the modification of lysine and, hence, includes lipidation of ovalbumin (Kosters et al., 2003; Wierenga et al., 2003), lysozyme (Liu et al., 2000), and gelatin (Aewsiri et al., 2010, 2011a, 2011b). Other modifications which can be quantified using the OPA assay are succinylation (Kosters et al., 2003; Wierenga et al., 2005), methylation (Kosters et al., 2003), glycosylation (Broersen et al., 2004; Kosters et al., 2003), and thiolation of proteins (Broersen et al., 2006).

The second frequently used method to quantify available amino groups upon modification involving the chemical TNBS was developed by Okuyama and Satake (1960) and Satake and colleagues (1960). This method was first employed to study free amino acid groups in trypsin and chymotrypsin inhibitors (Haynes et al., 1967) and for routine screening of protein concentrates for animal feeds (Hall et al., 1973). The chemical TNBS reacts with high preference to free amino groups resulting in the formation of trinitrophenyl derivatives. The reaction product can be quantified spectrophotometrically at 335 nm. A disadvantage of this method is that TNBS also reacts with free sulfhydryl groups, albeit at a slower rate than with amino groups and to form a labile product (Kotaki et al., 1964). The TNBS assay has been used for various types of protein modification including fatty acid incorporation (Andersson et al., 1971; Ibrahim et al., 1991, 1993), glycosylation of proteins as a result of the Maillard reaction (Sun et al., 2004), succinylation (El-Adawy, 2000; Schwenke et al., 1998; Zhao et al., 2004a, 2004b), and acetylation of faba bean legumin (Krause et al., 1996).

The compound 2,2-dihydroxyindane-1,3-dione (ninhydrin) reacts with ε-amino groups and ammonium ions into a blue-purple Schiff base product called Ruhemann's purple, that can be colorimetrically detected at a wavelength of 440 nm (Yemm & Cocking, 1955; Schilling et al., 1963; Samejima et al., 1971). The ninhydrin assay has been used to quantify degrees of succinylation of soy protein (Franzen & Kinsella, 1976a), and acylation of flax protein isolates (Wanasundara & Shahidi, 1997).

Comparison of TNBS, OPA or ninhydrin to determine α-amino groups in pea protein isolates and hydrolysates thereof lead to the conclusion that TNBS and OPA produced comparable results while ninhydrin detected only half of the amino groups that were detected by TNBS and OPA (Panasiuk et al., 1998).

3.8.2 Sulfhydryl groups

Two commonly used assays are available to evaluate the successful conjugation or blockage of sulfhydryl groups in proteins. The assay which was developed first by Ellman (1959) involves the reaction of 5,5'-dithiobis(2-nitrobenzoic acid) (DTNB) or Ellman's reagent with free sulfhydryl groups yielding colored 3-carboxylato-4-nitrothiophenolate (CNT). Spectrophotometric absorbance intensity at 412 nm provides for a direct measure of the concentration of CNT in solution and cysteine is commonly used as calibration standard. Ellman's assay has been used before to determine the extent of *N*-ethylmaleimide modification of β-lactoglobulin A (Kitabatake et al., 2001; Wada & Kitabatake, 2001), thiolation of ovalbumin (Broersen et al., 2006). or acylation of soy protein sulfhydryl groups (Franzen & Kinsella, 1976a). However, the sizeable DTNB at 400 dalton may not be able to detect sulfhydryl groups which are buried inside the folded structure of intact proteins or aggregated proteins (reviewed by Visschers & de Jongh, 2005). A useful alternative for the Ellman's assay has been developed by Owusu Apenten and colleagues (2003) for β-lactoglobulin and bovine serum albumin. Chemical reactivity of thiol groups is an absolute requirement to enable covalent cross-linking through disulfide bond formation (Hillier et al., 1980). This assay, called the sulfhydryl-disulfide exchange (SEI) index, therefore provides a direct measure of the chemical reactivity of thiol groups as it determines the conversion of substrate in time and kinetically relates the conversion to that of fully exposed thiol groups (Owusu Apenten et al., 2003). Chemical activity of introduced sulfhydryl groups by modification of ovalbumin using the SEI index has been verified for example by Wierenga and colleagues (2006).

3.8.3 Carboxylic acid groups

The carboxylic acid content of proteins is primarily investigated using the compound 2-ethyl-5-phenylisoxazolium-3'-sulfonate or Woodward's reagent K (Woodward & Olofson, 1961; Woodward et al., 1961; Sinha & Brewer, 1985). The activity of Woodward's reagent K is the result of a multistep process. First, Woodward's reagent K is converted into ketoketenimine at neutral pH. The intermediate compound is then either further disintegrated to form ketoamide or interacts with carboxylic acid groups of a protein. The latter interaction results in the formation of an enol ester (Pétra, 1971) which absorbs at 340 nm with a molar extinction coefficient of 7000 M^{-1} cm^{-1} (Sinha & Brewer, 1985). At a later stage, Kosters and de Jongh (2003) revised the extinction coefficient of the product to 3150 M^{-1} cm^{-1} at 269 nm to improve specificity of the reaction and eliminate the substantial contribution of side reactions with other nucleophiles in proteins (Llamas et al., 1986), and histidine and cysteine (Bustos et al., 1996; Johnson & Dekker, 1996) to the absorbance at 269 nm. This revised version of the assay employing Woodward's reagent K to estimate the number of carboxylic acid groups of chemically modified proteins has been used by a number of researchers. Wierenga and colleagues (2005) used Woodward's reagent K to estimate the degree of modification of succinylated ovalbumin to study the relation between protein net charge and adsorption to air-water interfaces. Similar net charge variation induced by succinylation was used to investigate colloidal versus conformational stability of ovalbumin to aggregation (Broersen et al., 2007a). To investigate the stability of ovalbumin, the protein was modified by succinylation, methylation, glycosylation, and lipophilization and the degrees of modification were validated using Woodward's reagent K and the OPA assay (paragraph 3.8.1).

3.8.4 Conjugated groups

An alternative route to obtain information on the degree of protein modification is to selectively probe the conjugated group. This can be achieved for example by incorporating an isotopically labeled reagent or inclusion of a chromophore or fluorophore which can then be quantified by read-out of fluorescence intensity using a fluorimeter or simple absorbance measurements using a standard spectrophotometer. Raman spectroscopy was shown to provide direct insight into degrees of succinylation and acetylation of a range of proteins originating from soy, egg white or whey by distinct contributions of the conjugated groups at 1737 cm^{-1} and 1420 cm^{-1} (Zhao et al., 2004a, 2004b). The peak intensities at these wavelengths could be directly converted to obtain information on the degree of modification. Degrees of palmitoylation of soybean glycinin (Haque et al., 1982) and α_{s1}-casein (Haque & Kito, 1983a) have been determined using gas liquid chromatography.

Chromatography has been further explored in the shape of cation exchange chromatography to validate the degree of methylation of ovalbumin which was found to provide comparable read-outs as the revised version of Woodward's reagent K method to determine carboxylic acid groups (Kosters & de Jongh, 2003). All these techniques provide insight in the ensemble average degree of modification. Chemical engineering inherently implies the rise of heterogeneous species of proteins. Mass spectrometry, often employed as Matrix-assisted laser desorption/ionization-time of flight (MALDI-tof) mass spectrometry, has proven a powerful method to specifically obtain insight into the distribution of the modification reactions. This method has been employed to derive information on modification distributions of *N*-ethylmaleimide modified β-lactoglobulin A (Wada & Kitabatake, 2001), and glycosylation of β-lactoglobulin (Broersen et al., 2004; van Teeffelen et al., 2005). These studies demonstrated that degrees of modification obtained upon chemical engineering of proteins are rather broad and show a Gauss distribution profile rather than a single well-defined modification degree (Broersen et al., 2004; van Teeffelen et al., 2005). Some attempts have been made to isolate modified protein fractions with more defined degrees of modification, for example by using ion exchange chromatography of succinylated ovalbumin (Wierenga et al., 2005).

4. Application potential of food protein modification

From the above it is clear that in the past decades a lot of effort has been spend on better understanding and controlling protein behavior and protein-based microstructure formation by making use of chemical engineering approaches. But how much impact have these insights had on the development of new food applications? There are a number of well-known product categories where engineered protein functionality has led to improved product properties. In the early seventies Unilever produced new lines of margarines that showed better performance in aspects like spreadability, prolonged storage stability and during baking, caused by acetylation of milk proteins leading to better fat emulsification (Evans & Irons, 1970). Also for mayonnaises and salad dressings modification of egg yolk proteins (via *N*-succinylation) provided improved product quality (Evans & Irons 1970). Another example of an application of improved protein functionality is that of the use of succinylation to improve the solubility/dispersability properties of soy proteins in the extraction and refinery process (Melnychyn & Stapley, 1973) It is interesting to evaluate via a patent-literature screening how frequent the wide variety of technological possibilities as described in section 3 to better control protein behavior have led to unique market-propositions.

4.1 Current application of protein modifications in food/feed-related products

Figure 1 illustrates a landscape representation of patents (worldwide), filed in the last decade, in the area of food and feed where protein modification has played a crucial role in deriving a new type of product functionality. The height of the contour indicates the activity in that particular area. The distance between patents reflects their commonality. In total only 445 relevant patents (grouped in 157 families) can be found. In comparison, a search on any biobased-product (including products with protein-based technical polymers as in coatings, paints, paper, etc.) revealed more than 8600 patents. Clearly the role of protein engineering to derive new food product specifications are very limited, especially in view of the potential commercial impact. The observation that the patent-families found are spread rather constant over the plot illustrates that these patents are not directly linked to each other in terms of engineering approach or application area. When evaluating the patent filings of the last ten years five areas can be distinguished with a relatively high patenting activity and these will be discussed below in more detail. These are numbered 1 to 5 in the figure.

1. *Refinery of seed storage proteins (12 patents).* There is an increasing interest to use readily abundant and relatively cheap seed storage proteins as nutritional component in food and feed. The major difficulty of this protein-source is to obtain functional proteins after the refinery steps. Loss of functionality occurs especially when the protein is used as powdered ingredient, because of difficulties in resolubilization and unpredictable caking of the powder during storage/transport. A number of patents have been filed that use mild engineering tools, like Maillardation, or enzyme-treatments to preserve functional proteins during refinery steps (e.g. patent EP1370157B1: "Highly soluble, high molecular weight soy protein"). Interestingly, all patents filed in this area pay attention to the in principle reversibility of the modification applied.
2. *Nano-particles (23 patents).* To better direct properties of protein-based nano-particles enzymatic introduction of lipidic groups like small fatty acids or PEGylation has been employed to encapsulate bioactives, typically as microemulsions (e.g. patent US20070154907A1: "Microemulsions as precursors to solid nanoparticles"). These applications are considered food-grade or have passed medical-ethical approval in their testing/and or usage. There are no indications that these application have found their way into the food/feed product market.
3. *Nutritional availability in feeds (13 patents).* A considerable number of patents can be found in the application area to increase the nutritional availability of amino acids in feed (14 patents). Most patent-positions are dealing with destabilizing protein-structures to promote their digestibility and their (proposed) nutritional uptake. Typical patented approaches are de-amidation and Maillardation (e.g. patent WO2004020977A3: "De-amidation to promote nutritional value of rumen in feed"). No patents can be found that link nutritional value and protein modification in foods.
4. *Protein-based emulsion-stabilizers (two times 7 patents).* A few patents are found where specific protein modification is used to improve emulsification of (oil in water) products. Especially enzymatic glycosylation and lipidation approaches have been used (e.g. patent US7126042B1: "Recombinant oleosins from cacao and their use as flavoring or emulsifying agents"). In view of the small scale examples provided in these patents, it is not likely that these inventions have been implemented in a commercial product.
5. *Edible coatings (12 patents).* A whole family of patents is present on enzymatic protein cross-linking (mainly by transglutaminase) in relation to the production of edible coatings. Modifications typically act on a microstructural level, and not so much on the protein molecular level, to strengthen the spatial network formed (e.g. patent EP963704B1: "Food containing proteinaceous material treated with a transglutaminase and an oxidoreductase"). As these interventions in a product do not necessarily need to be labeled on the product, it is difficult to evaluate whether they have resulted in product development.

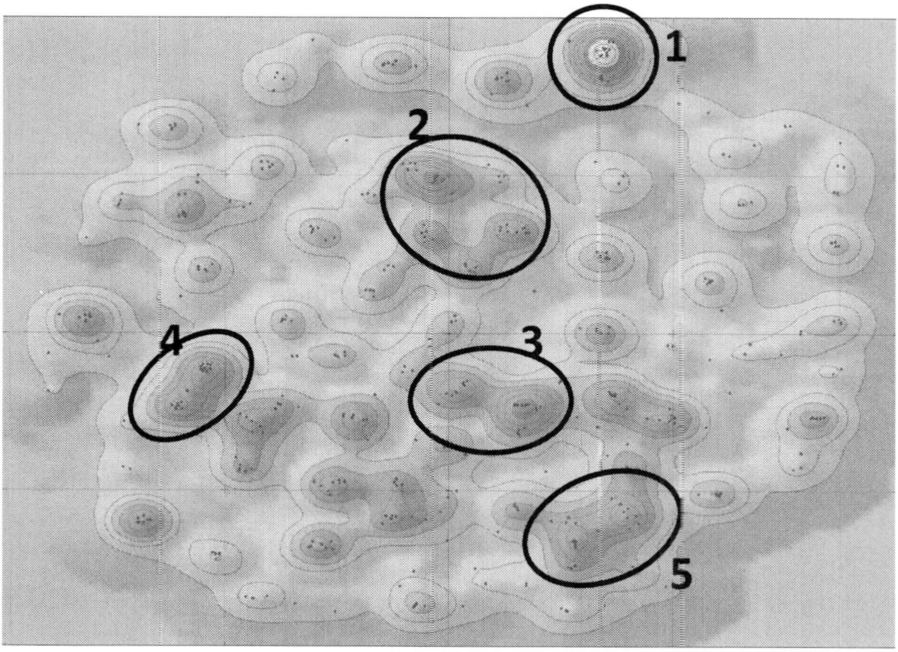

Fig. 1. Landscape representation of patent (families) filed between 2000 and 2011 where engineering/modification of protein functionality has been used to derive new material/product characteristics. The small black dots indicate the position of a patent-family. The numbered circles represent the areas with the highest activity in patents. The numbers are explained in the body text.

4.2 Food legislation

Whereas in paragraph 3 it was demonstrated how active and progressing the understanding of protein functionality in complex systems has been in the past decades, it is striking to see how minor the contributions to new applications these insights apparently have been. To understand this better one needs to take into account that the technological developments coincided with an increasing level of legislation. In the United States this is embodied in the Food and Drug Administration (FDA). The European Food Safety Authority (EFSA) is its European counterpart. The FDA is responsible for protecting and promoting public health through the regulation and supervision of food safety. It does so by formulating acts that set the boundaries for implementation of new ingredients, processes or compositions related to foods. There are a few acts that have had a strong impact on the food sector. In 1990 the Nutrition Labeling and Education Act was launched. This required food products to be labeled in terms of composition, allowing traceability of its ingredients to their source. This act also amended that all nutrient content claims (e.g. 'high fiber', 'low fat', etc.) would meet the standards set by the FDA. In the end it meant that every engineered protein would require a new label and would need to be recognized and approved first by the FDA.

Exceptions were those modified ingredients that could be considered as 'occurring from a natural process' or that were 'reversibly modified' (so temporary). The FDA Modernization Act of 1997 was designed to reduce the time for the approval of new pharmaceutical drugs, but also had an impact on food technology by the acknowledgment of the advancement of technological, trade, and public health complexities. Basically, a 'new' food ingredient needed to be seen and evaluated in the complex role it had considering its production up to its digestion in the food. This act was further refined by the Amendments Act of 2007, leading to much sharper defined criteria in what was considered as safe in food products. Recently, in 2010, The Food Safety Modernization Act was signed. Sections of this act require food producers to enable tracking and tracing of all ingredients used. The use of engineered proteins (either genetically or chemically) requires a separate approval for market-clearance. Summarizing, the FDA (and EFSA) have acknowledged that (future) food production requires innovations at the ingredient level, but also via processing routes, and they are in principle open to protein engineering routes. At the same time it enforces that functionally improved ingredients are checked along the full chain from refinery to nutritional value and human health within the complexity of the product.

4.3 Potentials in the area of food product sustainability

With increasing world-population and welfare the demand for protein as food-nutritional component is rising sharply. Also the identification of proteins as building blocks in non-food applications in view of a more sustainable economy, has led to increasing pressure on innovations in production, refinery, and application of proteins from wider sources than in the current economy is provided. For the development of a vision on food quality and especially the role of nutritional impact needs to be seen in the context of the basic requirements set not only by consumer demands, but also by participating industries. This can be presented by a so-called Pyramid of Food Innovation, as shown in figure 2. On top one finds the foods that need to be developed in a most sustainable way; to achieve this prospect one needs to comply with lower levels of restrictions, limitations and concomitant scientific challenges.

Food *safety* forms the most fundamental aspect here. Terminology like for example 'nanotechnology' cannot count on consumer acceptance and also the inclusion of genetically modified ingredients gives rise within Europe to hesitance in applicability by food-producing companies. Especially for traceability and chemical characterization the demands become exceedingly higher. This is the level though where new protein engineering routes could be contributing most. A major bottleneck today in reformulation-strategies is the occurrence of (sensory) differentiations relative to the original product. Exceptions are reduced sugar or salt products, but from a marketing-technical perspective moderations of structuring components are preferably performed within the frame as novel food. A continuously on-going drive to elucidate structure-texture-product acceptance relations is essential, just like innovations in the area of optimized processing tools to deliver products that are *acceptable* with retained food safety. Typically the efforts on applying protein modifications are focusing in deriving food structures more efficiently or to provide new functional building blocks to create new food structures. In an economically global society there is a strong pressure on both ingredient-prices and commercial acceptable processing. Industrial entrepreneurship is essential and the availability of second-line ingredients for products hampers the implementation of new food production strategies. Especially within

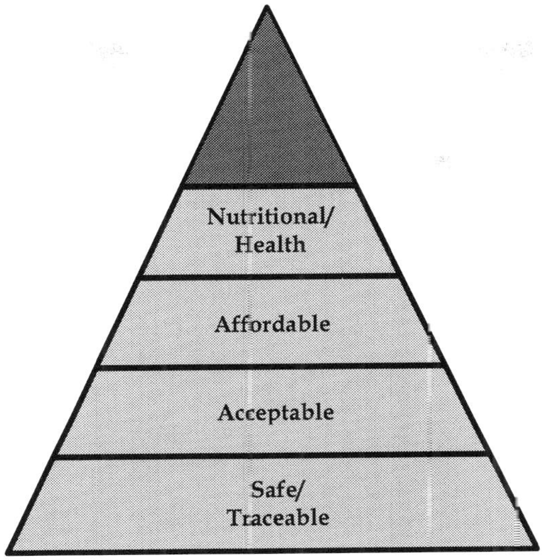

Fig. 2. The Pyramid of Food Innovation

the EU there is an increasing attention to social inequality issues. From a governmental regulatory point of view there is increasing support for open innovations and shared responsibilities to produce products of good quality that are *affordable* to all social classes. It is unclear what the role of optimized protein behavior by engineering approaches can be. Consumers are susceptible to additional *health*-aspects when safe and acceptable good quality products are available for reasonable prices. In this information technology era consumers are capable to evaluate the added value of products far better, setting higher demands for health marks of food products. The use of added value to products, like bioactives, has not led to major winners in the food sector yet. The desire (and need) to innovate in food product developments could be fed by the wealth of information on how product functionality relates to microstructural morphologies and how these in turn are dictated by molecular properties of specific proteins. On the other hand there are robust approaches in the field of protein engineering that allows us to direct protein behaviour and their propensities to (self)assemble into spatial networks. Still, one has to recognize that these two aspects have not come together yet to contribute to a more sustainable production and use of proteins in foods. This will be further discussed in section 5.

4.4 Considerations of chemical modification of proteins

While genetic modification of proteins generally results in a homogeneous product, chemical engineering approaches to modify proteins are well-known to result in heterogeneity. A good example is the application of the Maillard reaction to β-lactoglobulin (Broersen et al., 2004) or fish parvalbumin (de Jongh et al., 2011). On average one may find that for example 6 of the 12 available lysine residues has become glycosylated, where mass analysis will demonstrate that a significant population of protein molecules is present with 5 or 7 sugar groups, some molecules may contain 4 or 8 groups and even traces of molecules with 3 and 9 groups are present. This is commonly the result of variation in reactivity between the different amino groups present in a specific protein structure: i.e. some amino acids are more exposed and hence are more prone to rapid modification. Buried amino acids do not present a straight forward target for modification as most chemicals are unable to access the folded or aggregated protein structure. Another effect which likely contributes to the extent of heterogeneity is the location of target amino acids in the primary sequence. For example, modification of an amino acid located next to a negatively charged amino acid by succinic anhydride will be hampered as a result of steric hindrance. Some attempts have been made to fractionate protein molecules with different degrees of modification. Five ovalbumin preparations with different degrees of modification have been purified using ion exchange chromatography (Wierenga et al., 2005). The heterogeneous nature of chemically modified proteins has marked implications on the acceptance of these novel ingredients under current food law. Chemical heterogeneity would imply that each obtained modified species would have to be tested separately on toxicity in order to become regarded as safe to use as novel food ingredient.

4.5 Health risks

In this paragraph we will discuss potential health implications induced by chemically modified proteins used in food products including allergenic response, the presence of anti-nutritional factors linked to proteins and the development of potential toxic compounds as a side-effect of the modification reaction. Concerns can be raised regarding the impact of protein modification on the potential health risks from products containing modified protein ingredients. Bernstein and colleagues (2003) identified three possible modes for novel food ingredients to result in adverse health effects. These include toxicity, impaired nutrition and food allergy.

4.5.1 Toxicity

One of the known toxic compounds to be formed upon chemical engineering of proteins is acrylamide, which results from the Maillard reaction between reducing sugars and asparagine or methionine (Mottram et al., 2002; Stadler et al., 2002). The high solubility of this compound induces rapid absorption and metabolism in the body (reviewed in Dearfield et al., 1988). Following absorption, acrylamide can bind to DNA aiding a genotoxic and carcinogenic response which has been demonstrated in animals (Rudén, 2004). The occurrence of carcinogenic acrylamide in foods has been related to frying and cooking of food products (Rosen & Hellenäs, 2002; Tareke et al., 2002). Two communications published in Nature in 2002 demonstrated the requirement of the reaction between asparagine or methionine and a reducing sugar to intermediate formation of the dicarbonyl reactant followed by Strecker degradation (Mottram et al., 2002; Stadler et al., 2002). Particularly plant proteins are rich in asparagine suggesting that glycosylation of proteins from plant-origin using the Maillard reaction should be carefully considered in terms of the known toxic effects of acrylamide. Other toxic side effects of chemical modification of proteins have not been reported.

4.5.2 Impaired nutrition

Chemical engineering can theoretically have far-reaching consequences for the nutritional value of proteins. In a limited number of cases protein engineering has been intentionally employed to modify the nutritional quality, for example to increase the nutritional value of plant proteins (Liao et al., 2010). However, the main objective of most studies employing protein engineering is to investigate the consequences of the modification procedure for a range of functional properties. Only a small number of studies investigates nutritional aspects of chemical protein engineering which is mostly regarded as a convenient or inconvenient side effect of the modification procedure. Nutritional aspects covered in literature are exclusively based on *in vitro* studies and cover protein digestibility, availability of essential amino acids or the presence of anti-nutritional factors.

4.5.2.1 Anti-nutritional factors

Anti-nutritional factors are related to reduced protein digestibility and amino acid availability (reviewed by Gilani et al., 2005; Salunkhe et al., 1982) and are commonly present in large concentrations in plant products (Kay, 1979; Liener, 1980). One of the studies in this field reports on the nutritional quality of mung bean isolate following the exposure to varying concentrations of acetic or succinic anhydride to induce acylation (El-Adawy, 2000). The concentrations of anti-nutritional factors tannin, phytic acid, and trypsin inhibitor showed a significant loss with increasing degrees of modification suggesting that this type of modification can positively impact the effect of anti-nutritional factors. The concentration of trypsin inhibitor even decreased with 70% of the original level of trypsin inhibitor in unmodified protein. The introduction of bulky and/or negatively charged side groups was postulated to affect the extent of protein-tannin (El-Adawy, 2000), protein-mineral-phytic acid (Dua et al., 1996; El-Adawy, 2000), or protein-phenol (Loomis (1974) interactions. Loomis (1974) further showed that the flour and protein production processes provide for optimal conditions for the conversion of polyphenols into quinone oxidation products which, in turn, may bind covalently with sulfhydryl groups of cysteine and ε-amino groups of lysine and N-terminal amino groups. Further support was provided by Dua and colleagues (1996) who showed that acylation and methylation of rapeseed meal and its water-soluble fraction resulted in loss of anti-nutritional factors polyphenol, glucosinolates and phytic acid. However, the methylation procedure employed by Dua and colleagues (1996) resulted in very limited degrees of modification compared to the succinylation and acetylation process suggesting that other factors than the chemical conjugation itself may play a role in the loss of anti-nutritional factors upon chemical modification. As El-Adawy (2000) comments, the extensive dialysis of the protein following the acylation procedure may well be primarily responsible for the loss of water-soluble anti-nutritional factors upon modification. No studies are known to date that report on the loss of anti-nutritional factors upon dialysis.

4.5.2.2 *In vitro* digestibility

In vitro digestibility of modified proteins is often assayed through exposure of the proteins to a single or a mixture of enzymes including trypsin and pancreatin (Salgó et al., 1984), a combination of trypsin, chymotrypsin and peptidase (Hsu et al., 1977), or pepsin-pancreatin mixtures (Haque et al., 1982) to simulate (post)gastrointestinal digestion of food proteins. The small increase in digestibility of acylated mung bean protein isolate reported by El-Adawy (2000) was primarily correlated to the concurrent loss of tannin; tannins have been shown to play an important role in the reduction of protein digestibility (Barroga et al., 1985). Alternative factors proposed to induce increased digestibility of modified proteins include the improved access of sites susceptible to enzymatic cleavage as a result of the dissociation of quaternary complexes of proteins or partial unfolding induced by the modification procedure (Achouri & Zhang, 2001). Protein unfolding as a result of modification has been shown for soy protein hydrolysate using techniques to study the secondary and tertiary structure content of the protein upon succinylation (Achouri & Zhang, 2001). However, the acylation of cotton seed flour did not improve *in vitro* protein digestibility (Rahma & Narasinga Rao, 1983) suggesting that results in this area are somewhat controversial and perhaps other factors play a role (see paragraph 4.5.2.1). *In vitro* protein digestibility using a multienzyme system containing trypsin, chymotrypsin and peptidase was not impaired for pea proteins upon acetylation (Johnson & Brekke, 1983) while Ma (1984) reported increased digestibility for acylated pea protein, similar to low degrees of succinylation of soy protein hydrolysate (Achouri & Zhang, 2001). Loss of *in vitro* digestibility has also been reported upon succinylation of a variety of proteins, particularly affecting the release of lysine (Matoba & Doi, 1979; Siu & Thompson, 1982; Wanasundara & Shahidi, 1997) or lipophilization of soy bean glycinin with palmitic acid (Haque et al., 1982). Data on *in vivo* digestibility of ingested proteins upon modification have not been reported.

4.5.2.3 Availability of essential amino acids

A commonly used target for protein modification are the lysine or cysteine residues, lysine being classified as an essential amino acid, i.e. this amino acid cannot be synthesized *de novo* by humans and should therefore be ingested. Extensive modification of these amino acids can therefore result in a lower availability. Few studies report on the impact of lysine or cysteine modification on the availability of these amino acids, usually assayed through amino acid analysis. Overall only small decreases in lysine were reported upon succinylation of soy protein hydrolysate (Achouri & Zhang, 2001). Similar findings were reported for acylated soy proteins (Franzen & Kinsella, 1976a), acylated sunflower proteins (Kabirullah & Wills, 1982), and succinylation of oat proteins (Ma & Wood, 1987). It is not clear whether these reported effects of chemical engineering inducing the loss of lysine availability ultimately result in a noticeable and substantial loss of nutritional quality for the human population.

4.5.3 Food allergy and intolerance

An estimated 3-4% of the children and 1-2% of adults in the industrialized world exert an allergenic response to one or more ingested food proteins (Baral & Hourihane, 2005; Jansen et al., 1994). Not all ingested proteins behave as allergens but proteins implicated in allergenic response often share features such as unusual resilience against heat, acid or protease digestion, propensity to bind to lipids and are glycosylated to some degree (Lehrer et al., 2002; Metcalfe et al., 1996). Also posttranslational modifications including *N*-glycosylation, and hydroxylation of proline residues have been postulated to affect IgE reactivity to Phl p 1 present in timothy grass pollen (Petersen et al., 1998). The precise mechanism of the effects of glycosylation on the allergenic response are not clear. Attempts have been made to develop and evaluate algorithms which predict protein allergenic response based on sequence homology (Aalberse & Stapel, 2001; Jameson & Wolf, 1988), structural identity, and evolutionary relationship (Jenkins et al., 2007), albeit with limited predicting power. The Food and Agriculture Organization (FAO) of the United Nations together with the World Health Organization (WHO) developed a decision tree to assess the allergenic potential of genetically engineered food ingredients based on sequence homology followed by serum screens (FAO/WHO 2001, reviewed by Bernstein et al., 2003). Based on these findings and the fact that to date there is no indication for a common sequence motif of linear IgE epitopes, it can be concluded that the underlying mechanisms of immunological sensitization to food proteins remain elusive. To be able to assess the potential of chemically modified food proteins to induce allergenic reactions implies extensive knowledge of the underlying mechanical aspects of the allergic response. No publications are known which report on the allergenic effect in humans of modified ingested food proteins. In the pharmaceutical field PEGylation is used to modulate drug delivery of proteins or peptides. An early paper in the late 1970s reported that, upon covalent attachment of methoxypolyethylene glycols of 1900 daltons or 5000 daltons to bovine liver catalase, no evidence was observed of a modulated immune response following repeated injection of the modified enzymes into mice (Abuchowski et al., 1977).

5. Future prospects

There is a clear ambition to implement fundamental insights obtained on protein behavior in complex systems more effectively in sustainable food production in the future. There is also the need, considering that one should become more flexible in protein sources and more effective in utilizing the proteins structuring and nutritional potential. The technology to enable this is largely present, as demonstrated by the diversity of literature presented in paragraph 3. The mechanistic insight that has been derived in underlying principles on the relation between (engineered) protein functionality is impressive. And still, the development of new applications using chemical protein engineering to make a better usage of protein functionality or to promote protein source exchangeability is poor. There are a number of reasons to identify. The use of food protein modification has been limited to the domain of food chemistry or food physics. There is no significant literature available where the fine-tuning of microstructure formation as part of product development has been studied at a sensory or (human) physiological level. Compared to genetic modification, chemically engineered proteins pose an inherent heterogeneity. Instead of a single ingredient with an altered functionality, the FDA needs to consider all levels of heterogeneity in their approvals. This is an enormous laborious task as it requires fractionation of derived materials and sufficient stability of the formed products. Moreover, many attempts to reduce heterogeneity in the protein material will lead to non-food grade materials that cannot be studied in relation to e.g. sensory or human physiological aspects. The food safety authorities have amended (see paragraph 4.2) that the protein needs to be seen in the complex context of the food product. This is not just chemical analysis, but also includes product stability, consumer acceptance (sensory) and human physiological data. Especially in these latter two domains there is no public literature readily available to help these authorities to establish objective views on tolerance. This is where the scientific community has a highly needed role in bringing different disciplines together to produce literature relevant for authorities to base their legislations on.

6. References

Aalberse, R.C. & Stapel, S.O. (2001). Structure of food allergens in relation to allergenicity. *Pediatr. Allergy Immunol.* 12: 10-14.

Abuchowski, A.; McCoy, J.R.; Palczuk, N.C.; van Es, T. & Davis, F.F. (1977). Effect of covalent attachment of polyethylene glycol on immunogenicity and circulating life of bovine liver catalase. *J. Biol. Chem.* 252: 3582-3586.

Achouri, A. & Zhang, W. (2001). Effect of succinylation on the physicochemical properties of soy protein hydrolysate. *Food Res. Int.* 34: 507-514.

Aewsiri, T.; Benjakul, S.; Visessanguan W.; Wierenga, P.A. & Gruppen, H. (2010). Improvement of foaming properties of cuttlefish skin gelatin by modification with *N*-hydroxysuccinimide esters of fatty acid. *Food Hydrocoll.* In press.

Aewsiri, T.; Benjakul, S.; Visessanguan, W.; Wierenga, P.A. & Gruppen, H. (2011a). Emulsifying property and antioxidative activity of cuttlefish skin gelatin modified with oxidized linoleic acid and oxidized tannic acid. *Food Bioprocess. Technol.* In press.

Aewsiri, T.; Benjakul, S.; Visessanguan, W.; Wierenga, P.A. & Gruppen, H. (2011b). Surface activity and molecular characteristics of cuttlefish skin gelatin modified by oxidized linoleic acid. *Int. J. Biol. Macromol.* In press.

Alting, A.C.; Hamer, R.J.; de Kruif, C.G. & Visschers, R.W. (2003). Cold-set globular protein gels: interactions, structure and rheology as a function of protein concentration. *J. Agric. Food Chem.* 51: 3150-3156.

Andersson, L.-O.; Brandt, J. & Johansson, S. (1971). The use of trinitrobenezenesulfonic acid in studies on the binding of fatty acid anions to bovine serum albumin. *Arch. Biochem. Biophys.* 146: 428-440.

Anfinsen, C.B. (1973). Principles that govern the folding of protein chains. *Science* 181: 223-230.

Anson, M.L. (1940). The reactions of iodine and iodoacetamide with native egg albumin. *J. Gen. Physiol.* 23: 321-331.

Antoniewski, M.N.; Barringer, S.A.; Knipe, L. & Zerby, H. (2007). The effect of a gelatin coating on the shelf-life of fresh meat. *J. Food Sci.* 72: E382-E387.

Arnaudov, L.N.; de Vries, R.; Ippel, H. & van Mierlo, C.P. (2003). Multiple steps during the formation of β-lactoglobulin fibrils. *Biomacromolecules* 4: 1614-1622.

Arnaudov, L.N. & de Vries, R. (2005). Thermally induced fibrillar aggregation of hen egg white lysozyme. *Biophys. J.* 88: 515-526.

Arnaudov, L.N. & de Vries, R. (2006). Strong impact of ionic strength on the kinetics of fibrilar aggregation of bovine β-lactoglobulin. *Biomacromolecules* 7: 3490-3498.

Arntfield, S.D.; Murray, E.D. & Ismond, M.A.H. (1991). Role of disulfide bonds in determining the rheological and microstructural properties of heat-induced protein networks from ovalbumin and vicilin. *J. Agric. Food Chem.* 39: 1378-1385.

Arvanitoyannis, I.S. (1999). Totally and partially biodegradable polymer blends based on natural and synthetic macromolecules: preparation, physical properties, and potential as food packaging materials. *Polym. Rev.* 39: 205-271.

Audic, J.L. & Chaufer, B. (2005). Influence of plasticizers and crosslinking on the properties of biodegradable films made from sodium caseinate. *Eur. Polym. J.* 41: 1934-1942.

Bai, J.; Alleyne, V.; Hagenmaier, R.D.; Mattheis, J.P. & Baldwin, E.A. (2003). Formulation of zein coatings for apples (*Malus domestica* Borkh). *Postharv. Biol. Technol.* 28: 259-268.

Baldursdottir, S.G.; Fullerton, M.S.; Nielsen, S.H. & Jorgensen, L. (2010). Adsorption of proteins at the oil/water interface - observation of protein adsorption by interfacial shear stress measurements. *Colloids Surf. B Biointerfaces* 79: 41-46.

Baldwin, R.L. (2008). The search for folding intermediates and the mechanism of protein folding. *Annu. Rev. Biophys.* 37: 1-21.

Baldwin, E.A. (1994). Edible coatings for fresh fruits and vegetables: past, present, and future. In: Edible Coatings and Films to Improve Food Quality. Ed. By Krochta, J.M., Bandwin, E.A., Nisperos-Carriedo, M.O., Technomic Publishing, Lancaster PA, pp. 25-64.

Baral, V.R. & Hourihane, J.O. (2005). Food allergy in children. *Postgrad. Med. J.* 81: 693-701.

Barbut, S. & Foegeding, E.A. (1993). Ca^{2+}-induced gelation of preheated whey protein solutions. *J. Food Sci.* 58: 867-871.

Barroga, C.F.; Laurena, A.C. & Mendoza, E.M.T. (1985). Polyphenol in mung bean (*Vigna radiata* L. Wilczek) determination and removal. *J. Agric. Food Chem.* 33: 1006-1009.

Bauer, H.H.; Aebi, U.; Häner, M.; Hermann, R.; Müller, M.; Arvinte, T. & Merkle, H.P. (1995). Architecture and polymorphism of fibrillar supramolecular assemblies produced by *in vitro* aggregation of human calcitonin. *J. Struct. Biol.* 115: 1-15.

Baumketner, A.; Jewett, A. & Shea, J.E. (2003). Effects of confinement in chaperonin assisted protein folding: rate enhancement by decreasing the roughness of the folding energy landscape. *J. Mol. Biol.* 332: 701-713.

Bernstein, J.A.; Bernstein, I.L.; Bucchini, L.; Goldman, L.R.; Hamilton, R.G.; Lehrer, S.; Rubin, C. & Sampson, H.A. (2003). Clinical and laboratory investigation of allergy to genetically modified foods. *Environ. Health Persp.* 111: 1114-1121.

Betz, S.F. (1993). Disulfide bonds and the stability of globular proteins. *Protein Sci.* 2: 1551-1558.

Borgia, A.; Williams, P.M. & Clarke, J. (2008). Single-molecule studies of protein folding. *Annu. Rev. Biochem.* 77: 101-125.

Bowes, J.H. & Cater, C.W. (1968). The interaction of aldehydes with collagen. *Biochim. Biophys. Acta* 168: 341-352.

Broersen, K.; Weijers, M.; de Groot, J.; Hamer, R.J. & de Jongh, H.H. (2007a). Effect of protein charge on the generation of aggregation-prone conformers. *Biomacromolecules* 8: 1648-1656.

Broersen, K.; Elshof, M.; de Groot, J.; Voragen, A.G.J.; Hamer, R.J. & de Jongh, H.H.J. (2007b). Aggregation of β-lactoglobulin regulated by glucosylation. *J. Agric. Food Chem.* 55: 2431-2437.

Broersen, K.; Voragen, A.G.; Hamer, R.J. & de Jongh, H.H (2004). Glycoforms of beta-lactoglobulin with improved thermostability and preserved structural packing. *Biotechnol. Bioeng.* 86: 78-87.

Broersen, K.; van Teeffelen, A.M.; Vries, A.; Voragen, A.G.; Hamer, R.J. & de Jongh, H.H. (2006). Do sulfhydryl groups affect aggregation and gelation properties of ovalbumin? *J. Agric. Food Chem.* 54: 5166-5174.

Bromley, E.H.C.; Krebs, M.R.H. & Donald, A.M. (2006). Mechanisms of structure formation in particulate gels of β-lactoglobulin formed near the isoelectric point. *Eur. Phys. J. E.* 21: 145-152.

Buchner, S.; Kinnear, M.; Crouch, I.J.; Taylor, J. & Minnaar, A. (2011). Effect of kafirin protein coating on sensory quality and shelf-life of 'Packham's Triumph' pears during ripening. *J. Sci. Food Agric.* In press.

Buchner, G.S.; Murphy, R.D.; Buchete, N.V. & Kubelka, J. (2011). Dynamics of protein folding: probing the kinetic network of folding-unfolding transitions with experiment and theory. *Biochim. Biophys. Acta* 1814: 1001-1020.

Bustos, P.; Gajardo, M.I.; Gómez, C.; Hughes, G.; Cardemil, E. & Jabalquinto, A.M. (1996). Woodward's reagent K reacts with histidine and Cysteine residues in *Escherichia coli* and *Saccharomyces cerevisiae* phosphoenolpyruvate carboxykinases. *J. Protein Chem.* 15: 467-472.

Calamai, M.; Taddei, N.; Stefani, M., Ramponi, G. & Chiti, F. (2003). Relative influence of hydrophobicity and net charge in the aggregation of two homologous proteins. *Biochemistry* 42: 15078-15083.

Canet, D.; Sunde, M.; Last, A.M.; Miranker, A.; Spencer, A.; Robinson, C.V. & Dobson, C.M. (1999). Mechanistic studies of the folding of human lysozyme and the origin of amyloidogenic behavior of its disease-related variants. *Biochemistry* 38: 6419-6427.

Cao-Hoang, L.; Chaine, A.; Grégoire, L. & Waché, Y. (2010). Potential of nisin-incorporated sodium caseinate films to control *Listeria* in artificially contaminated cheese. *Food Microbiol.* 27: 940-944.

Chang, T.-S. & Sun, S.F. (1978). Structural studies on the succinylated bovine serum albumin. *Int. J. Peptide Protein Res.* 11: 65-72.

Charissou, A.; Ait-Ameur, L. & Birlouez-Aragon, I. (2007). Kinetics of formation of three indicators of the Maillard reaction in model cookies: influence of baking temperature and type of sugar. *J. Agric. Food Chem.* 55: 4532-4539.

Chibnall, A.C. (1942). Bakerian lecture: amino-acid analysis and the structure of proteins. *Proc. Roy. Soc. B* 131: 136-160.

Chiti, F. & Dobson, C.M. (2006). Protein misfolding, functional amyloid, and human disease. *Annu. Rev. Biochem.* 75: 333-366.

Chiti, F.; Stefani, M.; Taddei, N.; Ramponi, G. & Dobson, C.M. (2003). Rationalization of the effects of mutations on peptide and protein aggregation rates. *Nature* 424: 805-808.

Chiti, F.; Taddei, N.; Bucciantini, M.; White, P.; Ramponi, G. & Dobson, C.M. (2000). Mutational analysis of the propensity for amyloid formation by a globular protein. *EMBO J.* 19: 1441-1449.

Church, F.C.; Swaisgood, H.E.; Porter, D.H. & Catignani, G.L. (1983). Spectrophotometric assay using *o*-phthaldialdehyde for determination of proteolysis in milk and isolated milk proteins. J. Dairy Sci. 66: 1219-1227.

Cieplak, M. (2004). Cooperativity and contact order in protein folding. *Phys. Rev. E Stat. Nonlin. Soft Matter Phys.* 69: 031907.

Corzo-Martínez, M.; Soria, A.C.; Villamiel, M.; Olano, A.; Harte, F.M. & Moreno, F.J. (2011). Effect of glycation on sodium caseinate-stabilized emulsions obtained by ultrasound. *J. Dairy Sci.* 94: 51-58.

Creighton, T.E. (1988). Disulphide bonds and protein stability. *Bioessays* 8: 57-63.

Cucu, T.; Platteau, C.; Taverniers, I.; Devreese, B.; de Loose, M. & de Meulenaer, B. (2011). ELISA detection of hazelnut proteins: effect of protein glycation in the presence or absence of wheat proteins. *Food Addit. Contam. Part A Chem. Anal. Control Expo. Risk Assess.* 28: 1-10.

Damodaran, V.B. & Fee, C. (2010). Protein PEGylation: an overview of chemistry and process considerations. *Eur. Pharm. Rev* 15: 18-26.

Dearfield, K.L.; Abernathy, C.O.; Ottley, M.S.; Brantner, J.H. & Hayes, P.F. (1988). Acrylamide: its metabolism, developmental and reproductive effects, genotoxicity, and carcinogenicity. *Mutat. Res.* 195 45-77.

De Groot, J.; Kosters, H.A. & de Jongh, H.H. (2007). Deglycosylation of ovalbumin prohibits formation of a heat-stable conformer. *Biotechnol. Bioeng.* 97: 735-741.

De Jongh, H.H.J.; Kosters, H.A.; Kudryashova, E.; Meinders, M.B.J.; Trofimova, D. & Wierenga, P.A. (2004). Protein adsorption at air-water interfaces; a combination of details. *Biopolymers* 74: 131-135.

De Jongh, H.H.J. & Wierenga, P.A. (2006). Assessing the extent of protein intermolecular interactions at air-water interfaces using spectroscopic techniques. *Biopolymers* 82: 384-389.

De Jongh, H.H.; Taylor, S.L. & Koppelman, S.J. (2011). Controlling the aggregation propensity and thereby digestibility of allergens by Maillardation as illustrated for cod fish parvalbumin. *J. Biosci. Bioeng.* 111: 204-211.

Del Rosario Moreira, M.; Pereda, M.; Marcovich, N.E. & Roura, S.I. (2011). Antimicrobial effectiveness of bioactive packaging materials from edible chitosan and casein polymers: assessment on carrot, cheese, and salami. *J. Food Sci.* 76: M54-M63.

DeSantis, G. & Jones, J.B. (1999). Chemical modification of enzymes for enhanced functionality. *Curr. Opin. Biotechnol.* 10: 324-330.

Dickinson, E. (1992). An introduction to food colloids. Oxford, UK: Oxford University Press.

Djagny, K.B.; Wang, Z. & Xu, S. (2001). Conformational changes and some functional characteristics of gelatin esterified with fatty acid. *J. Agric. Food Chem.* 49: 2987-2991.

Dobson, C.M. (1999). Protein misfolding, evolution and disease. *Trends Biochem. Sci.* 24: 329-332.

Doi, E. (1993). Gels and gelling of globular proteins. *Trends Food Sci. Technol.* 4: 1-5.

Dondero, M.; Figueroa, V.; Morales, X. & Curotto, E. (2006). Transglutaminase effects on gelation capacity of thermally induced beef protein gels. *Food Chemistry* 99, 546-554.

Dua, S.; Mahajan, A. & Mahajan, A. (1996). Improvement of functional properties of rapeseed (*Brassica campestris* Var. Toria) preparations by chemical modification. *J. Agric. Food Chem.* 44: 706-710.

DuBay, K.F.; Pawar, A.P.; Chiti, F.; Zurdo. J.; Dobson, C.M. & Vendrusculo, M. (2004). Prediction of the absolute aggregation rates of amyloidogenic polypeptide chains. *J. Mol. Biol.* 341: 1317-1326.

El-Adawy, T.A. (2000). Functional properties and nutritional quality of acetylated and succinylated mung bean protein isolate. *Food Chem.* 70: 83-91.

Ellman, G.L. (1959). Tissue sulfhydryl groups. *Arch. Biochem. Biophys.* 82: 70-77.

Emmambux, M.N.; Stading, M. & Taylor, J.R.N. (2004). Sorghum kafirin film property modification with hydrolysable and condensed tannins. *J. Cereal Sci.* 40: 127-135.

Englander, S.W.; Mayne, L. & Krishna, M.M. (2007). Protein folding and misfolding: mechanisms and principles. *Q. Rev. Biophys.* 40: 287-326.

Evans, M.T.A. & Irons, L. (1970). N-succinylated egg yolk proteins. German Patent 1 951 247.

FAO/WHO (2001). Evaluation of allergenicity of genetically modified foods. Report of a Joint FAO/WHO Expert Consultation of Allergenicity of Foods Derived from Biotechnology. Available: ftp://ftp.fao.org/es/esn/food/allergygm.pdf [accessed 08 August 2011].

Farouk, M.M.; Price, J.F. & Salih, A.M. (1990). Effect of an edible film overwrap on exudation and lipid oxidation in beef round steak. *J. Food Sci.* 55: 1510-1512, 1563.

Feeney, R.E.; Blankenhorn, G. & Dixon, H.B. (1975). Carbonyl-amine reactions in protein chemistry. *Adv. Protein Chem.* 29: 135-203.

Feeney, R.E.; Yamasaki, R.B. & Geoghegan, K.F. (1982). Chapter 1: Chemical modification of proteins: an overview. In 'Modification of Proteins', Advances in Chemistry; American Chemical Society, Washington DC. pp 3-55.

Fernandez-Escamilla, A.M.; Rousseau, F.; Schymkowitz, J. & Serrano, L. (2004). Prediction of sequence-dependent and mutational effects on the aggregation of peptides and proteins. *Nat. Biotechnol.* 22: 1302-1306.

Ferreira, S.T.; De Felice, F.G. & Chapeaurouge, A. (2006). Metastable, partially folded states in the productive folding and in the misfolding and amyloid aggregation of proteins. *Cell Biochem. Biophys.* 44: 539-548.

Ferreon, A.C. & Deniz, A.A. (2011). Protein folding at single-molecule resolution. *Biochim. Biophys. Acta* 1814: 1021-1029.

Fields, R. (1971). The measurement of amino groups in proteins and peptides. *Biochem. J.* 124: 581-590.

Fraenkel-Conrat, H. (1957). Methods for investigating the essential groups for enzyme activity. *Methods Enzymol.* 4: 247-269.

Finley, J.W. (1975). Deamidated gluten: a potential fortifier for fruit juices. *J. Food Sci.* 40: 1283-1285.

Francis, G.E.; Fisher, D.; Delgado, C.; Malik, F.; Gardiner, A. & Neale, D. (1998). PEGylation of cytokines and other therapeutic proteins and peptides: the importance of biological optimisation of coupling techniques. *Int. J. Hematol.* 68: 1-18.

Franzen, K.L. & Kinsella, J.E. (1976a). Functional properties of succinylated and acetylated soy protein. *J. Agric. Food Chem.* 24: 788-795.

Franzen, K.L. & Kinsella, J.E. (1976b). Functional properties of succinylated and acetylated leaf proteins. *J. Agric. Food Chem.* 24: 914-919.

Frister, H.; Meisel, H. & Schlimme, E. (1988). OPA method modified by use of *N,N*-dimethyl-2-mercaptoethylammonium chloride as thiol component. *Fresenius'Z. Anal. Chem.* 330: 631-633.

Gerbanowski, A.; Rabiller, C.; Larré, C. & Guéguen, J. (1999a). Grafting of aliphatic and aromatic probes to bovine serum albumin: influence on its structural and physicochemical characteristics. *J. Protein Chem.* 18: 325-335.

Gerbanowski, A.; Malabat, C.; Rabiller, C. & Guéguen, J. (1999b). Grafting of aliphatic and aromatic probes on rapeseed 2S and 12S proteins: influence on their structural and physicochemical properties. *J. Agric. Food Chem.* 47: 5218-5226.

Gilani, G.S.; Cockell, K.A. & Sepehr, E. (2005). Effects of antinutritional factors on protein digestibility and amino acid availability in foods. *J. AOAC Int* 88: 967-987.

Gillgren, T. & Stading, M. (2008). Mechanical and barrier properties of avenin, kafirin and zein films. *Food Biophys.* 3: 287-294.

Graham, D.E. & Phillips, M.C. (1979a). Proteins at liquid interfaces. I. Kinetics of adsorption and surface denaturation. *J. Colloid Interface Sci.* 70: 403-414.

Graham, D.E. & Phillips, M.C. (1979b). Proteins at liquid interfaces. II. Adsorption isotherms. *J. Colloid Interface Sci.* 70: 415-426.

Graham, D.E. & Phillips, M.C. (1979c). Proteins at liquid interfaces. III. Molecular structures at adsorbed films. *J. Colloid Interface Sci.* 70: 427-439.

Graña-Montes, R.; de Groot, N.S.; Castillo, V.; Sancho, J.; Velazquez-Campoy, A. & Ventura, S. (2011). Contribution of disulfide bonds to stability, folding and amyloid fibril formation: the PI3-SH3 domain case. *Antioxid. Redox Signal.* In press.

Grigoriev, D.O.; Derkatch, S.; Krägel, J. & Miller, R. (2007). Relationship between structure and rheological properties of mixed BSA/Tween 80 adsorption layers at the air/water interface. *Food Hydrocoll.* 21: 823-830.

Groninger, H.S. Jr. (1973). Preparation and properties of succinylated fish myofibrillar protein. *J. Agric. Food Chem.* 21: 978-981.

Gruener, L. & Ismond, M.A.H. (1997). Effects of acetylation and succinylation on the physicochemical properties of the canola 12S globulin. *Food Chem.* 60: 357-363.

Gueguen, J.; Bollecker, S.; Schwenke, K.D. & Raab, B. (1990). Effect of succinylation on some physicochemical and functional properties of the 12S storage protein from rapeseed (*Brassica napus* L.). *J. Agric. Food Chem.* 38: 61-69.

Gurin, S. & Clarke, H.T. (1934). Allocation of the free amino groups in proteins and peptides. *J. Biol. Chem.* 107: 395-419.

Habeeb, A.F.S.A. & Atassi, M.Z. (1969). Enzymic and immunochemical properties of lysozyme II. Conformation, immunochemistry and enzymic activity of a derivative modified at tryptophan. *Immunohistochemistry* 6: 555-556.

Hall, R.J.; Trinder, N. & Givens, D.I. (1973). Observations on the use of 2,4,6-trinitrobenzenesulphonic acid for the determination of available lysine in animal protein concentrates. *Analyst* 98: 673-686.

Halling, P.J. (1981). Protein-stabilized foams and emulsions. *Crit. Rev. Food Sci. Nutr.* 15: 155-203.

Hamada, J.S. & Marshall, W.E. (1989). Preparation and functional properties of enzymatically deamidated soy proteins. *J. Food Sci.* 54: 598-601.

Hamano, Y. (2011). Occurrence, biosynthesis, biodegradation, and industrial and medical applications of a naturally occurring ε-poly-L-lysine. *Biosci. Biotechnol. Biochem.* 75: 1226-1233.

Haque, Z.; Matoba, T. & Kito, M. (1982). Incorporation of fatty acid into food protein: palmitoyl soybean glycinin. *J. Agric. Food Chem.* 30: 481-486.

Haque, Z. & Kito, M. (1983a). Lipophilization of α_{s1}-casein. 1. Covalent attachment of palmitoyl residue. *J. Agric. Food Chem.* 31: 1225-1230.

Haque, Z. & Kito, M. (1983b). Lipophilization of α_{s1}-casein. 2. Conformational and functional effects. *J. Agric. Food Chem* 31: 1231-1237.

Hardy, P.M.; Nicholls, A.C. & Rydon, H.N. (1969). The nature of gluteraldehyde in aqueous solution. *J. Chem. Soc. Chem. Commun* 10: 565-566.

Harper, J.D.; Wong, S.S.; Lieber, C.M. & Lansbury, P.T. Jr. (1997). Atomic force microscopy imaging of seeded fibril formation and fibril branching by the Alzheimer's disease amyloid-β protein. *Chem. Biol.* 4: 951-959.

Hart, G.W. (1992). Glycosylation. *Curr. Opin. Cell Biol.* 4: 1017-1723.

Havard, C. & Harmony, M.X. (1869). Improved process of preserving meat, fowls, fish. June 8, U.S. Patent 90: 944.

Hayakawa, S. & Nakai, S. (1985). Contribution of the hydrophobicity, net charge and sulfhydryl groups to thermal properties of ovalbumin. *Can. Inst. Food Sci. Technol. J.* 18: 290-295.

Haynes, R.; Osuga, D.T. & Feeney, R.E. (1967). Modification of amino groups in inhibitors of proteolytic enzymes. *Biochemistry* 6: 541-547.

He, X.-H.; Shaw, P.-C. & Tam, S.-C. (1999). Reducing the immunogenicity and improving the inv vivo activity of trichosanthin by site-directed PEGylation. *Life Sci.* 65: 355-368.

Helenius, A.; Trombetta, E.S.; Hebert, D.N. & Simons, J.F. (1997) Calnexin, calreticulin and the folding of glycoproteins. *Trends Cell Biology* 7: 193-200.

Hillier, R.M.; Lyster, R.L.J. & Cheeseman, G.C. (1980). Gelation of reconstituted whey powders by heat. *J. Sci. Food Agric.* 31: 1152-1157.

Hoagland, P.D. (1968). Acylated β-caseins. Effect of alkyl group size on calcium ion sensitivity and on aggregation. *Biochemistry* 7: 2542-2546.

Hoffmann, M.A.M. & van Mil, P.J.J.M. (1997). Heat-induced aggregation of β-lactoglobulin: role of the free thiol group and disulfide bonds. *J. Agric. Food Chem.* 45: 2942-2948.

Hopwood, D. (1969). A comparison of the crosslinking abilities of gluteraldehyde, formaldehyde and α-hydroxyadipaldehyde with bovine serum albumin and casein. *Histochemie* 17: 151-161.

Hopwood, D., Callen, C.R. & McCabe, M. (1970). The reactions between gluteraldehyde and various proteins. An investigation of their kinetics. *Histochem. J.* 2: 137-150.

Hsu, H.W.; Vavak, D.L.; Satterlee, L.S. & Miller, G.A. (1977). A multienzyme technique for estimating protein digestibility. *J. Food Sci.* 42: 1269-1273.

Huggins, C. & Jensen, E.V. (1949). Thermal coagulation of serum proteins; the effects of iodoacetate, iodoacetamide, and thiol compounds on coagulation. *J. Biol. Chem.* 179: 645-654.

Hunter, J.R.; Carbonell, R.G. & Kilpatrick, P.K. (1991). Coadsorption and exchange of lysozyme/β-casein mixtures at the air/water interface. *J. Colloid Interface Sci.* 143: 37-53.

Hurle, M.R.; Helms, L.R.; Li, L.; Chan, W. & Wetzel, R. (1994). A role for destabilizing amino acid replacements in light-chain amyloidosis. *Proc. Natl. Acad. Sci. U.S.A.* 91: 5446-5450.

Ibrahim, H.R.; Kato, A. & Kobayashi, K. (1991). Antimicrobial effects of lysozyme against gram-negative bacteria due to covalent binding of palmitic acid. *J. Agric. Food Chem.* 39: 2077-2082.

Ibrahim, H.R.; Kobayashi, K. & Kato, A. (1993). Length of hydrocarbon chain and antimicrobial action to gram-negative bacteria of fatty acylated lysozyme. *J. Agric. Food Chem.* 41: 1164-1168.

Izzo, H.V.; Lincoln, M.D. & Ho, C.-T. (1993). The effect of temperature, feed moisture, and pH on protein deamidation in an extruded wheat flour. *J. Agric. Food Chem.* 41: 199-202.

Jameson, B.A. & Wolf, H. (1988). The antigenic index: a novel algorithm for predicting antigenic determinants. *Comp. Appl. Biosci.* 4: 181-186.

Jansen, J.J.N.; Kardinaal, A.F.M.; Huijbers, G.; Vlieg-Boerstra, B.J.; Martens, B.P.M. & Ockhuizen, T. (1994). Prevalence of food allergy and intolerance in the adult Dutch population. *J. Allergy Clin. Immun.* 93: 446-456.

Jarrett, J.T. & Lansbury, P.T. Jr. (1993). Seeding "one-dimensional crystallization" of amyloid: a pathogenic mechanism in Alzheimer's disease and scrapie? *Cell* 73: 1055-1058.

Jenkins, J.A.; Breiteneder, H. & Mills, E.N.C. (2007). Evolutionary distance from human homologs reflects allergenicity of animal food proteins. *J. Allergy Clin. Immunol.* 120: 1399-1405.

Jiménez, J.; Nettleton, E.J.; Bouchard, M.; Robinson, C.V.; Dobson, C.M. & Saibil, H. (2002). The protofilament structure of insulin amyloid fibrils. *Proc. Natl. Acad. Sci. U.S.A.* 99: 9196-9201.

Jiménez, J.; Guijarro, J.I.; Orlova, E.; Zurdo, J.; Dobson, C.M.; Sunde, M & Saibil, H.R. (1999). Cryo-electron microscopy structure of an SH3 amyloid fibril and model of the molecular packing. *EMBO J.* 18: 815-821.

Johnson, E.A. & Brekke, J. (1983). Functional properties of acylated pea protein isolates. *J. Food Sci* 48: 722-725.

Johnson, A.R. & Dekker, E.E. (1996). Woodward's reagent K inactivation of *Escherichia coli* L-threonine dehydrogenase: increased absorbance at 340-350 nm is due to modification of cysteine and histidine residues, not aspartate or glutamate carboxyl groups. *Protein Sci.* 5: 382-390.

Jonas, A. & Weber, G. (1970). Partial modification of bovine serum albumin with dicarboxylic anhydrides. Physical properties of the modified species. *Biochemistry* 9: 4729-4735.

Ju, Z.Y. & Kilara, A. (1998). Aggregation induced by calcium chloride and subsequent thermal gelation of whey protein isolate. *J. Dairy Sci.* 81: 925-931.

Kabirullah, M. & Wills, R.B.H. (1982). Functional properties of acetylated and succinylated sunflower protein isolate. *J. Food Technol.* 17: 235-249.

Kaneko, R. & Kitabatake, N. (2001). Structure-sweetness relationship in thaumatin: importance of lysine residues. *Chem Senses* 26: 167-177.

Kato, A.; Ibrahim, H.; Takagi, T. & Kobayashi, K. (1990). Excellent gelation of egg white preheated in the dry state is due to the decreasing degree of aggregation. *J. Agric. Food Chem.* 38: 1868-1872.

Kato, A.; Nagase, Y.; Matsudomi, N. & Kobayashi, K. (1983) Determination of molecular-weight of soluble ovalbumin aggregates during heat denaturation using low-angle laser-light scattering technique. *Agric. Biol. Chem.* 47: 1829-1834.

Kay, D.E. (1979). Crop and product digest. No. 3. Food legumes. London, Tropical Products Institute, 435 p.

Kelly, J.W. (1998). The alternative conformations of amyloidogenic proteins and their multi-step assembly pathways. *Curr. Opin Struct. Biol.* 8: 101-106.

Kitabatake, N.; Wada, R. & Fujita, Y. (2001). Reversible conformational change in β-lactoglobulin A modified with *N*-ethylmaleimide and resistance to molecular aggregation on heating. *J. Agric. Food Chem.* 49: 4011-4018.

Klotz, I.M. & Heiney, R.E. (1962). Introduction of sulfhydryl groups into proteins using acetylmercaptosuccinic anhydride. *Arch. Biochem. Biophys.* 96: 605-612.

Korn, A.H.; Feairheller, S.H. & Filachione, E.M. (1972). Gluteraldehyde: nature of the reagent. *J. Mol. Biol.* 65: 525-529.

Kornfeld, R. & Kornfeld, S. (1985). Assembly of asparagines-linked oligosaccharides. *Annu. Rev. Biochem.* 54: 631-664.

Koseki, T.; Kitabatake, N. & Doi, E. (1989). Irreversible thermal denaturation and formation of linear aggregates of ovalbumin. *Food Hydrocoll.* 3: 123-134.

Kosters, H.A. & de Jongh, H.H. (2003). Spectrophotometric tool for the determination of the total carboxylate content in proteins; molar extinction coefficient of the enol ester from Woodward's reagent K reacted with protein carboxylates. *Anal. Chem.* 75: 2512-2516.

Kosters, H.A.; Broersen, K.; de Groot, J.; Simons, J.W.; Wierenga, P. & de Jongh, H.H. (2003). Chemical processing as a tool to generate ovalbumin variants with changed stability. *Biotechnol. Bioeng.* 84: 61-70.

Kotaki, A.; Harada, M. & Yagi, K. (1964). Reaction between sulfhydryl compounds and 2,4,6-trinitrobenzene-1-sulfonic acid. *J. Biochem.* 55: 553-561.

Krause, J.-P.; Mothes, R. & Schwenke, K.D. (1996). Some physicochemical and interfacial properties of native and acetylated legumin from faba beans (*Vicia faba* L.). *J. Agric. Food Chem.* 44: 429-437.

Krebs, M.R.H.; Devlin, G.L. & Donald, A.M. (2009). Amyloid fibril-like structure underlies the aggregate structure across the pH range for β-lactoglobulin. *Biophys. J.* 96: 5013-5019.

Kudryashova, E.V.; Meinders, M.B.J.; Visser, A.J.W.G.; van Hoek, A. & de Jongh, H.H.J (2003). Structural properties and rotational dynamics of egg white ovalbumin adsorbed at the air/water interaface. *Eur. J. Biophys.* 32: 553-562.

Kudryashova, E.V.; Visser, A.J.W.G., & de Jongh, H.H.J. (2005). Reversible self-association of ovalbumin at air-water interfaces and the consequences for the exerted surface pressure. Protein Science 14: 483-493.

Lakkis, J. & Villota, R. (1992). Effect of acylation on substructural properties of proteins: a study using fluorescence and circular dichroism. *J. Agric. Food Chem.* 40: 553-560.

Land, A. & Braakman, I. (2001). Folding of the human immunodeficiency virus type 1 envelope glycoprotein in the endoplasmatic reticulum. *Biochimie* 83: 783-790.

Langton, M. & Hermansson, A.-M. (1992). Fine-stranded and particulate gels of β-lactoglobulin and whey protein at varying pH. Food Hydrocoll. 5: 523-539.

Le, T.T.; Bhandari, B. & Deeth, H.C. (2011). Chemical and physical changes in milk protein concentrate (MPC80) powder during storage. *J. Agric. Food Chem.* 59: 5465-5473.

Le Floch-Fouéré, C.; Pezennec,S.; Pézolet, M.; Rioux-Dubé, J.F.; Renault, A. & Beaufils, S. (2011). Unexpected differences in the behavior of ovotransferrin at the air-water interface at pH 6.5 and 8.0. *J. Colloid Interface Sci.* 356: 614-623.

Lehrer, S.B.; Ayuso, R. & Reese, G. (2002). Current understanding of food allergens. *Ann. N. Y. Acad. Sci.* 964: 69-85.

Liao, L.; Zhao, M.; Ren, J.; Zhao, H.; Cui, C. & Hu, X. (2010). Effect of acetic acid deamidation-induced modification on functional and nutritional properties and conformation of wheat gluten. *J. Sci. Food Agric.* 90: 409-417.

Liener, I.E. (1980). Toxic constituents of plant foodstuffs. New York and London, Academic Press, 502 p.

Liu, S.-T.; Sugimoto, T.; Azakami, H. & Kato, A. (2000). Lipophilization of lysozyme by short and middle chain fatty acids. *J. Agric. Food Chem.* 48: 265-269.

Llamas, K.; Ownes, M.; Blakeley, L. & Zerner, B. (1986). *N*-ethyl-5-phenylisoxazolium-3'-sulfonate (Woodward's reagent K) as a reagent for nucleophilic side chains of proteins. *J. Am. Chem. Soc.* 108: 5543-5548.

Lomakin, A.; Chung, D.S.; Benedek, G.B.; Kirscher, D.A. & Teplow, D.B. (1996). On the nucleation and growth of amyloid β-protein fibrils: detection of nuclei and quantitation of rate constants. *Proc. Natl. Acad. Sci. U.S.A.* 93: 1125-1129.

Lomakin, A.; Teplow, D.B.; Kirschner, D.A. & Benedek, G.B. (1997). Kinetic theory of fibrillogenesis of amyloid β-protein. *Proc. Natl. Acad. Sci. U.S.A.* 94: 7942-7947.

Longares, A.; Monahan, F.J.; O'Riordan, E.D. & O'Sullivan, M. (2005). Physical properties of edible films made from mixtures of sodium caseinate and WPI. *Int. Dairy J.* 15: 1255-1260.

Lönnerdal, B. (2002). Expression of human milk proteins in plants. *J. Am. Coll. Nutr.* 21: 218–221

Loomis, W.D. (1974). Overcoming problems of phenolics and quinones in the isolation of plant enzymes organelles. In: *Methods in Enzymology*; Fleischer, S., Packer, L., Eds.: Academic Press: New York, p. 528.

Losso, J.N. & Nakai, S. (2002). Stabilization of oil-in-water emulsions by β-lactoglobulin-polyethylene glycol conjugates. *J. Agric. Food Chem.* 50: 1207-1212.

Lubelski, J.; Rink, R.; Khusainov, R.; Moll, G.N. & Kuipers, O.P. (2008). Biosynthesis, immunity, regulation, mode of action and engineering of the model lantibiotic nisin. *Cell. Mol. Life Sci.* 65: 455-476.

Luey, J.-K.; McGuire, J. & Sproull, R.D. (1991). The effect of pH and NaCl concentration on adsorption of β-lactoglobulin at hydrophilic and hydrophobic silicon surfaces. *J. Colloid Interface Sci.* 143: 489-500.

Luheshi, L.M. & Dobson, C.M. (2009). Bridging the gap: from protein misfolding to protein misfolding diseases. *FEBS Lett.* 583: 2581-2586.

Ma, C.-Y. (1984). Functional properties of acylated oat protein. *J. Food Sci.* 49: 1128-1131.

Ma, C.-Y. & Wood, D.F. (1987). Functional properties of oat proteins modified by acylation, trypsin hydrolysis or linoleate treatment. *J. Am. Oil Chem. Soc.* 64: 1726-1731.

MacRitchie, F. (1978). Proteins at interfaces. *Adv. Protein Chem.* 32: 283-326.

Malaki Nik, A.; Wright, A.J. & Corredig, M. (2010). Interfacial design of protein-stabilized emulsions for optimal delivery of nutrients. *Food Funct.* 1: 141-148.

Marcon, G.; Plakoutsi, G.; Canale, C.; Relini, A.; Taddei, N.; Dobson, C.M.; Ramponi, G. & Chiti, F. (2005). Amyloid formation from HypF-N under conditions in which the protein is initially in its native state. *J. Mol. Biol.* 347: 323-335.

Margoshes, B.A. (1990). Correlation of protein sulfhydryls with the strength of heat-formed egg white gels. *J. Food Sci.* 55: 1753.

Martin, A.H.; Meinders, M.B.J.; Bos, M.A.; Cohen Stuart, M.A. & van Vliet, T. (2003). Conformational aspects of proteins at the air/water interface studied by infrared reflection-absorption spectroscopy. *Langmuir* 19: 2922-2928.

Martins, J.T.; Cerqueira, M.A.; Souza, B.W.; Carmo Avides, M. & Vicente, A.A. (2010). Shelf life extension of ricotta cheese using coatings of galactomannans from nonconventional sources incorporating nisin against *Listeria monocytogenes*. *J. Agric. Food Chem.* 58: 1884-1891.

Masuda, T.; Ueno, Y. & Kitabatake, N. (2001). Sweetness and enzymatic activity of lysozyme. *J. Agric. Food Chem.* 49: 4937-4941.

Masuda, T.; Ide, N. & Kitabatake, N. (2005a). Structure-sweetness relationship in egg white lysozyme: role of lysine and arginine on the elicitation of lysozyme sweetness. *Chem. Senses* 30: 667-681.

Masuda, T.; Ide, N. & Kitabatake, N. (2005b). Effects of chemical modification of lysine residues on the sweetness of lysozyme. *Chem. Senses* 30: 253-264.

Matoba, T. & Doi, E. (1979). In vitro digestibility of succinylated protein by pepsin and pancreatic proteases. *J. Food Sci.* 44: 537-539.

Matsudomi, N.; Kato, A. & Kobayashi, K. (1982). Conformation and surface properties of deamidated gluten. *Agric. Biol. Chem.* 46: 1583-1586.

Matsudomi, N.; Sasaki, T.; Kato, A. & Kobayashi, K. (1985). Conformational changes and functional properties of acid-modified soy protein. *Agric. Biol. Chem.* 49: 1251-1256.

Maurer-Stroh, S.; Debulpaep, M.; Kuemmerer, N.; Lopez de la Paz, M.; Martins, I.C.; Reumers, J.; Morris, K.L.; Copland, A.; Serpell, L.C.; Serrano, L.; Schymkowitz, J.W.H. & Rousseau, F. (2010). Exploring the sequence determinants of amyloid structure using position-specific scoring matrices. *Nature Methods* 7: 237-242.

Marquardt, T. & Helenius, A. (1992). Misfolding and aggregation of newly synthesized proteins in the endoplasmatic reticulum. *J. Cell Biol.* 117: 505-513.

McClellan, S.J. & Franses, E.I. (2003). Effect of concentration and denaturation on adsorption and surface tension of bovine serum albumin. *Colloids Surf, B Biointerfaces* 28: 63-75.

McClements, J. (2005). Food emulsions: principles, practice, and techniques. 2nd Ed. Boca Raton, FL: CRC Press.

McMahon, D.J.; Adams, S.L. & McManus, W.R. (2009). Hardening of high-protein nutrition bars and sugar/polyol-protein phase separation. *J. Food Sci.* 74: E312-E321.

Means, G.E. & Feeney, R.E. (1971). Chemical modifications of proteins. Holden-Day, San Francisco.

Meinders, M.B.J.; van den Bosch, G.G.M. & de Jongh, H.H.J. (2001). Adsorption properties of proteins at and near the air/water interface from IRRAS spectra of protein solutions. *Eur. Biophys. J.* 30: 256-267.

Melnychyn, P. & Stapley, R. (1973). Acylated protein from coffee whitener formulations. U.S. Patent 3,764,711.

Mendis, E.; Rajapakse, N. & Kim, S. (2005). Antioxidant properties of a radical-scavenging peptide purified from enzymatically prepared fish skin gelatin hydrolysate. *J. Agric. Food Chem.* 53: 581-587.

Metcalfe, D.D.; Astwood, J.D.; Townsend, R.; Sampson, H.A.; Taylor, S.L. & Fuchs, R.L. (1996). Assessment of the allergenic potential of foods derived from genetically engineered crop plants. *Crit. Rev. Fd. Sci. Nutr.* 36: S165-S186.

Migneault, I.; Dartiguenave, C.; Bertrand, M.J. & Waldron, K.C. (2004). Gluteraldehyde: behavior in aqueous solution, reaction with proteins, and application to enzyme crosslinking. *BioTechniques* 37: 790-802.

Mine, Y. (1992). Sulfhydryl groups changes in heat-induced soluble egg white aggregates in relation to molecular size. *J. Food Sci.* 57: 254-255.

Ming, D. & Hellekant, G. (1994). Brazzein, a new high-potency thermostable sweet protein from *Pentadiplandra brazzeana* B. *FEBS Lett.* 355: 106-108.

Mirmoghatadaie, L.; Kadivar, M. & Shahedi, M. (2009). Effects of succinylation and deamidation on functional properties of oat protein isolate. *Food Chem.* 114: 127-131.

Moore, S. & Stein, W.H. (1948). Photometric ninhydrin method for use in the chromatography of amino acids. *J. Biol. Chem.* 176: 367-388.

Morel, B.; Varela, L. & Conejero-Lara, F. (2010). The thermodynamic stability of amyloid fibrils studied by differential scanning calorimetry. *J. Phys. Chem. B* 114: 4010-4019.

Morris, J.A. & Cagan, R.H. (1972). Purification of monellin, the sweet principle of *Dioscoreophyllum cumminsii*. *Biochem. Biophys. Acta* 261: 114-122

Mossine, V.V.; Glinsky, G.V. & Feather, M.S. (1994). The preparation and characterization of some Amadori compounds (1-amino-1-deoxy-D-fructose derivatives) derived from a series of aliphatic omega-amino acids. *Carbohydr. Res.* 262: 257-270.

Mottram, D.S.; Wedzicha, B.L. & Dodson, A.T. (2002). Acrylamide is formed in the Maillard reaction *Nature* 419: 448.

Moure, A.; Sineiro, J.; Domínguez, H. & Parajó, J.C. (2006). Functionality of oilseed protein products: a review. *Food Res. Int.* 39: 945-963.

Nakai, S. (1983). Structure-function relationships of food proteins with an emphasis on the importance of protein hydrophobicity. *J. Agric. Food Chem.* 31: 676-683.

Nakai, S. & Li-Chan, E. (1988). Hydrophobic interaction in food systems. CRC Press: Boca Raton, FL.

Narayana, K. & Rao, N. (1991). Effect of acetylation and succinylation on the physicochemical properties of winged bean (Psophocarpus tetragonolobus) proteins. *J. Agric. Food Chem.* 39: 259-261.

Niestijl Jansen J.J.; Kardinaal, A.F.M.; Huijbers, G.; Vlieg-Boerstra, B.J.; Martens, B.P.M. & Ockhuizen, T. (1994). Prevalence of food allergy and intolerance in the adult Dutch population. *J. Allergy Clin. Immunol.* 93: 446-456.

Nimni, M.E., Cheung, D.; Strates, B.; Kodama, M. & Sheikh, K. (1987). Chemically modified collagen: a natural biomaterial for tissue replacement. *J. Biomed. Mater. Res.* 21: 741-771.

Nizet, V. (2006). Antimicrobial peptide resistance mechanisms of human bacterial pathogens. *Curr. Issues Mol. Biol.* 8: 11-26.

Ohta, K.; Masuda, T.; Ide, N. & Kitabatake, N. (2008). Critical molecular regions for elicitation of the sweetness of sweet-tasting protein, thaumatin I. *FEBS J.* 275: 3644-3652.

O'Kane, F.E.; Happe, R.P.; Vereijken, J.M.; Gruppen, H. & van Boekel, M.A.J.S. (2004). Heat-induced gelation of pea legumin: comparison with soy glycinin. *J. Agric. Food Chem.* 52: 5071-5078.

Okuyama, T. & Satake, K. (1960). On the preparation and properties of 2,4,6-trinitrophenyl-amino acids and –peptides. *J. Biochem.* 47: 454-466.

Onuchic, J.N.; Luthey-Schulten, Z. & Wolynes, P.G. (1997). Theory of protein folding: the energy landscape perspective. *Ann. Rev. Phys. Chem.* 48: 545-600.

Owusu Apenten, R.K.; Chee, C. & Hwee, O.P. (2003). Evaluation of a sulphydryl-disulphide exchange index (SEI) for whey proteins – beta-lactoglobulin and bovine serum albumin. *Food Chem.* 83: 541-545.

Pabit, S.A.; Roder, H. & Hagen, S.J. (2004). Internal friction controls the speed of protein folding from a compact configuration. *Biochemistry* 43: 12532-12538.

Paik, W. & Kim, S. (1972). Effect of methylation on susceptibility of proteins to proteolytic enzymes. *Biochemistry* 11: 2589-2594.

Panasiuk, R.; Amarowicz, R.; Kostyra, H. & Sijtsma, L. (1998). Determination of α-amino nitrogen in pea protein hydrolysates: a comparison of three analytical methods. *Food Chem.* 62: 363-367.

Paparcone, R. & Buehler, M.J. (2011). Failure of Aβ(1-40) amyloid fibrils under tensile loading. *Biomaterials* 32: 3367-3374.

Paparcone, R.; Keten, S. & Buehler, M.J. (2010). Atomistic simulation of nanomechanical properties of Alzheimer's Aβ(1-40) amyloid fibrils under compressive and tensile loading. *J. Biomechanics* 43: 1196-1201.

Park, H.J.; Chinnan, M.S. & Shewfelt, R.L. (1994). Edible corn-zein film coatings to extend storage life of tomatoes. *J. Food Process Preserv.* 18: 317-331.

Park, H.J. (1999). Development of advanced edible coatings for fruits. *Trends Food Sci. Technol.* 10: 254-260.

Patel, K. & Borchardt, R.T. (1990a). Chemical pathways of peptide degradation. II. Kinetics of deamidation of an asparaginyl residue in a model hexapeptide. *Pharm. Res.* 7: 703-711.

Patel, K. & Borchardt, R.T. (1990b). Chemical pathways of peptide degradation. III. Effect of primary sequence on the pathways of deamidation of asparaginyl residues in hexapeptides. *Pharm. Res.* 7: 787-793.

Patil, S.M.; Mehta, A.; Jha, S. & Alexandrescu, A.T. (2011). Heterogeneous amylin fibril growth mechanisms imaged by total internal reflection fluorescence microscopy. *Biochemistry* 50: 2808-2819.

Payne, P.I.; Holt, L.M.; Jackson, E.A. & Law, C.N. (1984). Wheat storage proteins: their genetics and their potential for manipulation by plant breeding. *Phil. Trans. R. Soc. Lond. B* 304: 359-371.

Pedroche, J.; Yust, M.M.; Lqari, H.; Giron-Calle, J.; Alaiz, M.; Vioque, J. & Millan, F. (2004). *Brassica carinata* protein isolates: chemical composition, protein characterization and improvement of functional properties by protein hydrolysis. *Food Chem.* 88: 337-346.

Pellegrino, L.; van Boekel, M.A.J.S.; Gruppen, H.; Resmini, P. & Pagani, M.A. (1999). Heat-induced aggregation and covalent linkages in β-casein model systems. *Int. Dairy J.* 9: 255-260.

Pereda, M.; Aranguren, M.I. & Marcovich, N.E. (2008). Characterization of chitosan/caseinate films. *J. Appl. Polymer Sci.* 107: 1080-1090.

Pereda, M.; Aranguren, M.I. & Marcovich, N.E. (2009). Water vapor absorption and permeability of films based on chitosan and sodium caseinate. *J. Appl. Polym. Sci.* 111: 2777-2784.

Petersen, A.; Schramm, G.; Schlaak, M. & Becker, W.M. (1998). Post-translational modifications influence IgE reactivity to the major allergen Phl p 1 of timothy grass pollen. *Clin. Exp. Allergy* 28: 315-321.

Pétra, P.H. (1971). Modification of carboxyl groups in bovine carboxypeptidase A. I. Inactivation of the enzyme by *N*-ethyl-5-phenylisoxazolium-3'-sulfonate (Woodward's reagent K). *Biochemistry* 10: 3163-3170.

Plaxco, K.W.; Simons, K.T.; Ruczinski, I. & Baker, D. (2000). Topology, stability, sequence, and length: defining the determinants of two-state protein folding kinetics. *Biochemistry* 39: 11177-11183.

Plaxco, K.W.; Simons, K.T. & Baker, D. (1998). Contact order, transition state placement and the refolding rates of single protein domains. *J. Mol. Biol.* 277: 985-994.

Plotkin, S.S. & Onuchic, J.N. (2002). Understanding protein folding with energy landscape theory. Part I: Basic concepts. *Quart. Rev. Biophys.* 35: 111-167.

Pogaku, R.; Eng Seng, C.; Boonbeng, L. & Kallu, U. (2007). Whey protein isolate-starch system – a critical review. *Intl. J. Food Eng.* 3: 104-119.

Ponce, A.G.; Roura, S.I.; Del Valle, C.E. & Moreira, M. (2008). Antimicrobial and antioxidant activities of edible coatings enriched with natural plant extracts: in vitro and in vivo studies. *Postharvest. Biol. Technol.* 49: 294-300.

Promeyrat, A.; Bax, M.L.; Traoré, S.; Abry, L.; Santé-Lhoutellier, V. & Gatellier, Ph. (2010). Changed dynamics in myofibrillar protein aggregation as a consequence of heating time and temperature. *Meat Sci.* 85: 625-631.

Qiu, L. & Hagen, S.J. (2004). A limiting speed for protein folding at low solvent viscosity. *J. Am. Chem. Soc.* 126: 3398-3399.

Quintanilla-Guerrero, F.; Duarte-Vázquez, M.A.; Tinoco, R.; Gómez-Suárez, M.; García-Almendárez, B.E.; Vazquez-Duhalt, R. & Regalado, C. (2008). Chemical modification of turnip peroxidase with methoxypolyethylene glycol enhances activity and stability for phenol removal using the immobilized enzyme. *J. Agric. Food Chem.* 56: 8058-8065.

Quintas, A.; Saraiva M.J. & Brito, R.M. (1997). The amyloidogenic potential of transthyretin variants correlates with their tendency to aggregate in solution. *FEBS Lett.* 418: 297-300.

Quiocho, F.A. & Richards, F.M. (1966). The enzymic behaviour of carboxypeptidase-A in the solid state. *Biochemistry* 5: 4062-4076.

Rahma, E.H. & Narasinga Rao (1983). Effect of acetylation and succinylation of cottonseed flour on its functional properties. *J. Agric. Food Chem.* 31: 351-355.

Ramírez-Alvarado, M.; Merkel, J.S. & Regan, L. (2000). A systematic exploration of the influence of protein stability on amyloid fibril formation in vitro. *Proc. Natl. Acad. Sci. U.S.A.* 97: 8979-8984.

Ramírez-Jiménez, A.; Guerra-Hernández, E. & García-Villanova, B. (2000). Browning indicators in bread. *J. Agric. Food Chem.* 48: 4176-4181.

Richard, F.M. & Knowles, J.R. (1968). Gluteraldehyde as a protein cross-linkage reagent. *J. Mol. Biol* 37: 231-233.

Riha, W.E.; Izzo, H.V.; Zhang, J. & Ho, C.T. (1996). Nonenzymatic deamidation of food proteins. *Crit. Rev. Food Sci. Nutr.* 6: 225-255.

Robinson, A.B. & Rudd, C. (1974). Deamidation of glutaminyl and asparaginyl residues in peptides and proteins. *Curr. Top. Cell. Regul.* 8: 247-295.

Rodríguez Patino, J.M.; Carrera Sánchez, C. & Rodríguez Niño, M.R. (2008). Implications of interfacial characteristics of food foaming agents in foam formulations. *Adv. Colloid Interf. Sci.* 140: 95-113.

Rondeau, P.; Navarra, G.; Cacciabaudo, F.; Leone, M.; Bourdon, E. & Militello, V. (2010). Thermal aggregation of glycated bovine serum albumin. *Biochim. Biophys. Acta* 1804: 789-798.

Rondeau, P.; Armenta, S.; Caillens, H.; Chesne, S. & Bourdon, E. (2007). Assessment of temperature effects on β-aggregation of native and glycated albumin by FTIR spectroscopy and PAGE: relations between structural changes and antioxidant properties. *Arch. Biochem. Biophys.* 460: 141-150.

Röper, H.; Röper, S.; Heyns, K. & Meyer, B. (1983). N.M.R. spectroscopy of N-(1-deoxy-D-fructos-1-yl)-L amino acids ("fructose-amino acids"). *Carbohydr. Res.* 116: 183-195.

Rosén, J. & Hellenäs, K.E. (2002). Analysis of acrylamide in cooked foods by liquid chromatography tandem mass spectrometry. *Analyst* 127: 880-882.

Roth, M. (1971). Fluorescence reaction for amino acids. *Anal. Chem.* 43: 880-882.

Rudd, P.M.; Colominas, C.; Royle, L.; Murphy, N.; Hart, E.; Merry, A.H.; Hebestreit, H.F. & Dwek, R.A. (2001). A high-performance liquid chromatography based strategy for rapid, sensitive sequencing of N-linked oligosaccharide modifications to proteins in sodium dodecyl sulphate polyacrylamide electrophoresis gel bands. *Proteomics* 1: 285-294.

Rudén, C. (2004). Acrylamide and cancer risk - expert risk assessments and the public debate. *Food Chem. Tox.* 42: 335-349.

Sala, G.; de Wijk, R.A.; van de Velde, F. & van Aken G.A. (2008). Matrix properties affect the sensory perception of emulsion-filled gels. *Food Hydrocolloids* 22: 353-363.

Salgó, A.; Granzler, K. & Jecsai, J. (1984). Simple enzymatic methods for prediction of plant protein digestibility. In: R. Lásztity, M. Hidvé. Proc. Int. Assoc. Cereal Chem. Symp. (pp. 311-321). Budapest: Akadémiai Kiadó.

Salunkhe, D.K.; Jadhav, S.J.; Kadam, S.S. & Chavan, J.K. (1982). Chemical, biochemical, and biological significance of polyphenols in cereals and legumes. *Crit. Rev. Food Sci. Nutr.* 17: 277-305.

Samejima, K.; Dairman, W. & Underfriend, S. (1971). Condensation of ninhydrin with aldehydes and primary amines to yield highly fluorescent ternary products. *Anal. Biochem.* 42: 222-247.

Sanchez-Ruiz, J.M. (2011). Probing free-energy surfaces with differential scanning calorimetry. *Annu. Rev. Phys. Chem.* 62: 231-255.

Sancho, A.I.;. Rigby, N.M.; Zuidmeer, L.; Asero, R.; Mistrello, G.; Amato, S.; González-Mancebo, E.; Fernández-Rivas, M.; van Ree, R. & Mills, E.N.C. (2005). The effect of thermal processing on the IgE reactivity of the non-specific lipid transfer protein from apple, Mal d 3. *Allergy* 60: 1262-1268.

Sanger, F. (1945). The free amino groups of insulin. *Biochem. J.* 39: 507-515.

Santé-Lhoutellier, V.; Astruc, T.; Marinova, P.; Greve, E. & Gatellier, P. (2008). Effect of meat cooking on physicochemical state and in vitro digestibility of myofibrillar proteins. *J. Agric. Food Chem.* 56: 1488-1494.

Satake, K.; Okuyama, T.; Ohashi, M. & Shinoda, T. (1960). The spectrophotometric determination of amine, amino acid and peptide with 2,4,6-trinitrobenzene-1-sulfonic acid. *J. Biochem.* 47: 654.

Sawyer, W.H. (1968). Heat denaturation of bovine β-lactoglobulins and relevance of disulfide aggregation. *J. Dairy Sci.* 51: 323-329.

Schilling, E.D.; Burchill, P.I. & Clayton, R.A. (1963). Anomalous reactions of ninhydrin. *Anal. Biochem.* 5: 1-6.

Schwenke, K.D.; Knopfe, C.; Mikheeva, L.M. & Grinberg, V.Y. (1998). Structural changes of legumin from Faba beans (*Vicia faba L.*) by succinylation. *J. Agric. Food Chem.* 46: 2080-2086.

Sheih, I.C.; Wu, T.K. & Fang, T.J. (2009). Antioxidant properties of a new antioxidative peptide from algae protein waste hydrolysate in different oxidation systems. *Bioresour. Technol.* 100: 3419-3425.

Shental-Bechor, D. & Levy, Y. (2008). Effect of glycosylation on protein folding: a close look at thermodynamic stabilization. *Proc. Natl. Acad. Sci. U.S.A.* 105: 8256-8261.

Shih, F.F. (1990). Deamidation during treatment of soy protein with protease. *J. Food Sci.* 55: 127-129.

Shimada, K. & Cheftel, J.C. (1989). Sulfhydryl group/disulfide bond interchange reactions during heat-induced gelation of whey protein isolate. *J. Agric. Food Chem.* 3: 161-168.

Shirahama, H.; Lyklema, J. & Norde, W. (1990). Comparative protein adsorption in model systems. *J. Colloid Interf. Sci.* 139: 177-187.

Siepen, J.A. & Westhead, D.R. (2002). The fibril_one on-line database: mutations, experimental conditions, and trends associated with amyloid fibril formation. *Protein Sci.* 11: 1862-1866.

Silva Freitas, D. & Abrahão-Neto, J. (2010). Batch purification of high-purity lysozyme from egg white and characterization of the enzyme modified by PEGylation. *Pharm. Biol.* 48: 554-562.

Sinha, U. & Brewer, J.M. (1985). A spectrophotometric method for quantitation of carboxyl group modification of proteins using Woodward's reagent K. *Anal. Biochem.* 151: 327-333.

Siu, M. & Thompson, L.U. (1982). In vitro and in vivo digestibilities of succinylated cheese protein concentrates. *J. Agric. Food Chem.* 30: 743-747.

Skraup, Z.H. & Kaas, K. (1906). Über die Einwirkung von salpetriger Säure auf Ovalbumin. *Justus Liebigs. Ann. Chemie* 351: 379-389.

Smythe, C.V. (1936). The reaction of iodoacetate and of iodoacetamide with various sulfhydryl groups, with urease, and with yeast preparations. *J. Biol. Chem.* 114: 601-612.

Soderling, T.R. (1975). Regulation of glycogen synthetase. Specificity and stoichiometry of phosphorylation of the skeletal muscle enzyme by cyclic 3':5'-AMP-dependent protein kinase. *J. Biol. Chem.* 250: 5407-5412.

Song, Y.; Azakami, H.; Hamasu, M. & Kato, A. (2001). In vivo glycosylation suppresses the aggregation of amyloidogenic hen egg white lysozymes expressed in yeast. *FEBS Lett.* 491: 63-66.

Sorci, M.; Silkworth, W.; Gehan, T. & Belfort, G. (2011). Evaluating nuclei concentration in amyloid fibrillation reactions using back-calculation approach. *PLoS One* 6: e20072.

Stadler, R.H.; Blank, I.; Varga, N.; Robert, F.; Hau, J.; Guy, P.A.; Robert, M.-C. & Riediker, S. (2002). Acrylamide from Maillard reaction products. *Nature* 419: 449.

Straub, J.E. & Thirumalai, D. (2011). Toward a molecular theory of early and late events in monomer to amyloid fibril formation. *Annu. Rev. Phys. Chem.* 62: 437-463.

Sun, Y.; Hayakawa, S. & Izumori, K. (2004). Modification of ovalbumin with a rare ketohexose through the Maillard reaction: effect on protein structure and gel properties. *J. Agric. Food Chem.* 52: 1293-1299.

Sun, Y. & Hayakawa, S. (2002). Heat-induced gels of egg white/ovalbumins from five avian species thermal aggregation, molecular forces involved, and rheological properties. *J. Agric. Food Chem.* 50: 1636-1642.

Suttiprasit, P.; Krisdhasima, V. & McGuire, J. (1992). The surface activity of α-lactalbumin, β-lactoglobulin and bovine serum albumin. I. Surface tension measurements with single-component and mixed solutions. *J. Colloid Interface Sci.* 154: 316-326.

Tareke, E.; Rydberg, P.; Karlsson, P.; Eriksson, S. & Törnqvist, M. (2002). Analysis of acrylamide, a carcinogen formed in heated foodstuffs. *J. Agric. Food Chem.* 50: 4998-5006.

Tawfik, D.S. (2002). Chapter 63: Side chain selective chemical modifications of proteins. In: The Protein Protocols Handbook, 2nd Edition, Edited by J.M. Walker, Humana Press Inc. Totowa, NJ, pp 465-467.

Teplow, D.B.; Lazo, N.D.; Bitan, G.; Bernstein, S.; Wyttenbach, T.; Bowers, M.T.; Baumketner, A.; Shea, J.E.; Urbanc, B.; Cruz, L.; Borreguero, J. & Stanley, H.E. (2006). Elucidating amyloid-β protein folding and assembly: a multidisciplinary approach. *Acc. Chem. Res.* 39: 635-645.

Thomas, M.E.C.; Scher, J.; Desobry-Banon, S. & Desobry, S. (2004). Milk powders ageing: effect on physical and functional properties. *Crit. Rev. Food Sci. Nutr.* 44: 297-322.

Tyler-Cross, R. & Schirch, V. (1991). Effects of amino acid sequence, buffers, and ionic strength on the rate and mechanism of deamidation of asparagines residues in small peptides. *J. Biol. Chem.* 266: 22549-22556.

Van den Berg, L.; Carolas, A.L.; van Vliet, T.; van der Linden, E.; van Boekel, M.A.J.S. & van de Velde, F. (2008). Energy storage controls crumbly perception in whey proteins/polysaccharide mixed gels. *Food Hydrocolloids* 22: 1404-1417.

Van der Wel, H. & Loeve, K. (1972). Isolation and characterization of thaumatin I and II, the sweet-tasting proteins from *Thaumatococcus danielli* Benth. *Eur. J. Biochem.* 31: 221-225.

Van Teeffelen, A.M.M.; Broersen, K. & de Jongh, H.H.J. (2005). Glucosylation of β-lactoglobulin lowers the heat capacity change of unfolding; a unique way to affect protein thermodynamics. *Protein Sci.* 14: 2187-2194.

van Vliet, T. & Walstra, P. (1995). Large deformation and fracture behaviour of gels. *Faraday Discussions* 101: 359-370.

Veerman, C.; Ruis, H.; Sagis, L.M.C.; van der Linden, E. (2002). Effect of electrostatic interactions on the percolation concentration of fibrillar β-lactoglobulin gels. *Biomacromolecules* 3: 869-873.

Visschers, R.W. & de Jongh, H.H.J. (2005). Disulphide bond formation in food protein aggregation and gelation. *Biotechnol. Adv.* 23: 75-80.

Vioque, J.; Sánchez-Vioque, R.; Clemente, A.; Pedroche, J.; Bautista, J. & Millán, F.J. (1999). Production and characterization of an extensive rapeseed protein hydrolysate. *J. Am. Oil Chem. Soc.* 76: 819-823.

Wada, R. & Kitabatake, N. (2001). β-Lactoglobulin A with *N*-ethylmaleimide-modified sulfhydryl residue, polymerized through intermolecular disulfide bridge on heating in the presence of dithiothreitol. *J. Agric. Food Chem.* 49: 4971-4976.

Walder, R. & Schwartz, D.K. (2010). Single molecule observations of multiple protein populations at the oil-water interface. *Langmuir* 26: 13364-13367.

Walsh, D.M.; Lomakin, A.; Benedek, G.B.; Condron, M.M. & Teplow, D.B. (1997). Amyloid β-protein fibrillogenesis. Detection of a protofibrillar intermediate. *J. Biol. Chem.* 272: 22364-22372.

Wanasundara, P.K.J.P.D. & Shahidi, F. (1997). Functional properties of acylated flax protein isolates. *J. Agric. Food Chem.* 45: 2431-2441.

Wang, C.; Eufemi, M.; Turano, C. & Giartosio, A. (1996). Influence of the carbohydrate moiety on the stability of glycoproteins. *Biochemistry* 35: 7299-7307.

Waniska, R.D. & Kinsella, J.E. (1979). Foaming properties of proteins: evaluation of a column aeration apparatus using ovalbumin. *J. Food Sci.* 44: 1398-1411.

Waniska, R.D. & Kinsella, J.E. (1985). Surface properties of β-lactoglobulin: adsorption and rearrangement during film formation. *J. Agric. Food Chem.* 33: 1143-1148.

Weijers, M.; Broersen, K.; Barneveld, P.A.; Cohen Stuart, M.A.; Hamer, R.J.; de Jongh, H.H. & Visschers, R.W. (2008). Net charge affects morphology and visual properties of ovalbumin aggregates. *Biomacromolecules* 9: 3165-3172.

Wierenga, P.A.; Kosters, H.; Egmond, M.R., Voragen, A.G.J. & de Jongh, H.H.J. (2006). Importance of physical vs. chemical interactions in surface shear rheology. *Adv. Colloid Interface Sci.* 119: 131-139.

Wierenga, P.A.; Meinders, M.B.J.; Egmond, M.R.; Voragen, F.A.G.J. & de Jongh, H.H.J. (2003). Protein exposed hydrophobicity reduces the kinetic barrier for adsorption of ovalbumin to the air-water interface. *Langmuir* 19: 8964-8970.

Wierenga, P.A.; Meinders, M.B.; Egmond, M.R.; Voragen, A.G. & de Jongh, H.H. (2005). Quantitative description of the relation between protein net charge and protein adsorption to air-water interface. *J. Phys. Chem. B* 109: 16946-16952.

Wilde, P.J. (2000). Interfaces: their role in foam and emulsion behaviour. *Curr. Opin. Colloid Interface Sci.* 5: 176-181.

Wilde, P.J.; Mackie, A.; Husband, F.; Gunning, P. & Morris, V. (2004). Proteins and emulsifiers at liquid interfaces. *Adv. Colloid Interface Sci.* 108-109: 63-71.

Wimley, W.C. (2010). Describing the mechanism of antimicrobial peptide action with the interfacial activity model. *ACS Chem. Biol.* 5: 905-917.

Wolynes, P.G. (2005). Recent successes of the energy landscape theory of protein folding and function. *Quart. Rev. Biophys.* 38: 405-410.

Woodward, R.B. & Olofson, R.A. (1961). The reaction of isoxazolium salts with bases. *J. Am. Chem. Soc.* 83: 1007-1009.

Woodward, R.B.; Olofson, R.A. & Mayer, H. (1961). A new synthesis of peptides. *J. Am. Chem. Soc.* 83: 1010-1012.

Wright, H.T. & Urry, D.W. (1991). Nonenzymatic deamidation of asparaginyl and glutaminyl residues in proteins. *Crit. Rev. Biochem. Mol. Biol.* 26: 41-52.

Xu, L.J.; Sheldon, B.W.; Larick, D.K. & Carawan, R.E. (2002). Recovery and utilization of useful by-products from egg processing wastewater by electrocoagulation. *Poult. Sci.* 81: 785-792.

Xu, Z.; Paparcone, R. & Buehler, M.J. (2010) Alzheimer's Aβ(1-40) amyloid fibrils feature size-dependent mechanical properties. *Biophys. J.* 96: 2053-2062

Yemm, E.W. & Cocking, E.C. (1955). The determination of amino acids with ninhydrin. *Analyst* 80: 209-213.

Zamora, R. & Hildalgo, F.J. (2006). Coordinate contribution of lipid oxidation and Maillard reaction to the nonenzymatic food browning. *Crit. Rev Food Sci. Nutr.* 45: 49-59.

Zavodsky, M.; Chen, C.W.; Huang, J.K.; Zolkiewski, M.; Wen, L. & Krishnamoorthi, R. (2001). Disulfide bond effects on protein stability: designed variants of *Cucurbita maxima* trypsin inhibitor-V. *Protein Sci.* 10: 149-160.

Zhang, J.; Lee, T.C. & Ho, C.-T. (1993). Comparative study on kinetics of nonenzymatic deamidation of soy protein and egg white lysozyme. *J. Agric. Food Chem.* 41: 2286-2290.

Zhao, Y.; Ma, C.-Y.; Yuen, S.-N. & Phillips, D.L. (2004a). Study of succinylated food proteins by raman spectroscopy. *J. Agric. Food Chem.* 52: 1815-1823.

Zhao, Y.; Ma, C.-Y.; Yuen, S.-N. & Phillips, D.L. (2004b). Study of acetylated food proteins by raman spectroscopy. *J. Food Sci.* 69: 206-213.

Zhou, P.; Liu, X. & Labuza, T.P. (2008). Effects of moisture-induced whey protein aggregation on protein conformation, the state of water molecules, and the microstructure and texture of high-protein-containing matrix. *J. Agric. Food Chem.* 56: 4535-4540.

Zidar, J. & Merzel, F. (2011). Probing amyloid-beta fibril stability by increasing ionic strengths. *J. Phys. Chem. B* 115: 2075-2081.

Zinoviadou, K.G.; Koutsoumanis, K.P. & Biliaderis, C.G. (2009). Physico-chemical properties of whey protein isolate films containing oregano oil and their antimicrobial action against spoilage flora of fresh beef. *Meat Sci.* 82: 338-345.

Part 2

Catalysis and Reaction Engineering

6

Rational Asymmetric Catalyst Design, Intensification and Modeling

Zeeshan Nawaz[1], Faisal Baksh[1],
Ramzan Naveed[2] and Abdullah Alqahtani[1]
[1]*Chemical Technology Development,*
Saudi Basic Industries Corporation (SABIC), Riyadh
[2]*Department of Chemical Engineering,*
University of Engineering and Technology, Lahore
[1]*Kingdom of Saudi Arabia*
[2]*Pakistan*

1. Introduction

The development of new catalysts for a variety of chemical processes, molecular-level fundamental understanding of how it works and knowledge of surface science, chemistry, materials science, process modeling, process systems engineering, etc. is needed. Catalyst design and kinetic modeling have long been based on chemical intuition, i.e., the combination of a large empirical database and qualitative concepts of chemical reaction engineering and surface science. Recently the first principle kinetic modeling has become an important tool to investigate catalytic reactions and catalyst structures for superior operational benefits by integrating first principle guided exploration and experimental data. Nanocatalysis design is the building blocks of the concept of micro-level controlled reaction engineering, and the size and shape dependant material properties are the key enabling factor of the emerging technology. The qualitative difference is infect the material properties those changes in 1-10nm scale. Drastic enhancement in capabilities in nanomaterial synthesis with increasing control in size and shape were observed in last two decades but the rational catalyst design criteria is still not mature. Consideration of reaction brings intrinsic complexity in nanomaterial topology development (for a catalyst design) and emphasize on systematic multiscale simulation method to design and forecast.

In the field of heterogeneous catalysis, desired topological improvements on the basis of experimental information is well known and tedious [1]. Advances in surface science enable data manipulation of individual atoms on the catalyst surface with experiments provides initial guess towards systematic design. Number of model based catalyst design strategies were reported in open literature like use of qualitative reasoning and knowledge-based systems [2, 3], computational models [4-6] and detailed microkinetic modeling for catalytic systems [7]. In order to get the job done, compositional analysis of physical, chemical, and/or biological properties and validation of appropriate molecular structure (robust forward model linked with material description and knowledge extraction from experiment). Reaction modeling systems, optimization, and artificial intelligence based tools were extensively developed in automated environment to avoid reaction engineering expertise in building robust kinetic models. These tools allow rigorous analysis of multiple reaction mechanisms in the light of experimental data. Katare et al. [6] demonstrated kinetic model development for propane aromatization on H-ZSM-5 zeolite catalyst with a proof-of-concept in his studies.

Similarly, organometalic catalyst has interests in the broad areas of homogeneous catalysis, and catalyst modeling and mechanism elucidation. The main thrust of the field is in catalyst design and investigation of mechanistic aspects includes organic synthetic, organometallic and inorganic chemistry, molecular modeling and reaction monitoring. Fundamental research provide a theoretical basis and looks at the catalyst modeling using both experimental and theoretical techniques, providing information on active species and likely transition states. Ultimately helps in designing real experimental systems to be modeled with reliability and confidence. Mechanistic investigations using spectroscopic and kinetic methods help elucidate catalytic pathways and possible catalyst decomposition routes. Assessment and optimization of catalyst performance factors effecting catalyst behavior are important. The study of heterocyclic carbenes and synthesis of mixed carbene-donor ligands (transition metal complexes) were under extensive consideration. The structure, stoichiometric reaction behaviour and catalytic potential of the new complexes are the key goals to elaborate ways of manipulating the chemistry of the complexes in catalysis. The synthetic work is coupled with theoretical studies on orbital behavior and effects on the energetics of decomposition reactions. Synthesis of new polydentate ligands and their transition metal complexes offer prospects for the formation of new alkene polymerisation, copolymerisation and oligomerisation catalysts. Ligands with phosphorus, nitrogen and/or oxygen donors are the main classes of ligands and their selection depends on the metal being investigated. Therefore, methodology for catalyst discovery targets suitable strategy based on the conceptual tools of surface mechanism, molecular, bio and solid state chemistry.

2. Reaction modeling suite

The model building procedure needs rapid screening of reaction mechanistic hypothesis that could explain the experimental data. Number of user driven tools were available those facilitates knowledge archiving and retrieval to an automated reaction kinetic modeling. These modeling suites develop reaction mechanism and directly construct kinetic model. The architectural philosophy behind cumbersome mathematical modeling is shown in figure 1. Modelling can provide reaction energies (ΔH_r), energy barriers (ΔE_a), location of intermediates, etc. and it can explore catalyst composition, poisons, promoter's effect, efficiencies of catalysts using alternative routes, etc. In homogeneous catalyst design control of activity, ligand type, concentration of co-catalyst, molecular weight, tacticity (depends upon equilibrium), etc are also important.

The automated kinetic models development software tools of reaction networking, parameter optimization and overall kinetic modeling etc., are available in bulk like Reactor, Reaction Modeling Suite (RMS), KINAL A, Aspen Custom Modeler, gPROMS, Presto Kinetics, GREGPAK, Matlab, MLAB, ParaMetra, Scientist, Eurokin, polymath, Chemkin, Mitsubishi, MKM, IBM CKS, NetGen, XMG, Dynetica, Forcite, COMPASS, Sorption, CASTEP, DMol^3, Gepasi, Athena Visual Studio parameter estimation tool, E-cell, DBsolve,

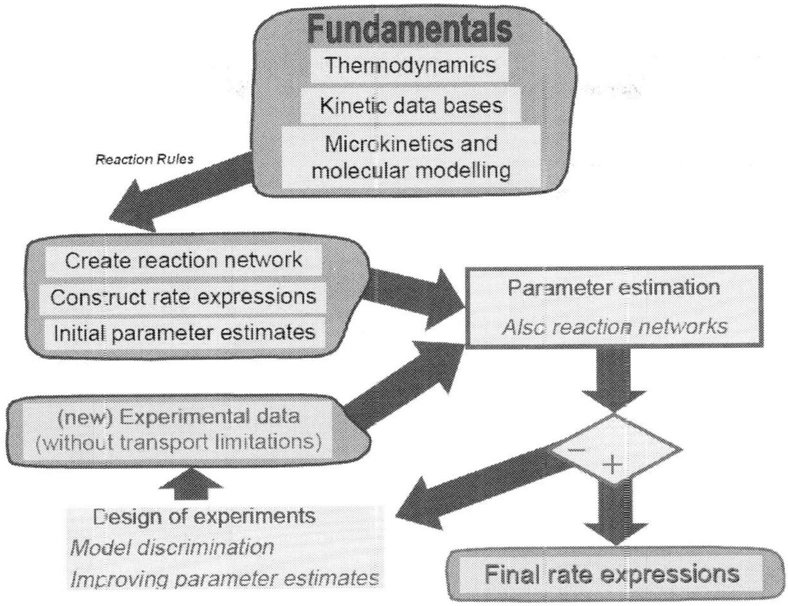

Fig. 1. Inventory of model building process

etc. For catalyst design Katare et al. developed a predictive model for surface reactions studies and claims to be rational and robust [6]. The key steps involved in this model building are generation of the simplest plausible reaction mechanism, translation of the reaction mechanism to a computationally tractable mathematical model, solving the model to estimate the parameters in light of high throughput and/or insufficient experimental data, refining the model to better fit the data by altering the mechanism, suggesting new experiments that could help in discriminate among multiple models. These software tools needed state of the art features for specific purposes [6]:

Reaction mechanism generation and networking

- Accurate representation of the large number of elementary reactions, overall reactions and species.
- Robust compiler that fallows the generic reaction rules to generate network generator.
- Pruning the reaction mechanism to get it in simpler and sophisticated.
- Reduce the number of thermodynamic and/or kinetic parameters to optimize based on thermokinetic and experimental data manipulation.

Kinetic parameter estimation

- Robust solvers that can handle large number of differential-algebraic equations and validate parameter estimation techniques.
- Evaluation of the multiple solutions for the parameters that explain the experimental data also.

Results and forecasting

- Discrepancies identification and optimization between the key features of model and experimental data that leads to design of experiment and questions on generated mechanism.
- Explain robustness of the model development with boundaries and shows statical analysis.

The most critical and important task in building a robust model is development of kinetic model and its also the most-time-consuming (excepting experimental or pilotplant campaigns). Without a satisfactory reaction mechanism and kinetics, a model may be applicable to narrow ranges of conditions and dangerous to use for predictions. The open literature information may not exactly match your conditions or catalyst formulation, therefore, formulation or selection of reaction-mechanism and rate expressions is best suggested. Establishing the reaction kinetics involves selection of rate expressions, and determination of rate parameters. Rate expressions involves four parameters; pre-exponential factor, activation energies, reaction order to each component and adsorption constants. The pre-exponential factors must be determined from the actual experimental data for catalyst or reaction conditions. This chapter give pathway to model information for catalyst design and intensification from state of the art reaction kinetics data generation, to the developing of computer-based knowledge organization for catalyst development, with brief review said techniques and models.

3. Thermodynamical consistency

In mechanism development and rate parameter optimization, thermodynamic consistency at both the enthalpic and entropic levels was often overlooked. An inconsistency in enthalpies gives incorrect solutions to the energy conservation equation, which lead towards ambiguous predictions of heat exchange, conversion, selectivity, etc.; while entropic inconsistency explains fundamental inconsistency (i.e. the pre-exponential factors). As there is no specific criterion for thermodynamic consistency evaluation and it distorts underlying equilibrium constant, only few published mechanisms are thermodynamically consistent [9-12]. In general, for any i-th reaction in a, the following equations form the basis of the enthalpic and entropic constraints [13]. For thermodynamically consistent mechanism the ratio of equilibrium constants (determined by gas phase thermochemistry) should be closer to 1 and for rigorous account of temperature variations data its necessary.

$$\Delta H_i = E_i^f - E_i^b = \sum_{j=1}^{L} c_{ij}\left(E_j^f - E_j^f\right), \qquad i = L+1, I$$

$$A_i^f \,/\, A_i^b = e^{\Delta S_i / R}$$

$$\frac{\Delta S_i}{R} = \ln(A_i^f \,/\, A_i^b) = \ln\prod_{j=1}^{L}(A_j^f \,/\, A_j^b)^{c_{ij}}\ , \quad i = L+1, I,$$

where f stands for the forward, b for backward reaction steps, A is the pre-exponential factor, E is the activation energy, R is the universal gas constant, ΔH is enthalpy change and

ΔS is entropy change of the reaction, respectively. For a linearly independent set of reactions network, if L is the size of reaction mechanism (unique in size) having I number of reactions, then $L \leq I$. The linear dependent reactions constraints can be expressed as c_{ij}, those are the coefficients of i-th linearly decomposed reaction onto the reaction basis.

4. Computer aided catalyst design

Computer aided catalyst design and intensification is an integrated approach in catalysis research and development, as chemical engineering is currently re-emerging under the label process intensification. The concept of process intensification is catalysis is new and not well established in the catalyst design industry. It is particularly well suited for the design and development of new catalyst. The incorporation of appropriately designed micro-structured catalytic having good control over the surface chemistry of the reaction is presented. Modelling of reactions at the active sites of catalysts also performed some times using density functional theory, and we will not discuss this here.

5. Reaction modeling suite

Reaction Modeling Suite (RMS) was developed by Purdue University, USA [6]. The quantitative kinetic description of a reaction mechanism involves specification of a differential and algebraic equation for each reaction component. RMS can develop reaction mechanism that typically involves more than 10 differential equations for a simple reaction and for complex mechanisms it can be 100 or more differential equations. Each set of differential equations includes a number of kinetic parameters that must be regressed from experimental data. It is a tedious, time-consuming task to develop code and solve the appropriate set of differential equation, where there is a real possibility of mistakes due to the complexity. The key steps involved in this process are reaction mechanism form chemistry rules, conversion of mechanism to a tractable mathematical model in computer language, integration of differential/algebraic set of equations with the suitable initial conditions, optimum model parameters estimation and finally analysis of reliability of the model parameters.

Each of the above steps requires considerable expertise, time and effort. Moreover, there are typically a number of potential kinetic models that need to be evaluated, and the model building/solving tasks must then be repeated for each new physical model. RMS presents the appropriate tools so that chemical experts can focus on chemistry without be overly burdened with the implementation of the required numerical methods. Purdue University, has successfully developed a set of systems tools, called the Reaction Modeling Suite (RMS) for rapid generation of complex kinetic models and evaluates these models with large, diverse sets of experimental data. These system tools are essential if model building and evaluation are to keep a pace commensurate with the rate of data production from high throughput experimentation [6]. In RMS, Chemistry Compiler takes a set of chemistry rules (i.e. any kinetic model is defined by a set of reaction rules), ensures that the rule set is chemically consistent, and finally generates the appropriate kinetic expressions. Next, an Equation Generator takes the kinetic expressions and develops the appropriate differential and algebraic equations in a form that can be solved numerically. Parameter Optimization is the next step, where the best set of parameters for a given model needed to describe the experimental data was determined i.e. non-trival for the large, nonlinear, coupled sets of

DAEs associated with complex kinetic mechanisms. Finally, a Statistical Analyzer has been developed to quantitatively assess the reliability of the model parameters with the ability to use nonlinear statistics. The RMS tools enable the nearly automatic analysis of multiple reaction mechanism and kinetics that assist catalyst design.

6. Catalyst intensification and optimization

The following is an overview of the catalyst modeling and optimization features necessary in developing attractive suite. Catalysts can be modeled in several different ways, with increasing levels of complexity by considering pellet balances, pellet factor effectiveness factor using is basic activity profiles, etc. With out pellet balances optimization is a very basic modelling level, it is not necessary to make any extra calculations due to the presence of a catalyst by expressing reaction kinetics in terms of composition of the catalyst. The catalyst composition can then be made an optimization variable. In the case of plug flow reactors, the composition can be varied along the length of the reactor, and viewed using the catalyst profile. Pellet factor calculation involves balances between component concentrations in the bulk, and within the catalyst pellet. The activity distribution inside the pellet is represented by a "Dirac-delta" function (infinite concentration at a single location). Although this cannot be achieved in practice, the Dirac-delta function can be closely approximated by a narrow step distribution of active material around the optimal location of the delta function without noticeable loss in performance.

Catalyst effectiveness factor calculations may involve Thiele modulus and effectiveness factor for the pellet. The effectiveness factor is a function of temperature, pressure and catalyst properties such as shape, size and diffusivity of components. When using this method the catalyst properties become optimizable parameters, and optimized profiles can be obtained. This method is sub-divided into basic activity profile as represented below:

- Uniform - The active material is uniformly distributed through the pellet.
- Egg-shell - The active material is concentrated at the pellet surface, and rarefied towards the centre.
- Egg-yolk - The active material is concentrated at the pellet centre, and rarefied towards the surface.
- Peaked - The active material is concentrated at a location partway between the centre and surface, and rarefied towards the centre and surface.
- Layered - The active material is distributed in a layer of thickness T, and location L from the pellet centre, where T and L are expressed in fractions of the pellet diameter.

The activity profiles are represented as infinitely variable profiles and by definition, when using the effectiveness factor method, the intrapellet mass balances are solved to determine the conentration profile throughout the catalyst pellet. There are also three further levels of complexity which can be selected to improve the model at the expense of computational time:

- Interphase mass balances causes the concentration profile between the pellet surface and the bulk material to be calculated.
- Interphase heat balances calculates the temperature profile between the pellet surface and the bulk material
- Intrapellet heat balances calculates the temperature profile within the catalyst pellet

Catalyst deactivation is defined as the steady degradation of the performance of the catalyst over time. Deactivation can be modeled by describing a deactivation correlation which relates the catalyst performance to its age. The deactivation factor varies from 1.0 (fresh catalyst) to 0.0 (fully depleted catalyst). If catalyst deactivation is significant it is important to consider its effect when optimising the catalyst properties. Properties which are optimal when the catalyst is fresh may become highly non-optimal by the time catalyst has degraded. By considering catalyst deactivation during optimization it is possible to design a catalyst that, while non-optimal initially, becomes optimal when its performance is averaged out over its lifespan. An additional factor which may be considered is the possibility of modifying the reactor temperature over time, to help minimize the impact of the catalyst deactivation.

Center of Process Integration, University of Manchester, United Kingdom provides a unique software suite have all above features named "REACTOR". Figure 2, 3 and 4 shows its user interface for optimizing catalyst properties. It also develops rate expressions and optimized the reactor at desired operating conditions. Heterogeneous catalyst's fluid-particle interface is modeled in mass and energy balance below

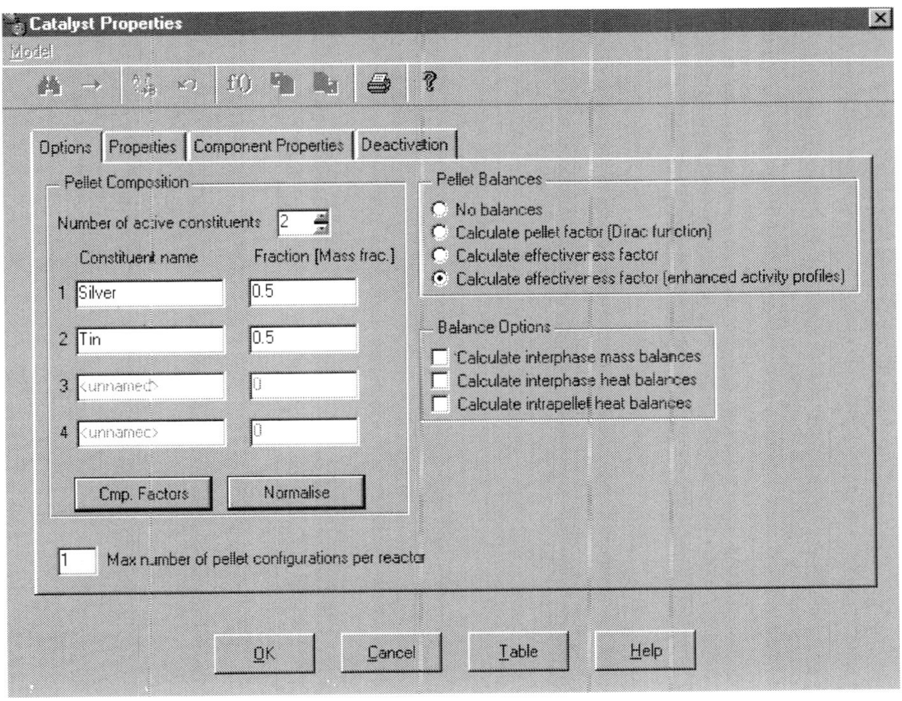

Fig. 2. Catalyst properties options in REACTOR

Mass balance:

$$k_{g,i}\left(C_{s,i} - C_{j,i}\right) = -D_{e,i}\frac{dC_i}{dz} = \eta\sum\left(V_{i,j} - R_{s,j}\right)\rho p S' \tag{1}$$

Energy balance:

$$h\left(T_s - T_g\right) = -\lambda\frac{dT}{dz} = \eta\sum\left(\Delta H_j - R_{s,j}\right)\rho p S' \tag{2}$$

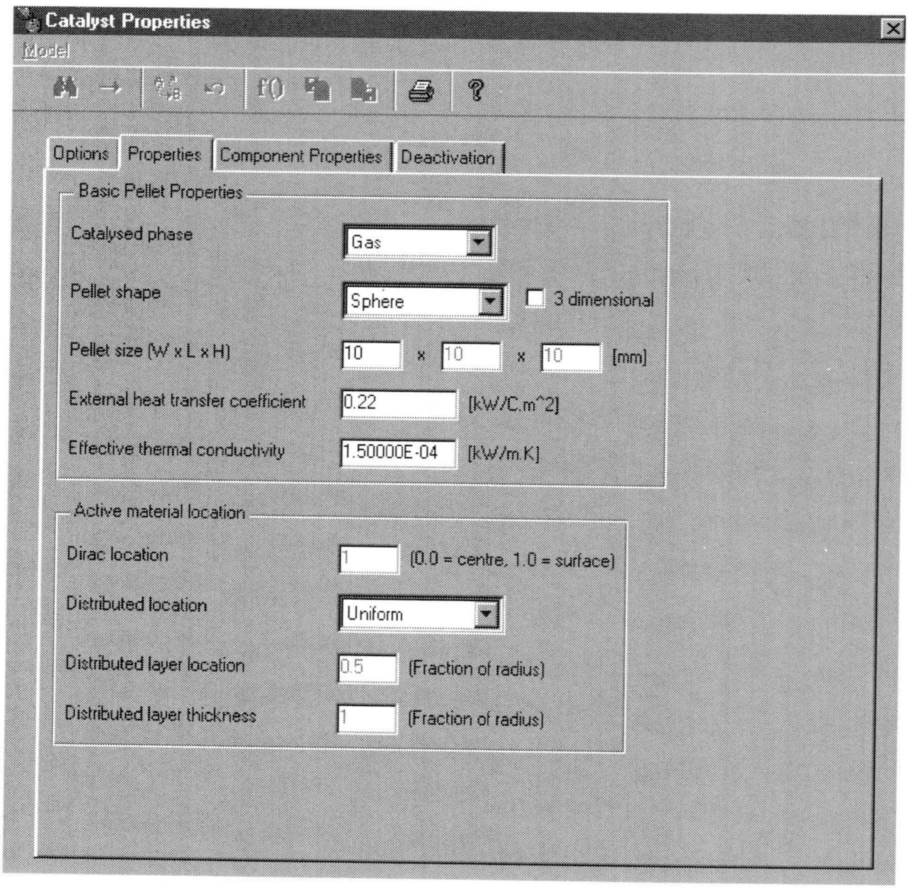

Fig. 3. Catalyst physical properties options in REACTOR

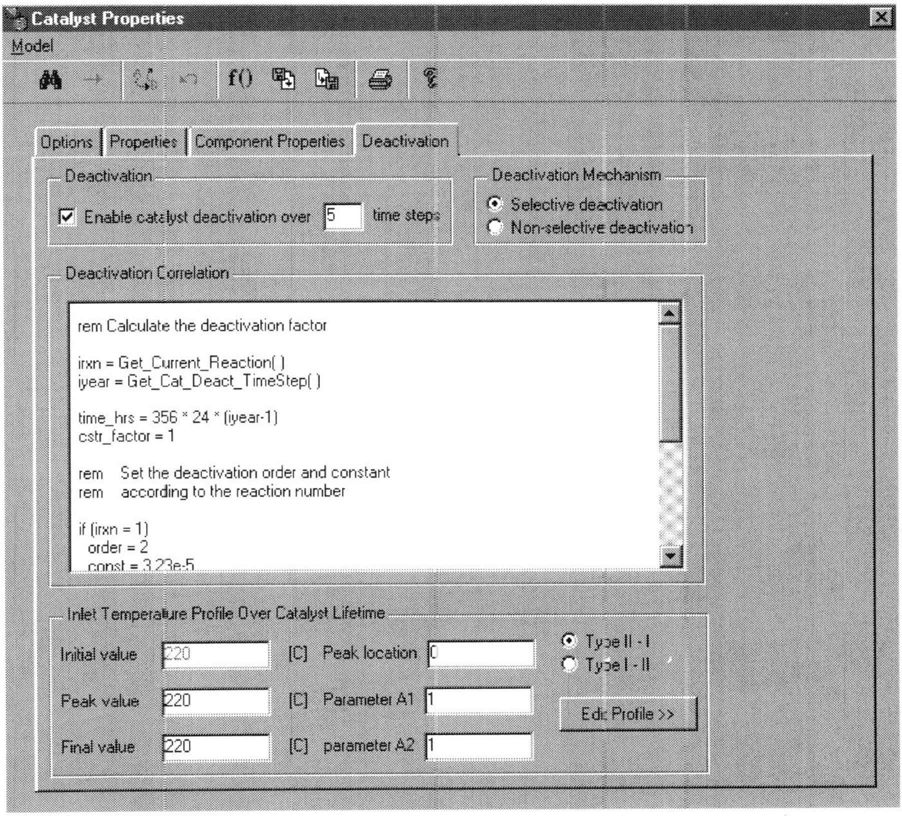

Fig. 4. Catalyst deactivation modeling window in REACTOR

7. Microchannel design and modeling

Microchannel has distinct identical channels and shape with characteristic dimensions in the order of micrometers and surface area-to-volume ratios ca. 50-100 times higher than those of their conventional counterparts. Due to the presence of high surface areas, sub-millimeter dimensions and metallic substrates usage, heat can be distributed quickly and uniformly distributed along the channel walls of the catalyst. High heat integration opertunities are available to drive endothermic and exothermic reactions independently and/or in parallel. The representative units can be modeled by incorporating two-dimensional continuity, momentum, energy conservation and species mass transport equations for the fluid and catalytic wall phase, and energy conservation equation for the metallic/catalytic wall phase. Simultaneous solution of these equations may carry out using finite elements methods. The model is then used to figure out the effects of various configurationally parameters such as wall thickness, type of wall material and the presence of micro-baffles on temperature distribution. The results forecast the parameters affects the degree of heat flow between channels and provide useful insights for the design of microchannels/catalysts.

8. Thermoanalytical technique

Thermoanalytical techniques were considered as transient response techniques, those relate characteristic properties of solid catalysts to its temperature programmed heating. Exchanges of matter and/or energy between and/or on surfaces of the solid catalysts provide means to follow physical or chemical transformations. The response obtained against certain temperature (thermogram) reflects the nature of the system and suitable experimental conditions. This analysis is used as a tool for quantitative and qualitative analysis for evaluating the influence of different activity factors. The most common thermoanalytical techniques are listed in Table 1. Number of methods exists in the open literature for extraction of kinetics information from thermoanalytical data [14]. Particularly in the heterogeneous catalysis, these techniques are powerful tool for investigating the influence of composition, preparation method and pretreatment on the reactivity of the surface or bulk with gases.

Thermal analysis techniques	Characteristic factor monitored
TPx Temperature-programmed reaction Temperature-programmed desorption (TPD)	Gas composition at the reactor outlet Characterisation of adsorptive properties of materials, surface acidity, total adsorption and desorption capacity with temperature, temperatures of rate maxima, metal surface area and dispersion, surface energetic heterogeneity, binding states and energies of adsorbed molecules, mechanism and kinetics of adsorption and desorption
Temperature-programmed reduction (TPR)	Characterisation of redox properties of materials, consumption of reducing agent with temperature range, temperatures of rate maxima, valence states of metal atoms in zeolites and metal oxides, metal oxide and support interaction, indication of alloy formation in bimetallic catalysts
Temperature-programmed oxidation (TPO)	Characterisation of coke species in deactivated catalysts and redox properties of metal oxides, total coke content in deactivated catalysts, mechanism and kinetics of oxidation reactions, deactivation kinetic mechanism
Thermogravimetry (TG)	Weight of sample
Differential thermogravimetry (DTG)	Rate of weight changes
Thermomechanical analysis (TMA), Dilatometry	Specific volume of solid sample
Thermomagnetic analysis (TMA)	Magnetic susceptibility
DMC Differential microcalorimetry	Enthalpy difference between sample and reference

Table 1. Thermal analysis techniques and their uses

The suitable model can account for heterogeneity in terms of chemisorption by introducing a sufficient number of different adsorption states. The readsorption formulated and tested against experimental data by describing the intrinsic dynamics of an adsorption state as a quasiequilibrium adsorption/desorption between the gas phase and the surface. Conceptually it is potentially useful for the elucidation of fundamental reaction mechanistic information for design and intensification of catalyst. The transient response methods place similar requirements on experimentation as conventional methods of kinetic analysis. In slow transient input, all or part of the elementary reactions may proceed in quasiequilibrium and their individual rates then can not properly elucidated. In general this is based on nonlinear regression analysis, and omits assumptions of pseudo-steady state.

In our extensive studies of catalyst intensification we conduct series of kinetic analysis for TPx experiments in phenomenological way [15-17]. Modelling of TPD data is based on adsorption and desorption theories, while TPR and TPO call for gas-solid surface reaction mechanisms, which may include topochemical characteristics. Further the rate of surface reaction divided into intrinsic rate of reaction per area of reaction interface and change in the reaction interface in reaction [18]. In general kinetic analysis of thermoanalytical data where there is a linear temperature rise (β) represented in the rate equation, where, rate of reaction is proportional to Arrhenius temperature dependence, A is pre-exponential factor, E is activation energy, and these are function of degree of conversion, f(α).

$$\beta \frac{d\alpha}{dT} = A exp\left[\frac{-E}{RT}\right] f(\alpha) \tag{3}$$

The transforming eq. 3 by taking natural logarithms on both sides gives

$$ln\left[\beta \frac{d\alpha}{dT}\right] = ln\left(Af(\alpha)\right) - \frac{E}{RT} \tag{4}$$

The plot of ln(βdα/dT) versus 1/T results straight line at a selected conversion level, the slope of the line gives -E/R at certain conversion level. This also did not explain reaction mechanism but by varing activation energy E=E(α) results from the Friedman analysis explains that eq. 3 with a normal physico-chemical interpretation will not describes the reaction kinetics.

9. Design and intensification of Pt-Sn-based alkane dehydrogenation catalyst

The control of chemical reaction pathways at molecular level presents undoubtedly the most important scientific challenge on the way to fully sustainable, thermodynamically efficient chemical processes. Integrated molecular reaction control is one of the major breakthroughs in achieving sustainable and efficient chemical processes design. Numerous fundamental works on this topic have been reported in literature and it is now the key challenge to advance them in an interdisciplinary way towards the industrial catalysis. The molecular alignment and geometry of collisions as well as selectivity of desired product was tackled by catalyst topology. In this example we discusses Pt-Sn based catalyst design and intensification for light alkane dehydrogenation to light alkenes. Constraints of this reaction are very complex due to endothermic limitation and endothermicity.

Bringing more molecules at the energy levels exceeding the activation energy threshold by conductive heating but it offers only a macroscopic control over the reaction and is thermodynamically inefficient. There are two broader approaches to control stereochemistry of reaction. First consists of immobilizing the molecules during reaction in confined nano-spaces by imposing "hard walls"of structures or of other molecules, or to force reacting species to assume certain position inside the offered structure. The second approach consists in acting upon moving molecules with spatially oriented external fields. Methods for controlling molecular alignment and orientation via nano-structural confinement invite shape-selective and imprinted catalyst. Shape-selective zeolites (having distinct sized nano-cage) discriminate between reactants and products and between tran-sition states molecules within the pores [19, 20].

Pt-Sn-based catalysts supported on amorphous supports or zeolites, for propane dehydrogenation have been discussed in many studies [15-29]. The selection of Sn as a promoter has been explained in terms of geometric and/or electronic effects [30]. In the geometric effect, Sn decreases the size of platinum ensembles, which reduces hydrogenolysis and coking reactions. Sn also modifies the electronic density of Pt, either by positive charge transfer from the Snn+ species or the different electronic structures in Pt-Sn alloys [31]. The deactivation of Pt-based catalyst during propane dehydrogenation is fatal, and mainly due to the aggregation of Pt particles and carbon deposition [31, 32]. However, the mechanistic understanding is still controversial and the reaction results were not conclusive, especially on stability and selectivity. In early days much focus has been given to Pt-Sn-based catalysts supported on Al_2O_3 for propane dehydrogenation. But the lifetime of these catalysts is too short. Therefore, there is a growing interest in developing zeolite supported catalysts, e.g. Pt-Sn/ZSM-5 zeolite for propane dehydrogenation, because of longer activation times and higher conversions [15-17]. Many studies have improved the performance of Pt-Sn/ZSM-5 catalysts by adding a more metallic promoter (particularly from alkali or alkaline earth metals) like Na, Fe, Zn, etc. [33, 34] and by increasing the Si/Al ratio of the ZSM-5 support [16, 34]. However, ZSM-5 supported catalysts were detrimentally affected by frequent regeneration with steam [35] and have lower propylene selectivity [16]. ZSM-5 supported catalysts can further take part in cracking and their large pore size fails to create shape selectivity effect for propylene. Moreover comparing the effects of supports and additional promoters, it was observed that the performance of propane dehydrogenation reaction significantly dependent on support. In particular it was noted that the control of intermediates products to desired product over the bi-functional catalyst was significantly support dependent.

Detailed TPx experimentation and kinetic studies with continuous experimental validity took us towards the innovative design of Pt-Sn-based zeolite supported catalyst. In this case, competitive reactions present a picture of a nano-snooker game. Unfortunately, in practicing industrial reactors has very limited degree of control of molecular-level events, therefore the catalyst should be robust enough. Extensive analysis suggests SAPO-34 as support (almost inert to dehydrogenation), with good shape selective opportunities and hydrothermally stable. The silico-aluminophosphate zeolite SAPO-34 is a microporous sieve with a chabasite-like structure (structural code CHA), has an extremely good shape selective effect for propylene [36-38]. Their Bronsted acid sites are responsible for their activities and based on two distinct hydroxyls (OH-bridges to Al and Si) on the structural framework. The structure, acidity and catalytic properties of SAPO-34 depend on the number and distribution of Si in the framework [39, 40]. Its Si based acid sites make it resistant to hydrothermal treatment and gives superiority over ZSM-5, where dealumination occur frequently [33, 34]. These observations suggest that the weak acidic support SAPO-34, which can sophisticatedly control the stereo-chemistry and gives shape selectivity effect for propylene in addition to being a bi-functional catalyst, Pt-Sn/SAPO-34.

The influences of the support on the reduction properties of Pt-Sn/SAPO-34 catalyst were analyzed by H_2-TPR and results are shown in Fig. 5. The reduction peaks at higher temperatures (above 450 °C) indicated that the Pt was well interacted with the support in the presence of Sn [5, 6, 14, 34]. The peak height and width increases with the increase in Sn loading, between 400 to 600 °C that may be due to joint reduction and valuable interaction between Pt and Sn on the SAPO-34 [24]. At higher temperatures towards Sn^o species forms. But the oxidation state of Sn (+2) in a bimetallic Pt-Sn catalyst was important, while the Pt-Sn alloy formation results in permanent deactivation [16, 24, 30]. The Pt attached to SnO is believed to be active for dehydrogenation reaction. The reduction peaks at the higher temperature indicated that the Pt had a strongly interaction with the SAPO-34 incomparative to other supports. The H_2 reduction results from Pt-Sn/SAPO-34 displayed sharp reduction peaks that confirmed the joint reduction and valuable interaction between Pt and Sn on the SAPO-34.

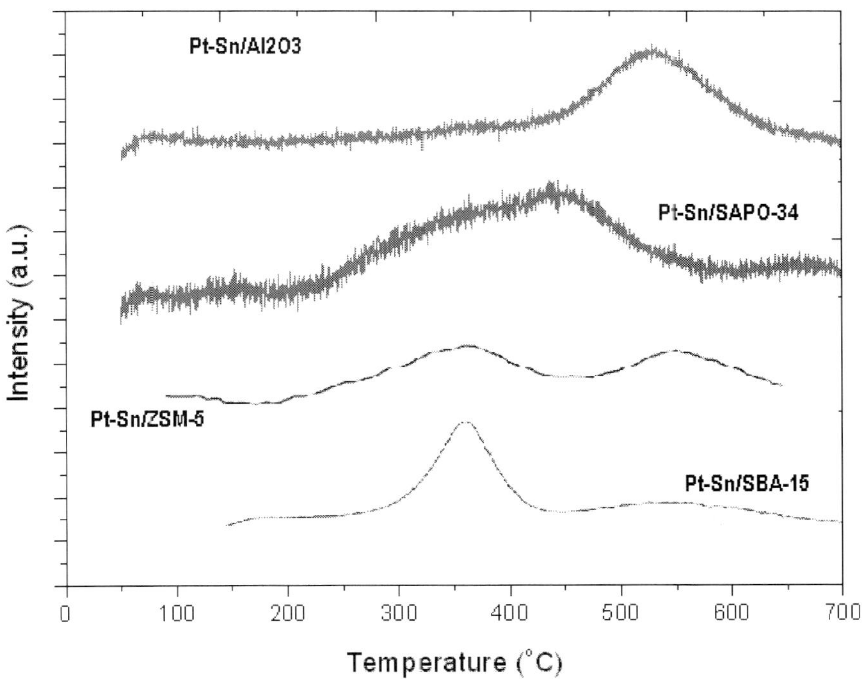

Fig. 5. H_2-TPR profiles of PtSn based different support catalysts of identical Pt and Sn composition.

Coke formation during propane dehydrogenation is the inherent factors that adversely affect catalyst performance. The amount of coke formed over different catalysts (after 8 hr reaction) was analyzed by TPO. The results are shown in Fig. 6. The typical TPO profile shows two successive peaks, indicating coke deposited on the metallic sites and support, while the peak height represents the intensity of deposition. The peaks between 300-400 °C and 550 - 650 °C corresponds to coke deposited on metallic site and support, respectively. It was noted that the strong electronic attachment over SAPO-34 facilitated the transfer of carbon deposits on metal sites to support. It's only possible when the metals are well interacted with the support. Moreover Sn modifies Pt ensembles by decreasing surface area and reduces carbon–metal bonds/interaction, and allows the intermediates to mobile easily to support [41].

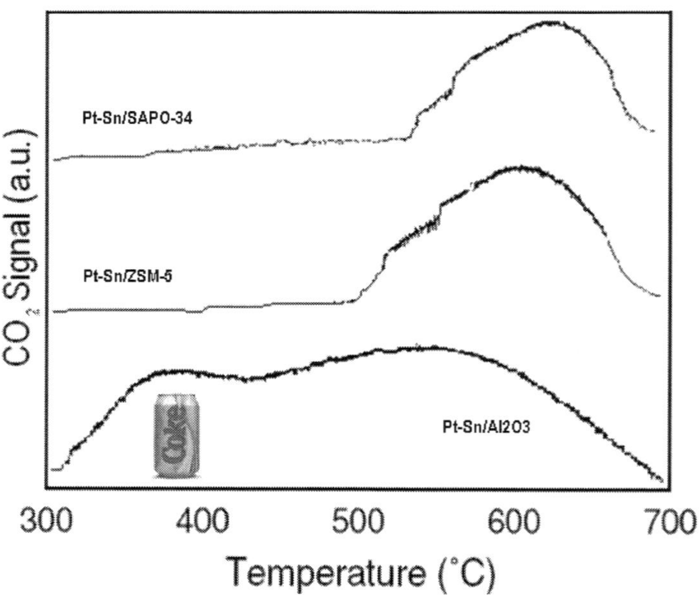

Fig. 6. TPO profiles of coke deposits on different supported PtSn-based catalysts after 8 hr propane dehydrogenation reaction under identical operating conditions.

The superiority of novel support SAPO-34 performance was experimentally verified (results are shown in Fig. 7 and 8), and it was observed that Pt-Sn/SAPO-34 catalyst controls the stereo-chemistry of propane dehydrogenation reaction much better than PtSn-based catalysts supported on ZSM-5 and Al_2O_3.

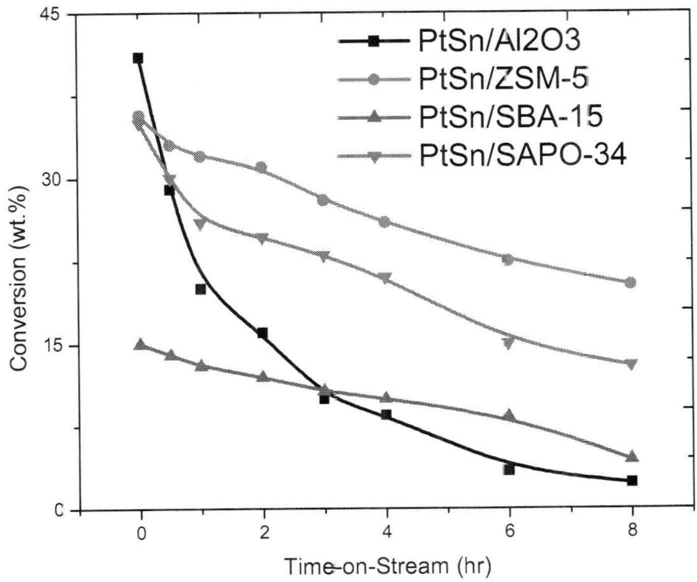

Fig. 7. Catalysts activity at 586 °C for propane conversion

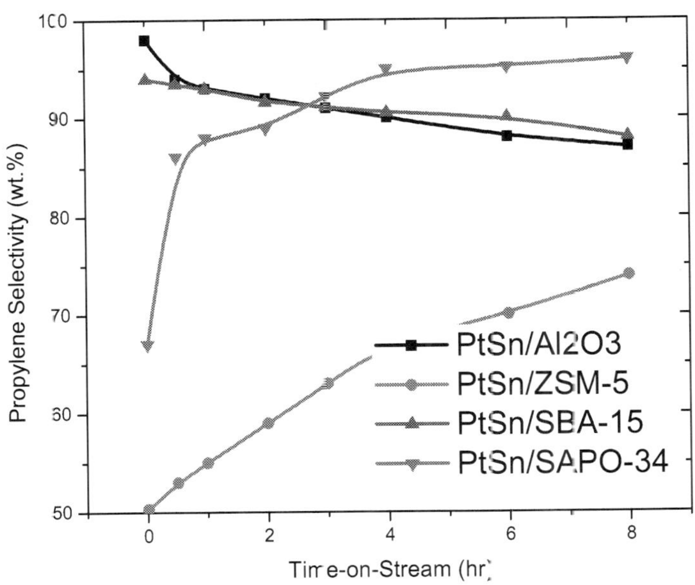

Fig. 8. Catalysts selectivity of propylene over different supported PtSn-based catalysts

10. Conclusion

Tight time and cost constraints force catalysis research to continuously reduce their experimental efforts during catalyst development and to facilitate the application of model-based strategy. Nevertheless, the effort of setting up a sophisticated model or criteria in coupling with novel TPx analysis for the design and intensification of catalysts is still higher. To overcome bottlenecks in catalyst modelling, considerable effort has to be spent with the systematization of kinetic models, formalization of reaction pathways, and development of knowledge-based software tools. This type of contribution claimed by number of software developers as discussed, while for the catalyst design we must need TPx analysis to achieve a robust design. It will be shown, how informal textual or metal-support information defining requirements on catalyst for a specified use. To find the optimal design under these uncertainties, a stochastic optimization method can be employed and verified by experimentation. In this work, the optimal properties of a catalyst for direct dehydrogenation of propane to propylene is presented, and uncertainties associated with the reactions and their parameters are modelled to design sustainable catalyst.

11. References

[1] Jones, C. W.; Tsuji, C.; Davis, M. E. Organic-Functionalized Molecular Sieves as Shape-Selective Catalysts. Nature 1998, 393, 52-54.

[2] Banares-Alcantara, R.; Westerberg, A. W.; Ko, E. I.; Rychener, M. D. Decadesa Hybrid Expert System for Catalyst Selection-I. Expert System Consideration. Comput. Chem. Eng. 1987, 11 (3), 265-277.

[3] Banares-Alcantara, R.; Ko, E. I.; Westerberg, A. W.; Rychener, M. D., Decadesa Hybrid Expert System for Catalyst Selection-Ii. Final Architecture and Results. Comput. Chem. Eng. 1988, 12 (9-10), 923-938.

[4] Van Santen, R. A. Theory, Spectroscopy and Kinetics of Zeolite Catalyzed Reactions. Catal. Today 1999, 50, 511-515.

[5] Jacobsen, J. H. C.; Dahl, S.; Clausen, B. S.; Bahn, S.; Logadottir, A.; Norskov, J. K. Catalyst Design by Interpolation in the Periodic Table: Bimetallic Ammonia Synthesis Catalysts. J. Am. Chem. Soc. 2001, 123, 8404-8405.

[6] Santhoji Katare, James M. Caruthers, W. Nicholas Delgass, and Venkat Venkatasubramanian, An Intelligent System for Reaction Kinetic Modeling and Catalyst Design, Ind. Eng. Chem. Res. 2004, 43, 3484-3512.

[7] Dumesic, J. A.; Rudd, D. F.; Aparicio, L. M.; Rekoske, J. E.; Trevino, A. A. The Microkinetics of Heterogeneous Catalysis; American Chemical Society: Washington, DC, 1993; p 316.

[8] J.M. Caruthers, J.A. Lauterbach, K.T. Thomson, V. Venkatasubramanian, C.M. Snively, A. Bhan, S. Katare, and G. Oskarsdottir, Catalyst design: knowledge extraction from high-throughput experimentation, Journal of Catalysis 216 (2003) 98–109.

[9] Dumesic, J. A.; Rudd, D. F.; Aparicio, L. M.; Rekoske, J. E.; Trevino, A. A., The Microkinetics of Heterogeneous Catalysis, American Chemical Society: Washington, D.C., 1993.

[10] Stoltze, P., Prog. Surf. Sci., 2000, 65, 65-150.

[11] Schmitz, G., J. Chem. Phys., 2000, 112, 10714-10717.

[12] Aghalayam, P.; Park, Y. K.; Vlachos, D. G., AIChE J., 2000, 46, 2017-2029, Chem. Eng. Progress, Oct. issue, 2000.

[13] Ashish B. Mhadeshwar, Hai Wang, and Dionisios G. Vlachos, THERMODYNAMIC CONSISTENCY IN MICROKINETIC DEVELOPMENT OF SURFACE REACTION MECHANISMS, Prepr. Pap.-Am. Chem. Soc., Div. Fuel Chem. 2003, 48(2), 514.

[14] Sestak, J., Satava V., Wendlandt, W. W., The study of heterogeneous processes by thermal analysis, Thermochim. Acta 7 (1973) 333-556.

[15] Zeeshan Nawaz and Fei Wei, Pt-Sn-based catalyst's intensification using Al2O3-SAPO-34 as a support for propane dehydrogenation to propylene, Journal of Industrial and Engineering Chemistry, 17 (2011) 389-393.

[16] Zeeshan Nawaz, Tang Xiaoping, Qiang Zhang, Wang Dezheng and Fei Wei, A highly selective Pt-Sn/SAPO-34 catalyst for propane dehydrogenation to propylene, Catalysis Communications, 10 (2009) 1925-1930.

[17] Zeeshan Nawaz, Xiaoping Tang, Yao Wang and Fei Wei, Parametric characterization and influence of Tin on the performance of Pt-Sn/SAPO-34 catalyst for selective propane dehydrogenation to propylene, Industrial & Engineering Chemistry Research, 49 (2010) 1274–1280.

[18] Haber, J., Kinetics and mechanism of the reduction of VI-b transition metal oxides and their oxysalts, J. Less-Comm. Met. 54 (1977) 243-261.

[19] M.E. Davis, Katz A, W.R. Ahmad, Rational catalyst design via imprinted nanostructured materials, Chem. Mater. 8 (1996) 1820–1839.

[20] M. Stöcker, Gas phase catalysis by zeolites, Microp. Mesop. Mater. 82 (2005) 257–292.

[21] H. Lieske, J. Volter, J. Catal. 90 (1984) 96-105.

[22] G.D. Angel, A. Bonilla, Y. Pena, J. Navarrete, J.L.G. Fierro, D.R. Acosta, J. Catal., 219 (2003) 63-73.

[23] T. Waku, J.A. Biscardi, E. Iglesia, Chem. Commun., (2003) 1764-1772.

[24] F.B. Passos, M. Schmal, M.A. Vannice, J. Catal. 160 (1996) 106-117.

[25] P.L.D. Cola, R. Gläser, J. Weitkamp, Appl. Catal. A. Gen. 306 (2006) 85-97.

[26] M.S. Kumar, D. Chen, J.C. Walmsley, A. Holmen, Catal. Commun. 9 (2008) 747-750.

[27] S.M. Stagg, C.A. Quenrini, W.E. Alvarez, D.E. Resasco, J. Catal. 168 (1997) 75-94.

[28] Y. Zhang, Y. Zhou, A. Qiu, Y. Wang, Y. Xu, P. Wu, Catal. Comm. 7 (2006) 860–866.

[29] G. Aguilar-Rios, P. Salas, M.A. Valenzuela, H. Armendariz, J.A. Wang, Salmones, Catal. Lett. 60 (1999) 21-25.

[30] J. Llorca, N. Homs, J. Leon, J. Sales, J.L.G. Fierro, P. Ramirez, Appl. Catal. A. Gen. 189 (1999) 77-86.

[31] C. Kappenstein, M. Guierin, K. Liaziar, K. Matusek, Z. Paial, J. Chem. Faraday Trans. 94 (1998) 2463-2474.

[32] M.S. Kumar, D. Chen, A. Holmen, J.C. Walmsley, Catal. Today. 142 (2009) 17-23.

[33] Y. Zhang, Y. Zhou, A. Qiu, Y. Wang, Y. Xu, P. Wu, Acta Phys.-Chim. Sin., 22 (2006) 672−678.

[34] Y. Zhang, Y. Zhou, H. Liu, Y. Wang, Y. Xu, P. Wu, Appl. Catal. A. Gen. 333 (2007) 202-210.

[35] Y. Zhang, Y. Zhou, K. Yang, Y. Li, Y. Wang, Y. Xu, P. Wu, Micropor. Mesopor. Mater. 96 (2006) 245–254.

[36] M.A. Anderson, B. Sulikowski, B.J. Barrie, J. Klinowski, J. Phys. Chem. 94 (1990) 2730-2734.

[37] Y. Xu, C.P. Grey, J.M. Thomas, A.K. Cheetham, Catal. Lett. 4 (1990) 251-260.
[38] J. Chen, P.A.Wright, J.M. Thomas, S. Natarajan, L. Marchese, S.M. Bradley, G. Sankar, C.R.A. Catlow, P.L. Gai-Boyes, R.P. Townsend, C.M. Lok, J. Phys. Chem. 98 (1994) 10216-10224.
[39] G. Sastre, D.W. Lewis, C. Richard, A. Catlow, J. Phys. Chem. B 101 (1997) 5249-5262.
[40] E. Dumitriu, A. Azzouz, V. Hulea, Micropor. Mesopor. Mater. 10 (1997) 1-8.
[41] F.N. Thomas, V. Hannelore, A.L. Johannes, J. Catal., 157 (1995) 388-395.

7

Preparation, Catalytic Properties and Recycling Capabilities Jacobsen's Catalyst

Jairo Cubillos
Department of Chemistry Sciences,
Pedagogical and Technological University of Colombia, Tunja
Colombia

1. Introduction

Metal salen complexes, such as the Jacobsen's catalyst, have attracted much interest in the last few decades because of their unique catalytic activity, especially as olefin epoxidation catalysts, in the presence of terminal oxidants like iodosylbenzene (PhIO), sodium hypochlorite (NaOCl), and *meta*-chloroperoxybenzoic acid (*m*-CPBA) (Gladasi, 2007). Salen ligands of Jacobsen's catalyst bind manganese ions through four atoms, two nitrogen and two oxygen atoms. This tetradentate-binding motif is reminiscent of the porphyrin framework in the heme-based oxidative enzymes (Canali & Sherrington, 1999). Nonetheless, salen derivatives are more easily synthesized than porphyrins and their structures are more easily manipulated to create an asymmetric environment around the active metal site (Canali & Sherrington, 1999). However, this homogeneous catalyst cannot be separated from the reaction media and, subsequently, cannot be recycled. Moreover, it suffers deactivation in homogeneous phase by either formation of dimeric μ-oxo-manganese (IV) species, which are inactive in the olefin epoxidation or oxidative degradation of the salen ligand through the imine group (Figure 1) (Xia et al., 2005). The conventional ways to solve these problems are to immobilize Jacobsen-type catalysts onto solid supports. The last decade has witnessed an intense research effort to heterogenise Jacobsen-type catalysts, and in general chiral manganese(III) salen complexes, using several types of supports in order to make them recyclable as well as economical (Murzin et al., 2005). Reports on the heterogenization of Jacobsen-type catalysts have been centered on their covalent binding to organic polymers (Clapham et al., 2001) and on their encapsulation, entrapment, adsorption and covalent attachment to porous inorganic supports, such as zeolites, Si-MCM-41, Al-MCM-41 and clays (Cubillos, 2010; Dasa, 2006; Kureshy, 2006), and also on activated carbon (Mahata et al., 2007). Unfortunately, the catalytic activity of the recovered catalyst decreases during the catalytic tests of reuse. It is found that isolating Mn(III) salen complexes onto a solid support increases the catalyst stability (Baleizaõ & García, 2006) by suppressing the formation of inactive dimeric μ-oxo manganese(IV) species. However, the deactivation route by ligand oxidation cannot be avoided by anchoring the catalyst to a solid matrix, since it depends on the oxidation conditions. On the other hand, the immobilized catalysts usually lead to partial loss of activity and/or enantioselectivity as compared to their analogous homogeneous catalysts (Baleizaõ & García, 2006). It is well known that in homogeneous phase, the catalyst acquires an appropriate geometric configuration that promotes both, the

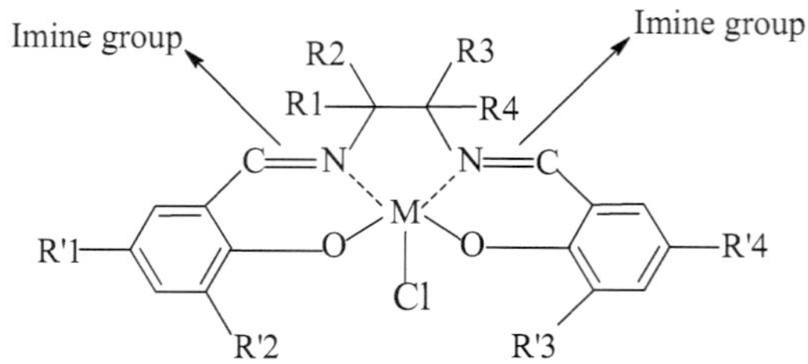

M = Mn, Cr, Co, Cu, V, Ru, Fe, Al, Ti

R1, R2, R3, R4 = H, alkyl or phenyl group

R'1, R'2, R'3, R'4 = H, nitro, alkyl, aryl or alcoxy group

Fig. 1. Chemical structure of metal salen complexes.

oxygen transfer from oxo-Mn(V) active species to the double bond of olefin and chiral induction (Adam et al., 2000). In contrast, when the catalyst is not free to move due to the influence of the solid support, an inappropriate geometrical configuration can be obtained (Fan et al., 2002). In addition to that, structural modification of either the catalyst or the solid support during the immobilization process, generally lead to diminished catalytic activity (Mastrorilli & Nobile, 2004). The most common immobilization methods by covalent bond modify one or two tert-butyl groups of the salen ligand. It is known that the tert-butyl groups are very important, since they define the optimal trajectory of the incoming olefin towards the oxo-Mn(III) active species (Linde et al., 2005). In summary, compared with the homogeneous counterparts, some of the immobilized complexes often suffer from various disadvantages, such as poor activity, leaching of the active species into the reaction medium, and low substrate accessibility (Baleizaõ & García, 2006). Therefore, for industrial practical merit and academic interest of homogeneously catalyzed reactions, the development of an efficient strategy for catalyst recovery is still challenging.

An alternative convenient and economical strategy is to adjust the solubility of homogeneous catalysts by varying the reaction conditions, resulting in the direct separation of the catalysts during the reaction (Kureshy et al., 2001). Recently, I and my colleagues described that using dimethyldioxirane (DMD) as oxidizing agent for the epoxidation of R-(+)-limonene and *cis*-ethyl cinnamate with Jacobsen-type catalysts could facilitate product isolation and catalyst recovery by segregating the catalysts into a solid phase during the reaction (Cubillos et al., 2009). Limonene di-epoxide was the main product from limonene, whereas an epoxide was obtained with 78% e.e in the case of *cis*-ethyl cinnamate. Moreover, it was reported that catalyst stability towards the oxidative degradation could be enhanced in the reaction medium, using DMD as oxidizing agent (Cubillos et al., 2009). On the other hand, it was found that the stereogenic center of the pure enantiomerically Jacobsen's catalyst did not influence the catalytic activity for the epoxidation of R-(+)-limonene (Cubillos et al., 2010).

Here I report the preparation and the catalytic properties as well as recycling capabilities of pure enantiomerically Jacobsen's catalysts (R,R-Jacobsen and S,S-Jacobsen) and its racemic form for epoxidation of R-(+)-limonene and *cis*-ethyl cinnamate with *in situ* generated DMD as oxidant. In order to compare the activity, recovery and reuse of the catalysts with DMD, other oxidizing agents (NaOCl and *m*-CPBA) most commonly used, were employed for R-(+)-limonene epoxidation with R,R-Jacobsen as catalyst. The influence of catalyst optical configuration on catalytic performance for this reaction is also revised.

2. Experimental section

The main reagents used for the synthesis of Jacobsen-type catalysts were 2,4-di-tert-butylphenol, tin(IV) chloride (99%, Aldrich®), paraformaldehyde (99.5%, Aldrich®), L-(+)-tartaric acid (99.5%, Aldrich®), D-(–)-tartaric acid (99, Aldrich®), *cis/trans* 1,2-diaminocyclohexane (99%, Aldrich®), potassium carbonate (99.995%, Aldrich®), manganese acetate tetrahydrate (99%, Aldrich®) and lithium chloride (99.99%, Aldrich®). All these materials were directly used as received. The solvents were absolute ethanol, chloroform, acetone and dichloromethane. 3,5-di-tert-butyl salicylaldehyde was prepared from the formylation of 2,4-di-tert-butylphenol using paraformaldehyde as reactant and tin(IV) chloride as catalyst (Deng & Jacobsen, 1992). (R,R)-1,2-diammoniumcyclohexane and (S,S)-1,2-diammoniumcyclohexane mono-(+)-tartrate salt were prepared from the racemic resolution of *cis/trans* 1,2-diaminocyclohexane with L-(+)-tartaric acid and D-(–)-tartaric acid, respectively (Deng & Jacobsen, 1992). Other reagents included R-(+)-limonene (97%, Aldrich®), ethyl phenylpropiolate (98%, Aldrich®), Lindlar's catalyst (5 wt.% on calcium carbonate, Aldrich®), Oxone® ($2KHSO_5 \cdot KHSO_4$, K_2SO_4, Aldrich®), manganese(II) acetate tetrahydrate (99%, Aldrich®), and sodium bicarbonate (Merck), aqueous sodium hypochlorite (NaOCl, 7% active chlorine basis, Carlo Erba), 4-phenylpyridine N-oxide (98%, Aldrich®), *meta*-chloroperoxybenzoic acid (*m*-CPBA, 77%, Aldrich), 4-methylmorpholine N-oxide (97%, Aldrich®). *Cis*-ethyl cinnamate was prepared from the partial hydrogenation of ethyl phenylpropiolate using the Lindlar's catalyst (Larrow & Jacobsen, 1994).

Different salen ligands R,R-Jacobsen, S,S-Jacobsen and racemic were prepared according to the methods previously described (Deng & Jacobsen, 1992), by condensing the appropriate diamine (3.5 mmol) and 3,5-di-tert-butyl salicylaldehyde (7.0 mmol), for 2 h, using ethanol as a solvent. Upon reaction, completion of a yellow–orange precipitate appeared. These precipitates were filtered and dried under vacuum. The obtained solids were characterized by FT-IR (Bahramian et al., 2006). The Jacobsen's catalysts R,R-Jacobsen, S,S-Jacobsen and Jacobsen racemic (Figure 2), were prepared as previously described (Deng & Jacobsen, 1992). Ethanolic solutions of 3.0 mmol of different ligands, and 3.3 mmol of manganese (II) acetate tetrahydrate were refluxed, for 1 h. During reflux, a change in the color of the solution from yellow–orange to dark-brown was observed. The manganese complexes were recrystallized in heptane and dried under vacuum. The FT-IR and DR UV–Vis spectral bands of all complexes were found to be identical to those reported (Bahramian, 2006; Chaube, 2005).

R,R-Jacobsen, S,S-Jacobsen and Jacobsen racemic catalysts were then used as catalysts for the epoxidation of R-(+)-limonene (1) and *cis*-ethyl cinnamate (2). In a standard procedure, 1.0 mmol of substrate, 1.2 mmol of sodium bicarbonate and 0.05 mmol of catalyst were dissolved in 4 mL of acetone. A buffer solution (aqueous $NaHCO_3$, 5 % wt) was added to bring the pH in the range between 8.0-8.5 (mixture A). In another vessel, 2.0 mmol of $KHSO_5$ (Oxone®) was dissolved in 4 mL of water (mixture B). While mixture A was being stirred, mixture B was slowly added, keeping the pH in the range 8.0 - 8.5 using $NaHCO_3$ solution (5 % w/w aqueous). When mixture B was completely added, stirring was stopped and the

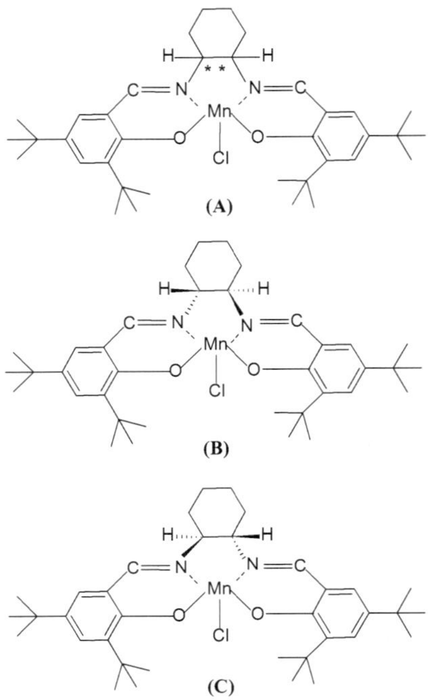

Fig. 2. Chemical structure of Jacobsen type catalysts. A: Racemic Jacobsen, B: R,R-Jacobsen, C: S,S-Jacobsen.

formed solids separated by filtration and/or centrifugation. The liquid phase was extracted with dicloromethane in a separation funnel and analyzed by GC. The solid phase was washed with sufficient water up to reach a constant weight in the obtained residue. This dark brown residue (catalyst) was easily dissolved in acetone and thus ready for recycling. Also, the chemical identity of this residue was analyzed by FTIR.

An Agilent Technologies 7890A gas chromatograph (GC), equipped with a DB-1 capillary column (50 m long, 0.32 mm ID and 1.20 mm film thickness) and a FID detector was used for the analysis of solvent purity, olefin and oxidation products. Ultra high pure helium was used as carrier gas (30 mL/min). The injection port temperature was kept at 300 °C. For separation of R-(+)-limonene the column temperature was programmed between 80 and 140 °C while for *cis*-ethyl cinnamate it was kept isothermal at 140 °C. The area normalization method was used to determine conversion, selectivity and relative yield. The enantiomeric excess (ee) for the single epoxide derived from *cis*-ethyl cinnamate epoxidation was determined by GC using a chiral capillary column, i.e. Betadex GTA (60 m long, 250 mm ID and 0.25 mm film thickness). In this case, a commercially available 3-ethyl-phenylglycidate (*cis/trans* = 10/90, Aldrich) racemic mixture was used. In the case of the chiral epoxides derived from R-(+)-limonene, 1,2-limonene oxide (97%, mixture of *cis* and *trans*, Aldrich) was used. Limonene diepoxide was prepared by oxidation of R-(+)-limonene (1.0 mmol) oxide using *m*-CPBA (5.0 mmol) as oxidizing agent and confirmed by GC-MS. The optical configuration was assigned by comparing the chromatogram of our products with those of *cis*-ethyl cinnamate isomers available in literature (Steiner et al., 2002). Also, isolated yields of the major epoxide product for either substrate were calculated. Thus, the major epoxide originating from the *cis*-ethyl cinnamate oxidation was purified by short-path distillation (110 °C and 0.5 mmHg), while in the case of R-(+)-limonene, diepoxide was collected at 140 °C and 0.5 mmHg.

In order to compare the catalytic activity, recovery and reuse of catalysts with DMD, other oxidizing agents such as NaOCl and *m*-CPBA were explored for R-(+)-limonene epoxidation according to reported conventional methods (Wang et al., 2006), using the same molar ratio of substrate/oxidant (0.5 mmol/mmol), substrate/catalyst (20 mmol/mmol), reaction time (30 min) and reaction temperature (25 °C) as DMD.

3. Results and discussion

Figure 3 shows the FTIR spectra of the salen ligands and their corresponding catalysts. The salen ligands show a characteristic band around 1620 cm^{-1}, which is associated with the vibrations of the imine group (HC=N) (Bahramian et al., 2006). In the catalyst samples, this band is displaced towards lower wavelengths (1600-1590 cm^{-1}) as the first evidence of the formation of the organometallic complex. Additionally, characteristic bands at 1530 (C-O), 550 (Mn-O) and 480 cm^{-1} (Mn-N) are also associated with the complexation of manganese by the salen ligand (Bahramian et al., 2006).

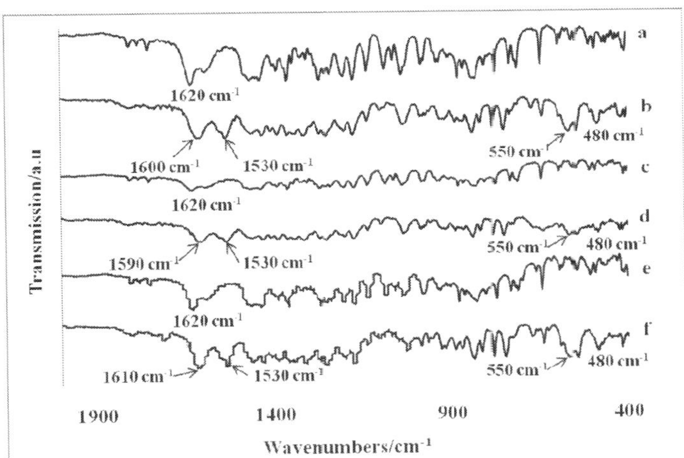

Fig. 3. FT-IR Spectra of (R,R) Jacobsen's ligand (a), (R,R) Jacobsen's catalyst (b), (S,S) Jacobsen's ligand (c), (S,S) Jacobsen's catalyst (d), Jacobsen's racemic ligand (e), Jacobsen's racemic catalyst (f).

Figure 4 shows DR UV-vis spectra of the salen ligands and their corresponding catalysts. The salen ligands exhibit absorption bands at 265 nm and 335 nm. These bands are attributed to $\pi \rightarrow \pi^*$ transitions. The band at 265 nm has been assigned to the benzene ring and the one at 335 nm, to the imino groups (Chaube et al., 2005). The imino $\pi \rightarrow \pi^*$ transitions in the Mn salen complexes is shifted to larger wavelengths due to metal coordination,

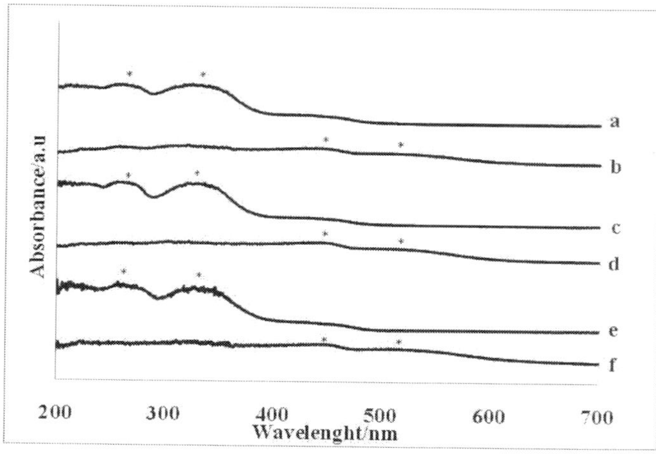

Fig. 4. DR UV-vis Spectra of (R,R) Jacobsen's ligand (a), (R,R) Jacobsen's catalyst (b), (S,S) Jacobsen's ligand (c), (S,S) Jacobsen's catalyst (d), Jacobsen's racemic ligand (e), Jacobsen's racemic catalyst (f).

confirming the formation of Mn(III) salen complex (Chaube et al., 2005). UV–Vis and FT-IR spectra revealed that the salen ligands are unaffected and are not decomposed upon coordination of the organo-funtional groups with manganese (Chaube et al., 2005).

R,R-Jacobsen, S,S-Jacobsen and Jacobsen racemic catalysts were examined in the liquid phase oxidation reaction of *cis*-ethyl cinnamate and R-(+)-limonene using *in situ* generated DMD as oxidizing agent. Different oxidation products were obtained, depending on the substrate type. Figures 5 and 6 outline the test reactions. Di-epoxides appear as the main products from R-(+)-limonene oxidation (Figure 5), while a single epoxide ((2R,3R)-*cis* ethyl-3-phenylglycidate) was produced from *cis*-ethyl cinnamate oxidation (Figure 6). Similarly to the previously reported experiments of olefins oxidation (Cubillos, 2009, 2010), the Jacobsen type catalyst was easily separated from the obtained solid phase at the end of reaction.

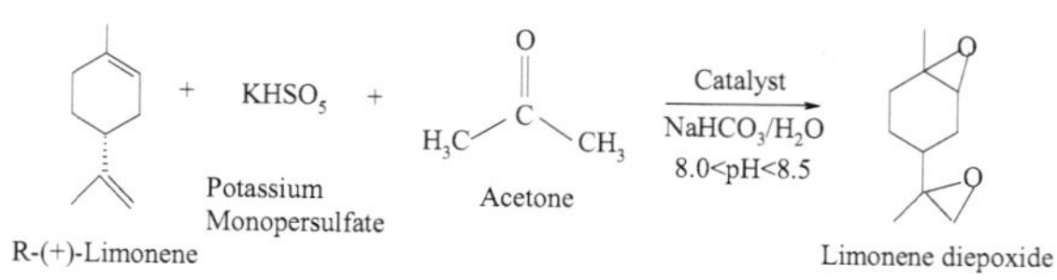

Fig. 5. Selective oxidation of R-(+)-limonene (1).

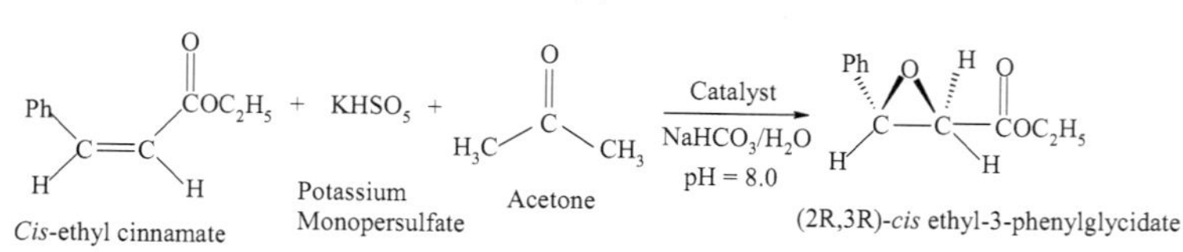

Fig. 6. Enantioselective oxidation of *cis*-ethyl cinnamate (2).

The results of catalytic activity and recyclability are collected in Tables 1 and 2. In the absence of catalyst, reaction was observed for R-(+)-limonene (conversion 53 % and relative rates of reaction 0.145 mol/Lt×h). In this case, 79% selectivity to limonene diepoxide with 46% relative yield and 30% isolated yield was achieved. 1,2-limonene oxide was obtained as secondary product with very low selectivities (Table 1, entry 1,). In contrast, no reaction was observed using *cis*-ethyl cinnamate as substrate (Tables 1 and 2, entry 8). It is known that unfunctionalized olefins are more reactive than its functionalized counterparts when Jacobsen type catalysts are used (Porter & Skidmore, 2000).

Entry	Substrate	Catalyst	Conversion (%)	Epoxide selectivity (%)	Diepoxide selectivity (%)	Relative yield[h] (%)
1	(1)[a]	None	58	18[e]	79[g]	46[i]
2	(1)[a]	I	100	5[e]	93[g]	93[i]
3	(1)[a]	II	100	3[e]	93[g]	93[i]
4	(1)[a]	III	100	4[e]	95[g]	95[i]
5	(1)[b]	I	99	7[e]	90[g]	89[i]
6	(1)[c]	I	98	5[e]	87[g]	85[i]
7	(1)[d]	I	98	8[e]	90[g]	88[i]
8	(2)[a]	None	0	-	-	0
9	(2)[a]	II	18	88[f]	-	16[j]
10	(2)[a]	III	20	84[f]	-	17[j]
11	(2)[a]	I	15	90[f]	-	14[j]
12	(2)[b]	II	12	82[f]	-	10[j]
13	(2)[c]	II	8	83[f]	-	7[j]
14	(2)[d]	II	6	82[f]	-	5[j]

[a] Reaction conditions: substrate = 1.0 mmol; $KHSO_5$ = 2.0 mmol (from Oxone®), catalyst = 0.05 mmol; acetone = 4.0 mL; water = 4.0 mL; reaction time = 30 min. [b] First reuse. [c] Second reuse. [d] Third reuse. [e] Selectivity to 1,2-limonene oxide. [f] Selectivity to ethyl-3- phenylglycidate (sum of *cis* and *trans*). [g] Sum of the four diepoxides of R-(+)-limonene. [h] Relative yield = Conversion×Selectivity/10,000. [i] Relative yield to the diepoxides. [j] Relative yield to ethyl-3-phenylglycidate (sum of *cis* and *trans*).

Table 1. Results of catalytic activity and reutilization

Similar yields (relative yields 93-95% and isolated yields 56-60%) to limonene diepoxide were reached with all catalysts, while in the absence of catalyst 46% and 30% respectively, were obtained. This finding suggests that the chiral center of enantiomerically pure (R,R and S,S-Jacobsen) appear to have little or no influence on the catalytic activity; rather the state of coordination given by the salen ligand to the manganese appears to be crucial. Here, the contribution of the catalyst is proven once again by an increase of both R-(+)-limonene conversion (or relative rates of reaction) and selectivity to limonene diepoxide (sum of the four diepoxides of R-(+)-limonene). None of the four diepoxides of R-(+)-limonene predominated. In contrast, the catalytic oxidation of *cis*-ethyl cinnamate offers the possibility

Entry	Substrate	Catalyst	Relative rate of reaction[e] (mol/lt×h)	Enantiomeric excess (%)	Isolated yield (%)
1	**(1)**[a]	None	0.145	-	30[h]
2	**(1)**[a]	I	0.25	-	56[h]
3	**(1)**[a]	II	0.25	-	57[h]
4	**(1)**[a]	III	0.25	-	60[h]
5	**(1)**[b]	I	0.25	-	52[h]
6	**(1)**[c]	I	0.245	-	53[h]
7	**(1)**[d]	I	0.245	-	50[h]
8	**(2)**[a]	None	0	-	0
9	**(2)**[a]	II	0.045	78[f]	11[i]
10	**(2)**[a]	III	0.050	55[g]	9[i]
11	**(2)**[a]	I	0.038	0	7[i]
12	**(2)**[b]	II	0.030	76[f]	7[i]
13	**(2)**[c]	II	0.020	74[f]	5[i]
14	**(2)**[d]	II	0.015	73[f]	4[i]

[a] Reaction conditions: substrate = 1.0 mmol; $KHSO_5$ = 2.0 mmol (from Oxone®);catalyst = 0.05 mmol; acetone = 4.0 mL; water = 4.0 mL. Reaction time = 30 min. [b] First reuse. [c] Second reuse. [d] Third reuse. [e] Relative rate of reaction = Conversion×initial mol/reaction volume×reaction time. [f] Enantiomeric excess of (2R,3R)-cis-ethyl-3-phenylglycidate over (2S,3S)-cis-ethyl-3-phenylglycidate. [g] Enantiomeric excess of (2S,3S)-*cis*-ethyl-3-phenylglycidate over (2R,3R)-*cis*-ethyl-3-phenylglycidate. [h] Isolated yield to R-(+)-limonene diepoxides. [i] Isolated yield to ethyl-3-phenylglycidate (sum of *cis* and *trans*).

Table 2. Results of catalytic activity and reaction rates

to epoxidize the unique C=C double bond located in its chemical structure. In order to perform enantioselective epoxidation, a pure enantiomerically catalyst is required, which can be reached either using R,R-Jacobsen or S,S-Jacobsen. In general, good selectivities to ethyl-3-phenylglycidate (sum of *cis* and *trans*, 84-90%) and good enantiomeric excesses to (2R,3R)-*cis*-ethyl-3-phenylglycidate (78%) were obtained with R,R-Jacobsen. In contrast, lower yields were reached (relative yields 16 and isolated yields 11). Additionally, it is worth to note that the product stereochemistry is strongly dependent on the absolute configuration of catalyst. Thus, *cis*-ethylcinnmate with R,R-Jacobsen gives(2R,3R)-*cis*-ethyl-3-phenylglycidate (Table 2, entry 9), whereas R-(+)-limonene with S,S-Jacobsen gives (2S,2S)-*cis*-ethyl-3-phenylglycidate (Table 2, entry 10). This shows clearly the specificity of a pure enantiomeric catalyst for inducing the preferential formation of the observed product.

Table 3 shows the results of catalytic activity and catalyst recovery for R-(+)-limonene epoxidation using the R,R-Jacobsen catalyst and three oxidants: DMD, NaOCl and *m*-CPBA. For these experiments, Figure 7 shows the spectra of the fresh and used catalyst after reaction with either oxidant. As listed in Table 3, the largest conversion was obtained with DMD (100%), although different selectivities were obtained. Thus, 1,2-limonene oxide was the major product when NaOCl and *m*-CPBA were used as oxidizing agents, obtaining

Entry	Oxidizing agent	Conversion (%)	selectivity (%)	Process of catalyst recovery
1	None	0	-	-
2	NaOCl[a]	65	75[d]	Vacuum distillation
3	*m*-CPBA[b]	35	95[d]	Vacuum distillation
4	DMD[c]	100	93[e]	Centrifugation and filtration

[a] R-(+)-limonene = 1.0 mmol; NaOCl = 2.0 mmol (0.05 M aqueous Na2HPO4 solution) ; 4-phenyl pyridine N-oxide = 0.4 mmol; R,R-Jacobsen = 0.05 mmol; dichloromethane = 4 mL Reaction time = 30 min. [b] R-(+)-limonene = 1.0 mmol; *m*-CPBA= 2.0 mmol; 4-methylmorpholine N-oxide = 5 mmol; R,R-Jacobsen = 0.05 mmol; dichloromethane = 4 mL; Reaction time = 30 min. [c] R-(+)-limonene = 1.0 mmol; KHSO5 = 2.0 mmol (from Oxone®); catalyst = 0.05 mmol; acetone = 4.0 mL; water = 4.0 mL; Reaction time = 30 min. [d] Selectivity to 1,2-limonene oxide; [e] Selectivity to R-(+)-limonene diepoxides (sum of the four diepoxides of R-(+)-limonene).

Table 3. R-(+)- limonene epoxidation using R,R-Jacobsen as catalyst . Effect of the oxidizing agent.

about 95% with *m*-CPBA. In the case of DMD, R-(+)-limonene diepoxides, were the major products. These differences in selectivities can be associated to the easy segregation of the catalyst. Presumably, this phenomenon creates active sites in the precipitated solid that promotes double epoxidation. It has been reported that the catalyst is not separable by means of physical-mechanical methods, when NaOCl and *m*-CPBA were used as oxidizing agents (Abdi et al., 2004). In these cases, a distillation process under vacuum (160 °C and 0.08 MPa) was required in order to separate the catalyst from reaction products, whereas with DMD the catalyst was isolated by physical-mechanical separation methods. On the other hand, figure 7 reveals that the catalyst exhibited the best stability to the oxidative degradation when *in situ* generated DMD was used as the oxidizing agent, since the main band associated to the Mn(III) salen complexes (1530 cm^{-1}) is still present in the FTIR spectrum, where DMD was used as oxidizing agent. Other important bands associated to the Mn-O (550 cm^{-1}) and Mn-N (480 cm^{-1}) stretching vibrations are slightly displaced towards a lower wavelength.

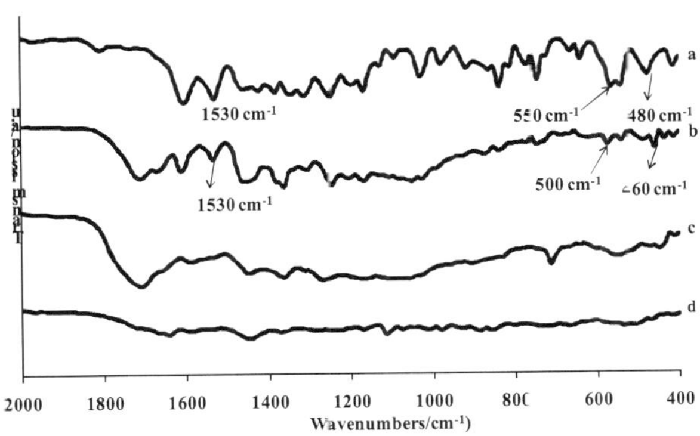

Fig. 7. FT-IR Spectra of the (R,R) Jacobsen's catalyst used with different oxidizing agents: fresh catalyst (a), DMD (b), NaOCl (c) and m-CPBA (d).

Finally, catalyst reuse was explored for both reactions. As can be observed in Tables 1 and 2, catalysts experienced a slight decrease in their initial catalytic activity through three consecutive runs in both reactions. The catalyst was segregated into a solid phase, while the reaction products remained in the liquid phase. This allowed the easy separation of catalyst and reaction products. On the other hand, Figure 8 shows the spectra of the catalyst used in three cycles. Clearly, it is observed that the main bands associated to the Mn(III) salen complexes are retained in the used catalysts. It indicates that the catalyst is very stable to the reaction conditions during three consecutive runs, whereas with oxidizing agents like NaOCl and *m*-CPBA the catalyst was deactivated in the first run (Figure 7). Therefore, the slight loss of catalytic activity with DMD is associated to physical loss of the catalyst during the isolation process rather than to oxidative degradation.

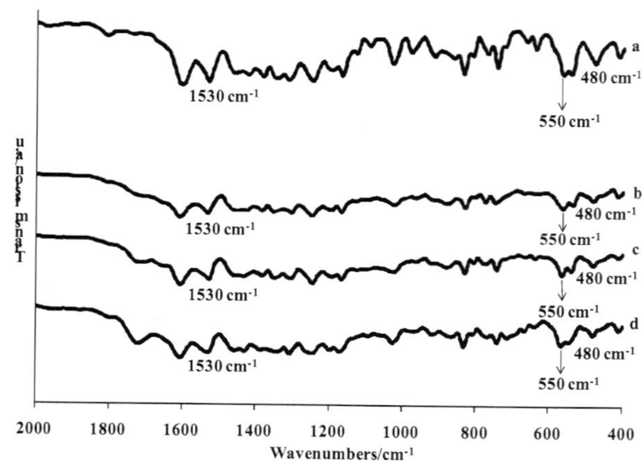

Fig. 8. FT-IR Spectra of the (R,R) Jacobsen's catalyst used in various run: fresh catalyst (a), second run (b) third run (c) fourth run.

4. Conclusion

Diepoxides were the main products from the oxidation of R-(+)-limonene, whereas a monoepoxide is obtained in the case of the catalytic oxidation of *cis*-ethyl cinnamate. In the latter case, a pure enantiomerically catalyst is required, while the Jacobsen racemic catalyst was sufficient in the case of the catalytic oxidation of R-(+)-limonene. Given that the catalytic activity of the three catalysts is very similar for R-(+)-limonene epoxidation and considering that the unique difference among the three catalysts is its stereogenic center located in the bond C1-C2 of the 1,2-diamino cyclohexane component, I conclude that the catalytic activity is not dependent on the stereogenic center. The catalyst could be recycled three times without appreciable loss of its initial activity.

5. References

Abdi, S. H. R.; Kureshy, R. I.; Khan, N. H. & Jasra, R. V (2004). Asymmetric epoxidation of non-functionalized alkenes using transition metal complexes. *Catalysis Surveys from Asia*, Vol.8, No.3, (September 2004), pp. 187-197, ISSN 1574-9266.

Adam, W.; Mock-Knoblauch, C.; Saha-Moller, C. R. & Herderich, M (2000). Are MnIV species involved in Mn(salen)-catalyzed Jacobsen–Katsuki epoxidations? a mechanistic elucidation of their formation and reaction modes by EPR spectroscopy, mass-spectral analysis, and product studies: chlorination versus oxygen transfer. *J. Am. Chem. Soc*, Vol.122, No.40, (September 2000), pp. 9685–9691, ISSN 0002-7863.

Bahramian, B.; Mirkhani, V.; Moghadam, M. & Amin, A. H (2006). Water-soluble manganese(III) salen complex as a mild and selective catalyst for oxidation of alcohols. *Appl. Catal. A: Gen*, Vol.315, (November 2006), pp. 52-57, ISSN 0926-960X.

C. Baleizaõ. & H. García (2006). Chiral salen complexes: an overview to recoverable and reusable homogeneous and heterogeneous catalysts. *Chem. Rev*, Vol.106, No.9, (September 2006), pp. 3987-4043, ISSN 0009-2665.

Canali, C. & Sherrington, D (1999). Utilisation of homogeneous and supported chiral metal(salen) complexes in asymmetric catalysis. *Chem. Soc. Rev*, Vol.28, No.2, (1999), pp. 85-93, ISSN 0306-0012.

Chaube, V. D.; Shylesh, S. & Singh, A. P (2005). Synthesis, characterization and catalytic activity of Mn(III)- and Co(II)-salen complexes immobilized mesoporous alumina. *J. Mol. Catal. A: Chem*, Vol.241, No.1-2, (November 2005). pp. 79-87, ISSN: 1381-1169.

Clapham, B.; Reger, T. S. & Janda K. D (2001). Polymer-supported Catalysis in Synthetic Organic Chemistry. Tetrahedron, Vol.57, No.22, (May 2001), pp. 4637-4662, ISSN 0040-4020.

Cubillos, J.; Montilla, I. & Montes de Correa, C (2009). Easy separation and reutilization of the Jacobsen's catalyst in olefin oxidation. *Appl. Catal. A: Gen*, Vol. 366, No.2, (September 2009). pp. 348–352, ISSN 0926-960X.

Cubillos, J.; Vargas, M.; Reyes, J.; Villa, A. & Montes de Correa, C (2010). Effect of the substrate and catalyst chirality on the diastereoselective epoxidation of limonene using Jacobsen-type catalysts. *Chirality*, Vol.22, No.4, (June 2009), pp. 403-410, ISSN 1520-636X.

Cubillos, J.; Vasquéz, S. & Montes de Correa, C (2010). Salen manganese (III) complexes as catalysts for R-(+)-limonene oxidation. *Appl. Catal. A: Gen*, Vol. 373, No. 1-2. (January 2010). pp. 57–65, ISSN 0926-960X.

Dasa, P.; Kuźniarska-Biernacka, I.; Silva, A. R.; Carvalho, A. P.; Pires, J. & Freire, C (2006). Encapsulation of chiral Mn(III) salen complexes into aluminium pillared clays: Application as heterogeneous catalysts in the epoxidation of styrene. *J. Mol. Catal. A: Chem*. Vol.248, No.1-2, (January 2006), pp. 135-143, ISSN 1381-1169.

Deng, L. & Jacobsen, E. N (1992). A practical, highly enantioselective synthesis of the Taxol side chain via asymmetric catalysis. *J. Org. Chem*, Vol. 57, No.15, (July 1992), pp. 4320–4323, ISSN 0022-3263.

Fan, Q. H.; Li, Y. M. & Chan A. S. C (2002). Recoverable catalysts for asymmetric organic synthesis. *Chem. Rev.*, Vol.102, No.10, (September 2002). pp. 3385-3466, ISSN 0009-2665.

Gladiali, S. (2007) Guidelines and Methodologies in Asymmetric Synthesis and Catalysis. *C. R. Chimie*, Vol.10, No.3, (March 2007), pp. 220-231, ISSN 1631-0748.

Kureshy, R. I.; Ahmad, I.; Khan, N. H.; Abdi, S. H. R.; Pathak, K. & Jasra R. V (2006). Chiral Mn(III) salen complexes covalently bonded on modified MCM-41 and SBA-15 as efficient catalysts for enantioselective epoxidation of nonfunctionalized alkenes. *J. Catal*, Vol.238, No.1, (January 2006), pp. 134-141, ISSN 0021-9517.

Kureshy, R. I.; Khan, N-u. H.; Abdi, S. H. R.; Patel, S. T. & Jasra, R. V (2001). Dimeric chiral Mn(III) Schiff base complex-catalysed enantioselective epoxidation of non-functionalised alkenes. *Tetrahedron Lett*, Vol.42, No.15. (April 2001). pp. 2915-2918, ISSN 0040-4039.

Larrow, J. F. & Jacobsen, E. N (1994). A Practical process for the large-scale preparation of (R,R)-N,N'-bis(3,5-di-tert-butylsalicylidene)-1,2-cyclohexanediaminomanganese (III) chloride, a highly enantioselective epoxidation catalyst. *J. Org. Chem*, Vol.59, No.7, (April 1994), pp. 1939–1940, ISSN 0022-3263.

Linde, C.; Anderlund, F. M. & Åkermark, B (2005). The effect of phenolates in the (salen)Mn-catalyzed epoxidation reactions. *Tetrahedron Lett*, Vol.46, No.33, (August 2005), pp. 5597-5600, ISSN 0040-4039.

Mahata, N.; Silva, A. R.; Pereira, M. F.; Freire, C.; de Castro, B. & Figueiredo, J. L (2007). Anchoring of a [Mn(salen)Cl] complex onto mesoporous carbon xerogels. *J Colloid Interface Sci*. Vol.311, No.1, (July 2007), pp. 152-158, ISSN 0021-9797.

Mastrorilli, P. & Nobile, C. F (2004). Supported catalysts from polymerizable transition metal complexes. *Coord. Chem. Rev*. Vol.248, No.3-4, (February 2004), pp. 377-395, ISSN 0010-8545.

Murzin, D. Y.; Mäki-Arvela, P.; Toukoniitty, E. & T. Salmi (2005). Asymmetric heterogeneous catalysis: science and engineering. *Catal. Rev., Sci. Eng*, Vol.47, No.2, (February 2007), pp. 175-256, ISSN 1520-5703.

Porter, M. J. & Skidmore, J (2000). Asymmetric epoxidation of electron-deficient olefins. *Chem. Commun*, No.14, pp. 1215-1225, ISSN 1359-7345.

Steiner, D.; Ivison, L.; Goralski, C. T.; Appell, R. B.; Gojkovic, J. R. & Singaram, B (2002). A facile and efficient method for the kinetic separation of commercially available *cis*- and *trans*-limonene epoxide. *Tetrahedron asymmetry*, Vol.13, No.21, (October 2002), pp. 2359-2363, ISSN: 0957-4166.

Wang, D.; Wang, M.; Zhang, R.; Wang, X.; Gao, A.; Ma, J. & Sun, L (2006). Asymmetric epoxidation of styrene and chromenes catalysed by dimeric chiral (pyrrolidine salen)Mn(III) complexes. *Appl. Catal. A. Gen*, Vol.315, (November 2006), pp. 120-127, ISSN 0926-960X.

Xia, Q. H.; Ge, H. Q. ; Ye, C. P.; Liu, Z. M. & Su, K. X (2005). Advances in homogeneous and heterogeneous catalytic asymmetric epoxidation. *Chem. Rev*, Vol.105, No.5, (2005), pp. 1603-1662, ISSN 0009-2665.

8

Carbohydrate-Based Surfactants: Structure-Activity Relationships

Hary Razafindralambo, Christophe Blecker and Michel Paquot
University of Liege
Belgium

1. Introduction

Carbohydrate-based surfactants (CBS) are, today, among the most important classes of amphiphilic compounds (Dembintsky, 2004; Queneau et al, 2008; Ruiz, 2009). Their structure is the result of the saccharide and lipid combination, naturally biosynthesized within living cells, or synthetically prepared by sequential reactions using carbohydrate and fatty materials, through one or several bonds. The growing interests of such compounds arise from many reasons of fundamental, practical, economical, and environmental orders (Razafindralambo et al, 2009, 2011a; Hill, K. & LeHen-Ferrenbach, 2009; Kitamoto et al, 2009). First, they can be easily prepared from the most abundant renewable vegetable raw materials (cellulose, pectin, hemicellulose, starch, etc.) in a wide range of structure and geometry by modular synthesis thanks to the presence of numerous reactive hydroxyl groups. Second, such a structural diversity makes them, on the one hand, as excellent models for getting insight into the surfactant mechanisms in modifying interfacial properties, which control the formation and the stability of colloidal systems such as micelles, vesicles, foams, emulsions, and suspensions (Razafindralambo et al, 2011b). One the other hand, numerous properties and functionalities would be expected from such a quasi-unlimited number of various compounds that can find specific applications in different industrial areas. Third, their compatibility to the environment, for instance, a higher biodegradability and lower toxicity, is an excellent criterion for their uses as alternatives to surfactants from petrochemical sources. Owing to the two former reasons, a systematic investigation of structure-activity relationships, which has rarely been carried out in the past, appears valuable for increasing knowledge on the impact of each CBS structural entity on their activities-functionalities, and ultimately, for achieving successfully a rational design in selecting and combining suitable compounds for further developments and applications. In the present chapter, we report the results of dynamic and equilibrium surface properties of homologous and analogous series of uronic acid derivatives, and evidence, consequently, the impact of different structural entities on their fundamental properties at the air-water interface.

2. Scope of the contribution

The aim of the present contribution is: (1) to review CBS in terms of structural classification based on their molecular size (mono-, oligo-, polymeric surfactants), geometry (standard, bipolar or bolaform, and gemini surfactants), and the nature of the polar headgroup (charged or not, cyclic or not), the apolar tail (number and length of alkyl chain), and the linker (amide, ester, ...) and/or the spacer; (2) to present systematically results on structure-activity relationships of uronic acid derivatives (UADs), a particular class of carbohydrate-based surfactants. These concern the impact of each structural entity including the polar headgroup (stereochemistry), apolar tail (chain length, number, and unsaturation), and linkage/spacer, on the performance of UADs to change surface properties, and possibly, to form and stabilize colloidal systems.

3. Classification of carbohydrate-based surfactants

Carbohydrate-based surfactants, currently known as glycolipids (Chester, 1997; Hato et al., 1999), are constituted by a saccharide unit (mono-, di-, oligo-, or polysaccharide) linked to a hydrophobic part of one, two, or multi-hydrocarbon chains by a single, or several bonds. These may be an ester, thioester, ether, amine, or/and amide group (Stubenrauch, 2001). Generally, glycolipids may be classified according to their amphiphilic structure, which depends on the polar headgroup, the apolar tail, but also the linkage between these main entities. Based on these three structural parameters, glycolipids are grouped in different categories, as presented in Table 1.

<table>
<tr><th>Glycolipids</th><th>Hydrophilic part</th><th>Hydrophobic part</th><th>Linkers</th></tr>
<tr><td>Monocatenary
Bicatenary
Multicatenary</td><td>One headgroup</td><td>One tail
Two tails
Multi tails</td><td rowspan="7">Ester, thioester, Ether, amine, amide</td></tr>
<tr><td>Glycoglycerolipids</td><td></td><td>Acylglycerol</td></tr>
<tr><td>Glycosphingolipids</td><td></td><td>Sphingoides</td></tr>
<tr><td>Bolaforms</td><td>Two headgroups</td><td>One spacer</td></tr>
<tr><td>Geminis</td><td></td><td>One spacer/two tails</td></tr>
<tr><td>Alkylpolyglucosides</td><td>Multi-headgroups</td><td>One or multi-tails</td></tr>
</table>

Table 1. Main classes of glycolipids, based on the amphiphilic structure and geometry

Besides the nature ionic, nonionic, and amphoteric of the polar headgroup, as for all surfactants, glycolipids can also be classified in small, medium, and large compounds, in relation to their molecular mass and size. According to these criteria, three main classes belonging to glycolipid-based monomeric, oligomeric, and polymeric surfactants may be distinguished. In addition to structural criteria, glycolipids may have two origins, those from microbial fermentation, crops, and animals, which are natural products, and those synthetized by chemical, enzymatic, and chemo-enzymatic routes, globally considered as synthetic compounds. In the following section, we try to illustrate by a few representative examples the most important classes of glycolipids with their chemical structure on the basis of their source, molecular mass, and geometry while considering the other classification aspects.

3.1 Natural glycolipids

3.1.1 Microbial biosurfactants

Bioglycolipids	Structure	Specificities (headgroup, tail, linkage)
	Low molecular mass	
• Rhamnolipids	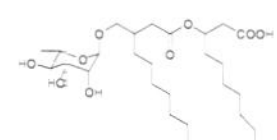	nonionic, mono/di, ester, (Dembitsky, 2005)
• Sophorolipids		nonionic, mono, ester, (Dembitsky, 2005)
• Trehalolipids		nonionic, di, ester, (Dembitsky, 2005)
• Mannosylerythriol lipid	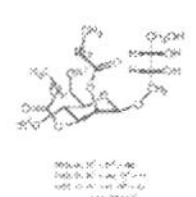	nonionic, mono, ester, (Kitamoto et al. 2009)
• Monogalactosyldiacylglycerol	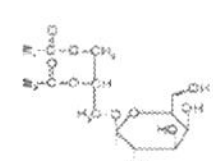	nonionic, di, ester, (Infante et al. 1997)
	High molecular mass	
• Cyclo-oligosaccharides		nonionic, mono, ester-amide, (Dembitsky, 2005)
• Emulsan (Polymeric biosurfactant)	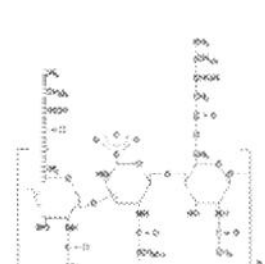	nonionic, multi, ester/amide, (Desai et Banat, 1997)

3.1.2 Plant biosurfactants

Bioglycolipids	Structure	Specificities (headgroup, tail, linkage)
• Butanolide glycosides		nonionic, multi, ester, (Dembitsky, 2005)
• Sucrose ester	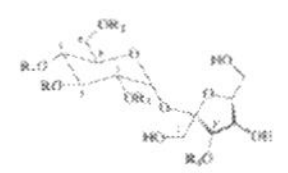	nonionic, multi, ester, (Dembitsky, 2005)

3.1.3 Animal biosurfactants

Bioglycolipids	Structure	Specificities (headgroup, tail, linkage)
• Cerebrosides	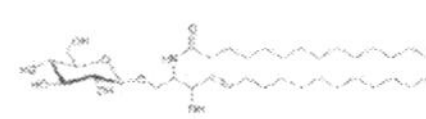	nonionic, di, amide, (Dembitsky, 2005)
• Gangliosides	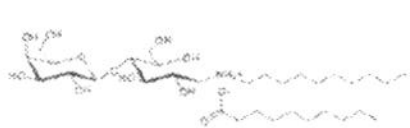	cationic, di, amine, (Dembitsky, 2005)

3.2 Synthetic glycolipids

Glycolipids	Structure	Specificities (headgroup, tail, linkage)
	Monomeric	
• Sorbitan monoester	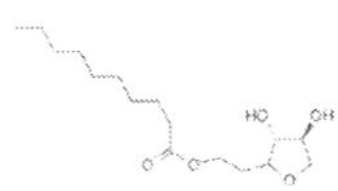	nonionic, mono, ester, (Hill &LeHen- Ferrenbach, 2009)
• Isosorbide derivative	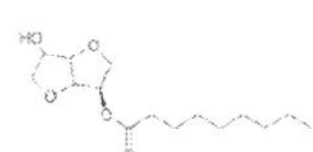	nonionic, mono, ester (Hill &LeHen- Ferrenbach, 2009)
• Uronic acid derivatives: *a. Monocatenary*	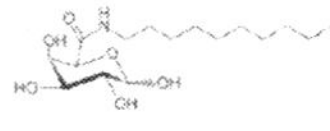	nonionic, mono, amide (Laurent et al. 2011)
b. Bicatenary		nonionic, di, ester-ether (Richel et al. 2010)
• Alkylglucoside		nonionic, mono, ether (Vulfson, 1990)

Glycolipids	Structure	Specificities (headgroup, tail, linkage)
Bolaform		
• α,ω-diamino-(oxa) [Spacer: -$(CH_2)_n$-]	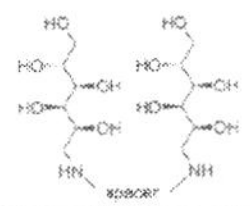	nonionic, bi, amide (Wagenaar&Engberts, 2007)
Gemini		
• Glu(n)-2-Glu(n)	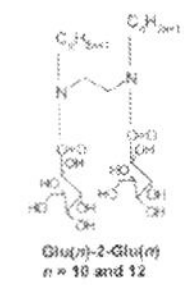	nonionic, bi, amides (Sakai, 2008)
Oligomeric/polymeric		
• Alkylpolyglucoside		nonionic, mono, ether (Queneau et al. 2008)
• Inulin carbamate	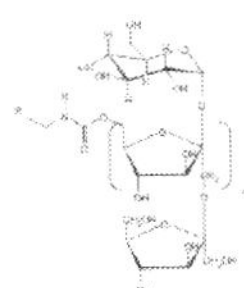	nonionic, mono, ester (Tadros, 2004)

4. Uronic acid derivatives

Uronic Acid-Derivative Surfactants (UADs) are a particular class of monomeric glycolipids (Fig.1). Their polar headgroup represents a considerable part of carbohydrate components, widely distributed in natural plant polysaccharides (Langguth & Benet, 1992). In addition, uronic acids are the result of the primary alcohol oxydation into a carboxylic group within a monosaccharide like glucose and galactose. Thus, they contain, in their structure, both hydroxyl and carboxylic groups that are highly reactive, explaining their potential as a basic unit for generating closely related surfactant compounds. The polar headgroup configurations according to the stereochemistry of OH groups and geometry (cyclic or not, bipolarity), the hydrophobic tail (number and length of alkyl chain), and the type of linker (ester, ether and amide, etc) are among the main variables in their structural entities. Therefore, they represent a set of ideal compounds for investigating structure-surface activity relationships.

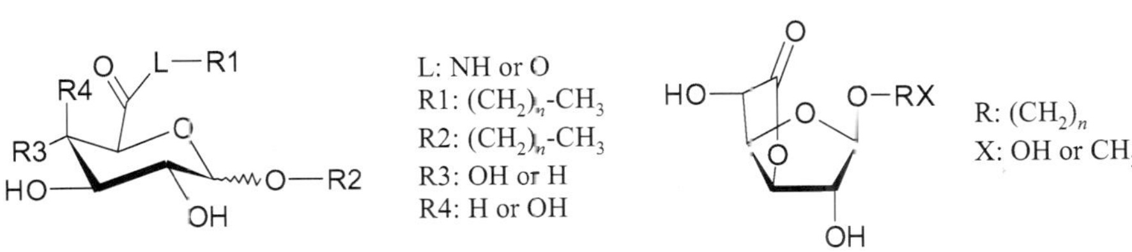

Fig. 1. General structure of uronic acid (left) and uronolactone (right) derivatives

Among several activities that CBS are able to develop (Ruiz, 2009), those related to the modification of surface and interface fundamental properties are undeniably the most important to evaluate and understand. Through several research works during last decades, it is established that fundamental properties such as dynamic and equilibrium surface tensions play a key role in the formation and stability of colloidal systems like foams, emulsions, and suspensions (Razafindralambo et al. 2011b). In this chapter, we are only focused on the relationships between chemical structure of UADs and their surface-activities at the air-water interface. UADs are a set of glucuronic and galacturonic acid-based surfactants varying in the polar headgroup configuration, including a cyclic (lactone bond) or non-cyclic structure and a α- or β- anomeric form, the hydrophobic tail chain length (C8 to C14) and number (mono- and bicatenary), the presence of a double bond, as well as an OH group at the terminal carbon, and in the type of linkage, ester, amide in C6, or ether in C1. These compounds have been synthesized by chemical or enzymatic routes, purified, and their chemical structure has been confirmed by means of various spectroscopic techniques (Nuclear Magnetic Resonance, Infrared, and Mass Spectrometry).

5. Surface-active properties

Surface-active properties of any surfactants can be measured through their capacity, either to reduce the surface tension of aqueous solution (γ_o) to any value (γ), or to increase the surface pressure ($\pi = \gamma_o - \gamma$) of the air-water interface. Measurements can be performed in both dynamic and static/quasi-static modes. Based on the solubility of the coumpound to characterize, one of the two following approaches must be used. The Gibbs'approach, measuring γ by adsorption, i.e. by migration of surfactant to the surface, is more convenient for short chain derivative surfactants (high solubility in water) whereas the Langmuir approach, determining π after spreading and compressing surfactant molecules at the surface, is rather appropriate for long chain derivative surfactants (low solubility in water). Surface activities of uronic acid derivatives may be evaluated in both dynamic and static/quasi-satic modes using a series of complementary techniques at room temperature comprised between 20-25°C (Razafindralambo et al., 1995; 2009; 2011).

5.1 Dynamic surface properties

In dynamic mode, surface tensions (γ_{dyn}) are measured using a drop volume tensiometer (TVT1, Lauda, Germany) by determining the critical volume (maximum pending size) of a series of drop created at different growing rates from 3 to 120 s (Razafindralambo et al., 2004).

5.2 Equilibrium surface properties

5.2.1 Gibbs approach

In static and quasi-static modes, the equilibrium surface tensions (γ_e) are measured by means of Wilhelmy plate (Tensimat N3, Prolabo, France), drop volume (TVT1, Lauda, Germany), or pending drop (Tracker, IT-Concept, France)-based tensiometers as detailed in previous papers (Razafindralamboet al., 1995; 2009). These methods consist of measuring continuously γ_e as a function of time, for surfactant solutions at different concentrations. The equilibrium surface tension (γ_e) can be determined as the fitting value of γ from representative experimental data, using appropriate mathematical models. All physical parameters such as critical micelle concentration (CMC), surface tension at the CMC (γ_{cmc}), surface excess (Γ_{max}), and minimum area per molecule(A_{min}), can be calculated using Gibbs equations for nonionic surfactants :

$$\Gamma_{max} = -\frac{1}{RT}\left(\frac{d\gamma}{d\ln C_o}\right) \tag{1}$$

and

$$A_{cmc} = \left(\frac{1}{\Gamma_{max} N}\right) \tag{2}$$

where C_o is the bulk solution concentration; R = 8.314 $J.K^{-1}.m^{-2}$, the gas constant; T, the temperature in Kelvin, and *N*, the Avogadro's number.

5.2.2 Langmuir approach

By Langmuir approach, a small volume of surfactant organic solution is spread drop wise onto the clean milli-Q water surface by means of Hamilton micro-syringe using a reproducibility adapter. After a given laps time, allowing the solvent evaporation, the surface pressure (π) of the monolayer is determined by measuring the horizontal force per unit of length, for a delimited area, during the compression. With an automatic Langmuir film Waage LFW2 3"5 (Lauda, Königshofen, Germany), a total trough area of 927 cm^2 is available. This is progressively reduced by moving a single barrier made in Teflon, and the force per length unit detected at a float gives the surface pressure of the monolayer at a constant temperature. For characterizing uronic acid derivatives, mainly bicatenary surfactants, the research of optimized experimental conditions is required (Razafindralambo et al. 2011).

6. Structure-activity relationships

6.1 Polar headgroup effect

6.1.1 Stereochemistry

Anomeric forms

The effect of the octyl glucuronolactone anomeric form is observed for the equilibrium surface properties (Table 2). The anomer α CMC is about the half of that of the anomer β, all other related parameters being quite similar. γ_{cmc} and A_{min} are quite similar, indicating that the anomeric form does not affect the molecular arrangement at the air-water interface, but changes that in the bulk water. In other words, the anomer α self-aggregates readily in water while adopting the same close-packed configuration at the air-water interface than the anomer β. Therefore, the anomer α appears more hydrophobic than the anomer β, which is attributed to the difference in the steric effect of their polar headgroup. This interacts stronger with water molecules for the anomer β, delaying its self-aggregation process. Concerning their dynamic surface properties, the anomer β appears slightly more efficient than the anomer α (data not shown), the difference being less important than that observed for equilibrium surface properties.

	Υ_{cmc} [mN/m]	CMC [mM]	A_{min} [Å^2/mol]
α-C_8- Glucuronolactone	32.8	1.3	43
β-C_8- Glucuronolactone	35.2	2.4	40

Table 2. Physical parameters of octyl glucuronolactone with α and β anomers

4-hydroxyl group (4-OH) axial or equatorial position effect

The second parameter related to the polar headgroup stereochemistry is the axial or equatorial position of the hydroxyl group of the fourth carbon within bicatenary derivatives of galacturonic and glucuronic acids. This stereochemistry effect impacts their configuration and behaviour at the air-water interface, as observed with the (π-A) isotherms (Fig.2). The glucuronic acid derivative (GlcA-$C_{14/14}$) is more expanded when it is spread at the air-water around 20°C, and shows a transition phase under compression, which does not occur for the galacturonic acid derivative (GalA-$C_{14/14}$). This difference is in agreement with film morphologies and thicknesses, and is also supported by molecular models (Razafindralambo et al., 2011). In contrast, the 4-OH axial or equatorial position has no effect when the molecules are vertically oriented, i.e. within a film at a condensed state. Such results have been attributed to the configuration of two alkyl chains, which is in "open" or "close" structure according to the film is in expanded or in condensed state.

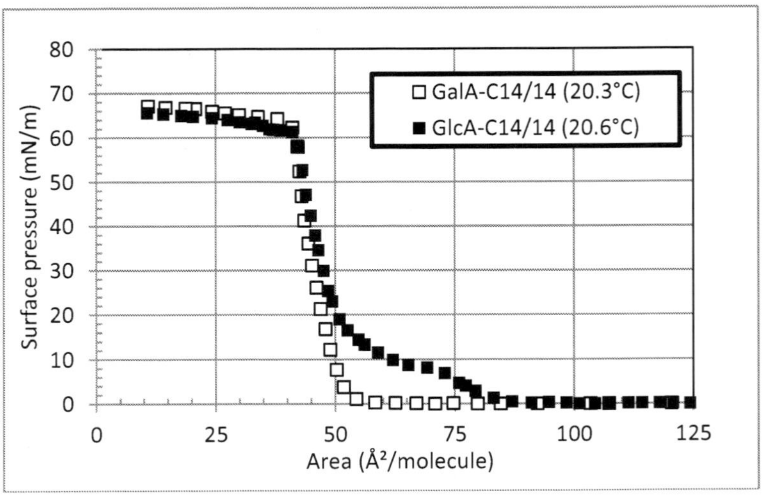

Fig. 2. (π-A) isotherms of GalA-$C_{14/14}$ and GlcA-$C_{14/14}$ spread at the air-water interface 20°C

6.1.2 Impact of the bipolarity: (OH group ending the alkyl chain)

Adding an OH group at the end of the alkyl chain changes the geometry of uronic acid derivatives, which becomes a bipolar asymmetric. Such a modification impacts some dynamic (Fig.3) and equilibrium surface properties (Table 3) of glucuronic acid derivatives.

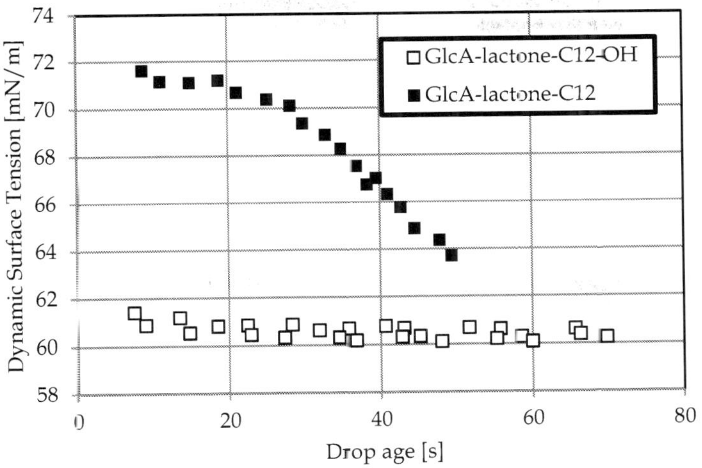

Fig. 3. Dynamic surface tension vs. time of dodecyl glucuronolactones (0.2 g/L), with and without a hydroxyl group at the end, measured by TVT1 at 20°C

	γ_{cmc} [mN/m]	CMC [mM]	A_{min} [Å^2/mol]
OH-C_{12}-Glucuronolactone	43.9	0.7	65
C_{12}-Glucuronolactone	47.2	0.6	34

Table 3. Physical parameters of dodecyl glucuronolactones ending with, and without hydroxyl group

It increases the area occupied per molecule (cross-sectional area), and the performance in the adsorption time by reducing faster the dynamic surface tension. This difference is easy to understand, and may be explained by the fact that the two polar headgroups are directed into the aqueous phase, leaving the hydrophobic spacer in the gaseous phase, and forming, as a consequence, a "convex" configuration. In contrast, it has no significant effect on the equilibrium (long time) surface properties, based on their CMC and γ_{cmc} values.

6.2 Hydrophobic effect

6.2.1 Alkyl chain length

The impact of the alkyl chain length on all of interfacial properties can be evaluated with a homologous series of monocatenary and bicatenary uronic acid derivatives. As for all monomeric surfactants, whatever their nature, the general trend is respected with uronic acid derivatives, that is, the longer the alkyl chain length, the lower the CMC of the uronate and uronamide derivatives (Blecker et al., 2002; Laurent et al., 2011). About dynamic surface properties, an optimum chain length of eleven carbon atom gives the best performance, regarding the adsorption time and the maximum reduction of the surface tension. Concerning bicatenary derivatives, the comparison of GalA-$C_{10/10}$ and GalA-$C_{14/14}$ monolayer properties shows that the shorter the two alkyl chains ($C_{10/10}$), the more expanded and compressible the spread molecules, and the less stable the molecular film (Fig.4). Such behaviour of the shorter chain derivatives is attributed to stronger interactions with the subphase, which is comparable with the case of the dimyristoyl-phosphatidylcholine (DMPC), an ionic phospholipid having the same alkyl chains than that of GalA-$C_{14/14}$ (data not shown).

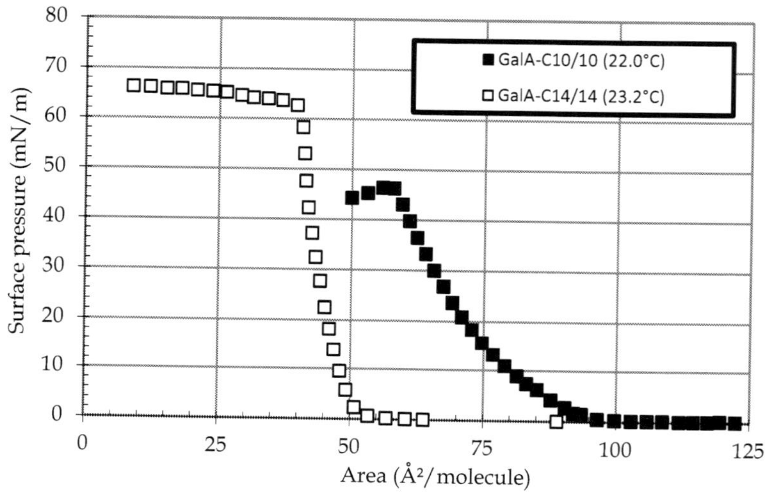

Fig. 4. (π-A) isotherms of GalA-$C_{10/10}$ and GlcA-$C_{14/14}$ spread at the air-water interface

6.2.2 Alkyl chain number (Monocatenary vs. bicatenary)

The alkyl chain number of uronic acid derivatives also appears crucial on their surface properties characterized by the Langmuir approach. When a second alkyl chain is added to the polar headgroup, that is, the compound becomes bicatenary, the mechanical properties of the film at the air-water interface are improved, regarding its collapse pressure (Fig.5). It is to be noted, from this figure, that the inclusion of a cyclic ester bond into the polar headgroup affects slightly the behaviour of the monolayer.

6.2.3 Unsaturation

By including a double bond at the end of the alkyl chain, no significant effect has been observed on the equilibrium surface properties, the γ_{cmc}, CMC, and A_{min} values being similar and equal to 34.2 ± 1.1 mN/m, 0.12 ± 0.01 g/L, and 39.0 ± 2.0 Å²/mol., respectively. In contrast, it impacts extensively the dynamic surface properties of the undecanoyl glucuronate (Fig.6). The derivative with an unsaturated chain migrates faster at the air-water interface, and reduces readily the dynamic surface tension, compared to the saturated one. It is attributed to the "shortening effect" of the double bond inclusion (Milkereit et al. 2005), reducing the hydrophobicity of the surfactant, and therefore, its adsorption time.

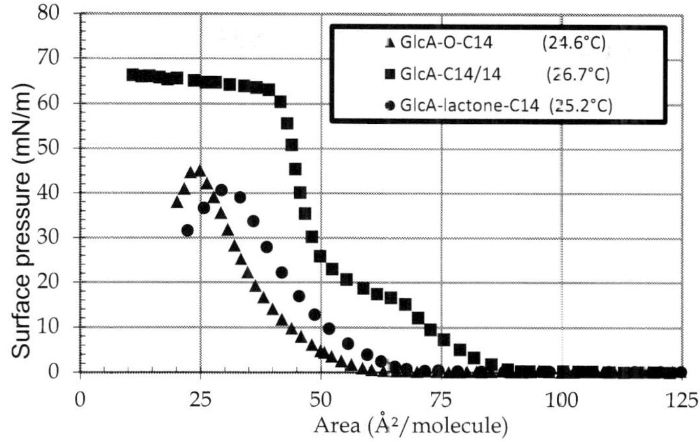

Fig. 5. (π-A) isotherms of GlcA-O-C_{14}, GlcA-lactone-C_{14}, and GlcA-$C_{14/14}$ spread at the air-water interface 25°C

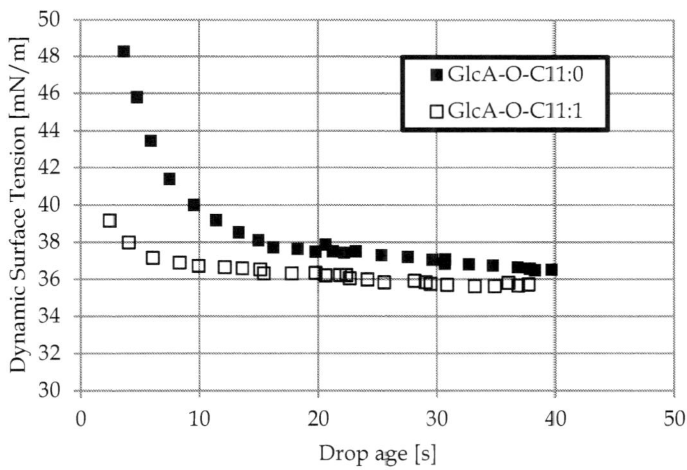

Fig. 6. Dynamic surface tension vs. time of undecanyl (C11:0) and undecenyl (C11:1)-glucuronolactones (0.1 g/L) measured by TVT1 at 20°C.

6.3 Linkage effect

6.3.1 Ester bond direction

One of the spectacular effects of the linkage between the polar headgroup and the hydrophobic tail is the change of the ester bond orientation within the octyl sugar derivatives. By comparing the dynamic and equilibrium surface properties of glucose octanoate (C8-**CO**-O-Glc) and octyl glucuronate (C8-O-**CO**-GlcA), significant differences are observed (Table 4). When the carbonyl group is closed to the polar headgroup constituted by glucuronic acid (C8-O-**CO**-GlcA), the surfactant is more hydrophobic, and the polar head group is larger. The consequence is the increase of dynamic properties, and the decrease of the CMC (10.7 mM vs. 19.1 mM). When the carbonyl group is close to the alkyl chain (C8-**CO**-O-Glc), there is an "interruption" of the hydrophobic tail, which reduces its hydrophobicity (Razafindralambo et al. 2009). The molecular area becomes smaller and the CMC is higher. On the other hand, the adsorption time rate characterized by t* and v_{max} are increased, reducing its performance on dynamic surface properties.

Parameters	Glucose octanoate [Glc-O-C_8]	Octylglucuronate [GlcA-O-C_8]
Dynamics (1) :		
t* (10^{-3} s)	36.1	3.4
n	0.14	0.56
v_{max}(mN/m/s)	0.024	0.0081
Equilibrium :		
Υ_{cmc} (mN/m)	25.1	28.0
CMC (mM)	19.1	10.7
A_{min} (Å^2/mol)	37	44

Table 4. Physical parameters of glucose octanoate and octyl glucuronate; (1) Compared at 1.63 mM (25°C).

6.3.2 Ester vs. amide bond

By comparing the surface properties of three derivatives with an octyl chain, the impact of the type of the linkage bond can be evidenced, despite the difference in the ratio alpha/beta of anomeric forms (Table 5). It appears that the uronamide is the most effective in terms of equilibrium surface tensions. This result may be attributed to the difference in the molecular area occupied by the two octyl derivatives. The amide bond has a smaller bulk size than the ester ones. In comparison with a unique form of octyl glucoside, the glucuronamide is more effective. Even though the linkage is not at the same position, on the C1 for glucoside, and on the C6 for glucuronamide, the impact may be significant, considering the similarity in the molecular mass, and the number of the OH group available.

Physical parameters	C8-glucuronate [C6-Ester linkage] α/ β=1.5	C8-glucuronamide [C6-Amide linkage] α/ β=0.6	C8-glucoside (25°C) [C1-Ether linkage] α ; β [Boyd et al, 2000]
Υ_{cmc} (mN/m)	28.0	24.0	35.2 ; 31.3
CMC (mM)	10.7	3.3	12.0; 18.2
A_{min} (Å^2/mol)	44 ; 45(1)	25 ; 35(1)	42 ; 42(2)

Table 5. Physical parameters of glucuronic acid and glucose derivatives containing different linkages at the air-water interface. (1) By modeling; (2) by linear fit

7. Conclusion

In the present chapter, we contribute to a better understanding of the structure-surface activity relationships of uronic acid derivatives, a promising class of carbohydrate-based surfactants. Each structural element impacts either their dynamic performances, measured over a short period range, or their equilibrium activities, evaluated after a longer period. Besides scientific interests of such fundamental information, the approach also leads to the identification of some suitable structures for practical performances in forming and stabilizing colloidal systems like foams, emulsions, and suspensions, which are encountered virtually in all soft manufactured products.

8. Acknowledgments

This work carried out in the framework of the "TECHNOSE" excellence Program was supported by Belgian Walloon Region. The authors thank Drs Aurore Richel, Gaëtan Richard, Pascal Laurent for the syntheses and characterization of uronic acid derivatives, and Mr. Alexandre Schandeler for technical assistance.

9. References

Blecker, C.; Piccicuto, S.; Lognay, G.; Deroanne, C.; Marlier, M. & Paquot, M. (2002). Enzymatically prepared *n*-alkyl esters of glucuronic acid: the effect of hydrophobic chain length on surface properties.*Journal of colloid and interface science*, 247, 424-428.

Boyd, B. J.; Drummond, C.; Krodkiewska, I. & Grieser, F. (2000). How Chain Length, Headgroup Polymerization, and Anomeric Configuration Govern the Thermotropic and Lyotropic Liquid Crystalline Phase Behavior and the Air-Water Interfacial Adsorption of Glucose-Based Surfactants. *Langmuir, 16*, 7359-7367

Chester, A. (1997). Nomenclature of glycolipides. *Pure and Applied Chemistry*, 69, 2475-2487.

Dembitsky, V. (2005). Astonishing diversity of natural surfactants: 3. Carotenoid glycosides and isoprenoid glycolipids. *Lipids*, 40, 535-557.

Dembitsky, V. (2004). Astonishing diversity of natural surfactants: 1. Glycosides of fatty acids and alcohols. *Lipids*, 39, 933-953.

Dembitsky, V. (2005). Astonishing diversity of natural surfactants: 4. Fatty acid amide glycosides, their analogs and derivatives. *Lipids*, 40, 641-660.

Desai, J.& Banat, I. (1997). Microbial production of surfactants and their commercial potential. *Microbiology and Molecular Biology Reviews*, 61, 47-64.

Hato, M.; Minamikawa, H. ; Tamada, K. ; Baba, T. & Tanabe, Y. (1999). Self-assembly of synthetic glycolipid/water systems. *Advances in Colloid and Interface Science*, 80, 233-270.

Hill, K. & LeHen-Ferrenbach, C. (2009). Sugar-Based Surfactants for Consumer Products and Technical Applications, In: *Sugar-based surfactants fundamentals and Applications*, C.C. Ruiz (Ed.), 1-20, CRC Press, ISBN 978-1-4200-5166-7.

Infante, M. ; Pinazo, A.& Seguer, J. (1997). Non-conventional surfactants from amino acids and glycolipids: Structure, preparation and properties. *Colloids and Surfaces A: Physicochemical and Engineering Aspects*, 123-124, 49-70.

Kitamoto, D.; Morita, T. ; Fukuoka, T. ; Konishi, M. & Imura, T. (2009). Self-assembling properties of glycolipid biosurfactants and their potential applications. *Current Opinion in Colloid & Interface Science*14, 315-328.

Laurent, P.; Razafindralambo, H. ; Wathelet, B. ; Blecker, C.; Wathelet, J. & Paquot, M. (2011). Synthesis and Surface-Active Properties of Uronic Amide Derivatives, Surfactants from Renewable Organic Raw Materials.*Journal of Surfactants and Detergents,* 14, 51-63.

Milkereit, G.; Garamus,V.; Veermans, K.; Willumeit, R. &Vill,V. (2005). Structures of micelles formed by synthetic alkyl glycosideswith unsaturated alkyl chains. *Journal of Colloid and Interface Science,* 284, 704–713

Queneau, Y.; Chambert, S. ; Besset, C. & Cheaib, R. (2008). Recent progress in the synthesis of carbohydrate-based amphiphilic materials: the examples of sucrose and isomaltulose. *Carbohydrate Research,* 343, 1999-2009.

Razafindralambo, H.; Richel, A.; Wathelet, B.; Blecker, C.; Wathelet, J.; Brasseur, R.; Lins, L.; Miñones, J. & Paquot, M. (2011a). Monolayer properties of uronic acid bicatenary derivatives at the air-water interface: effect of hydroxyl group stereochemistry evidenced by experimental and computational approaches. *Physical chemistry chemical physics: PCCP,* 13, 15291–15298.

Razafindralambo,H.; Blecker, C.& Paquot, M.(2011b). Screening of Basic Properties of Amphiphilic Molecular Structures for Colloidal System Formation and Stability: The Case of Carbohydrate-Based Surfactants. In *Amphiphiles: Molecular Assembly and Applications,* Nagarajan, R., Ed. ACS: Washigton; Vol. 1070, p53-66.

Razafindralambo, H.; Blecker, C.; Mezdour, S.; Deroanne, C.; Crowet, J.; Brasseur, R.; Lins, L.; Paquot, M. (2009). Impacts of the Carbonyl Group Location of Ester Bond on Interfacial Properties of Sugar-Based Surfactants: Experimental and Computational Evidences. *The journal of physical chemistry. B,* 113, 8872-8877.

Razafindralambo, H.; Thonart, P. & Paquot, M. (2004). Dynamic and equilibrium surface tensions of surfactin aqueous solutions.*Journal of Surfactants and Detergents, 7* (1), 41-46.

Razafindralambo, H.; Blecker, C.; Delhaye, S.& Paquot, M. (1995). Application of the Quasi-Static Mode of the Drop Volume Technique to the Determination of Fundamental Surfactant Properties. *Journal of colloid and interface science,* 174, 373-377.

Richel, A.; Laurent, P.; Wathelet, B.; Wathelet, J. & Paquot, M. (2010). Microwave-assisted synthesis of D-glucuronic acid derivatives using cost-effective solid acid catalysts. *Tetrahedron letters,* 51, 1356-1360.

Ruiz, C. (2009). Sugar-Based Surfactants Fundamentals and Applications. CRC Press. ISBN978-1-4200-5166-7.

Sakai, K.; Tamura, M.; Umezawa, S.; Takamatsu, Y.; Torigoe, K.; Yoshimura, T.; Esumi, K.; Sakai, H. & Abe, M. (2008). Adsorption characteristics of sugar-based monomeric and gemini surfactants at the silica/aqueous solution interface. *Colloids and Surfaces A: Physicochemical and Engineering Aspects,* 328, 100-106.

Stubenrauch, C. (2001). Sugar surfactants - Aggregation, interfacial, and adsorption phenomena. *Current Opinion in Colloid and Interface Science,* 6, 160-170.

Tadros, T.; Vandamme, A.; Levecke, B.; Booten, K. & Stevens, C. (2004). Stabilization of emulsions using polymeric surfactants based on inulin. *Advances in Colloid and Interface Science,* 108-109, 207-226.

Vulfson, E.; Patel, R. & Law, B. (1990). Alkyl-α-glucoside synthesis in a water-organic two-phase system. *Biotechnology Letters,* 12, 397-402.

Wagenaar, A.& Engberts, J. (2007). Synthesis of nonionic reduced-sugar based bola amphiphiles and gemini surfactants with an α,ω-diamino-(oxa) alkyl spacer. *Tetrahedron,* 63, 10622-10629.

9

CO_2 Biomitigation and Biofuel Production Using Microalgae: Photobioreactors Developments and Future Directions

Hussein Znad, Gita Naderi, H.M. Ang and M.O. Tade
Curtin University
Australia

1. Introduction

Increased concerns about global warming and greenhouse gas emissions as well as the exhaustion of easily accessible fossil fuel resources are calling for effective carbon dioxide (CO_2) mitigation technologies and clean and renewable energy sources. One of the major gases leading to global warming is carbon dioxide. CO_2 makes up 68% of the estimated total greenhouse gas emissions (Harrington & Foster, 1999).

There have been several approaches proposed for managing the levels of CO_2 emitted into the atmosphere, including ocean sequestration such as deep ocean injection or increasing the amount of CO_2 dissolved in the ocean. Another proposed form of sequestration is to sequester the CO_2 into terrestrial ecosystems (Halmann, 1993). The short term options of sequestration by direct injection into geologic or oceanic sinks are recognized as methods to reduce the CO_2 levels but do not address issues of sustainability (Stewart & Hessami, 2005). Carbon sequestration can also be accomplished through chemical approaches; some problems with these approaches are that they must be safe for the environment, stable for long- term storage, and cost - competitive to other sequestration options. Other technologies have been considered, such as chemical absorption, membrane separation, cryogenic fractionation and adsorption using molecular sieves, but they are even less energy efficient as to be considered economically viable (Stewart & Hessami, 2005).

One of the most understudied methods for CO_2 mitigation is the use of biological processes (via microalgae) in a direct CO_2 to biomass conversion from point source emissions of CO_2 in engineered systems such as photobioreactors. Microalgal biofixation of carbon dioxide (CO_2) in photobioreactors has recently gained renewed interest as a promising strategy for CO_2 mitigation. The use of photobioreactors for microalgal CO_2 sequestration offers the principal advantages of increased microalgae productivity, owing to controlled environmental conditions, and optimized space/volume utilization and, thus, more efficient use of costly land. In fact, the photosynthetic solution when scaled up would present a far superior and sustainable solution under both environmental and economic considerations.

Fig. 1 shows the importance of the microalgae photobioreactor and its general applications, microalgae used to capture waste CO_2 utilizing the nutrients in wastewater and natural

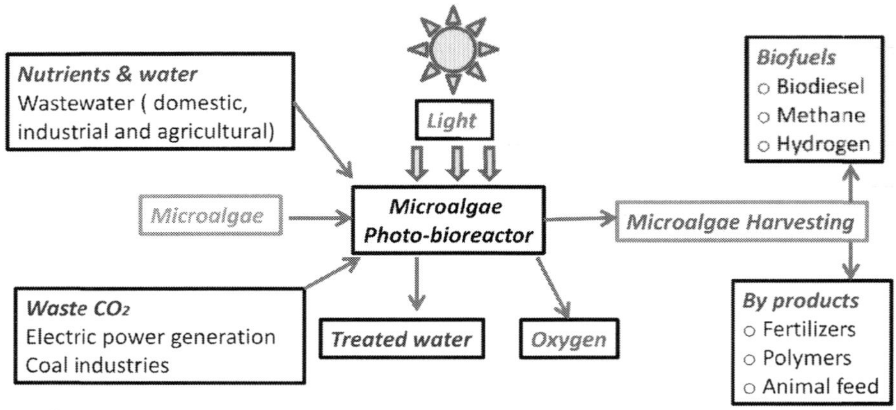

Fig. 1. Microalgae photobioreactors applications

solar light. The microalgae biomass produced can be used for biofuel production (such as biodiesel and methane) and other by products (such as animal feeds and polymers).

Broadly, photo-bioreactors can be classified as open (pond) systems or closed systems. Considering all the limitations and shortcomings of the pond systems, most researchers, had oriented their research works towards the development of an unconventional way for microalgae culture, which should be fully closed and compact with high surface-to-volume ratio and all the growth factors be optimized.

With these desired characteristics as the main goals, research on tubular and airlift photo-bioreactors were the right orientation and some forms of designs had in certain aspects succeeded when used in the lab scale. However, few of these forms could be really applicable in the pilot production scale, due to serious obstacles of operational problems and growth limitations. Amongst them, were primarily the oxygen build-up in the growth medium, photoinhibition, light saturation effect and the overheating inside the tube walls by the intensive solar radiation when operating in summer seasons especially in the midday light hours. Besides, the poor circulation of the growth medium causes the algal staining on the inner walls of the tubes, gave eventually an uneconomic results. Over the years, several solutions have been proposed to overcome these fundamental limitations to productivity. However, these systems are complicated to scale up and may be suitable for small-scale cultivation. Moreover, there is little knowledge about the feasibility of photobioreactors scale-up and developments.

The developmental state of the photobioreactor technology for CO_2 mitigation and biofuel production will be reviewed in this chapter, focusing on several essential issues, such as effective and efficient provision of light; supply of carbon dioxide while minimizing losses; removal of photo-synthetically generated oxygen that may inhibit metabolism or otherwise damage the culture if allowed to accumulate; sensible scalability of the photobioreactor technology; harvesting the microalgae biomass and biofuel production. The theoretical background of microalgae cultivation will be summarized in this chapter as well. The chapter will present possibly new insights that could be gained in the future for the potential commercial exploitation of microalgae for CO_2 biomitigation and biofuel production.

2. Mechanism of the photosynthesis and biophotolysis

Photoautotrophic microorganisms like eukaryotic green microalgae, possess chlorophyll and other pigments to capture sunlight energy and use photosynthetic systems (PSII and PSI) to carry out plant-like oxygenic photosynthesis (Kruse et al. 2005). The pigments in PSII (P680) absorb the photons with a wavelength shorter than 680 nm, generating a strong oxidant capable of splitting water into protons (H^+), electrons (e^-) and O_2 as shown in Fig. 2.

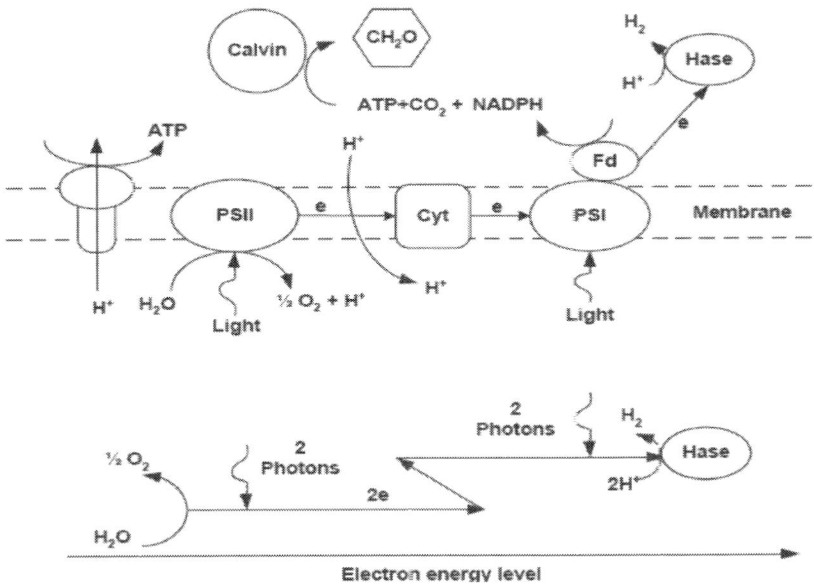

Fig. 2. Schematic mechanisms of photosynthesis and biophotolysis of photoautotrophic microbes (Amos, 2004).

The electrons or reducing equivalents are transferred through a series of electron carriers and cytochrome complex to PSI. The pigments in PSI (P700) absorb the photons with a wavelength under 700 nm, which further raises the energy level of the electrons to reduce the oxidized ferredoxin (Fd) and/or nicotinamide adenine dinucleotide phosphate (NADP+) into their reduced forms. The proton gradient formed across the cellular (or thylakoid) membrane drives adenosine triphosphate (ATP) production via ATP synthase. CO_2 is reduced with ATP and NADPH via a reductive pentose phosphate pathway or Calvin cycle for cell growth.

The excess reduced carbon is stored inside the cells as carbohydrates (CH_2O) and/or lipids. The type of carbohydrate product produced depends on the type of strain being used. The reducing power (Fd) could also be directed to hydrogenase (Hase) for hydrogen evolution.

3. Microalgae and microalgae cultivation

Algae are a large and diverse group of simple, typically autotrophic organisms, ranging from unicellular to multicultural forms. The largest and most complex marine forms are called seaweeds. They are photosynthetic, like plants, and simple because they lack the many distinct organs found in land plants. Some unicellular species rely entirely on external energy sources and have limited or no photosynthetic apparatus. All algae have photosynthetic machinery ultimately derived from the cyanobacteria, and so produce oxygen as a byproduct of photosynthesis.

Strain	Protein	Carbohydrates	Lipids	Nucleic acid
Scenedesmus obliquus	50-56	10-17	12-14	3-6
Scenedesmus quadricauda	47	-	1.9	-
Scenedesmus dimorphus	8-18	21-52	16-40	-
Chlamydomonas rheinhardii	48	17	21	-
Chlorella vulgaris	51-58	12-17	14-22	4-5
Chlorella pyrenoidosa	57	26	2	-
Spirogyra sp.	6-20	33-64	11-21	-
Dunaliella bioculata	49	4	8	-
Dunaliella salina	57	32	6	-
Euglena gracilis	39-61	14-18	14-20	-
Prymnesium parvum	28-45	25-33	22-38	1-2
Tetraselmis maculata	52	15	3	-
Porphyridium cruentum	28-39	40-57	9-14	-
Spirulina platensis	46-63	8-14	4--9	2-5
Spirulina maxima	60-71	13-16	6-7	3-4.5
Synechoccus sp.	63	15	11	5
Anabaena cylindrica	43 - 56	25-30	4-7	-

Table 1. Chemical composition of algae expressed on a dry matter basis (%) (Becker,1994)

Algae can be classified into two types based on their sizes, microalgae and macroalgae. Microalgae are microscopic photosynthetic organisms (less than 2 mm in diameter). However, macroalgae, these organisms that are found in both marine and freshwater environments. Biologists have categorized microalgae in a variety of classes, mainly distinguished by their pigmentation, life cycle and basic cellular structure (Amos, 2004). The most frequently cited microalgae as carrying one or more of the desirable features for efficient and economical combination of CO_2 biofixation, wastewater treatment and lipid synthesis toward biofuel production are:

1. The diatoms (Bacillariophyceae). These algae dominate the phytoplankton of the oceans, but are also found in fresh and brackish water. Approximately 100,000 species are known to exist. Diatoms contain polymerized silica (Si) in their cell walls. All cells store carbon in a variety of forms. Diatoms store carbon in the form of natural oils or as a polymer of carbohydrates known as chyrsolaminarin.
2. The green algae (Chlorophyceae). These are also quite abundant, especially in freshwater. They can occur as single cells or as colonies. Green algae are the evolutionary progenitors of modern plants. The main storage compound for green algae is starch, though oils can be produced under certain conditions.
3. The blue-green algae (Cyanophyceae). Much closer to bacteria in structure and organization, these algae play an important role in fixing nitrogen from the atmosphere. There are approximately 2,000 known species found in a variety of habitats.
4. The golden algae (Chrysophyceae). This group of algae is similar to the diatoms in pigmentation and biochemical composition. They have more complex pigment systems, and can appear yellow, brown or orange in color. Approximately 1,000 species are known to exist, primarily in freshwater systems. The golden algae produce natural oils and carbohydrates as storage compounds.

All algae primary comprise of the following, in varying proportions (Table 1): proteins, carbohydrates, fats and nucleic acids. While the percentages vary with the type of algae, there are algae types that are comprised of up to 40% of their overall mass by fatty acids that could be extracted and converted into biodiesel.

4. Photo-bioreactor for carbon dioxide sequestration

Photobioreactors for microalgae cultivation can be classified as open systems or closed systems. Open systems are ponds, constructed on the large open areas, in rows with growth medium exposed to environment and sunlight. Closed systems are those where growth medium enclosed from the environment. Open systems have many disadvantages over closed system, for instance they are hard to control and, contamination from external environment is high and could cause the microalgae mutate (Camacho Rubio et al., 1999). Closed systems are easy to monitor, less chances of contamination, better mass transfer (varies based on the type of bioreactor), occupy less space for the same algal growth. Closed systems can be classified as tubular photobioreactors; stirred photobioreactors; flat plate photobioreactors; hollow fiber membrane photobioreactors; airlift and sparged bubble column photobioreactors. Unfortunately, none of the these bioreactor configurations is able to control effectively all process parameters that are required for maximum CO_2 biofixation, microalgal growth and metabolic rates, particularly at large scale production. Below is a brief description for the most widely used photobioreactors for CO_2 biofixation and biofuel production:

4.1 Tubular photo-bioreactors

Tubular photo-bioreactors consist of long thin tubes arranged in different geometrical patterns (helical, straight tubes) to optimize irradiance from a point light source (sun). Generally liquid growth medium is circulated in these tubes by air bubbling and by injection of air into one end of the system and degassed at the other end. Construction, light regime, mass transfer and scale up issues of these photo-bioreactors have been discussed (Molina Grima, 2000). Experiments have showed that a large-scale tubular photo-bioreactor has failed and the main reason attributed for its failure was the large dissolved oxygen in the system (Molina Grima, 2000). Hence a system should not at any stage be over saturated with oxygen as this would cause algae to shutdown photosynthesis and growth (Camacho Rubio et al., 1999). It was also reported that tubular photo-bioreactors are difficult to build and maintain, and have limited scalability.

4.2 Mechanically stirred photo-bioreactors

Mechanically stirred photo-bioreactors use baffles to stir the growth medium to attain a mass transfer of air/CO_2 into liquid. A drawback of the stirred medium is if stirred vigorously the algae cell wall would be damaged by the high fluid shear forces (Molina Grima et al., 1996). If it is stirred slowly, eddy currents will not be established that move the algae toward the light source thereby decreasing the efficiency of light available for the photosynthetic process and also reducing mass transfer of nutrients from the air/CO_2 to the liquid in the systems.

4.3 Airlift photo-bioreactors

Airlift photo-bioreactors are basically a column divided into two parts, air/CO_2 is bubble through only one side of the partition which causes a liquid current pattern to develop with the air bubble side called the riser and other part called the downcomer (Sánchez Mirón et al., 2000). These bioreactors are extensively investigated for fermentation process and wastewater treatment (Znad et al. 2004, Znad et al. 2006) but have not been looked at as a replacement for the popular tubular photo-bioreactors until recent times (Sánchez Mirón et al., 2000). The airlift photo-bioreactor characterized by; high mass transfer, good mixing with low shear stress, low energy consumption, high potentials for scalability, easy to sterilize, readily tempered, good for immobilization of algae, reduced photo-inhibition. However, the main limitations are; the small illumination surface area and decrease of illumination surface area upon scale-up. It has become clear that biological carbon sequestration and hydrogen production technologies have been poorly studied in the airlift photo-bioreactors and are in their infancy of development.

4.4. Bubble column photo-bioreactors

Bubble column bioreactors are vertical columns either cylindrical or rectangular filled with growth medium and air is bubble through a sparged system installed at the bottom. These systems have the highest gas hold ups rates which means they have the best mass transfer compared to other systems (Miron et al. 2000; Kommareddy & Anderson , 2003). A modified version of these bubble column bioreactors is porous membrane reactors, which have efficient aeration, give smaller bubbles, and pressure drop across the membrane is low compared with other rigid sparged bubble column reactors. These characteristics are achievable at high gas flow rates, with low energy costs (Poulsen & Iversen, 1997).

5. Factors affecting the photobioreactor performance

The key parameters that affect the photobioreactor performance, i.e., the growth of the microalgae in the photo-bioreactor, are the effective and efficient provision of light, carbon dioxide level, photo-synthetically generated oxygen, Gas transfer, mixing rates, temperature, pH and nutrient requirements.

5.1 Light provision

Light is the basic energy source for phototrophic microorganisms. The intensity and utilization efficiency of the light supplied are thus of crucial importance in microalgal bioreactors. Light intensity decreases deeper within the culture medium, especially in high-density cultures; hence, the issue of optical depth, which measures the proportion of radiation absorbed or scattered along a path through a partially transparent medium, should be considered in microalgal bioreactor design (Kumar et al., 2010).

Both sunlight and artificial light have been used via outer surface exposure as well as inner volume exposure, through the placement of lighting devices (e.g. LEDs or optical fibers) inside the bioreactor itself (Suh & Lee, 2003). The photosynthetically active radiance is normally assumed to be 43-45% in the wavelength range of 400–700 nm (Laws et al. 1987). The light intensity available to microalgae in high-density cultures is significantly attenuated by mutual shading; to maximize light absorbance and minimize light attenuation, bioreactors should be designed with a high surface area-to-volume ratio, coupled with a short light path (Richmond et al. 2003).

Good microalgal growth rates have been reported (Hu et al 1998) under a light intensity of 4000 μmol m^{-2} s^{-1}; this intensity is twice the solar flux in a medium latitude spot at midday during summer. However, a strong species-dependence exists that should be taken into account. By contrast, light above a saturation point causes light inhibition, which can be counterbalanced by exposing microalgal cells to very short cyclic periods of light and darkness (Pulz, 2001).

The ratio of light to dark (or low-intensity light) periods in a cycle is crucial for microalgal productivity (Munoz & Guieysse, 2006). Similar overall numbers of moles of photons do not necessarily produce equal growth rates of (or CO_2 assimilation by) microalgae. When the light/dark cycle period approaches the photosynthetic unit turnover time (equal to the dark reaction time, estimated to lie within 1–15 ms), maximum photosynthetic efficiencies can be achieved (Richmond et al. 2003). Moreover, compared with periodic darkness, periods of low light intensity significantly increase growth, CO_2 assimilation and lipid productivity in microalgae for a given whole light level (Cuello et al. 2008). This type of lighting design can be achieved via artificial light, such as hybrid lighting systems (Muhs, 2000). Different lamps generate distinct spectra, and different microalgal species possess dissimilar absorption optima; therefore, each individual case should be studied before deciding on the set point of this important operational parameter. Variation of the exponential growth rates of *Phorphyridium cruentum* have been recorded (Suh & Lee, 2003) with variable radiation energies and light spectra, concluding that blue light (400–500 nm) increases cell growth and polysaccharide production.

In terms of artificially illuminated bioreactors, the need for small reactor diameters to increase the illuminated surface area per unit volume of culture can be circumvented through provision of internal illumination. High biomass yields are more crucial in the case of artificially illuminated reactors, because the light provided adds to the overall operational cost of the underlying process. Such costs can be kept below acceptable thresholds via in situ growth-monitoring and associated online control of the intensity of light supplied.

5.2 Carbon uptake

Biological CO_2 fixation can be carried out by higher plants and microalgae, yet the latter possess a greater ability to fix CO_2 (Li et al. 2008; Chisti, 2007; Tredici 2010). Usual sources of CO_2 for microalgae include atmospheric CO_2; CO_2 from industrial exhaust gases (e.g. flue gas and flaring gas); and CO_2 chemically fixed in the form of soluble carbonates (e.g. $NaHCO_3$ and Na_2CO_3) (Kumar et al., 2010). The tolerance of various microalgal species to the concentration of CO_2 is variable; however, the CO_2 concentration in the gaseous phase does not necessarily reflect the CO_2 concentration to which the microalga is exposed during dynamic liquid suspension, which depends on the pH and the CO_2 concentration gradient created by the resistance to mass transfer. Under heterotrophic or mixotrophic conditions, some microalgal species can metabolize a variety of organic compounds, including sugars, molasses and acetic acid, as well as compounds present in wastewater and petroleum (Becker, 1994). Atmospheric CO_2 levels (0.0387% (v/v)) are not sufficient to support the high microalgal growth rates and productivities needed for full-scale biofuel production.

Waste gases from combustion processes, however, typically contain >15% (v/v) CO_2; this percentage indicates, in principle, that combustion processes will provide sufficient amounts of CO_2 for large-scale production of microalgae (Doucha et al. 2005). Owing to the cost of upstream separation of CO_2 gas, direct utilization of power plant flue gas has been considered in microalgal biofuel production systems (Lackne, 2003). Flue gases that contain CO_2 at concentrations ranging from 5 to 15% (v/v) have indeed been introduced directly into ponds and bioreactors of various configurations that contain several microalgal species (Kumar et al., 2010).

5.3 Oxygen generated

Another specific issue of microalgal bioreactors is the accumulation of photosynthetically generated oxygen that may inhibit metabolism or otherwise damage the culture if allowed to accumulate, especially when the rate of photosynthesis, which often correlates with the rate of CO_2 transfer, is high (as typical in horizontal tubular reactors) (Kumar et al., 2010). Most solutions to this problem rely on the use of a degasser (or gas exchange unit), where dissolved oxygen can be released (Morita et al. 2000). However, to attain effective separation between the gas and liquid phases, the path through the degasser should be such that the smallest bubbles have sufficient time to disengage from the liquid.

In tubular bioreactors, connections between tubes can incorporate a tube specifically for oxygen degassing, or a layer of parallel tubes connected by two manifolds: the lower manifold is used to inject air into the culture, and the higher one acts as the degasser (Kumar et al., 2010). Nevertheless, microalgal productivities were lower than expected in these tubular systems, possibly because of build-up of dissolved oxygen during high light intensity periods and along the bioreactor path between manifolds. In systems with exhaust gas recirculation, dissolved oxygen accumulation can be avoided by bubbling exhaust gas through a sodium sulfite solution before its return to the bioreactor (Cien-Fernandez et al. 2005).

Unfortunately, the efficiencies of most techniques used to date for dissolved oxygen removal from microalgal cultures are still not satisfactory. As a result, the classical bubbling mode of operation has been employed to avoid the costlier need for degassing devices. The use of several small bioreactors instead of one large unit also alleviates this problem. Scale-up is indeed easier for facilities that use many small reactors in parallel, even though investment costs might be higher than with fewer large equipment units. Continued research is needed to accurately match the amount of CO_2 supplied to the actual uptake requirement of the metabolizing microalgae, as well as the amount of O_2 removed to the actual amount of O_2 produced.

5.4 Gas transfer

Gases introduced into bioreactors serve a number of purposes in microalgal cultivation, including: supply of CO_2 as sources of carbon for biomass primary and secondary metabolism; provision of internal mixing, which avoids nutrient concentration gradients; promotion of exposure of all cells to light (especially in high density cultures), while minimizing self-shading and phototoxicity; control of pH by assuring dissolution of CO_2 and avoiding gradients thereof; and stripping of accumulated dissolved oxygen, hence reducing its toxicity to microalgae ((Kumar et al., 2010, Pulz, 2001).

Among the various alternatives, bubbling CO_2-enriched air into the bottom of the bioreactor with bubble diffusers has been the most frequently used approach. Moderate overall transfer efficiencies (13–20%) can be achieved by this mode of gas delivery (Carvalho et al. 2006); however, associated drawbacks are loss of CO_2 to the atmosphere, biofouling of diffusers, and poor mass transfer rates owing to a relatively low interfacial specific surface area. Better overall efficiencies are expected for hollow-fiber membrane bioreactors in which the slightly lower mass transfer coefficients that arise from a less turbulent local hydrodynamic pattern are compensated by the much larger area per unit volume available for mass transfer. In addition, the area of mass transfer is well defined, and the pressure on the gas side can be controlled so as to supply only the required amount of CO_2, hence permitting more accurate control of the transfer rate and a dramatic reduction in the amount of CO_2 lost to the atmosphere (Carvalho & Malcata, 2001).

5.5 Mixing rates

Mixing is a key parameter for acceptable performance of microalgal bioreactors. Low mixing rates hamper gaseous mass transfer and might even permit biomass settling. In either case, poor mixing leads to emergence of stagnant zones, where light and nutrients are insufficiently available and anoxic/anaerobic conditions will thus prevail, which results in a decrease of productivity (Kumar et al., 2010). Culture viability might also be compromised by production and accumulation of toxic compounds in stagnant zones (Becker, 1994). Conversely, high mixing rates can cause shear damage to cells (Carlsson et al. (2007), besides requiring a large energy input.

The most common methods of mixing in microalgal bioreactors are pumping, mechanical stirring and gas injection. Pumping offers fair mixing efficiency, but low gas transfer rates; the associated hydrodynamic stress increases with the rotation speed of the pumps, or the number of passes of the microalgal suspension through the pump units (Jaouen et al. 1999).

Mechanical stirring has been reported to provide good mixing efficiency and gas transfer; however, it is likely to produce significant hydrodynamic stress (Tredici, 2003), which can be managed via adequate use of baffles to create a controlled turbulence pattern. Gas injection (bubbling) produces lower hydrodynamic stress, while providing good gas transfer and reasonable mixing efficiency (Richmond & Cheng-Wu, 2001); however, cell damage in sparged cultures increases as the biomass concentration increases, because exponentially higher degrees of stirring are needed to maintain a high-density culture at a predefined level of mixing (Pulz, 2001). One approach to minimize this problem is to maintain a low gas input per nozzle, so as to reduce shear stress and consequent cell damage (Barbosa et al. 2003).

5.6 Temperature effects

Temperature is one of the major factors that regulate cellular, morphological and physiological responses of microalgae: higher temperatures generally accelerate the metabolic rates of microalgae, whereas low temperatures lead to inhibition of microalgal growth (Munoz & Guieysse, 2006). The optimal temperature varies among microalgal species (Ono & Cuello, 2003); however, optimal temperatures are also influenced by other environmental parameters, such as light intensity. Optimal growth temperatures of 15–26 ^{o}C have been reported for some species, with maximum cell densities obtained at 23 ^{o}C. Only daytime higher temperatures were observed to have clearly favorable effects on microalgal growth rates due to photosynthesis, except when the night temperature was as low as 7 ^{o}C (Tamiya, 1957).

5.7 pH effects

Most microalgal species are favored by neutral pH, whereas some species are tolerant to higher pH (e.g. Spirulina platensis at pH 9) (Hu et al. 1998) or lower pH (e.g. Chlorococcum littorale at pH 4) (Kodama et al. 1993). There is a complex relationship between CO_2 concentration and pH in microalgal bioreactor systems, owing to the underlying chemical equilibria among such chemical species as CO_2, H_2CO_3, HCO_3^- and CO_3^{2-}. Increasing CO_2 concentrations can lead to higher biomass productivity, but will also decrease pH, which can have an adverse effect upon microalgal physiology. By contrast, microalgae have been shown to cause a rise in pH to 10–11 in open ponds because of CO_2 uptake (Oswald, 1988). This increase in pH can be beneficial for inactivation of pathogens in microalgal wastewater treatment, but can also inhibit microalgal growth. Similarly, the speciation of NH_3 and NH_4^+ in microalgal bioreactors is strongly dependent on pH - NH_3 uncouples electron transport in the microalgal photosystem and competes with water molecules in oxidation reactions, thus leading to release of O_2 (Hu et al. 1998).

5.8 Nutrient requirements

In addition to the carbon, nitrogen is the most important element that is required for microalgal nutrition (Becker, 1994) and, as a constituent of both nucleic acids and proteins, nitrogen is directly associated with the primary metabolism of microalgae. Fast-growing microalgal species prefer ammonium rather than nitrate as a primary nitrogen source (Green & Durnford, 1996); intermittent nitrate feeding, however, will enhance microalgal growth if a medium that lacks nitrate is used (Jin et al. 2006). Under partial nitrogen deprivation, microalgae grow at lower rates, but produce significantly more lipids, which are reserve compounds synthesized under stress conditions, even at the expense of lower productivities (Lardon et al. 2009).

Phosphorus is the third most important nutrient for microalgal growth, and should be supplied to significant excess as phosphates because not all phosphorus compounds are bioavailable (e.g. those combined with metal ions) (Kumar et al. 2009). In the case of marine microalgae, seawater supplemented with commercial nitrate and phosphate fertilizers is commonly used for production of microalgae (Green & Durnford, 1996). Nevertheless, trace species, such as metals (Mg, Ca, Mn, Zn, Cu and Mb) and vitamins, are typically added for effective cultivation (Becker, 1994).

6. Microalgae harvesting and conversion to fuels

6.1 Microalgae biomass harvesting

Harvesting of the microalgae biomass, i.e., concentrating microscopic algal cells from the dilute solutions of the algal mass culture, is an essential step to secure high-quality effluents and to prevent cell washout (Richmond et al. 2003, Muncz & Guieysse, 2006). The main difficulties encountered in harvesting microalgae arise from the relatively low biomass concentration in conventional bioreactors, coupled with the small size of its constituent microalgal cells. Harvesting typically contributes to 20–30% of the total cost of microalgal biomass production (Carlsson et al. 2007). The major techniques presently applied in the harvesting of microalgae include coagulation, flocculation, sedimentation, centrifugation, foam fractionation, ultrasonic separation, flotation, membrane filtration, and electrophoresis techniques (Carlsson et al. 2007; Kumar et al., 2010; Uduman et al. 2010).

Selection of the harvesting method mainly depends on the properties of microalgae, such as density, size, the value of the desired products. Microalgae harvesting can generally be divided into a two-step process, bulk harvesting, to separate microalgal biomass from the bulk suspension, in this method, the total solid mater can reach 2–7% using flocculation, flotation, or gravity sedimentation; and the second step is thickening, to concentrate the slurry, using filtration and centrifugation. This step needs more energy than bulk harvesting (Brennan & Owende, 2010).

Microalgal cell immobilization has been proposed to circumvent the harvesting issue, but large-scale applications are limited. Further investigation is clearly needed to optimize operating conditions and design new processes (Mallick, 2002).

Following biomass harvest by centrifugation or filtration, microalgal paste traditionally consists of 90% (w/w) water, which meets the requirements for anaerobic digestion. However, it is necessary to reduce this value to a maximum of 50% (w/w) water for efficient oil extraction (Kumar et al., 2010). Despite its energy-intensive nature, drying has often been the dewatering process of choice.

Almost 90% of the energy required for biodiesel production is indeed accounted for by harvesting and dewatering of biomass, besides lipid extraction itself (Lardon et al. 2009). In addition to lipid extraction for biodiesel production, a novel process that gasifies biomass to methane and concentrated CO_2 has recently been proposed (Stucki et al. 2009) for improved overall energy efficiency.

Most microalgae exhibit the phenomenon of bioflocculation, which is the spontaneous aggregation of algal cells into large flocs. These flocs will then settle rather rapidly. The process yet not fully understood and need more investigation. It depends on the elaboration of polymers by the algal cells that makes the cells stick together (Benemann & Pedroni, 2008). Sufficient experience exists to suggest that bioflocculation, possibly in combination with centrifugation, could achieve the cost goals for efficient CO_2 biofixation and biofuel production. Further study and development of this process remains a central problem, next to productivity and controlled cultivation of specific algal species in the designed photobioreactor (Benemann & Pedroni, 2008).

6.2 Microalgae biomass conversion to fuels

For economic and environmental reasons the demand for liquid energy from renewable resources will have an ascending trend in the coming year. The advantages of biomass include that it is biodegradable, sustainable and also causes less pollution when compared with fuels being used. Microalgae with high lipid content produces higher biodiesel than commercially used oilseed crops (rapeseed, soybean oil) utilizing less amount of water (Sheehan et al. 1998).

Converting the harvested biomass to a biofuel considered the least difficulty step. The high water content of the harvested biomass makes drying or any thermochemical conversion process (e.g. combustion, gasification, pyrolysis) impractical, and an even more critical problem is the high nitrogen content of algal biomass. Any thermochemical processing would result in unacceptable NOx generation and loss of this valuable nutrient and resource. Thus, microalgae biomass fuel conversion processes are dependent on fermentations to produce methane or ethanol, or the metabolism of the algae themselves, to produce oils and hydrocarbons, useable for conversion to biodiesel, or to evolve hydrogen. Methane production from microalgae biomass is technically and economically feasible, but still requires some research and development to improve yields and overall efficiency.

Compared to anaerobic digestion, very little work has been done on ethanol fermentations of algal biomass. The reason is that ethanol fermentations, typically carried out by yeast, are restricted to sugars, starches and similar easily degraded carbohydrates. Microalgae typically contain only about 20% or less of such carbohydrates, present as starch in green algae and glycogen in cyanobacteria. For practical ethanol production, an algal biomass with very high fermentable carbohydrate content, preferably over 60% on a dry weight basis, is required. Such high starch or glycogen accumulation is only observed under conditions of nitrogen limitation, where cell growth is reduced and much or most of the photosynthetically-fixed CO_2 is diverted to storage reserves. Thus the issue is whether it is possible to optimize for both high carbohydrate content and high productivity (e.g. CO_2 fixation) using nitrogen limitation (Benemann & Pedroni, 2008).

7. Conclusion

Microalgae have attracted a great deal of attention for CO_2 fixation and biofuel production because they can convert CO_2 into biomass via photosynthesis at much higher rates than conventional biofuel crops.

Several challenges, addressed in this chapter, for microalgal based CO_2 sequestration and biofuel production remain. Most studies reported to date have been performed on the bench-scale, and were conducted under strictly controlled conditions. As a result, little is known about the feasibility of the photobioreactor scale-up. Factors, such as supply of adequate amounts of CO_2, nutrients and light to microalgal cells, should be investigated and optimized at large scale.

Co-digestion of microalgae with wastewater sludge for biogas production should also be considered, because this strategy could be integrated into the existing wastewater infrastructure.

Microalga-based CO_2 fixation and biofuel production can be more sustainable by coupling microalgal biomass production with existing power generation and wastewater treatment infrastructures. Microalgae can utilize low-quality water, such as agricultural runoff or municipal, industrial or agricultural wastewaters, as a source of water for the growth medium as well as a source of nitrogen, phosphorus and minor nutrients (Becker, 1994). Hence, an additional economic and environmental incentive exists as a result of the decreased cost of water and chemicals required for the formulation of the growth medium, while providing a pathway for wastewater treatment (Kumar et al., 2010; Mallick, 2002; Demirbas, 2004). Such a coupling technology need to be further investigated. A number of crucial research gaps remain that must be overcome to achieve full-scale operation such as improved algal growth and nutrient uptake rates; integration of biosystems with waste gas, wastewater and water reclamation systems; improved gas transfer and mixing; and improved algal harvesting and dewatering.

Harvesting, dewatering and lipid extraction from microalgal biomass are still challenging issues because they consume large amounts of energy – mainly because of the small cell size and relatively low biomass levels of microalgal cultures. Therefore, more efficient and economic harvesting technology should be developed to enhance the commercial viability of microalgae biofuels industry (Kumar et al., 2010).

A key challenge for microalgal biodiesel production is the use of microalgal species that can maintain a high growth rate in addition to a high metabolic rate, thus leading to significant lipid yields. This major challenge can be duly addressed via extensive bio-prospecting or target oriented genetic engineering – currently such approaches starting to appear as promising approaches (Kumar et al., 2010).

Finally, it seems there is a lack of fundamental information needed to rationally optimize the performance of existing bioreactors. Novel bioreactor configurations and designs are also needed that promote microalgal growth and CO_2 biofixation, characterized by volumetric productivities at least one order of magnitude above those of conventional open pond systems.

8. References

Acién Fernández, FG.; Fernández Sevilla, JM.; Egorova-Zachernyuk, TA.; Molina Grima, E. (2005). Cost-effective production of 13C, 15N stable isotope-labelled biomass from phototrophic microalgae for various biotechnological applications. *Biomol. Eng.* 22, 193–200

Amos, W.A. (2004). Updated cost analysis of photobiological hydrogen production from chlamydomonas reinhardtii green algae, NREL/MP-560-35593, National Renewable Energy Laborotary, Golden, Colorado, USA.

Barbosa, MJ.; Albrecht, M.; Wijffels, RH. (2003). Hydrodynamic stress and lethal events in sparged microalgae culture. *Biotechnol. Bioeng*. 83, 112-120.

Becker, E.W. (1994). In "Microalgae: biotechnology and microbiology" Ed. Baddiley, J. et al., 178 (1994) Cambridge Univ. Press, Cambridge, New York

Benemann, J.R. & Pedroni, P.M. (2008). Biofixation of Fossil CO_2 by Microalgae for Greenhouse Gas Abatement. *Treccani Encylopedia of Hydrocarbons*, Volume III, p 837 -861.

Brennan, L. & Owende, P. 2010. Biofuels from microalgae - a review of technologies for production, processing, and extractions of biofuels and co-products. *Renew. Sust. Energ. Rev*. 14, 557-577.

Camacho Rubio, F.; Acie´n Ferna´ndez, F.G.; Sa´nchez Pe´rez, J.A.; Garcıa Camacho, F. & Molina Grima, E. (1999). Prediction of dissolved oxygen and carbon dioxide concentration profiles in tubular photobioreactors for microalgal culture. *Biotechnol. Bioeng* 62, 71-86.

Carlsson, A. S.; Beilen, J. V.; Möller, R.; Clayton, D. (2007). Micro and Micro-algae: Utility for Industrial Applications, University of York.

Carvalho AP.; Meireles, LA.; Malcata FX. (2006). Microalgal reactors: a review of enclosed systems design and performances. *Biotechnol. Prog*. 22, 1490-1506.

Carvalho, A.P. & Malcata, F.X. (2001). Transfer of carbon dioxide within cultures of microalgae: plain bubbling versus hollow-fiber modules. *Biotechnol. Prog*. 17, 265-272

Chisti, Y. (2007). Biodiesel from microalgae. *Biotechnol. Adv*. 25, 294- 306.

Cuello, J.L., et al. (2008). Hybrid solar and electric lighting (HYSEL) for space light support. *Presentation at Carbon Recycling Forum* 2008, Arizona

Demirbas, A. (2004). Current technologies for the thermo-conversion of biomass into fuels and chemicals. *Energy Source* 26, 715-730

Doucha, J.; Straka, F.; Lívanský, K. (2005). Utilization of flue gas for cultivation of microalgae (Chlorella sp.) in an outdoor open thin-layer photobioreactor. *J. Appl. Phycol*. 17, 403-412

Green, B.R. & Durnford, D.G. (1996). The chlorophyll-carotenoid proteins of oxygenic photosynthesis. *Ann. Rev. Plant Physiol. Plant Mol. Biol*. 47, 685-714.

Halmann, M. (1993). Chemical fixation of carbon dioxide: methods for recycling CO_2 into useful products. *CRC Press*, Ann Arbor, MI.

Harrington, L. Foster, R. (1999). Australian residential building sector greenhouse gas emissions 1990-2010. Final Report, *Energy Efficient Strategies*. Australian Greenhouse Office

Hu, Q., Zarmi, Y. and Richmond, A. (1998). Combined effects of light intensity, light path and culture density on output rate of Spirulina platensis (Cyanobacteria). *Eur. J. Phycol*. 33, 165-171.

Jaouen, P.; Vandanjon, L.; Quéméneur, F. (1999). The shear stress of microalgal cell suspensions (Tetraselmis suecica) in tangential flow filtration systems: the role of pumps. *Biores. Technol*. 68, 149-154.

Jin, HF.; Lim, BR.; Lee K. (2006). Influence of nitrate feeding on carbon dioxide fixation by microalgae. *J. Environ. Sci. Health A Tox. Hazard Subst. Environ. Eng.* 41, 2813–2824.

Kodama, M.; Ikemoto, H.; Miyachi, S. (1993). A new species of highly CO_2-tolerant fastgrowing marine microalga suitable for high-density culture. *J. Marine Biotechnol.* 1, 21–25.

Kommareddy, A. & Anderson G. (2003). Mixing and Eddie Currents in a Modified Bubble Column Reactor . Paper Number: RRV03-0043 An ASAE Meeting Presentation. Written for presentation at the 2003 CSAE/ASAE Annual Intersectional Meeting Sponsored by the Red River Section of ASAE Quality Inn & Suites 301 3rd Avenue North Fargo, North Dakota, USA October 3-4, 2003

Kruse, O.; Rupprecht, J.; Mussgnug, J R.; Dismukes, G.C. & Hankamer, B. (2005). *Photochemical & Photobiological Sciences*, 4 , 957.

Kumar, A.; Ergas, S.; Yuan, X.; Sahu, A.; Zhang, Q.; Dewulf, J., Xavier Malcata, F. & Herman van Langenhove. (2010). Enhanced CO_2 fixation and biofuel production via microalgae: recent developments and future directions. *Trends in Biotechnology* 28, 371–380

Kumar, A.; Yuan, X.; Sahu, A. K.; Dewulf, J.; Ergas, S. J.; Van Langenhove, H. (2010). Hollow fiber membrane photo-bioreactor for CO_2 sequestration from combustion gas coupled with wastewater treatment: a process engineering approach. *J. Chem. Technol. Biotechnol.* 85, 387-394

Lackne, K.S. (2003), Climate Change: A guide to CO_2 sequestration. *Science* 300, 1677–1678.

Lardon, L.; Hélias, A.; Sialve, B.; Steyer, J.P.; Bernard, O. (2009). Life-cycle assessment of biodiesel production from microalgae. *Environ. Sci. Technol.* 43, 6475–6481.

Laws, E.A.; Taguchi, S.; Hirata, J.; Pang, L. (1986). Continued studies of high algal productivities in a shallow flume. *Biomass* 11, 39–50.

Li Y.; Horsman M.; Wu N.; Lan CQ.; Dubois-Calero N. (2008). Biofuels from microalgae. *Biotech. Prog.* 24, 815–820.

Mallick, K. (2002). Biotechnological potential of immobilized algae for wastewater N, P and metal removal: a review. *Biometals* 15, 377–390.

Miron, A. S.; Camacho, F. G.; Gomez, A. C.; Grima, E. M. & Chisti, Y. (2000). Bubble-Column and Airlift Photobioreactors for Algal Culture. *American Institute of Chemical Engineers Journal.* 46(9): 1872-1887.

Molina Grima, E.; Acién Fernández, F.G.; Garcıa Camacho, F.; Camacho Rubio, F. & Chisti, Y. (2000). Scale-up of tubular photobioreactors, *J Appl Phycol* 12:355–68

Molina Grima, E.; Fernandez Sevilla, J. M.; Sanchez Perez, J. A. & Garcıa Camacho, F. (1996). A study on simultaneous photolimitation and photoinhibition in dense microalgal cultures taking into account incident and averaged irradiances. *J. Biotechnol.* 45, 59–69

Morita, M.; Watanabe Y.; Saiki H. (2000). Investigation of photobioreactor design for enhancing the photosynthetic productivity of microalgae. *Biotechnol. Bioeng.* 69, 693–698.

Muhs, J. (2000). Design and analysis of hybrid solar lighting and fullspectrum solar energy systems. Proceedings of the International Solar Energy Conference, *Solar Engineering*, pp. 229–237.

Munoz, R. & Guieysse, B. (2006). Algal-bacterial processes for the treatment of hazardous contaminants: a review. *Water Res.* 40, 2799– 2815

Ono, E. & Cuello, J. L. (2003). Selection of optimal microalgae species for CO_2 sequestration. Proceedings of Second Annual Conference on Carbon Sequestration. Alexandria, VA

Oswald, W.J. (1988). Large-scale systems (engineering aspects). In Micro-algal Biotechnology (Borowitzka, M.A. and Borowitzka, L.J., eds), pp. 357-394, Cambridge University Press.

Poulsen, B. R. & J. J. L. Iversen. (1997). Characterization of Gas Transfer and Mixing in a Bubble Column Equipped with a Rubber Membrane Diffuser. *Biotechnology and Bioengineering*. 58(6): 633-641.

Pulz, O. (2001). Photobioreactors: production systems for phototrophic microorganisms. *Appl. Microbiol. Biotechnol.* 57, 287–293.

Richmond, A. & Cheng-Wu, Z. (2001). Optimization of a flat plate glass reactor for mass production of Nannochloropsis sp. outdoors. *J. Biotechnol.* 85, 259–269

Richmond, A. et al. (2003). Efficient use of strong light for high photosynthetic productivity: interrelationships between the optical path, the population density and the cell growth inhibition. *Biomol. Eng.* 20, 229–236.

Sánchez Mirón, A.; Garc´ıa Camacho, F.; Contreras Gómez, A.; Molina Grima, E. & Chisti, Y.(2000). Bubble column and airlift photobioreactors for algal culture. *AIChE J* 46:1872–87

Sheehan, J.; Dunahay, T.; Benemann, J.; Roessler, P. (1998). *A look back at the US Department of Energy's aquatic species program - Biodiesel from algae,* Golden (CO), National Renewable Energy Laboratory, July, NERL/TP 580-24190.

Stewart, C. & Hessami, M. A. (2005). Energy Conversion and Management, 46, 403-420.

Stucki, S.; Vogel, F.; Ludwig, C.; Haiduc, A. G.; Brandenberger, M. (2009). Catalytic gasification of algae in supercritical water for biofuel production and carbon capture. *Energy Environ. Sci.* 2, 535–541.

Suh, I.S. & Lee, S.B. (2003). A light distribution model for an internally radiating photobioreactor. *Biotechnol. Bioeng.* 82, 180–189.

Tamiya, H. (1957). Mass culture of algae. *Ann. Rev. Plant Physiol.* 8, 309–334.

Tredici, M. R. (2003). Closed photobioreactors: basic and applied aspects. In Proceedings of Marine Biotechnology: Basics and Applications, p. 1, Matalascan as, Spain

Tredici, M.R. (2010). Photobiology of microalgae mass cultures: understanding the tools for the next green revolution. *Biofuels* 1, 143–162.

Uduman, N.; Qi, Y.; Danquah, M.K.; Forde, G.M. & Hoadley, A. 2010. Dewatering of microalgal cultures: a major bottleneck to algae-based fuels. *J. Renew. Sust. Energ.* 2, 012701.

Znad, H.; Báleš, V. & Kawase, K. (2004). Modeling and scale up of airlift bioreactor. *Computers & Chemical Engineering*, 28, 2765-2777.

Znad, H.; Kasahara, N. & Kawase, Y. (2006). Biological decomposition of herbicides (EPTC) by activated sludge in a slurry bioreactor. *Process Biochemistry* 41, 1124–1128.

10

Production of Biodiesel from Microalgae

Marc Veillette, Mostafa Chamoumi, Josiane Nikiema, Nathalie Faucheux and Michèle Heitz*
Chemical Engineering and Biotechnological Engineering Department, Université de Sherbrooke
Canada

1. Introduction

Since the early 70s, several major energy crises have forced the scientific community to find alternative sources of power. In July 2008, the price of crude petroleum reached 145$US/barrel, the highest price ever achieved in 30 years (BP, 2011; Ervin & Associated, 2011). In July 2011, the price of crude petroleum was still relatively high at 108$US/barrel (Ervin & Associated, 2011). Moreover, the 2008 world economic crisis encouraged the United States government to develop biofuels in order to blend them with petroleum fuels without engine modifications or distribution process changes (Bindraban *et al.*, 2009; The White House, 2010). Furthermore, in Canada, by 2012, the objective of adding biodiesel into transportation diesel and heating fuels is 2% (v/v) (Natural Ressources Canada, 2011b).

Consequently, in July 2007, a nine-year investment of 1.5 billion CAD$ (the ecoENERGY for Biofuels Initiative) was announced by the Canadian government to stimulate the production of biofuels in Canada (Natural Ressources Canada, 2011a).

In recent years, most of industrial biodiesels are made from oil (triglycerides) of raw materials (rapeseed, sunflower, soybean, etc.). With the intent to change their physicochemical properties similar to petroleum-based diesel, triglycerides are transesterifed into fatty acid alkyl esters, which can be used in a conventional engine without modifications (Knothe, 2010). On the ecological side, in addition to the ability of oleaginous plants to reduce pollutant emissions of greenhouse gases (GHG) by their capacity to trap and use the carbon dioxide (CO_2), using biodiesel also reduces net emissions of pollutants. Typically, the addition of 20% (v/v) of soybean-based biodiesel in petrodiesel reduces emissions of carbon monoxide (CO), CO_2, particulate matter (PM) and hydrocarbons (HC) by 11%, 15.5%, 10% and 21%, respectively (Sheehan *et al.*, 1998a; United States Environmental Protection Agency, 2002).

The raw materials are also necessary to feed humans and animals. A large demand for raw materials to produce biodiesel could thus increase their price. Moreover, the culture of conventional vegetable material requires an important amount of water, chemical fertilizers and pesticides, which have a negative impact on the environment (Smith *et al.*, 2009).

* Corresponding Author

To overcome these problems, researchers are currently exploring a new way of producing biodiesel using microalgae. Indeed, during their growth, photoautotrophic microalgae metabolize inorganic carbon (CO_2) through photosynthesis (Chen *et al.*, 2011). According to Cadoret & Bernard (2008), microalgae have also the capacity to absorb other pollutants such as phosphates and nitrates. Furthermore, microalgae can accumulate a high amount of fatty acids and have a culture yield per hectare at least 10 times higher than any oily plants (Chisti, 2007).

In this paper, the different sources of energy such as oil, conventional and unconventional biofuels like microalgae based biodiesel will be discussed. Then, the different aspects of the microalgae valorization into biodiesel will be considered. The culture of microalgae, the extraction of lipids, the transesterification process and the biodiesel characteristics will be especially discussed.

1.1 Background

Finding new energy resources to compensate the decrease of the world petroleum reserves is an important challenge. The estimation of the world crude oil reserves is a difficult task because it is influenced by political, economic and technological factors (Pirog, 2005). The proven oil reserves represent the amount of petroleum that could be obtained from deposits already discovered with actual technological and economic conditions (Institut Français du Pétrole, 2005; Pirog, 2005). They also include the oil discovered, but that cannot be recovered with current technologies. In 2009, these reserves were estimated to be about 1,376 billion barrels (BP, 2011), which corresponds to a reserve that will last for 35 years (i.e. until 2045) (Shafiee & Topal, 2009).

According to a predictive model from Huber (1956), the proven reserves of oil should have reached a roof value by 2000, as the production of petroleum should begin to fall. In contrast, more recent data show that between 1989 and 2008, the proven oil reserves seem to have increased from 1006 to 1333 billion barrels (BP, 2011). On the other hand, while the world oil consumption was about 86 million barrels/day in 2006, it would reach 107 million barrels/day by 2030. The transportation sector would be responsible for 80% of this increase (U.S. Energy Information Administration, 2009) and would consume 76% of the world oil production by 2030 (International Energy Agency, 2008).

Fossil fuel dependency, mainly in the transportation sector, has encouraged research on biofuels. A recent study shows that microalgae biofuels have the potential to replace 17% of oil imports in the United States used as transportation fuel by 2022 (Wigmosta *et al.*, 2011). Moreover, following the BP oil spill in the Atlantic Ocean, the United States administration is considering reducing his oil imports by 1/3 by 2021 using, among others, biofuels (N. Banerjee, 2011).

2. Biofuels as a main alternative

2.1 Conventional biofuels

Intending to replace oil-based fuels, many studies have been conducted on non-renewable and renewable energy alternatives.

The researches on non-renewable energy are mainly based on coal and natural gas. In 2009, 89% of the energy produced in Canada (15 exaJoules) was coming from non-renewable sources (Statistics Canada, 2011). For example, coal can be converted into syngas (CO, H_2) by gasification (Naveed *et al.*, 2010), and then into fuel by Fischer-Tropsch reaction, methanol or synthetic natural gas by catalytic processes and hydrogen by water/gas shift reaction (Dry, 1996; Longwell *et al.*, 1995).

Renewable energy sources used in the world in 2006 were mainly biomass and waste (58%), hydropower (31%) and others (12%), including wind, geothermal and solar (International Energy Agency, 2008). Most of renewable energies in Canada was produced by hydroelectricity (89%), while biomass was responsible for 6% of the 81 GWatt of renewable energy produced in 2008 (Nyboer & Grove, 2010). In United States, the energy generated from biomass is expected to increase by 300% between 2009 and 2030 to reach 153 billion of 12 kWatt-h (U.S. Energy Information Administration, 2011). Biomass is a source of energy, which can be used to produce 1st, 2nd and 3rd generation biofuels.

2.1.1 First generation

First generation biofuels correspond to those issued from food-based crops (Antizar-Ladislao & Turrion-Gomez, 2008). They mainly correspond to ethanol-based fuels obtained from the fermentation of sugars (corn, beet, sugar cane, etc.), vegetable oil-based fuels (raw oil, biodiesel and renewable diesel produced from catalytic hydrodeoxygenation) (Knothe, 2010; Natural Ressources Canada, 2011b) from oleaginous plants (colza, palm, canola, etc.) and biogas emitted from raw material or landfills (Naik *et al.*, 2010).

However, on many levels (environmental, societal), the fact that food resources could be used to produce biofuels shows several limits, as this would create land pollution, a lack of agricultural land (world hunger) and deforestation (Goldemberg & Guardabassi, 2009; National Research Council, 2007). For example, in some European countries such as France, the arable lands available for cultivation of oleaginous plants used for 1st generation biofuels production will not be able to support the biofuels demand by 2015, except by saturating the lands in fallow, which would create soil impoverishment problems (Bordet *et al.*, 2006).

2.1.2 Second and third generation

Second generation biofuels are the cellulosic-based biofuels obtained from non-food crops materials (wood, leaves, straw, etc.). These biofuels include bioalcohols, bio-oil, 2,5-dimethylfuran (BioDMF), biohydrogen, Fischer-Tropsch diesel, wood diesel (Fatih Demirbas, 2009; Román-Leshkov *et al.*, 2007).

Third generation biofuels are microorganisms (yeast, fungi) biofuels and algae-based fuels like vegetable oils, bio-oil, jet-fuels, biohydrogen, biodiesel, renewable diesel and many others (Fatih Demirbas, 2009; Nigam & Singh, 2011).

Second and 3rd generation biofuels are better than 1st generation biofuels for sustainable development as they are carbon neutral, or they reduce atmospheric CO_2 as they are carbon negative (Naik et al., 2010).. For example, 1st generation biodiesel (like soybean) only induces a net reduction of GHG emissions by 41% (Hill *et al.*, 2006). In comparison, for each ton of microalgal biomass produced, some authors estimate that 1.8 tons of CO_2 would be consumed (180% reduction) (Chisti, 2007).

2.2 Microalgae as a source of biofuels

Microalgae can generate diverse biofuels, which are mainly: biomethane produced by anaerobic digestion (Sialve *et al.*, 2009; Spolaore *et al.*, 2006), biohydrogen by photobiological process (Fedorov *et al.*, 2005; Kapdan & Kargi, 2006), bioethanol by fermentation (Choi *et al.*, 2010; Dexter & Pengcheng, 2009), liquid oil by thermal liquefaction (A. Banerjee *et al.*, 2002; Miao & Wu, 2004; Miao *et al.* 2004; Sawayama *et al.*, 1995) and biodiesel (M. B. Johnson & Wen, 2009; Koberg *et al.*, 2011; Nagle & Lemke, 1990; H. Xu *et al.*, 2006).

Even if industrial scale biofuels from microalgae remain at an early stage, they remain a sustainable solution as a transportation fuel.

2.3 Microalgae for biodiesel

Some microalgae species like *Botryococcus braunii* or *Schizochytrium* sp. can contain up to 80% of their dry weight of lipids (Deng *et al.*, 2009). These species can produce a yield of lipids by acre up to 770 times higher than oleaginous plants (colza, sunflower, etc.) and their high scale production allows to consider developing high-yield biodiesel (Chisti, 2007). Another advantage of using microalgae to produce biodiesel is that microalgae can double from 1 to 3 times in 24 hours (Khan *et al.*, 2009). Consequently, the microalgae biomass can be harvested more than once a year. Microalgae can potentially be used also in food, cosmetic, fertilizing and many other industries (Jacob-Lopes & Teixeira Franco, 2010).

2.3.1 Sustainable development issues

Environmental issues

Microalgae growth is made naturally from sunlight photosynthesis in diverse media (Deng *et al.*, 2009). Replacing 1st generation biodiesel with biodiesel from microalgae would reduce the carbon footprint of the process. In fact, producing microalgal biodiesel requires no mechanical seeding of grains, no watering or "harmful" chemical products spreading and no harvesting using heavy engines powered by fossil fuels.

The microalgae production can add value to the gases emitted by coal fired power plants (or other processes) (de Morais & Costa, 2007) as they are able to absorb CO_2, nitrogen dioxide (NO_2) and sulphur dioxide (SO_2), which are important nutrients for their growth (Malińska & Zabochnicka-Świątek, 2010). Moreover, some studies have shown that some algae can be grown on municipal *(Chlorella* sp.*)* and industrial wastewaters (*Chlamydomonas globosa, Chlorella minutissima or Scenedesmus bijuga)* since they also use compounds like phosphorus (P) and metals (Al, Fe, Mg, Mn, Zn, etc.) (Chinnasamy *et al.*, 2010; Wang *et al.*, 2010).

Economic and social issues

The production of 1st generation biofuels is causing a substantial rise of the world food prices. In order to evaluate the monthly variations for different international food commodities, The FAO Food Price Index can be used. From 2000 to 2011, The FAO Food Price Index (FFPI) and The FAO Cereal Price Index increased from 88 to 238 and from 87 to

245, respectively (FAO, 2011b). In the long term, producing biofuels on arable land could contribute to increasing world hunger. The increase of cereal prices could have an impact on the cost of 1st generation biodiesel production as the FAO Oils/Fats Price Index from 2000 to 2011 also increased from 78 to 278 (FAO, 2011b). In North America, the best lands could be used for biodiesel production, because the harvested area for soybean, which is the most used oil for biodiesel production, represents 32% of the 99.5 million ha harvested worldwide (FAO, 2011a).

As microalgae industrial culture does not directly compete with food and wood production, it can represent a great potential economic development. In fact, the use of microalgae for biofuel production would permit to reduce deforestation and preserving the forest heritage. Moreover, the development of valorization of microalgae could favour the energetic autonomy of all countries, including developing countries. Thus, the industrial production of microalgae could be considered as a sustainable solution to energetic, environmental and food problematic.

2.3.2 Types of microalgae

Microalgae can be found in a large range of places where light and water are present including ocean, lake, soils, ice, rivers, etc. (Deng *et al.*, 2009). Microalgae demonstrate a great biodiversity (between 200 000 and several millions of species) (Natrah *et al.*, 2007), which can be divided into categories depending on their pigmentation, biological structure and metabolism.

Size classification

Microalgae are small organisms, which can be divided into 4 size categories as the microplankton (20 to 1000 µm), the nanoplankton (2 to 100 µm), the ultraplankton (0.5 to 15 µm) and the picoplancton (0.2 to 2 µm) (Callieri & Stockner, 2002; Gopinathan, 2004). Their small size allows them to do an effective photosynthesis, converting light energy with CO_2 dissolved in water to produce lipids, carbon hydrates, proteins, etc..

Taxonomic groups

Microalgae can be classified into 4 main taxonomic groups: diatoms (*Bacillariophyceae*), green algae (*Chlorophyceae*), cyanobacteria or blue green algae (*Cyanophyceaes*) and golden algae (*Chrysophyceae*). However, there are 6 other groups of microalgae composed of yellow-green (*Xanthophyceae*), golden algae (*Chrysophyceae*), red algae (*Rhodophyceae*), brown algae (*Phaeophyceae*), dinoflagellates (*Dinophyceae*), *Prasinophyceae* and *Eustigmatophyceae* (Williams & Laurens, 2010). However, these species are not equally interesting for biodiesel production. As an example, green algae taxonomic group includes the most promising species of microalgae like *Botryococcus* sp., *Dunaliella* sp. and *Chlorella* sp. (Garofalo, 2010) for biodiesel production.

Metabolism classification

Microalgae can be separated into 4 main types of metabolism called photoautotrophic, heterotrophic, mixotrophic and photoheterotrophic (Chen *et al.*, 2011).

In photoautotrophic microalgae metabolisms, light (source of energy) converts inorganic carbon (CO_2) and water to biomass by photosynthesis reaction (Cadoret & Bernard, 2008).

Even if photoautotrophic microalgae contain high level of lipids, their biomass productivity in photobioreactors and open ponds is generally lower than heterotrophic microalgae, between 0.117 and 1.54 kg/m^3/day (Chisti, 2007).

Heterotrophic microalgae need organic carbon as a source of carbon and energy. Their production is carried out in closed bioreactors (fermentors). They are more promising than the photoautotrophic species for the production of biodiesel (Martek, 2008; Xiong *et al.*, 2008; H. Xu *et al.*, 2006). For example, Xiong *et al.* (2008) showed that the biomass productivity obtained from *Chlorella* sp. grew under heterotrophic conditions could reach 7.4 kg/m^3/day with lipid content ranging from 50 to 58% (g lipid/g dry weight). However, they do not achieve the main goal of producing microalgae, i.e. mitigation of the emissions of CO_2.

Some microalgae species also have mixotrophic metabolism as they grow in light or dark using both inorganic and organic carbon sources. Growing *Chlorella vulgaris* in the dark, Liang *et al.* (2009) changed CO_2 contained in air for glucose (as a source of carbon) and observed an increase of maximal biomass productivity from 10 to 151 mg/L/day. When they used light and glucose as a source of carbon, the biomass productivity increased to 254 mg/L/day. The use of microalgae with mixotrophic metabolism is relatively rare for biodiesel production (Chen *et al.*, 2011).

Photoheterotrophic metabolism means that microalgae need light as a source of energy and a source of organic carbon (Chen *et al.*, 2011). For example, using glycerol as a source of carbon and a light intensity of 35 $\mu E/m^2/s^1$, Yang *et al.* (2011) obtained an increase in *Chlorella minutissima* (UTEX2341) biomass concentration from 1.2 to 8.2 g/L after 15 days of culture. As these microalgae need an inexpensive source of organic carbon and are dependant of the sunlight periods, photoheterotrophic metabolism seems less interesting for biodiesel production.

2.3.3 Strain promising

Even if microalgae are a large group of microorganisms, not all species are suitable to produce biodiesel. The factors to consider were stated by Grobbelaar (2000): "1-Carbon dioxide tolerance and uptake, 2-Temperature tolerance, 3-Stability for cultivation in specific bioreactors," 4-Secondary valuable products, "5-Specific growth requirements and competing algal, 6-Vulnerability to infection and herbivotory potential, 7-Excretion of autoinhibitor, 8-Harvesting and down-stream processing and 9-Manipulation potential of genetic engineering".

2.3.4 Lipids yield of microalgae

Intending to increase lipid production, Huntley & Redalje (2007) proposed a method of growth for microalgae in a photobioreactor (PBR) which allows reaching high lipids yield followed by 2 days of nitrogen deficiency in a pond. In general, nutrient deprivation can lead to an increase in lipid content, but not for all species of microalgae. For example, the microalgae *Navicula* (NAVIC1) had the highest lipid content, which raised in exponential phase from 22 to 49% (g lipid/g dry weight) in silicon (Si) deficiency and increased to 58% (g lipid/g dry weight) when nitrogen was limited. On the other hand, nutrient limitation (nitrogen or Si) limitation had less or no significant effect on lipid content of microalgae *Amphora* (AMPHO1) and *Cyclotella* (CYCLO2), respectively (Sheehan *et al.*, 1998b). Other studies found that P deprivation could have a positive effect on lipid content (Khozin-Goldberg & Cohen, 2006; Xin *et al.*, 2010). For example, increasing the P concentration from 0.14 to 0.37 mg/L, Xin et al. (2010) observed a raise of microalgae concentration from 0.14 to 0.37 g/L, while the lipid content decreased from 53 to 23.5% (g lipid/g dry weight).

Finally, an osmotic shock might also stimulate the lipids production. For example, Takagi *et al.* (2006) enhanced the sodium chloride (NaCl) concentration from 3.5 to 7 g/L (0.5 to 1 mol/L) and found an increase in lipid production from 60 to 67% (g lipid/g dry weight). However, these physicochemical treatments could also favour the synthesis of polar lipids like phospholipids or glycolipids associated with cell walls of the microalgae (Cadoret & Bernard, 2008); such lipids are less interesting for biodiesel production (Nagle & Lemke, 1990).

3. Production of biodiesel

3.1 Culture of microalgae

The large scale production of microalgae is generally performed with solar energy (photoautotrophic metabolism) in open ponds (raceways), closed systems (photobioreactors) or fermentors.

Open ponds are generally circular with nested loops and are 30 cm deep (Chisti, 2007). However, ponds can have several non neglectable disadvantages. Indeed, as they are open, evaporation and contaminants (protozoa, bacteria or other microalgae) could affect the microalgae productivity (Blanco *et al.*, 2007).

Photobioreactors are continuous culture systems which can achieve concentration of microalgae up to 6.7 g/L (Bai *et al.*, 2011; Chisti, 2007; Ranjbar *et al.*, 2008) in fresh or sea water. Different models of photobioreactors (indoor or outdoor) have been developed including tubular, flate plate, airlift, bubble column and stirred tank (L. Xu *et al.*, 2009). Even if the closed photobioreactor has a higher harvesting efficiency (more biomass) and a good control on culture parameters (temperature, pH, CO_2 concentration etc.) (Suh & Lee, 2003), its capital costs remain higher (around 10 times) than those of open ponds (Carvalho *et al.*, 2006). However, the combination of ponds and photobioreactors can be profitable because microalgae can be grown in open ponds while reducing contamination by undesired species (Huntley & Redalje, 2008). In this culture process, the first step of microalgae production is conducted in a controlled temperature (e.g. by a sea water bath (16-18°C)) photobioreactor. Microalgae are then transferred into an open pond for a 5 days second culture step (Huntley & Redalje, 2007; Huntley & Redalje, 2008).

Fermentors are mainly used to produce heterotrophic microalgae using an organic source of carbon such as glucose, fructose, galactose acetate, glycerol and acetic acid (Cantin, 2010). These bioreactors can reach high biomass concentration (150 g/L) without rheological problems (Wu & Shi, 2008). However, heterotrophic production costs of microalgae in fermentors remains relatively high (Wei *et al.*, 2009).

3.2 Harvesting of microalgae

Microalgae biomass could be harvested by centrifugation, flocculation, gravity sedimentation, filtration, screening, flotation or by electrophoresis techniques (Chen *et al.*,

2011). As microalgae are floating in pond at a concentration less than 0.5 g/L, the harvesting costs can represent 20 to 30% of the industrial microalgae biomass production cost (Carlsson *et al.*, 2007) of 2.95 and 3.80 $US/kg biomass (photobioreactor and raceway, respectively) (Chisti, 2007).

3.3 Extraction techniques

In order to produce biodiesel from microalgae lipids, the later must be priory extracted. The main lipid extraction techniques are the use of chemical solvents, supercritical CO_2, physicochemical, biochemical and direct transesterification.

3.3.1 Chemical solvents extraction

Chemical solvents method is by far the most commonly used, but less effective when microalgae are still wet (Samorì *et al.*, 2010). Consequently, for laboratory scale extraction of lipids, freeze-drying (J. Lee *et al.*, 2010) is a popular method, but spray-drying (Koberg *et al.*, 2011), oven-drying (Cooney *et al.*, 2009) or vacuum-evaporation (Umdu *et al.*, 2009) have also been used to dry microalgae. However, drying microalgae prior to lipid extraction could require 2.5 times more energy than a process without drying, which makes a process using a prior drying unprofitable (negative balance) (Lardon *et al.*, 2009).

In laboratory scale studies, even if chloroform-methanol blends have been extensively used with high extraction yields up to 83% (g lipid/g dry weight) (Yaguchi *et al.*, 1997), less polar solvent like hexane are often preferred because of their lower toxicity and affinity for non-lipid contaminants (less polar) (Halim *et al.*, 2010). As an example, hexane was used to obtained lipids content up to 55% (g lipid/g dry weight) from a heterotrophic microalgae, *Chlorella protothecoides* (Miao & Wu, 2006). For microalgae lipid extraction on an industrial scale, Soxhlet extraction is not recommended due to high energy requirement (Halim *et al.*, 2010).

Other less toxic solvents like alcohols (ethanol, octanol) or 1,8-diazabicyclo-[5.4.0]-undec-7-ene (DBU) have been tested but the yield of fatty acid methyl ester (FAME) obtained was up to 5 times lower than with n-hexane extraction (Samorì *et al.*, 2010) even if the hydrocarbon (lipid) yield was more than twice higher.

3.3.2 Supercritical carbon dioxide extraction

Supercritical CO_2 (Dhepe *et al.*, 2003) has the advantages of being not toxic, easy to recover and usable at low temperatures (less than 40°C) (Andrich *et al.*, 2005). However, this technique requires expensive equipments (Perrut, 2000) and a huge amount of energy to reach high pressures (Tan & Lee, 2011). Few studies used supercritical CO_2 extraction to recover microalgae lipids and transformed them into biodiesel (Halim *et al.*, 2010) even if some studies obtained lipid content up to 26% (g lipid/g dry weight) from *Nannocloropsis* sp. (Andrich *et al.*, 2005). Using supercritical CO_2 extraction at operating temperature of 60°C and pressure of 30 MPa to extract lipids from *Chlorococcum* sp. microalgae, Halim *et al.* (2010) obtained a higher extraction yield of lipids with supercritical CO_2 than hexane Soxhlet extraction (5.8 and 3.2% (g lipid/g dry weight), respectively). Moreover, using supercritical CO_2 extraction with wet microalgae, Halim *et al.* (2010) obtained a maximum yield of lipids of 7.1% (g lipid/g dry weight) for the same experimental conditions, which was a relatively low lipid yield compared to other species such as *Botryococcus* sp. (28.6% g lipid/g dry weight) (J. Lee *et al.*, 2010). Consequently, in opposition to chemical solvent extraction, supercritical CO_2 lipid extraction can be stimulated by the presence of water in the blend of microalgae.

3.3.3 Physicochemical extraction

Some physicochemical techniques like microwave, autoclaving, osmotic shock, bead-beating, homogenization, freeze-drying, French press, grinding and sonication can be used for microalgae cell disrupting in order to recover lipids (Cooney *et al.*, 2009; J. Lee *et al.*, 2010; S. Lee *et al.*, 1998). Using microwave or bead-beating seems to be the most promising techniques to increase the lipid yield. As an example, J. Lee *et al.* (2010) increased the lipid extraction yield of *Botryococcus* sp. microalgae in water phase from 7.7 to 28.6% (g lipid/g dry weight) using a 5 min microwave pretreatment.

3.3.4 Biochemical extraction

Few studies have used biochemical extraction to extract lipids from microalgae. Using a 72 h cellulase hydrolysis pretreatment of the *Chlorella* sp. microalgae, Fu *et al.* (2010) have obtained a hydrolysis yield of sugars of 70% (concentration reducing sugar/concentration total sugar), although the lipids yield has increased only from 52 to 54% (g lipid/g dry weight).

3.3.5 Direct (in situ) transesterification

Direct transesterification is a process that blends the microalgae with an alcohol and a catalyst without prior extraction. Number of acid catalysts have been investigated for heterotrophic microalgae biomass including hydrochloric (HCl) or sulphuric acid (H_2SO_4) but acetyl chloride (CH_3COCl) remains the catalyst producing the higher FAME yield of 56% (g FAME /g dry weight) (Cooney *et al.*, 2009). A less polar solvent, like hexane or chloroform, can be added to increase the yield of biodiesel production (M. B. Johnson & Wen, 2009). Direct transesterification using a heterogeneous catalyst could be more effective coupled with microwaves heating. As an example, using microwave with direct transesterification of *Nannochloropsis* in presence of a heterogeneous catalyst (SrO), Koberg *et al.* (2011) reported an increase in the FAME yield from 7 to 37% (g FAME /g dry weight). However, direct transesterification requires a dry biomass, increasing the cost of harvesting.

3.4 Transesterification

The direct use of crude vegetable oils in diesel engines is envisageable, but could lead to numerous technical problems. For example, their characteristics (high viscosity, high density, difficulty to vaporize in cold conditions) cause deposits in the combustion chamber, with a risk of fouling and an increase in most emissions (Basha *et al.*, 2009). These drawbacks can be mitigated, but not without some modifications of the diesel engine (Altin *et al.*, 2001). To overcome all these inconveniences, the transformation of microalgae lipids in corresponding esters is essential.

3.4.1 Reaction

In the transesterification process, a catalyst and an alcohol are added to a blend of microalgae lipids. The reaction reduces the molecular weight, the viscosity and increases the volatility of microalgae lipids. Different parameters can influence the yield of transesterification like the ratio of alcohol-oil, catalyst types and concentration, reaction time, temperature and agitation rate.

3.4.2 Microalgae lipid content

The yield of the transesterification reaction depends on the nature of lipids. For example, monoglycerides like palmitic acid (C16:0) produced a FAME yield of 93% (g FAME / g lipid) while triglycerides like triolein had a yield of 88% (g FAME / g lipid). Furthermore, phospholipids and glycolipids gave a lower yield of 54 to 65% (g FAME / g lipid) and 47 to 56% (g FAME / g lipid), respectively (Nagle & Lemke, 1990). The nature of lipids is an important data for biodiesel production because some microalgae can contain up to 93% (g / g lipid) of phospholipids and glycolipids (Williams & Laurens, 2010). Moreover, some microalgae can also contain lipids such as unsaponifiable lipids carotenoids and other elements (chlorophyll) which are considered as by-products (Bai *et al.*, 2011).

3.4.3 Alcohol

Methanol is the most commonly used alcohol because of its low price. However, other alcohols such as ethanol or butanol can also be employed (Chisti, 2007). In traditional alkali-based catalyst transesterification of vegetable oil, the most used methanol to oil molar ratio for transesterification is 6:1 (Marchetti *et al.*, 2007) even if the stoichiometric value is 3:1 for triglycerides (Berriosa & Skelton, 2008). For microalgae lipids transesterification, the optimal methanol to oil ratio is higher. For example, performing a direct transesterification during 8h at 25°C, Ehimen *et al.* (2010) obtained a decrease in the specific gravity (SG) of the biodiesel from 0.8887 to 0.8849 when the molar ratio methanol to oil was increased from 105:1 to 524:1.

3.4.4 Catalyst

Catalysts used for transesterification of microalgae lipids are mainly homogenous or heterogeneous. Another method of transesterification using methanol in the supercritical state (without catalyst) has been developed, but the cost of this technology renders its use impossible to date (Tan & Lee, 2011).

Homogenous alkaline catalysts used for transesterification of vegetable oils mainly include sodium or potassium hydroxide (NaOH or KOH) and sodium or potassium methoxide (CH_3ONa or CH_3OK)) while homogenous acid-catalysts includes H_2SO_4, HCl and sulphonic acids ($R\text{-}SO_3H$) (Helwani *et al.*, 2009). In industrial vegetable oil biodiesel, homogenous alkali-catalysed transesterification is commonly used because homogenous acid-catalyzed transesterification is around 4000 times slower (Chisti, 2007) and these catalysts (NaOH or KOH) are relatively less expensive (Helwani *et al.*, 2009).

However, acidic catalyst is preferable for homogenous transesterification of microalgae lipids because the biodiesel yield obtained by Nagle & Lemke (1990) was more than 50 times higher using HCl than NaOH (68 and 1.3% (g FAME/g lipids), respectively) as a catalyst for the same conditions of transesterification (0.1h, 70°C). The most used acid catalysts are CH_3COCl (Cooney *et al.*, 2009), HCl (Tran *et al.*, 2009), or H_2SO_4 (Miao & Wu, 2006). Using 4 concentrations of H_2SO_4 (0.56, 1.13, 1.35 or 2.25 mol/L) and a temperature of 90°C, Miao & Wu (2006) found that adding 2.25 mol/L catalyst gave lowest SG (0.863) but a slightly lowest yield of biodiesel (38%, g biodiesel/g lipid).

Different homogenous acid-alkaline (H_2SO_4-CH_3OK, KOH-HCl) catalysts can also be used to transesterify lipids from microalgae (Halim *et al.*, 2010; J. Lee *et al.*, 2010). Using a 1st transesterification step of 2 hours with H_2SO_4 at 50°C and a 2nd step of 2 hours with CH_3OK at 55°C, Halim *et al.* (2010) obtained a maximum biodiesel yield of 44% (g FAME/g lipid). The purpose to add an alkaline catalyst is mainly to increase the FAME yield (transmethylate acylglycerols and phospholipids) (Halim *et al.*, 2010).

Heterogeneous catalyst used for transesterification of triglycerides can be acid, alkali or enzymatic (immobilized enzymes) (Helwani *et al.*, 2009).

Transesterification studies testing heterogeneous catalysts were conducted with *Nannochloropsis* microalgae using alkaline earth oxide base catalyst such as strontium oxide (SrO) (Koberg *et al.*, 2011), calcium oxide (CaO) or magnesium oxide (MgO) (Umdu *et al.*, 2009). As example, performing a direct transesterification, Koberg *et al.* (2011) reported biodiesel yields up to 99.9% (g biodiesel/g lipid) for a time of reaction of 2 min at 60°C for ratio SrO:microalgae of 30% (g catalyst/g microalgae).

Few studies used enzymatic catalysts to transesterify lipids from microalgae. Generally, the most common used enzyme is lipase. A biodiesel yield of 98% (g FAME/g lipid) is reported for heterotrophic microalgae *Chlorella protothecoides* lipids transesterified using lipase (*Candida* sp. *99-125*) immobilized on macroporous resins at a concentration of 30% (g/g lipid, 12000 U/g lipid) at 38°C during 12h (Li *et al.*, 2007; Xiong *et al.*, 2008). However, enzymatic biodiesel production is often seen expensive due to the cost of the catalyst from 236 to 836 $US/ton biodiesel (Sotoft *et al.*, 2010). In comparison, the price of acid catalysts like H_2SO_4 (98%, w/w) is as low as 100 $US/ton H_2SO_4 (Tao & Aden, 2009) for a price of catalyst for biodiesel production from 1.4 to 5.3 $US/ton biodiesel (1 and 5% (w/w) catalyst, respectively) (Lotero *et al.*, 2005).

3.4.5 Reaction time, temperature and stirring

Increasing the reaction time has a positive effect on the SG of the biodiesel produced. For example, performing a direct transesterification of *Chlorella* microalgae at 30°C with H_2SO_4 as catalyst, Ehimen *et al.* (2010) found that the SG decreased from 0.914 to 0.884 when the reaction time was increased from 0.25 to 12h.

The temperature seems to have less effect on the microalgae biodiesel production than reaction time except for high temperatures. For example, Miao & Wu (2006) used H_2SO_4 catalyst (2.25 mol/L) and found similar biodiesel yield of 56 and 58% (g biodiesel/g lipid) at temperatures of 30 and 50°C, respectively. At 90°C, the biodiesel yield dropped of about 38% (g biodiesel/g lipid).

Stirring can have a positive effect on the biodiesel quality. For example, Ehimen *et al.* (2010) observed a decrease of the SG from 0.9032 (stirring at 500 rpm) to 0.8831 (no stirring).

3.4.6 Purification of biodiesel and by-products

Microalgae biodiesel and by-products must be separated for increasing the biodiesel production. The main separation processes used hot water (50°C) (Li *et al.*, 2007), organic solvents such as hexane (Halim *et al.*, 2010; Wiltshire *et al.*, 2000) and water-organic solvent for a liquid-liquid separation (Couto *et al.*, 2010; Lewis *et al.*, 2000; Samorì *et al.*, 2010). When using a non-polar co-solvent for transesterification of lipids, only water is added to separate biodiesel from the by-products (M. B. Johnson & Wen, 2009).

To our present knowledge, there is no study on the purification of biodiesel from microalgae. Based on 1st generation biodiesel (Leung *et al.*, 2010), three mains means of purification on biodiesel could be applied to microalgae biodiesel purification: "1-water washing 2-dry washing 3-membrane extraction." Based on vegetable oil biodiesel production (Berriosa & Skelton, 2008) and microalgae composition, the main by-products could be unreacted lipids, water, alcohol, chlorophyll, metals and glycerol.

Among the by-products obtained from the biodiesel production, glycerol is the most interesting. Glycerol worldwide consumption remains relatively constant, in recent years, with a consumption of 600 kton/year. Twenty-six percent of glycerol consumption was associated with pharmaceutical, cosmetic and soap industries (Bondioli, 2003). From 2004 to 2011, massive biofuel production created a problem of overproduced glycerol and the price of crude glycerol (80% pure) decreased from 110 to 7.5 $US/ton (The Jacobsen, 2011; Yazdani & Gonzalez, 2007). Gained glycerol can be transformed into added-value products using many paths including chemical, thermochemical or biological conversion.

Among the chemical added-value products, glycerol can be oxidized or reduced to many compounds like propylene glycol, propionic acid, acrylic acid, propanol, isopropanol, allyl alcohol and acrolein but only some of these products are interesting in terms of market or profitability (D. T. Johnson & Taconi, 2007).

Glycerol can be also converted into Fischer-Tropsch fuel at low temperature (225-300°C) by catalytic processes (Soares *et al.*, 2006) or transformed into hydrogen (H_2) by catalytic (generally nickel, platinum or ruthenium) or non-catalytic reforming (Vaidya & Rodrigues, 2009).

Biological conversion of glycerol includes fermentation into alcohols (ethanol, butanol, 1,3-propanediol) and other products like H_2, formate, propionic or succinic acid (Yazdani & Gonzalez, 2007).

Anaerobic digestion of by-products is another possible way to make biodiesel from microalgae cost effective if the lipid content of the microalgae does not exceed 40% (g lipid/g dry weight) (Sialve *et al.*, 2009). For example, Ehimen *et al.* (2010) used anaerobic digestion of microalgae residues issued from a direct transesterification with a constant loading rate of 5 kg volatile solids (VS)/m^3, temperatures and carbon-to-nitrogen (C/N) mass ratio varying from 25 to 40°C ant 5.4 to 24, respectively. For a temperature of 40°C and a C/N mass ratio of 8.53, a maximum methane (CH_4) concentration of 69% (v/v) with a specific CH_4 yield of 0.308 m^3 CH_4/kg VS was obtained.

4. Characteristics of biodiesel

The physicochemical properties of biodiesel are nearly similar to diesel fuel. The most important properties for biodiesel are cetane number, heat of combustion, viscosity, oxidative stability, cold flow properties and lubricity (Knothe *et al.*, 2005). Table 1 presents the main properties of microalgae biodiesel compared with diesel and 1st generation biodiesel.

1. The cetane number is an indicator of quality of ignition of a fuel which increases with the number of carbon and decreases with the number of unsaturated carbon bounds (Hart Energy Consulting, 2007). Consequently, a higher unsaturated biodiesel like microalgae biodiesel would have a lower cetane number. Based on our present knowledge, no measurement of the cetane number of microalgae biodiesel has been performed. However, some studies approximated the cetane numbers of many species based on their FAME content and found cetane numbers ranging from 39 to 54 (Stansell *et al.*, 2011), while cetane number of petrodiesel fuel are at least between 47 and 51 (ASTM Standard D6751-10, 2010; Knothe, 2006).
2. The heat of combustion shows if a biodiesel is suitable to burn in a diesel engine. The heat of combustion increases with the length of the carbon chain (Knothe, 2005a). Using lipids extracted from heterotrophic microalgae in the presence of H_2SO_4 in methanol, Miao & Wu (2006) obtained a biodiesel with a heat of combustion of 35.4 MJ/L which is in the range of diesel fuel (36-38 MJ/L).
3. As the cetane number, the viscosity increases with the number of carbon and decreases with the degree of unsaturation (Knothe, 2005b). A higher kinematic viscosity would create engine problems like engine deposits (Knothe & Steidley, 2005a). Transesterification favours a decrease in the viscosity of the oil at values usually between 4 to 6 mm^2/s (40°C) (National Renewable Energy Laboratory, 2009).
4. Oxidation of biodiesel could happen when the FAME are in contact with oxygen and are transformed into hydrogenoperoxides, aldehydes, acids and other oxygenates, which might form deposits (Knothe, 2005). Oxidation of the biodiesel increases as a function of the degree of unsaturation (Hart Energy Consulting, 2007). Oxidation stability of microalgae lipids is therefore a real problem (Stansell *et al.*, 2011) that can be overcome by adding antioxidants if the biodiesel blend is stored more than a few months (National Renewable Energy Laboratory, 2009)
5. Cold flow properties are important parameters for biodiesel production for northern countries like Canada and could be measured by cloud and pour points. The decrease of temperature could lead to the formation of visible crystals ($d \geq 0.5$ µm) in the biodiesel at a limit called cloud point (Knothe, 2005). Cloud point temperature decreases with the mole fraction of unsaturated compounds and slightly increases with the length of the carbon chain (Imahara *et al.*, 2006). Pour point is defined as the temperature at which biodiesel does not flow anymore. Usually, cloud and pour points increase as a function of the molar ratio of biodiesel in diesel fuel from 0 to 100% (National Renewable Energy Laboratory, 2009). A higher level of polyunsaturated compounds in microalgae biodiesel could be a benefit in terms of cold properties (cold and poor points) for a blend microalgae biodiesel/petrodiesel in cold climates.
6. The definition of lubricity for a fuel is "the ability to reduce friction between solid surfaces in relative motion" (Chevron Corporation, 2007; Shumacher, 2005). The lubricity of diesel fuel is influenced by the viscosity, the acidity, the water content and the sulphur compounds (Seregin *et al.*, 1975). Even with additives, the measured friction (no unit reported) of biodiesel (0.114 and 0.117) is lower than the one of petrodiesel

Fuel property	Microalgae biodiesel	References	Diesel	References	Biodiesel	References
Heating value (MJ/L)	35.4	(Miao & Wu, 2006)	36 to 38	(National Renewable Energy Laboratory, 2009)	32 to 36	(Demirbas, 2008; Srivastava & Prasad, 2000)
Cetane Number	39-54[b]	(Stansell *et al.*, 2011)	40 to 55	(Lapuerta *et al.*, 2008)	45 to 70	(Bozbas, 2008; Srivastava & Prasad, 2000)
Kinematic viscosity (at 40°C, mm^2/s)	3.87-5.2	(M. B. Johnson & Wen, 2009; Miao & Wu, 2006)	1.9 to 3.8	(Bozbas, 2008)	2.8 to 5.7	(Demirbas, 2008; Srivastava & Prasad, 2000)
Specific gravity (15.5°C)	0.864	(Miao & Wu, 2006)	0.81 to 0.86	(Lapuerta *et al.*, 2008)	0.86 to 0.89	(Alptekin & Canakci, 2008; Srivastava & Prasad, 2000)
Sulfur (wt.%)	0.0069	(M. B. Johnson & Wen, 2009)	0.01 to 0.04	(Bozbas, 2008)	0.0000 to 0.0024	(Bozbas, 2008)
Flash point (°C)	115-204	(M. B. Johnson & Wen, 2009; Miao & Wu, 2006)	60 to 80	(National Renewable Energy Laboratory, 2009)	96 to 190	(Srivastava & Prasad, 2000)
Cloud point (°C)	Not reported		-20 to 0	(Lapuerta *et al.*, 2008)	-3 to 15	(National Renewable Energy Laboratory, 2009; Srivastava & Prasad, 2000)
Pour point (°C)			-35 to -15	(National Renewable Energy Laboratory, 2009)	-5 to 10	(Tyson, 2001)
Oxidation stability (h)					3 max[c]	(ASTM Standard D6751-10, 2010)
Lubricity (25°C)			0.509 to 0.238[d]	(Knothe & Steidley, 2005b)	0.114[d]	(Knothe & Steidley, 2005b)

a Calculated

Table 1. Properties of microalgae biodiesel, diesel and biodiesel

(0.238 and 0.210) for 25 and 60°C (Knothe & Steidley, 2005b). Consequently, a benefit of adding biodiesel in conventional low sulphur diesel fuel is to improve lubricity (Muñoz *et al.*, 2011). For microalgae biodiesel, no lubricity study, to our knowledge, was reported from the literature.

5. Standards

European (EN 14214) and American standard (ASTM D6751-10) differ on some levels for biodiesel. For example, EN 14214 stipulates that, polyunsaturated (≥ 4 bonds) ester content and linoleic ester content must be less than 1% (mol/mol) and 12% (mol/mol), respectively (Knothe, 2010). These restrictions are important for biodiesel production from microalgae as the following ester composition is often rich in polyunsaturated content and over these levels (M. B. Johnson & Wen, 2009; Koberg *et al.*, 2011). Some other parameters of EN 14214 are not present in ASTM D6751-10 like esters content, density, iodine value and fatty acid contamination.

EN 14214 is more restrictive for biodiesel from microalgae, as the standard of cetane number is higher (51 vs 47), the maximum viscosity is lower (5 vs 6 mm^2/s) and the oxidative stability must be higher (6 vs 3 h). Consequently, the ASTM D6751 standard seems to be more adapted to biodiesel from microalgae.

Despite the fact that many studies produced biodiesel from microalgae, few studies ensure that the latter satisfies the official standards (M. B. Johnson & Wen, 2009; Miao & Wu, 2006). Producing biodiesel from *Schizochytrium limacinum*, Johnson & Wen (2009) reported a biodiesel that failed to meet the Grade S15 ASTM standard of water and sediments content of 0.05% (v/v) max (0.1% (v/v)) and sulfur content of 15 ppm max (69 ppm) but satisfied the ASTM acid number of 0.5 mg KOH/g max (0.11 mg KOH/g), free glycerine of 0.020% (w/w) max (0.003% w/w), total glycerine of 0.240% (w/w) max (0.097% w/w), corrosiveness to copper of no. 3 max (1a), kinematic viscosity (40°C) between 1.9-6.0 (3.87 mm^2/s) and flash point (closed cup) of 93°C min (204°C). In another study, Miao & Wu (2006) satisfied to meet the Grade S15 ASTM standard of flash point (closed cup) (115°C), viscosity (5.2 mm^2/s, 40°C) and acid number (0.374 mg KOH/g). However, based on our present knowledge, methanol content (EN 14110), sulphated ash (D874), cloud point (D2500), cold soak filterability, P content (D4951), distillation temperature (D1160), sodium and potassium (EN 14538), oxidative stability (EN 15751) have not been measured for microalgae biodiesel, as microalgae biodiesel is relatively new.

6. Conclusion

With the increase of the price of crude oil in the late 00s, blending biodiesel with petrodiesel appears a sustainable solution to reduce the dependency on oil producing countries. For now, the 1st generation of biofuels currently used could have economic, environmental and social negative consequences. To overcome these problems, producing biodiesel from microalgae lipids seems to be a sustainable solution as microalgae could be used to reduce the CO_2 emissions from coal power plants or wastewater pollution. Researchers are working to engineer super lipids producing microalgae strain in order to increase the yield of biodiesel. Producing biodiesel by transesterification of lipids remains relatively costly compared to 1st generation biodiesel. Moreover, the polyunsaturated content of these biodiesels could be profitable for cold properties, but seems to cause problems with the cetane number and the oxidative stability to meet the American standard (ASTM). Thus, more efforts must be made to reduce the process costs and to increase the biodiesel quality.

7. Acknowledgment

The authors are grateful to le Fond Québécois de la Recherche sur la Nature et les Technologies (FQRNT) for the grant to Michèle Heitz and Nathalie Faucheux for the research program in partnership contributing to the reduction of greenhouse gases.

The author for correpondence is Michèle Heitz, a full professor at the Chemical Engineering and Biotechnological Engineering Department at Université de Sherbrooke, 2500 boulevard de l'Université, Sherbrooke, QC, J1K 2R1, Canada (phone: +1-819-821-8000 ext 62827; fax: +1-819-821-7955; e-mail: Michele.Heitz@USherbrooke.ca).

8. References

Alptekin, E. & Canakci, M. (2008). Determination of the density and the viscosities of biodiesel-diesel fuel blends. *Renewable Energy,* Vol.33, No.12, pp. 2623-2630

Altin, R.; Çetinkaya, S. & Yücesu, H.S. (2001). Potential of using vegetable oil fuels as fuel for diesel engines. *Energy Conversion and Management,* Vol.42, No.5, pp. 529-538

Andrich, G.; Nesti, U.; Venturi, F.; Zinnai, A. & Fiorentini, R. (2005). Supercritical fluid extraction of bioactive lipids from the microalga *Nannochloropsis* sp.. *European Journal of Lipid Science and Technology,* Vol.107, No.6, pp. 381-386

Antizar-Ladislao, B. & Turrion-Gomez, J.L. (2008). Second-generation biofuels and local bioenergy systems. *Biofuels, Bioproducts and Biorefining,* Vol.2, No.5, pp. 455-469

ASTM Standard D6751-10, Standard specification for biodiesel fuel blend stock (B100) for middle distillate fuels, Available from www.astm.org

Bai, M.; Cheng, C.; Wan, H. & Lin, Y. (2011). Microalgal pigments potential as byproducts in lipid production. *Journal of the Taiwan Institute of Chemical Engineers,* Vol.42, No.5, pp. 783-786

Banerjee, A.; Sharma, R.; Chisti, Y. & Banerjee, U.C. (2002). *Botryococcus braunii*: a renewable source of hydrocarbons and other chemicals. *Critical Reviews in Biotechnology,* Vol.22, No.3, (2002), pp. 245-279

Banerjee, N. (2011). *U.S. to reduce oil imports by a third by 2021, Obama says. Los Angeles Times,* Available from http://articles.latimes.com/2011/mar/31/nation/la-na-obama-energy-20110331/2

Basha, S.A.; Gopal, K.R. & Jebaraj, S. (2009). A review on biodiesel production, combustion, emissions and performance. *Renewable and Sustainable Energy Reviews,* Vol.13, No.6-7, pp. 1628-1634

Berriosa, M. & Skelton, R.L. (2008). Comparison of purification methods for biodiesel. *Chemical Engineering Journal,* Vol.144, pp. 459-465

Bindraban, P.S.; Bulte, E.H. & Conijn, S.G. (2009). Can large-scale biofuels production be sustainable by 2020? *Agricultural Systems,* Vol.101, pp. 197-199

Blanco, A.M.; Moreno, J.; Del Campo, J.A.; Rivas, J. & Guerrero, M.G. (2007). Outdoor cultivation of lutein-rich cells of *Muriellopsis* sp. in open ponds. *Applied Microbiology and Biotechnology,* Vol.73, No.6, pp. 1259-1266

Bondioli, P. (2003). From oil seeds to industrial products present and near future of oleochemistry. *Italian Journal of Agronomy*, Vol.7, No.2, pp. 129-135, ISSN 1125-4718

Bordet, J.; Michez, J. & Gilot, A. (2006). *Mise en oeuvre du plan biocarburant au regard de la protection de la ressource en eau.* Ministère de l'agriculture, de l'alimentation, de la pêche, de la ruralité et de l'aménagement du territoire, pp. 1-39, Available from http://agriculture.gouv.fr/IMG/pdf/plan_biocarburants.pdf

Bozbas, K. (2008). Biodiesel as an alternative motor fuel: Production and policies in the European Union. *Renewable and Sustainable Energy Reviews,* Vol.12, No.2, pp. 542-552

BP (2011). BP Statistical Review of World Energy, 07.132011, Available from http://www.bp.com/liveassets/bp_internet/globalbp/globalbp_uk_english/reports_and_publications/statistical_energy_review_2011/STAGING/local_assets/pdf/statistical_review_of_world_energy_full_report_2011.pdf

Cadoret, J. & Bernard, O. (2008). La production de biocarburant lipidique avec des microalgues : promesses et défis. *Journal de la Société de Biologie,* Vol.202, No.3, pp. 201-211

Callieri, C. & Stockner, J.G. (2002). Freshwater autotrophic picoplankton: A review. *Journal of Limnology,* Vol.61, No.1, pp. 1-14

Cantin, I. (2010). *La production de biodiesel à partir des microalgues ayant un métabolisme hétérotrophe.* Maîtrise en environnement (M.Env), Université de Sherbrooke, Sherbrooke, Québec, Canada, pp. 1-82

Carlsson, A.S.; van Beilen, J.B.; Moller, R. & Clayton, D. (2007). *Micro- and macro-algae: utility for industrial applications.* CPL Press, Newbury, UK, pp. 1-86

Carvalho, A.P.; Meireles, L.A. & Malcata, F.X. (2006). Microalgal reactors: A review of enclosed system designs and performances. *Biotechnology Progress,* Vol.22, pp. 1490-1506

Chen, C.; Yeh, K.; Aisyah, R.; Lee, D. & Chang, J. (2011). Cultivation, photobioreactor design and harvesting of microalgae for biodiesel production: A critical review. *Bioresource Technology,* Vol.102, No.1, (1), pp. 71-81, ISSN 0960-8524

Chevron Corporation (2007). *Diesel Fuels Technical Review.* Chevron Products Company, San Ramon, CA, USA, pp. 1-116, Available from http://www.chevron.com/products/prodserv/fuels/documents/Diesel_Fuel_Tech_Review.pdf

Chinnasamy, S.; Bhatnagar, A.; Hunt, R.W. & Das, K.C. (2010). Microalgae cultivation in a wastewater dominated by carpet mill effluents for biofuel applications. *Bioresource Technology,* Vol.101, No.9, pp. 3097-3105

Chisti, Y. (2007). Biodiesel from microalgae. *Biotechnology Advances,* Vol.25, No.3, pp. 294-306

Choi, S.P.; Nguyen, M.T. & Sim, S.J. (2010). Enzymatic pretreatment of *Chlamydomonas reinhardtii* biomass for ethanol production. *Bioresource Technology,* Vol.101, pp. 5330-5336

Cooney, M.; Young, G. & Nagle, N. (2009). Extraction of bio-oils from microalgae. *Separation & Purification Reviews,* Vol.38, pp. 291-325

Couto, R.M.; Simões, P.C.; Reis, A.; Da Silva, T.L.; Martins, V.H. & Sánchez-Vicente, Y. (2010). Supercritical fluid extraction of lipids from the heterotrophic microalga *Crypthecodinium cohnii. Engineering in Life Sciences,* Vol.10, No.2, pp. 158-164 de Morais, M.G. & Costa, J.A.V. (2007). Isolation and selection of microalgae from coal fired thermoelectric power plant for biofixation of carbon dioxide. *Energy Conversion and Management*, Vol.48, No.7, pp. 2169-2173

Demirbas, A. (2008). Relationships derived from physical properties of vegetable oil and biodiesel fuels. *Fuel*, Vol.87, No.8-9, pp. 1743-1748

Deng, X.; Li, Y. & Fei, X. (2009). Microalgae: A promising feedstock for biodiesel. *African Journal of Microbiology Research*, Vol.3, No.13, pp. 1008-1014

Dexter, J. & Pengcheng, F. (2009). Metabolic engineering of cyanobacteria for ethanol production. *Energy & Environmental Science*, Vol.2, pp. 857-864

Dhepe, P.L.; Fukuoka, A. & Ichikawa, M. (2003). Novel fabrication and catalysis of nano-structured Rh and RhPt alloy particles occluded in ordered mesoporous silica templates using supercritical carbon dioxide. *Physical Chemistry Chemical Physics*, Vol.5, No.24, pp. 5565-5573

Dry, M.E. (1996). Practical and theoretical aspects of the catalytic Fischer-Tropsch process. *Applied Catalysis A: General*, Vol.138, pp. 319-344

Ehimen, E.A.; Sun, Z.F. & Carrington, C.G. (2010). Variables affecting the in situ transesterification of microalgae lipids. *Fuel*, Vol.89, No.3, pp. 677-684, ISSN 0016-2361

Ehimen, E.A.; Sun, Z.F.; Carrington, C.G.; Birch, E.J. & Eaton-Rye, J.J. (2010). Anaerobic digestion of microalgae residues resulting from the biodiesel production process. *Applied Energy*, Vol.88, No.10, pp. 3454-3463, ISSN 0306-2619

Ervin, M.J. & Associated (2011). Petroleum Price Data, In: kentmarketingservices.com, 23.07.2011, Available from
http://www.kentmarketingservices.com/dnn/ PetroleumPriceData.aspx

FAO (2011a). FAOSTAT: Crops, 07.142011, Available from http://faostat.fao.org

FAO (2011b). FAOSTAT: Food Price Index, In: FAO, 23.05.2011, Available from
http://www.fao.org/worldfoodsituation/wfs-home/foodpricesindex/en/

Fatih Demirbas, M. (2009). Biorefineries for biofuel upgrading: A critical review. *Applied Energy*, Vol.86, No.SUPPL. 1, pp. S151-S161

Fedorov, A.S.; Kosourov, S.; Ghirardi, M.L. & Seibert, M. (2005). Continuous hydrogen photoproduction by *Chlamydomonas reinhardtii*: Using a novel two-stage, sulfate-limited chemostat system. *Applied Biochemistry and Biotechnology - Part A Enzyme Engineering and Biotechnology*, Vol.121, No.1-3, pp. 403-412

Fu, C.; Hung, T.; Chen, J.; Su, C. & Wu, W. (2010). Hydrolysis of microalgae cell walls for production of reducing sugar and lipid extraction. *Bioresource Technology*, Vol.101, No.22, (November 2010), pp. 8750-8754

Garofalo, R. (2010). Algae and aquatic biomass for a sustainable production of 2nd generation biofuels, 07.15.2011, Available from http://www.aquaculture.ugent.be

Goldemberg, J. & Guardabassi, P. (2009). Are biofuels a feasible option? *Energy Policy*, Vol.37, No.1, pp. 10-14

Gopinathan, C.P. (2004). Marine microalgae, Proceedings of Ocean Life Food and Medicine ExpoChennai, India, February 2004

Grobbelaar, J.U. (2000). Physiological and technological considerations for optimising mass algalcultures. *Journal of Applied Phycology*, Vol.12, pp. 201-206

Halim, R.; Gladman, B.; Danquah, M.K. & Webley, P.A. (2010). Oil extraction from microalgae for biodiesel production. *Bioresource Technology*, Vol.102, No.1, pp. 178-185

Hart Energy Consulting (2007). *Establishment of the Guidelines for the Development of Biodiesel Standards in the APPEC Region*. Asia Pacific Economic Cooperation, pp. 1-136

Helwani, Z.; Othman, M.R.; Aziz, N.; Kim, J. & Fernando, W.J.N. (2009). Solid heterogeneous catalysts for transesterification of triglycerides with methanol: A review *Applied Catalysis*, Vol.363, pp. 1-10

Hill, J.; Nelson, E.; Tilman, D.; Polasky, S. & Tiffany, D. (2006). Environmental, economic, and energetic costs and benefits of biodiesel and ethanol biofuels. *PNAS*, Vol.103, No.30, pp. 11206-11210

Hubbert, K. (1956). Nuclear energy and the fossil fuels, 01.07.2010, Available from http://www.hubbertpeak.com/hubbert/1956/1956.pdf

Huntley, M.E. & Redalje, D.G. (2007). CO_2 mitigation and renewable oil from photosynthetic microbes: a new appraisal. *Mitigation et Adaptation Strategies for Global Change*, Vol.12, pp. 573-608

Huntley, M.E. & Redalje, D.C. (2008). *Continous-batch hybrid process for production of oil and other useful products from photosynthetic microbes*, volume US 2008/0118964, n° 7770322, p. 1-9. Available from http://www.freepatentsonline.com/7770322.html

Imahara, H.; Minami, E. & Saka, S. (2006). Thermodynamic study on cloud point of biodiesel with its fatty acid composition. *Fuel*, Vol.85, No.12-13, pp. 1666-1670

Institut Français du Pétrole (2005). Réserves de pétrole : des données évolutives en fonction de la technique et de l'économie, In: Institut français du pétrole, 15.06.2010, Available from http://www.ifp.fr/content/download/57554/1264970/version/1/ file/IFP-ConfPresse-31mai05_Reserves_6_DonneesEvolutivesSurLesReserves.pdf

International Energy Agency (2008). *World Energy outlook 2008*. OECD/IEA, Paris, pp. 1-569, Available from http://www.iea.org/textbase/nppdf/free/2008/weo2008.pdf

Jacob-Lopes, E. & Teixeira Franco, T. (2010). Microalgae-based systems for carbon dioxide sequestration and industrial biorefineries, In: *Biomass*, Momba, M. & Bux, F., pp. 135-146 978-953-307-113-8, Sciyo, Croatia

Johnson, M.B. & Wen, Z. (2009). Production of biodiesel fuel from the microalga *Schizochytrium limacinum* by direct transesterification of algal biomass. *Energy Fuels*, Vol.23, pp. 5179-5183

Johnson, D.T. & Taconi, K.A. (2007). The glycerin glut: Options for the value-added conversion of crude glycerol resulting from biodiesel production. *Environmental Progress*, Vol.26, No.4, pp. 338-348

Kapdan, I.K. & Kargi, F. (2006). Bio-hydrogen production from waste materials. *Enzyme Microbiological Technology*, Vol.38, pp. 569-582

Khan, S.A.; Hussain, M.Z.; Prasad, S. & Banerjee, U. (2009). Prospects of biodiesel production from microalgae in India. *Renewable & Sustainable Energy Reviews*, Vol.13, No.9, pp. 2361-2372

Khozin-Goldberg, I. & Cohen, Z. (2006). The effect of phosphate starvation on the lipid and fatty acid composition of the fresh water eustigmatophyte *Monodus subterraneus*. *Phytochemistry*, Vol.67, No.7, pp. 696-701

Knothe, G. (2005). Oxidative stability of biodiesel, In: *The biodiesel handbook,* Knothe, G., Krahl, J. & Gerpen, J. V., pp. 122-126 AOCS Press, Campaign, Illinois, USA

Knothe, G.; Gerpen, J.V. & Krahl, J. (2005). *The biodiesel handbook,* AOCS Press, Campaign, Illinois, USA

Knothe, G. (2005a). Cetane numbers–heat of combustion–why vegetable oils and their derivatives are suitable as a diesel fuel, In: *The biodiesel handbook,* Knothe, G., Krahl, J. & Gerpen, J. V., pp. 72-80 AOCS Press, Campaign, Illinois, USA

Knothe, G. (2005b). Viscosity of Biodiesel, In: *The biodiesel handbook,* Knothe, G., Krahl, J. & Gerpen, J. V., pp. 81-82 AOCS Press, Campaign, Illinois, USA

Knothe, G. (2006). Analyzing biodiesel: standards and other methods. *Journal of the American Oil Chemists' Society,* Vol.83, No.10, pp. 823-833

Knothe, G. (2010). Biodiesel and renewable diesel: A comparison. *Progress in Energy and Combustion Science,* Vol.36, No.3, (2010), pp. 364-373

Knothe, G. & Steidley, K.R. (2005a). Kinematic viscosity of biodiesel fuel components and related compounds. Influence of compound structure and comparison to petrodiesel fuel components. *Fuel,* Vol.84, No.9, pp. 1059-1065

Knothe, G. & Steidley, K.R. (2005b). Lubricity of components of biodiesel and petrodiesel. The origin of biodiesel lubricity. *Energy and Fuels,* Vol.19, No.3, pp. 1192-1200

Koberg, M.; Cohen, M.; Ben-Amotz, A. & Gedanken, A. (2011). Bio-diesel production directly from the microalgae biomass of *Nannochloropsis* by microwave and ultrasound radiation. *Bioresource Technology,* Vol.102, pp. 4265-4269

Lapuerta, M.; Armas, O. & Rodríguez-Fernández, J. (2008). Effect of biodiesel fuels on diesel engine emissions. *Progress in Energy and Combustion Science,* Vol.34, No.2, pp. 198-223

Lardon, L.; Hélias, A.; Sialve, B.; Steyer, J. & Bernard, O. (2009). Life-cycle assessment of biodiesel production from microalgae. *Environmental Science and Technology,* Vol.43, No.17, pp. 6475-6481

Lee, J.; Yoo, C.; Jun, S.; Ahn, C. & Oh, H. (2010). Comparison of several methods for effective lipid extraction from microalgae. *Bioresource Technology,* Vol.101, pp. 575-577

Lee, S.; Yoon, B. & Oh, H. (1998). Rapid method for the determination of lipid from the green alga *Botryococcus braunii. Biotechnology Techniques,* Vol.12, No.7, pp. 553-556

Leung, D.Y.C.; Wu, X. & Leung, M.K.H. (2010). A review on biodiesel production using catalyzed transesterification. *Applied Energy,* Vol.87, pp. 1083-1095

Lewis, T.; Nichols, P.D. & McMeekina, T.A. (2000). Evaluation of extraction methods for recovery of fatty acids from lipid-producing microheterotrophs. *Journal of Microbiological Methods,* Vol.42, pp. 107-116

Li, X.; Xu, H. & Wu, Q. (2007). Large-scale biodiesel production from microalga *Chlorella protothecoides* through heterotrophic cultivation in bioreactors. *Biotechnology and Bioengineering,* Vol.98, No.4, pp. 764-771

Liang, Y.; Sarkany, N. & Cui, Y. (2009). Biomass and lipid productivities of *Chorella vulgaris* under autotrophic, heterotrophic and mixotrophic growth conditions. *Biotechnology Letters,* Vol.31, pp. 1043-1049

Longwell, J.P.; Rubint, E.S. & Wilson, J. (1995). Coal: Energy for the future. *Progress in Energy and Combustion Science,* Vol.21, pp. 269-360

Lotero, E.; Liu, Y.; Lopez, D.E.; Suwannakarn, K.; Bruce, D.A. & Goodwin Jr., J.G. (2005). Synthesis of biodiesel via acid catalysis. *Industrial and Engineering Chemistry Research,* Vol.44, No.14, pp. 5353-5363

Malińska, K. & Zabochnicka-Świątek, M. (2010). Biosystems for air protection, In: *Air Pollution,* Villanyi, V., pp. 177-194, InTech, 978-953-307-143-5,

Marchetti, J.M.; Miguel, V.U. & Errazu, A.F. (2007). Possible methods for biodiesel production. *Renewable and Sustainable Energy Reviews,* Vol.11, No.6, pp. 1300-1311

Martek (2008). Martek Biosciences Corporation, In: Martek, 17.06.2010, Available from http://aboutmartek.martek.com/researchanddevelopment/

Miao, X. & Wu, Q. (2004). High yield bio-oil production from fast pyrolysis by metaboliccontrolling of *Chlorella protothecoides. Journal of Biotechnology,* Vol.110, pp. 85-93

Miao, X.; Wu, Q. & Yang, C. (2004). Fast pyrolysis of microalgae to produce renewable fuels. *Journal of Applied Pyrolysis,* Vol.71, No.2004, pp. 855-863

Miao, X. & Wu, Q. (2006). Biodiesel production from heterotrophic microalgal oil. *Bioresource Technology,* Vol.97, No.6, pp. 841-846

Muñoz, M.; Moreno, F.; Monné, C.; Morea, J. & Terradillos, J. (2011). Biodiesel improves lubricity of new low sulphur diesel fuels. *Renewable Energy,* Vol.36, No.11, pp. 2918-2924

Nagle, N. & Lemke, P. (1990). Production of methyl-ester fuel from microalgae. *Applied Biochemistry and Biotechnology,* Vol.24, No.24-25, pp. 355-361

Naik, S.N.; Goud, V.V.; Rout, P.K. & Dalai, A.K. (2010). Production of first and second generation biofuels: A comprehensive review. *Renewable and Sustainable Energy Reviews,* Vol.14, No.2, pp. 578-597

National Renewable Energy Laboratory (2009). *Biodiesel handling and use guide.* pp. 1-56

National Research Council (2007). Water Implications of Biofuels Production, In: National Research Council, 01.07.2010, Available from http://nationalacademies.org/wstb

Natrah, F.M.I.; Yusoff, F.M.; Shariff, M.; Abas, F. & Mariana, N.S (2007). Screening of Malaysian indigenous microalgae for antioxidant properties and nutritional value. *Journal of Applied Phycology,* Vol.19, No.6, pp. 711-718

Natural Ressources Canada (2011a). Government of Canada calls on industry to participate in new biofuels initiative, In: Natural Ressources Canada, 25.05.2011, Available from http://www.nrcan.gc.ca/media/newcom/2007/2007124-eng.php

Natural Ressources Canada (2011b). Initiative de démonstration nationale sur le diesel renouvelable, In: Natural Ressources Canada, 23.05.2011, Available from http://oee.nrcan.gc.ca/transports/carburants-remplacement/programmes/idndr/rapport-2010/chap1.cfm?attr=16

Naveed, S.; Ramzan, N.; Ashraf, A.; Usman, J. & Shabbir, H.A., S. (2010). Modelling and simulation of coal gasification plant. *Journal of Pakistan Institute of Chemical Engineers,* Vol.38, pp. 1-6

Nigam, P.S. & Singh, A. (2011). Production of liquid biofuels from renewable resources. *Progress in Energy and Combustion Science,* Vol.37, No.1, (2), pp. 52-68, ISSN 0360-1285

Nyboer, J. & Grove, S. (2010). *A review of renewable energy in Canada, 2008.* pp. 1-25, Available from http://www.cieedac.sfu.ca

Perrut, M. (2000). Supercritical fluid applications: Industrial developments and economic issues. *Industrial and Engineering Chemistry Research,* Vol.39, No.12, pp. 4531-4535

Pirog, R. (2005). *CRD Report for Congress: World oil demand and its effect on oil prices.* Congressional Research Service - The Library of Congress, pp. 1-23

Ranjbar, R.; Inoue, R.; Shiraishi, H.; Katsuda, T. & Katoh, S. (2008). High efficiency production of astaxanthin by autotrophic cultivation of *Haematococcus pluvialis* in a bubble column photobioreactor. *Biochemical Engineering Journal,* Vol.39, No.3, pp. 575-580

Román-Leshkov, Y.; Barrett, C.J.; Liu, Z.Y. & Dumesic, J.A. (2007). Production of dimethylfuran for liquid fuels from biomass-derived carbohydrates. *Nature,* Vol.447, No.7147, pp. 982-985

Samorì, C.; Torri, C.; Samorì, G.; Fabbri, D.; Galletti, P.; Guerrini, F.; Pistocchi, R. & Tagliavini, E. (2010). Extraction of hydrocarbons from microalga *Botryococcus braunii* with switchable solvents. *Bioresource Technology,* Vol.101, No.9, pp. 3274-3279

Sawayama, S.; Inoue, S.; Dote, Y. & Yokoyama, S. (1995). CO_2 fixation and oil production through microalga. *Energy Conversion and Management,* Vol.36, pp. 721-731

Seregin, E.P.; Gureev, A.A.; Bugai, V.T.; Makarov, A.A.; Sarantidi, P.G. & Skovorodin, G.B. (1975). Lubricity of diesel fuels. *Chemistry and Technology of Fuels and Oils,* Vol.11, No.5, pp. 360-363

Shafiee, S. & Topal, E. (2009). When will fossil fuel reserves be diminished? *Energy Policy,* Vol.37, pp. 181-189

Sheehan, J.; Combreco, V.; Duffield, J.; Graboski, M. & Shapouri, H. (1998a). *An overview of biodiesel and petroleum diesel life cycles.* National Renewable Energy Laboratory, Golden, Colorado, USA, pp. 1-47, Available from http://www.nrel.gov/docs /legosti/fy98/24772.pdf

Sheehan, J.; Dunahay, T.; Benemann, J. & Roessler, P. (1998b). *A look back at the U.S. Department of Energy's aquatic species program – biodiesel from algae.* U.S. Department of Energy's Office of Fuels Development, pp. 1-325, Available from http://www.nrel.gov/docs/legosti/fy98/24190.pdf

Shumacher, L. (2005). Biodiesel Lubricity, In: *The biodiesel handbook,* Knothe, G., Krahl, J. & Gerpen, J. V., pp. 127-133 AOCS Press, Campaign, Illinois, USA

Sialve, B.; Bernet, N. & Bernard, O. (2009). Anaerobic digestion of microalgae as a necessary step to make microalgal biodiesel sustainable. *Biotechnology Advances,* Vol.27, pp. 409-416

Smith, V.H.; Sturm, B.S.M.; deNoyelles, F.J. & Billings, S.A. (2009). The ecology of algal biodiesel production. *Trends in Ecology and Evolution,* Vol.25, No.6, pp. 301-309

Soares, R.R.; Simonetti, D.A. & Dumesic, J.A. (2006). Glycerol as a source for fuels and chemicals by low-temperature catalytic processing. *Angewandte Chemie - International Edition,* Vol.45, No.24, pp. 3982-3985

Sotoft, L.F.; Rong, B.; Christensen, K.V. & Norddahl, B. (2010). Process simulation and economical evaluation of enzymatic biodiesel production plant. *Bioresource Technology,* Vol.101, No.14, pp. 5266-5274, ISSN 0960-8524

Spolaore, P.; Joannis-Cassan, C.; Duran, E. & Isambert, A. (2006). Commercial applications of microalgae. *Journal of Bioscience and Bioengineering,* Vol.101, No.2, pp. 87-96

Srivastava, A. & Prasad, R. (2000). Triglycerides-based diesel fuels. *Renewable & Sustainable Energy Reviews,* Vol.4, No.2, pp. 111-133

Stansell, G.R.; Gray, V.M. & Sym, S.D. (2011). Microalgal fatty acid composition: implications for biodiesel quality. *Journal of Applied Phycology,* , ISSN 0921-8971

Statistics Canada (2011). *Energy statistics handbook: First quarter 2011.* pp. 1-170, Available from http://www.statcan.gc.ca/pub/57-601-x/57-601-x2011001-eng.pdf

Suh, I.S. & Lee, C. (2003). Photobioreactor engineering: Design and performance. *Biotechnology and Bioprocess Engineering,* Vol.8, No.6, pp. 313-321

Takagi, M.; Karseno & Yoshida, T. (2006). Effect of salt concentration on intracellular accumulation of lipids and triacylglyceride in marine microalgae *Dunaliella* cell. *Journal of Bioscience and Bioengineering,* Vol.101, pp. 223-226

Tan, K.T. & Lee, K.T. (2011). A review on supercritical fluids (SCF) technology in sustainable biodiesel production: Potential and challenges. *Renewable and Sustainable Energy Reviews,* Vol.15, No.5, pp. 2452-2456

Tao, L. & Aden, A. (2009). The economics of current and future biofuels. *In Vitro Cellular and Developmental Biology - Plant,* Vol.45, No.3, pp. 199-217

The Jacobsen (2011). Renewable fuels: The biodiesel bulletin, 07.04 2011, Available from http://www.thejacobsen.com

The White House (2010). Obama announces steps to boost biofuels, clean coal, In: Office of the Press Secretary, 23.05.2011, Available from http://www.whitehouse.gov/the-press-office/obama-announces-steps-boost-biofuels-clean-coal

Tran, H.; Hong, S. & Lee, C. (2009). Evaluation of extraction methods for recovery of fatty acids from *Botryococcus braunii* LB 572 and *Synechocystis* sp. PCC 6803. *Biotechnology and Bioprocess Engineering,* Vol.14, No.2, pp. 187-192

Tyson, S.K. (2001). *Biodiesel handling and use guidelines* pp. 1-22, Available from http://www.angelfire.com/ks3/go_diesel/files042803/biodiesel_handling.pdf

U.S. Energy Information Administration (2009). World proved reserves of oil and natural gas, most recent estimates, In: U.S. Energy Information Administration, 19.03.2011, Available from http://www.eia.doe.gov/international/reserves.html

U.S. Energy Information Administration (2011). *Annual energy outlook - 2011.* pp. 1-246, Available from http://www.eia.doe.gov/forecasts/aeo/pdf/0383(2011).pdf

Umdu, E.S.; Tuncer, M. & Seker, E. (2009). Transesterification of *Nannochloropsis oculata* microalga's lipid to biodiesel on Al_2O_3 supported CaO and MgO catalysts. *Bioresource Technology,* Vol.100, No.11, (6), pp. 2828-2831, ISSN 0960-8524

United States Environmental Protection Agency (2002). *A comprehensive analysis of biodiesel impacts on exhaust emissions.* pp. 1-118

Vaidya, P.D. & Rodrigues, A.E. (2009). Glycerol reforming for hydrogen production: A review. *Chemical Engineering and Technology,* Vol.32, No.10, pp. 1463-1469

Wang, L.; Min, M.; Li, Y.; Chen, P.; Chen, Y.; Liu, Y.; Wang, Y. & Ruan, R. (2010). Cultivation of green algae *Chlorella* sp. in different wastewaters from municipal wastewater treatment plant. *Applied Biochemistry and Biotechnology,* Vol.162, No.4, pp. 1174-1186

Wei, A.; Zhang, X.; Wei, D.; Chen, G.; Wu, Q. & Yang, S. (2009). Effects of cassava starch hydrolysate on cell growth and lipid accumulation of the heterotrophic microalgae

Chlorella protothecoides. Journal of Industrial Microbiology and Biotechnology, Vol.36, No.11, pp. 1383-1389

Wigmosta, M.S.; Coleman, A.M.; Skaggs, R.J.; Huesemann, M.H. & Lane, L.J. (2011). National microalgae biofuel production potential and resource demand. *Water Ressources Research,* Vol.47, No.W00H04, pp. 1-13

Williams, P.J. & Laurens, L.M. (2010). Microalgae as biodiesel and biomass feedstocks: Review and analysis of the biochemistry, energetics and economics. *Energy & Environmental Science,* Vol.3, pp. 554-590

Wiltshire, K.H.; Boersma, M.; Möller, A. & Buhtz, H. (2000). Extraction of pigments and fatty acids from the green alga *Scenedesmus obliquus* (Chlorophyceae). *Aquatic Ecology,* Vol.34, No.2, pp. 119-126

Wu, Z. & Shi, X. (2008). Rheological properties of *Chlorella pyrenoidosa* culture grown heterotrophically in a fermentor. *Journal of Applied Phycology,* Vol.20, No.3, pp. 279-282

Xin, L.; Hong-ying, H.; Ke, G. & Ying-xue, S. (2010). Effects of different nitrogen and phosphorus concentrations on the growth, nutrient uptake, and lipid accumulation of a freshwater microalga *Scenedesmus* sp. *Bioresource Technology,* Vol.101, No.14, pp. 5494-5500

Xiong, W.; Li, X.; Xiang, J. & Wu, Q. (2008). High-density fermentation of microalga *Chlorella protothecoides* in bioreactor for microbio-diesel production. *Applied Microbiology and Biotechnology,* Vol.78, pp. 29-36

Xu, H.; Miao, X. & Wu, Q. (2006). High quality biodiesel production from a microalga *Chlorella protothecoides* by heterotrophic growth in fermenters. *Journal of Biotechnology,* Vol.126, pp. 499-507

Xu, L.; Weathers, P.J.; Xiong, X. & Liu, C. (2009). Microalgal bioreactors: Challenges and opportunities. *Engineering in Life Sciences,* Vol.9, No.3, pp. 178-189

Yaguchi, T.; Tanaka, S.; Yokochi, T.; Nakahara, T. & Higashihara, T. (1997). Production of high yields of docosahexaenoic acid by *Schizochytrium* sp. strain SR21. *JAOCS, Journal of the American Oil Chemists' Society,* Vol.74, No.11, pp. 1431-1434

Yang, J.; Rasa, E.; Tantayotai, P.; Scow, K.M.; Yuan, H. & Hristova, K.R. (2011). Mathematical model of *Chlorella minutissima* UTEX2341 growth and lipid production under photoheterotrophic fermentation conditions. *Bioresource Technology,* Vol.102, No.3, pp. 3077-3082

Yazdani, S.S. & Gonzalez, R. (2007). Anaerobic fermentation of glycerol: a path to economic viability for the biofuels industry. *Current Opinion in Biotechnology,* Vol.18, No.3, pp. 213-219

11

Sulfonation/Sulfation Processing Technology for Anionic Surfactant Manufacture

Jesús Alfonso Torres Ortega
Universidad de La Salle
Colombia

1. Introduction

In 2008, global production of surfactants was 13 million metric tons reaching a turnover of US$24,33 million at 2009, which means an increment of 2% from the previous year. Moreover, it is projected a strong growth ca. 2,8% annually till 2012 and between 3,5 - 4% thereafter (Resnik et al., 2010). Sulfonation plants are scattered around the globe in production units with capacities varying from 3.000 to 50.000 tons/year, mainly of anionic surfactants. At least 800 sulfonation plants are estimated to be currently in operation around the World. However, about 20% of the global production (2.500.000 tons/year of sulfonated anionic surfactants) is concentrated in the United States, Western Europe and Japan (Acmite Market Intelligence, 2010).

Anionic surfactants are the key component in a detergent formulation. A molecule of anionic surfactant is composed of a lipophilic oil soluble "tail" (typically an organic molecule C_{12}-C_{14}) and a hydrophilic water soluble "head" (such as SO_3^-). Mixtures of organic molecules, either form non-renewable resources, such as crude oil or from renewable sources, such as vegetable oils, are currently used as raw materials for household detergents. The cleaning process performed by anionic surfactants (active detergents) is described in the following way (de Groot, 1991):

i. Wetting of the substrate and dirt due to reduction of surface tension;
ii. Remotion of dirt from substrate;
iii. Retaining the dirt in a stable solution or suspension.

Sulfonation is the term that identifies an electrophilic chemical reaction where a sulfonic group SO_3H is incorporated into a molecule with the capacity to donate electrons. The product of this chemical reaction is recognized as sulfonic acid if the electron donor molecule is a carbon. Sulfuric anhydride reacts easily with delocalized electronic densities as those present in aromatics groups or alkenes in general. These reactions produce a variety of products, including derivate polysulfones. On the other hand, the sulfating process involves the incorporation of the SO_3H molecules to an oxygen atom in an organic molecule to form C−O−S bonds and the sulfate group (Figure 1). Sulfates acids can be easily hydrolyzed, and for this reason an immediate neutralization is required after the sulfate group is formed (Foster, 1997). Although sulfonation and sulfating processes are employed industrially to obtain a wide range of products from hair dyes to pesticides and organic intermediates, their main applications are in the production of anionic surfactants (Foster, 2004).

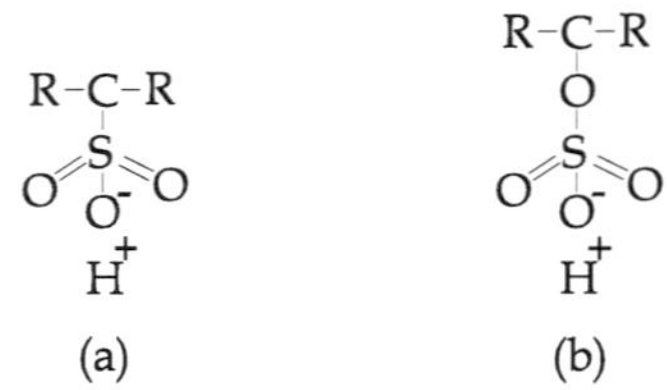

Fig. 1. Functional groups: (a) sulfonate $-SO_3H$ and (b) sulfate $-OSO_3H$

2. Main anionic surfactants

2.1 Linear alkyl benzene sulfonate (LABS)

Linear alkylbenzene is the most common organic feedstock employee in the detergent industry (Figure 2). LABS of low molecular weight (230 - 245) lay in the category of anionic surfactants most used in all ranges of household detergent formulations. Dishwashing liquids are prepared from LABS of low molecular weight in combination with other anionic surfactant as Lauryl Ether Sulfate (LES) promoting high detergency, foam stability, degreasing capacity, and high stability in hard water (Zhu et al., 1998). Common concentrations of active detergents in liquid products are: LABS 10-15% (30%), Primary alcohol sulfate/LES 3-5% (10%), where values in brackets are the maximum for concentrated products (Table 1). LABS of high molecular weight (245-260) are the anionic surfactants more used in all ranges of household detergents formulation, but especially in heavy duty laundry products, sometimes in combination with nonionics alcohol sulfates from tallow and soaps (Mungray & Kumar, 2009).

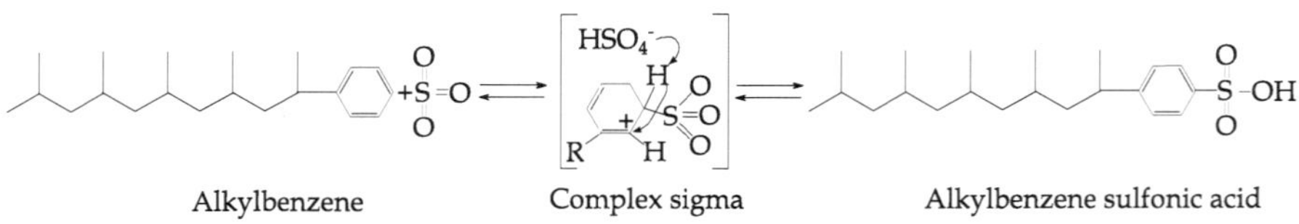

Fig. 2. Sulfonation of alkylbenzene (adapted from Foster, 1997)

Heavy powders detergents (no soapy)	High foam	Low foam
LABS, high molecular weight (245-260)	20 - 30%	5 - 10%
Tallow Alcohol Sulfate (TAS)		2 - 5 %
Nonionics		2 - 5 %
"Tallow" soap		2 - 5 %

Table 1. Heavy powders detergents used in all ranges of household detergents formulation (de Groot, 1991)

2.2 Primary alcohol sulfates (PAS)

PAS are categorized in different groups regarding the number carbon that compose them: The so called lauryl alcohol sulfates C_{12}-C_{14}, the "tallow" alcohol sulfates (TAS) C_{16}-C_{18}, and the broader cut C_{10}-C_{18} alcohol sulfate comprising coconut fatty alcohol sulfates. The broad cut (C_{10}-C_{18}) alcohol sulfates presents cost/performance equilibrium in terms of detergency, solubility and foaming properties. This product can partially or totally substitute other anionic surfactants either in liquid or powder detergent formulations with adequate biodegradability and low "defatting action", which is important for human tissue and delicate natural or synthetic fibers. The narrow cut (C_{12}-C_{14}) alcohol sulfates find their main application in a wide range of personal care products such as shampoos, bubble bath products, tooth pastes, dishwashing liquid, delicate products for laundry wash. The C_{16}-C_{18} alcohol sulfates ("tallow") are used as sodium salts in the formulation of heavy duty laundry products for hand and machine washing. Their detergency power is up to 10% higher than LABS in a wide range of detergent formulations (de Groot, 1991). Furthermore, TAS shows controlled foam, which is important mainly at high temperatures, still keeping the advantage of softness in the wash of sensitive natural and synthetic fibers (Rosen, 2005). The physical detergency and biodegradability of primary alcohols can be affected by the carbon chain length distribution. Therefore, each new supply may require testing to determine whether the desired properties in the chosen application can be achieved. The mechanism for alcohol sulfation is thought to be similar to that for linear alkylbenzene sulfonation with $\Delta H = -150$ kJ/mol (Figure 3).

$$ROH + 2SO_3 \xrightarrow[\text{(fast)}]{} \underset{\text{alkyl hydrogen pyrosulfate}}{ROSO_2OSO_3H}$$

$$ROSO_2OSO_3H + ROH \xrightarrow[\text{(less fast)}]{} \underset{\text{alkyl hydrogen sulfate}}{2ROSO_3H}$$

Fig. 3. Mechanism of alcohol sulfation (adapted from Roberts, 1998)

2.3 Alcohol ether sulfates (AES)

Primary alcohol ethoxylates are made by the addition of ethylene oxide to a primary alcohol in the presence of an alkaline catalyst (Boskamp & Houghton, 1996). The addition of the second ethylene oxide molecule to the alcohol is kinetically favored in comparison with the addition of the first ethylene oxide; hence the product of ethoxylation contains a distribution of ethylene oxide chain lengths attached to the alcohol along with the starting alcohol itself. Consequently the physical, detergency and biodegradation characteristics are affected not only by the carbon chain length distribution as is the case for primary alcohols, but also by the ethylene oxide distribution which in turn can be supplier depend (de Groot, 1991). The most common alcohol ethoxylates found as feedstocks for sulfation have an average of 2 to 3 molecules of ethylene oxide (2EO or 3EO).

During the sulfating of alcohol ethoxylates the by-products 1,4-dioxane may be formed (Figure 4). Although the formation of 1,4-dioxane is governed predominantly by sulfation and neutralization conditions and by the chemical composition of the feedstock, other factors such as the quality of the raw material also contribute. These factors must be considered during the store and handling of the alcohol ethoxylate feedstock.

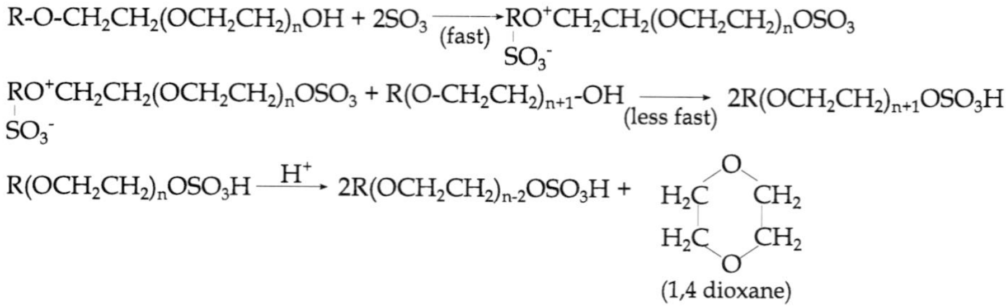

Fig. 4. Reaction scheme of ethoxylated alcohol sulfation (adapted from de Groot, 1991)

LES (C_{12}-$C_{14/15}$ 2-3 ethylene oxide) can be considered as the most efficient anionic surfactant in terms of: superior detergency power, good tolerance for water hardness, and mildness on hands and fibers. The application therefore is wide: from household to personal care and cosmetic product. Unfortunately, sulfated alcohol ether sulfates show a limited stability to hydrolysis at high temperatures, and this restricts their use in heavy duty laundry powders, where high temperatures occur in the spray drying process of powder manufacture.

The high stability to calcium ions permits formulation of liquid detergents with limited or no addition of water "softeners" even in case of use in hard water (Matthijs et al., 1999). The optimum compromise of ethylene oxide addition to keep adequate foam levels and solubility/mildness ratio vary from 2 to 3 moles per mole of fatty alcohol. The most important worldwide application of AES 2-3 ethylene oxide (EO) are in dish washing liquid detergent, generally combined with LABS and in shampoos/bubble baths (Table 2).

	Liquids dish wash detergent	Shampoos/bubble baths
LES (C_{12}-$C_{14/15}$ 2-3 EO)	5 – 10 %	10 – 30 %
LABS (low molecular weight)		15- 20%
Coconut Ethanol Amides (CEA)	2 – 3 %	2 – 3 %
Hydrotopes (Sodium Tolunene Sulfonate – Sodium Xylene Sulfonate) alcohol solvent		1 – 3 %
Other actives (i.e. amphoteric/nonionic)		5 – 10 %
H_2O, perfume, color, preservatives	Balance	Balance

Table 2. Applications worldwide of AES 2-3EO combined with LABS and in shampoos/bubble baths (de Groot, 1991)

2.4 Alfa-olefins sulfonates (AOS)

Alfa-olefins are a potential replacement for alkylbenzenes in detergent applications. Olefin sulfonation is highly exothermic with ΔH = -210 kJ/mol (Roberts, 2001). The neutralized product of alfa-olefin sulfonation requires hydrolysis to remove the sultones, which are skin sensitizers (Figure 5). Their exploitation, however, is largely limited to the Far East, Centre on Japan, at present. Commercial supplies of alfa-olefins are produced by the

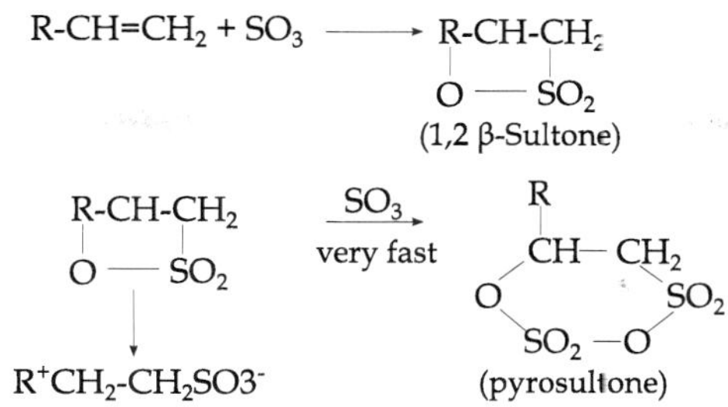

Fig. 5. Reactions of alfa-olefin sulfonation (adapted from de Groot, 1991)

oligomerisation of ethylene. The physical, detergency and biodegradation characteristics of alfa-olefins are affected by the carbon chain length distribution and therefore each new supply may require testing to determine whether the desired properties for the new chosen application can be achieved. The Lion Corporation, Japan, is one of the principal producers and users of alfa-olefin sulfonates. In addition to fabric washing powders, they also market fabric washing liquid, shampoos, tooth paste and foam bath products containing this active. In the USA, Minnetonka has utilized AOS in hand cleaners/liquid soaps. AOS is a potential replacement for alkyl benzene sulfonates in dish wash detergent liquids formulations with performance peaking at C_{16} chain length (de Groot., 1991).

2.5 Fatty acid methyl esters sulfonates (FAMES)

FAMES are called to be the main feedstock for detergent formulating in the future due to their applicability in detergent formulations (Ingegar & Martin, 2001; Johansson & Svensson, 2001; Roberts & Garrett, 2000; Satsuki, 1998). Moreover, when it is derived from palm oil presents special biodegradable properties that place them over the surfactants derived from petrochemicals compounds. To date, the application of FAMES is under development in various detergent products, and their presence on the market is still highly restricted. The typical cut of FAMES (C_{16}-C_{18}) shows interesting surface activity (about 90% compared to LABS), high detergent, dispersing and emulsifying power in hard water, high lime soap dispersion and moderate foam levels. FAMES show high stability to pH and temperature hydrolysis. Therefore, they can be easily spray dyer and/or incorporated in detergent bars. Methyl ester sulfonates have a wide range of application and important biological properties. As aggregated value the FAMES can be used in cosmetics, as auxiliary agents in the production of fibers, plastics, and rubber, and in leather manufacture (Cohen et al., 2008; de Groot, 1991; Roberts et al., 2008; Stein & Baumann, 1975).

3. Sulfonation process used for the manufacturing of anionic surfactants

Sulfonation reactions can be carried out in different configurations, either liquid-liquid contact, or gas-liquid contact reactors, and a diversity of sulfonating reagents can be applied for the sulfonation process, such as: Sulfuric acid, SO_3 from stabilized liquid SO_3, SO_3 from sulfur burning and subsequent conversion of the SO_2 formed, SO_3 from boiling concentrated oleum and chlorosulfonic acid. However some reasons why SO_3/air in gas-liquid contactor (sulfonator) is becoming the predominant process for the manufacture of anionic surfactants are (Foster, 1997):

i. Adaptability: All types of organic feedstocks, like alkylbenzenes, primary alcohols, alcohol ethers, alfa-olefins and fatty acid methyl esters, can be successfully transformed to high-quality sulfonate/sulfate active detergents using SO_3/air as sulfonating reagent. Sulfonating reagents like sulfuric acid and oleum are less desirable because only alkylbenzene feedstocks can be converted to high-quality alkylbenzene sulfonic acids.
ii. Security: Concentrated sulfuric acid, liquid SO_3, and oleum (20 or 65%) are hazardous to be handled, transported, and storage. Sulfur, either in liquid or solid form, although less dangerous option as initial material for the manufacture of SO_3, is still risky.
iii. Price: SO_3 obtained directly from the sulfur combustion is the most economical option among all the others options mentioned above regarding transport, handle and storage.
iv. Availability: Liquid SO_3, 65% and 20% oleum and even sulfuric acid are not produced everywhere. Even close to sulfuric acid plants, it is not guaranteed the availability of all the gamma of oleum concentrations.

Several studies have been done about absorption along with exothermic reaction in a Falling Film Reactor - FFR (Mann & Moyes, 1977; Villadsen & Nielsen, 1986), particularly for dodecylbenzene and tridecylbenzene sulfonation. However, due to de complexity of processes taking place inside the FFR has not been completely elucidated, being of special interest today. The SO_3-sulfonation is carried out in tubular reactors where the organic matter (liquid) wets the wall of the tubes while a gas stream containing the sulfonating reagent flows in co-current with the organic matter to avoid over-sulfonation (MacArthur et al., 1999). The simplest FFR configuration can be described as a two concentric tubes arranged in a vertical way (Figure 6).

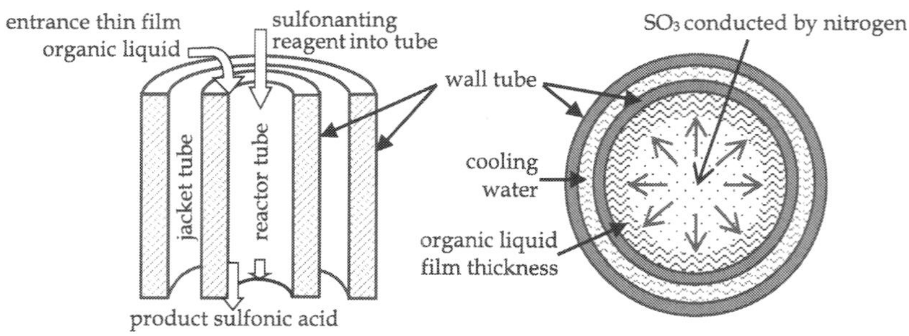

Fig. 6. Sketch of falling film reactor

Organic matter forms a thin film covering the inner wall of the inner tube. The film descends from the top of the reactor in laminar flow forming an annulus for whose interior a gas stream flows in turbulent regimen. In the first reaction section the concentrated sulfonate reagent get in touch with fresh organic matter. The reaction rate is high as well as the amount of heat released (150 - 170 kJ/mole). A coolant stream flows by the external wall of the inner tube in parallel with the reactant streams. As long as the reaction advances the viscosity of the liquid phase increases (ca. 100 times the initial value). The depletion of reactants reduces the reaction rate and the increase of viscosity slow down the mass transfer process in the film. In this point the co-current coolant, this has already removed a huge amount of heat from the first reactor zone, works as a heating current that controls the viscosity of the film.

Figure 7 shows a diagram of film SO_3-sulfonation along with an additional step (bleaching) than could be required depending of the feedstock and characteristic of the final product. Depending on the type of organic feedstock and consequent organic acid, further reaction steps may be required before neutralization. Sulfonic acids of LABS are one of those materials that no require an aging step to reach full conversion. Moreover, a hydrolysis or stabilization step is required to convert anhydrides form during the sulfonation process. Alcohol and alcohol ethoxylate sulfonic acids, as well as FAMES, must be neutralized immediately after a delayed aging to avoid undesired by-products formed in side reactions.

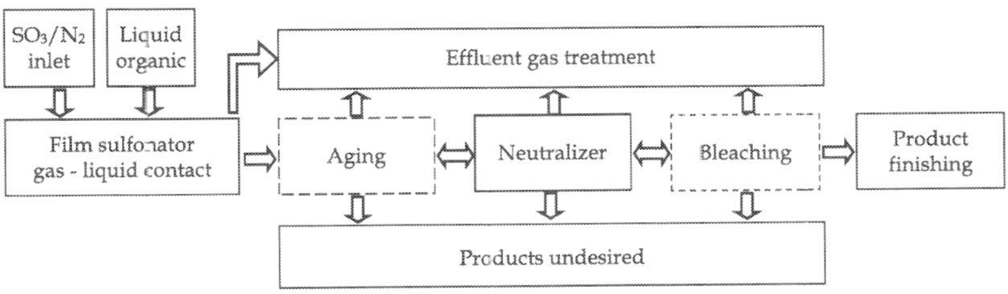

Fig. 7. Process diagram for film SO_3-sulfonation

After aging and hydrolysis a stable product is obtained, then the neutralization stage can be carried out with many alkaline chemicals like caustic, ammonia and sodium carbonate. Neutralization with diluted caustic is recognized as instantaneous and highly exothermic it may form gel at high temperatures or undesired reactions may occur if micro-dispersion of organic acid in the diluted caustic phase fails. Various loop-type reactors, consisting of a circulation pump, homogenizer (where the acid is introduced in the circulating alkaline paste), and heat exchanger, are used for the complex neutralization step (Foster, 1997).

4. Phenomenological description of film sulfonation

Organic liquid flow through of the reactor wall in laminar regimen, the high flow of the gas phase by gravitational effects intensifies the formation of random waves all along the gas-liquid interface. Depending on the flow rate of organic liquid and gas stream the thickness of the film can increase or decrease up to twice its average value in the zone where the waves are present (Díaz, 2009). This induced turbulence affects the local values of concentration and temperature in the regions where appears, hence altering the mass transfer and temperature profiles in the film. Mathematical models which describe the sulfonation of tridecylbenzene in FFRs have been developed by Akanksha et al. (2007), Davis et al., (1979), and Johnson & Crynes, (1974), while Dabir et al. (1996), Gutiérrez et al. (1988) and Talens (1999) focused on dodecylbenzene sulfonation. Nevertheless, these models have been subject of debate due to the assumption that either the chemical reaction is limited to the gas liquid interface, the mass transfer of the sulfonating reagent in the gas phase is the rate determining step, and/or the flow profiles in the film are neatly laminar, neglecting the effects of the waves formed at the gas-liquid interface.

Recently Torres et al. (2009b) proposed a model for the methyl esters sulfonation that is appropriate for both laminar and turbulent films and it considers effects of wavy film flow by using eddy diffusivity parameter. The eddy diffusivity models proposed by Lamourelle & Sandall (1972) for the outer region and modified by van Driest (1956) for the region near the wall were used. Effects of interfacial drag at the gas-liquid interface and the gas-phase heat and mass transfer resistance have also been considered in the proposed model. The model takes into account the variations of physical properties with temperature and predicts conversion profiles, gas-liquid interface temperature in the axial direction, and average liquid film thickness along the reactor length. Knowledge of temperature distribution along with the reactor is important for the product quality control, since for highly exothermic reactions under certain conditions can produce degradation of the products. The equations described in the following section account for the mass, momentum and heat transfer. In the development of these equations was considered the turbulent diffusivity for mass transfer coupled with chemical reaction, according to the theory of Yih & Seagrave (1978), and with heat transfer according with Yih & Liu (1983).

Finally some additional assumptions were made for the mathematical model:

i. No entrainments of liquid droplets into gas or of gas bubbles into the liquid film occur;
ii. Fully developed film (entrance and exit effects to reactor are neglected);
iii. The liquid film is symmetric with respect to the reactor axis.

According with these assumptions the mathematical model is showed in the following sections.

4.1 Mass balance

Only three components are considered in the liquid phase: organic liquid, acid product and sulfonating reagent, therefore two microscopic balances are sufficient to determine the concentration profiles (Figure 8), where y varies from $y = 0$ (at the wall surface) to $y = \delta$ (at the liquid surface).

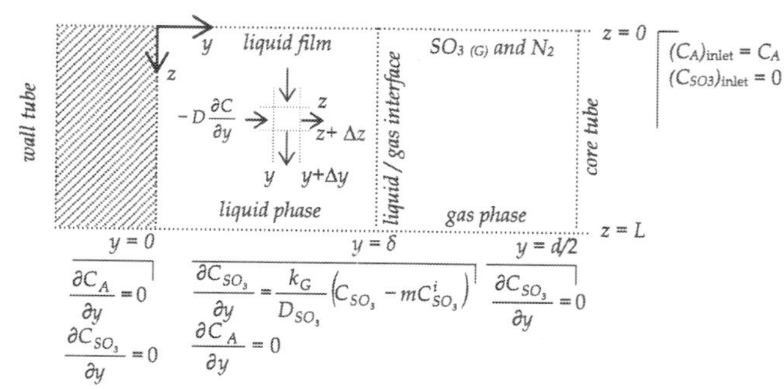

Fig. 8. Mass balance on finite volume includes the boundary conditions at the solid wall and liquid/gas interface

It is assumed that the mass balance for $SO_{3(G)}$ absorbing by the liquid (equation 1) can also be applied to reagent in the liquid phase where reaction occurs, then equation 2 is the steady state mass balance on the absorbing species A in liquid phase.

$$v_z \frac{\partial C_{SO_3}}{\partial z} = \frac{\partial}{\partial y}\left[(D_{SO_3} + D_T)\frac{\partial C_{SO_3}}{\partial y}\right] - r \quad \delta \le y \le \frac{d}{2} \tag{1}$$

$$v_z \frac{\partial C_A}{\partial z} = \frac{\partial}{\partial y}\left[(D_A + D_T)\frac{\partial C_A}{\partial y}\right] - r \quad 0 \le y \le \delta \tag{2}$$

As discussed by Knaggs, (2004), even if the liquid film is turbulent and does wavy flow then turbulent diffusivity cannot be neglected, this and turbulent viscosity in the liquid phase can be taken of work suggested by Yih & Liu (1983).

$$\frac{D_T}{v_T} = -0,5 + 0,5\left\{1 + 0,64(y^{+2})\frac{\tau}{\tau_w} \times \left[1 - \exp\left(\frac{-y^+(\tau/\tau_w)^{0,5}}{A^+}\right)\right]^2 f\right\}^{0,5} \tag{3}$$

$$\frac{\tau_w}{\tau} = 1 - \left(\frac{\tau_L}{\tau_G + \tau_L}\right)^3 \left(\frac{y^+}{\delta^+}\right) \tag{4}$$

Turbulent Schmidt number is evaluated from the Cebeci's modification of the van Driest model and is further modified as:

$$Sc_T = \frac{v_T}{D_T} = \frac{1 - \exp(-y^+(\tau/\tau_w)^{0,5}/A^+)}{1 - \exp(-y^+(\tau/\tau_w)^{0,5}/E^+)} \tag{5}$$

$$B^+ = Sc^{-0,5}\sum_{i=1}^{5} C_i \left(\log_{10} Sc\right)^{i-1} \tag{6}$$

with $A^+ = 25{,}1$; $C_1 = 34{,}96$; $C_2 = 28{,}97$; $C_3 = 13.95$; $C_4 = 6{,}33$ and $C_5 = -1{,}186$. For non-volatile liquids such as methyl stearate, the vapor pressure is zero at working temperatures. At the interface, it is assumed that Henry and Raoult's laws are applicable to determine the SO_3 solubility. The Henry constant m, is determined from the SO_3 vapor pressure:

$$N_{SO_3}^G = k_G\left(C_{SO_3}^G - mC_{SO_3}^i\right) \tag{7}$$

$$\frac{k_G}{u} = 0{,}8Sc^{-0{,}704} \quad \text{(McCready \& Hanratty, 1984)} \tag{8}$$

where the turbulent velocity is defined as:

$$u = \left(\frac{\tau_G}{\rho_G}\right)^{0,5} \tag{9}$$

4.2 Momentum balance

Axial liquid velocity v_z, can be found from the momentum equation after neglecting the pressure gradient and axial terms (Figure 9). The flow profile of the liquid falling is

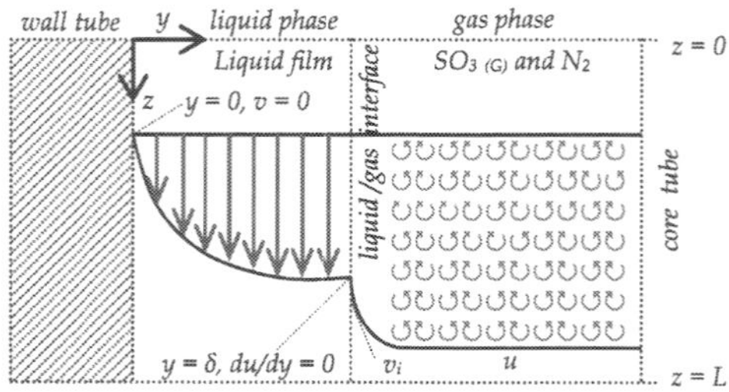

Fig. 9. Schematic representation of velocity profiles in laminar and turbulent regimes for both liquid and gas phase

predominantly laminar, while SO_3 flow is clearly turbulent and consequently the SO_3 is absorbed at the gas/liquid interface. In equation 10 for co-currents systems J is +1 and for counter current systems J is -1.

$$v_z = \frac{\rho_L g}{\mu_L}\left[y\delta - \frac{y^2}{2}\right] + J\frac{\tau_G y}{\mu_L} \tag{10}$$

For high gas flow where the shear force predominates over the gravitational force, the linear velocity distribution is:

$$v_z = \frac{\tau_G y}{\mu_L} \tag{11}$$

Calculation of τ_G based on the relations proposed by Riazi & Faghri (1986) shows that when the gas flow is turbulent, the effects of the interfacial drag cannot be neglected; τ_G can also be verified by means of experimental pressure drop data. In this way a set of parameters can be introduced and adjusted to minimize the deviation from a data set for sulfonation:

$$\Gamma = \frac{\rho_L g}{3\mu_L}\delta^3 - \frac{\tau_G}{2\mu_L}\delta^2 \tag{12}$$

$$\tau_G = C_f \rho_G u^2 \tag{13}$$

$$\frac{1}{C_f^2} = -4Log\left\{\frac{\varphi\delta}{3,7d} - \frac{5.02}{\mathrm{Re}_G}Log\left(\frac{\varphi\delta}{3,7d} + \frac{13}{\mathrm{Re}_G}\right)\right\} \quad \text{(Talens, 1999)} \tag{14}$$

$ln(\varphi) = 3,59 - 5,14\, v_{iL}$ (if $v_{iL} < 0,175$ ms^{-1}) or $ln(\varphi) = 20,55\, v_{iL} - 0,93$ (if $v_{iL} \geq 0,175$ ms^{-1}) (15)

The initial value of the film thickness δ can be obtained through equation 16, after this value is calculated by iteration:

$$\delta = \left(\frac{3\Gamma\mu}{g\rho_L}\right)^{\frac{1}{3}} \tag{16}$$

4.3 Heat balance

For exothermic reactions such as sulfonation a large amount of heat may be released, the boundary conditions showed at Figure 10 is applying for the energy balance (equation 17).

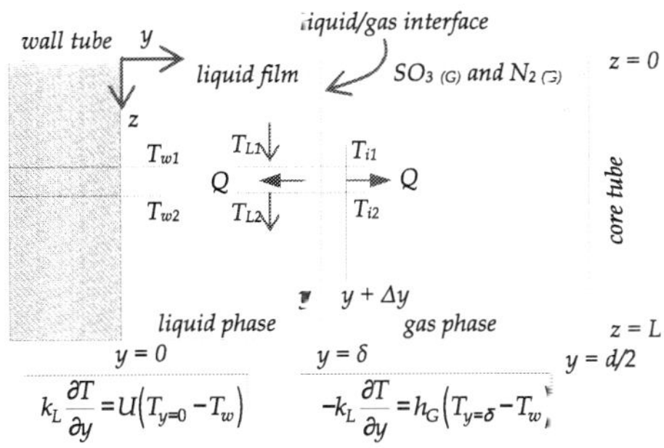

Fig. 10. Schematic representation of model for a segment heat balance

$$v_z\frac{\partial\rho cT}{\partial z}=-\frac{\partial}{\partial y}\left[-k_L\frac{\partial T}{\partial y}\right]+(\Delta H)r \tag{17}$$

$$\frac{1}{U}=\frac{1}{k_w\frac{d_{lm}}{d_{ex}}}+\frac{1}{h_w\frac{d_{ex}}{d_{in}}} \tag{18}$$

Heat transport equations follow the Prandtl analogy and are equivalent to those used for mass transfer.

$$\frac{h_G}{u}=0{,}8Sc^{-0{,}704} \tag{19}$$

5. Main parameters of film SO_3-sulfonation

An experimental set was developed to study the effect of follows factors: (i) mole ratio between SO_3 and organic liquid, (ii) wall temperature and, (iii) volumetric percentage of SO_3 in the phase gaseous. The variables representing the quality of the sulfonated product are: active matter, unsulfonated matter, acid value and color (Ahmad et al., 2007; Inagaki, 2001). The experimental matrix is presented in Table 3 and detailed information about the analysis is presented below (Torres et al., 2008b).

Conditions	value
SO_3/N_2 inlet (gaseous sulfonate mixture), % vol/vol	3 - 7
SO_3/N_2 temperature inlet, °C	50 - 60
SO_3 /organic liquid, mole ratio	1:1 - 1,2:1

Table 3. Operating conditions used for the methyl ester sulfonation (Torres et al., 2009a)

5.1 Active matter

Increase of active matter in product was proportional to the increase of SO_3/organic liquid mole ratio as well as the increase of SO_3 percentage in the gas stream. Slope changes observed with respect to the temperature are probably due to side reactions occurring at higher temperatures; the formations of undesired matters decrease active matter (Figure 11).

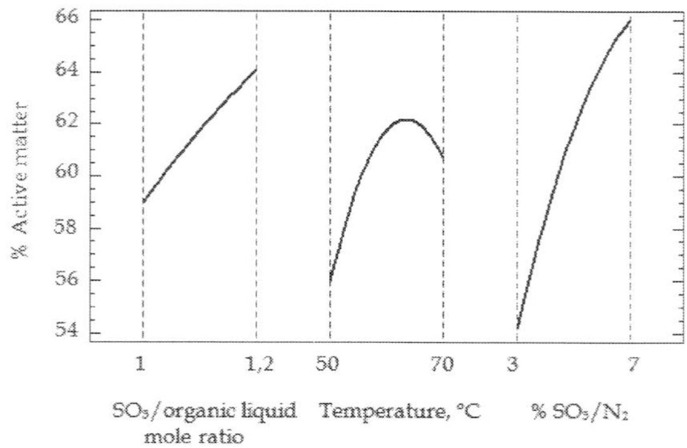

Fig. 11. Impact of the operation conditions on the degree of sulfonation

5.2 Unsulfonated matter

The impact of the experimental factors is initially inverse compared with the effect obtained with the active matter; however the SO_3/organic liquid mole ratios beyond 1,1 produce an increase in the quantified unsulfonated matter. This change can be explained by over sulfonation of reactant and formation of side products. This assumption is consistent with the effects of temperature on the percentage of non-sulfonated matter (Figure 12).

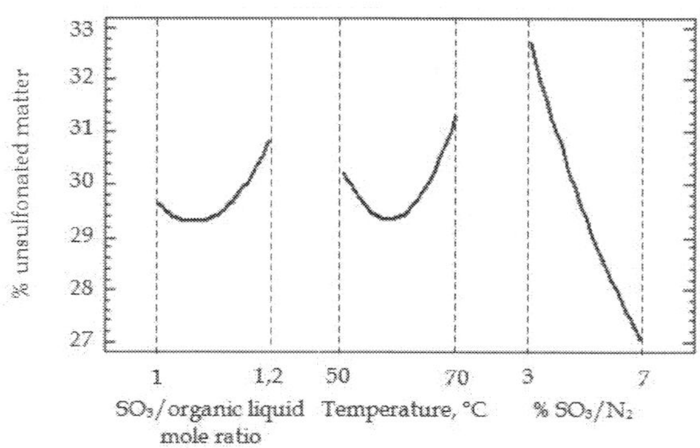

Fig. 12. Effects of the experimental factors on the unsulfonated matter

Over-sulfonated products do not have the same characteristics of the washing active substance and therefore are not identified as active matter but yet as free oil (unsulfonated).

Figure 12 proves that the unsulfonated matter percentage decreases the same as the sulfonating reagent in the gas mixture increases.

5.3 Acid value

Figure 13 shows the effect of the conditions process on acid value. Increase in the sulfur trioxide has a positive effect in agreement with expectations; also the increase of temperature in process enlarges the acidity in the product. Change in the slope can be explained by kinetics effects favored by temperature rising. Both the increase in the mole ratio and sulfur trioxide in the sulfonating mixture can be explained by the effectiveness of the reaction because an excess of SO_3 promotes the consumption of the same reactants for the generation of over-sulfonated matter.

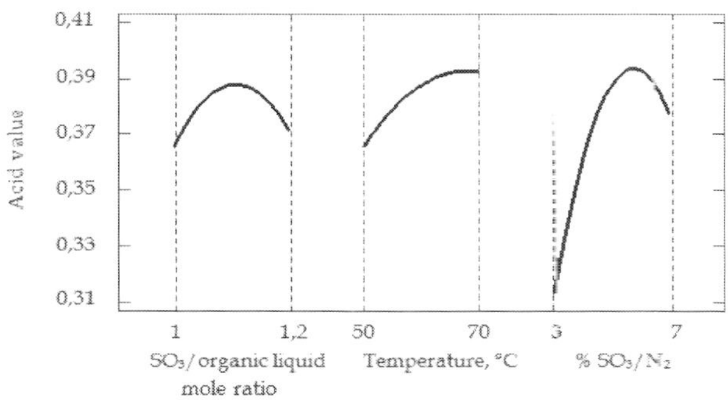

Fig. 13. Acid values obtain by changes of mole ratio, temperature and SO_3 inlet

5.4 Coloration

Figure 14 presents the trends for the coloration in the sulfonated product which intensifies the values of concentration of SO_3 some that the mole ratio and SO_3 content in the gas stream. All variables show a direct influence to the increase in color which is associated with

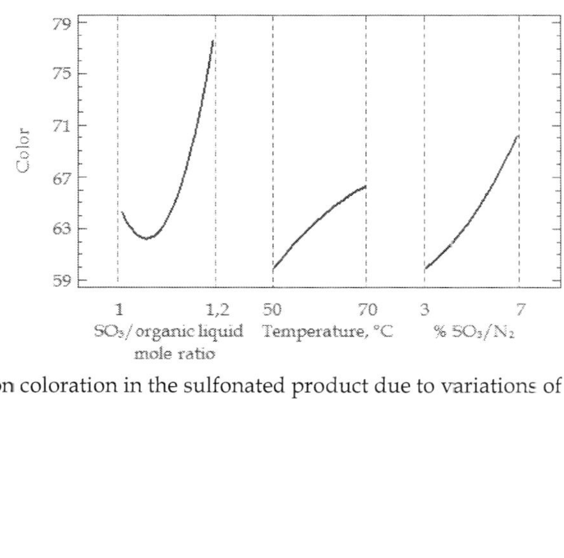

Fig. 14. Impact on coloration in the sulfonated product due to variations of process

higher sulfonation degrees. Although the coloration is identified mostly as an esthetic factor for the commercialization of sulfonates, higher colorations can also be a qualitative indicator of over-sulfonation. Anionic surfactants in aqueous solution have colors ranging from yellow to reddish orange (Inagaki, 2001).

6. Film SO_3-sulfonation applied

6.1 Sulfonation of methyl ester with SO_3

Use of methyl esters (ME) in the industry detergent, although under investigation and development since more than 25 years, has not yet expanded to high levels, mainly because the following reason:

i. Controversial forecasts about availability of petrochemical feedstocks with related cost comparison via vs. natural sources;
ii. Viability of sufficient quality of sulfonation grade methyl esters;
iii. The process to produce high quality α-sulfonated methyl ester (SME) is generally more complex than that for alkyl benzene sulfonates;
iv. Application know-how is not yet completely availed and low FAMES solubility involves some restrictions in application, notably concerning the use in liquid detergent products and low temperature washing cycles.

Complex chemistry is not yet fully elucidated, but may be summarized as is shown in Figure 15 (Morales & Martínez, 2009).

$$SO_3 + H_2SO_4 \longrightarrow H_2SO_3^+ + HSO_4^- \quad \text{(i)}$$

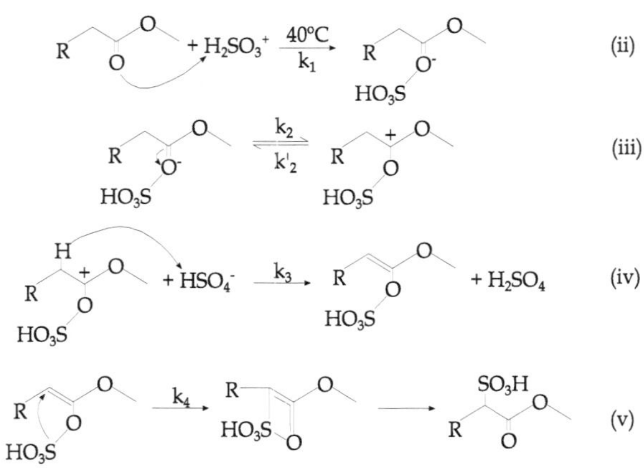

Fig. 15. Mechanism of methyl ester sulfonation in sulfonator (Torres et al., 2009b)

The methyl ester molecule is initially di-sulfonated in a relatively fast reaction accompanied with a high amount of heat released (Roberts, 2003). There is a third reaction stage considerably slower than the previous ones, where an SO_3 group is liberated (on aging). For some researchers, this SO_3 group just released would be especially active and therefore capable of directly sulfonating another methyl ester molecule in an alfa position. The diffusivities of reagents are estimated from the Wilke-Chang equations and diffusivities in the mixture are estimated through the Vignes equation (Vignes, 1966). Generally, in methyl ester sulfonation the amount of intermediate III in the final product varies from 10-20% (Foster, 2004), but this amount can be reduced by a long and heated digestion (aging stage).

An experimental apparatus showed in Figure 16 was utilized by Torres (2009) for researching the parameters of film SO_3-sulfonation of methyl ester derived from hydrogenated stearin palm, this apparatus was designed by Chemical Engineering Laboratory from Universidad Nacional de Colombia (Bogotá, Colombia).

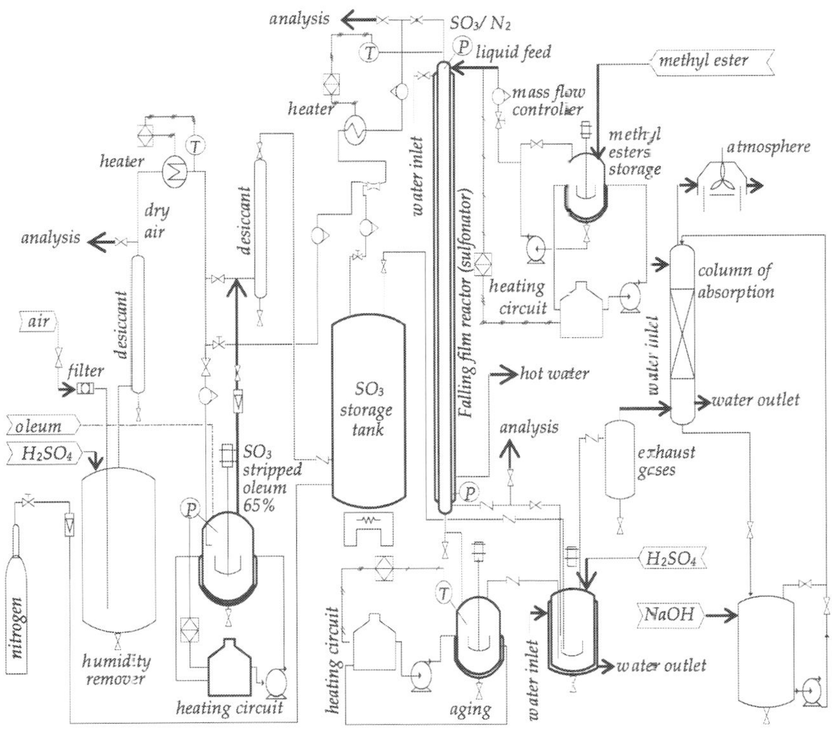

Fig. 16. Experimental apparatus for methyl ester sulfonation using a falling film reactor and SO_3 stripped from 65% oleum with dried air process

High degree of sulfonation is obtained in the aging step controlling simultaneously the temperature and residence time. At higher temperatures it is feasible to obtain higher conversion levels, whereas at low temperatures (below 80°C) the time required to reach high conversions is considerably long. These reactions are highly exothermic in order of 150 - 170 kJ/mole (including 25 kJ/mole of the absorption heat of gaseous SO_3). A kinetic model has been development by Roberts (2007) based on the proposal that two major intermediates are involved in aging (Figure 17).

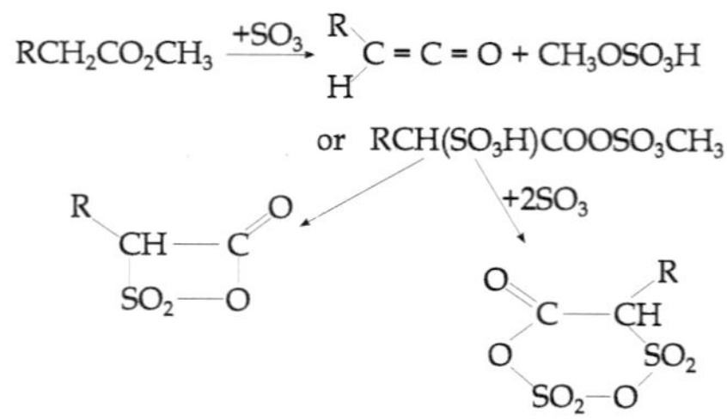

Fig. 17. Reactions in the aging step (adapted from Roberts, 2001)

$$k = Ae^{\left(-\frac{B}{T}\right)} \tag{20}$$

	k_f, (s^{-1})	k_s, (s^{-1})
LogA	12,10	11,52
B	12,060	12,130

Table 4. Values for equation 20 on aging stage (Roberts, 2008)

The overall conversion as function of time and mole ratio M of SO_3/ME is given by:

$$\%\ conversion = 100M\left(\frac{1}{M_{100}}\right) - 0.25e^{-k_f t} - 0.167e^{-k_s t} \tag{21}$$

M_{100} is mole ratio for a conversion at 100%, after a delayed aging M_{100} = 1,2. These equations, for aging in a batch reactor system or in a plug flow systems, can be used as guidelines when setting initial conditions before fine-tuning plant operation to meet a required specification (Roberts, 1998). Methyl esters are less active than aromatic compounds to sulfonating due to the less electronic density of the aliphatic chains. The methyl ester sulfonation include a neutralization step to obtain monosodium salts of α-sulfo methyl esters as desire products (Kapur et al. 1978). If neutralization is immediate disodium salt is formed (see Figure 18a). However, if neutralization of the acid is delayed, the sulfo ester monodisodium salt is obtained as final product (see Figure 18b).

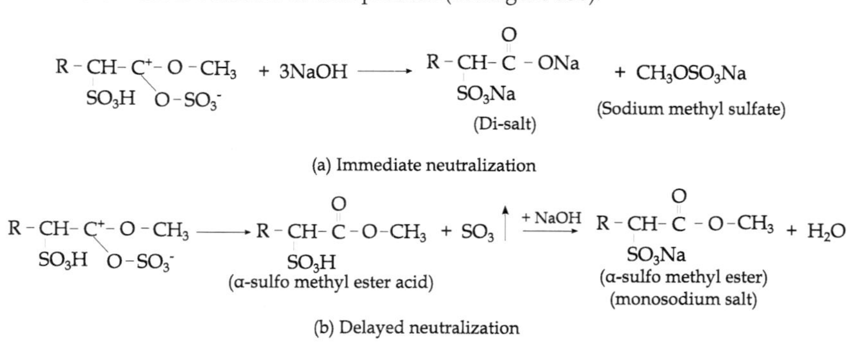

Fig. 18. Neutralization chemistry of SME (Torres et al., 2009b)

Generally, feedstock for the manufacture of FAMES containing unsaturated fatty acids and these has been attributed to the formation of polysulfone in the double bonds (Yamada & Matsutani, 1996). Unsaturated in methyl ester make it an olefin with a carboxyl methyl group at the end of the chain. Olefins are more rapidly sulfonated by SO_3 also unsaturated bound produces oversulfonation and oxidation of the olefin which competes with the saturated ester obtain product more colored, however the color can be improved by bleaching. Unsaturated make it an olefin with a carboxyl methyl group at the end of the chain. Olefins are more rapidly sulfonated also unsaturated bound produces oversulfonation and oxidation of the olefin which competes with the saturated ester obtain product more colored, however the color can be improved by bleaching (Figure 19).

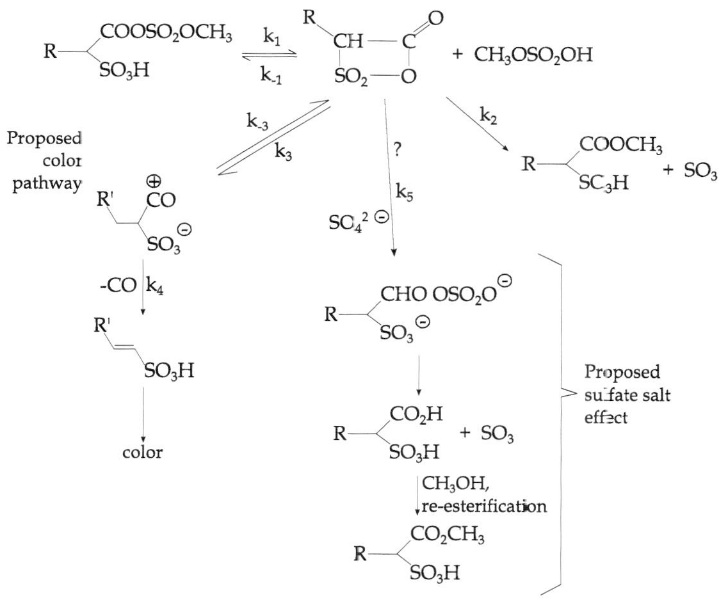

$$k_{-1} >> k_1,\ k_{-1} >> k_2,\ k_4k_3/k_{-3} << k_2,\ k_5 >> k_4k_3/k_{-3}$$

Fig. 19. Reaction scheme for the coloration in aging step (adapted from Roberts et al., 2008)

Mechanism proposed by Roberts (2007) suggests a reversible formation of β-dioxide cycle and CH_3SO_3H this β-anhydride reacts opening its cycle, sintering itself, and losing a carbon monoxide to become an alkene sulfonic acid. This is formed mainly in reactions of sulfonation of alfa-olefins, these are very intensive in color when aged in the acid form (Clippinger, 1964)

6.2 Validation of model

The input variables more important for the conversion are: the length and diameter reactor, flow of liquid reactant, mole ratio between SO_3 and organic liquid, in this case methyl stearate derived of hydrogenated stearin from palm oil (Narváez et al., 2005; Torres et al., 2005), and amount sulfur trioxide in the gas phase (SO_3/N_2), finally the temperature of the process. This mathematical model permits to calculate the profiles of interfacial liquid temperature, liquid film density, liquid viscosity for any column height and longitudinal profiles of conversion. The proposed model may be suitable for use in design and operation of industrial film reactors. To ensure convergence of the system of equations then transformation of the equations proposed by Agrawal & Peckover (1980) was chosen following the same development by Talens (1999). The set of equations resulting from the mass, momentum and heat transfer is solved numerically. Figure 20 shows schematic view from the top of a reactor: the liquid is evenly distributed around the wall, and the gas mixture is injected through the center of the column. The interfacial temperature is affected by the SO_3 amount in the gas mix. It is clear an increase of interface temperature result of the SO_3 excess in the gas flow. The temperature of the reagents is a key control variable to avoid undesirable side impact that damage the product mainly by strong coloration.

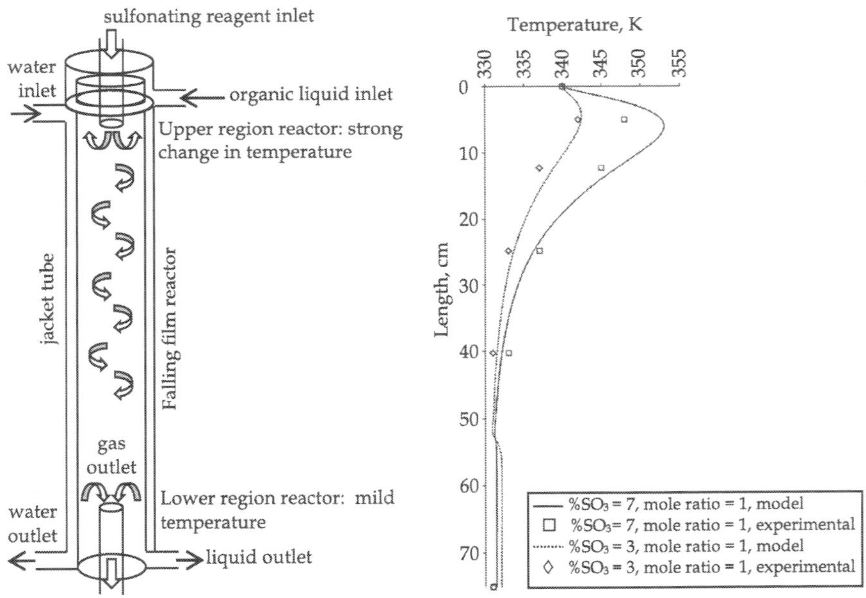

Fig. 20. An example of interfacial temperature profiles fall in the reactor longitudinal

Other example of the results provided by the model for longitudinal conversion profile from top of the reactor (expressed as percentage of active matter) is shown in Figure 21(a). The input values of the model are: SO_3/N_2 percentage at 5%, SO_3/methyl stearate mole ratio at 1; T_G, T_L, and T_w at 343 K, 333 K and 313 K, respectively. This figure shows schematically the fast conversion region at the top of the reactor (associated with gas phase control) and slow conversion region at the bottom (linked with liquid phase control). The washing active substance was determinate using a two titration technique with Hyamine 1622 as the titrant reagent and methylene blue as indicator (Tsubochi et al., 1979; Milwdsky & Gabriel, 1982;

Battaglini et al., 1986; Schambil & Schwuger, 1990). The progress of the reaction is decisive for sulfonation degree expressed as amount active matter.

Physicals and chemicals properties used in model are listed in Table 5.

Parameters	Correlation
Kinetic in sulfonator, kmol/m³s (Torres et al., 2008a)	$r = -kC_{SO_3}C_{ME}$ (22) $k = 1,14\times10^{19} e^{-\left(\frac{14.350}{T}\right)}$ (23)
Diffusivity, m²/s (Wilke & Chang, 1955)	$D_{SME\text{-}ME} = 3,12\times10^{-11}\frac{T_L}{\mu_L^{2/3}}$ (24) $D_{ME\text{-}SME} = 6,288\times10^{-11}\frac{T_L}{\mu_L^{2/3}}$ (25) $D_{SO3} = 2,031\times10^{-10}\frac{T_L}{1000\mu_L^{2/3}}$ (26)
Gas thermal conductivity, J/msK (Davis et al., 1979)	$k_G = 0,0279$
Heat capacity of liquid, J/kmolK (Broström, 1975; Reid et al., 1987)	$c_L = 507,300 - 101,010x$ (27)
Heat capacity of gas mix, J/kmolK (Same at nitrogen)	$c_G = 29,82$
Liquid mixture density, kg/m³ (Talens & Gutiérrez, 1995; Broström, 1975)	$\rho_L = 980 + 192x - 0,66T_L$ (28)
Liquid thermal conductivity, J/msK (Davis et al., 1979)	$k_L = 0,276$
Surface tension, N/m (de Groot, 1991)	$\sigma = 0,046$
Viscosity of methyl esters mix, kg/ms (Torres & Sánchez, 2008)	$\mu_{ME} = 8\times10^{-6} e^{\frac{1949,6}{273+T_L}}$ (29)
Viscosity of methyl esters sulfonic acid, kg/ms (Talens & Gutiérrez, 1995)	$\mu_{SME} = 1,36\times10^{-8} e^{\frac{5700}{T_L+5,88}}$ (30)
Viscosity of gas mix, kg/ms (Same at nitrogen)	$\mu_G = 1,9\times10^{-5}$
Viscosity of liquid mix, kg/ms (Broström, 1975; Talens & Gutiérrez, 1995)	a) $0 < x \le 0,25$: (31) $\mu_L = 2,72\times10^{-6} e^{\frac{2980}{T_L+3,86x}}$ b) $0,25 < x \le 0,6$: (32) $\mu_L = 3,7\times10^{-9} e^{\frac{4850}{T_L+1,035x}} + 0,003$ c) $0,6 < x < 1$: (33) $\mu_L = 3,7\times10^{-9} e^{\frac{4850}{T_L+5,22x}} + 0,003$
All thermal conductivity, J/msK (Davis et al , 1979)	$k_w = 16,3$

Table 5. Correlation to estimate heat and mass transfer coefficients and properties of organic liquid and gas mixture (Torres & Sánchez, 2009)

The model was constructed to predict the sharp increase in the conversion that takes place in the first stage inside of the falling film reactor. It confirms that the mass transfer is initially controlled by the resistance in the gas phase. After due to several factors the resistance occurs in the liquid phase. Figure 21(a) presents the conversion profile in the film reactor from the model, experimental results showed conversions lower than those predicted by the reactor model in upper reactor region. This is due that the model assumes a fully developed flow and entrance effects of the streams to the reactor are neglected. However, the model is able to predict adequately conversions downstream for longer lengths. In the bottom of reactor is a small jump in the conversion predicted by the model, perhaps due to kinetic effects that reach importance by the consumption of reactants.

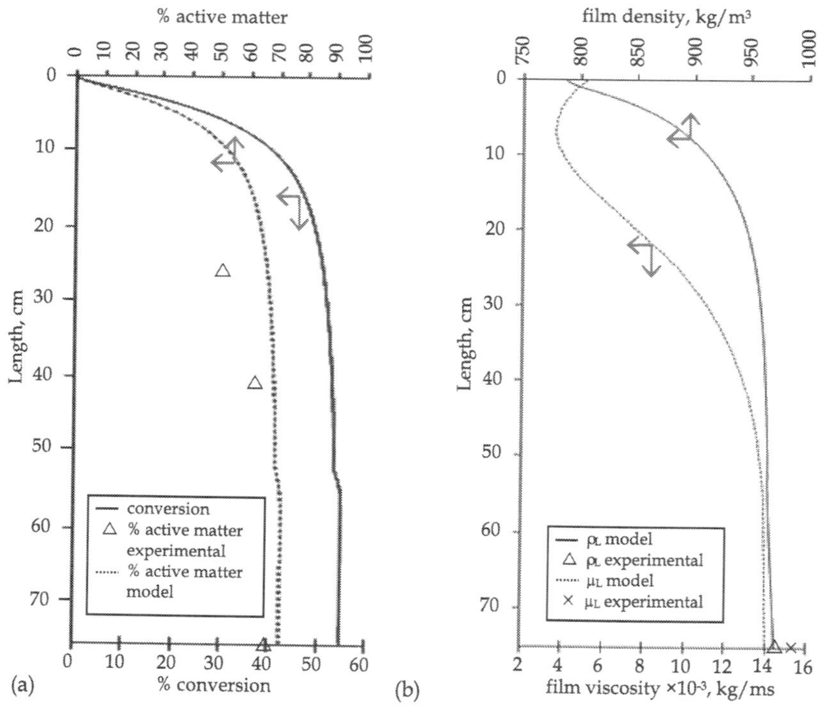

Fig. 21. (a) Longitudinal conversion profile for SME and (b) density and viscosity model estimated by the model

The most important outlet data obtained by solving the mathematical model are conversion, density and viscosity of the product. The density and viscosity of the effluent, downstream in the reactor film, estimated by the mathematical model is similar to that obtained experimentally, as shown in the Figure 21(b). The increase of temperature produces a decrease in viscosity enhancing the solubility of SO_3 in the liquid, and causing a decrease in film thickness. These variations are the result of abrupt change in composition and release of heat in the initial part of the reactor, which produce an increasing of the temperature in this area. Same phenomena occur with the film thickness. The jump in conversion takes place in the top reactor, the temperature rises considerably and reduces the viscosity of the liquid, even canceling the effect of viscosity then in the bottom reactor increases composition and the interfacial velocity. Subsequently, the reduced generation heat and descent of the temperature is increase the viscosity again.

7. Conclusions

Transfer rates in the gas phase are affected by changes in the tubular reactor. Increases uncontrolled in the gas flow could drag some liquid into the gas phase. Therefore the gas velocity has to be set at the point where no liquid drops can be pulled to the gas phase. Temperature is a critical parameter in the quality control of the sulfonated products. Although inlet streams' temperature should be adjusted above room to enhance the reaction and avoid the solidification of the organic matter, an adequate control is required due to the high release of heat attributed to the sulfonation reaction. The SO_3/organic liquid mole ratio requires rigorous control. Excess of SO_3 enhance side reactions and extended reaction times will also enhance side reactions.

The comparison obtained for this same process with petrochemicals compounds indicates that the model could be applied to any film sulfonation but adjusting the parameters and specific conditions, such as the physicochemical properties of the compounds used, since the sulfonation process described in this work is one of the more complicated cases. Although some of the physical and chemical properties of mixture are obtained of a similar form, these should be tested and approach to achieve convergence of the model; these yielded the best results in the mathematical model of falling film reactor.

The model predicts two distinct transfer areas. The first is characterized by an abrupt increase in conversion and temperature, in which the controlling step depends initially of the gas phase and in accordance with the extent of the sulfonation reaction, the viscosity fluid increases, the film thickness is also higher and the film velocity decreases, then the liquid phase becomes the controlling stage with a mild increase of the temperature and conversion. The mathematical model proposed for a film SO_3-sulfonation fits adequately the trend of experimental results, so it is now possible to make a prediction on the conversion in a falling film reactor, because the profiles of temperature, density, viscosity and conversion are consistent with experimental results that satisfy the conditions to minimize the strictest mathematical calculations mistakes due to the usage of numerical solutions.

8. Notation

A	pre-exponential factor, s^{-1}
A^+	a van Driest constant
B^+	a van Driest parameter
C	concentration, kmol/m^3; van Driest constants
c	heat capacity, J/(kmolK)
C_f	friction factor, dimensionless
D	diffusivity, m^2/s
d	reactor diameter, m
f	damping factor, $f = e^{-1.66(1-\tau/\tau_w)}$
g	acceleration due to gravity, m/s^2

ΔH	reaction enthalpy, J/kmol
h	heat transfer coefficient, J/(m^{2}sK)
k	reaction rate constant, m^3/kmols; thermal conductivity, J/msK; mass transfer coefficient, kmol/m^{2}s
L	reactor length, m
M	mole ratio SO_3/ME
m	Henry constant, (kmol of SO_3/m^3 of gas)/(kmol of SO_3/m^3 of liquid)
N	mass flux of gaseous reactant, kmol/m^{2}s
P	pressure, atm
Q	heat of reaction, J/mol
r	reaction rate
Re	Reynolds number, dimensionless: $Re_G = \rho u(d - 2\delta)/\mu$; $Re_L = 4\Gamma/\mu$.
Sc	Schmidt number, dimensionless
T	temperature, K
U	global heat transfer coefficient, J/(m^{2}sK)
u	turbulence characteristic velocity of gas, m/s
x	conversion expressed as molar fraction of the sulfonic acid, dimensionless
v	axial velocity of liquid film, m/s
y	transversal coordinate (from wall toward the liquid free surface)
y^+	non-dimensional distance to the wall: $y(\tau_w g/\rho)^{1/2}/\nu$
z	axial coordinate

Greek letters

Γ	volumetric flow rate of the liquid per unit wetted perimeter, m^2/s
δ	film thickness, m
δ^+	dimensionless film thickness, $\delta^+ = \delta u/\nu$
φ	roughness enhancement factor, dimensionless
μ	liquid viscosity, kg/ms
ν	kinematic viscosity, m^2/s
ρ	liquid density, kg/m^3
τ	interfacial shear stress, N/m^2
σ	Surface tension, N/m

Subscripts

A	absorbing specie
ex	exterior
G	gas phase
i	interface
in	interior
L	liquid phase
lm	logarithmic mean
T	turbulent
w	wall

9. Acknowledgment

Gratefully acknowledge at Dr. Federico I. Talens Alesson from University of Nottingham (UK), Dr. David W. Roberts from Liverpool John Moores University (UK) and Dr. Icilio

Adami from Ballestra S.p.A. (Italy) by communications received. Same wish to thank COLCIENCIAS (Departamento Administrativo de Ciencia, Tecnología e Innovación) for providing financial support. The experimental work presented here was finished at 2009 in the Chemical Engineering Laboratory from Universidad Nacional de Colombia (Colombia), under the direction from Professor Francisco J. Sánchez C. Dr. Paulo C. Narváez R., Dr. Oscar Y. Suárez P. and MSc. Luis A. Díaz A. assisted with the experiments.

10. References

Acmite Market Intelligence. (2010). *Market report. World Surfactant Market.* (Market report Ratingen, Germany: Acmite Market Intelligence, Retrieved from http://www.acmite.com/brochure/Brochure-Surfactant-MarketReport.pdf

Agrawal, A. K., & Peckover, R. S. (1980). Nonuniform grid generation for boundary-layer problems. *Comput. Phys. Comm.*, Vol 19, No. 2, pp.171-178, ISSN 0010-4655

Ahmad, S., Siwayanan P., Murad Z. A., Aziz H. A., & Soi H. S. (2007). Beyond biodiesel, Methyl esters as the route for the production of surfactants feedstock. *INFORM,* Vol.18, pp.216-221, ISSN 0897-8026

Akanksha, Pant, K. K., & Srivastava, V. K. (2007). Modeling of sulphonation of tridecylbenzene in a falling film reactor. *Math. Comp. Model.*, Vol.46, No. 9-10, pp.1332-1344, ISSN 0895-7177

Battaglini, G., Larson-Zobus, J., & Baker, T. G. (1986). Analytical methods for alpha sulfo methyl tallowate. *JAOCS*, Vol.63, No. 8, pp.1073-1077, ISSN 1558-9331

Boskamp, J. V. & Houghton, M. P. (1996). Detergent compositions containing primary alcohol sulfate, ethoxylated nonionic surfactant, and zeolite builder. *Zeolites,* Vol. 17, Issues 5-6, pp.525, ISSN 0144-2449

Broström, A. (1975). A mathematical model for simulating the sulphonation of dodecylbenzene with gaseous sulphur trioxide in an industrial reactor of votator type. *Transactions of the Institution of Chemical Engineers,* Vol.53, pp.29-33, ISSN 0046-9858

Clippinger, E. (1964). Reactions of alpha-olefins. *Ind. Eng. Chem. Proc. D. D.*, Vol.3, No. 1, pp.3-7, ISSN 0196-4305

Cohen, L., Soto, F., Melgarejo, A., & Roberts, D. W. (2008). Performance of Φ-sulfo fatty methyl ester sulfonate versus linear alkylbenzene sulfonate, secondary alkane sulfonate and α-sulfo fatty methyl ester sulfonate. *J. Surfactants Deterg.*, Vol.11, No. 3, pp.181-186, ISSN 1558-9293

Dabir, B., Riazi, M. R., & Davoudirad, H. R. (1996). Modelling of falling film reactors. *Chem. Eng. Sci.*, Vol.51, No. 11, pp.2553-2558, ISSN 0009-2509

Davis, J. E., Van Ouwerkerk, M., & Venkatesh, S. (1979). An analysis of the falling film gas-liquid reactor. *Chem. Eng. Sci.*, Vol.34, No. 4, pp.539-550, ISSN 0009-2509

de Groot, W. H. (1991). *Sulphonation technology in the detergent industry.* Kluwer Academic Publisher, ISBN 0-7923-1202-3, Dordrecht, The Netherlands.

Díaz, L. (2009). Hydrodynamics analysis of a falling film reactor for the sulfonation of methyl esters derived from palm oil. *M.Sc. Thesis,* Universidad Nacional de Colombia, Bogotá, Colombia

Foster, N. C. (2004). Manufacture of methyl ester sulfonates and other derivates. *Soaps, Detergents, Oleochemicals and Personal Care Products,* In: Spitz L. (Ed.), AOCS Publishing, ISBN 9781893997769, Seattle, USA

Foster, N. C. (1997). *Sulfonation and sulfation processes.* The Chemithon Corporation, Retrieved from: http://www.chemithon.com/Resources/pdfs/Technical_papers/Sulfo%20and%20Sulfa%201.pdf

Gutiérrez, G. J., Mans, T. C., & Costa, L. J. (1988). Improved mathematical model for a falling sulfonation reactor. *Ind. Eng. Chem. Res.*, Vol.27, No. 9, pp.1701-1707, ISSN 1520-5045

Inagaki, T. (2001). Development of α-Sulfo fatty acid esters. *Proceedings World Conference on Oleochemicals,* ISBN 0-935315-34-9, Kuala Lumpur, Malaysia, October 8th -12th, 2000

Ingegard, J. & Martin, S. (2001). Surfactants based on fatty acids and other natural hydrophobes. *J. Colloid Interf. Sci.*, Vol.6, pp. 178-188, ISSN 0021-9797

Johansson, I., & Svensson, M. (2001). Surfactants based on fatty acids and other natural hydrophobes. *Curr. Opin. Colloid In.*, Vol.6, No. 2, pp.178-188, ISSN 0196-4305

Johnson, G. R., & Crynes, B. L. (1974). Modelling of a thin-film sulfur trioxide sulfonation reactor. *Ind. Eng. Chem. Proc. D. D.*, Vol.13, No. 1, pp.6-14, ISSN 0196-4305

Kapur, B. L., Solomon, J. M. & Bluestein, B. R. (1978) Summary of the technology for the manufacture of higher alpha-sulfo fatty acid esters. *JAOCS*, Vol.55, pp. 549-557, ISSN 1558-9331

Knaggs, E. (2004). Comment on "Optimization of the linear alkyl benzene sulfonation process for surfactant manufacture" and "Sulfonation technology for anionic surfactant manufacture": Falling film SO_3 sulfonation - Laminar or turbulent flow controversy. *Org. Proc. Res. Dev.*, Vol.8, No. 6, pp.1079-1080, ISSN 1520-586X

Lamourelle, A., & Sandall, O. (1972). Gas absorption into a turbulent liquid. *Chem. Eng. Sci.*, Vol.27, pp.1035-1043, ISSN 0009-2509

MacArthur, B. W., Brooks, B., Sheats W.B., & Foster, N. C. (1999). In: *Proceedings of the 1999 PORIM International Palm Oil Congress, "Emerging Technologies and Opportunities in the Next Millenium"*, pp. 229-250, ISBN 978-967-961-075-8, Kuala Lumpur, Malaysia

Mann, R., & Moyes, H. (1977). Exothermic gas absorption with chemical reaction. *AIChE J.*, Vol.23, No. 1, pp.17-23, ISSN 0001-1541

Matthijs, E., Holt, M.S., Kiewiet, A., Rijs, G.B.J. (1999). Environmental monitoring for linear alkylbenzene sulfonate, alcohol ethoxylate, alcohol ethoxy sulfate, alcohol sulfate, and soap. *Environ. Toxicol. Chem.* Vol.18, No. 11, pp.2634–2644, ISSN 1552-8618

McCready, M. J., & Hanratty, T. J. (1984). A comparison of turbulent mass transfer at gas-liquid and solid-liquid interfaces. *Gas transfer at water surfaces.* In: Brutsaert W., & Jirka G. H. (Eds.), pp. 283-292, Reidel Publishing Company, ISBN 9027716978, Dordrecht, Netherlands

Milwdsky, B. M., & Gabriel, D. M. (1982). *Detergents Analysis – A Handbook for cost-effective quality control,* John Wiley & Sons, Inc., ISBN 0-7114-5735-2, London, UK.

Morales, G. M., & Martínez, R. R. (2009). Molecular and multiscale modeling: New tools in the analysis of chemical engineering phenomena. *CT&F*, Vol. 3, No. 5, pp. 205-224, ISSN 0122-5383

Mungray, A. K., & Kumar, P. (2009). Fate of linear alkylbenzene sulfonates in the environment: A review. *Int. Biodet. Biodeg.*, Vol. 63, No. 8, pp. 981–987, ISSN 0964-8305

Narváez, R. P. C., Torres, J. A., Sánchez, C. F. J., Ponce de León, Q. L. F. (2005). Gas chromatography determination of fatty acid alkyl esters (methyl and ethyl) in the presence of mono-, di- and triglycerides. *Ing. Investig.*, Vol. 57, No. 1, pp.58-62, ISSN 0120-5609

Reid, R. C., Prausnitz,J. M., & Poling,B. E. (1987). *The properties of gases and liquids* (4th ed.) McGraw-Hill Book Company. ISBN 0-07-051799-1, NY, USA

Resnik, G. O., Vishwanath, P., Pynn, M. A., Sitnik, J. M., Todd, J. J., Wu, J., . . . Jarrell, K. A. (2010). Use of sustainable chemistry to produce an acyl amino acid surfactant. *Appl. Microbiol. Biot.*, Vol.85, No. 5, pp.1387-1397, ISSN 0175-7598

Riazi, M. R., & Faghri, A. (1986). Effect of the interfacial drag on gas absorption with chemical reaction in a vertical tube. *AIChE J.*, Vol.32, No. 4, pp.696-699, ISSN 0001-1541

Roberts, D. W. (1998). Sulfonation technology for anionic surfactant manufacture. *Org. Proc. Res. Dev.*, Vol. 2, *No.3*, pp. 194–202, ISSN 1520-586X

Roberts, D. W. (2001). Manufacture of anionic surfactants. *Oleochemical Manufacture and Applications*. In: Gustone F. D., & R. J. Hamilton (Eds.), pp. 55-73, Sheffield Academic Press, ISBN 9780849397851, UK

Roberts, D. W. (2003). Optimization of linear alkyl benzene sulfonation process for surfactant manufacture. *Org. Process Res. Dev.*, Vol.7, No. 2, pp.172-184, ISSN 1520-586X

Roberts, D. W. (2007). The origin of colour formation in methyl ester sulfonation. *Communication Journal Com. Esp. Det.*, Vol.37, pp.153-159, ISSN 0017-3495

Roberts, D. W., Giusti, L., & Forcella, A. (2008). Chemistry of methyl ester sulfonates. *INFORM*, Vol.19, No. Suppl. 5, pp.2-9, ISSN 0897-8026

Roberts, D. W., & Garrett, P. R. (2000). Ester sulphonates derived from non-methyl esters. In: *Proceedings of the 1999 PORIM International Palm Oil Congress, "Emerging Technologies and Opportunities in the Next Millenium"*, pp. 209-215, ISBN 978-967-961-075-8, Kuala Lumpur, Malaysia, 1999

Rosen, M. R. (2005). *Delivery System Handbook for Personal Care and Cosmetic Products: Technology, Applications, and Formulations*. William Andrew Publishing, ISBN: 0-8155-1504-9, Norwich, N.Y., USA

Satsuki, T. (1998). Methyl ester sulfonates, new products and applications in surfactant technology. In: *Annual surfactants review*, Karsa D. R. (Ed.), pp. 123-155, Sheffield Academic Press, ISBN 978-084-939-738-7, UK

Schambil, F., & Schwuger, M. J. (1990). Physico-chemical properties of alpha-sulpho fatty acid methyl esters and alpha-sulpho fatty acid di salts. *Tenside Surfact. Det.*, Vol.37, No. 6, pp.380-385, ISSN 0932-3414

Stein,W., & Baumann, H. (1975). a–sulfonated fatty acids and esters: manufacturing process, properties, and applications. *JAOCS*. Vol.52, No. 9, pp.323-329, ISSN 1558-9331

Talens, F. I., & Gutiérrez, J. M. (1995). Estimation of the viscosity and density of lauryl alcohol/laurylsulphuric acid mixtures as a function of the temperature and laurylsulphuric and molar fraction. *Chem. Eng. Res. Des.*, Vol.73A, pp.206-207, ISSN 0263-8762

Talens, F. I. (1999). The modelling of falling film chemical reactor. *Chem. Eng. Sci.*, Vol.54, No. 12, pp.1871-1881, ISSN 0009-2509

Talens, F. I. (2000). The effect of supercritical gas velocity on wavy films and its use in enhancing the performance of falling film reactors. *Chem. Eng. Technol.*, Vol.23, No.7, pp.629-632, ISSN 0930-7516

Torres, J. A. (2009). Production of alfa-sulfo methylstearate in a falling film reactor. *PhD Thesis*, Universidad Nacional de Colombia, Bogotá, Colombia

Torres, J. A., Díaz, L., & Sánchez, F. J. (2008a). Kinetic study of the sulfonation of methyl esters derived of palm stearin. *Corpoica Cienc. Tecnol. Agropecu.*, Vol.9, No. 2, pp.88-95, ISSN 0122-8706

Torres, J. A., Díaz, L., & Sánchez, F. J. (2009a). Falling film reactor for methyl ester sulphonation with gaseous sulphur trioxide. *Ing. Investig.*, Vol.29, No. 3, pp.48-53, ISSN 0120-5609

Torres, J. A., Morales, G. M., Suárez, P. O. Y., & Sánchez, F. J. (2009b). Mathematical model of a falling film reactor for methyl ester sulfonation. *Chem. Prod. Process Model.* Vol. 4, Iss. 5, Art. 12, ISSN 1934-2659

Torres, J. A., Narváez, P. C., Suárez, O. Y., & Sánchez, F. J. (2008b). Sulfonation of methyl esters derived from palm oil. *Corpoica Cienc. Tecnol. Agropecu.*, Vol.9, No. 2, pp.77-87, ISSN 0122-8706

Torres, J. A., & Sánchez, F. J. (2008). Physicochemical properties of the liquid mixture of methyl stearate/sulfo acid methyl stearate. *Publ. Invest.*, Vol.2, No. 1, pp.97-118, ISSN 1900-6608

Torres, J. A., & Sánchez, F. J. (2009). Small scale SO_3-sulfonation of methyl esters. *Program Book 8th World Congress of Chemical Engineering*, ISBN 0-920804-44-6, Montréal, Canada, August 23th - 27th, 2009

Torres, J. A., Sánchez. C. F. J, Narváez, R. P. C., Ponce de León, Q. L. F. (2005). Production of methyl stearate derived from stearin hydrogenated derived of palm oil. *Revista de Investigaciones de la UNAD*, Vol.4, No. 2, pp. 150-162, ISSN 0124-793X.

Tsubochi, M., Yamasaki, N., & Matsuoka, K. (1979). Determination of anionic surfactants by two-phase titration with tetrabromophenolphthalein ethyl ester as indicator. *JAOCS*, Vol.56, No. 11, pp.921-923, ISSN 1558-9331

van Driest, E. R. (1956). On turbulent flow near a Wall. *Journal of the Aeronautical Sciences*, Vol.23, pp.1007-1011, ISSN 0095-9812

Vignes, A. (1966). Diffusion in Binary Solutions. Variation of Diffusion Coefficient with Composition. *Ind. Eng. Chem. Fundam.*, Vol. 5, pp.189-199, ISSN 0196-4313

Villadsen, J., & Nielsen, H. P. (1986). Models for strongly exothermic absorption and reaction in falling films. *Chem. Eng. Sci.*, Vol.41, No. 6, pp.1655-1671, ISSN 0009-2509

Wilke, C. R., & Chang, P. (1955). Correlation of diffusion coefficients in dilute solutions. *AIChE J.*, Vol.1, No. 2, pp.264-270, ISSN 0001-1541

Yamada K, & Matsutani S. (1996). Analysis of the dark colored impurities in sulfonates fatty acid methyl ester. *JAOCS*, Vol.73, pp.121-125, ISSN 1558-9331

Yih, S. M., & Seagrave, R. C. (1978). An analytical solution for mass transfer in turbulent falling films with or without chemical reaction. *Chem. Eng. Sci.*, Vol.33, No. 12, pp.1581-1587, ISSN 0009-2509

Yih, S. M., & Liu, J. L. (1983). Prediction of heat transfer in turbulent falling liquid films with or without interfacial shear. *AIChE J.*, Vol.29, No. 6, pp.903-909, ISSN 0001-1541

Zhu, Y. P., Rosen, M. J., Morral, S. W. & Tolls, J. (1998) Surface properties of linear alkyl benzene sulfonate in hard river water. *J. Surfartants Deterg.* Vol. 1, No. 2, pp.187-193, ISSN 1558-9293

12

Pollutant Formation in Combustion Processes

Grzegorz Wielgosiński
Technical University of Lodz,
Faculty of Process and Environmental Engineering
Poland

1. Introduction

Each combustion process is a source of various emissions. During combustion, are formed not only carbon dioxide and water, but still a lot of other products of combustion and incomplete combustion. Knowledge of the mechanisms and the pathways of formation allow the use of so-called primary methods of reducing emissions and thereby reduce emissions to the atmosphere.

2. Various fuels composition

From the chemical point of view combustion is the oxidation reaction of organic compounds. Organic compounds - a variety of hydrocarbons which have in the molecule the atoms of carbon (*C*) and hydrogen (*H*). The oxidation reaction of hydrocarbons is accompanied by the emission of large amount of heat - that is why the reaction is exothermic. Hydrocarbons and their derivatives containing other atoms in the molecule, such as sulfur (*S*), nitrogen (*N*), oxygen (*O*), chlorine (*Cl*), etc. make a flammable substance which can be called fuel only when it meets certain qualitative conditions. Fuels can be divided according to several methods. One of them is the division according to the state of matter - into solid, liquid and gas fuels. Another method is the division according to the origin - natural fuels (e.g. fossil fuels) and synthetic fuels produced by processing natural fuels.

Natural solid fuels include mainly coal, lignite, peat, wood, etc. Natural liquid fuels include first of all oil from which numerous synthetic fuels such as petrol, kerosene, diesel, fuel oil, etc. are produced. In a group of natural gas fuels natural gas is crucial, while synthetic gas fuels include blast furnace gas, generator gas, water gas, city gas or extracted gas fractions such as methane, propane or butane as well as their mixtures.

Another type of fuel are wastes and produced from them the so-called alternative fuel which is known as refuse derived fuel (RDF) or secondary recovered fuel (SRF). Generally, fuel, in addition to the information about its form (physical state), is first of all characterized by elemental composition i.e. the contents of basic elements such as carbon (*C*), hydrogen (*H*), sulfur (*S*), nitrogen (*N*), oxygen (*O*), chlorine (*Cl*), etc. Apart form the contents of these elements, the content of water (*W*) and non-flammable substance (A), also known as ash, in the fuel are extremely important. Typical elemental composition of selected solid, liquid and gas fuels is shown in Table 1.

Fuel	*C*	*H*	*O*	*N*	*S*	*Cl*	*W*	*A*
hydrogen	-	100	-	-	-	-	-	-
acetylene	92,3	7,7	-	-	-	-	-	-
methane	75	25	-	-	-	-	-	-
propane	81,8	18,2	-	-	-	-	-	-
butane	83	17	-	-	-	-	-	-
natural gas	69,4	22,9	0,7	7,0	0,05	-	-	-
blast furnace gas	41,6	19,4	9,0	2,2	-	-	-	-
coke oven gas	15,0	50,0	23,0	12,0	-	-	8,5	-
petrol	85	15	-	-	0,05	-	-	-
diesel	87	13	-	-	0,05	-	-	-
heating oil	86	13	-	-	0,2	-	-	-
coal	72-83	3,4-5,3	1,8-12,5	1,0-1,2	0,1-6,5	0,1-1	3-20	3-30
lignite	25-77	2-6,5	1,5-20	0,1-0,5	0,2-6,5	0,1-1	10-60	3-30
peat	40	5	25	2	1	0,1	20	7
coke	80-90	0,3-1,5	1,0-2,0	1,2-2,2	0,5-1,0	0,5	1,5-8,5	2-17
wood	35-50	5-7	35-45	0,3-07	0,01-0,1	0,01	5-65	0,3-7

Table 1. Typical elemental composition of selected solid, liquid and gas fuels [in %]

3. Lower and upper heat value

The elemental composition determines another very important parameter which characterizes fuel i.e. heat of combustion so called upper heat value. Heat of combustion (H_o) - is the amount of heat that is generated during complete and perfect burning of unit weight or unit volume of the analyzed substance in constant volume, wherein:

- temperature and pressure are constant,
- non-volume work is equal to zero,
- products of combustion are present at the initial temperature,
- water contained in exhaust gas is in liquid state,
- nitrogen in the products is in the form of molecules (as N_2),
- sulfur contained in the fuel after combustion is in the form of SO_2

The term complete combustion means the process in which all the fuel (organic matter) will be oxidized (burned) and perfect combustion occurs when all the combustion products are non-flammable. Data on typical values of combustion heat of the selected fuels are shown in Table 2.

Heat of combustion (H_o) can be easily estimated on the basis of the elemental composition. The approximate value for the heat of combustion of solid and liquid fuels (upper heat value) can be calculated from the formula (Recknagel et al., 1994):

$$H_o = 34{,}8 \cdot C + 93{,}9 \cdot H + 10{,}5 \cdot S + 6{,}3 \cdot N - 10{,}8 \cdot O \qquad (1)$$

where:

H_o - combustion heat [MJ/kg]
C - carbon content in the fuel [kg/kg],
H - hydrogen content in the fuel [kg/kg],
S - sulfur content in the fuel [kg/kg],
N - nitrogen content in the fuel [kg/kg],
O - oxygen content in the fuel [kg/kg].

Fuel	Heat of combustion (upper heal value)	Heat value (lower heat value)
hydrogen	147,8	119,9
acetylene	49,9	48,2
methane	55,5	50,1
propane	50,3	46,4
butane	49,5	45,7
natural gas	41,1	37,5
blast furnace gas	4,1	4,0
coke oven gas	19,7	17,5
petrol	46,1	42,7
diesel	44,7	41,8
heating oil	44,8	42,7
coal		19-31
lignite		6-14
peat		12-16
coke		28-32
wood		10-18

Table 2. Combustion heat and calorific value of fuel (in MJ/kg)

Heat of combustion can be determined experimentally in an adiabatic calorimeter. However, fuel usually contains also water (moisture) whose presence and evaporation significantly reduces the amount of heat generated during the combustion of the fuel unit. In order to determine the amount of heat that can be effectively generated in the combustion process the concept of heat (calorific) value (H_u), called the lower heat value, was introduced.

The calorific (heat) value is the amount of heat emitted during combustion of mass unit or volume unit of fuel at its complete and perfect combustion, assuming that the steam contained in the exhaust gas does not condense, even though the temperature of the exhaust gases reach the initial temperature of the fuel. Knowing the heat of combustion the calorific value can be calculated from the formula (Recknagel et al., 1994):

$$H_u = H_o - (9 \cdot H + W) \cdot r_w \quad (2)$$

where:

H_u - calorific value [MJ/kg]
W - water content in the fuel [kg/kg],
r_w - water evaporation heat [MJ/kg].

4. Oxygen excess

Every combustion process is conducted with an excess of oxygen relative to stoichiometric amount resulting from the oxidation reaction of individual components of the fuel. The measure of this excess is so called coefficient of excess air (λ) defined as a ratio of the actual amount of air supplied to the combustion process to the theoretical amount of air required to carry out the complete and perfect combustion - the amount resulting from the stoichiometry of combustion (oxidation) reaction. If in the combustion process coefficient λ is equal to 1 then exactly the same amount of air that is required by the stoichiometry of the oxidation reaction is supplied to the process. If the coefficient λ amounts to less than 1 the combustion process is incomplete. As mentioned above, each actual combustion process is conducted with an excess of air, hence the coefficient λ is usually more than 1. Practically, its value depends on the type of fuel burned (gaseous and liquid fuels require less excess air than solid fuels) and design of the equipment in which combustion is carried out.

If the thermal process takes place without the presence of oxygen it is the process of pyrolysis. Pyrolysis - is the thermal decomposition of flammable substance without oxygen. This is an endothermic process that requires supplying heat from the outside. Pyrolysis process usually takes place at temperatures of 450-900°C in which the organic part of the fuel is converted into: the gas phase (pyrolysis gas), the liquid phase (pyrolysis oils) and solid phase (pyrolysis coke). Pyrolysis is sometimes also called degassing or carbonization. Depending on the temperature we distinguish: smouldering (below 600°C) and coking (above 600°C).

The composition and quantity of pyrolysis products depend on the type of fuel, their physicochemical properties and the temperature of the process. During the process of pyrolysis the mass of fuel is converted into (Thome-Kozmiensky, 1994):

- pyrolysis gas containing mainly hydrogen, methane, ethane and their homologues, carbon monoxide and carbon dioxide as well as other compounds such as: hydrogen sulfide, ammonia, hydrogen chloride and hydrogen fluoride
- pyrolytic coke - a solid phase containing carbon and metals as well as other inert substances
- liquid phase containing a mixture of oils, tars and water as well as dissolved in it simple aldehydes, alcohols and organic acids.

The process of pyrolysis of hydrocarbons can be described by the following simplified chemical equation:

$$C_xH_yO_z \xrightarrow[heat]{} CO_2 + H_2O + CO + CH_4 + C_2H_6 + CH_2O + ... + C_mH_n + C - q \tag{3}$$

However, if the thermal process takes place with limited access of air (oxygen) - when the coefficient λ is less than 1, often fuel gasification process occurs. Gasification is a total conversion of solid fuels into gas fuels under the influence of gasifying medium e.g. air or steam or both. During the flow through the glowing layers of solid fuel intended for the gasification a reaction of oxygen or water with carbon takes place in result of which carbon monoxide (CO) and hydrogen (H_2) are formed as well as small amounts of methane (CH_4), carbon dioxide (CO_2), nitrogen (N_2) and water steam (H_2O). The process of gasification of solid hydrocarbons can be described by the following simplified chemical equations (Thome-Kozmiensky, 1994):

$$C+\frac{1}{2}O_2 \longrightarrow CO+q \tag{4}$$

$$C+O_2 \longrightarrow CO_2+q \tag{5}$$

$$C+2H_2O \longrightarrow CO_2+2H_2-q \tag{6}$$

$$C+H_2O \longrightarrow CO+H_2-q \tag{7}$$

$$C+2H_2 \longrightarrow CH_4+q \tag{8}$$

$$CO+H_2O \longrightarrow CO_2+H_2+q \tag{9}$$

$$CO+3H_2 \longrightarrow CH_4+H_2O+q \tag{10}$$

$$C+CO_2 \longrightarrow 2CO-q \tag{11}$$

which are accompanied at the same time by thermal decomposition according to the simplified reaction equation:

$$C_xH_yO_z \xrightarrow[heat]{} CO_2+H_2O+CO+CH_4+C_2H_6+CH_2O+...-C_mH_n+C-q \tag{12}$$

Some of the mentioned above chemical reactions are exothermic (+q - proceeding with heat emission) and some endothermic (-q - requiring supplying heat from the outside).

5. Volume of flue gases from combustion processes

The stream of flue gases from the combustion process consists of a stream resulting from the oxidation reaction of flue gases and a stream of excess air supplied to the combustion process (with excess air coefficient λ greater than 1). Knowing the elemental composition of the fuel or only its calorific value it is possible to estimate both stoichiometric amount of air needed for combustion process and the amount of flue gases generated in the process.

For liquid and solid fuels, the amount of air needed for combustion reaction (stoichiometric) can be calculated from the formula (Recknagel et al., 1994):

$$V_T = 8,88 \cdot C + 26,44 \cdot H + 3,32 \cdot S - 3,33 \cdot O \tag{13}$$

where:

V_T - minimal (stoichiometric) amount of air for combustion converted to standard conditions (T = 273K, P = 1013hPa) [m^3/kg]

The volume of flue gases can be calculated on the basis of known elemental composition of the fuel from the formula (Recknagel et al., 1994):

$$V_S = 1,85 \cdot C + 11,11 \cdot H + 0,68 \cdot S + 0,8 \cdot N + 1,24 \cdot W \tag{14}$$

where:

V_S - volume of flue gas from the combustion process converted to standard conditions [m^3/kg]
N – nitrogen content in the fuel [kg/kg],

In case the elemental composition of fuel is not known the minimum amount of combustion air and exhaust gas volume can be calculated from approximate formulas taking into account only the calorific value of fuel (Recknagel et al., 1994):

- for solid fuel:

$$V_T = 0,241 \cdot H_u + 0,5 \tag{15}$$

$$V_S = 0,212 \cdot H_u + 1,65 \tag{16}$$

- for liquid fuel:

$$V_T = 0,203 \cdot H_u + 2,0 \tag{17}$$

$$V_S = 0,265 \cdot H_u \tag{18}$$

The total gas flow (after taking into account the excess air coefficient λ) under normal conditions can be calculated from the formula:

$$V_C = (\lambda - 1) \cdot V_T + V_S \tag{19}$$

Multiplying the value of the total flue gas stream V_C by the amount of fuel burned per unit of time we can calculate the volumetric flow of flue gases from combustion process converted to standard conditions:

$$V_N = B \cdot V_C \tag{20}$$

where:

V_N - volumetric flow of flue gases from the combustion process converted to standard conditions [m^3/h]
B – fuel consumption [kg/h],

Taking into account the current temperature (T) and flue gases pressure (P) it is possible to estimate flue gas stream in the real conditions from the formula:

$$V = V_N \cdot \frac{T + 273}{273} \cdot \frac{1013}{P} \tag{21}$$

where:

T – flue gas temperature [°C]
P – flue gas pressure [hPa]

6. Combustion as an autothermal process

Heat emitted in a chemical reaction of oxidation in the combustion process is used for heating of flue gases from the process (combustion products and excess air supplied to the combustion process), solid secondary products of the combustion process (slag and ashes) and the equipment in which the combustion process is conducted (plus the obvious loss to the environment) up to the process temperature. This is illustrated in Figure 1.

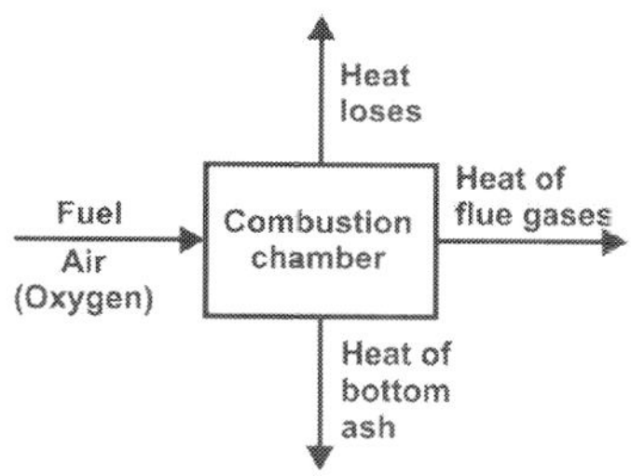

Fig. 1. Simplified heat balance in the combustion chamber

It is obvious that, if the generated amount of heat is too small, the combustion process does not run properly, its temperature is too low and the flue gases do not reach the required temperature. If, however, fuel of high calorific value is burned the amount of heat is sufficient to ensure proper combustion temperature and temperature of flue gases. This is particularly important in the case of waste incineration, where the law (Directive 2000/76/EC on the incineration of waste) requires that the exhaust gases have a minimum temperature of 850°C and remain in this temperature for at least 2 seconds when the waste fuel contains less than 1% chlorine. When the chlorine content in waste exceeds 1% it is required to achieve a minimum temperature of 1100°C. If the required temperature of flue gas (and the combustion process) is reached in the device, in which the combustion process is run, it is called autothermal combustion process, but when due to too low calorific value of fuel the required temperatures cannot be reached, combustion is not an autothermal process and requires additional fuel of higher calorific value.

7. Combustion as a set of chemical reactions

As mentioned at the beginning from the chemical point of view combustion is the reaction of oxidation of hydrocarbons. Combustion of gaseous fuel is the simplest combustion, in which simple reaction of hydrocarbons oxidation takes place according to the simplified reaction scheme:

$$C_xH_y + \left(x + \frac{y}{4}\right)O_2 \longrightarrow x\,CO_2 + \frac{y}{2}H_2O \tag{22}$$

$$C_xH_y + \left(\frac{x}{2} + \frac{y}{4}\right)O_2 \longrightarrow x\,CO + \frac{y}{2}H_2O \tag{23}$$

and the basic products of oxidation are carbon dioxide and water. In case of insufficient oxygenation of the combustion zone the additional reaction product is carbon monoxide. Flammable gas usually mixes well with air, causing a kinetic character of the combustion i.e. controlled by the rate of chemical oxidation reaction. In rare cases, when the flammable gas is mixed with air, the process can be controlled by diffusion of oxygen to the reaction zone. With inadequate oxygenation of the reaction zone the product of the combustion process is carbon monoxide, and elemental carbon (soot), because the reaction rate of water formation is higher than the rate of carbon oxidation.

Combustion of liquid fuels is a more complicated process, because to ensure the oxidation (combustion) it is necessary to evaporate the liquid, because its vapor is burned. Evaporation is promoted by the liquid spray, however, it is very difficult to obtain a homogeneous mixture of liquid vapor and air, hence the combustion of liquid fuels has rarely kinetic character, usually this is a diffusion combustion. Organic liquids generally have a more complex chemical structure than gases, and hence more often we are dealing with the formation of products of incomplete combustion, some of them being other organic compounds:

$$C_xH_y + \frac{3}{2}O_2 \longrightarrow C_{x-1}H_{y-2} + CO_2 + H_2O \tag{24}$$

of course simpler than the compound that is subject to burning.

In extreme cases there can be a situation when due to shortage of air in result of the combustion process the decomposition of organic compounds contained in the fuel will be incomplete and not all carbon will be oxidized. Consequently, in addition to gaseous products from the combustion process a solid product will be also formed - elemental carbon (C) or soot:

$$C_xH_y + \left(x - \frac{1}{2} + \frac{y}{4}\right)O_2 \longrightarrow C + (x-1)\,CO_2 + \frac{y}{2}H_2O \tag{25}$$

The process of solid fuel combustion runs in even more complicated way. It is a multidirectional and multi-stage process, a combination of combustion (oxidation), gasification, thermal decomposition, including pyrolysis (lack of oxygen process). One should be aware that direct heterogeneous reaction between gaseous oxygen and solid hydrocarbon is problematic and because of the hetero phases it must be slow. Thus, direct combustion of solid phase fuel runs in a minimal degree. The real combustion process is much more complicated. In the initial period, at sufficiently high temperatures degassing processes (pyrolysis) begin and flammable gases are emitted from the solid fuel. They are relatively easily oxidized (burned) in the gas phase generating heat flux heating the solid phase and accelerating the release of flammable gases. These processes run as long as fuel in a given element does not run out of hydrogen. The solid phase in result of degassing is enriched in the process of carbonization in carbon hence degassing process (the release of flammable gases) is accompanied by solid phase carbonization. At this point a change in the mechanism of the process takes place and the process of gasification of solid fuel begins (using air and/or steam), which results in carbonization phase being converted into flammable gases (carbon monoxide and hydrogen), which then in the gas phase are oxidized (burning). In the gas phase the oxidation reaction dominates, however, at the same time a number of other reactions take place - such as synthesis, cyclization, etc.

Apart from that we should not forget about other reactions running in parallel such as Boudouard reaction in which carbon dioxide, previously formed during the combustion, at the temperature in the range 400-950°C can be reduced with the participation of carbon to less oxidized form - a flammable carbon monoxide according to the equation:

$$CO_2 + C \longrightarrow 2\,CO \tag{26}$$

which can participate in further reactions e.g. oxidation.

It is assumed that depending on the temperature in the combustion process of solid fuels several important steps can be defined (Thome-Kozmiensky, 1994):

- 100–200°C - thermal drying process, separation of water (physical process),
- 250°C - deoxidation reductive processes, reduction, decomposition of sulfuric acid esters, separation of bound moisture and carbon dioxide, depolymerization, the beginning of the release of hydrogen sulfide,
- 340°C - formation of aliphatic compounds, including unsaturated, the beginning of release of methane and other aliphatic compounds,
- 380°C - beginning of carbonization in the smouldering process,
- 400°C - beginning of the formation of the compounds of carbon with oxygen and nitrogen,
- 400–600°C - transformation of bituminuous substances into smouldering oil or tar,
- 600°C - cracking of bituminous substances in the direction of thermally persistent substances (gases, hydrocarbons of short chain structure), formation of new compounds (benzene derivatives in result of cyclization of unsaturated aliphatic compounds),
- > 600°C - further course of the hypothetical reaction of dimerization of butylene formation, dehydration to butadiene, the formation of cyclohexane C_6H_{12}, thermal aromatization to benzene and other higher aromatic compounds.

8. Complete and uncompleted combustion

However, we should be aware that the actual combustion process is far from ideal - that is complete and perfect combustion. The combustion process (thermal decomposition and oxidation) of many organic compounds (in particular those contained in the waste) is not a perfect process that produces only carbon dioxide, carbon monoxide and water. In this process usually a large number of intermediate products of decomposition and oxidation are formed which then are not further decomposed. It would seem that in the drastic combustion conditions, at about 1000°C any organic material must be burned. Unfortunately this is not true. Many chemical compounds, often flammable, are not subject to complete destruction during process of combustion. Some of the organic compounds are generated in result of many secondary reactions running in the combustion zone and outside it. Simplified diagram of the formation of pollutants in combustion processes is shown in Figure 2.

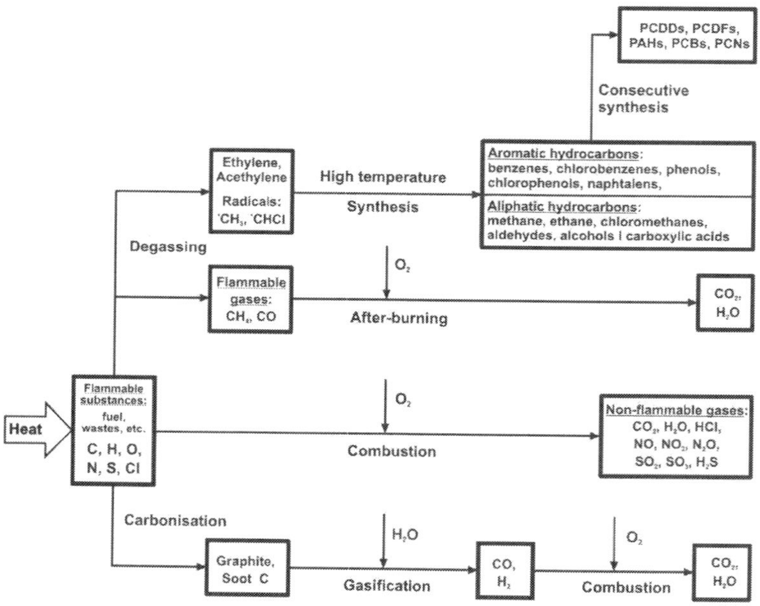

Fig. 2. Simplified diagram of the formation of pollutants in the combustion process of solid flammable substance (fuel, waste, etc.)

In the process of fuel combustion there are four main pathways. The first pathway - direct oxidation reactions. If we treat fuel as a collection of various chemicals consisting of carbon, hydrogen, oxygen, nitrogen, sulfur, chlorine, etc. so in result of direct oxidation typical flammable gases will appear such as carbon dioxide (CO_2), water (H_2O) in accordance with reaction equation (22) and also sulfur dioxide (SO_2), nitrogen oxides (NO, NO_2, N_2O) or hydrogen chloride (HCl). At the same time the combustion process will be conducted according to other two pathways. In the process of thermal decomposition flammable gases will be emitted that contain for example carbon monoxide (CO), methane (CH_4) and other simple hydrocarbons, which are relatively easily oxidized to carbon dioxide and water in the gas phase. Thermal decomposition process impoverishes the solid phase of fuel in hydrogen and other volatile substances, causing its carbonization.

On the third pathway carbonized solid phase is gasified with water and/or air to carbon monoxide and hydrogen which in the gas phase are easily oxidized to carbon dioxide and water. The last fourth pathway of the course of combustion processes is a secondary synthesis path. In result of decomposition and gasification processes in the gas phase simple hydrocarbons and hydrocarbon radicals are formed. Since virtually all fuels contain trace admixtures of chlorine (carbon even sometimes even about 1%!) in the gas phase there are also simple chlorinated hydrocarbons (including unsaturated - for example, acetylene C_2H_6) and chlorinated hydrocarbon radicals. They are extremely reactive and in the gas phase appear a number of synthesis reactions, including cyclization. In this way chlorinated and non-chlorinated aliphatic hydrocarbons are formed (methane, ethane, chloromethanes, chloroethanes etc., aldehydes: e.g. formaldehyde and acetaldehyde, alcohols: e.g. methanol, simple carboxylic acids: e.g. formic acid and acetic acid) and aromatic hydrocarbons (benzenes, chlorobenzenes, phenols and chlorophenols, naphthalenes and many others). The latter take part in the synthesis of polychlorinated dibenzo-p-dioxins (PCDDs), polychlorinated dibenzofurans (PCDFs), polychlorinated biphenyls (PCBs), polychlorinated naphthalenes (PCNs) and polycyclic aromatic hydrocarbons (PAHs) (Wielgosiński, 2009).

9. Formation of sulphur oxides

Practically all fuels solid, liquid and gas contain some amounts of sulphur (Table 3).

Kind of fuel	Sulphur content
Gas fuels: LNG, CNG, LPG	below 0,01%
Diesel oil	below 0,10%
Petrol	below 0,05%
Heating oil	0,2 - 0,3%
Wood	0,02 - 0,04%
Coal (Poland)	0,5 - 2,5%
Coal (World)	0,1 - 4,5%
Lignite (Poland)	0,4 - 1,2%
Lignite (World)	0,4 - 6,5%
Coke	0,2 - 1,0%
Municipal waste	0,2 - 0,6%

Table 3. Sulphur content in different fuels

Sulphur in the combustion process is relatively easily oxidized to sulphur dioxide (SO_2) in accordance with the simplified reaction scheme.

$$S + O_2 \xrightarrow[combustion]{} SO_2 \tag{27}$$

The rate of this reaction at temperatures, under which the combustion process is run, is very high. On the basis of many research results it can be assumed that approximately 90-95% of the sulfur contained in fuel in the combustion process will be oxidized to SO_2. The presence in the burned material of the non-combustible fraction containing some metals (e.g. vanadium - *V*) will result in partial catalytic oxidation of the formed sulfur dioxide to sulfur trioxide (SO_3) in accordance with reaction equation:

$$SO_2 + \frac{1}{2}O_2 \xrightarrow[catalyst-e.g.vanadium]{} SO_3 \tag{28}$$

Coal as a chemical element, like hydrogen, has reductive properties. In high temperature accompanying the combustion process, in the conditions of limited access of oxygen, in the presence of hydrogen in the gasification reaction, sulfur contained in the fuel can be reduced to hydrogen sulphide according to reaction equation:

$$S + H_2 \xrightarrow[carbon]{} H_2S \tag{29}$$

Hydrogen sulphide formed during the reaction (30) is a flammable gas and at a later stage may be oxidized to sulphur dioxide and water according to reaction equation:

$$H_2S + \frac{3}{2}O_2 \longrightarrow SO_2 + H_2O \tag{30}$$

It should be clearly stated that in the flue gases from combustion process both SO_3 and H_2S may be present in trace amounts (well below 1% of the total emissions of sulfur compounds) and sulphur dioxide - SO_2 will be the main pollutant that is emitted. It is the so called raw material pollutant whose quantity in the exhaust almost in 100% corresponds to the amount of sulphur introduced to the combustion process. This allows relatively accurate calculation of sulphur emissions from combustion processes.

10. Formation of nitrogen oxides

The studies of the combustion processes lead to the conclusion that the formation of nitrogen oxides (*NOx*) is observed for many fuels. This applies primarily to nitric oxide (*NO*), nitrogen dioxide (NO_2) and nitrous oxide (N_2O). Of course there are known many other chemical compounds which are a combination of oxygen and nitrogen, however, in the flue gas from combustion they are in fact absent. The basic nitrogen oxide formed in the combustion process is nitrogen monoxide - *NO*. The extensive literature on the subject gives the information about three mechanisms of the formation of this compound in thermal processes. They are:

- thermal mechanism of Zeldovich,
- fuel mechanism,
- Fenimore prompt mechanism (radical).

Thermal mechanism was discovered and described for the first time by Zeldovich in the forties of the last century (Zeldovich, 1946). It includes the direct synthesis of nitric oxide from oxygen and nitrogen contained in air at high temperature. In the first stage there is decomposition of molecular oxygen to two molecules of active atomic oxygen in result of contact with high-energy inert molecule *M*, acting as a catalyst (it may be e.g. a molecule of hot metallic wall of the combustion chamber, burner, etc.):

$$O_2 + M \longrightarrow O + O + M \tag{31}$$

At high temperature molecular oxygen is decomposed into highly active form of atomic oxygen. Atomic oxygen molecules attack the nitrogen molecules and nitric oxide is formed, while formed in this reaction active nitrogen atoms, by attacking oxygen molecules, also form nitric oxide and active atomic oxygen. So it is a classic example of chain reaction, for which the first reaction (30) is the stage of initiation while the next two reactions (31) and (32) are the propagation stage:

$$O + N_2 \longrightarrow NO + N \tag{32}$$

$$N + O2 \longrightarrow NO + O \tag{33}$$

in fuel-rich mixtures where the concentration of hydroxyl radicals is significant, greater than the concentration of hydrogen and oxygen atoms ($OH >> H > O$) the following reaction can be considered as the last stage of termination:

$$N + OH \longrightarrow NO + H \tag{34}$$

An additional source of nitric oxide formation may be the following reactions in accordance with the mechanism described by Bozzelli (Bozzelli et at., 1994):

$$H + N_2 \longleftrightarrow N_2H \tag{35}$$

$$N_2H + O \longleftrightarrow NO + NH \tag{36}$$

The rates of formation of nitrogen oxides in the thermal mechanism are relatively high but only at high temperatures. This mechanism becomes negligible at temperatures above 1400°C.

Fuel mechanism is directly related to nitrogen content in fuels. As shown in Table 1 most solid and liquid fuels contain nitrogen. So the source of nitrogen in this mechanism is fuel while the source of oxygen is air introduced to the combustion process. The formation of nitric oxide in this mechanism is quite long and it goes through a number of succeeding-parallel reactions (Bowman et al., 1982, Miller & Bowman, 1989) shown in Figure 3.

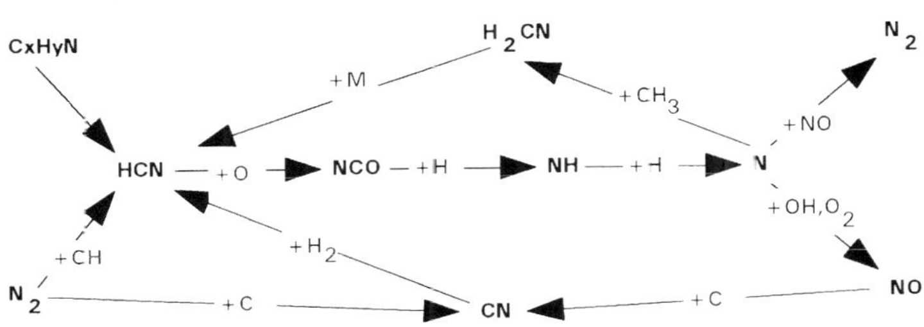

Fig. 3. Diagram of nitric acid formation according to fuel mechanism

In the first stage from organic matter, subject to combustion process (hydrocarbons containing nitrogen), hydrogen cyanide is released which by further oxidation and hydrogenation gives first radical *NCO*, next *NH* and then free atomic nitrogen (*N*). At this point there are three possible further courses of reactions - oxidation reaction in which nitric oxide is formed, the reaction with previously formed nitric oxide in result of which neutral molecular nitrogen is formed and the reaction with hydrocarbon radical returning the reaction to the beginning. Thus, only one course leads to the formation of nitric oxide while the other two effectively reduce its quantity - hence you can see that there are technical possibilities of influencing on the amount of produced nitrogen oxides - through the organization of the combustion process.

Prompt mechanism was first described by Fenimore in 1971 (Fenimore, 1971). He noted that in the early stage of flame, in the presence of numerous hydrocarbon radicals, occurs a synthesis of nitric oxide from oxygen and nitrogen introduced into the combustion process. Since *NO* formation reactions occur in the early stage of flame this mechanism was called the prompt mechanism. The prompt mechanism is complementary to the fuel mechanism and some chemical reactions follow the same path. Generally it can be assumed that the following reactions proceed in this mechanism:

$$CH + N_2 \longrightarrow HCN + N \tag{37}$$

$$CH_2 + N_2 \longrightarrow HCN + NH \tag{38}$$

$$HCN + O \longrightarrow NCO + H \tag{39}$$

$$NCO + O \longrightarrow NO + CO \tag{40}$$

The rate of formation of nitric oxide in this mechanism is very high, but the amount of formed *NO* according to this mechanism is relatively small.

Generally, during combustion of fuels, especially solid fuels, the greatest amount of nitrogen oxide is produced by the fuel mechanism (sometimes even 80-90%). Thermal mechanism begins to play a role only after exceeding the temperature of 1400°C, so the contribution of this mechanism usually does not exceed 10-20%. The contribution of the prompt mechanism is usually at the level of 1-5%. Of course the contribution of thermal mechanism increases with the increase of temperature.

In recent years, it was noted that during low-temperature combustion of fuel-poor mixtures with low excess air there appears an additional mechanism of nitric oxide formation. It is connected with the described by Malte and co-workers (Steele et al., 1995) synthesis of nitrous oxide in the combustion conditions. Most likely, this reaction occurs with the participation of active atoms of hydrogen, oxygen or carbon monoxide - a product of incomplete combustion according to the following equations:

$$N_2O + H \longrightarrow NO + NH \tag{41}$$

$$N_2O + O \longrightarrow 2\,NO \tag{42}$$

$$N_2O + CO \longrightarrow NO + NCO \tag{43}$$

Reactions of nitric oxide formation from nitrous oxide proceed relatively quickly at temperatures below 1200°C in the area of good oxygenation of the combustion zone. The nitrous oxide synthesis itself proceeds according to several possible paths, one of them requires, just as in the thermal mechanism, the contact with the high-energy inert molecule M:

$$O + N_2 + M \longrightarrow N_2O + M \tag{44}$$

$$NCO + NO \longrightarrow N_2O + CO \tag{45}$$

$$NH + NO \longrightarrow N_2O + H \tag{46}$$

$$2NO \longleftrightarrow N_2O + O \tag{47}$$

The reaction (45), (46) and (47) shows that the synthesis of N_2O can take place outside the combustion zone with the participation of nitric oxide previously formed in the presence of numerous radicals.

The studies of combustion products show above all the presence of nitric oxide (NO) and small quantities of nitrous oxide (N_2O). Nitrogen dioxide (NO_2), from the oxidation of NO formed during combustion, is a component of flue gas as well. The formation of NO_2 has been described by Miller and Bowman in 1989 (Miller & Bowman, 1989). As a result of diffusion of hydrogen radicals beyond flame zone to the area of lower temperatures (below 750°C) in the presence of excess oxygen there are formed, involving high-energy inert particles of M, HO_2 radicals which also in the same zone can react with nitric oxide according to the reaction:

$$NO + HO_2 \longrightarrow NO_2 + OH \tag{48}$$

In parallel to the synthesis reaction, nitrogen dioxide decomposition reactions take place:

$$NO_2 + O \longrightarrow NO + O_2 \tag{49}$$

$$NO_2 + H \longrightarrow NO + OH \tag{50}$$

Moreover, in the further zone of the installation, after heat recovery systems, at temperatures below 200°C occurs direct oxidation of NO to NO_2 by oxygen present in the flue gas:

$$NO + \frac{1}{2}O_2 \longrightarrow NO_2 \tag{51}$$

Under proper combustion conditions, the participation of NO_2 in the whole stream of emitted nitrogen oxides does not exceed 5-10%.

Nitrogen oxides (*NOx*) understood as the sum of *NO*, NO_2 and N_2O are pollutants whose rate of emission is minimally dependent on nitrogen content in the fuel. Their emissions is a direct result of plant construction and organization of the combustion process.

11. Particulate matter emission

The problem of particulate matter emissions practically applies only to solid fuel combustion process. It occurs in a minimal degree during the combustion of liquid fuels and practically does not occur during the combustion of gaseous fuels. The presence in the burned material of inorganic solid substance - non-flammable, most often called ash makes that in the combustion process solid residue in the form of slag and ash is formed. The flow of air through the combustion zone results in entrainment of fine solid particles and thus dust emission from combustion processes is formed. This emission is the higher, the better is the oxygenation of the combustion zone and the higher are air velocities observed in this zone. Solid particles, lifted in the flue gas stream from the combustion zone, have usually very small diameter and their main component is silicon dioxide (SiO_2 - silica). At the same time these particles are carriers of metals, elemental carbon (soot) as well as the adsorbed products of incomplete combustion and products of secondary synthesis outside the flame zone. In modern solid fuel combustion installations the release of solid particles is usually not more than 60% for the pulverized fuel and fluidal installations, and about 25-40% for the grate installations. The emission of solid particles, as dust from the combustion process is proportional to the content of non-flammable substances in the fuel and the factor of proportionality, the so called release factor, which is different for each installation and depends on its structure. So it depends partly on raw materials characteristics.

12. Metals emission

The emission of metals from the combustion process is very closely linked with the emission of solid particles (particulate matter) and therefore this problem does not exist during the combustion of gaseous fuels; it occurs to a small extent during combustion of liquid fuels and it applies fully to the combustion of solid fuels containing significant admixtures of non-flammable substance. Metals (mainly heavy metals), due to the speciation in the products of combustion, can be divided roughly into three groups:

- metals easily evaporated in the combustion zone and present mainly in emission to air,
- metals with high melting points and boiling points cumulative in solid combustion products (slag and grate ash)
- metals of intermediate properties for which there is division into gas and solid combustion products.

The first group includes above all mercury (*Hg*) and cadmium (*Cd*). Similar properties have selenium (*Se*). Most metals belong to the second group - they are mainly iron (*Fe*), magnesium (*Mg*), cobalt (*Co*), chromium (*Cr*), copper (*Cu*), manganese (*Mn*), molybdenum (*Mo*) and nickel (*Ni*). Similar properties also have aluminum (*Al*) and titanium (*Ti*). The third intermediate group, includes mainly arsenic (*As*), lead (*Pb*), tin (*Sn*) and zinc (*Zn*) as well as antimony (*Sb*). The detailed data on the speciation of the selected metals in the combustion process, based on the example of waste incineration plants (Belevi & Moench, 2000, Belevi & Langmaister 2000, Sukrut et al., 2002), are given in Table 4.

Metal	Solid phase [%]	Gas phase [%]
Antimony - Sb	25	75
Arsenic - As	68,6	31,4
Chromium - Cr	92	8
Tin - Sn	50,8	49,2
Zinc - Zn	52	48
Aluminium - Al	87,8	12,2
Cadmium - Cd	9,5	90,5
Cobalt - Co	90,1	9,9
Magnesium - Mg	91,9	8,1
Manganese - Mn	94	6
Copper - Cu	97,3	2,7
Molybdenum - Mo	91,7	8,3
Nickel - Ni	98,2	1,8
Lead - Pb	59	41
Mercury - Hg	0,7	99,3
Titanium - Ti	85,3	14,7
Iron - Fe	99,1	0,9

Table 4. Speciation of the selected metals in the combustion process

The situation is substantially changed, if in the burned area appear halides - especially chlorine and bromine. Melting point and boiling point of heavy metal salts (chlorides) is significantly lower than the melting point and boiling point of pure metal or its oxide.

According to Barton (Barton et al., 1991) and Niessen (Niessen, 2002) it is particularly evident in the case of nickel (*Ni*), thallium (*Tl*) and lead (*Pb*). This is illustrated by the data included in Table 5.

Metal	Volatility temperature [°C]	
	0% Cl	10% Cl
Antimony - Sb	660	660
Arsenic - As	32	32
Chromium - Cr	1613	1610
Cadmium - Cd	214	214
Nickel - Ni	1210	693
Lead - Pb	627	-15
Mercury - Hg	14	14
Selenium - Se	318	318
Thallium - Tl	721	138

Table 5. Volatility temperature (reaching resilience 0,1 Pa) for the selected metals in the absence and in 10% presence of chlorine.

So the emission of metals from combustion processes largely depends on the type of metal, temperature of the combustion process, but also on the presence of halides - chlorine and bromine which significantly increase the presence of some metals in flue gas.

13. Hydrogen chloride emission

Chlorine (*Cl*) is an element that is widespread in the environment. Its small, sometimes even trace amounts are found in each fuel. Of course, the least amount of chlorine (almost immeasurable amount) is in fuel gas, slightly more in liquid fuels, while in solid fuels its content may be as high as 2%. In the process of combustion, chlorine - like metals is subject to speciation. The largest part of chlorine (50-60%) is bound in the form of chlorides in the fly ash, a part is also bound in a similar way in slag and grate ashes. The remaining part of chlorine is released into the environment as emissions of hydrogen chloride (*HCl*), while only a small amount is released in form of molecular chlorine (Cl_2). With the increase of combustion temperature, the amount of chlorine released in the form of emissions increases and at the same time the amount of chlorine present in the slag and ashes decreases. Data on the speciation of chlorine in the combustion process at different temperatures according to the work of Liu (Liu et al., 2000) are shown in Table 6

Presence of chlorine	Temperature (°C)					
	600	650	700	750	800	900
HCl	7,6	10,7	10,1	19,9	27,4	43,7
Cl_2	0,5	0,2	0,4	0,6	0,8	1,3
Fly ash	50,9	47,5	56,3	53,2	62,4	51,9
Slag	40,5	41,3	32,8	25,6	8,5	1,2

Table 6. Speciation of chlorine in the combustion process depending on the temperature

The presence of chlorine in fly ash is of great importance in the formation of chloroorganic compounds outside the combustion zone. The presence of free chlorine which can participate in the chlorination and oxychlorination reactions outside the combustion zone is of similar importance. In the flue gas, outside the combustion zone, hydrogen chloride may undergo a catalytic decomposition (Deacon reaction) in accordance with the reaction equation (Griffin, 1986):

$$4\ HCl + O_2 \xrightarrow[catalyst - Cu, Fe, Al]{} 2\ Cl_2 + 2\ H_2O \tag{52}$$

In the case of sulfur dioxide (SO_2) present in flue gas, at the same time proceeds hydrogen chloride reproduction reaction (Lindbauer et al., 1994):

$$SO_2 + Cl_2 + H_2O \longrightarrow SO_3 + 2\ HCl \tag{53}$$

The balance in a chlorine - hydrogen chloride system is also affected by hydroxyl radicals and water present in the flue gas:

$$HCl + OH \longleftrightarrow Cl + H_2O \tag{54}$$

Thus the concentration of hydrogen chloride and free chlorine in the flue gas depends on many factors, including temperature and that is why their emission rates cannot be predicted on the basis of chlorine content in the fuel.

14. Products of incomplete combustion

The term products of incomplete combustion (PICs) means organic compounds introduced to combustion and formed during combustion that were not oxidized in the combustion zone as well as products of secondary synthesis outside the combustion zone. As already mentioned when discussing the combustion process paths shown in Figure No. 2, in the combustion zone and directly next to it, in the gas phase there are many hydrocarbon and chlorinated hydrocarbons radicals as well as simple aliphatic hydrocarbons, often unsaturated, which may participate in secondary, high-temperature synthesis reactions. Among these radicals there is also acetylene, which plays a key role in the later stages of synthesis.

The key here is a reaction first described by Aubrey and van Wazer (Aubrey & van Wazer, 1964), in which in high temperature aromatic compounds are formed from aliphatic compounds (acetylene):

$$9\,C_2Cl_6 \longleftrightarrow 12\,CCl_4 + C_6Cl_6 \qquad (55)$$

A special role is played by acetylene. It is always present in the flue gas from the combustion process and is a precursor of the formation of many chloro-aromatic compounds (Lenoir et al., 2001). It is subject in the first stage to chlorination to dichloroacetylene in result of ligand exchange reaction and then is subject either to cyclization to hexachlorobenzene or condensation to hexachlorobutadiene (Lenoir et al., 1998). Subsequent studies have shown that, for example from acetylene at high temperatures chlorobenzenes, chlorophenols and chloronaphthalenes can be formed.

In addition, the presence in the waste of organic substances of unsaturated character, that is containing in the molecule double or triple carbon-carbon bonds ($C=C$ or $C \equiv C$) causes that at a temperature of about 500-800°C, occurs the synthesis reaction of polycyclic aromatic hydrocarbons (Liow et al., 1997). Polycyclic aromatic hydrocarbons (PAHs) are among the most dangerous substances in the environment. Many of them, such as benzo(a)pyrene, benzo(a)anthracene, benzo(k)fluoranthene, dibenzo(a,h)anthracene, indeno(1,2,3-c,d)pyrene are classified by the International Agency for Research on Cancer (IARC) as substances with proven carcinogenic effect. They are emitted to the atmosphere practically from all combustion processes, not only from the waste incineration plants (Mastral et al., 2000).

The result of these reactions, which proceed outside the combustion zone, is the presence of pollutants in the flue gas, among others such as (Eduljee, 1994):

- hydrocarbons C_1 and C_2,
- acrylonitrile, acetonitrile,
- benzene, toluene, ethylbenzene, xylene,
- 1,2-dichlorobenzene, 1,4-dichlorobenzene, 1,2,4-trichlorobenzene, hexachlorobenzene,
- phenol, 2,4-dinitrophenol, 2,4-dichlorophenol, 2,4,5-trichlorophenol, pentachlorophenol,

- chloromethane, chloroform, methylene chloride, carbon tetrachloride,
- 1,1-dichloroethane, 1,2-dichloroethane, 1,1,1-trichloroethane, 1,1,2-trichloroethane, 1,1,2,2-tetrachloroethane,
- 1,1-dichloroethylene, trichloroethylene, tetrachloroethylene,
- formaldehyde, acetaldehyde, acetone, methyl-ethyl-ketone,
- vinyl chloride, diethyl phthalate,
- formic acid, acetic acid and many others.

Subsequent studies conducted by other researchers (Wienecke et al., 1995, Jay & Stieglitz, 1995, Mascolo et al., 1997, Trenholm 1998) showed the presence of more than 350 different types of chemical compounds (organic) in the flue gases from the waste incineration plants in concentrations above 5 $\mu g/m^3$. Similar results were obtained for burning of wood and other biomass. Unfortunately, there are no such test results for the combustion of coal, although it is expected that the situation is similar.

15. Formation of PCDD/Fs, PCBs, PCNs and PAHs

Polychlorinated dibenzo-p-dioxins (PCDDs), polychlorinated dibenzofurans (PCDFs), polychlorinated biphenyls (PCBs), polychlorinated naphthalenes (PCNS) and polycyclic aromatic hydrocarbons (PAHs) are also products of incomplete combustion formed outside the combustion zone. Their common feature is the negative biological impact - some of them have proven carcinogenic properties (PAHs), others are xantoestrogens (endocrine disrupters) that disrupt the hormonal balance of living organisms. In recent years many myths referring to them have appeared, however, most of them have nothing to do with reality.

According to the diagram in Figure 2 showing the combustion process of solid fuels, the crucial stage for the synthesis of PCDD/Fs, PCBs, PCNs and PAHs is the formation of hydrocarbon radicals, often including halogenated (mainly chlorinated) ones as well simple unsaturated hydrocarbons e.g. ethylene and acetylene. These reactions proceed at high temperatures and in the next stage their products undergo further reactions such as cyclization - the formation of aromatic often chlorinated compounds - including benzene, naphthalene, chlorobenzenes and chlorophenols. All these chemical compounds are formed in virtually any combustion process of solid and liquid fuels and also, although less frequently, gas fuels. If in the zone of respective temperatures appear chlorine, oxygen and organic matter chlorinated compounds are formed - such as PCDD / Fs, PCBs and PCNs. This phenomenon is observed not only for combustion process but also for most thermal processes running at temperatures of 200-700°C. The overall reaction is as follows:

$$R - Cl + O_2 \xrightarrow[combustion]{} PCDD/Fs,\ PCBs,\ PCNs \qquad (56)$$

On the basis of the critical analysis of numerous literature data, we can assume the following paths of appearance of dioxins (as well as PCBs and PCNs) in the flue gas stream from combustion:

- dioxins introduced to the system which did not participate in any chemical reactions or were subject to the reactions of partial chlorination and/or dechlorination of the molecule (Tosine et al., 1985),

- dioxins from the aromatic rings formation and rearrangement of substituents with the participation of chlorinated precursors such as PVC, PCBs, chlorinated benzenes, chlorinated phenols directly behind the combustion zone - this is a reaction that occurs in the gas phase at temperatures of 500-700°C (Ballschmiter et al., 1985),
- dioxins formed outside the combustion zone in result of the reaction of non-chlorinated organic compounds such as PAHs, lignin, lignite, etc. and then chlorination reaction as well as from various types of chlorinated organic precursors such as chlorophenols, chlorobenzenes, polychlorinated diphenyl ethers, polychlorinated biphenyls and other products, etc. as well as other products of incomplete combustion which may react with each other and on the surface of metals in the fly ash particles (Dickson & Karasek 1987), where they are subject to the reactions of chlorination and dechlorination of the molecule,
- dioxins formed outside the combustion zone of molecular carbon (carbonized residues of the combustion, unburned carbon in the particles of dust, soot and organic or inorganic chlorine in the presence of fly ash at a temperature of 200-400°C usually called *de novo* synthesis (Vogg & Stieglitz 1986).

The main difference in these mechanisms is the source of carbon and the temperature range in which the synthesis takes place. In the light of the research results it seems that the most important mechanism owing to which most dioxins are formed is the third mechanism (*de novo*), then the second and least dioxins are produced by the first mechanism (Huang & Buekens, 1995).

The analysis of the above mentioned paths of appearance of dioxins in the flue gas stream from thermal processes, including combustion of solid fuels shows that the first case is relatively unlikely. PCDD/Fs and PCBs are not chemicals of high thermal stability. In the combustion conditions (presence of oxygen, mixing, flow) practically most organic compounds, including dioxins, is decomposed at the temperature about 850°C. The temperature of 99.9% of the destruction of PCDD/Fs is around 700°C while the temperature of the destruction of other compounds that are precursors of the synthesis of dioxins can reach up to 950°C. However, in the case of lack of oxygen the limit of the decomposition of organic compounds may be increased to around 1000°C. It therefore seems unlikely that the dioxins contained in the material subject to combustion at temperatures reaching and exceeding 1000°C could not be subject to destruction, although in the case of poor construction of the combustion installation and the formation of cold zones in the combustion chamber this is not excluded.

The second path of the appearance of PCDD/Fs, PCBs, PCNs in flue gas is a high temperature synthesis which is a condensation of precursor molecules - benzene, naphthalene, phenanthrene, acetone, trichloroethanes, benzaldehyde, dibenzofuran, benzofuran, phenol, mono-, di-, trichlorophenoles, chlorobenzenes, tetrachlorethylene or polychlorinated diphenyl ethers. These precursors may react with each other both at 500-700°C in the gas phase and on the surface of fly ash at a temperature of 200-500°C. The key to the formation of dioxins in result of synthesis in the gas phase are the conditions for conducting the combustion process. Numerous studies have shown a very close relationship between the formation of chlorinated and non-chlorinated aromatic compounds in the combustion and afterburning chamber (at a temperature 650-900°C) and the parameters of combustion. It was also noted that by-products of gas fuel combustion - such as natural gas, acetylene, ethane and ethylene may be a sufficient source of carbon for the synthesis of dioxins.

Outside the combustion zone at temperatures below 500°C dioxins are formed as a result of a series of catalytic reactions occurring on the surface of dust containing metals. Usually it is assumed that dioxins are formed there by two mechanisms - the catalytic synthesis of precursors such as chlorobenzenes, chlorophenols, polychlorinated naphthalenes and polychlorinated biphenyls (similarly as in the gas phase) and catalytic synthesis of elemental carbon (soot) contained in dust particles or polycyclic aromatic hydrocarbons as well as gaseous chlorine and oxygen.

The first of these two mechanisms was first described in 1987 by Dickson (Dickson & Karasek 1987). Numerous studies conducted in subsequent years have shown that this reaction proceeds in principle as a condensation reaction of chlorobenzenes or chlorophenols or as a reaction of selective oxychlorination of polychlorinated naphthalenes and polychlorinated biphenyls. This reaction can be catalyzed by many metals – e.g. copper, titanium, manganese, cobalt, zinc but definitely copper has by far the largest catalytic ability. The characteristic of the condensation mechanism (high temperature in gas phase or on the surface of dust particles) is the fact that in this way are formed low chlorinated dioxins, containing from one to maximum 3-4 atoms of chlorine. So the problem of the origins of high chlorinated dioxins, containing from 4 to 8 atoms of chlorine per molecule, in flue gas remained unresolved. The third mechanism of dioxin formation offers the solution to this problem - the *de novo* synthesis.

The second mechanism of the formation of dioxins in the zone after the combustion chamber, in the area of flue gas cooling, was discovered in the late eighties of the last century by Stieglitz and co-workers, the so called mechanism of *de novo* synthesis (Vogg & Stieglitz 1986). The *de novo* synthesis is a slow heterogenic catalytic reaction in which dioxins are formed without the formation of gaseous intermediate products. It occurs within the temperature range 200-500°C from carbon particles contained in fly ash, through the formation and closure of benzene rings, which then react with oxygen and chlorine on the catalyst surface (Stieglitz & Vogg, 1987). The source of carbon in *de novo* synthesis is usually so called elemental carbon contained in fly ash particles, as a carbonized residue of thermal decomposition and oxidation in the combustion process. In practice it is impossible to achieve complete burnout of carbon contained in the fuel and small amounts of unburned carbon are present in flue gas in form of soot which is made up of elemental carbon and the polycyclic aromatic hydrocarbons - PAHs. The emission of dioxins from the combustion of pure wood suggests that the carbon source can also be large, complex molecules of organic compounds such as lignin and lignite. The chlorinating agent in *de novo* synthesis can be free, molecular chlorine, chlorinated organic compounds as well as volatile inorganic chlorine salts - such as NaCl or $FeCl_3$. However, the fundamental role in the formation of dioxins plays the concentration of chlorine free radicals in the zone of combustion reaction and cooling. Despite the obvious role of chlorine in *de novo* synthesis the data from over 1 900 research works on an industrial scale (Rigo & Handler, 1998) shows that there is no clear relationship between the amount of chlorine introduced in the fuel to the combustion process and the amount of dioxins emitted. There is, however, relationship between the amount of dioxins emitted and the concentration of chlorine or hydrogen chloride at the outlet of the combustion chamber.

The emission of dioxins depends primarily on the construction of the combustion unit and burning conditions. The concentration of chlorine in the flue gases significantly affects the profile of formed congeners PCDD and PCDF. The higher the concentration the greater the share of congeners with higher number of chlorine atoms. This suggests that *de novo* synthesis is the reaction of subsequent chlorination of individual congeners of dioxins and furans. Analyzing the above mechanism of *de novo* synthesis it can be stated that on the fly ash containing $CuCl_2$ there can also run the follow-chlorination of lower substituted dioxins generated in the gas phase as well as dechlorination of higher chlorinated dioxins. The optimal temperature assumed for *de novo* synthesis is about 350°C. Oxygen concentration also influences the course of the *de novo* synthesis. This is obvious if we consider that oxygen is present in the molecule of both dibenzo-p-dioxin and dibenzofuran. Studies have confirmed that the increase in oxygen concentration outside the combustion zone significantly increases the amount of generated dioxins, the increase in the concentration of hydrogen chloride and gives the same effect.

The catalyst in the *de novo* synthesis is primarily copper chloride. Copper has a very strong effect in supporting the formation and closing the rings such as chlorophenols. Subsequent studies have shown that transition metals catalyze both ring closing and the process of chlorination and *de novo* synthesis may be treated as an electrophilic substitution reaction that runs in two stages - the first stage includes the chlorination of carbon surface while the second stage is oxidation and decomposition in which dioxins and furans are formed. Similar to the catalytic properties of copper, in the *de novo* synthesis similar properties also have aluminum, iron, magnesium, manganese, cobalt as well as sodium, potassium and zinc. Other metals - such as nickel does not exhibit catalytic activity or they decrease the amount of generated dioxins e.g. chromium, wolfram or vanadium. But there is no doubt that the best catalyst for *de novo* synthesis is copper. In reality, however, on the surface of fly ash there are many metals, whose influence on the synthesis of dioxins is different - some of them catalyze the *de novo* synthesis or the chlorination process of low chlorinated dioxins formed earlier (in the synthesis in gas phase) - for example, copper, zinc, iron or aluminum, while others catalyze the decomposition of generated dioxins or their dechlorination – e.g. chromium, vanadium, tungsten or nickel.

After analyzing the emissions of dioxins from combustion processes and all the described paths of the presence of dioxins in the flue gas, especially the paths of their synthesis it should be noted that probably most dioxins are formed outside the combustion zone. According to Goldfarb, who analyzed the problem of dioxin emissions from Canadian incineration plants, this is most probably the vast majority (Goldfarb, 1986).

Summing up the problem of paths of dioxins formation in combustion processes we should first mention the decisive influence of the combustion process conditions on the quantity of produced dioxins and furans. Combustion process conditions include first of all temperature, residence time of gases from the combustion in the zone of high temperatures, oxygen concentration in the combustion zone and directly connected with it concentrations of carbon monoxide in the flue gas. Poor combustion conditions i.e. too low temperature and high concentration of carbon monoxide contribute to the formation of incomplete combustion products such as chlorobenzenes and chlorophenols which are precursors of dioxins and can undergo condensation reactions in both high and low temperature zone in the presence of metals catalyzing condensation reactions. Poor combustion conditions also favor the formation of polycyclic aromatic hydrocarbons (PAHs) and unburned coal residues in the dust entrained in the air which is an excellent raw material for *de novo* synthesis and this in turn means that you can prove the thesis that there is a direct impact of the combustion conditions on the amount of dioxins generated in all three known mechanisms described earlier.

When conducting combustion processes we are always dealing with two opposing reactions (reaction systems) - formation of dioxins as a result of various homo- and heterogenic reactions and decomposition of dioxins as a result of high temperature and catalytic reactions. Graphic illustration of this problem is shown in Figure 4. Most of the dioxins generated in waste incineration plants is formed in result of *de novo* synthesis in the temperature range 200-400°C with maximum temperatures around 300°C. This maximum is the result of synthesis and decomposition processes running in parallel.

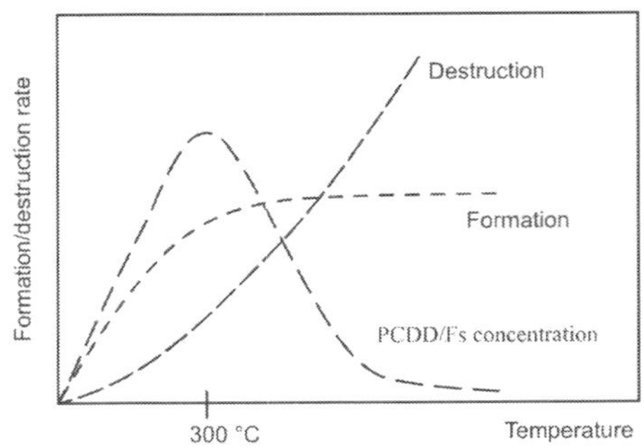

Fig. 4. Formation and decomposition of dioxins in a catalytic process on dust particles.

The process of formation of PCDD/Fs, PCBs, PCNS and PAHs during combustion depends on many factors and that is why it is very difficult to predict how much of these compounds will be formed in a specific process with fixed parameters. Momentary dioxin emissions results not only from the current combustion conditions. Due to the fact that probably most of the dioxins is produced by *de novo* synthesis, which is a very slow reaction, the formation of dioxins and their release continues long after the optimization of the parameters of the combustion process and minimizing the amount of generated dioxins. This is so called memory effect occurring in waste incineration plants. In the case of unstable operation of incinerators high concentrations of dioxins in the flue gas can be observed over long periods of time (Hunsinger et al., 2007).

16. Primary methods of emission reduction

Understanding the mechanisms of pollutants formation in the combustion process enables the development of such technologies and methods of combustion which ensure the smallest amount of generated pollutants. There are well known so called low emission combustion processes relating to carbon monoxide and nitrogen oxides - the use of the combustion zone, in stages, flue gas recirculation and the so-called reburning (introducing fuel to the periphery of flame zone) allows to reduce NO_x emissions by about 50-60% compared to the original process (Jarosiński, 1996, Hill & Smoot, 2000).

Such efforts to reduce emissions, which are an interference in the combustion process are called primary methods of reducing emissions. They become more and more important in recent years, because from the economic point of view, they are more cost effective (cheaper) than the secondary methods, called "end of pipe technologies". Their effectiveness is generally lower than that of flue gas cleaning methods, however, they significantly help to improve the work of the latter, allowing to obtain the total degree of purification of gases at a level above 99%.

In recent years the determination of the effect of parameters of the combustion process on the emission of metals, PAHs, dioxins and other organic substances from the combustion process became an important issue. Numerous literature reports clearly show that keeping good combustion process parameters significantly affects the reduction of emissions of organic substances (including PAHs and dioxins) to the atmosphere. A parameter which characterizes "good combustion conditions" very well is the concentration of carbon monoxide in the flue gas. Other parameters defining the "good combustion conditions" are above all so called "3T"- temperature, turbulence and time of flue gas residence in proper temperature. Detailed conditions are specified among others in the Directive on the incineration of waste (2000/76/EC) - they are as follows: flue gas residence time at 850°C should be not less than 2 seconds at high gas turbulence (for waste containing less than 1% of chlorine compounds, in the case of waste containing more than 1% of chlorine compounds the temperature should be higher than 1100°C).

It is essential, however, to ensure combustion conditions close to complete and perfect combustion, with minimized amount of formed carbon monoxide. Such conditions prevail in properly oxygenated combustion zone at an optimal concentration of oxygen (for coal combustion 6-8%, excess air coefficient about 1.4-1.8). In the case of waste incineration better oxygenation of the combustion zone and providing excess air coefficient of the order of 2,0-2,4 (oxygen concentration 10-13%) is required. The effect of oxygen concentration and combustion temperature on the concentration of carbon monoxide in the flue gas is illustrated in Figure 5 (Seeker, 2001).

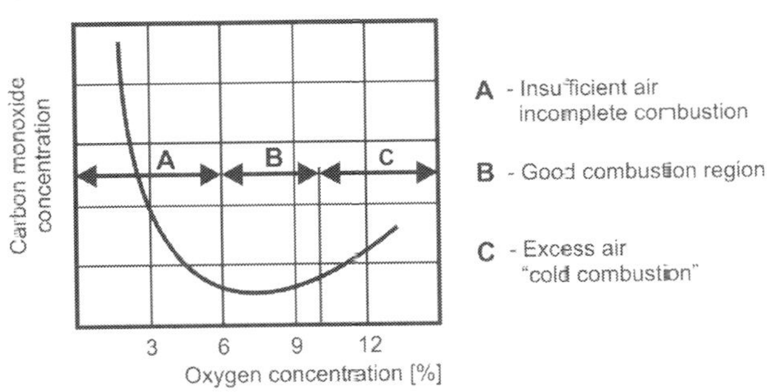

Fig. 5. The effect of oxygen concentration in the combustion zone on the carbon monoxide emission.

According to the research of Seeker (Seeker, 2001) the concentration of carbon monoxide, that is combustion conditions, has a great impact on the amount of dioxins produced in the combustion process and their emission. This is illustrated in Table 7.

Combustion parameters	Slag, bottom ash	Fly ash	Dust from electrostatic precipitator	Emission to atmosphere
T = 830 °C CO = 25 ppm	0,33 µg/Mg	0,14 µg/Mg	21,2 mg/Mg	11,2 µg/Mg
T = 850 °C CO = 1 ppm	0,35 µg/Mg	0,22 µg/Mg	21,2 mg/Mg	1,6 µg/Mg

Table 7. The effect of combustion conditions on the rate of PCDD/Fs emission to atmosphere and the concentration in secondary waste from the combustion process (per 1 Mg of incinerated waste).

The research conducted by Schmöckel (Schmöckel & Streit, 1994) of wood combustion process in plants with a capacity of 11 kW to 13.9 MW have shown a very close correlation between the concentration of some pollutants in the flue gas and the concentration of carbon monoxide, which clearly shows the impact of combustion conditions on the emission of organic pollutants. The results of these studies are shown in Figure 6.

Combustion parameters have also great impact on emissions of metals. This is due to the different volatility of metals and their speciation depending on the forms of occurrence. The presence of chlorine and hydrogen chloride in the combustion zone causes that different amounts of metals are in slag, fly ash, ash from de-dusting equipment or in gases emitted to the atmosphere (Chiang et al. 1997). There is also a clear effect of combustion temperature and oxygen concentration in different zones of combustion on the balance of metals in individual secondary products of the combustion process (Wunsh et al., 1995, Wey et al., 1998). For example, controlling the amount of primary air in the individual zones of the grate and the temperature can result in almost zero emission of some metals with almost 100% their content in the slag (Modigell & Liebig, 1999). The use of primary methods of reducing emissions makes that secondary purification of flue gas to the level required by law (e.g. Directive 2000/76/EC) is easier. The positive effect of the use of primary methods of reducing emissions on the concentration of pollutants in flue gas is shown in Table 8.

Pollution	Unit	Installation without the use of primary methods	Installation using the primary methods	Allowable concentration according to Directive 2000/76/EC
Dust	mg/m^3	2 000 – 10 000	1 000 – 1 500	10
Carbon monoxide	mg/m^3	50 – 80	10 – 15	50
Sum of organic compounds	mg/m^3	10 – 100	0,5 – 1	10
PCDD/Fs	ng TEQ/m^3	5 – 12	0,6 – 1,2	0,1

Table 8. The effect of the primary methods for reducing emissions of pollutants in waste incineration plants (Büttenberger & Hansen 1997).

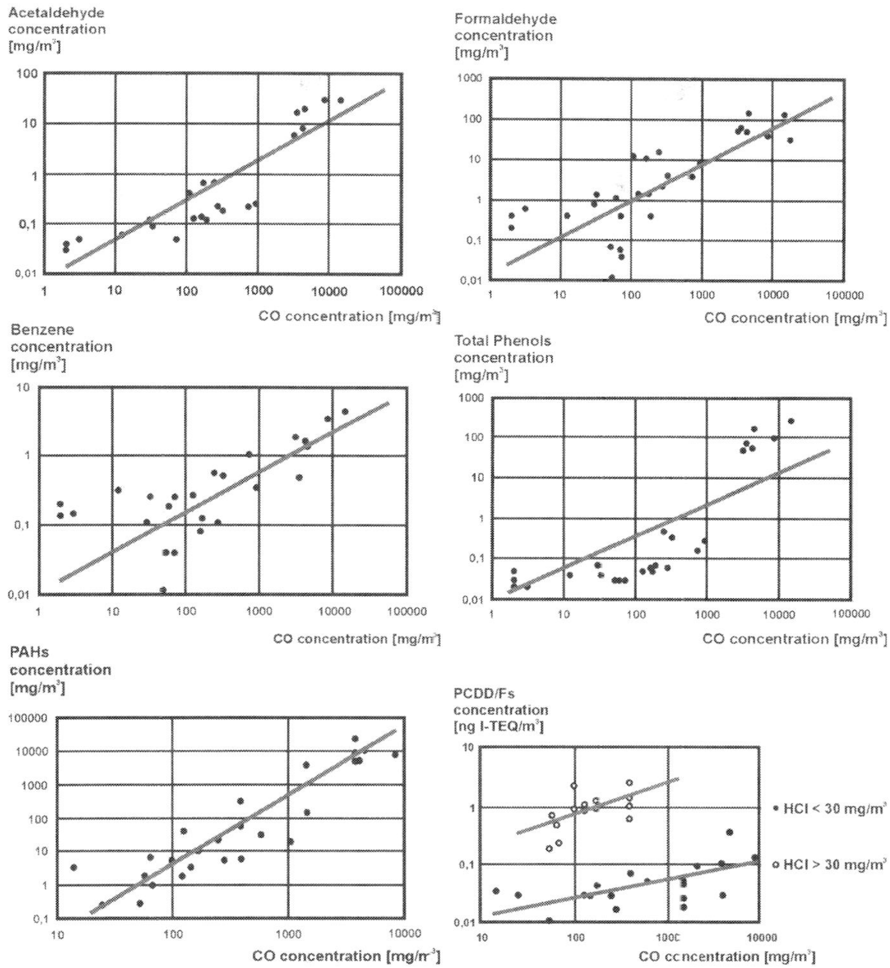

Fig. 6. The effect of combustion conditions (concentration of CO in flue gas) on the emission of some organic pollutants from combustion process (Schmöckel & Streit, 1994).

17. References

Aubrey, N. E., van Wazer, J. J. (1964) Equilibrium Rearrangement of Perchlorocarbon Compounds. *Journal of American Chemical Society*, Vol.86, pp.4380-4383

Ballschmiter, K., Zoller, W., Buchert, H., Clas Th. (1985) Correlation between substitution pattern and reaction pathway in the formation of polychlorodibenzofurans. *Fresenius Zeitschrift für Analistische Chemie*, Vol. 322, pp.587-594

Barton, R. G., Seeker, W. R., Bostian, H. E. (1991) The behavior of metals in municipal sludge incinerators. *Trans. Inst. Chem. Eng.* Vol.69, pp.29-36

Belevi, H., Langmeier, M. (2000) Factors determining the element behavior in municipal solid waste incinerators. Laboratory experiments. *Environ. Sci. Technol.* Vol.34, pp.2507-2512

Belevi, H., Moench, H. (2000) Factors determining the element behavior in municipal solid waste incinerators. Field studies. *Environ. Sci. Technol.* Vol.34, pp.2501-2506

Bowman, C.T., Hanson, R.K., Louge, M. Garman, A. (1982) Gas Phase Reaction Kinetics of NOx Formation from Fuel Nitrogen in Fossil Fuel Combustion. – *in: Proceedings of 1982 Joint Symposium on Stationary Symposium NOx Control*, Dallas, USA

Bozzelli, J. W., Chang, A. Y., Dean, A. M. (1994) Analysis of the reactions $H+N_2O$ and NH+NO: Pathways and rate constants over a wide range of temperature and pressure. *in: Proceedings of International Symposium on Combustion*, Vol.25, No.1, pp.965-974

Büttenberger, B., Hansen, H. (1997) Das EVT-Müllverbrennungssystem im Hinblick auf die neuesten gesetzlichen Anforderungen. *BWK/TÜV-Umwelt-Special*, October 1997, pp.26-38

Chiang, K.Y., Wang, K.S. Lin, F.L., Chu, W.T (1997) Chloride effects on the speciation and partitioning of heavy metals during the municipal solid waste incineration process. *The Science of the Total Environment*, Vol. 203, pp.129-140

Dickson, L. C., Karasek, F. W. (1997) Model studies of polychlorinated dibenzo-p-dioxin formation during municipal refuse incineration. - *Science*, Vol.237, pp.754-756

Eduljee, G. H. (1994) Organic micropollutant emissions from waste incineration. - in: Hester R. E., Harrison R. M. (ed.) - *„Waste incineration and the environment"*, Royal Society of Chemistry, London

Fenimore, C. P. (1971) Formation of nitric oxide in premixed hydrocarbon flames. – *in: Proceedings of XIIIth International Symposium on Combustion,* The Combustion Institute, Pittsburg, pp.373-384

Goldfarb, T. D. (1989) Evidence for post-furnace formation of PCDD and PCDF - implications form control. *Chemosphere*, Vol.18, pp.1051-1055

Griffin, R. D. (1986) A new theory of dioxin formation in municipal solid waste combustion. *Chemosphere*, Vol.15, pp.1987-1990

Hill, S.C., Smoot, L.D. (2000) Modeling of nitrogen oxides formation and destruction in combustion systems. *Progr. in Energy and Combustion Sci.* Vol.16, pp.417-458;

Huang, H., Buekens, A. (1995) On the mechanisms of dioxin formation in combustion processes. *Chemosphere*, Vol.31, pp.4099-4117

Hunsinger, H., Seifert, H., Jay, K. (2007) Control of PCDD/F formation under conditions of fluctuating combustion performance in MSWI. *Organohalogen Compounds*, Vol.69, pp.956-961

Jarosiński, J. (1996) Techniki czystego spalania (in polish). WNT, Warszawa;

Jay, K., Stieglitz, L. (1995) Identification and quantification of volatile organic components in emission of waste incineration plants. *Chemosphere*, Vol.30, pp.1249-1260

Lenoir, D., Wahrmeier, A., Schramm, K. W., Kaune, A., Zimmermann, R., Taylor, P. H., Sidhu, S. S. (1998) Thermal formation of polychlorinated dibenzodioxins and - furans: investigation of relevant pathways. *Environmental Engineering Science*, Vol.15, pp.37-478

Lenoir, D., Wahrmeier, A., Sidhu, S. S., Taylor, P. H. (2001) Formation and inhibition of chloroaromatic micropollutants formed in incineration processes. *Chemosphere*, Vol.43, pp.107-114

Lindbauer, R. L., Wurst, F., Prey, Th. (1994) PCDD/F emission control for MSWI by SO_3 addition. *Organohalogen Compounds*, Vol.19, pp.355-359

Liow, M. Ch., Lee, W. J., Chen, S. J., Wang, L. Ch., Chung, Ch. H., Chen, J. H. (1997) Emission polycyclic aromatic hydrocarbons from medical waste incinerators. *Journal of Aerosol Science*, Vol.28, pp.549-550

Liu, K., Pan, W-P., Riley, J. T. (2000) A study of chlorine behavior In a simulated fluidized bed combustion system. *Fuel*, Vol.79, pp.1115-1124

Mascolo, G., Spinos,a L., Lotito, V., Mininni, G., Bagnuolo, G. (1997) Lab-scale evaluations on formation odf products of incomplete combustion in hazardous waste incineration: influence of process variables. *Water Science and Technology*, Vol.36, pp.219-226

Mastral, A. M., Callen, M. S., Garcia, T. (2000) Toxic organic emissions from coal combustion. *Fuel Processing Technology*, Vol.67, pp.1-10

Miller, J. A., Bowman, C. T. (1989) Mechanism and modeling of nitrogen chemistry in combustion. *Progress in Energy and Combustion Science*, Vol.15, pp.287-338

Modigell, M., Liebig, D. (1999) Modeling and simulating the waste incineration process. - private communication

Nissen, W. R. (2002) Combustion and Incineration Processes. Marcel Dekker, New York,

Recknagel, H., Sprenger, E., Hönmann, W., Schramek, E. R. (1994) Taschenbuch für Heizung und Klimatechnik. - R. Oldenburg Verlag, München

Rigo, H. G., Handler, A. J. (1998) Is there a strong dioxin-chlorine link in commercial scale systems? *Chemosphere*, Vol.37, pp.2031-2046

Schmöckel, G., Streit, A. (1994) Emission organischer Stoffe bei der Holzfeuerung. Bayerisches Landsamt für Umweltschutz, Augsburg

Seeker, R. (2001) Current understanding of dioxin/furan formation and control. - private communication

Steele, R. C., Malte, P. C., Nicol, D. G., Kramlich, J. C. (1995) NO_x and N_2O in lean-premixed jet-stirred flames. *Combustion and Flame*, Vol.100, pp.440-449

Stieglitz, L., Vogg, H. (1987) On formation conditions of PCDD/PCDF in fly ash from municipal waste incinerators. *Chemosphere*, Vol.16, pp.1917-1922

Sukrut, S., Thipse, E., Dreizin, L. (2002) Metal partitioning in products of incineration of municipal solid waste. *Chemosphere*, Vol.46, pp.837-849

Thome-Kozmiensky, K. J. (1994) Thermische Abfallbehandlung. - EF-Verlag für Energie- und Umwelttechnik, Berlin

Tosine, H. M., Clemen,t E. R., Ozvacic, V., Wong, G. (1985) Levels of PCDD/PCDF and other chlorinated organics in municipal refuse. *Chemosphere*, Vol.14, pp.821-827

Trenholm, A. (1998) Identification of PICs in hazardous waste combustion emissions. *Waste Management*, Vol.18, pp.485-492

Vogg, H., Stieglitz, L. (1986) Thermal behavior of PCDD/PCDF in fly ash form municipal incinerators. *Chemosphere*, Vol.15, pp.1373-1378

Wey, M.Y., Su, J.L., Yan, M.H., Wei, M.C. (1998) The concentration distribution of heavy metals under different operation conditions. *The Science of the Total Environment*, Vol.212, pp.183-193

Wielgosiński, G. (2009) Dioxin Emission from Thermal Processes and Methods of Emission Reduction. (in polish) - Polish Academy of Science, Lodz

Wienecke, J., Kruse, H., Huckfeldt, U., Eickhoff, W., Wassermann, O. (1995) Organic compounds in the flue gas of a hazardous incinerator. *Chemosphere*, Vol.30, pp.907-913;

Wunsch, P., Matuschek, G., Kettrup, A. (1995) The influence of combustion bed temperature during waste incineration on slag quality. *Thermochimica Acta*, Vol.263, pp.95-100

Zeldovich Y. B. (1946) The Oxidation of Nitrogen in Combustion and Explosions. *Acta Physcochimica, Academy of Sciences U.S.S.R.*, Vol.21, 577-628

Part 3

Process Engineering

13

Systematic Framework for Multiobjective Optimization in Chemical Process Plant Design

Ramzan Naveed[1], Zeeshan Nawaz[2],
Werner Witt[3] and Shahid Naveed[1]
[1]*Department of Chemical Engineering,*
University of Engineering and Technology, Lahore
[2]*Chemical Technology Development, STCR,*
Saudi Basic Industries Corporation (SABIC)
[3]*Lehrstuhl Anlagen und Sicherheitstechnik,*
Brandenburgicshe Technische Universität, Cottbus
[1]*Pakistan*
[2]*Kingdom of Saudi Arabia*
[3]*Germany*

1. Introduction

For solving multiobjective decision making problems, a systematic and effective procedure is required. As far as the process or control system has to be modified process simulators like Aspen Plus™, Aspen Dynamics are widely used. But these simulators are not designed for investigation of other objectives as environment and safety. Due to complex and conflicting nature of multiobjective decision making an integrated optimization tool should be of value. In this chapter a systematic methodology based on independent modules and its different stages to deal this problem is presented in detail.

2. Proposed methodology

The methodology is built around several standard independent techniques. These techniques have been suitably modified/adopted and woven together in an integrated plate form. The main aim is to standardize the screening and selection of decisions during design/modification of chemical process plant and optimizing the process variables in order to generate a process with improved economics along with satisfaction of environmental and safety constraints. The methodology (see Figure 1) consists of four layers/stages:

- Generation of alternatives and problem definition;
- Analysis of alternatives i.e. generation of relevant data for comparison of Environmental, economic and safety objectives
- Multiobjective decision analysis/ optimization
- Design evaluation stage i.e. decision making from the pareto-surface of non-inferior solution or ranking of alternatives

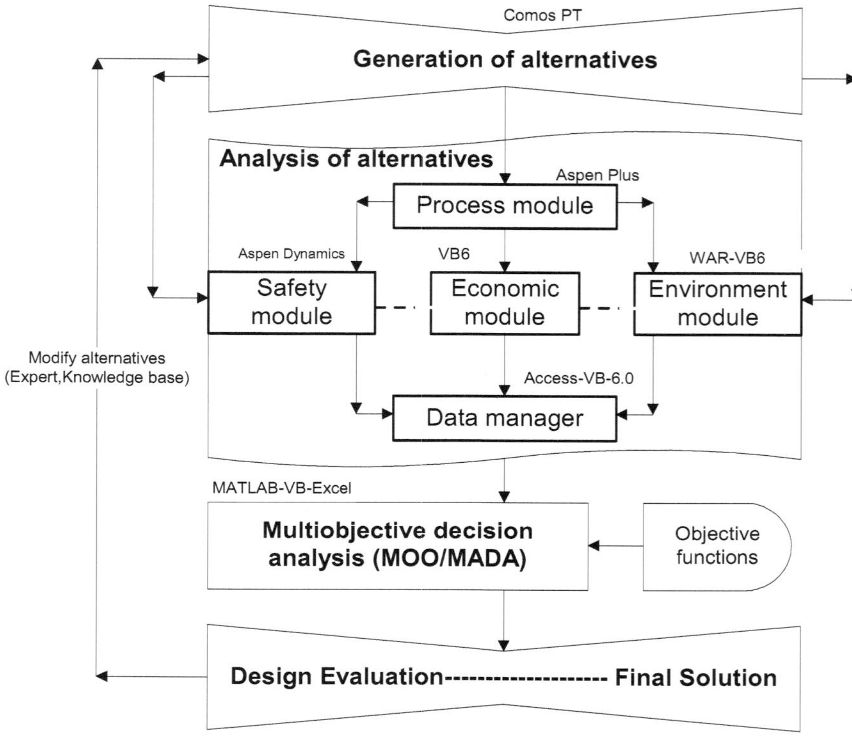

Fig. 1. Simplified block diagram of proposed methodology

2.1 Stage I: Generation of alternatives and problem definition

The first layer composed of following tasks:

Definition of the scope of the study,
Statement of key assumptions and the performance targets such as quality etc.,
Degree of freedom analysis,
Identification of the key design, control, and manipulated variables,
Definition of the system boundary,
Identification of constraints,
Choice of functional unit for all calculations,
Collection of relevant information about process and chemicals to be handled,
Generation of different alternatives either based on suggestion from independent departments or using the individual objective modules from stage II.

Seader et al. (1999)[6] has described rules for selection of process variables in the book "Process design principles-synthesis, analysis and evaluation". The data and information about the process and chemicals involved such as thermodynamic and kinetic data can be found from journal articles, patents or handbooks. Current chemical prices can be obtained from market reports if not available in main plant documentation or company central data base. In addition to these sources, some data related to quantification of environmental impacts and material safety data sheets of chemical are also collected from commercial data bases so that an impact assessment and safety analysis can be performed in subsequent design steps. Commercial computer aided tool like Comos[PT] can be used for plant documentation and to support stage-I of proposed methodology.

2.2 Stage II: Analysis of alternatives

This stage is composed of independent modules used to generate relevant information for evaluation of economic, safety and environmental performance objectives. These modules are:

- Process module
- Safety module [1,3]
- Economic module
- Environment module [2]
- and a data manager for managing the relevant information generated from these modules.

2.3 Process module

In the process module, an operation model of the process system has to be developed for evaluating alternatives. The configured simulation model has to be able to reproduce the selected results to an accepted degree of accuracy. This simulation model can be used for design and operation, revamping and debottlenecking of the process under study[7]. Three major integrated simulation systems widely used in the firms and companies for this purpose are Aspen technology (Aspen Plus, Aspen dynamics etc), Hyprotech (Hysys process, Hysys plant etc) and Simulation Sciences (Pro/II etc.). Aspen Plus™ 12.1 is used in this work for development of simulation model and linked in a visual basic platform for integration with safety, economic and environment modules. The most important results available from the process simulation model are material and energy balance information for both streams and units, rating performance of units and tables and graphs of physical properties. A brief description of Aspen Plus™ 12.1 and steps involved in development of the process simulation model is described here below.

Aspen Plus™

Aspen Plus™ supports both sequential modular and equation oriented computation strategy and allows the user to build and run a steady-state simulation model for a chemical process. It provides a flexible and productive engineering environment designed to maximize the results of engineering efforts, such as user interface mode manager, quick property analysis, rigorous and robust flowsheet modelling, interactive architecture, powerful model analysis tools and analysis and communication of results. Therefore, it lets the user to focus his/her energies on solving the engineering problems, not on how to use the software. It is not only good for process simulation but also allows to perform a wide range of other tasks such as estimating and regressing physical properties, generating custom graphical and tabular output results, sensitivity analysis, data-fitting plant data to simulation models, costing the plant, optimizing the process, and interfacing results to spreadsheets.

The development of a simulation model for a chemical process using Aspen Plus™ 12.1 involves the following steps (see details in table 1):

1. Define the process flowsheet configuration by specifying
   - Unit operations
   - Process streams flowing between the units
   - Unit operation models to describe each unit operation
2. Specify the chemical components,
3. Choose a thermodynamic model to represent the physical properties of the components and mixtures in the process,
4. Specify the component flow rates and thermodynamic conditions (i.e. temperature, pressure, or phase condition) of the feed streams,
5. Specify the operating conditions for the unit operations,

Step	Used to
Defining the flowsheet	Break down the desired process into its parts: feed streams, unit operations, and product streams
Specifying stream properties and units	Calculate the temperature, pressure, vapor fraction, molecular weight, enthalpy, entropy and density for the simulation streams
Entering components	From a databank that is full of common components
Estimating property parameters	Property Constant Estimate System (PCES) can estimate many of the property parameters required by physical property models
Specifying streams	Streams connect unit operation blocks in a flowsheet and carry material and energy flows from one block to another. For all process feed streams, we must specify flowrate, composition, and thermodynamic condition
Unit operation blocks	We choose unit operation models for flowsheet blocks when we define our simulation flowsheet

Table 1. Developmental process for an Aspen Plus™ simulation model

2.4 Safety module

Safety module is based on combination of conventional standard risk analysis techniques and process disturbance simulation. This module not only generates relevant information related to safety aspects for multiobjective decision analysis but also used for safety/risk analysis and optimization. The purpose of this module is to determine risk from operational disturbances and to develop effective risk reductions. It can be divided into the following steps (Figure 2):

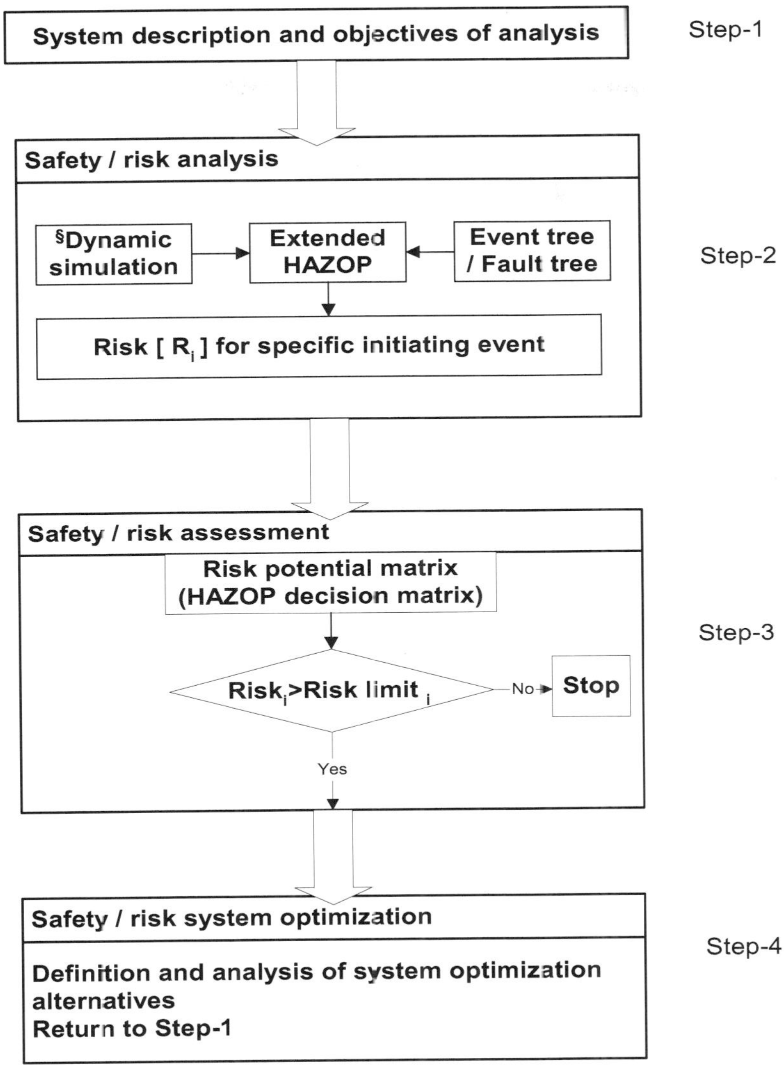

§ Simulation of process related malfunctions

i = { financial risk , environmental risk , human health risk }

Step 1: System description and objectives of analysis (before starting safety and risk analysis)
Step 2: Safety/risk analysis (identification of weak points via Extended HAZOP)
Step 3: Safety/risk assessment (categorization of risk via risk potential matrix (HAZOP decision matrix))
Step 4: Safety/risk system optimization

Fig. 2. Simplified block diagram of safety module

2.4.1 Step 1: (Before starting safety/risk analysis) - Description of system and objectives of analysis

For efficient safety/risk studies, the analyst must have an accurate description of the system to be investigated and a clear objective of the analysis study. Therefore, in this step the purpose, objectives, and scope of the study are clearly defined. The necessary information required for the study such as process flow diagrams, piping and instrumentation diagrams, plant layout schematics, material safety data sheets, equipment data sheets, operating instructions, start up and emergency shutdown procedures, and process limits, etc. is gathered from plant documentation. A team under a trained and experience leader with five to seven people including experts of the design and operation of the subject process may be formulated.

2.4.2 Step 2: Safety/risk analysis (Identification of weak points via extended HAZOP) - Extended HAZOP

Our intention is to identify weak points due to disturbances in operation, which may or may not be hazardous, in order to improve safety, operability, and/or profit at the same time. Extended HAZOP (HAZOP supported by dynamic simulation, event tree and fault tree techniques and HAZOP decision matrix) is used not only for identification of weak points but also for generation and analysis of optimization proposals [8-11]. Extended HAZOP differs from the standard HAZOP approach in following aspects:

i. Use of dynamic simulation:

In Extended HAZOP, the analysis of the influence of disturbances (failures) on the behaviour of the process is based on shortcut or simplified hand calculations or dynamic simulation. Aspen dynamics is used for this purpose.

ii. Classification of risk related consequences:

Each established consequence (hazard) has to be expressed by a consequence class (C). The plant specific scoring (from 0 (lowest) to 8 (highest)) chart is given in Table 2 (a & b) based on principle consequence analysis. For classification of consequences based on principle release estimates, accident consequence analysis techniques (models for calculation of toxic, fire and explosion effects) and plant location data (capital investment, population density etc.) have to be considered.

Illustrative Example 1

Figure 3 shows the plant lay out considered for developing plant specific consequence scoring chart. The area around the plant is open fields (rural condition). As weather conditions changed around the year, so certain assumptions are made to results in worse case conditions for consequence analysis. These include weather conditions and wind speed that result in smallest value of dispersion coefficients. Therefore, stability "F" and wind speed as low as possible (1.5 m/s) is selected. It is assumed that 10 workers are present (working 24 h each day), which are not distributed uniformly, on the land in area (100 m x 100 m) around the column under study. Acetone is selected as representative fluid for consequence analysis.

Acetone vapours released from the vent line at a rate of 1616 kg/h due to loss of cooling medium. It is assumed that released vapours form a cloud for 30 minutes before being

Effects	Class	Financial loss (€)	Class related consequences: examples
Function impairment	*	< 10	: Product quality lowering (brief)
	1	$10^1 - 10^2$	: Product quality lowering
	2	$10^2 - 10^3$	: Product quality lowering (long term)
Functional Loss	3	$10^3 - 10^4$	: Production disturbance (brief) Soil contamination Safe dispersion of material release from vent line
	4	$10^4 - 10^5$	: Production disturbance Material release from the piping Pump damage (pressure impacts)
	5	$10^5 - 10^6$	: Production disturbance (long term) Jet fire as result of release of material from vent line Pool fire (from pump leakage)
Safety and Environmental pollution	6	$10^6 - 10^7$	Fireballs due to catastrophic rupture of vapour product line
	7	$10^7 - 10^8$	Vapour cloud explosion (ignoring domino effect)
	8	$>10^8$	Vapour cloud explosion along with domino effect

Effects	Class	Community	Class related consequences: examples
Function impairment	*	No effect on people	: Product quality lowering (brief)
	1	Nuisance effect	: Product quality lowering
	2	Minor irritation effect to people & local news	: Product quality lowering (long term)
Functional Loss	3	Moderate irritation effect to people and non compliance to laws, local news	: Production disturbance (brief) Soil contamination Safe dispersion of material release from vent line
	4	Moderate irritation effects to people & environment, single injuries and regional news	: Production disturbance Material release from the piping Pump damage (pressure impacts)
	5	Significant effects to people and environment, > 1 injuries & regional news	: Production disturbance (long term) Jet fire as result of release of material from vent line Pool fire (from pump leakage)
Safety and Environmental pollution	6	Major effects to people and environement, multiple injuries, fatality likely, regional news	Fireballs due to catastrophic rupture of pipe or condenser (vapour product line)
	7	Severe effects to people and environment, fatality, regional news	Vapour cloud explosion (ignoring domino effect)
	8	Multiple fatalities and process shutdown certain, international news	Vapour cloud explosion along with domino effect

Environment and Health consequences

Table 2. Scoring chart for Consequence Financial consequences [3.4]

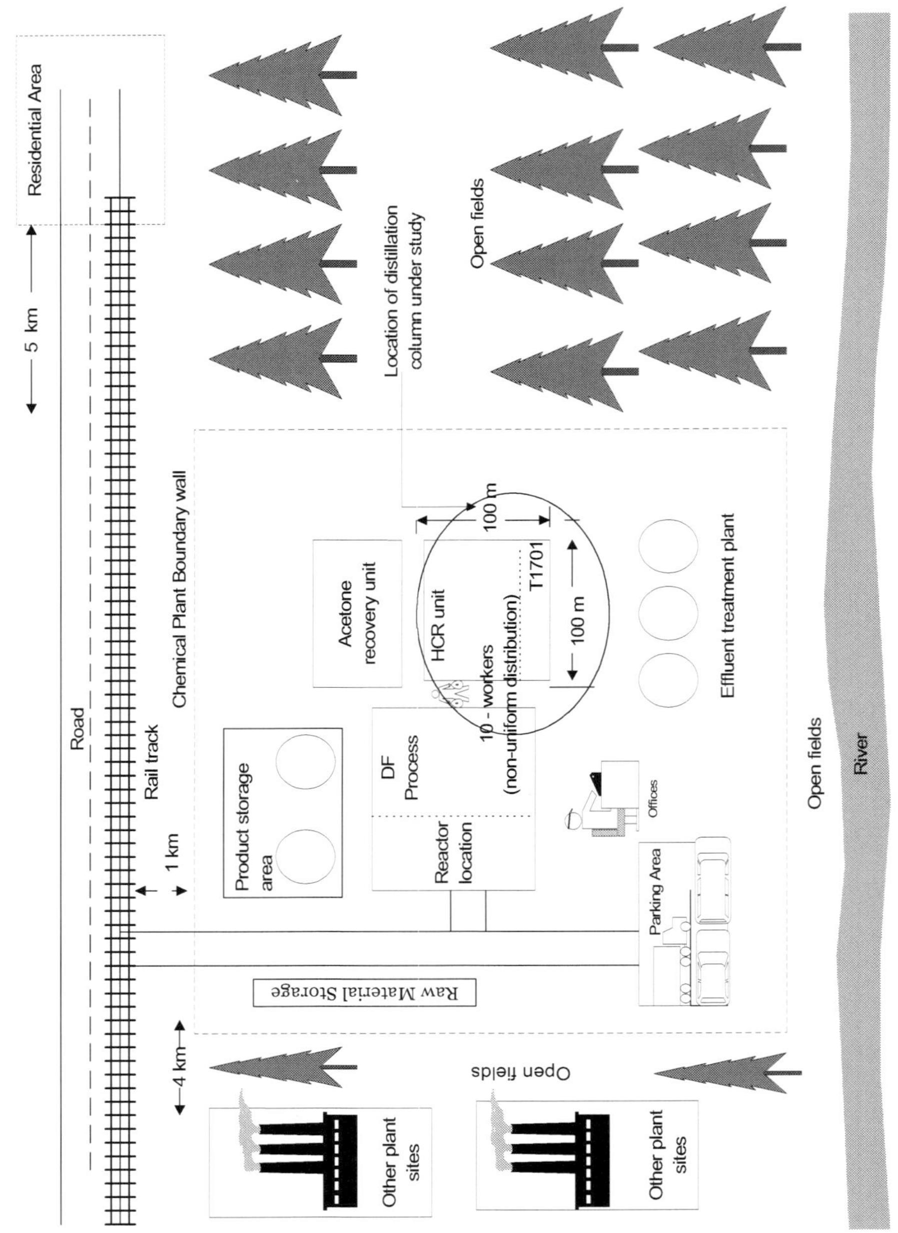

Fig. 3. Plant lay out for establishing consequence score chart

ignited and leads to vapour cloud explosion. The physical effects of this scenario or event is calculated as:

Weight of fuel in the cloud= M = 1616 / 2 = 308 kg

Then amount of TNT equivalent to the amount of this flammable material is

$$M_{TNT} = \alpha \cdot \frac{M \cdot H_c}{H_{TNT}}$$

Where $\alpha =$ explosion efficiency ~ 0.05 (Cameron 2005) $H_c =$ heat of combustion of fuel ~ 3.03 x 104 kJ/kg for acetone

$H_{TNT} =$ TNT blast energy ~ 5420 kJ/kg; so $M_{TNT} = 225.25$ kg

Then, using relation $Z = R/(M_{TNT})^{1/3}$ and figure 4, scaled distance and overpressure is estimated. Table 3 presents the results obtained.

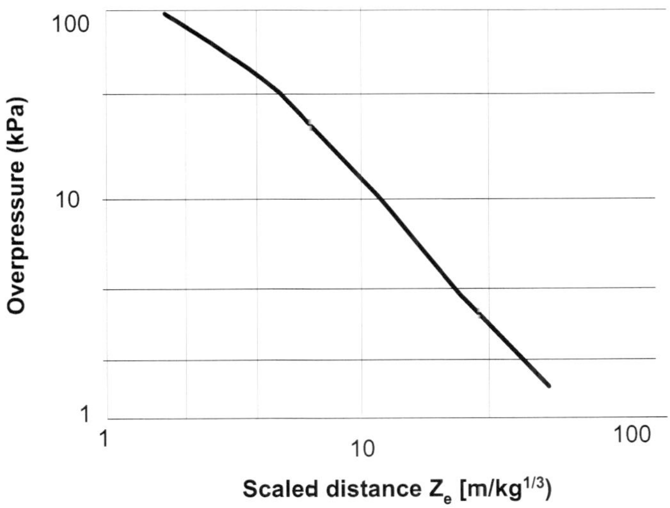

Fig. 4. Overpressure versus scaled distance for TNT explosions on flat surfaces (Tweeddale 2003, p. 115)

Distance , R M	Scaled distance, Z m / kg1/3	Overpressure, Δp kPa
10	1.64	90
20	3.28	40
50	8.21	20
100	16.40	7

Table 3. Results of physical effects of vapour cloud explosion

It is estimated that severe structural damage and 15 % chance of fatality outdoors or 50 % chance indoor will be experienced out to 20 m and almost complete destruction of all ordinary structures and 100 % chance of fatality indoors to 10 m distance.[8] (see Cameron 2005, p. 268).

iii. Classification of frequencies of risk related consequences:

The frequency of occurring for each possible consequence (hazard) has to be expressed by a frequency class, called (F) according to the scoring chart for frequency (Table 3.4). Definition of frequency class may be supported by Event Tree and/or Fault tree analysis techniques or Layer of protection analysis (LOPA) or historical databases.

For establishing frequency class: Estimation / calculation of frequency of vapour cloud explosion and fatality of person because of release of material due to catastrophic rupture of distillation column.

Frequency of catastrophic rupture of column = 10-6 (Taken from table 4)

Class	Frequency of occurring incident		
	Frequency 1/y	Comprehension	Examples based on general data bases
9	$<10^{-8}$	Very very small	Catastrophic rupture or leakage of pipe of diameter > 150 mm
8	10^{-8} - 10^{-7}	Very small	Catastrophic rupture of pipe of diameter ≤ 50 mm
7	10^{-7} - 10^{-6}	Small	Catastrophic rupture of fractionating system (excluding piping), storage tank rupture
6	10^{-6} - 10^{-5}	Less small	Pipe residual failure, 100 m full breach, Double wall tank leakage
5	10^{-5} - 10^{-4}	Moderate	Process vessel leakage of ≥ 1 mm diameter
4	10^{-4} - 10^{-3}	Less moderate	Pump leakage , Heat exchanger leakage
3	10^{-3} - 10^{-2}	Less high	Safety valve open spuriously, Large external fire
2	10^{-2} - 10^{-1}	High	Cooling water failure, BPCS instrument loop failure
1	10^{-1} – 10^{0}	Very high	Operator failure, Regulator failure , Solenoid valve failure
*	$>10^{0}$	Very very high	Power failure in developing countries, Operators failure under high stress

Table 4. Scoring chart for frequency [3.4]

Probability of ignition of released material = 0.10 (CCPs 2000, Borysiewich 2004)

Probability of VCE if released material ignited = 0.01 (CCPs 2000, Borysiewich 2004)

Probability of fatality of a person exposed to overpressure of 40 kPa due to VCE = 0.20 (Tweeddale 2003, p. 117 Figure 5-14)

Then, frequency of vapour cloud explosion = $10^{-6} \cdot 0.10 \cdot 0.01 = 10^{-9}$

So frequency class for this scenario = 9

Frequency of fatality of a person exposed to VCE = $10^{-9} \cdot 0.20 = 2.10^{-10}$

So frequency class for this scenario = 9

iv. Way of documenting the HAZOP results

The Extended HAZOP methodology worksheet for documenting the HAZOP team results is shown in Figure 5. Below consequence the physical effects and risk has to be documented first and next risk has to be classified using score charts (Table 3.2) related to financial, environment and health related consequences. The worst score of each risk has to be documented. For each risk related consequence, frequency class has also to be established.

Consequence, € / Frequency 1/y	C / F	<10	10^1 – 10^2	10^2 – 10^3	10^3 – 10^4	10^4 – 10^5	10^5 – 10^6	10^6 – 10^7	10^7 – 10^8	>10^8
		*	1	2	3	4	5	6	7	8
>10^0	*									
10^{-1} – 10^0	1									
10^{-2} - 10^{-1}	2									
10^{-3} - 10^{-2}	3									
10^{-4} - 10^{-3}	4									
10^{-5} - 10^{-4}	5									
10^{-6} - 10^{-5}	6									
10^{-7} - 10^{-6}	7									
10^{-8} - 10^{-7}	8									
<10^{-8}	9									

	Immediate action needed before further operation
	Action at next occasion after qualification of analysis for improving system
	Optional
	No further action needed

Fig. 5. Risk potential matrix (Extended HAZOP decision matrix)

Illustrative Example 2

Release of material to atmosphere from vent line or vapour line may disperse safely or has toxic effects or can lead to several outcomes such as flash fire, vapour cloud explosion and fire balls. So documenting consequence class in HAZOP work sheet, the score '8' of the most severe consequence will be documented.

<table>
<tr><td colspan="3">Plant/P&ID :
Equipment :
Volume :</td><td colspan="4">Process:

Function:</td><td colspan="3">Document:
Page :
Date :</td></tr>
<tr><td>No</td><td>Guide word / Process Parameter</td><td>Detection/ Safeguards</td><td>Possible causes</td><td>Conse-quences</td><td>FC</td><td>Recommended Actions</td><td>FC</td><td>Resp./ Ref.</td></tr>
<tr><td></td><td></td><td></td><td></td><td>Physical effects:

Risk related:</td><td></td><td></td><td></td><td></td></tr>
</table>

1- Short cut calculations 2-Dynamic simulation 3-deterministic models 4- Event tree 5- Fault tree
6- Historic data base

2.4.3 Step 3: Safety/risk assessment - Risk potential matrix (HAZOP decision matrix)

Figure 6 shows the risk potential matrix (HAZOP decision matrix) used for order of magnitude ranking of events. The rows of the matrix consider frequency class, while the columns show the consequence class. Each cell in the matrix represents a risk category. For the decision process, the matrix is divided into four risk category levels.

Risk level I --- red area --- scenario in this level is intolerable and immediate action (pant or process modification) is needed to reduce that risk category or more detailed quantified analysis has to be carried out in order to find arguments for wrong preliminary decisions.

Risk level II --- grey area --- scenario in this level is tolerable but not acceptable for long period of time so action at next schedule maintenance is needed to reduce that risk category.

Risk level III --- yellow area --- scenario in this level is acceptable and any action to reduce that risk category is optional.

Risk level IV--- green area --- scenario in this level needs no action.

Risk potential matrix (HAZOP decision matrix) may be also used for:

- Documentation of the status of the plant safety
- Selection and development of optimization proposals
- Importance of improvement
- Documentation of improvement achieved

The application of risk potential matrix (HAZOP decision matrix) in the Extended HAZOP is shown in figure 5. Arrows show the transformation of entries from the Extended HAZOP worksheet to the HAZOP decision matrix. The identity number (ID) of each scenario of the Extended HAZOP worksheet is placed in HAZOP decision matrix. Recommended actions for this scenario will be placed from Extended HAZOP sheet to the bottom of HAZOP decision matrix. First HAZOP decision matrix will shows the existing status and second HAZOP decision matrix shows the improved plant status after recommended actions.

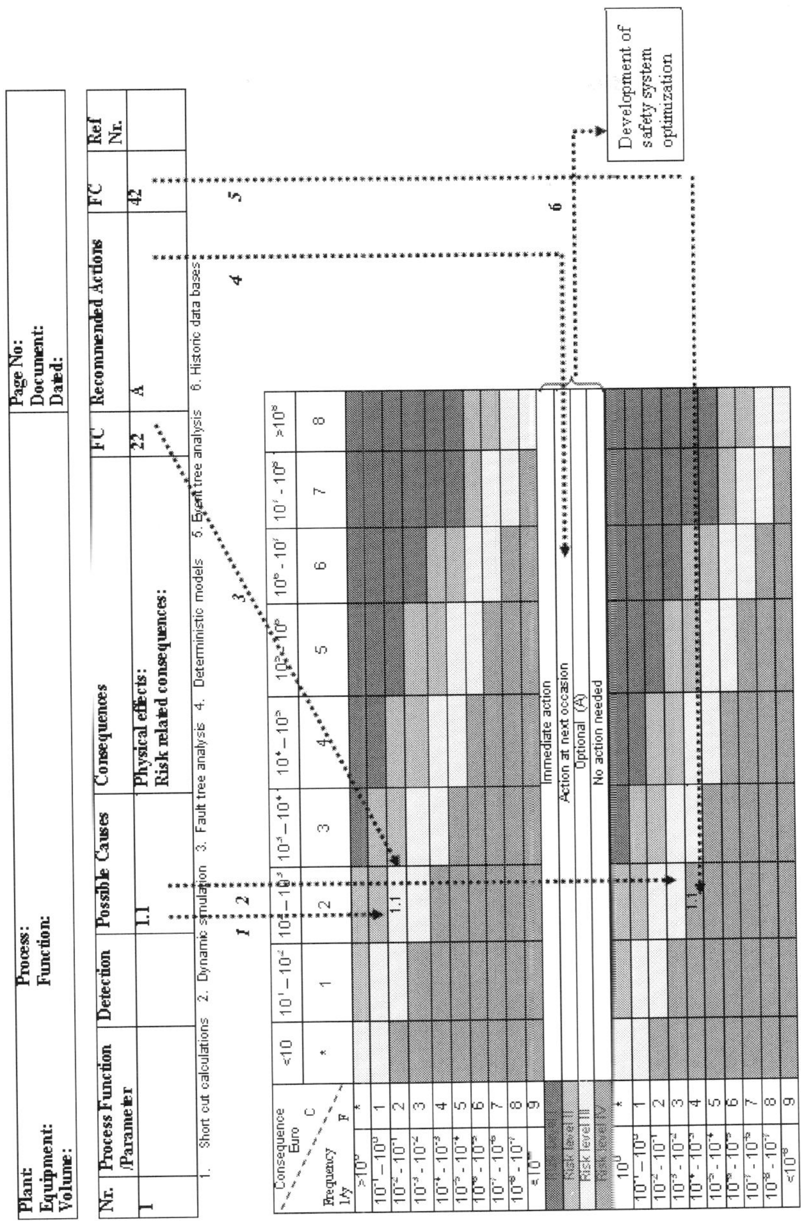

Fig. 6. Application of HAZOP decision matrix in Extended HAZOP

Similarly all results from Extended HAZOP worksheets are transferred to the HAZOP decision matrix. Keeping in view the risk target and depending on the scenario or recommended actions during the Extended HAZOP discussion, analysis team may reach a safety related modification proposal. Next, if safety/ risk optimization is in focus then weak points/scenarios with similar risk are clustered after analyzing HAZOP decision matrix and safety related optimization proposals are developed.

2.4.4 Step 4: Safety/risk system optimization (Development and analysis of optimization proposals)

In this step, safety related optimization proposals are generated and evaluated using dynamic simulation, Event tree analysis and/or Fault tree analysis. The optimization proposals can be developed at two levels:

- Simple optimization proposals e.g. addition of pressure alarm or change of location of sensor within the Extended HAZOP discussion
- Optimization proposals related to severe scenarios by evaluating risk potential matrix (HAZOP decision matrix)

The relevant information such as frequency and damage data will be transferred to economic module for safety related cost calculations and multiobjective decision making (if more than one alternatives developed).

2.5 Economic module

In all stages of design process, economic evaluation is crucial for the evaluation of process alternatives. Various objective functions are available in the literature of chemical engineering economics for economic evaluation of chemical processes. Some quite elegant objective functions, which incorporate the concept of the „time value of money", are net present value (NPV) and discounted cash flow. Business managers, accountants and economists prefer these methods because they are more accurate measures of profitability over an extended time period. However, application of these methods needs certain assumptions[12]. Total annualized cost (TAC) can be used as economic indicators/objective function for the evaluation of design alternatives and economic optimization.

Economic module developed in Visual Basic consists of two distinct sections. First section carries out standard cost calculations (i.e. Fixed capital investment (FC1) and operational cost (OC1)) and compute total annualized cost (TAC1) while second section carries out extended cost calculations i.e. process safety/risk related costs and computes the fixed capital investment related to safety system (FCISS), accident and incident damage related risk cost. Table 3.5 illustrates the difference of cost elements considered in standard practice of cost calculations of chemical process design and in this economic module. Figure 7 shows the simplified block diagram of economic module.

2.5.1 Standard cost calculations

Standard cost calculations involves fixed capital investment (FCI1) and operational cost (OC1). Fixed capital investment (FCI1) includes the cost of design and other engineering and construction supervision, all items of equipment and their installation, all piping,

Standard cost calculations	Economic module used in this work
• FCI1 = Fixed capital investment using either cost equations that have been derived by Ulrich or correlations developed by Guthrie depending on users choice • OC1= Operating cost (including both direct (e.g. raw material, utilities etc.) and indirect costs (e.g. taxes, overhead cost etc) • TAC1 = total annualized cost $= d \cdot (FCI1) + OC1$ normally d is taken 0.15-25 but can also be computed using depreciation calculation methods	• FCI1= Fixed capital investment using either cost equations that have been derived by Ulrich or correlations developed by Guthrie depending on users choice • FCI2= FCI1 + fixed capital investment related to safety system (FCISS) • OC1= Operating cost (including both direct (e.g raw material, utilities etc.) and indirect costs (e.g. taxes, overhead cost etc) • TAC1 = total annualized cost $= d \cdot (FCI1) + OC1$ • TAC2 = total annualized cost $= d \cdot (FCI2) + OC1$ normally d is taken 0.15-25 but can also be computed using depreciation calculation methods
	Extended cost calculations • RC1 = risk cost1= Asset risk cost + health risk cost + environmental risk cost • RC2 = risk cost2 = RC1+ production loss risk cost • RC3 = risk cost3 = process interruption cost • TRC = total risk cost = RC2+RC3 • ECC = Extended costs

Table 5. Elements of economic module and difference from standard cost calculations

instrumentation and control systems, buildings and structures, and auxiliary facilities such as utilities, land and civil engineering work. Several capital cost estimate methods ranging from order of magnitude estimate (ratio estimate) to detailed estimate (contractors estimate) are used for the estimation of installed cost of the process units in the chemical plant.

The most commonly used method that provides estimates within 20-30% of actual cost and widely used at design stage involve the usage of cost charts/correlations (Guthrie's article (1969) and book (1974), chapter 5 of Ulrich's 'A guide to chemical engineering process design and economics' (1984), 'Plant design and economics for chemical engineers' by Peters and Timmerhause (1991)) for estimating the purchase cost of major type of process equipment [13-15].

These cost charts / correlations were assembled in the 1960's or earlier and are projected to the date of installation using cost indices or escalation factors such as the chemical engineering plant cost index (published biweekly by chemical engineering magazine),

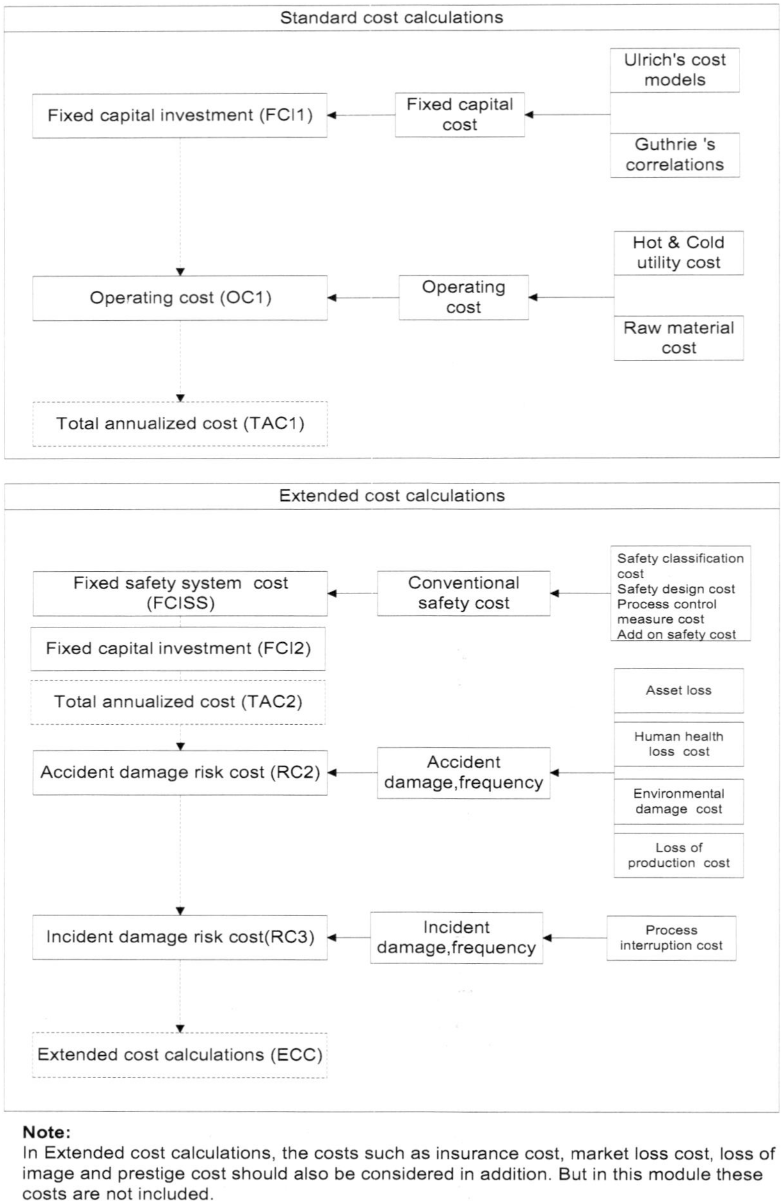

Fig. 7. Simplified block diagram of economic module

Marshall and Swift Index (also provided in chemical engineering magazine) and Nelson-Fabaar Index (from the oil and gas journal). For the comparison of process design alternatives, these study estimates for purchased cost of process units using cost charts or equations based on them are adequate. Given the purchase cost of a process unit, the installed cost is obtained by adding the cost of installation using factored-cost methods. For each piece of equipment Guthrie (1969, 1974) provides factors to estimate the direct cost of labor, as well as, indirect costs involved in the installation procedure. The cost elements that are included in the estimation of fixed capital investment are shown in figure 8.

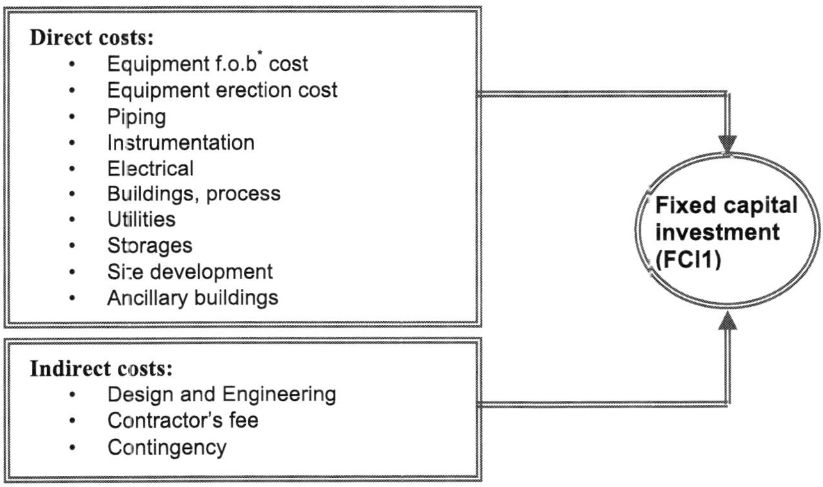

f.o.b ---- freight on board cost i.e. equipment purchase cost plus transport cost

Fig. 8. Typical cost elements for fixed capital investment

The operating cost (OC1) of a chemical plant is divided into two groups:

- Fixed operating cost
- Variable operating cost

The elements in fixed operating cost includes maintenance cost, operating labor cost, laboratory cost, supervision cost, plant overheads, capital charges, taxes, insurance, licence fees and royalty payments while the variable operating cost consists of raw material costs, miscellaneous operating material costs, utilities (services) and shipping and packaging. However this division of operating cost is somewhat arbitrary and depends on the accounting practice of a particular organization. The typical cost elements included in operating cost "OC1" are shown in figure 9.

However, from the existing process optimization point of view, energy cost and raw material costs are more important and often considered.

Economic module developed in this thesis using Visual basic computes fixed capital expenditure using either cost equations that have been derived by Ulrich or correlations

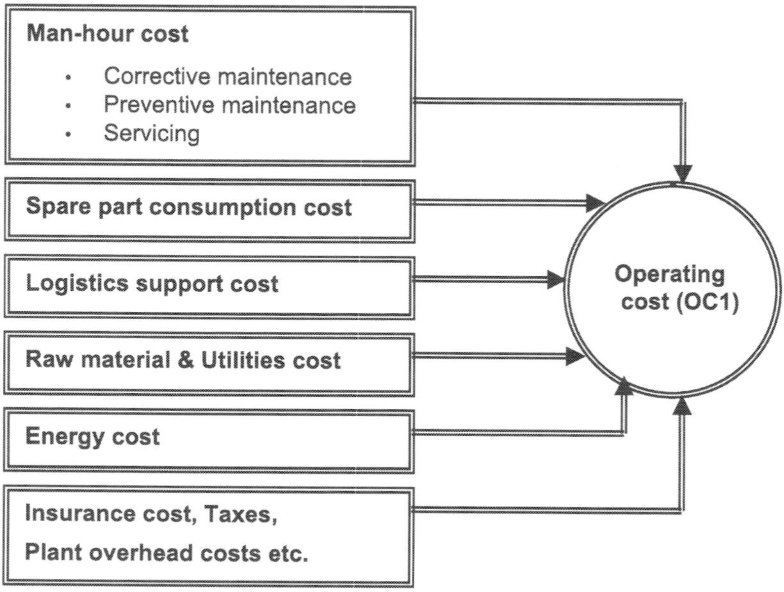

Fig. 9. Typical cost elements for operating cost (OC1)

developed by Guthrie depending on users choice, The significant operating cost for process optimization i.e. energy consumption cost (heating and cooling utilities cost) can also be calculated by using this module. Once the FCI1 and OC1 are calculated, then total annualized cost is obtained using the following equation:

$$\text{Total annualized cost (TAC1)} = \mathrm{d} \cdot (\mathrm{FCI1}) + \mathrm{OC1} \tag{1}$$

Here d is depreciation or capital recovery factor and normally taken between 0.15-0.25 but can also be computed using depreciation calculation methods e.g. double declining balance method.

2.5.2 Extended cost calculations

The second section of the economic module (see Figure 7) carries out Extended cost calculations, which considers the fixed capital investment related to safety system, and risk cost due to accident and incident damage.

i. Fixed capital investment related to Safety system:

The fixed investment related to safety system is calculated by the following equation:

$$FCISS = C_{SD} + \sum_{i=1}^{n} N_{SE,i} \cdot C_{SE,i} \tag{2}$$

Here, the first term C_{SD} is cost for safety design (i.e. cost related to safety classification, safety requirements and design specification, detailed design and engineering, factory acceptance test or pre-start up acceptance test and start up and correction). Table 3.6 gives the typical cost elements included in C_{SD} calculations. The second term $\sum_{i=1}^{n} N_{SE,i} \cdot C_{SE,i}$ is the sum of the purchase cost of safety equipment. Here CSE,i is the purchase cost of equipment "i" and NSE,i is the number (count) of that equipment. The costs for these devices are based on the recent detailed survey of available costs from various suppliers conducted by Khan and Annyotte (2004), however in the module the user has the possibility to enter the present market costs.

Cost element	Symbol
Safety classification cost e.g SIL determination cost	C_{SIL}
Safety requirements and design specifications (SRS) cost	C_{SRS}
Detailed design and engineering cost	C_{DE}
!Miscellaneous Cost:	C_{ME}
Initial training cost	C_{TC}
Factory acceptance test (FAT)/Installation/Pre-startup acceptance test (PSAT) cost	C_{FAT}
Startup and correction cost	C_{SCC}
$C_{SD} = C_{SIL} + C_{SRS} + C_{DE} + C_{ME} + C_{TC} + C_{FAT} + C_{SCC}$	

!power, wiring, junction boxes, operators interface cost

Table 6. Typical cost elements included in C_{SD} of safety system cost calculations

ii. Fixed capital investment (FCI2)

Then, extended fixed capital investment is calculated by adding FCISS to FCI1.

$$FCI2 = FCI1 + FCISS \tag{3}$$

iii. Total annualized cost (TAC2)

So, the extended total annualized cost will be calculated using extended fixed capital investment.

$$\mathrm{TAC2} = \mathrm{d} \cdot (\mathrm{FCI2}) + \mathrm{OC1} \tag{4}$$

Maintenance and repair cost of safety system should also be included in this calculation. But in this economic module these cost elements are not considered.

iv. Risk cost 1 (RC1)

Risk cost (RC1), which is the sum of property risk cost due to asset loss (PRC), health risk cost due to human health loss (HRC) and environmental risk cost due to environmental damage (ERC). The relations for calculation of these costs used in the module are:

Property risk cost due to asset lost (PRC) is the cost incurred due to lost of physical assets such as damage to property, loss of equipment due to accident/scenario and calculated by the equation below:

$$PRC = \sum_{i=1}^{n} \dot{F}_{A,i} \cdot A_{D,i} \cdot C_{A,i} \cdot t_{op} + \sum_{j=1}^{n} \dot{F}_{I,j} \cdot C_{D,j} \cdot t_{op} \tag{5}$$

$\dot{F}_{A,i}$ is frequency of occurring the hazardous accident, $A_{D,i}$ is damage area due to that accident, $C_{A,i}$ is the asset cost per unit area, t_{op} is total operation time, $\dot{F}_{I,j}$ is incident occurring frequency and $C_{D,j}$ is incident damage cost.

Health risk cost due to human health lost (HRC) is the cost of fatality and/or injury due to the accident scenario under study.

$$HRC = \sum_{i=1}^{n} \dot{F}_{A,i} \cdot N_{Peop,eff} \cdot C_{H,life} \cdot t_{op} \tag{6}$$

Here, $N_{peop,eff}$ is the number of person affected due to accident and is equal to $N_{peop,eff} = pop \cdot \wp$. Where POP is the population around the area of accident and $\wp$ is the population distribution factor ($\wp$ is 1 if population is uniform distributed (maximum value) and $\wp$ is 0.2 if population is localized and away from the area of accident (minimum value)) and $C_{H,life}$ is dollar value of human life or health. Though attempts to put value on human life have caused criticism and it changes from place to place. But a value for this can be obtained by dividing the annual gross national product by the annual number of births or by estimating how much money the person would have earned if not killed by the accident (Tweeddale 2003). A value for cost of loss of lives, marginal cost to avert the fatality, for the highest category of involuntariness risk 14 x 106 $ is used in this work (Passman, H.J. et al. 2003).

Environmental risk cost due to environmental damage is the cost incurred due to environmental damage.

$$ERC = \sum_{i=1}^{n} \dot{F}_{ED} \cdot A_{ED,i} \cdot C_{ED,i} \cdot t_{op} \tag{7}$$

Where $A_{ED,i}$ is the environmental damage area due to scenario "i", $\dot{F}_{ED}$ is the frequency of release of material to environment and $C_{ED,i}$ is the environmental damage cost per unit area.

so the sum of these three risk costs gives:

$$RC1 = PRC + HRC + ERC \tag{8}$$

v. Risk cost 2 (RC2)

Risk cost 2 (RC2), which is the sum of risk cost1 (RC1) and production loss risk cost (PLRC), accounts for accident damage risk cost. Here, production loss risk cost due to asset damage (PLRC) accounts for the cost due to the production loss because of accident and given by:

$$PLRC = \sum_{i=1}^{n} \dot{F}_{A,i} \cdot t_d \cdot \dot{C}_p \cdot t_{op} \tag{9}$$

Where PLRC is the production loss risk cost, td is the time lost due to accident and $\dot{C}_p$ is the production loss value in $/h. Thus Risk cost 2 (RC2) is

$$RC2 = RC1 + PLRC$$

vi. Risk cost 3 (RC3)

Risk cost 3 (RC3), which is sum of process interruption cost due the spurious trip of the safety system and process interruption cost because of safe shut down to avoid from accident, accounts for incident damage risk cost and calculated as follow:

$$RC3 = (\sum_{i=1}^{n} \dot{F}_S^{trip} \ t_{trip} + \dot{F}_R^{trip} \cdot t_{dR}) \cdot \dot{C}_p \cdot t_{op} \quad (10)$$

Here, $\dot{F}_S^{trip}$ is spurious trip frequency, $\dot{F}_R^{trip}$ is safe shut down frequency when trip system demand arises, ttrip is down time due to spurious trip and t_{dR} is down time to safe shut down when trip system demand arises.

vii. Total Risk cost (TRC)

Total Risk cost (TRC) is the sum of all risk costs:

$$TRC = RC2 + RC3 \quad (11)$$

Total risk cost can be annualized by dividing it with total operation time (t_{op}):

$$TAC_{risk} = TRC / t_{op}$$

viii. Extended Cost (ECC)

Extended cost calculations (ECC) is Life cycle related cost and calculated as follow :

$$ECC = FCI2 + PVC \quad (12)$$

Here, PVC is present value of the annual costs (OC1, TAC_{risk}) and calculated as follow:

$$PVC = (OC1 + TAC_{risk} - Insurance\ cost) \cdot \frac{1-(1+R)^{-t_{ly}}}{R}$$

R is the present interest rate and t_{ly} is the number of years (predicted life of system).

Besides, the cost elements mentioned above in Extended cost calculation section, the other elements such as warranty/insurance cost, lost of image and prestige cost, market lost cost should also be considered but quantification of these elements is still almost impossible.

2.6 Environment module

Environment module consists of four steps and introduced an environmental performance index (EPI1) for evaluation of environmental performance and environmental pollution index (EPI2) as environmental objective to be integrated along with economics. The environmental performance index (EPI1) is calculated by combining total PEI based on

WAR algorithm[16,17], resource depletion, energy conservation and fugitive emission rates while environmental pollution index (EPI2) is calculated by combining total PEI based on WAR algorithm and fugitive emissions because in this case other factors like resource depletion and energy consumption will be integrated in economic module or objective function. The Analytic hierarchy process (AHP) is used as multicriteria decision analysis tool for combining these different impacts and determination of weighting factors of individual impact categories in total PEI and later on in environmental performance index (EPI1) and environmental pollution index (EPI2) calculations. The module is developed using Microsoft Visual Basic 6.0 and WAR GUI (WAR graphical user interface) is integrated in the user plate form. The steps are:

Step I : Problem definition and data gathering

Step II : Individual impact categories calculation

Step III: Determination of weighting factors

Step IV: Environmental performance index calculation

Figure 10 shows the simplified block diagram of environment module and tasks to be performed.

2.6.1 Step I: Problem definition and data gathering

The primary task in step 1 is problem framing and scope definition. Information such as material and energy balance information, process conditions, process technology and nature of used materials/chemicals should be retrieved from process module. Process flow diagram is to be re-examined for identification of additional waste and emission streams. Collect additional data and information for environment evaluation to fill gaps. As sources of emissions such as fugitive emission sources, venting of equipment, periodic equipment cleaning, incomplete separations etc. are often missing in process so process is analyzed to identify these sources.

2.6.2 Step II: Individual impact categories calculation (Potential environmental impact calculations based on WAR algorithm)

The software WAR GUI (waste reduction algorithm graphical user interface) from the US Environmental Protection Agency is used to calculate individual potential environmental impacts. The generalized formula based on WAR algorithm for calculating individual PEI is given in equation 13.

$$PEI_L = (\dot{M}_b \cdot \sum_{k}^{Comps} x_{kb} \cdot \psi_{kL} + \dot{Q}_r \cdot \psi_L^E) / \dot{M}_p \qquad [Impact / kg\ product] \tag{13}$$

Where PEI_L is the potential environmental impact of category L, $\dot{M}_b$ is mass flow rate of base (effluent) stream, x_{kb} is the mass fraction of component k in the base stream, ψ_{kL} is the normalized impact score of chemical k for category L, $\dot{Q}_r$ is energy rate supplied for separation and ψ_L^E is the normalized impact score of category L due to energy. The sensitivity analysis results of individual potential environmental impact with respect to optimization variables should also be performed.

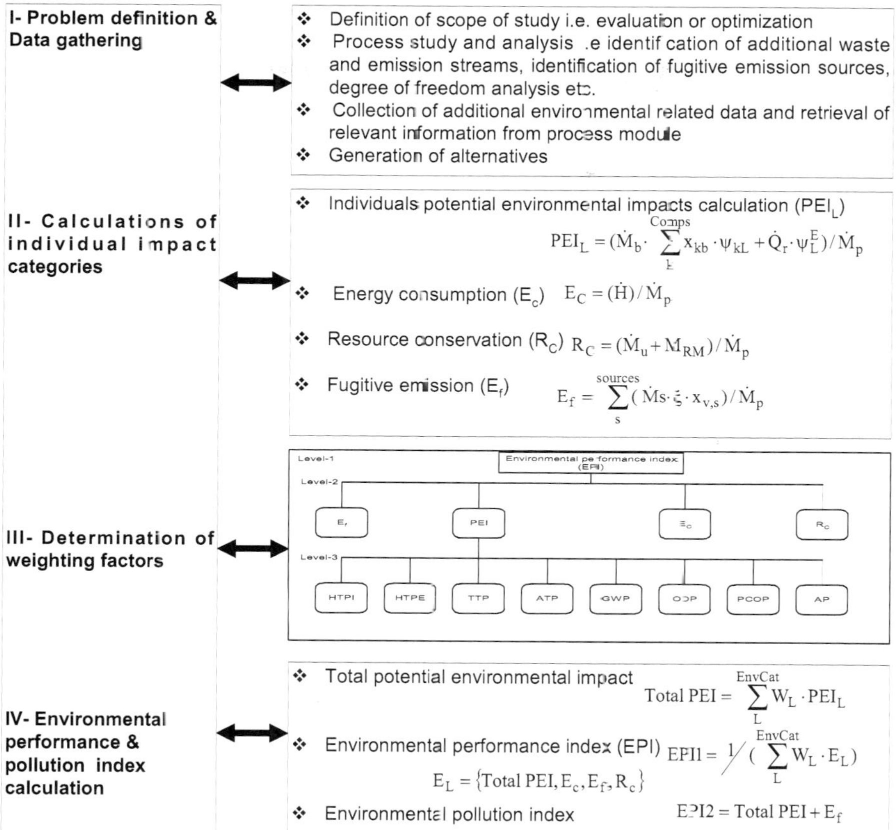

Fig. 10. Simplified block diagram of environment module

Energy consumption factor (E_C)

Energy consumption factor refers the total amount of energy consumed in the process per unit of product and is calculated as follow:

$$E_C = (\dot{H}) / \dot{M}_p \qquad \text{[kJ / kg product]} \tag{14}$$

Here $\dot{H} = \dot{M}_{steam}\ \hat{h}_{steam} + E_E$ where $\dot{M}_{steam}$ is the mass flow rate of steam [kg/h], $\hat{h}_{steam}$ is the enthalpy of steam per kg [KJ/kg], E_E is electrical energy consumed per unit time [KJ / h] and $\dot{M}_p$ is product rate [kg/h].

The sensitivity analysis of this factor with respect to optimization variables should also be performed.

Resource conservation factor (R_C)

The resource consumption refers all needed raw materials and utilities used and given by:

$$R_C = (\dot{M}_u + \dot{M}_{RM}) / \dot{M}_p \qquad \text{[kg / kg product]} \tag{15}$$

Where R_C is the resource conservation factor, $\dot{M}_U$ is utilities consumption rate, $\dot{M}_{RM}$ is raw material consumption rate.

Fugitive emission factor (E_f)

Fugitive emissions are unplanned or unmanaged, continuous or intermittent releases from unsealed sources such as storage tank vents, valves, pump seals, flanges, compressors, sampling connections, open ended lines etc and any other non point air emissions. These sources are large in number and difficult to identify. These emission rates depends on factors such as the age and quality of components, specific inspection and maintenance procedures, equipment design and standards of installation, specific process temperatures and pressures, number and type of sources and operational management commitment[18]. However, four basic approaches for estimating emissions from equipment leaks in a specific processing unit, in order of increasing refinement, in use are:

- Average emission factor approach
- Screening ranges approach
- EPA correlation approach
- Unit-specific correlation approach

All these approaches require some data collection, data analysis and/or statistical evaluation. On the other hand, using fundamental design / engineering calculations for accurate fugitive emission estimations for each source present in the process industry are difficult due to:

- large number and type of fugitive emission sources
- dependence of emission rates on other factors along with design and operating conditions e.g. installation standards, inspection and maintenance procedure etc.

As focus in this work is to integrate fuggitve emissions into environmental performance evaluation and optimization objectives so average emission factor approach giving a bit over estimates are used. Average emission factors for estimating fugitive emissions from fugitive sources found in synthetic organic chemical manufacturing industries operations (SOCMI) obtained from the US Environmental Protection Agency L & E Databases are used. The relation used in this work for calculation of fugitive emissions is:

$$E_f = \sum_{s}^{sources} (\dot{M}s \cdot \xi \cdot x_{v,s}) / \dot{M}_p \ \text{[kg/ kg product]} \tag{16}$$

Here E_f is fugitive emission factor per unit of product, $\dot{M}_s$ mass flow rate through the source 's', ξ is average emission factor and $x_{v,s}$ is mass fraction of volatile component through source 's' and $\dot{M}_p$ is product rate. It is assumed $x_{v,s}$ for the process fluids through fugitive sources such as pump seals, valves, flanges and connection is equal to 1, i.e. fluids are composed entirely of volatile compounds.

2.6.3 Step III: Determination of weighting factors (Application of multicriteria decision analysis technique)

The integration of these individual impact categories into one index is a hierarchical multicriteria decision analysis problem. The analytic hierarchy process (AHP) is used for this purpose[19] and a computer programme for it is developed in VB 6.0. In this stage, first a hierarchical structure of the problem, which is structured hierarchically similar to a flow chart, is constructed. The overall objective is placed at the top while the criteria and sub-criteria are placed below. For example, as shown in figure 11, the overall objective Environmental performance index (EPI1) is placed at the top (level 1), then below (level 2) are criterias Total PEI, Ef , Ec and RC and after this (level 3) sub-criterias as HTPI, HTPE, TTP, ATP, GWP, ODP, PCOP and AP. After this using the numerical scale given in table 2.6, two pairwise comparison matrices (see Table 7 and 8) are constructed for determination of weights for aggregation of individual impact categories of WAR to total PEI and for determination of weights of total PEI, Ef, Ec and RC to Environmental performance index (EPI1).

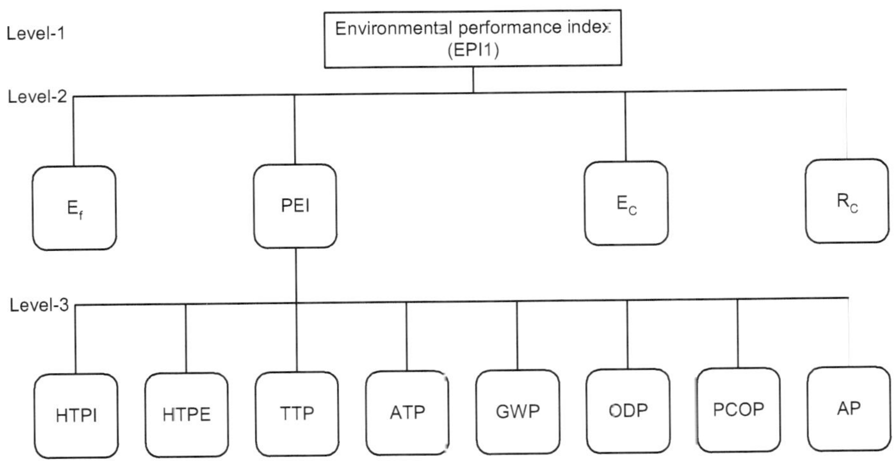

Fig. 11. Hierarchical structuring of multicriteria decision analysis problem for integrating individual environmental impacts

The right hand upper diagonal information in both matrices is to be provided by the decision maker giving the relative importance of the two criteria using the numerical scale of table 2.6 while the left hand lower diagonal is the reciprocal of the right hand upper diagonal. Once these pair wise comparison matrices are constructed, then developed computer programme using the AHP method, determines the weighting of individual impact categories. The level of inconsistency of decision makers input is checked by

Pairwise comparison matrix								
	HTPI	**HTPE**	**TTP**	**ATP**	**GWP**	**ODP**	**PCOP**	**AP**
HTPI	1	A12	A13	A14	A15	A16	A17	A18
HTPE		1	A23	A24	A25	A26	A27	A28
TTP			1	A34	A35	A36	A37	A38
ATP				1	A45	A46	A47	A48
GWP					1	A56	A57	A58
ODP						1	A67	A68
PCOP							1	A78
AP								1
W_L	W1	W2	W3	W4	W5	W6	W7	W8

Table 7. Pairwise comparison matrix for individual impact categories at level 3

Pairwise comparison matrix				
	PEI	**R_C**	**E_C**	**E_f**
PEI	1	A12	A13	A14
R_C		1	A23	A24
E_C			1	A34
E_f				1
W_L	W1	W2	W3	W4

Table 8. Pairwise comparison matrix for individual impact categories at level 2

consistency ratio before giving the output. Consistency ratio less than 0.1 is good and for ratios greater than 0.1, the input to pair wise matrix should be re-evaluated.

2.6.4 Step IV: Environmental performance & pollution index calculation

In the final step, first Total PEI is determined by multiplying each impact category values with its relevant weighting factor WL as given below:

$$Total\ PEI = \sum_{L}^{EnvCat} W_L \cdot PEI_L \tag{17}$$

After calculating Total PEI, Environmental performance index (EPI1) is determined for each alternative by multiplying the values of Total PEI, Ef, EC and RC with its relevant weighting factor WL (table 8) as given below:

$$EPI1 = 1/(\sum_{L}^{EnvCat} W_L \cdot E_L) \quad (18)$$

Where $E_L = \{Total\ PEI, E_c, E_f, R_c\}$

and environmental pollution index (EPI2) is calculated as follow

$$EPI2 = Total\ PEI + E_f \quad (19)$$

The higher value of environmental performance index (EPI1) shows that the process is environmentally better and vice versa. While the higher value of environmental pollution index (EPI2) shows that the environmental performance of process is worse.

2.7 Data manager

The relevant information generated from process module, safety module and environment module for each alternative is transferred to data manager. This information is used to formulate process diagnostic tables and multiobjective decision-making problem formulation. These tables consist of mass input/output table, energy input/output table, capital and utility annual expense summary, environmental impact summary and frequency of occurance of an event and their consequence categories and safety cost.

2.8 Stage III: Multiobjective decision analysis/ optimization

The purpose of this layer/stage is to set up multiobjective decision making/optimization among these conflicting objectives. The aim is to find out the trade-off surface for each alternative and /or complete ranking of alternatives. The calculation loop used for it is shown in figure 12.

In each independent performance module i.e. economic, environment and safety module, relevant information is generated and transferred automatically or manually to data manager for each alternative generated or under study. Before transferring the values of performance objective functions, each objective function is optimized within their independent module such as:

- Process/Economic optimization of each alternative is carried out using SQP optimization algorithm build within Aspen Plus™.
- The lower and upper limits for Environmental objective functions are calculated using environmental module from the material and energy balance information from process model.
- Safety/risk aspects are optimized in the safety module and information such as hazard occurance frequency, safety cost data (fixed safety system cost, accident and/or incident damage risk cost) for each alternative is transferred to the data manger.

Depending on the case under study or objectives of the study, graphical tool box of MatLab and/or multiobjective optimization technique (goal programming) or multiattribute decision analysis technique (PROMETHEE and/or AHP) is used for multiobjective decision analysis. The Data Manager is linked with MatLab 7.0 via Excel link Toolbox in the

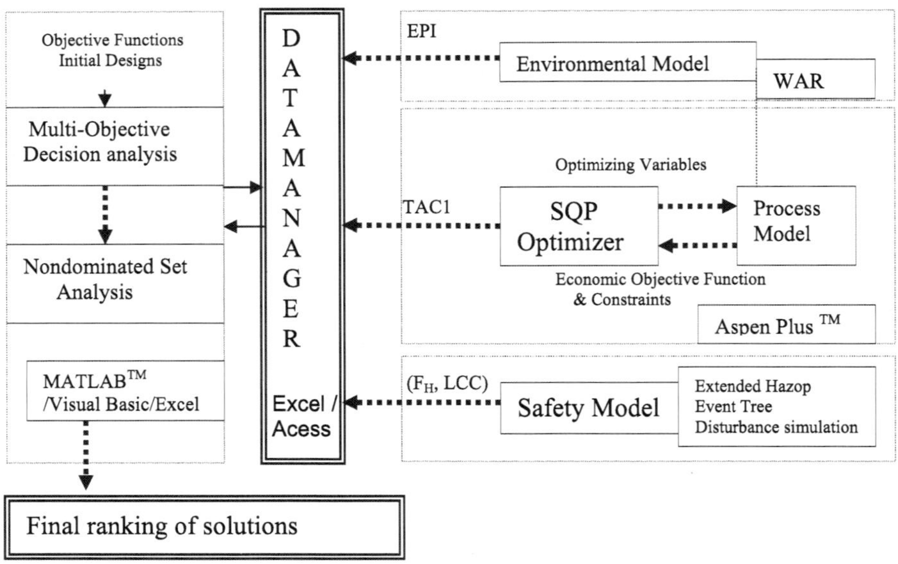

Fig. 12. Calculation loop for multiobjective optimization

integrated interface. Aspen plusTM is linked in the integrated interface via Visual basic 6.0 and Microsoft Excel. AHP technique is also programmed for the cases under study in Visual basic 6.0. The computer realization of these links in the integrated interface is explained in chapter four.

2.9 Stage IV: Design evaluation

The purpose of this layer/stage is to select the best alternative and/or find the complete ranking of alternatives under study based on the results of third stage/layer of the developed methodology. Pareto approach (non dominated analysis) or PROMETHEE is used for this purpose.

3. References

[1] Ramzan, N., Naveed, S., Feroze, N. and Witt, W.; "Multicriteria decision analysis for safety and economic achievement using PROMETHEE: A case study" Process Safety Progress (A Journal of Americal Institute of Chemical Engineering), 28(1), 68-83 (2009).

[2] Ramzan, N., Degenkolbe, S. and Witt, W.; "Evaluating and improving environmental performance of HC's recovery system: A case study of distillation unit", Chemical Engineering Journal (Journal published by Elsevier), 40(1-3), 201-213(2008). ISSN: 1385-8947.

[3] Ramzan, N., Compart, F. and Witt, W ; "Methodology for the generation and evaluation of safety system alternatives based on extended Hazop" Process Safety Progress (A Journal of Americal Institute of Chemical Engineering), 26(1),35-42 (2007).

[4] Ramzan, N., Compart, F. and Witt, W.; "Application of extended Hazop and event tree analysis for investigating operational failures and safety optimization of distillation column unit" Process Safety Progress (A Journal of Americal Institute of Chemical Engineering), 26(3),248-257 (2007).

[5] Ramzan, N., Witt, W.; "Multiobjective optimization in distillation unit: A case study", The Canadian Journal of Chemical Engineering, 84(5), 604-613(2006).

[6] Seader et al. (1999), Process design principles-Synthesis, Analysis, and Evaluation, John Wiley & Sons Inc. New York, 338-370.

[7] Nawaz, Z. Mahmood, Z. (2006) Importance Modeling, Simulation and Optimization in Chemical process Design, The Pakistan Engineer (Journal of Institute of Engineers Pakistan), 38-39.

[8] Cameron, I., Raman, R. (2005), Process Systems Risk Management, Elsevier Academic Press, NY, ISBN 0-12-156932-2

[9] Crowl, D.A., Louvar, J.F. (1999), Chemical Process Safety: Fundamentals with applications," Prentice Hall, New York.

[10] Kletz, T.A. (1997), Hazop-past and future, Reliability engineering and system safety, 55, 263-266.

[11] Lees, F.P. (1996), Loss prevention in CPI, Butterworth's, London, UK.

[12] Peters, M. S. and K.D. Timmerhaus (1991), Plant Design and Economics for Chemical Engineers, Ed.2nd, McGraw-Hill, New York, 90-145.

[13] Douglas, J. (1998), Conceptual design of chemical processes, McGraw Hill Inc.

[14] Guthrie, K.M. (1969), Data and techniques for preliminary capital cost estimating, Chem. Eng., 114-1421.

[15] Guthrie, K.M. (1974), Process plant estimating, evaluation and control, Craftsman, Solano Beach, CA.

[16] Cabezas, H. Bare, C. & Mallick, K. (1999), Pollution prevention with chemical process simulators: the generalized waste reduction (WAR) algorithm-full version. Computers and Chemical Engineering, 23,623-634.

[17] Cabezas, H. Bare, C. & Mallick, K. (1997), Pollution prevention with chemical process simulators: the generalized waste reduction (WAR) algorithm, Computers and Chemical Engineering, 21s, s305-s310.

[18] Dimian, A. C. (2003), Integrated design and simulation of chemical processes, 1st Eds., Elsevier Netherlands, 1-30,113-134.

[19] Dev, P.K. (2004), Analytic hierarchy process helps evaluate project in Indian oil pipelines industry, International journal of operation and production management, 24(6), 588-604

14

CFD Modelling of Fluidized Bed with Immersed Tubes

A.M.S. Costa, F.C. Colman, P.R. Paraiso and L.M.M. Jorge
Universidade Estadual de Maringa
Brazil

1. Introduction

Fluidized beds are widely used in combustion and chemical industries. The immersed tubes are usually used for enhancement of heat transfer or control of temperature in fluidized beds. By his turn, tubes subjected to the solid particle impact may suffer severe erosion wear. Many investigations have been devoted to erosion in tubes immersed in fluidized beds on the various influencing factors (cf. Lyczkowski and Bouillard, 2002). As pointed by Achim et al. (2002), the factors can be classified as particle characteristics, mechanical design and operating conditions.

Some previous experimental studies have focused on bubble and particle behaviors (Kobayashi et al., 2000, Ozawa et al., 2002), tube attrition, erosion or wastage (Bouillard and Lyczkowski, 1991; Lee and Wang, 1995; Fan et al., 1998; Wiman, 1994), heat transfer (Wong and Seville, 2006, Wiman and Almstedt, 1997) and gas flow regimes (Wang et. al, 2002).

Previous numerical studies were also performed using different CFD codes. Recently He et al. (2009, 2004), using the K-FIX code adapted to body fitted coordinates investigated the hydrodynamics of bubbling fluidized beds with one to four immersed tubes. The erosion rates predicted using the monolayer kinetic energy dissipation model were compared against the experimental values of Wiman (1994) for the two tube arrangement. The numerical values were three magnitudes lower than the experimental ones. Also employing a eulerian-eulerian model and the GEMINI numerical code, Gustavsson and Almstedt (2000, 1999) performed numerical computations and comparison against experimental results (Enwald et. al., 1999). As reported for those authors, fairly good qualitative agreement between the experimental and numerical erosion results were obtained, and the contributions to the erosion from the different fluid dynamics phenomena near the tube were identified.

In the present study, we revisit the phenomena of the immersed tubes in a gas fluidized bed with different immersed tube arrangements employing the eulerian-eulerian two fluid model and the MFIX code. The purposes of the numerical simulations are to compare and explore some effects not previously investigated in the above-mentioned references.

2. Two fluid and erosion model

The mathematical model is based on the assumption that the phases can be mathematically described as interpenetrating continua; the point variables are averaged over a region that is large compared with the particle spacing but much smaller than the flow domain (see Anderson, 1967). A short summary of the equations solved by the numerical code in this study are presented next. Refer to Benyahia et al. (2006) and Syamlal et al. (1993) for more detailment.

The continuity equations for the fluid and solid phase are given by :

$$\frac{\partial}{\partial t}(\varepsilon_f \rho_f) + \nabla \bullet (\varepsilon_f \rho_f \vec{v}_f) = 0 \tag{1}$$

$$\frac{\partial}{\partial t}(\varepsilon_s \rho_s) + \nabla \bullet (\varepsilon_s \rho_s \vec{v}_s) = 0 \tag{2}$$

In the previous equations ε_f, ε_s, ρ_f, ρ_s, $\vec{v}_f$ and $\vec{v}_s$ are the volumetric fraction, density and velocity field for the fluid and solids phases.

The momentum equations for the fluid and solid phases are given by:

$$\frac{\partial}{\partial t}(\varepsilon_f \rho_f \vec{v}_f) + \nabla \bullet (\varepsilon_f \rho_f \vec{v}_f \vec{v}_f) = \nabla \bullet \overline{\overline{S}}_f + \varepsilon_f \rho_f \vec{g} - \vec{I}_{fs} \tag{3}$$

$$\frac{\partial}{\partial t}(\varepsilon_s \rho_s \vec{v}_s) + \nabla \bullet (\varepsilon_s \rho_s \vec{v}_s \vec{v}_s) = \nabla \bullet \overline{\overline{S}}_s + \varepsilon_s \rho_s \vec{g} + \vec{I}_{fs} \tag{4}$$

$\overline{\overline{S}}_f$ $\overline{\overline{S}}_s$ are the stress tensors for the fluid and solid phase. It is assumed newtonian behavior for the fluid and solid phases, i.e.,

$$\overline{\overline{S}} = \left(-P + \lambda\ \nabla \bullet \vec{v}\right)\overline{\overline{I}} + 2\mu\ S_{ij} \equiv -p\ \overline{\overline{I}} + \overline{\overline{\tau}} \qquad S_{ij} = \frac{1}{2}\left[\nabla \vec{v} + (\nabla \vec{v})^T\right] - \frac{1}{3}\nabla \bullet \vec{v} \tag{5a,b}$$

In the above equation P, λ, μ are the pressure, bulk and dynamic viscosity, respectively.

In addition, the solid phase behavior is divided between a plastic regime (also named as slow shearing frictional regime) and a viscous regime (also named as rapidly shearing regime). The constitutive relations for the plastic regime are related to the soil mechanics theory. Here they are representated as:

$$p_s^P = f_1(\varepsilon^*, \varepsilon_f). \qquad \mu_s^P = f_2(\varepsilon^*, \varepsilon_f, \phi) \tag{6}$$

In the above equation ε^* is the packed bed void fraction and ϕ is the angle of internal friction.

A detailing of functions f_1 to f_4 and f_9 can be obtained in Benyahia (2008).

On the other hand, the viscous regime behavior for the solid phase is ruled by two gas kinetic theory related parameters (e, Θ).

$$p_s^v = f_3(\varepsilon_s, \rho_s, d_p, \Theta, e) \qquad \mu_s^v = f_4(\varepsilon_s, \rho_s, d_p, \Theta^{1/2}, e) \tag{7}$$

The solid stress model outlined by Eqs. (6) and (7) will be quoted here as the standard model. Additionally, a general formulation for the solids phase stress tensor that admits a transition between the two regimes is given by :

$$\overline{\overline{S}}_s = \begin{cases} \phi(\varepsilon_f)\overline{\overline{S}}_s^{v} + \left[1-\phi(\varepsilon_f)\right]\overline{\overline{S}}_s^{p} & \text{if } \varepsilon_f < \varepsilon^* + \delta \\ \overline{\overline{S}}_s^{v} & \text{if } \varepsilon_f \geq \varepsilon^* + \delta \end{cases} \tag{8}$$

According to Pannala et al.(2009), two diferent formulations for the weighting parameter "ϕ" can be employed :

$$\phi(\varepsilon) = \frac{1}{1+\upsilon^{\frac{\varepsilon-\varepsilon^*}{2\delta\varepsilon^*}}} \qquad \phi(\varepsilon) = \frac{\mathrm{Tanh}\left(\frac{\pi\left(\varepsilon-\varepsilon^*\right)}{\delta\varepsilon^*}\right)+1}{2} \tag{9a,b}$$

In the above equation the void fraction range δ and the shape factor υ are smaller values less than unity. It must be emphasized that when δ goes to zero and ϕ equals to unity, the "switch" model as proposed by Syamlal et al. (1993) based on the Schaeffer (1987) can be recovered. The models based on eqs. (9a) and (9b) will be referred in the numerical simulations as BLEND S and BLEND T, respectivelly.

On the other hand, the Srivastava and Sundaresan (2003), also called "Princeton model", can be placed on the basis of Eq. (9)

Also in equations (4) and (5) $\vec{I}_{fs}$ is the momentum interaction term between the solid and fluid phases, given by

$$\vec{I}_{fs} = -\varepsilon_s \nabla P_f - \beta\left(\vec{v}_s - \vec{v}_f\right) \tag{10}$$

There is a number of correlations for the drag coefficient β (Eqs. 11 to 16). The first of the correlations for the drag coefficient is based on Wen and Yu (1965) work. The Gidaspow drag coefficient is a combination between the Wen Yu correlation and the correlation from Ergun (1952). The Gidaspow blended drag correlation allows controlling the transition from the Wen and Yu, and Ergun based correlations. In this correlation the χ blending function was originally proposed by Lathowers and Bellan (2000) and the value of parameter C controls the degree of transition. From Eq. (14), the correlation proposed by Syamlal and O'Brien (1993) carries the advantage of adjustable parameters C_1 and d_1 for different minimum fluidization conditions. The correlations given in Eq. (15) and Eq. (16) are based on Lattice-Boltzmann simulations. For detailments of these last drag correlations refer to the works by Benyahia et al. (2006) and Wang et al. (2010).

$$\beta_{\text{Wen-Yu}} = \frac{3}{4}C_D \frac{\rho_f \varepsilon_f \varepsilon_s \left|\vec{v}_f - \vec{v}_s\right|}{d_p}\varepsilon_f^{-2.65}$$

$$C_D = \begin{cases} \frac{24}{\mathrm{Re}}\left(1+0.15\,\mathrm{Re}^{0.687}\right) & \mathrm{Re}<1000 \\ 0.44 & \mathrm{Re}\geq 1000 \end{cases} \qquad \mathrm{Re} = \frac{\rho_f \varepsilon_f \left|\vec{v}_f - \vec{v}_s\right| d_p}{\mu_f} \tag{11}$$

$$\beta_{\text{Gidaspow}} = \begin{cases} \beta_{\text{Wen-Yu}} & \varepsilon_f > 0.8 \\ \beta_{\text{Ergun}} = \dfrac{150\,\varepsilon_s(1-\varepsilon_s)\mu_f}{\varepsilon_f d_p^2} + \dfrac{1.75\,\rho_f\varepsilon_s\left|\vec{v}_f - \vec{v}_s\right|}{d_p} & \varepsilon_f \le 0.8 \end{cases} \tag{12}$$

$$\beta_{\text{Gidaspow-blended}} = \chi\,\beta_{\text{Wen-Yu}} + (1-\chi)\beta_{\text{Ergun}} \quad \chi = \frac{\tan^{-1}\left(C\,(\varepsilon_f - 0.8)\right)}{\pi} + 0.5 \tag{13}$$

$$\begin{aligned} &\beta_{\text{Syamlal-OBrien}} = \frac{3}{4}\frac{\rho_f\varepsilon_f\varepsilon_s}{V_r^2 d_p}\left(0.63 + 4.8\sqrt{\frac{V_r}{\text{Re}}}\right)^2\left|\vec{v}_f - \vec{v}_s\right| \\ &V_r = 0.5A - 0.03\text{Re} + 0.5\times\sqrt{(0.06\text{Re})^2 + 0.12\text{Re}\,(2B - A) + A^2} \\ &A = \varepsilon_f^{4.14} \quad B = \begin{cases} C_1\varepsilon_f^{1.28} & \varepsilon_f \le 0.85 \\ \varepsilon_f^{d_1} & \varepsilon_f > 0.85 \end{cases} \end{aligned} \tag{14}$$

$$\beta_{\text{Hill-Koch-Ladd}} = 18\mu_f(1-\varepsilon_s)^2\varepsilon_s\frac{F}{d_p^2} \qquad F = f_9(F_0, F_1, F_2, F_3) \tag{15}$$

$$\beta_{\text{Beestra}} = 180\frac{\mu_f\varepsilon_s^2}{d_p^2} + 18\frac{\mu_f\varepsilon_f^3\varepsilon_s\left(1 + 1.5\sqrt{\varepsilon_s}\right)}{d_p^2} + 0.31\frac{\mu_f\varepsilon_s\,\text{Re}}{\varepsilon_f d_p^2}\frac{\left[\varepsilon_f^{-1} + 3\varepsilon_f\varepsilon_s + 8.4\text{Re}^{-0.343}\right]}{[1 + 10^{3\varepsilon_s}\,\text{Re}^{-0.5-2\varepsilon_s}} \tag{16}$$

For closing the model, a transport equation for the granular energy Θ provides a way of determine the pressure and viscosity for the solid phase during the viscous regime. Equation (17) is a transport equation for the granular energy Θ. Its solution provides a way of determine the pressure and viscosity for the solid phase during the viscous regime. The terms κ_s γ and ϕ_{gs} are the granular energy conductivity, dissipation and exchange, respectively.

$$\frac{3}{2}\left[\frac{\partial}{\partial t}\varepsilon_s\rho_s\Theta + \nabla\bullet\rho_s\vec{v}_s\Theta\right] = \overline{\overline{S}}_s : \nabla\vec{v}_s - \nabla\bullet(\kappa_s\nabla\Theta) - \gamma + \phi_{gs} \tag{17}$$

$$\begin{aligned} \kappa_s &= f_5\left(\varepsilon_s, \rho_s, d_p, \Theta^{1/2}, e, \beta\right) \\ \gamma &= f_6\left(\varepsilon_s, \rho_s, d_p, \Theta^{3/2}, e\right) \\ \phi_{gs} &= f_7\left(\varepsilon_s, \rho_s, d_p, \Theta, \left|\vec{v}_f - \vec{v}_s\right|, \beta\right) \end{aligned} \tag{18}$$

In the algebraic approach, instead solving the full equation (17) , the granular energy is obtained by equating the first term on the right hand side with the dissipation term.

The model where Eqs. (5) to (8) and (17) are solved is the kinetic theory model, termed here as KTGF. Conversely, in the constant solids viscosity model (CVM) the solids pressure is defined as in Eq. (6) and the solids viscosity in either plastic and viscous regimes is set equal to a constant.

For erosion calculations in this work we use the monolayer energy dissipation model (Lyczkowski and Bouillard, 2002). In that model the kinetic energy dissipation rate for the solids phase in the vicinity of stationary immersed surfaces is related to erosion rate in m/s by multiplication with an appropriate constant. This constant is function of surface hardness, elasticity of collision and diameter of particles hitting the surface. The kinetic energy dissipation rate Φ_s in W/m³ for the solids phase is given by :

$$\Phi_s = \left[\varepsilon_s \overline{\overline{\tau}}_s : \nabla \vec{v}_s + \beta \frac{\vec{v}_s^{\,2}}{2} \right] \tag{19}$$

3. Numerical method

MFIX (Multiphase Flow with Interphase eXchanges) is an open source CFD code developed at the National Energy Technology Laboratory (NETL) for describing the hydrodynamics, heat transfer and chemical reactions in fluid-solids systems. It has been used for describing bubbling and circulating fluidized beds, spouted beds and gasifiers. MFIX calculations give transient data on the three-dimensional distribution of pressure, velocity, temperature, and species mass fractions.

The hydrodynamic model is solved using the finite volume approach with discretization on a staggered grid. A second order accurate discretization scheme was used and superbee scheme was adopted for discretization of the convective fluxes at cell faces for all equations in this work. With the governing equations discretized, a sequential iterative solver is used to calculate the field variables at each time step. The main numerical algorithm is an extension of SIMPLE. Modifications to this algorithm in MFIX include a partial elimination algorithm to reduce the strong coupling between the two phases due to the interphase transfer terms. Also, MFIX makes use of a solids volume fraction correction step instead of a solids pressure correction step which is thought to assist convergence in loosely packed regions. Finally, an adaptive time step is used to minimize computation time. See Syamlal (1998) for more details. The immersed obstacles were implemented using the cut-cell technique available in the code (Dietiker, 2009)

The numerical runs were based on experiments of Wiman (1994, 1997) in an air pressurized bed with horizontal tubes for four different tube-bank geometries. The T2 and I4 are portrayed in Fig. 1 whereas the S4 and S4D tubes geometry are given in Figure 3. Figure 2 details the domain for the numerical simulations and circumferential angle coordinate for erosion measurement. In Fig. 2a is detailed the mesh for the T2 arrangement This non-uniform stretched grid, follows Cebeci et al. (2005) approach, has fine spacing close to the surface of the tubes and coarse spacing away from the surface. For all the other arrangements the mesh was uniform as depicted in Fig. 2b. The mesh employed for the bed without tube, and the I4 arrangement was 60 × 260 cells, and for the S4 and S4D arrangement 60 × 340. The bed was operated at room temperature (24 C) at pressures between 0.1 and 1.6 MPa and at two different excess velocities: $U_{f1} - U_{mf} = 0.2$ m/s and $U_{f1} - U_{mf} = 0.6$ m/s. Here, U_{f1} is the superficial fluidization velocity based on the free bed cross-section. The voidage at minimum fluidization was 0.46 and the minimum fluidization velocities was 0.42 , 0.31 and 0.18 m/s, for 0.1, 0.4 and 1.6 MPa pressures, correspondingly. The particle diameter and density were 700 μm and 2600 kg/m³. The bubble parameters

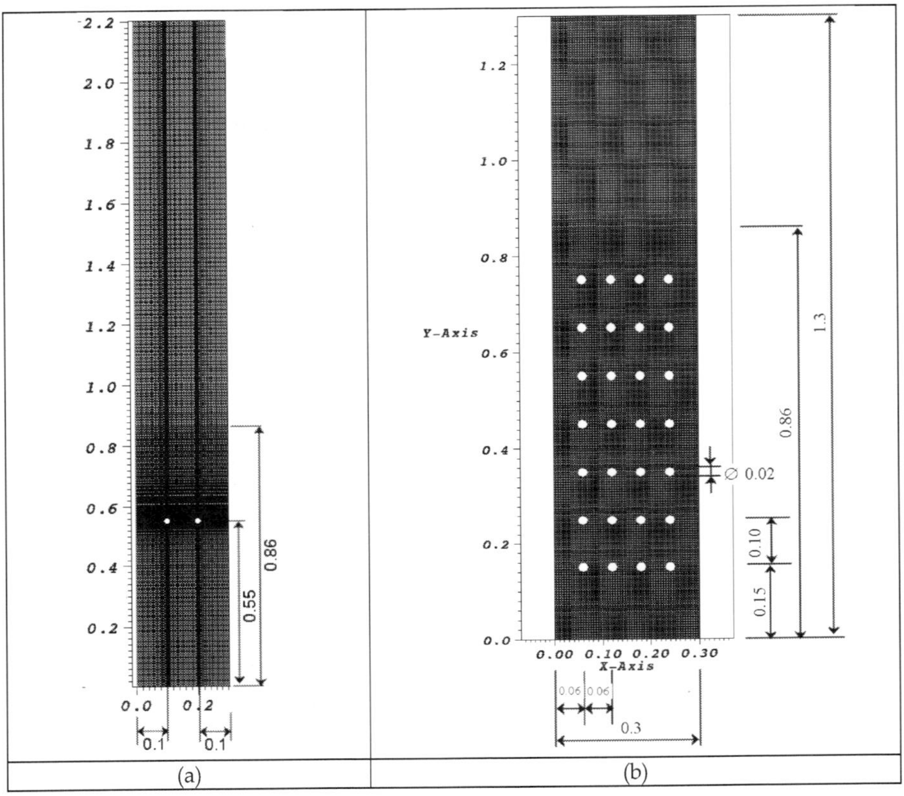

Fig. 1. Geometry and mesh for T2 and I4 tube arrangement.

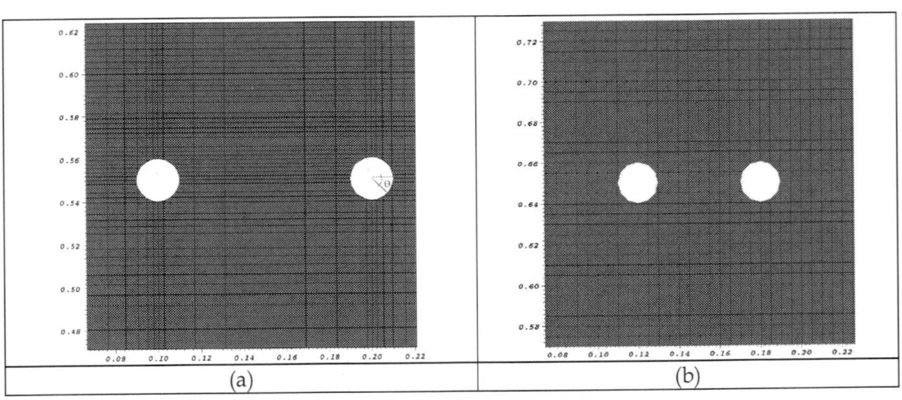

Fig. 2. Mesh detailment for (a) T2 arrangement (b) all the others arrangement

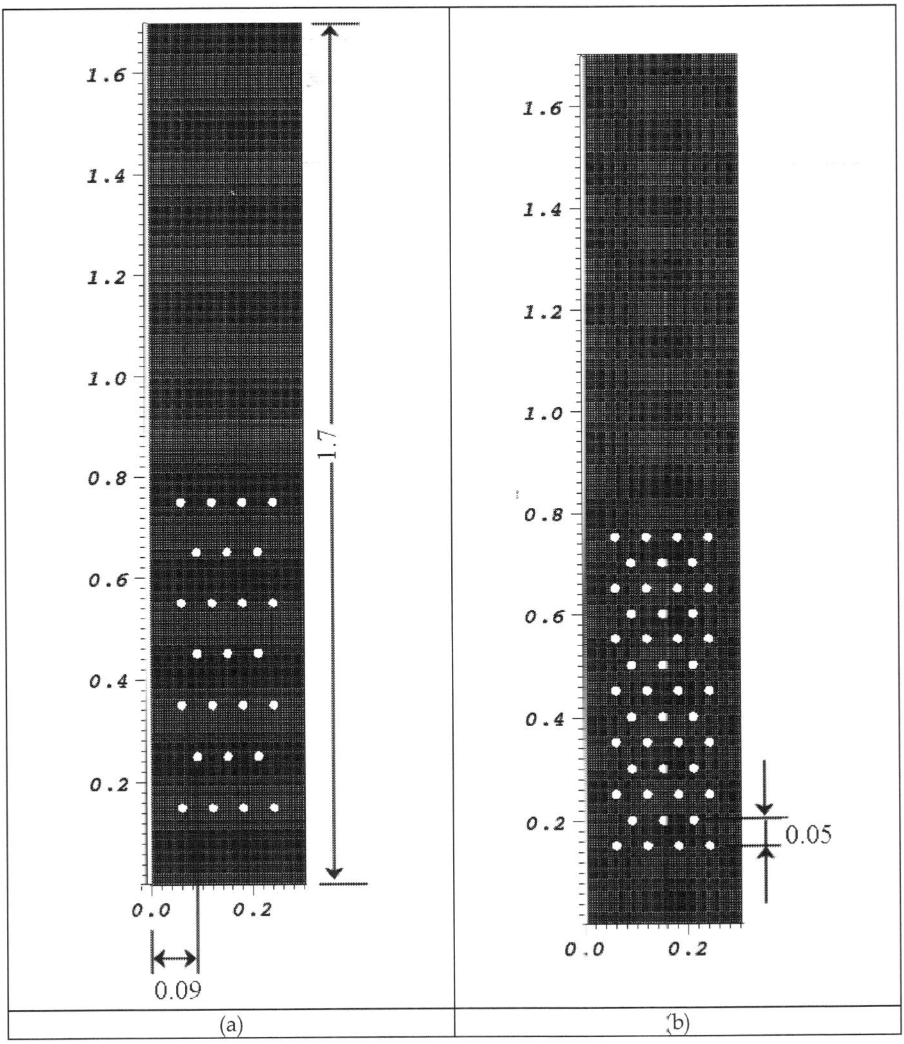

Fig. 3. Geometry and mesh for (a) S4 and (b) S4D tube arrangement

obtained from simulation were based on methodology described in Almstedt (1987), using numerical probes in the domain, centered at (0.15, 0.55) and separated from 15 mm. The target tube for erosion measurements is the one centered at (0.18, 0 55) for all the tube arrangements. More information about the experiments can be accessed from Wiman (1994, 1997).

In this work, the parameters for controlling the numerical solution (e.g., under-relaxation, sweep direction, linear equation solvers, number of iterations, residual tolerances) were kept as their default code values. Moreover, for setting up the mathematical model, when not otherwise specified the code default values were used. The computer used in the numerical simulations was a PC with OpenSuse linux and Intel Quad Core processor. The simulation time was 20 s.

For generating the numerical results, e employed the parameters listed above, referred here as baseline simulation. In addition, for the baseline simulation we employed the Syamlal-O´Brien drag model, the standard solid stress model, and slip and non-slip condition for solid and gas phase, correspondingly. The previous set of models will be referred in the result's section as baseline simulation models.

4. Results and discussion

Figure 4 is a sampling plot showing the instantaneous solids velocities and gas volumetric fraction fields following two different bubbles passage around the obstacle. Analysis of Fig. (4) shows the bubble splitting mechanism taking place and the characteristic time scale for the bubble passage. After the bubble passage, the solid wake has higher solid velocity magnitude around the obstacle. The solids upward movement following the bubble wake and wall downward movement is kept unchanged. Generally, for beds without internal obstacles, bubble size increases with bed height, particle size and superficial velocity. Analysis of CFD results shows that the presence of tubes was found to alter such general trends for bubble growth.

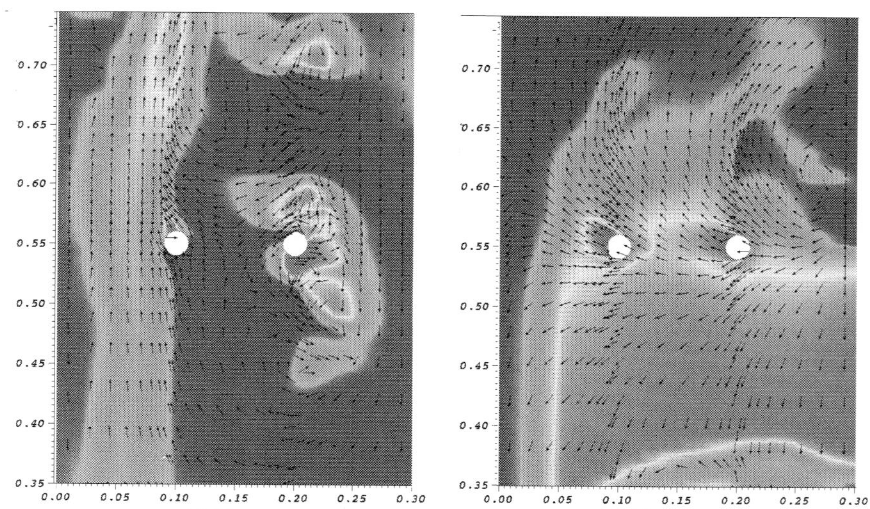

(c) 11 s (d) 12 s

Fig. 4. Detail of the instantaneous voidage and solids velocity vector field for the T2 arrangement

Figure 5 presents the time averaged kinetic energy dissipation as a function of circumferential position θ on the tube surface. As it would be seen from the majority of

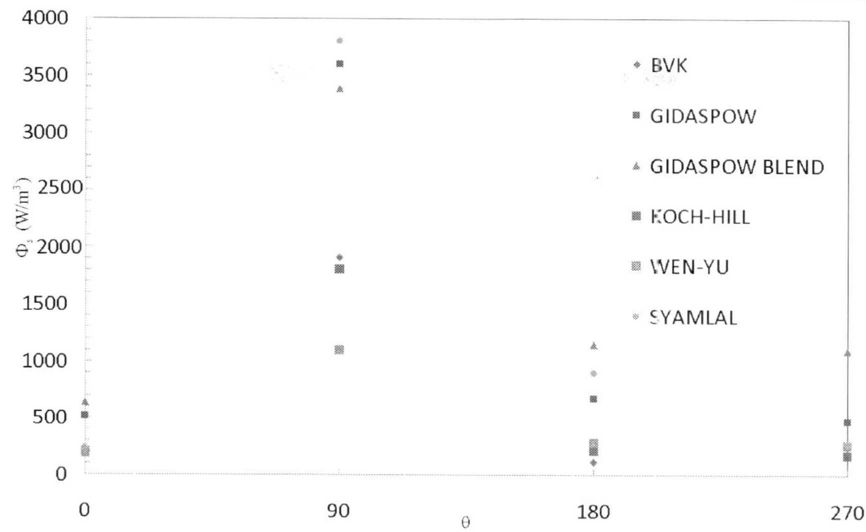

Fig. 5. Time averaged kinetic energy dissipation predicted for different gas-solid drag models – T2 arrangement

results the most severe dissipation occurs on the lower parts of the tube for an angle corresponding to 90 degrees. This fact is in agreement with experimental measured values of erosion from Wiman (1994). Analysis of results shows that, except for the BVK drag model, all the drag models predict the highest dissipation rate occurring at 90 degrees. The value predicted using the Syamlal-O'Brien drag model is the highest. There is a noticeable difference in the value of dissipation value between the Gidaspow and Gidaspow blend drag models, whereas the results by the Wen-Yu drag locate in the intermediate range. By his turn, the results by HYS and Koch-Hill models locates near the Wen-Yu values.

Also, analysis of transient simulated results shows the highest values of kinetic energy dissipation rate occur when the particle fraction suddenly changes from a low to a high value, which corresponds to the tube being hit by the wake of a bubble.

Figure 6 presents the result of kinetic energy dissipation for different solid stress models. As shown the peak values of the dissipation rate are close to 180 degrees for the model with (baseline) and without the blending function discussed in section 2. On the other hand, the maximum value predicted by the constant solids viscosity model (NO KTGF) locates around 180 degrees, and it is superior to the predicted by the solids kinetic energy theory.

Figure 7 presents the result of kinetic energy dissipation for different tube surface slip conditions. The baseline case considers the free slip condition for the solids and non-slip condition for the gas. The peak values do not change when considering the solids with the non-slip condition. On the other hand, when considering slip conditions for both phases the value decreased. The above results, suggests the free slip condition for the gas phase

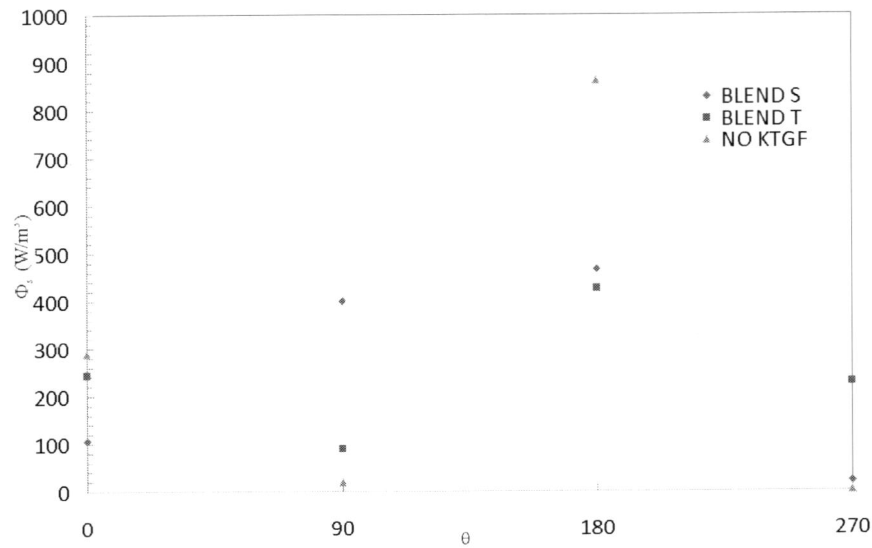

Fig. 6. Time averaged kinetic energy dissipation predicted for different solid stress models

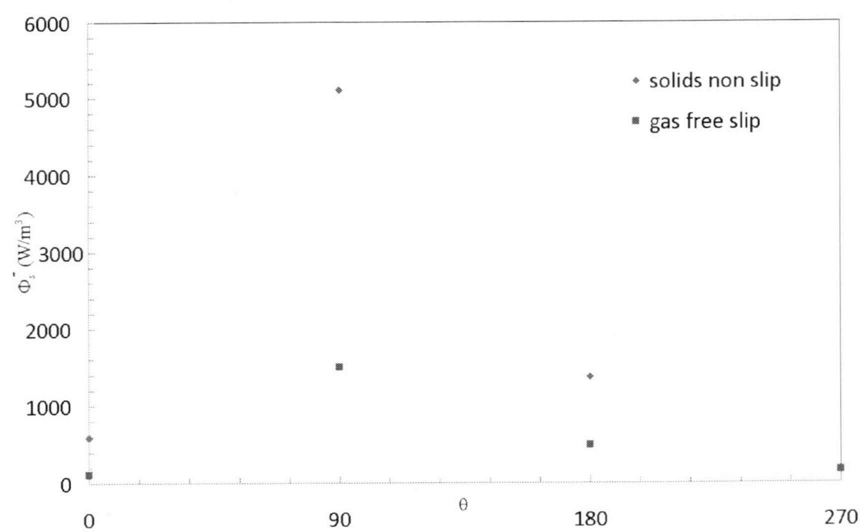

Fig. 7. Time averaged kinetic energy dissipation predicted for different tube surface slip conditions

on the surface of the tube plays an important role on the value of the kinetic energy dissipation.

Figures 8 and 9 shows a comparison between the numerically predicted time averaged values of the kinetic energy dissipation rate, using the baseline simulation models discussed in section 2, and those based on the experimental values for the T2 arrangement in the work of Gustavson and Almstedt (2000). A comparison of the numerical and experimental results for the two different operational pressures (c.f. Figs 8 and 9) shows some degree of discordance. However, the lack of minutely agreement with experimental values, the results can be compared for recognizing similar drifts. For instance, from the Fig 8, the higher simulated values of the dissipation rate occurring in the bottom position of the tube, i.e. for $\theta < 180^{\circ}$, is in agreement with the experimental counterpart. Similarly, the increase of dissipation rate with increased operational pressure for the simulated results is in agreement with the experimental values. Regarding the erosion and baseline models used for the simulation, some remarks towards better agreement with experimental values can also be done. According to the monolayer erosion model and its discussion above Eq. (19) some degree of uncertainty is associated to the multiplying constant, as the exact value of elasticity of collision is not known. It is also expected, that adjustments in the baseline simulation models, such as those outlined in Figs. (5) to (7) would produce better agreement.

Figure 10 is a sampling plot showing the instantaneous gas volumetric fraction fields for different tube arrangements. Analysis of Fig. (10) shows the influence of immersed obstacles on the bubble splitting mechanism taking place and the bubble passage pattern. Above the tube bank, the bubble appears to grow to size similar to the without tube geometry. For the geometry with tubes the bubble encompasses the obstacles but not at the full width of the bed. The interaction is stronger for the denser tube geometry.

Figures 11 and 12 shows a comparison between the numerically predicted values of bubble frequency, using the baseline simulation models discussed in section 2, and those based on the experimental measurements from the works of Almstedt (1987) and Wiman (1995). As shown in the Fig. 10, the calculated values of N_b are underestimated at higher pressures, while at low pressures, there is a quite good agreement between calculated and experimental results. This conclusion holds true both for the I4 and for the S4D tube arrangement. As in the experimental results there are no noticeable differences for the two tubes arrangements.

Figures 13 and 14 shows a comparison for the bubble frequency for the bed without tube and with the S4 arrangement. As shown in Fig. 13 for the bed without tubes the trend points to a frequency agreement between 0.4 and 0.6 MPa. Up this range the values differences increases as pressures increases up to 1.6 MPa. For the S4 arrangement the trend is similar to the I4 and S4D arrangement. The experimental results depicted in Figs. 11 to 14 suggest that the frequency increases with pressure both with and without tubes. For numerical results this holds true only for low pressures, i.e., 0.1 and 0.4 MPa. The numerical results suggest a maxima occurring between 0.4 and 1.6 MPa. On the other hand, the numerical results corroborate the experimental trend that the mean frequency is higher for the bed with tubes than for the freely bubbling bed.

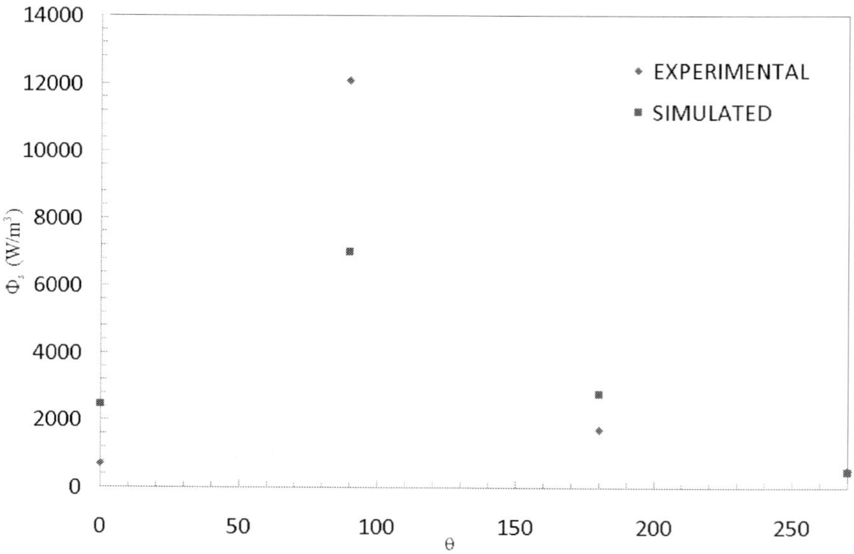

Fig. 8. Simulated kinetic energy dissipation at various circumferential angular positions in the surface of the tube and values from experimental results by Gustavsson and Almstedt, 2000. Operational pressures : 0.8 MPa ;

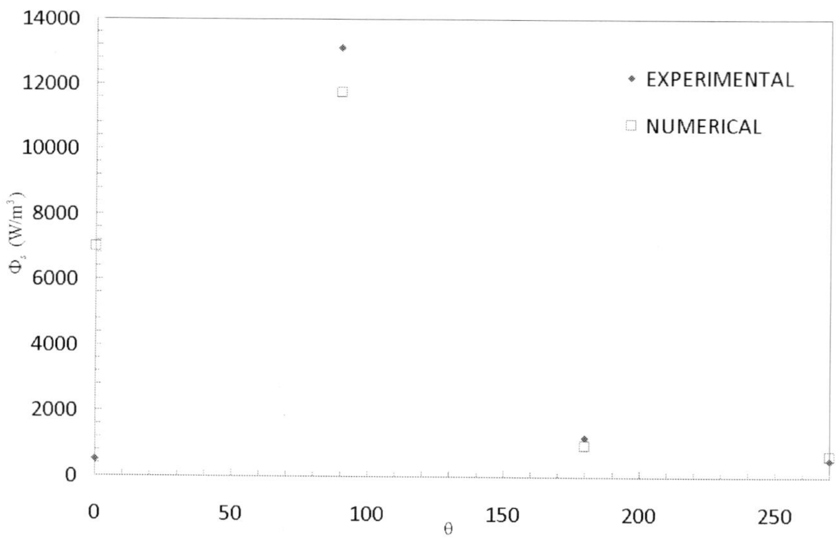

Fig. 9. Simulated kinetic energy dissipation at various circumferential angular positions in the surface of the tube and values from experimental results by Gustavsson and Almstedt, 2000. Operational pressures : 1.6 MPa

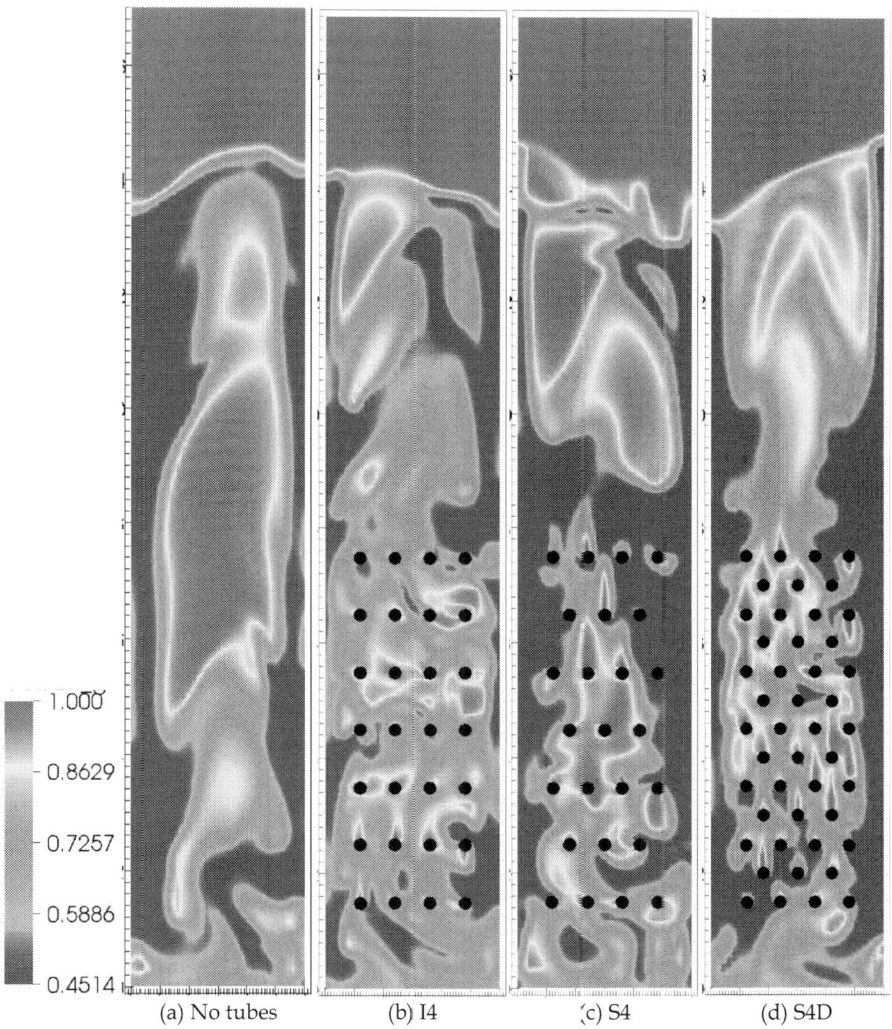

Fig. 10. Snapshots of voidage field at 4 s for different tube arrangements. P = 0.4 MPa , Uf = 0.6 m/s

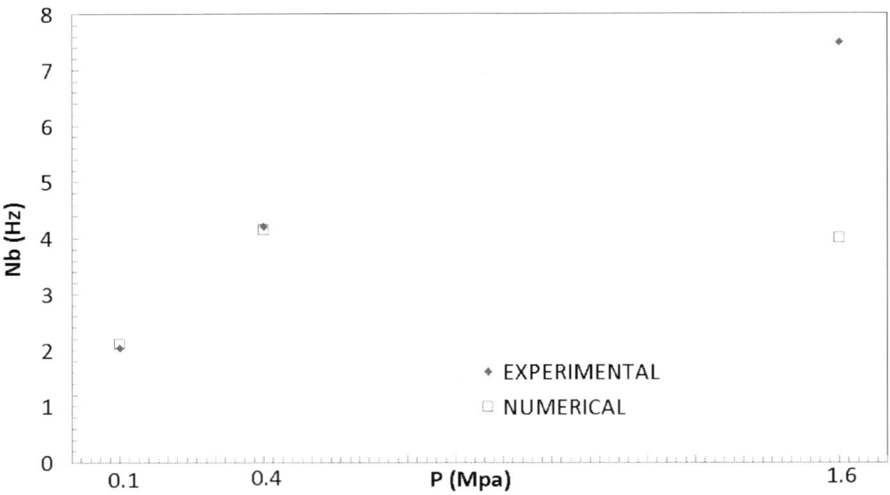

Fig. 11. Nb versus pressure, numerical X experimental: I4 arrangement, U_f = 0.6 m/s

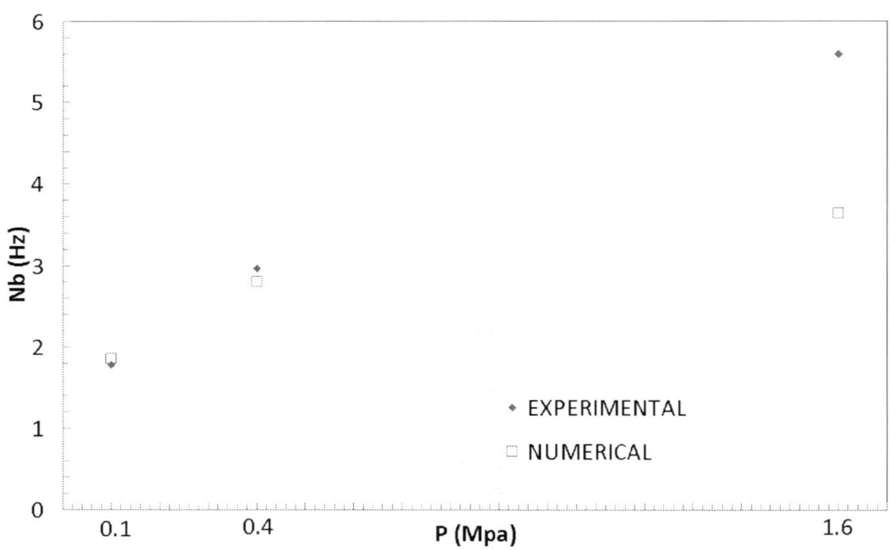

Fig. 12. Nb versus pressure, numerical X experimental: S4D arrangement, U_f = 0.6 m/s

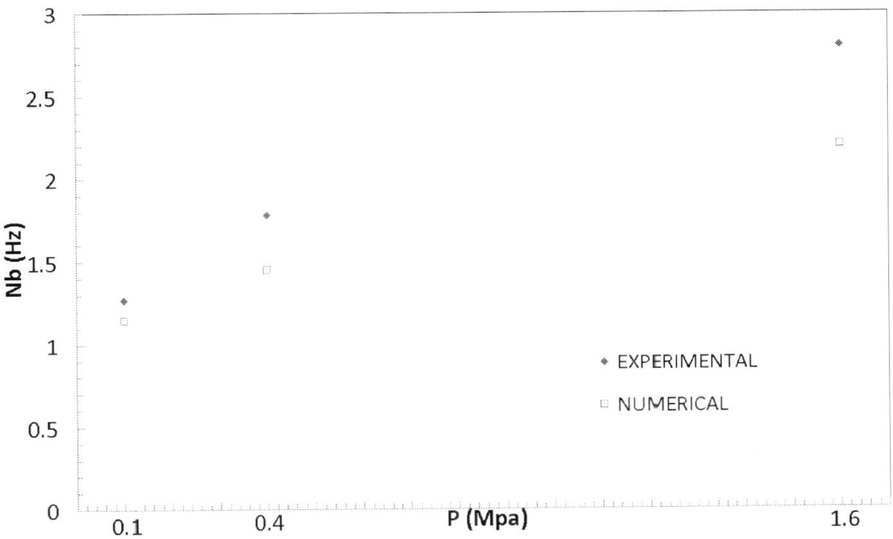

Fig. 13. Nb versus pressure, numerical X experimental: no tubes, U_f = 0.6 m/s

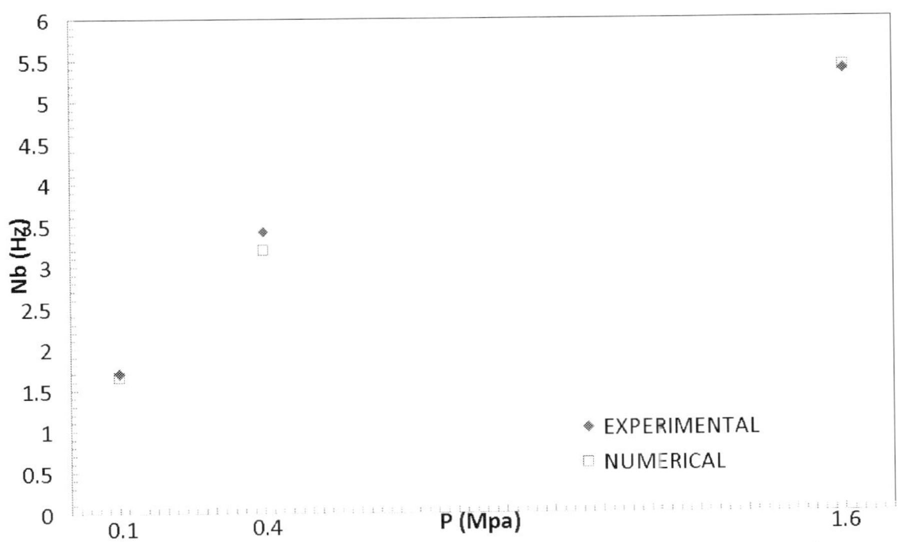

Fig. 14. Nb versus pressure, numerical X experimental: S4 arrangement, U_f = 0.6 m/s

Figure 15 to 18 shows a comparison for the mean bubble velocity. In all cases, the simulated results are underestimated in relation to the experiments. However, the trend observed for the experimental results with a maxima around 0.4 MPa is verified for the numerical results for all the tube arrangements. For the bed without tubes the experimental increase trend of Vb with pressure is valid for pressures higher than 0.4 MPa.

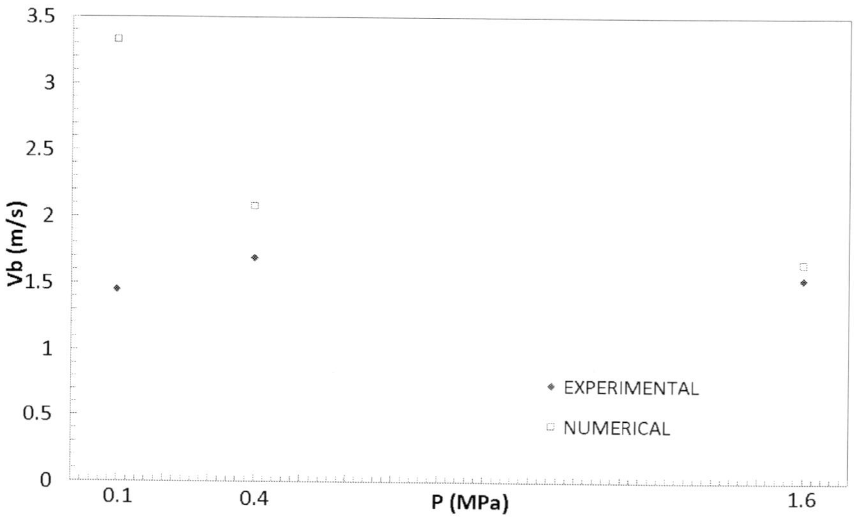

Fig. 15. Vb versus pressure, numerical X experimental: : I4 arrangement, U_f = 0.6 m/s

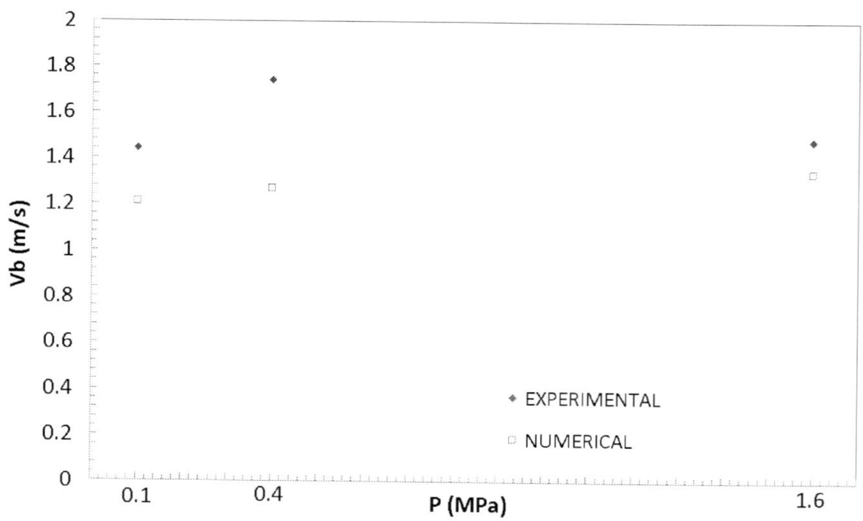

Fig. 16. Vb versus pressure, numerical X experimental: S4D arrangement, U_f = 0.6 m/s

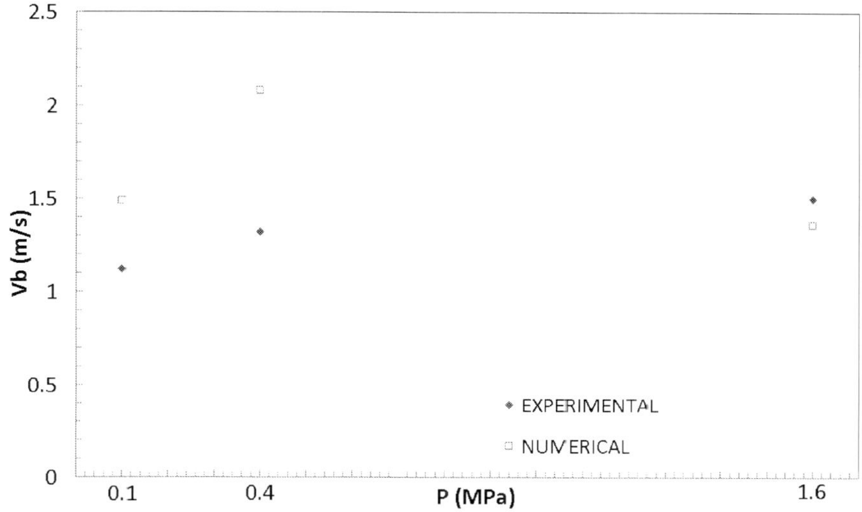

Fig. 17. Vb versus pressure, numerical X experimental: : no tubes, U_f = 0.6 m/s

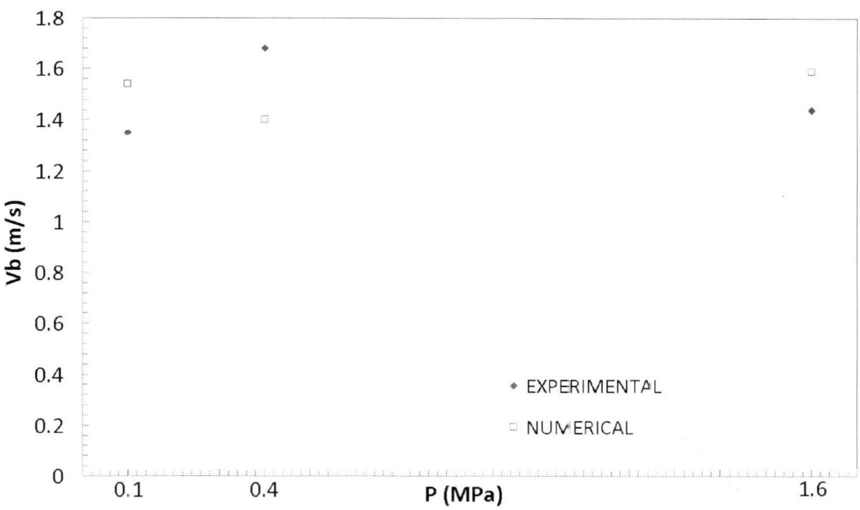

Fig. 18. Vb versus pressure, numerical X experimental: : S4 arrangement, U_f = 0.6 m/s

Figures 19 and 20 shows a comparison for the bubble frequency for U_f = 0.2 m/s. Comparison with results in Figs. 12 and 14 shows that the numerical results, although smaller are closer to the experimental. The tendency holds true for both S4 and S4D arrangements. Comparison with the numerical results for the S4D and S4 arrangements

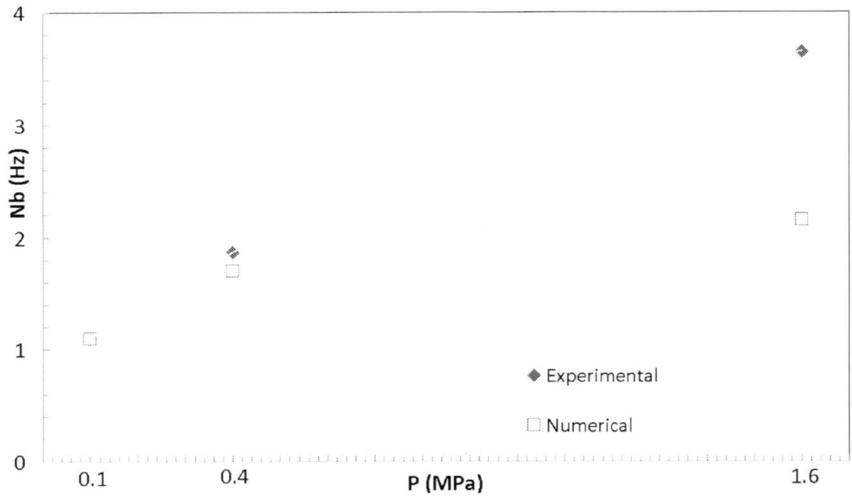

Fig. 19. Nb versus pressure, numerical X experimental: : S4 arrangement, U_f = 0.2 m/s

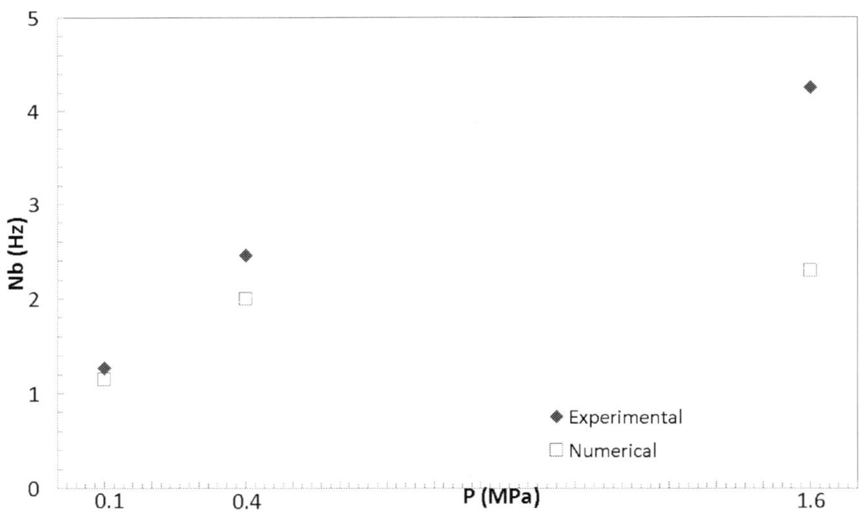

Fig. 20. Nb versus pressure, numerical X experimental: S4D arrangement, U_f = 0.2 m/s

given in Fig. 6 and 7, also shows that Nb increases with increasing excess velocity. The last, is the same drift verified for the experimental values.

Figures 21 and 22 present the results for the time averaged kinetic energy dissipation as a function of circumferential position θ on the tube surface for the S4 arrangement for two distinct pressures. As it would be seen the highest experimental results are between 90 and

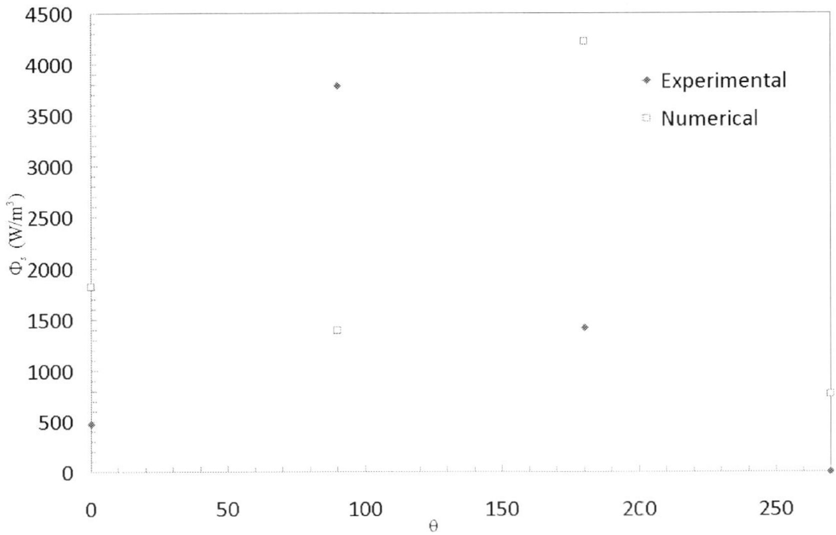

Fig. 21. Time averaged kinetic energy dissipation predicted for different tube arrangements at two different operating pressures: : S4 arrangement, P = 0.1 MPa

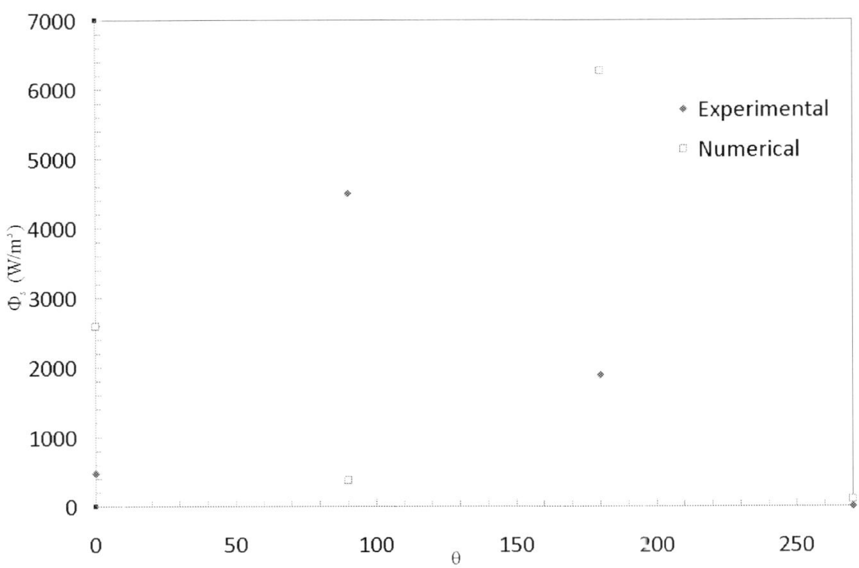

Fig. 22. Time averaged kinetic energy dissipation predicted for different tube arrangements at two different operating pressures : : S4 arrangement, P = 1_6 MPa

240 degrees, while for the numerical are between 30 and 150. The numerical values are over predicted in the range 0 to 100 degrees and above 240 degrees. In the range from 100 to 240 degrees the numerical are below the experimental. Similar trends are verified for the 1.6 MPa pressure, with the experimental curves less sensitive to pressure variation . By his turn the numerical values, show more sensitivity to pressure, although with the same magnitude order.

5. Conclusion

In this work was investigated numerically the hydrodynamics of two dimensional beds with immersed tubes. The simulations were based on an experimental bed with different tube bank geometries, operating pressures and gas excess velocities. The objective of this study was two fold: explore and investigate some effects not previously explored in the literature, to verify the feasibility of the MFIX code for such a kind of study.

The simulation's results were framed in terms of averaged solids kinetic energy dissipation rate and bubble parameters (frequency and mean velocity). A comparison between the numerical results for frequency and the experiments shows good agreement for low pressures. Also, some similar drifts were identified, e.g., greater frequency for greater excess velocities. By his turn, the bubble velocity numerical results agrees better with experiments for high pressures. For the bed with tubes, a maxima of bubble velocity for intermediate pressure is identifiable, this could not be verified for the numerical results. Our results points to significant influences in the predicted dissipation rates and consequently, in the erosion rate, when employing different drag models. The dissipation rate is also influenced by either the use of blending functions for the transition between the plastic and viscous regime of solids flow or the use of a constant viscosity model. The results are key sensitive to the slip condition for the gas phase in the surface of the tube. In the case of a free slip condition for the gas phase the lowest values of dissipation are obtained By his turn, for the solids kinetic dissipation energy, the range of angles that gives maximum values are shifted in relation to the experiments. Finally, remarks towards better agreement with experimental values can also be done for the energy dissipation model. Specifically, according to the monolayer erosion model and its discussion above Eq. (19) some degree of uncertainty is associated to the multiplying constant, as the exact value of elasticity of collision is not known.

6. References

Achim, D., Easton, A. K., Schwarz, M. P., Witt, P.J., Zakhari, A., 2002, "Tube erosion modelling in a fluidised bed", Applied Mathematical Modelling, Vol 26, pp. 191-201.

Almstedt, A. E., 1987, A study of bubble behaviour and gas distribution in pressurized fluidized beds burning coal. Thesis for the degree of Licentiate of Engineering, Chalmers University of Technology, Goteborg. Sweden.

Anderson, T. B., 1967, "A fluid mechanical description of fluidized beds: Equations of motion", Industrial Engineering Chemical Fundamentals, Vol 6, pp. 527-539.

Benyahia, S., 2008, "Validation study of two continuum granular frictional flow theories", Industrial Engineering Chemical Research, 47, 8926-8932.

Benyahia, S., Syamlal, M., O'Brien, T. J., "Summary of MFIX Equations 2005-4", 1 March 2006, Available from http://www.mfix.org/documentation/MfixEquations2005-4-1.pdf.

Benyahia, S., Syamlal, M., O'Brien, T. J., 2006, "Extension of Hill–Koch–Ladd drag correlation over all ranges of Reynolds number and solids volume fraction", Powder Technology, 162, 166-174.

Bouillard, J. X., Lyczkowski, R. W., 1991, "On the erosion of heat exchanger tube banks in fluidized-bed combustor", Powder Technology, Vol 68, pp. 37-51.

Cebeci, T., Shao, J. P., Karyeke, F., Laurendeau, E., 2005, Computational Fluid Dynamics for Engineers, Horizon Publishing Inc, USA, 402p.

Dietiker, J., "Cartesian Grid User Guide", 4 September 2009, Available from https //mfix.netl.doe.gov/documentation/Cartesian_grid_user_guide.pdf.

Enwald, H., Peirano, E., Almstedt, A. E., Leckner, B., 1999, "Simulation of the fluid dynamics of a bubbling fluidized bed: Experimental validation of the two-fluid model and evaluation of a parallel multiblock solver", Chemical Engineering Science, 54, 311-328.

Ergun, S., 1952, "Fluid-flow through packed columns", Chemical Engineering Progress, Vol 48, n. 2, pp. 91-94.

Fan, J. R., Sun, P., Chen, L. H., Cen, K. F., 1998, "Numerical investigation of a new protection method of the tube erosion by particle impingement", Wear, Vol 223, pp. 50-57.

Gustavsson, M., Almstedt, A. E., 1999, "Numerical simulation of fluid dynamics in fluidized beds with horizontal heat exchanger tubes", Chemical Engineering Science, 55, 857-866.

Gustavsson, M., Almstedt, A. E., 2000, "Two-fluid modelling of cooling-tube erosion in a fluidized bed", Chemical Engineering Science, 55, 867-879.

He, Y. R., Lu, H. L., Sun, Q. Q., Yang, L. D., Zhao, Y. H., Gidaspow, D., Bouillard, J., 2004, "Hydrodynamics of gas-solid flow around immersed tubes in bubbling fluidized beds", Powder Technology, 145, pp. 88-105.

He, Y. R., Zhan, W., Zhao, Y., Lu, H. Schlaberg, I., 2009, "Prediction on immersed tubes erosion using two-fluid model in a bubbling fluidized bed". Chemical Engineering Science, 64, pp. 3072-3082.

Kobayashi, N., Yamazaki, R., Mori, S., 2000, "A study on the behaviour of bubbles and solids in bubbling fluidized beds", Powder Technology, Vol 113, pp. 327-344.

Lathowers, D., Bellan, J., 2000, "Modeling of dense gas-solid reactive mixtures applied to biomass pyrolysis in a fluidized bed", Proceedings of the 2000 U.S. DOE Hydrogen Program Review, NREL/CP-570-28890. USA.

Lee, S. W., Wang, B. Q., 1995, "Effect of particle-tube collision frequency on material wastage on in-bed tubes in the bubbling fluidized bed combustor", Wear, Vol 184, pp. 223-229.

Lyczkowski, R. W., Bouillard, J. X., 2002, "State-of-the-art review of erosion modeling in fluid/solid systems", Progress in Energy and Combustion Science, Vol 28, pp. 543-602.

Ozawa, M., Umekawa, H., Furui, S., Hayashi, K., Takenaka, N., 2002, "Bubble behavior and void fraction fluctuation in vertical tube banks immersed in a gas-solid fluidized-bed model", Experimental Thermal and Fluid Science, Vol 26, pp. 643-652.

Pannala, S., Daw, C. S., Finney, C. E. A., Benyahia, S., Syamlal, M., O´Brien, T. J., "Modelling the collisional-plastic stress transition for bin discharge of granular material", 2009, Powders and Grains 2009 - Proceeding of the 6th International Conference on Micromechanics of Granular Media, pp. 657-660.

Schaeffer, D. G., 1987, "Instability in the evolution equations describing incompressible granular flow", Journal Differential Equations, v. 66, pp. 19-50.

Siravastava, A., Sundaresan, S., 2003, "Analysis of a frictional–kinetic model for gas–particle flow", Powder Technology, v. 129, pp. 72-85.

Syamlal, M., 1998, "MFIX Documentation, Numerical Techniques", Technical Note, DOE/MC-31346-5824, NTIS/DE98002029, National Technical Information Service, Springfield, VA, USA.

Syamlal, M., Rogers, W. A., O'Brien, T. J., 1993, "MFIX Documentation, Theory Guide", Technical Note, DOE/METC-94/1004, NTIS/DE94000087, National Technical Information Service, Springfield, VA, USA.

Wang, J., van der Hoef, M. A., Kuipers, J. A. M., 2010,"CFD study of the minimum bubbling velocity of Geldart A particles in gas-fluidized beds", Chemical Engineering Science, 65, pp. 3772-3785.

Wen, C. Y., Yu, Y. H., 1966, "Mechanics of Fluidization", Chemical Engineering Progress Symposium Series, Vol 62, n. 62, pp. 100-111.

Wiman, J., 1994, An experimental study of hydrodynamics and tube erosion in a pressurized fluidized with horizontal tubes. Thesis for the degree of Licentiate of Engineering, Chalmers University of Technology, Goteborg. Sweden.

Wiman, J., Almstedt, A. E., 1997, "Hydrodynamics, erosion and heat transfer in a pressurized fluidized bed: influence of pressure, fluidization velocity, particle size and tube bank geometry", Chemical Engineering Science, Vol 52, pp. 2677-2695.

Wong, Y. S., Seville, J. P. K., 2006, "Single-particle motion and heat transfer in fluidized beds", AIChe Journal, Vol 52, pp. 4099-4109.

15

Optimal Synthesis of Multi-Effect Evaporation Systems of Solutions with a High Boiling Point Rise

Jaime Alfonzo Irahola
Universidad Nacional de Jujuy
Argentina

1. Introduction

In the past, optimization had been studied only for typical flowpatterns like forward and backward feed. To make the choice between them, simple rules were applied based on the viscosity and the temperature of the initial dilute solution (T_F). Thus, forward feed was usually favored for the evaporation of low-viscous hot solutions featuring a temperature $T_F > T_p$, where T_P is the desired temperature of the final product. By doing so, the liquid heating load is largely cut down. In turn, backward feed was recommended for heavy-viscous cold liquors.

Moreover, a few contributions to the optimal synthesis of multiple effect evaporator systems (MEES) have so far been published. Most of the previous papers was focused on the analysis rather than the synthesis of evaporation systems. They generally assumed that important structural variables like vapor and liquid flowpatterns and the number of effects are all known data though they drastically change the performance of a MEES.

Nishitani and Kunugita (1979) first presented a multiobjective problem formulation to determine the optimal flowpattern of a multiple-effect evaporator system. However, they did not consider stream mixing/splitting. In addition, the solution method performed one-by-one the simulation of the MEES for all possible flowpatterns. More recently, Hillebrand and Westerberg (1988) developed a simple model to explicitly compute the utility consumption for multiple-effect evaporator systems exchanging sensible heat with outside streams. In turn, Westerberg and Hillebrand (1988) introduced the concept of "heat shunt" to derive the best liquid flowpattern in a heuristic way. Nonetheless, major assumptions like constant boiling point elevation, no liquid bypassing and negligible heat of mixing somewhat limit the applicability of their findings.

To resolve the problem presented here has been used a mathematical model rigorous previously developed (Irahola & Cerdá, 1994). It considers the possibility of optimizing the variables that you want to. This has allowed that the model is used to solve various types of problems, namely: simulation, optimization, optimal synthesis and optimum partial reengineering restrictive of multi effect evaporation system (MEES) (Irahola, 2008). The mathematical model developed is the MINLP and solved using commercial software. The approach was successfully applied to three industrial problems. Depending on the feed and the product temperature, the optimal configuration uses a distinct liquid flowpattern that often differs from the conventional forward and backward feed and leads to reasonable savings (Irahola & Cerdá, 1996)

Among the results should be noted that: the splitting of the flow of live steam can be a better alternative than the traditional cascade of steam; the best fixed cost curve is not always a monotonous increasing; the correct distribution of the areas of the effects of the MEES, the appropriate operating conditions and the correct choice of liquid and vapor flowpatterns, are the determining factors in the optimal design of the MEES.

Perhaps the greatest disadvantage found using the method proposed here to solve the formulated mathematical model (MINLP type) for optimal synthesis of the MEES is the presence of numerous local optimal what makes it difficult to obtain optimal Global.

2. Evaporation of an aqueous solution of caustic soda

Among the solutions of industrial interest that present a high increase in boiling point are sodium hydroxide solutions (caustic soda). The concentration of these substances by evaporation, presents significant disadvantages due to the characteristics of the caustic solutions, namely:

- Have a high boiling point elevation (BPE) which implies a great loss in the temperature difference available.
- Concentrated solutions are highly viscous, which severely reduces the rate of heat transfer in natural circulation evaporators.
- They can have detrimental effects on steel, causing what is called caustic fragility. In addition, they may require removal of large amounts of salt when the solution is concentrated.

Since the transfer of heat (U) of liquor film coefficient, depending on the speed of the caustic solution through tubes (among other variables), usually, seeks a high speed in order to obtain a large coefficient (Kern, 1999).

According to the literature, it has taken as standard for the concentration of caustic soda, a evaporation system of two or three effects operate in backward feeed (Kern, 1999). In this study, found that the structure in counterflow or backward feed, obviously presents a high performance, but is not the best. In order to confirm what was said, is going to solve a problem.

3. Optimal synthesis of a multi-effect evaporation system for the concentration of caustic soda

3.1 Problem

Find the optimal MEES to concentrate 30040 lb/h (13626 kg/h) of an aqueous solution of sodium hydroxide from 10 to 50% by weight. The type of used evaporator is long vertical tube with forced circulation. Available in the plant: live steam boiler to 63.69 Psia (4.48 kg/cm2). The allowable minimum absolute pressure in an effect is of 1.942 Psia (0.1365 kg/cm2) (Geankoplis, 1983).

3.2 Mathematical model

The scope of the rigorous mathematical model is limited by the following assumptions adopted in the formulation:

a. In each effect, the vapor and liquid phases are in equilibrium.
b. A solid phase never arises in any effect.
c. The impact of the hydrostatic head on the liquid boiling point is neglected.
d. There is no leakage or entrainment.
e. Heat losses from any effect need not be considered.
f. The steam always condenses completely.
g. Subcooling of the condensate is very small.
h. Flow of noncondensables is negligible.
i. The concentrated final product is withdrawn from a single effect which in turn does not transfer liquid to any other one.
j. If necessary, you can use a heat exchanger or condenser so that the product go out to the preset temperature (Tp).
k. Not consider any type of pump between the effects.

To solve the mathematical model and find the optimal design should be available before, the mathematical expressions for all dependent variables: enthalpy of steam (H), enthalpy of solution of soda caustic (h), latent heat of vaporization of the water (λ), overall heat transfer coefficient (U), temperature of the solution in the evaporator (T) and costs of forced circulation evaporator, barometric condenser multijet, surface condenser and heat exchanger. In general, useful information is available in graphics, which presented various authors cited in the bibliography, but there are no equations of those curves. These situations and other problems are resolved below. According to Standiford (1963), in forced circulation evaporators, film transfer coefficient (h) on the side of the liquid, can be calculated from the conventional Dittus-Boelter equation for forced circulation when there is no boiling.

$$\frac{hD}{k} = 0.0023\left(\frac{DG}{\mu}\right)^{0.8}\left(\mu\frac{Cp}{k}\right)^{0.4} \tag{1}$$

If there is this equation for two points and combine both equations, you can find the functionality of $\mathbf{h_1}$ with respect to another point (2) as:

$$h_1 = h_2\left(\frac{\mu_2}{\mu_1}\right)^{0.4}\left(\frac{C_{p1}}{C_{p2}}\right)^{0.4} \tag{2}$$

As the overall heat transfer coefficient U is practically determined by the film coefficient h fluid side, the above equation can be used to obtain a correlation for U. In Geankoplis for T=105 °F and X = 0.5, data is U = 400 (Btu/h ft^2 °F). T and X is obtained from a graph (Horvath, 1985) μ_2 = 22.84 centipoise. Then, for five values of concentration and six temperature values are obtained from graphics (Horvath, 1985) the values of Cp_1 and μ_1. With the data obtained can be calculated according to the above expression, the overall heat transfer coefficient U_1 with reference to U_2. Thus, U is plotted vs. X (Fig. 1) and U vs. T (Fig. 2). Finally, using the triple X, T, U_i, can be found by regression, the functionality of U in terms of concentration and temperature:

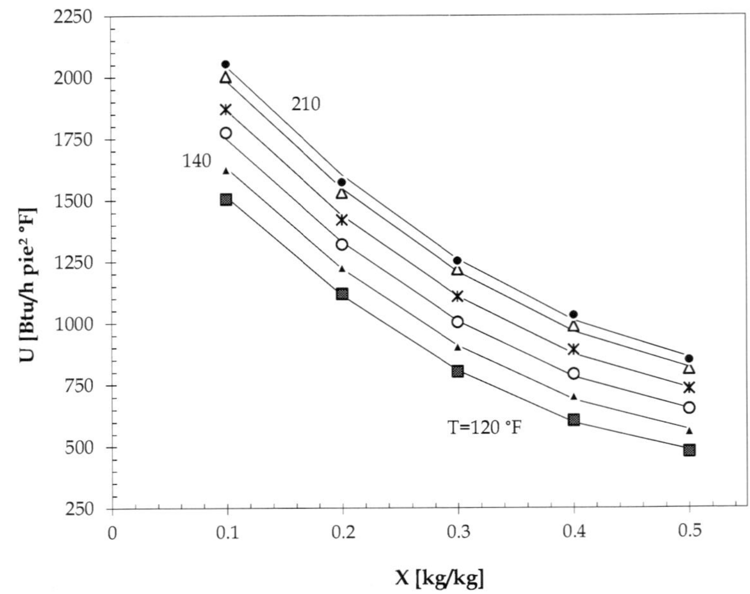

Fig. 1. Overall heat transfer coefficient correlation as a function of solute concentration.

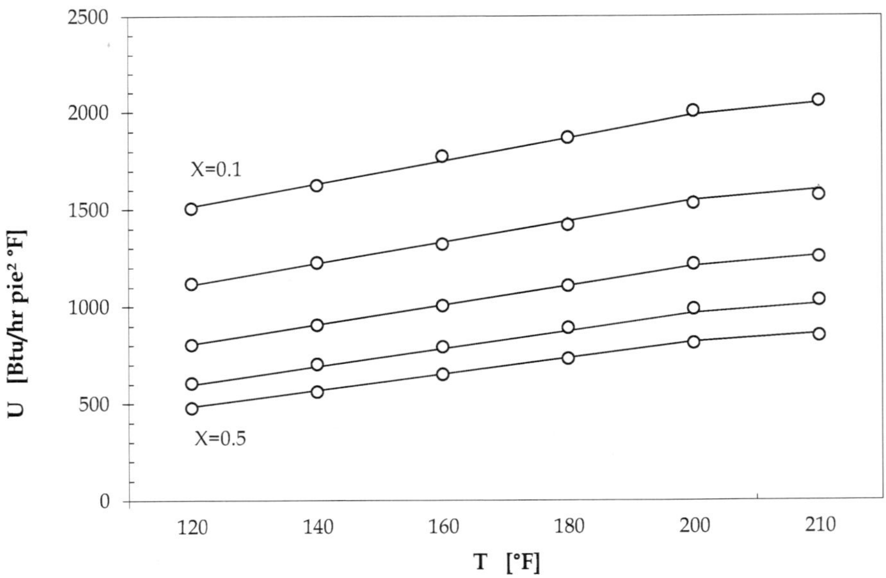

Fig. 2. Overall heat transfer coefficient correlation as a function of temperature.

$$U = 1254.780865 - 4954.6700\,X + 4832.059524\,X^2 + 6.321549\,T - 4.30974\,XT \left[\frac{Btu}{h\,ft_2\,°F}\right] \tag{3}$$

where: X: mass fraction and T [=] ° F

The regression function is chosen to achieve maximum correspondence with the data, but at the same time trying to maintain, if possible, the simplicity. Nevertheless, we could not avoid the bilinear term. Figures 1 and 2, you can appreciate the good fit of the correlation found for U.

Also, to represent the enthalpy of solution, depending on the concentration and temperature, it became a non-linear regression with data from Foust (1980) (Fig. 3):

$$h = -10.250000 - 319.591837\,X + 939\,795918\,X^2 + 0.963929\,T - 0.335714\,XT \left[\frac{Btu}{lb}\right] \tag{4}$$

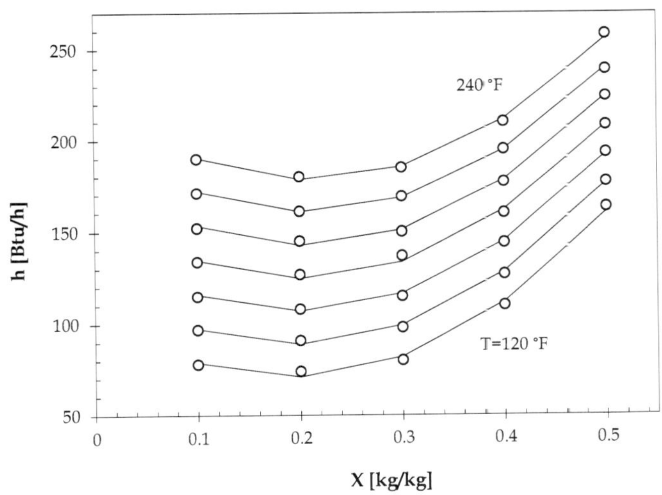

Fig. 3. Enthalpy of solution

The functional expression of the cost of barometric condenser multijet (C_{CBM}) is obtained by regression analysis of the chart presented by Peters (1991). Again, searching for mathematical expressions that minimize non-convexities of the mathematical model, we found that a quadratic expression represents excellent, the cost curve in the range of work desired (Fig. 4) (1992):

$$C_{CBM} = 4,753104 + 2,480885\,w + 0,281818\,w^2 \;\left(10^3\ USD\right) \tag{5}$$

Where: w [=] gpm. The flow rate of cooling water (w) required in the condenser is directly given by (Kern, 1999):

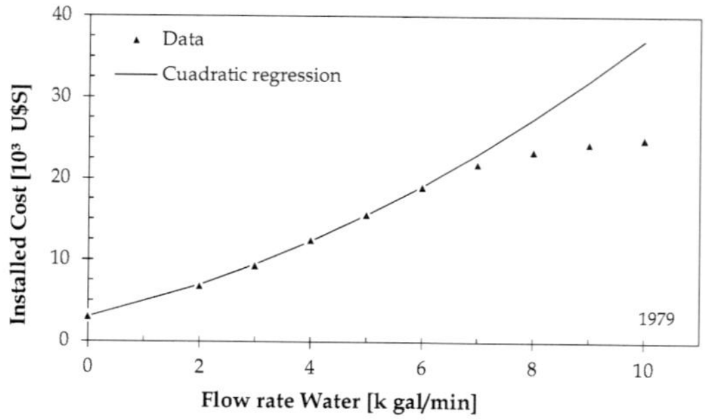

Fig. 4. Installed Cost of multijet barometric condenser.

$$w = \frac{Q}{500(T_s - T_w - t_a)}(\text{Gpm}) \tag{6}$$

where:

Q = heat load, Btu / h
T_s = temperature of saturated steam, °-F
T_w = temperature of cooling water, °-F
t_a = 15 °-F = degree of approximation at T_s

The cost of forced circulation evaporator is obtained based on information reported by Maloney, 2008:

$$CE_{FC} = 2420.5\ A^{0.7121}\ U\$S \tag{7}$$

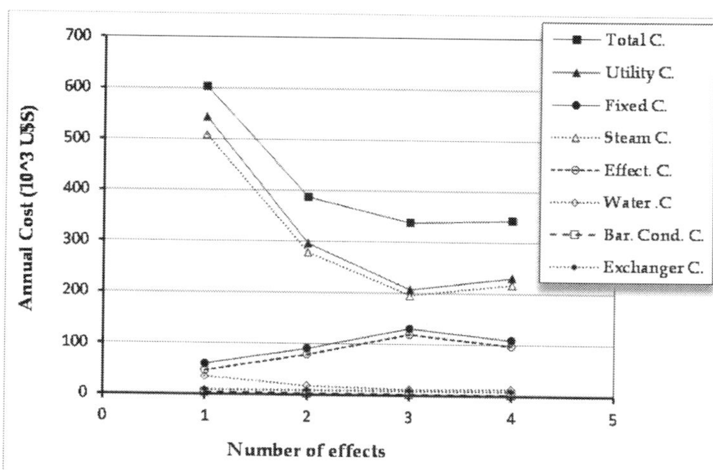

Fig. 5. Evolution of total operating costs.

The information of the data used in the resolution of the problem is presented in table 1. Also, the thermodynamic properties of the remaining functions and costs of services and equipment are presented.

In order to consider probable situations that could be presented in the industry, will study the cases in which the temperature of the weak solution (feed) is equal, higher or lower than the temperature of the strong solution (product). In the three cases, only change the values of the inlet of the weak solution temperature. The rest of the parametric conditions remain fixed.

<table>
<tr><td colspan="6">Feed and product conditions</td></tr>
<tr><td colspan="2">Item</td><td colspan="2">Feed</td><td colspan="2">Product</td></tr>
<tr><td colspan="2">Flow rate lb/h (Kg/h)</td><td>30040</td><td>(13626)</td><td>60080</td><td>(2725.2)</td></tr>
<tr><td colspan="2">Weight fraction</td><td colspan="2">0.1</td><td colspan="2">0.5</td></tr>
<tr><td rowspan="3">Temperature °F (°C)</td><td>Case I</td><td>180</td><td>(82.2)</td><td rowspan="3">130</td><td rowspan="3">(54.4)</td></tr>
<tr><td>Case II</td><td>130</td><td>(54.4)</td></tr>
<tr><td>Case III</td><td>80</td><td>(26.7)</td></tr>
<tr><td colspan="6">Operation conditions</td></tr>
<tr><td colspan="2"></td><td colspan="4">Temperature °F (°C)</td></tr>
<tr><td colspan="2">Steam</td><td colspan="4">296.6 (147.0)</td></tr>
<tr><td colspan="2">Cool water</td><td colspan="2">(min) 89.6 (32.0)</td><td colspan="2">(max) 107.6 (42.0)</td></tr>
<tr><td colspan="2">Effect</td><td colspan="2">(min) 125.0 (51.7)</td><td colspan="2">(max) 294.6 (145.9)</td></tr>
<tr><td colspan="2"></td><td colspan="4">Minimum allowable temperature difference
°F (°C)</td></tr>
<tr><td colspan="2">Condensers</td><td colspan="4" rowspan="2">18.0 (10)</td></tr>
<tr><td colspan="2">Heat Exchangers</td></tr>
<tr><td colspan="6">Thermodynamic properties</td></tr>
<tr><td colspan="2">Specific vapor enthalpy</td><td colspan="4">$H_i = 1075 + 0.3466\ Tv_i$ [Btu/lb]</td></tr>
<tr><td colspan="2">Solvent Latent Heat of Vaporization</td><td colspan="4">$\lambda_i = 1104 - 0.65\ Tv_i$ [Btu/lb]</td></tr>
<tr><td colspan="2">Operating temperature at effect i</td><td colspan="4">$T_i = (1 + 0.1419526\ X_i)Tv_i - 9.419608\ X_i + 271.3627\ X_i^2$ [°F]</td></tr>
<tr><td colspan="6">Costs of utilities and equipment</td></tr>
<tr><td colspan="2">Steam</td><td colspan="4">$2.922\ 10^{-6}$ (USD/Btu) 2.104E-2</td></tr>
<tr><td colspan="2">Cool water (305 K)</td><td colspan="4">$1.952\ 10^{-7}$ (USD/Btu) 1.405E-3</td></tr>
<tr><td colspan="2">Surface condenser</td><td colspan="4">$1092.83\ A^{0.65}$(USD)</td></tr>
<tr><td colspan="2">Heat exchanger</td><td colspan="4">$1144.16\ A^{0.65}$(USD)</td></tr>
</table>

Table 1. Data for example and, functional expression of thermodynamic properties and the costs of utilities and equipment.

4. Analysis and discussion

4.1 Case I. Feed temperature higher than the temperature of the product ($T_F>T_P$)

4.1.1 Comparative analysis of the optimal solution found

Adopted T_F = 180 °F (82.2 °C). Coinciding with the generally accepted criterion of optimality for the evaporation of caustic soda, has been found that the optimum number of effects of the MEES is equal to three (Fig. 5). However, a mixed structure {2,1,3} has been found in the path of the current liquid instead of backward feed. The feed stream enters to the effect 2 and then continuous countercurrent to the effect 1. Then go out and circulates in forward feed to the effect 3. (mixed liquid flowpattern). In this new structure that is presented (Fig. 6), we see significant increases in boiling point of the solution: 10.5 °C and 40 °C, in effects 1 and 3 respectively.

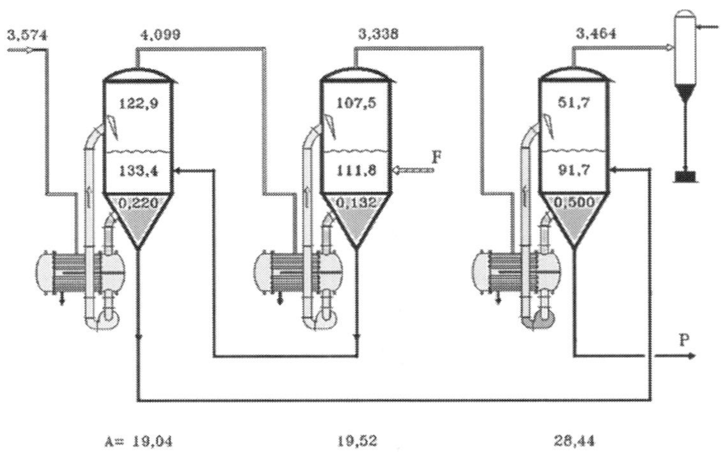

Fig. 6. Optimal configuration three-effect MEES. (TAC=337835 USD)

The optimal solution will be the one with the lowest total annual cost (TAC). From this point of view, the classical structure proposed as an optimum solution: evaporation system of three effects of equal area arranged backward feed (BFA), is 4.31% more expensive than the optimal solution (SO) (Fig. 7). More, even if it is allowed to in the structure backward feed, the effects have distinct areas (BF), do not get a better result that the optimal solution found. The difference in cost is 4.03%. In the figure 7, we also present results for a forward feed evaporation system. This structure, in its classic form forward feed with effects of equal area (FFA), is 7.91% more expensive. Which could corroborate, in some way, because in the past the BFA structure was preferred. If is allowed that the effects have different areas in the structure forward feed (FF) is very interesting the result obtained. The correct distribution of areas has led to a decrease in the total cost. The decrease is so great that, now, the FF structure is better than any of the above structures in backward feed. Its total annual cost (TAC) is only 2.95% greater than the of the optimum solution found (Fig. 7). In general, it appears that whatever the structure of MEES, the operating cost is significantly greater than the fixed cost (Fig. 8). It is approximately 65% of the total cost.

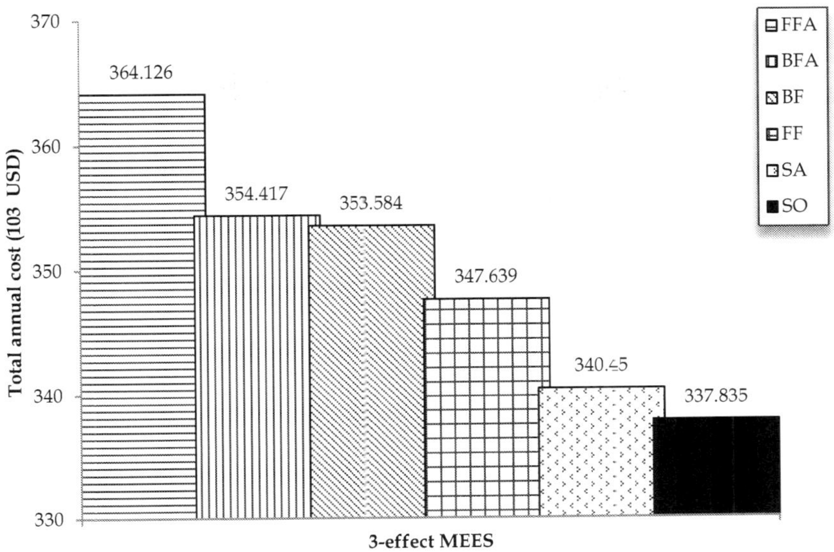

Fig. 7. Comparison with Typical Flowpatterns.

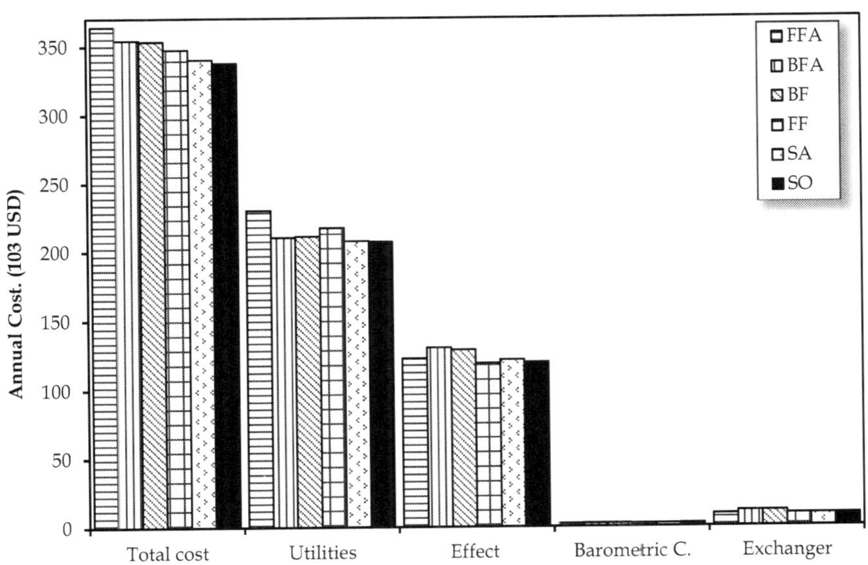

Fig. 8. Relative incidence of Operating and fixed Costs

4.1.2 Impact of the flow pattern

In this case, the trajectory of the liquid stream is the determining factor in the performance of a given MEES. The benefit achieved is even greater than obtained by allowing the effects having different areas with each other. That said, is based on the result of the structure SA. This has the same flow pattern that the optimal solution, but the effects of evaporation system are of equal area. The increase in cost is only 0.8% compared to MEES optimum (Fig. 7). From the practical point of view, the alternative SA may be the best option.

4.1.3 Profiles of the structural and parametric variables

As it will be seen later, only in this case it can be seen some regularity in the curves of the structural and parametric variables. Furthermore, after reaching the optimal point generally the next curve is anomalous with respect to the preceding ones. In the last effect occurs the maximum concentration jump (Δ X) of the solution (Fig. 9). At the same time, it has the maximum area as shown in the curves of the 1 to 4-effect optimal MEES (Fig. 10). The flow rate of steam produced in the effects is approximately the same. However, this does not apply to MEES with greater number of effects than the optimal. (Fig. 11). The profile of the temperature to the optimum MEES of a different number of effects, does not have a regular aspect. However, the maximum temperature jump occurs in the last effect (Fig. 12).

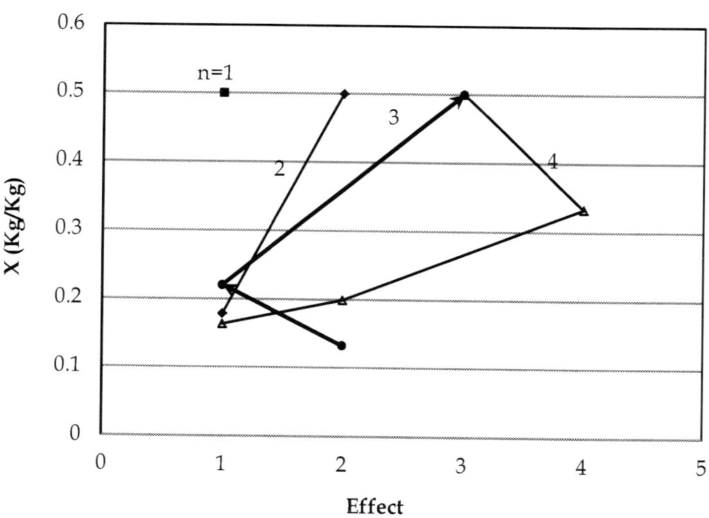

Fig. 9. Concentration profiles in the 1 to 4-effect MEES. (n: number of effects)

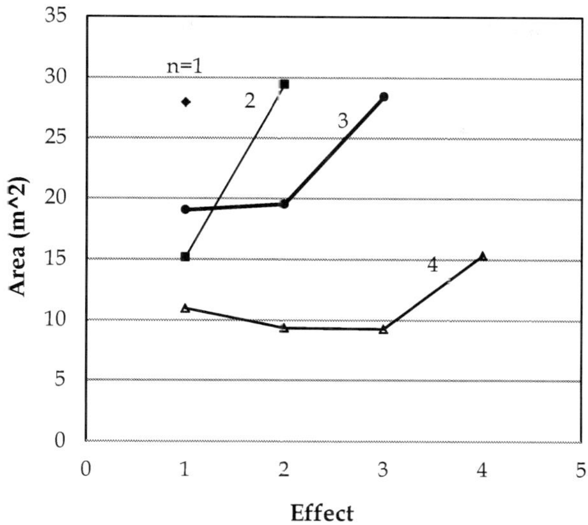

Fig. 10. Area profiles in the 1 to 4-effect MEES (n: number of effects).

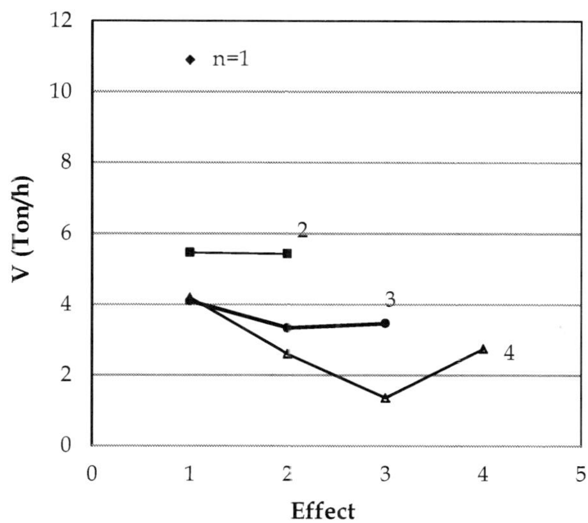

Fig. 11. Flow rate profiles in the 1 to 4-effect MEES (n: number of effects).

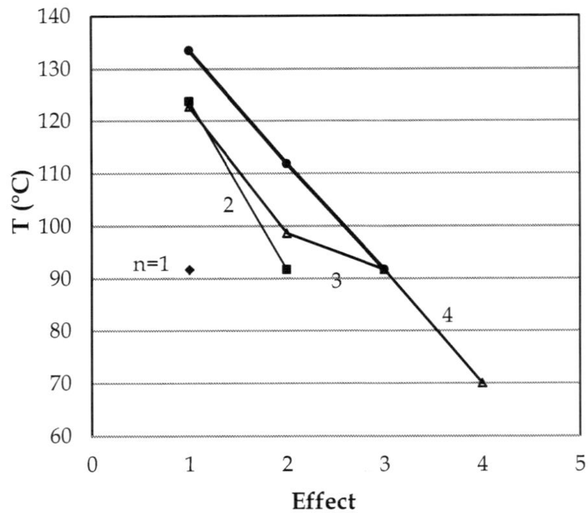

Fig. 12. Temperature profiles in the 1 to 4-effect MEES (n: number of effects).

4.2 Case II. Feed temperature equal than the temperature of the product ($T_F = T_P$)

4.2.1 Impact of the splitting of the live steam flow rate on the optimal solution

The result obtained when TF = TP, is different, not only to the found for the case I, but also with respect to the classical position. Found structure is highly innovative and simple in its conception.

If you look at the evolution of the total annual cost (TAC) curve of the 1, 2, 3, 4 and 5 effect Optimal MEES, it was found that the four-effect MEES is that of lower cost (optimal quasi-global) (Fig. 13). The flow pattern is backward feed and in this aspect, this result coincides with the classical motion, but not with the number of effects: proposed here a four-effect MEES, instead of three. However, this new proposal, would not be really the best alternative, if it was not associated to the new steam flow pattern proposed (Fig. 14). It emphasizes, splitting in the live steam flow pattern, it enters parallel to the effects 1 and 2; the by-passing effect 2 by the vapor stream from effect 1 and finally, the mixing of vapor streams from effects 1 and 2 for heating effect 3.

Against, this new trajectory of the flow of steam, first doubt that occurs, is the performance of this configuration against the unifilar cascade of high thermodynamic efficiency.

4.2.2 Comparative study of proposed flow pattern with respect to traditional configurations

For the purpose of explaining the improvement achieved, compares the structure backward feed with effects of different areas (BF) and the optimal solution (SO). Both structures have equal number of effects and same trajectory of the liquid flow and only differ in the

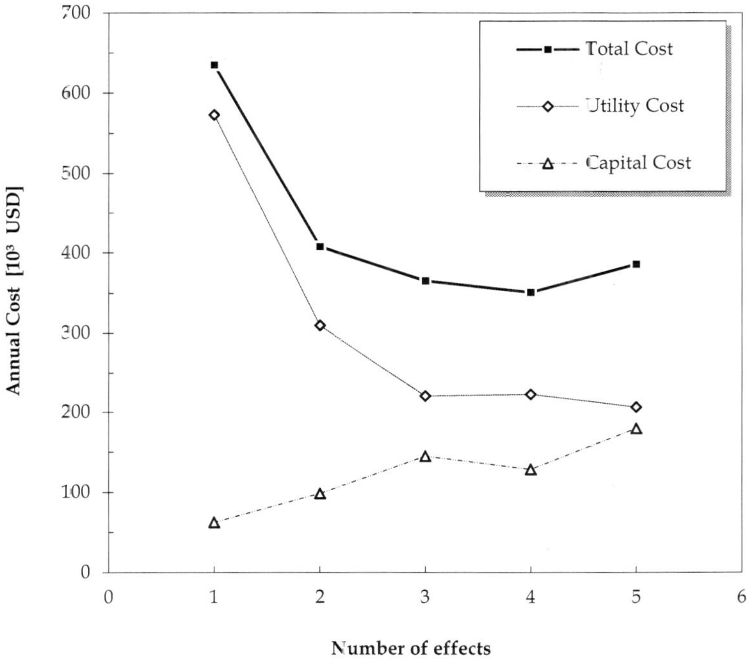

Fig. 13. Relative incidence of Operating and fixed Costs.

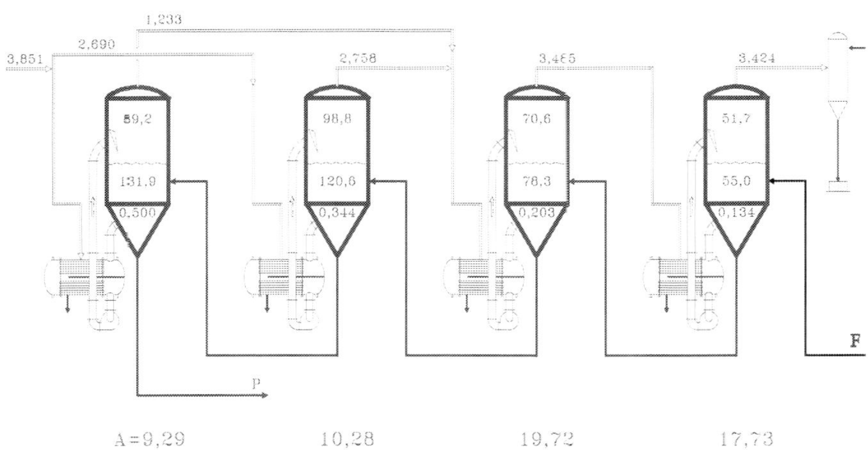

Fig. 14. Optimal configuration four-effect MEES. (TAC=351082 USD)

trajectory of the flow of steam. Two comparisons were made: one relating to the cost of auxiliary services and the other with respect to the cost of the effects, which is almost all of the fixed cost.

4.2.3 Energy efficiency and fixed cost of the traditional structure

The BF structure has a higher efficiency since the cost of auxiliary services is 174963 USD, 21.30% less than the cost for the **SO**. On the other hand, the cost of the effects is 193008 USD, i.e. 167,8% of the respective cost observed in SO. The net result of the comparison of the total costs, indicates that the BF structure is 8.9% more expensive than SO. With these results, following the classical position we can say, that BF 4 effects is not better than the optimum found (SO), because the MEES should be a structure BFA 3 effects, not four, which was used to compare. Therefore, will be then verified the validity of this rule, for the case study.

4.2.4 Difference in the number of effects due to non-traditional flow patterns

The optimal number of effects found by the mathematical model does not coincide with the optimal number for BFA and FFA traditional structures or even structures BF and FF. This explains why in the absence of a mathematical model to explore the multiple alternatives of design, the best answer to the problem was until now, a countercurrent system.

It was found that the optimum number of effects to structures backward feed and forward feed is 3. However, the developed model proposes the structure SO of 4 effects as the best solution. Therefore, to verify the quality of it, is advisable to compare the best results found for each structure.

4.2.5 The optimal solution compared to traditional structures

The total cost of the BFA MEES is 4.2% more than the optimal. Therefore, the structure and number of effects, traditionally proposed do not seem to be the most appropriate. Then, one might think that if you remove the restriction of equal area of the effects, could be improved, significantly, the current result. The results show that the BF structure of three effects is 4.1% more expensive than SO (Fig. 15). However, despite the difference in the number of effects, is convenient to analyze in more detail these recent results.

Structurally, BF and SO differ only in the flow pattern of steam. The cost of the auxiliary services of BF is 1.0% lower than the SO. On the other hand, the fixed cost is 13.4% greater determining to SO submit one minor TAC (Fig. 16).

4.2.6 Profiles of the structural and parametric variables

This case is characterized because the profiles of the process variables, for the various intermediate optimal MEES, they have no similarity among themselves. In particular, notes that the optimal solution presents the most discordant curve with respect to the others.

The temperature profile is irregular with temperature differences between effects non-uniform, being the most important jump located between 2 and 3 effect, following the drop of temperature effects 3 and 4, both heated with secondary steam. (Fig. 17). The greater temperature difference between the heating steam and the solution to evaporate, occurs in

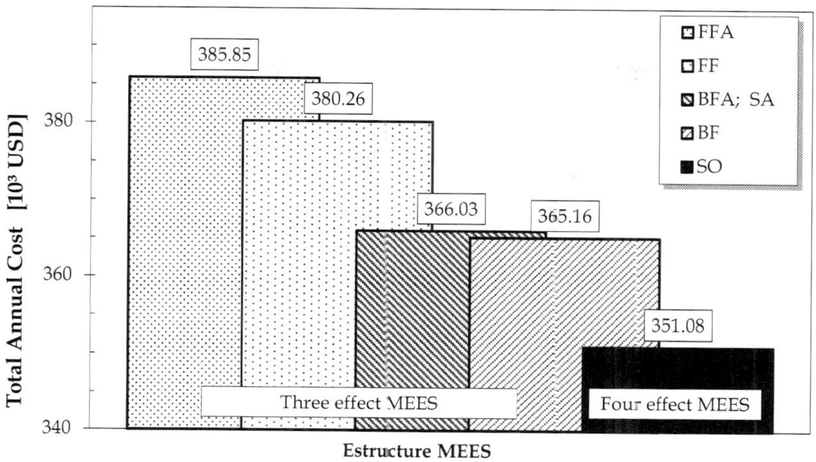

Fig. 15. Comparison with Typical Flowpatterns.

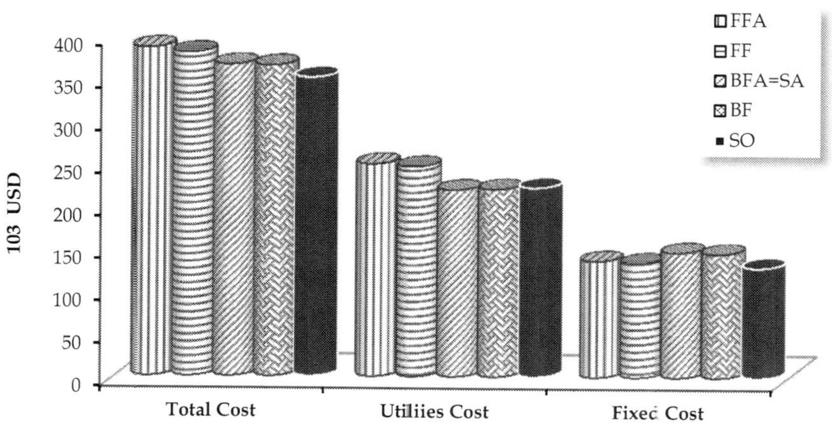

Fig. 16. Comparison with Typical Flowpatterns (Case II).

the effect 2 of the four effect MEES (Fig. 18), through the use of live steam in the effect. Thermal jumps that are achieved with the optimum structure are higher that in the triple and quintuple-effect MEES. In addition, it should be noted that the thermal jump in effect 2 is almost doubled with respect to other effects.

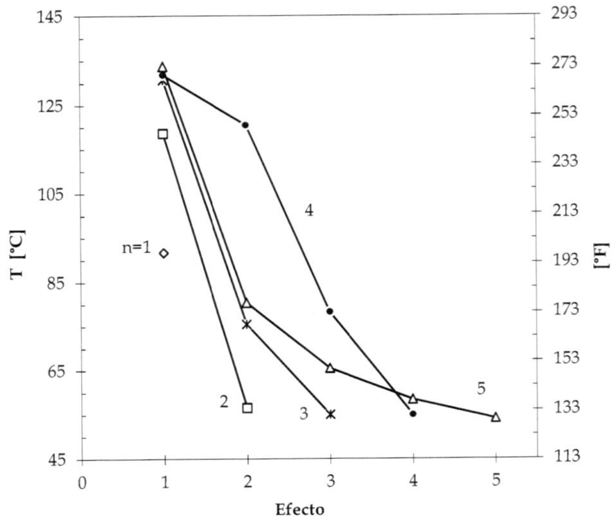

Fig. 17. Temperature profile optimal MEES for n=1 to 5 effects.

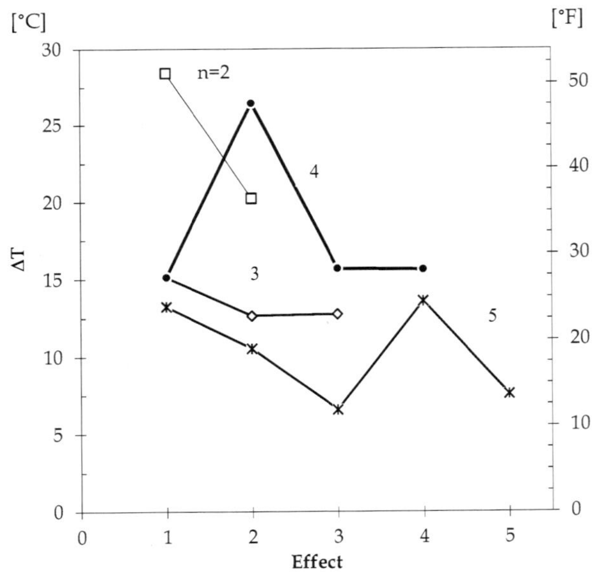

Fig. 18. Temperature difference profiles between the condensation and evaporation chambers in each effect for 2, 3, 4 and 5 effects.

Similarly, the profile of the area for each MEES, is far from being uniform, with the highest values located in the lower thermal effects. (Fig. 19). Should be mentioned, that the problem had been resolved for a fixed range area (100, 1000 ft²).

The flow rate of solvent evaporated in each effect is approximately the same in the double and triple-effect MEES (Fig. 20), where the chosen structure is backward feed. But in the optimal solution and after this, the values of vapor flow rate are far from each other. Although it can be seen that, in SO the curve is regular and decreasing with temperature

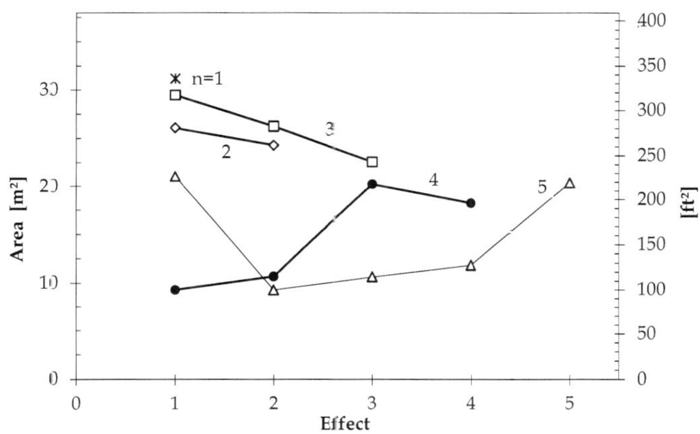

Fig. 19. Area profiles in the 1 to 5 effect MEES (n: number of effects)

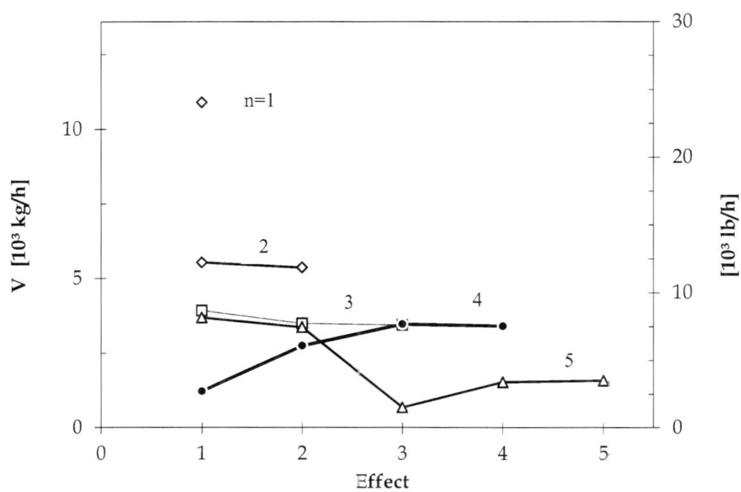

Fig. 20. Flow rate profiles in the 1 to 5-effect MEES (n: number of effects).

effects, i.e. the greater evaporated flow rate occurs in the effect 4. The profile of the global coefficient U is similar: lower in effect 1 and higher in the effect 4. It is important to clarify that, the values of the flow rate of steam produced in each effect are not directly indicative parameters. Yes it is a relative measure, for example, the percentage of the current liquid evaporates. Thus, the percentage amount of solvent evaporated, with respect to the flow rate of solvent that enters each effect, is maximum in the second effect (41%), followed in decreasing order, the third effect (34%), the first (31%) and finally the fourth (25%).

The concentration curve of the solution is monotone increasing, considering the path of the liquid stream (Fig. 21). The biggest jump in percentage concentration occurs in the second effect (69 %), then in descending order, the third effect (51 %), the first (45 %) and the fourth (34 %).

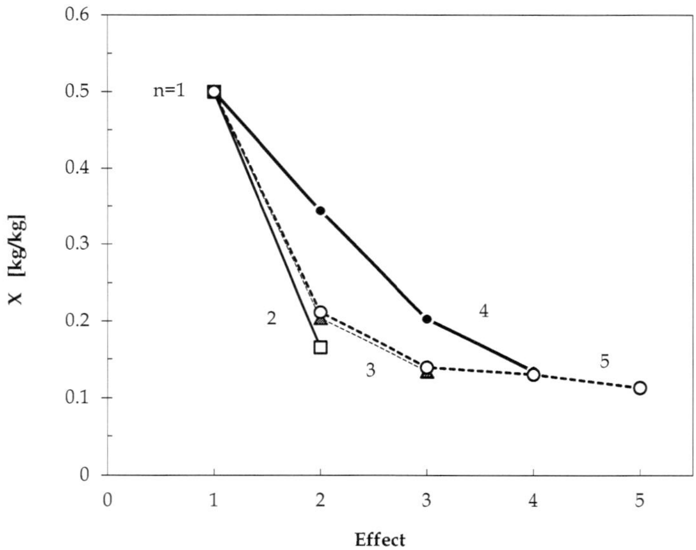

Fig. 21. Concentration profiles in the 1 to 5-effect MEES

The recent analysis of incremental concentration, the carried out for the steam produced in the effects, and the biggest jump thermal observed in effect 2 of the Optimal MEES show the importance of this effect. Its presence is the root cause of the improvement achieved in SO. This is achieved, thanks to the optimal design of the MEES, which allows an adequate relationship of the variables that define the system.

4.3 Case III. Temperature of the weak solution lower than strong solution ($T_F < T_P$)

4.3.1 Uniqueness of the cost curves

This case presents great similarity with the Case II. The optimal effect is the same (4) and intermediate structures found for n ≤ 4 are almost identical. The best configuration for the liquid stream, given by {4,3,2,1}, as in Case II, use live steam to heat the first two effects.

Evolution of cost curves show a singularity with respect to the previously analyzed cases. Now, to increase the number of effects to reduce the TAC, the consumption of live steam begins to increase (Fig. 22) rather than continue to decline with the addition of a new effect, and the cost of the effects decreasing, rather than continue to increase. Isn't traditional behavior: a curve monotone decreasing for the cost of the live steam and one monotone increasing for the cost of the effects. Here is not complied with this scheme because of the significant reduction of the U coefficient and the high increase of the boiling point (BPE) in the effects of greater concentration. It is now "pays" with live steam part of the savings in the areas of effects.

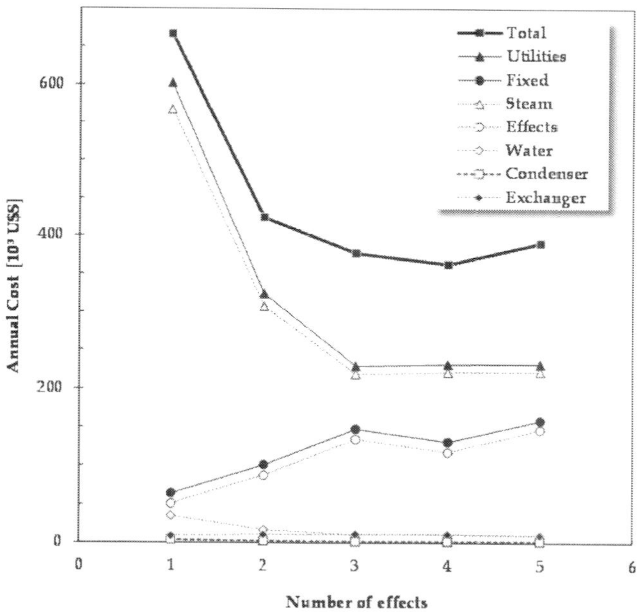

Fig. 22. Evolution of the TAC and its component terms with the number of effects (Case III).

4.3.2 Structure and distribution of temperature, concentration and heat transfer area

A structural analysis allows us to appreciate that the cuadruple-effect MEES results have added an effect, between the first and the second of the triple-effect MEES (Figs. 23 and 24, Table 2.).

Looking to reduce the total cost, the new effect requires one of the lowest values of area of heat transfer (Fig. 25). For this purpose, it operates with a large temperature difference; almost double that for the remaining effects (Fig. 26). On the other hand, contrary to expectations, their presence causes a slight increase in the consumption of live steam and allows at the same time, an increase in the thermal jump in the last two effects of, approximately, 6 °F (3.3 °C).

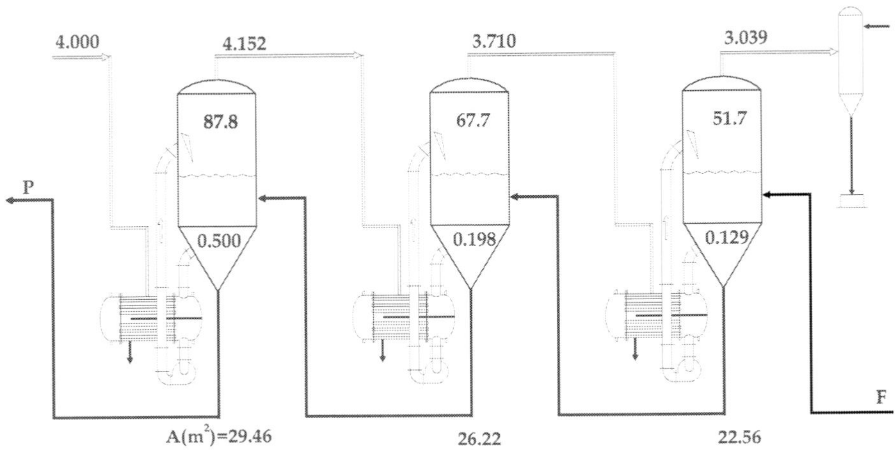

Fig. 23. Optimal Structure for three effects. (Local optimun).

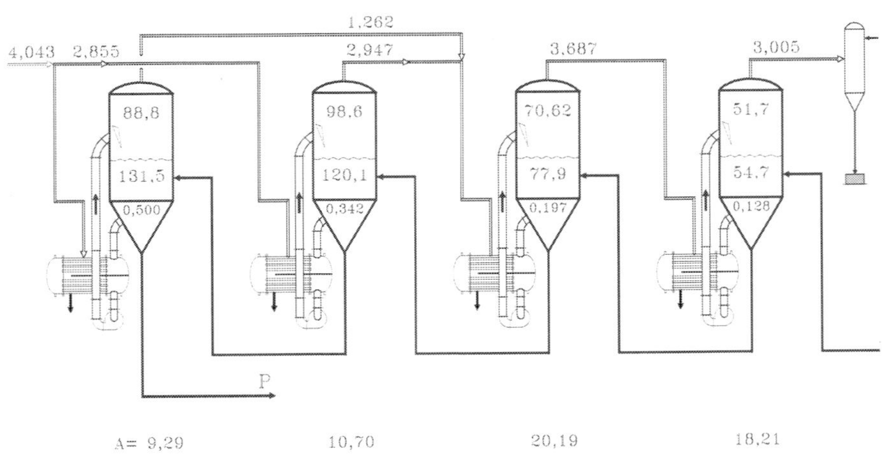

Fig. 24. Optimal Structure for four effects. (Optimal Solution).

Alternative Optimal Solutions (Case III)							
Equal area effects MEES: SA							
Effect	**T**	**Tv**	**Vs**	**L**	**V**	**X**	**F**
(i)	**[°C]**		**[10^3 kg/h]**			**[kg/kg]**	**[m^2]**
1	137.99	94.92	4.105	0	3.232	0.500	
2	128.93	107.28	0	0	2.979	0.340	15.33
3	80.62	73.39	0	0	3.665	0.195	
4	54.72	51.67	0	13.626	2.971	0.128	
Backward feed MEES. Optimal area effects: BF							
Effect	**T**	**Tv**	**Vs**	**L**	**V**	**X**	**F**
1	130.40	87.83	4.000	0	3.232	0.500	29.47
2	74.99	67.71	0	0	3.710	0.198	26.22
3	54.76	51.67	0	13.626	3.039	0.129	22.57
Backward feed MEES. Equal area effects: BFA							
Effect	**T**	**Tv**	**Vs**	**L**	**V**	**X**	**F**
1	128.39	85.95	3.984	0	4.126	0.500	
2	73.12	65.84	0	0	3.719	0.199	26.54
3	54.77	51.67	0	13.626	3.055	0.129	
Forward feed MEES. Optimal area effects: FF							
Effect	**T**	**Tv**	**Vs**	**L**	**V**	**X**	**F**
1	132.77	127.95	4.822	13.626	3.517	0.135	20.37
2	116.53	107.02	0	0	3.714	0.213	18.96
3	91.67	51.67	0	0	3.671	0.500	32.69
Forward feed MEES. Equal area effects: FFA							
Effect	**T**	**Tv**	**Vs**	**L**	**V**	**X**	**F**
1	135.12	130.25	4.870	13.626	3.530	0.135	
2	121.62	111.99	0	0	3.691	0.213	24.40
3	91.67	51.67	0	0	3.680	0.500	

Table 2. Results of alternative evaporation systems. (Vs: Live steam flow rate. V: Secondary Steam Flow rate)

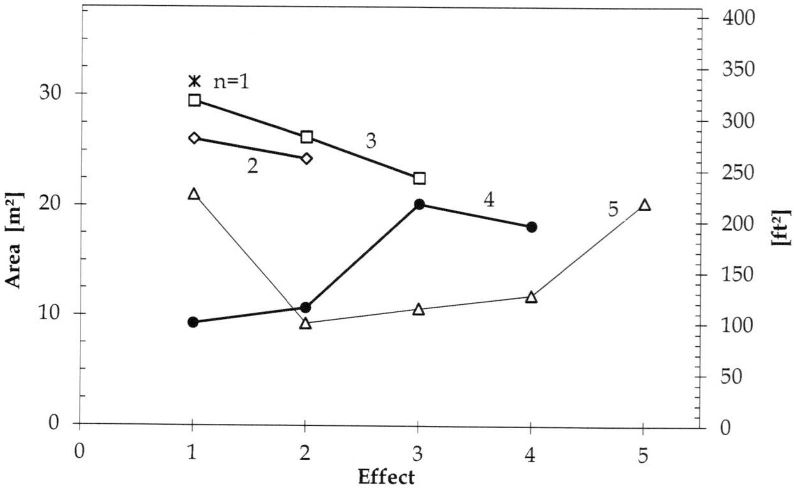

Fig. 25. Area profiles in the 1 to 5-effect MEES.

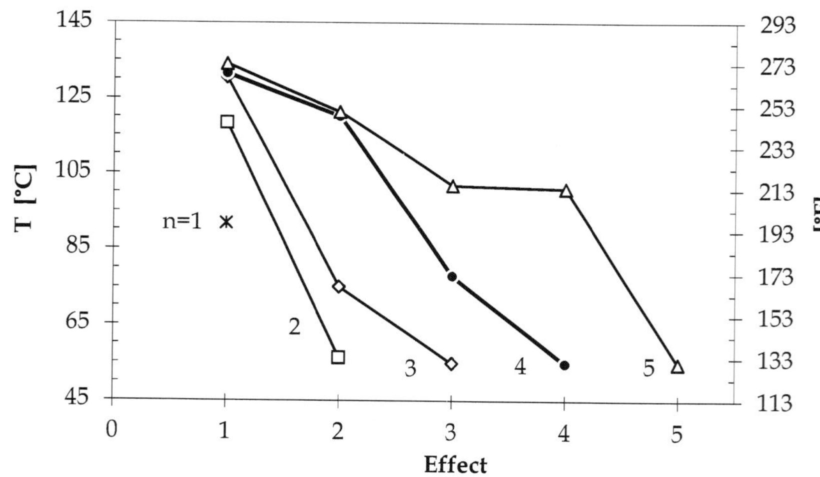

Fig. 26. Temperature profiles in the 1 to 5-effect MEES.

The net result is a drastic decrease in the area of thermal transfer, as you can see through the cost of the effects (Fig. 27). Moreover, the distribution of the heat transfer area curve changes dramatically, from being monotone decreasing for a triple-effect MEES to have a non-monotonic behavior in the Optimal MEES. Presents a maximum in the effect 3 and a minimum in the effect 1, where the product is removed and in which the coefficient U takes its smallest value (Fig. 25). This was achieved by lowering the flow rate of the solvent evaporated in the first two effects, especially in the first (Fig. 28).

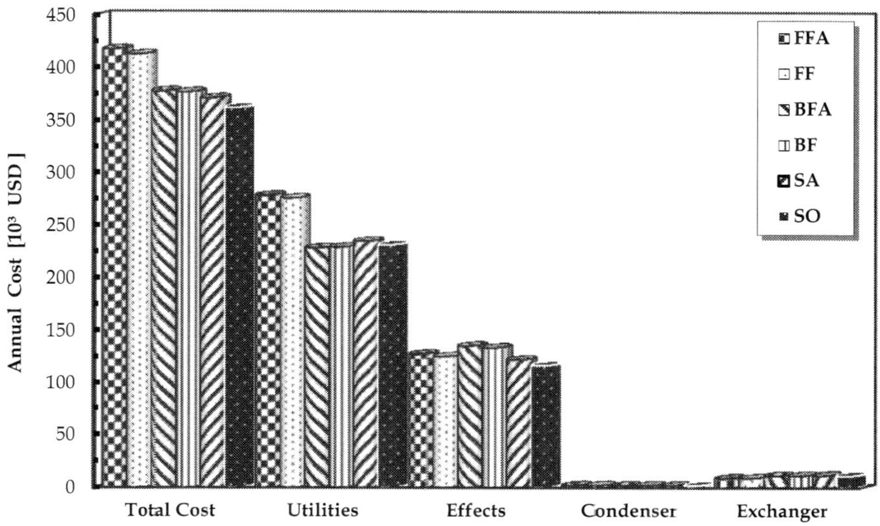

Fig. 27. Impact of the liquid flow pattern and the distribution of heat transfer area in different items of the total annual cost of the MEES (Case III).

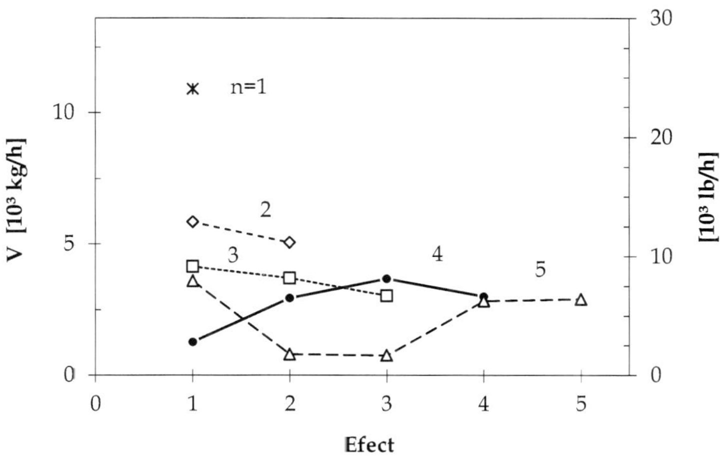

Fig. 28. Flow rate steam profiles in the 1 to 5 effect MEES.

Profile of concentration of the solution in the optimal MESS shows a non-uniform increase of concentration along the evaporator train (Fig. 29). Thus, following the path of the current liquid is seen a percentage increase of 28% in the effect 4, 53% in the effect 3, 73% in effect 2 and 46% in the first effect. Similarly as stated in Case II, the largest concentration jump

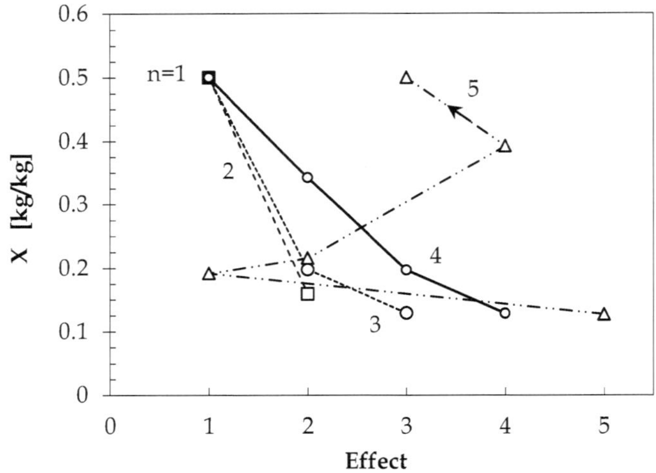

Fig. 29. Concentration profiles in the 1 to 5 effect MEES.

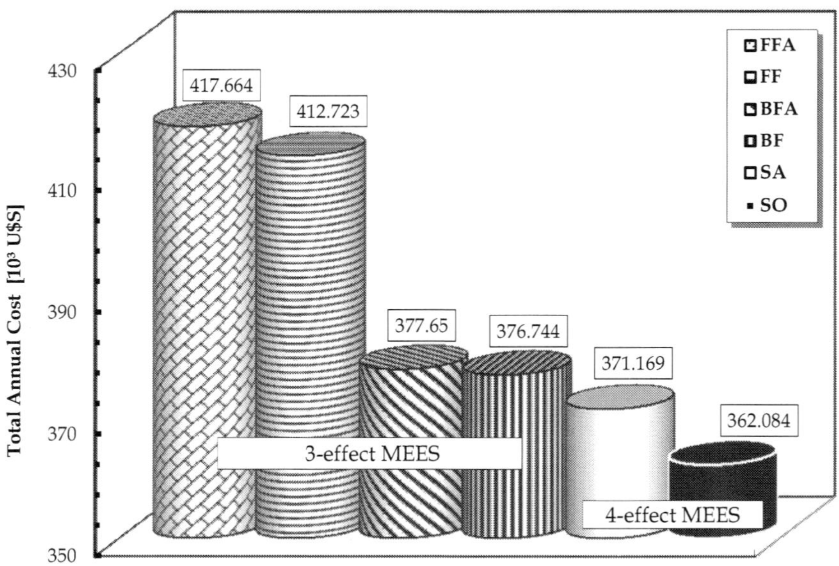

Fig. 30. Impact of the liquid and vapor flow pattern and area effects on the total annual cost of the MEES (Case III).

occurs in effect 2, as opposed to what was observed in Case I, where the largest jump occurs in the last effect, where in addition, the product leaves.

4.3.3 Comparison with traditional configurations

As in the case II, the optimum number of effects of optimal MEES is four and does not match the traditionally proposed by the classic bibliography: triple-effect countercurrent (BFA). If you seek the best solution between traditional structures with effects of equal area (Table 2), again confirms the superiority of the backward feed configuration and that three is the optimal number of effects. Comparatively, FFA is 11 4% more expensive that the structure BFA (Fig. 30). If in both structures allows you to optimize the distribution of the transfer area between the different effects, the improvement is not important. Of them, the best option (BF) is 4.1% more expensive than SO. However, the best configuration for a MEES whose only structural restriction is equality of areas of its effects (SA): is backward feed (BF).

5. Conclusion

In this paper, we solve the problem of designing a MEES, for the concentration of caustic soda, developing a rigorous mathematical model non-convex MINLP type and solving it using a mathematical optimizer.

Unlike previous papers, the new formulation proposal incorporates as decision variables: (a) the trajectories of the steam and liquid flows along the evaporator train whose correct choice determines, in the opinion of different authors, as has been demonstrated in the resolution of the example here presented, the level of operational costs and investment of the MEES. It is also considered (b) the number of effects of the MEES, as was proved in the results presented is another critical design decision and (c) heat transfer area in each effect, without resorting to the hypothesis of equal areas on the effects that in many cases substantially increases the total cost of investment. As an evaluation criterion of alternative designs included in the solutions space of the problem, we used the total annual cost of the system of evaporation, including fixed and operational costs.

Other important aspects, not usually treated and much less simultaneously with the search of the optimal flow pattern, were considered such as:

the rise of the boiling point of the solution and its dependency on temperature and concentration of the same, the variation of the overall heat transfer coefficient along the train following changes in the concentration and the temperature of each effect and the functional dependence of the heat of non-ideal solution with the concentration and temperature.

In addition, we studied other variants non-conventional design that arise by allowing: feed in parallel of the weak solution to two or more effects of the evaporator train, the entry of two or more liquid streams (even of different concentration) to a given effect and the derivation of a liquid solution branch around an effect to avoid its treatment in the same.

A serious drawback found in the resolution of the mathematical model is the presence of many stationary points (optimal local), and also the great influence of the initial point.

It has been found that in order to obtain the optimal design of lower total cost annual: (i) not always steam flow pattern should be the traditional unifilar cascade, could be useful to feed steam live in more than one effect, (ii) the heat transfer areas do not necessarily have to be equal, (iii) the fresh feed stream should not always come in the last effect evaporation train.

However, the synthesis of the optimal MEES only be achieved if you are optimized simultaneously structural, parametric, and operation variables.

6. Acknowledgment

I thank to SeCTER for their help in research.

I thank Liliana, my dear sister, for her love and permanent support throughout my life.

7. References

Foust, Alan S., *Principles of Unit Operations,* Second Ed. C. Wiley, New York, 1980.

Geankoplis, Ch. J., *Transport Processes and Unit Operations,* 2nd. Edition, 495-501, 1983.

Hillebrand Jr., J. B. & Westerberg, A.W., The Synthesis of Multiple-Effect Evaporator Systems Using Minimum Utility Insights - I. A Cascaded Heat Representation *Computers & Chemical Engineering,* Vol. 12, pp. 611-624, 1988.

Horvath, A. L., *Handbook of Aqueous Electrolyte Solutions, Physical Properties, Estimation and Correlation Methods,* Wiley, C. New York, 1985.

Irahola Ferreira, Jaime A., & Jaime Cerdá. Optimal Synthesis Of A Multiple-Effect Evaporator System. *Fifth International Symposium On Process Systems Engineering (PSE),* Korea, 1994.

Irahola Ferreira, Jaime A., & Jaime Cerdá. Síntesis Optima de un Sistema de Evaporación Multiefecto para la Concentración de Licor de Caña de Azúcar. *IX Simposio Internacional en Aplicaciones de Informática. Infonor'96.* Antofagasta. Chile. 1996.

Irahola Ferreira, Jaime A., Aplicación para Formular y Resolver Modelos Matemáticos de Sistemas de Evaporación Multiefecto. *Información Tecnológica.* Chile, 2008

Kern, D. Q. *Process Heat Transfer.* McGraw-Hill, New York .pp 375 - 452, 1999.

Maloney. James O. *Perry's Chemical Engineers Handbook,* 8th Edn. McGraw-Hill, New York 2008.

Nishitani, H. & Kunugita, E., The Optimal Flow-Pattern of Multiple Effect Evaporator Systems, *Computers & Chemical Engineering,* Vol. 3 , pp. 261-268, 1979.

Peters, Max Stone Klaus D. Timmerhaus. *Plantdesign and economics for chemical engineers.* McGraw-Hill, Inc. Singapore, ISBN 0-07-100871-3, 1991.

Standiford, Ferris C. Jr., W. L. Badger, Evaporation. *Chem. Engng,* 70, 158 – 176, 1963.

Westerberg, A.W. & Hillebrand Jr., J. B. The Synthesis of Multiple-Effect Evaporator Systems Using Minimum Utility Insights - II. Liquid Flow-Pattern Selection. *Computers & Chemical Engineering,* Vol. 12, pp. 625-636, 1988.

16

Optimization of Spouted Bed Scale-Up by Square-Based Multiple Unit Design

Giorgio Rovero, Massimo Curti and Giuliano Cavaglià
Politecnico di Torino
K&E Srl
Italy

1. Introduction

Among several configurations typical of gas-solids fluidization, spouted beds have demonstrated to be characterized by a number of advantages, namely a reduced pressure drop, a relatively lower gas flow rate, the possibility of handling particles coarser than the ones treated by bubbling fluidized beds. Additionally, significant segregation is prevented by the peculiar hydraulic structure.

Spouted beds appear to go through a revival, testified by a very recent and comprehensive book on the topic (Epstein & Grace, 2011). This renewed interest arises by implementing new concepts in scaling-up spouting contactors and devising potential applications to high temperature processes, noticeable examples being given by pyrolysis and gasification of biomass, kinetically controlled drying of moist seeds to guarantee the requested qualities and polymer upgrading processes.

2. Generalities on fluidization

Fluidization is a hydrodynamical regime in which a bed of solid particles is expanded and suspended by an upward fluid flow. This regime is established when the fluid velocity reaches a value corresponding to the minimum fluidization. The basic design of a fluidized unit is carried out by considering a vessel having a cross section of any shape (circular, squared or rectangular) with a perforated bottom which separates the volume holding the solids from the lower gas plenum.

Fluidized beds show a number of features which are summarized below:

- forces are in balance and there is no net force acting in the system;
- the solid particle bulk exhibits a liquid-like behaviour: the surface of the solids remains horizontal by tilting the vessel;
- if two or more vessels operating in a fluidization regime are connected, the solids reach an identical hydrostatic level;
- in the presence of a side opening under the bed surface, particles gush as a liquid flow;
- heterogeneous bodies may float or sink, depending on their actual density.

Fluidization, besides being influenced by the solid characteristics, depends on the physical properties of the fluid and its superficial velocity. When this parameter is very low, the fluid merely percolates through the particles and no movement is induced, this condition being defined as "static bed". By rising the flow rate, frictional forces between particles and fluid increases: when the upward component of force counterbalances the particle weight, the minimum condition to expand the bed is reached. When all the particles are suspended by the fluid, the bed can be considered in a state of "incipient fluidization" and the pressure drop through any bed section equalizes the weight of the fluid and solids in that section. By further increasing the velocity, some phenomena of instability such as "bubbling or turbulent fluidization" may occur, depending on the system geometry and particle properties. In a gas-solid system operated at high fluid velocity, gas bubbles tend to coalesce and grow in volume during their upward travel; if the bed is not wide enough, a gas bubble can take all the vessel cross section, then the solid particles are lifted as a piston, giving origin to the so-called "flat slugging". This undesired occurrence easily happens with coarse particles as well with cohesive powders. Finally, when a critical value is reached, the velocity of the gas is high enough to transport individually or in clusters the bed particles in a "pneumatic conveying" fashion. These hydraulic regimes are schematically shown in Figure 1.

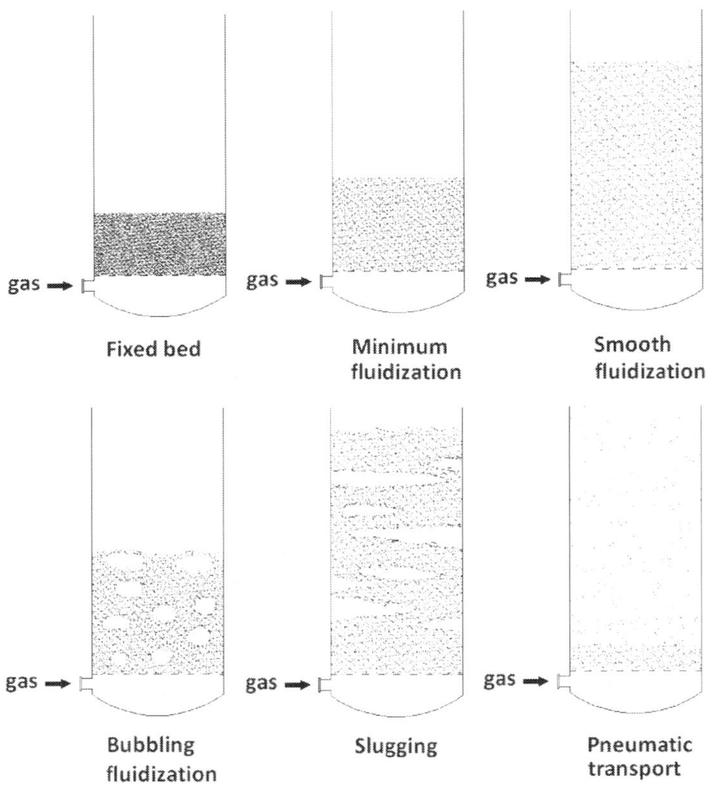

Fig. 1. Schematic representation of various fluidization regimes

In addition to properties of the fluidization medium, particle features play an essential role. A simple mapping was proposed (Geldart, 1973) to group particulate solid materials in four well-defined classes according to their hydrodynamic behaviour. To follow this categorization, Figure 2 shows the four different regions, proper of air/solids systems at ambient fluidization conditions and particles of the same shape, excluding pneumatic transport conditions.

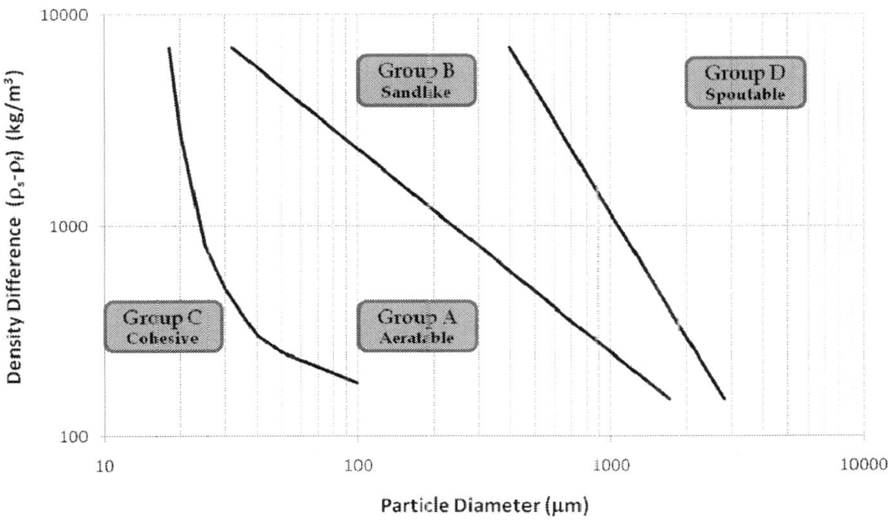

Fig. 2. Particle classification according to Geldart

The particles of the regions can be described as follows:

A type or "aeratable": solids with small diameter and density lower than about 1400 kg/m^3. The minimum fluidization velocity can be reached smoothly; then fine bubble fluidization occurs at higher gas velocities.

B type or "sandlike": these particles are coarser than the previous ones, ranging from 40 to 1000 μm and densities from 1400 to 4000 kg/m^3. A vigorous fluidization with large bubbles may be established.

C type or "cohesive": very fine powders, with a mean diameter generally lower than 50 μm. Strong interparticle forces render fluidization difficult.

D type or "spoutable": coarsest particles within a broad density range. These systems are characterized by a high permeability, which generates severe channelling and uneven gas distribution: the standard fluidization geometry should be modified to give origin to spouted systems.

3. Fluidization versus spouting

Fluidization is an operation characterized by several interesting peculiarities with desirable associated to non-optimal features. On one hand this technique guarantees a smooth and liquid-like flow of solid particulate materials that allows continuous and easily controlled operations. The good mixing of solids provides a large thermal flywheel and secures isothermal conditions throughout the reactor. If some minor wall-effects are neglected, the solids-to-fluid relationship is independent of the vessel size, so this operation can be easily scaled up and large size operations are possible.

On the other hand, fluidization may reach conditions of instability, such as bubbling or slugging, which usually represent a situation of inefficient contact between the two phases. Moreover, in high temperature operations or with sticky particles, solids sintering or agglomeration easily occur. Finally, but not less important, fines generation, erosion of vessel, internals and pipes is a serious problem caused by the random and intense movement of particles (i.e. particle or carry over exerted by the fluid.

Mass and heat transfer related to physical and chemical reactions are kinetically limited by surface area of large particle, either when the operation occurs in the fluid phase or on the solids. In these cases fluidized systems must operate with fine particulate materials; examples are given by heterogeneous catalysis or combustion/gasification of coal fines.

For reasons intrinsic to many processes (agricultural products upgrading, agglomeration, pelletization, etc.), large particle handling is required and fluidization does not represent an optimal technology. Material comminution to reach the size required by conventional fluidization is an additional negative aspect which increases the exergetic overall process cost. A very noticeable gas rate is required to reach fluidization of large particles, which often far exceeds the amount required for the physical or chemical operation considered. It should also be noted that fluidized systems are operated at a gas rate double or triple with respect to the minimum fluidization velocity to confer the system adequate mixing and avoid any dead zone. In conclusion, fluidization appears an interesting operation thanks to an easy scale-up, though its extensive feature (large gas flow rate need) as well as its intensive characteristics (random fluid-to-particle hydraulic interaction) counterbalance its favourable aspects to some extent.

A spouted bed can be realized by replacing the perforated plate distributor typical of a standard fluidized bed with a simple orifice, either located in the central position of a flat bottom or at the apex of a bottom cone, whose profile helps the solids circulation and avoids stagnant zones. Examples of non-axial orifices appear in the scientific literature, too. The fluidizing gas enters the system at a high velocity, generates a cavity which protrudes upward through the "spout", which, having an almost cylindrical shape, can be characterized by its spout diameter value D_S. When the gas flow rate is large enough, the spout reaches the bed surface and forms a "fountain" of particles in the freeboard. The fountain can be more or less developed depending on the gas rate and the overall system features. After falling on the bed surface, the solids continue their downward travel in the "annulus" surrounding the spout and reach different depths before being recaptured into the spout. The dual hydrodynamics, good mixing in spout and fountain and piston flow in the peripheral annulus with alternated high and slow interphase transfer, in the spout and in the annulus respectively, makes spouted beds unique reactors.

Consequently, spouted beds offer very peculiar features, which can be summarized as:

- very regular circulation of particles and absence of dead zones;

- reduced pressure drop and lower gas flow rate required to attain solids motion with respect to the minimum fluidization velocity, this result being possible as the gas transfers its momentum to a limited portion of solids constituting the whole bed;
- wide range of operating conditions starting from a value slightly exceeding the minimum spouting velocity ;
- possibility of handling coarse particles having a wide size range and morphology.

4. Spouted beds

The term "Spouted Bed" was coined with an early work carried out at the National Research Council of Canada by (Gishler & Mathur, 1954). A comprehensive book by (Mathur & Epstein, 1974) provided a systematic summary of the scientific work done in the years. Very recently (Epstein & Grace, 2011) the most advanced knowledge in the field has been updated.

Spouted beds were originally developed as an alternative method of drying moist seeds needing a prompt and effective processing. Sooner the interest in spouted beds grew and their application included coal gasification and combustion, pyrolysis of coal and oil shale, solid blending, nuclear particle coating, cooling and granulation as well as polymer crystallization and solid state polymerization processes. Fundamental studies were carried out to establish design correlations, the performance of spouted beds as chemical reactors, motion patterns and segregation of solids, gas distribution within the complex hydrodynamics.

Some additional improvement in the gas-to-solids contacting can be provided by independently aerating the annulus, thus generating the so-called "Spout-fluid Beds" (Chatterjee, 1970). Again, a perforated draft-tube can be placed to surround the spout, thus contributing in terms of stability and operational flexibility (Grbavčić et al., 1982).

Most studies were carried out in plain cylindrical geometries, either full sectional, half sectional or even in a reduced angular section of a cylinder in order to explore scale-up possibilities. In any case adding a flat transparent wall has been demonstrated to interfere to a moderate extent with the solids trajectory vectors within the annulus (Rovero et al., 1985) while does not affect the measurement of the fundamental parameters of spouted beds (U_{ms}, H_m, D_s). A limited number of examples consider multiple spouting, either in parallel or in series, squared and rectangular cross sections (Mathur & Epstein, 1974).

5. Design bases

A typical spouted bed scheme is given in Figure 3. This representation depicts the gas inlet and outlet, the upward movement of solids in the spout, their trajectories in the fountain and the subsequent descent in the annulus to reach the spout again at depths which depends on the path-lines originated by the landing position on the bed surface. The particle holdup can be loaded batchwise or, alternatively, in a continuous mode. The latter option depends on process requirements, nevertheless a continuous solids renewal in no way alters the above features. A proper solids feeding should minimize bypass towards the discharge port; in this view a direct feeding over the bed surface appears the best option to guarantee at least one circulation loop in the annulus to all the particles. A direct distributed feed over the fountain is required only in case of particles with high tendency to stick.

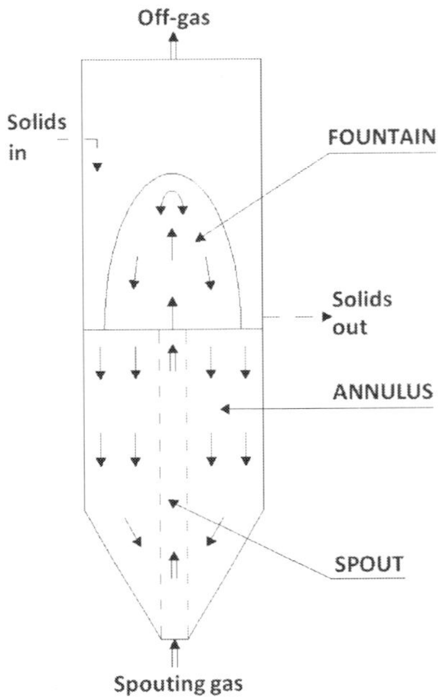

Fig. 3. Schematic of a spouted bed

The solids discharge from continuous operations is generally carried out with an overflow port, unless special process control is required. This case may be given by coating processes, where the total bed surface area should be controlled. In this case a submerged port preferentially discharges coarse material, due to local segregation mechanisms (Piccinini, 1980), while the entire spouted bed retains good mixing capacity, which can be regulated by the fountain action.

The gas flow distribution between spout and annulus is completely independent whether the solids are batch or continuously fed: part of the gas progressively percolates from the spout into the annulus by moving toward higher elevations in the bed of particles. In case of a bed of sufficient height the gas in the annulus may reach a superficial velocity close to the minimum fluidization velocity of the solids. In this event the annulus is prone to collapse into the spout, thus defining the "maximum spoutable bed depth, or H_m". This parameter represents one of the fundamental criteria to design a spouted bed unit (Mathur & Epstein, 1974); H_m depends on vessel geometry, fluid and particle properties.

The second fundamental parameter is given by the minimum rate of gas required to maintain the system spouting, the so-called "minimum spouting velocity, or U_{ms}". This operating factor can be either determined by an experimental procedure (as described in Figure 10) or can be calculated by the existing correlations.

A spouted bed, thanks to its flexibility, can be operated with a wide range of solids load to fill part or all the cone ("conical spouted beds"), or otherwise to engage also the upper portion of the vessel ("cylindrical spouted beds"). In both cases the conical included angle is in the range of 60 to 90°; by further diminishing the angle instability in solid circulation might occur, while increasing excessively the angle decreases solids circulation at the base A gross criterion that distinguishes conical and cylindrical spouted beds can be given considering the type of reaction to carry out: when solid phase undergoes a fast surface transformation, the optimum residence time of the gaseous phase is very short. This condition is satisfied by shallow beds as in the case of catalytic polymerization or coal gasification and pyrolysis. When the reaction is controlled by heat or mass transfer, the gas-solid contact must be adjusted with a deeper bed configuration as in drying, coating, solid phase polymerization, etc..

Thanks to this flexibility, the mean residence time of the solids in continuous operations can be regulated by optimizing the solids hold-up in a single vessel, which gives origin to a well mixed unit, or otherwise it is possible to conceive a cascade of several units to have a system approaching a plug flow. In the latter case square based units can have a number of advantages over a conventional cylindrical geometry. Specifically, the construction is cheaper, more compact and the heat dissipation toward the outside lower. Some scientific aspects remain open though: the design correlations that should be validated, the stability of multiple units proved both during the start-up and at steady state conditions and the possibility of fully predicting the solids residence time distribution as a function of geometry and number of stages.

6. Spouting regime

Stable spouting can be obtained by satisfying two hydrodynamic requirements: 1) the bed depth must be lower than the H_m value and 2) the gas flow rate has to exceed U_{ms}. From an initial condition of a static bed with a nil gas flow, by increasing the gas flow a certain pressure drop is built up through the bed of particles. The graph given in Figure 4 describes this hydrodynamic evolution, which implies a pressure drop/flow rate hysteresis between an increasing flow and the reverse situation. The hysteresis is caused by different packing conditions of the bed particles, that expand to attain a loose state once a spouting condition is reached. Starting from the static bed condition denoted by A, the pressure drop increases with the fluid velocity and reaches a maximum pressure drop (ΔP_M at B). With an additional increase of the gas velocity, the bed displays a moderate progressive expansion and a corresponding decrease of pressure drop to reach C. Finally an abrupt spouting leads to a sudden decrease of pressure drop which stabilizes at an nearly constant value (D), which is maintained in all the operating range of gas rate. This situation represents a stable spouting. In case of fluid velocity decrease, the pressure drop remains constant down to the spout collapse (E), which compacts the system to some extent and the pressure drop increases again to F, giving origin to the afore mentioned hysteresis. The minimum spouting velocity is recorded at E.

The whole system hydrodynamics is given by knowing ΔP_M, U_{ms}, ΔP_s and the U/U_{ms} ratio chosen for a stable spouting. A recent paper has compared data obtained in the mentioned 0.35 m side square-base unit to the existing literature correlations (Beltramo et al., 2009). The design data for the blower are ΔP_s and U, while the maximum pressure drop and the relative transitory flow rate can be easily generated by a side capacitive device, to be used at the start-up only. It is important knowing that the hydrodynamic transient can be as short as a few seconds, so that the capacitive device can be designed with a characteristic time shorter than a fraction of a minute. In this view, the timing of a spouting process onset is quantified in Figure 15.

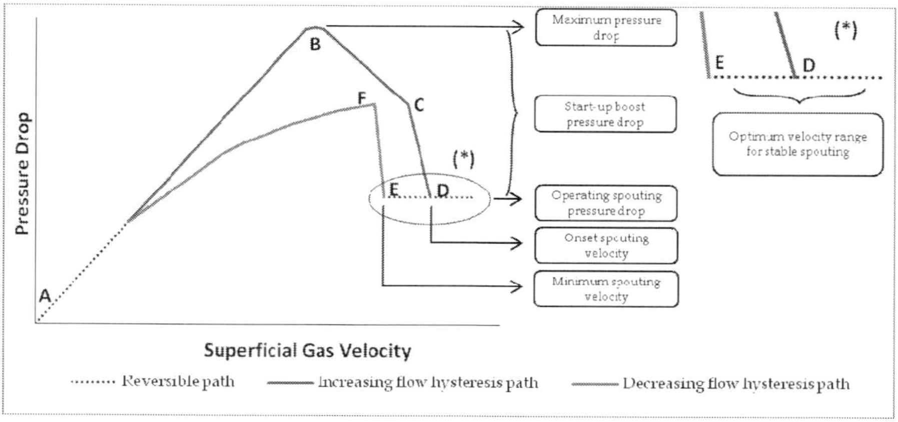

Fig. 4. Hydrodynamic diagram for spouting onset

Figure 5 displays a sequence of pictures that show the spout onset from the initial cavity generation (A) to the full spouting. The third picture qualitatively corresponds to the point B in Figure 4, the fourth picture shows the rapid sequence between C and D, while the last picture may describe any point in the interval E-D, or over.

Fig. 5. Photographic sequence of the evolution of a spouting process performed in a squared-based half sectional 0.2 m side unit

7. Scale-up

Due to the peculiarities of spouted beds, their application to industrial processes requires a sound experience and a clear vision of their hydrodynamics since scale-up from laboratory experience is required. A summary of industrial implementations appears in the recent book on spouted beds by (Epstein & Grace, 2011) together with general criteria, though the issue remains open.

Scaling-up a spouted and spout-fluid beds can be tackled according two approaches:

1. increasing the size of a single unit, or
2. repeating side by side several units.

Both routes must be discussed in terms of advantages and drawbacks. The first approach implies a simple geometry and mechanic construction: some doubts arise on the validity of the existing correlations and the overall hydrodynamics in the unit (gas distribution between spout and annulus, solids circulation, etc.). The use of the existing correlations up to a unit diameter (D_c) of about 0.6 m is generally thought fully safe. If a continuous operation is considered, this arrangement gives the solid particulate material a well-mixed behaviour with a broad distribution of particle residence time at the exit of the unit.

Conversely, if a sequence of multiple beds is realized, achieving a fully independence of the units becomes the fundamental goal. In other words a non-interfering system must be designed, so that it is up to the operator decide which unit to start-up or shut-down first according to process needs. Due to the complexity of a multiple system each unit must mandatorily replicate the foreseen behaviour of the basic component. In this case, the residence time distribution of the solids approaches closely a plug flow to meet most process requirements.

Design geometry and regulation criteria are thoroughly discussed in the continuation of this chapter. The design and the construction of a multiple unit implies a careful geometrical optimization to minimize heat loss, investment and operating costs, assure a straightforward start-up, guarantee stability and process performance. According to this key requirement, squared-based units could replace a standard cylindrical section geometry, according to an account presented in literature (Beltramo et al., 2009). A correlation between cylindrical and square-based units is also needed.

8. Experimentation on single and multiple square-based units

This chapter describes a systematic experimentation in a 0.13, 0.20 and 0.35 m side units to correlate the hydrodynamics of square-based spouted beds to the one of a corresponding cylindrical units and define the optimal geometrical configuration to assure solids circulation and transfer to downstream modules when multiple spouted beds in series are considered. The square-based experimental modules were made of wood with a frontal Perspex wall or of AISI 316 SS with a tempered glass window, depending on whether the apparatus had to be operated at room or higher temperatures. A 0.15 m ID cylindrical unit was also used for data comparison. Table 1 provides more geometrical details of the spouted beds. All bases, either frustum shaped or conical were characterized by an included angle of 60°. The spout orifice extended up of 1 mm over the base to improve solids circulation at the bed bottom, according to suggestion existing in the literature. The vessels were 1.5 m or 2 m high to allow the measurement of the maximum spoutable bed depth for all the solids tested.

The tests were carried out with several materials to cover a sufficient range of parameters, as they appear in Table 2.

Module type:	Section	Side (m)	Diameter (m)	Equivalent diameter (m)	Height of the unit (m)	Cone / Pyramid angle (deg)
L-13	squared	0.13	-	0.15	1.50	60
D-15	cylindrical	-	0.15	0.15	1.50	60
L-20	squared	0.20	-	0.23	2.00	60
L-35	squared	0.35	-	0.40	2.50	60
half L-20	rectangular	0.20	-	-	2.00	60

Table 1. Geometrical characteristics of spouted test apparatuses

Material	Equivalent mean diameter (mm)	Density (kg/m³)	Sphericity	Repose angle (deg)
PET chips	3.04	1336	0.87	35
Turnip seeds	1.50	1081	1.00	27
Corn	7.82	1186	0.80	27
Soya beans	7.23	1144	0.99	29
Sunflower seeds	6.16	696	0.87	37

Table 2. Physical properties of the particulate material used

Fig. 6. Materials used in the tests

The experimental equipment was composed of several units, whose pictures are given in Figure 7A and 7B. Depending on the flow rate required, air as spouting medium was either provided by two volumetric compressors (total flow rate of about 250 Nm³/hr) or from a blower (flow rate of 350 Nm³/hr). The air flow was cooled through a corrugated pipe heat exchanger to guarantee spouting at a constant room temperature. Two rotameters were used to meter the flow together with gauge pressure recording.

The experimental strategy was directed both to run batch experiments in a single vessel to assess the fundamental spouting parameters (H_m, U_{ms}, a stable U/U_{ms} ratio), as well as

Fig. 7A. See from left to right: a) 0.15 m ID cylindrical Perspex spouted bed vessel, b) 0.13 m side square-based wooden unit; equivalent to a), c) 0.20 m side square-based wooden unit and d) half-sectional 0.20x0.10 m² square-based wooden unit.

Fig. 7B. 0.35 m side square-based AISI 316 SS unit with tempered glass frontal window.

optimize the geometrical spouted bed features (details of the base, orifice diameter, particle traps at the gas exit, etc.). Additionally, cylindrical and square-based vessels were comparatively tested.

Continuously operating experiments were directed to design the geometry of internals required to guarantee easy start-up, spouting stability in a multiple stage unit, discharge facilities to minimize "off-spec" products, adequate solids transfer from the feeding port, through the inter-stage weir, to the final overflow discharge.

Figure 8A provides the schematic view of the spouted bed assemblage of D-15, L-13, L-20, half-L-20 and L-35 units for batchwise measurements of the fundamental operating

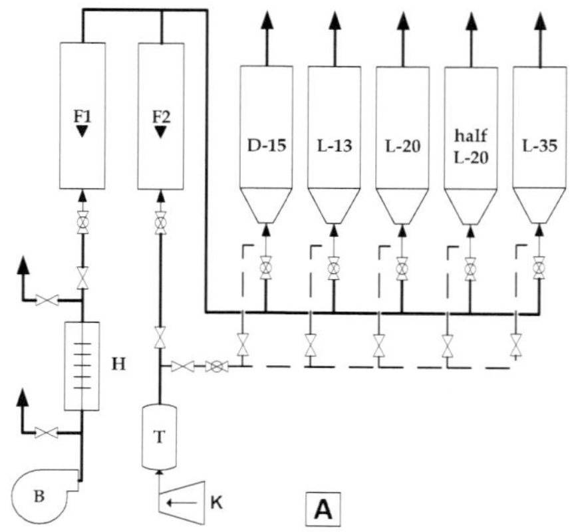

Fig. 8A. Schematic of the spouted bed assemblage for batch measurements of operating parameters (see legend in the below Fig. 8B)

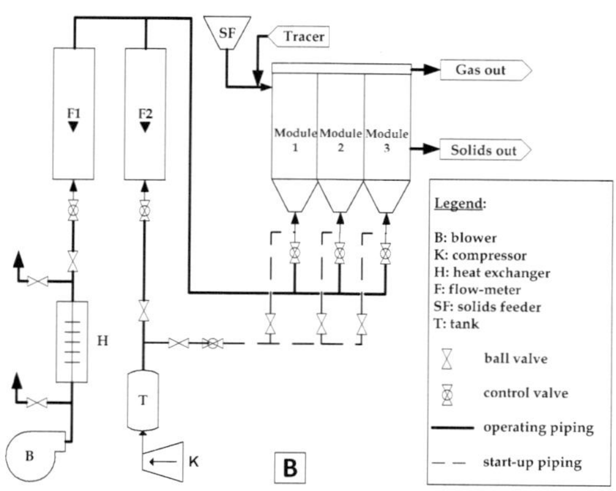

Fig. 8B. Schematic of the three module spouted bed for continuous hydrodynamic measurements

parameters characteristic (H_m, U_{ms}, D_s , mean particle velocity in the annulus, volume of spout and fountain). Figure 8B shows the scheme of the three cell equipment used to optimize geometry of internals, overall structure, effect of continuous solids feeding and residence time distribution of the solid phase.

8.1 Batch unit tests

The hydrodynamic behaviour of a square-based spouted beds was evaluated by exploring a wide range of conditions, from a bed depth corresponding to the frustum height to the maximum spoutable bed depth by operating with all the material shown in Figure 6. A reasonable ample scale-up factor (in excess of 7) was considered by running four square-based units of 0.13, 0.20 and 0.35 m side. Moreover the 0.13 m side unit was compared to the equivalent 0.15 m ID cylindrical unit to identify any difference in terms of U_{ms} and H_m and then validate the applicability of the existing correlations to the non-standard square-based geometry.

8.1.1 Maximum spoutable bed depth

The maximum spoutable bed depth is measured by progressively adding solid granulated material in the vessel and verifying that a stable spout could be formed. Some difficult transition may be encountered at $H \approx H_m$ to make the spouting process neatly evolve from an internal spout to an external well-formed one; in these cases some subjective uncertainty may be left in the measurements. In this case visualization in half-column can be useful (see the photographs of Fig.14).

Table 3 presents the experimental results obtained in two equivalent units, namely the cylindrical 0.15 m ID column (D-15) and the 0.13 m side square-based vessel (L-13); additionally, the larger L-20 vessel provided data useful for scale-up and validation of existing correlations. The test results show good and consistent agreement with the predictions given by literature equations (Malek & Lu, 1965; McNab & Bridgwater, 1977).

	D-15	L-13	L-20
Material	H_m, m	H_m, m	H_m, m
PET chips	0.69	0.65	1.26
Turnip seeds	0.84	0.85	1.44
Corn	0.41	0.46	0.98
Soya beans	0.37	0.36	0.65
Sunflower seeds	0.68	0.69	1.48

Table 3. Experimental values of the maximum spoutable bed depth H_m in the 0.13 m side square-based unit (L-13), in the 0.20 m side square-based unit (L-20) and in the 0.15 m ID cylindrical spouted bed (D-15).

Another interesting result derived from the comparison of the D-15 and L-13 units; the maximum spoutable bed depth values are very close for the same particles and identical orifice, so it allows us to assume that this fluid dynamic parameter does not depend on the cross section geometry. For this reason several correlations predicting H_m can be indifferently applied to both geometries. The herein below Malek-Lu (derived in SI units) and McNab-Bridgwater equations, among others, were compared with the experimental results, as it appears in Figure 9. The agreement can be defined fully satisfactory.

Malek-Lu equation:

$$\frac{H_M}{D_C} = 418 \cdot \left(\frac{D_C}{d_p}\right)^{0.75} \cdot \left(\frac{D_C}{d_i}\right)^{0.40} \cdot \left(\frac{\lambda^2}{\rho_s^{1.2}}\right) \tag{1}$$

McNab-Bridgwater equation:

$$H_M = \frac{D_C^2}{d_p} \cdot \left(\frac{D_C}{d_i}\right)^{2/3} \cdot \frac{700}{Ar}\left(\sqrt{1 + 35.9 \cdot 10^{-6} \cdot Ar} - 1\right)^2 \tag{2}$$

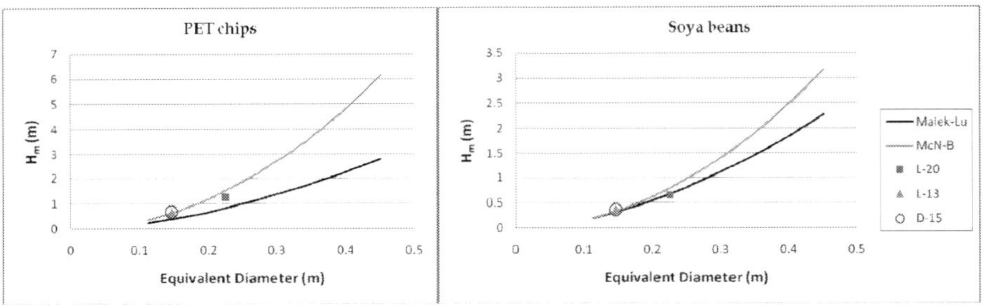

Fig. 9. Comparison of experimental data obtained in square-based and cylindrical units with literature correlations

8.1.2 Minimum spouting velocity

As given in the hydrodynamic diagram of Figure 4, the minimum spouting velocity U_{ms} represents the superficial velocity in the vessel below which spouting does not occur. Aiming to work in a stable situation, good practice suggests to moderately exceed this value by defining an operating spouting regime given by U/U_{ms} > about 1.05, which is the measure of a very modest excess of gas with respect to the minimum spouting condition. It is worthwhile noting that this value can be further reduced if the spouted bed approaches its maximum spoutable bed depth H_m, as the below Figure 10 diagram indicates. The onset spouting and minimum spouting velocities approach as H is closer to H_m. This occurrence also indicates that passing from a submerged to an external spouting is progressively easier as H_m is approached, since the bed of particles is already highly expanded by a very high gas rate (pictures on Figure 14 represent this case). Consistently, these operating conditions generate hydrodynamic diagrams with a much less pronounced hysteresis, as given in a recent paper (Beltramo et al., 2009). To predict the minimum spouting velocity in multiple cell systems an empirical correlation was proposed (Murthy and Singh, 1994). By observing carefully the type of plots appearing in Figure 10, it is possible to note that the maximum bed depth can be inferred from the overlapping of the two curves. This remarkable experimental statement (based on hydrodynamic data) may offer an interesting alternative

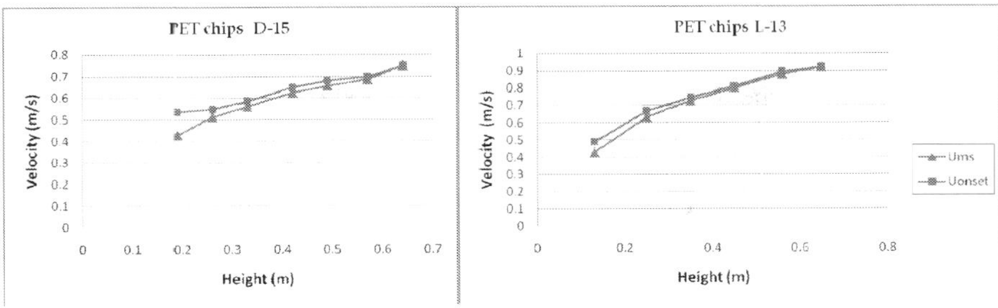

Fig. 10. Onset and minimum spouting velocities vs. bed depth

to direct observations, which are ambiguous in some instances due to bed instability as H_m is approached. Moreover careful extrapolation of the U_{ms} and U_{onset} curves to detect their intersection is a possible method to estimate H_m when the experimental conditions are not suitable for a complete direct measurement.

8.2 Continuously operating multiple units

A continuously operating unit was tested to devise a guideline to design, start-up, gain in stability and proper hydrodynamics in multiple square-based spouted beds.

A picture of the rig is present in Figure 11A, while a 3D scheme of the same unit is shown below, see Figure 11B.

Fig. 11A. Picture of the three-module experimental rig.

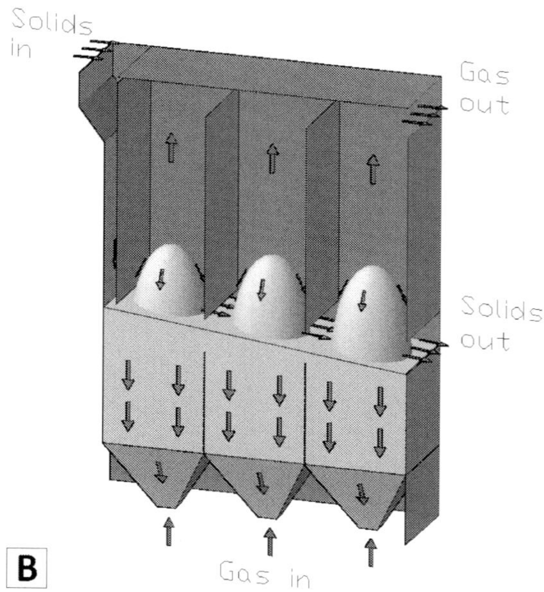

Fig. 11B. 3D diagram of the same continuously operating unit with a horizontal layout of stages

Designing a multiple spouted bed system does not represent a "black art", though may be more complex than other gas/solids contactors. In particular the following design aspects were focused in this project:

- diverting baffle at solids inlet,
- fountain height regulators,
- freeboard baffles between units,
- submerged baffles between units,
- overall layout of a multiple spouting unit

The following paragraphs illustrate the progressive tuning of a multiple stage spouted bed up to achieve safe know-how and run continuous operations.

8.2.1 Solids inlet design

Solid particles were stored in an elevated drum, from which they could flow by gravity, being metered by a rotary valve, a screw feeder or a simple calibrated orifice. The particles were distributed along one entire side of the spouted bed, addressed down by a vertical baffle protruding to a small distance from the bed free surface. This simple device avoided any solids bypass promoted by the fountain. Residence time distribution measurements quantified a bypass as high as about 20% of the total flow rate, in case of baffle absence. The elevation of the baffle over the bed may require some regulation, depending on solids feed rate.

8.2.2 Fountain height regulators

A spouted bed fountain can be defined underdeveloped, developed or overdeveloped, depending on whether its geometrical margins reach the side of the spouting vessel. Since the volumetric solids circulation through the spout/fountain system can be estimated to range to a few percents of the spouting gas flow rate, the fountain alone can pour very noticeable flows of material onto contiguous stages; then, its action should be limited by some mechanical device to restrain its hydrodynamic effect. Some devices, the so-called "Chinese hats" were presented in the sector literature (Mathur & Epstein, 1974) and tested in this unit. The experimental output was entirely disappointing, as these regulators failed in sufficiently defining the fountain shape, acted as a target for the particles propelled by the spout and interfered hydrodynamically with the gas flow in the freeboard, either when they were made of solid steel plates or wire mesh screen. To conclude, these devices are not advisable as internals in multistage spouted beds.

8.2.3 Freeboard baffles between stages

Each stage was segregated from its adjacent ones by side vertical baffles protruding down from the vessel top to a very short elevation over the bed free surface. The gap left had to assure solids flow only, depending on the continuous throughput rate, this gap between the submerged and the freeboard baffles has to be regulated to allow solids transfer by overflow, with restricted particle bounces from the fountain. Also bypass was minimized by this precaution. These flat and inexpensive devices were chosen to completely separate the freeboard into as many stage as the spouted bed design required. As a result, the action of each fountain (independently of its shape, thus gaining in spouting regime flexibility) was limited to its own stage. The use of these simple baffle repartition was observed to be fundamental for minimizing any interference between stages and enormously gain in stability.

8.2.4 Submerged baffles between stages

In principle, according to the fundamentals of fluidization, a multiple orifice spouted bed does not require a repartition between the annuluses. This consideration is also compatible with the particle vertical streamlines and the side-to-side homogeneous percolation of gas from a spout into the corresponding annulus. This assessment can be fully accepted when the system is operated batchwise and no net solids flow from one stage to the downstream one has to be steadily maintained. Practical reasons (easy start-up, spouting stability over time, independent gas flow rate regulation in each spouting module) have demonstrated that submerged baffles greatly help in defining the solids holdup in each stage. The separation of contiguous annular regions contributes in properly distributing the gas rate and giving origin to fully independent spouts. Conversely, if the holdup of solids is out of control, all the system stability may be affected.

8.2.5 Overall layout of a multiple spouting unit

In the recent past, in the frame of industrialization of a novel patented process for polyethylene terephthalate solid state polymerization (Cavaglià, 2003). a unit was conceived as a series of n-fluid beds (where n>5), operated either in turbulent fluidization or in spouting regime, where polyester beads with low intrinsic viscosity are heated-up and solid reacted in one equipment. A sextuple spouting demonstration unit was then built and operated as a prototype equipment for PET chips upgrading (Beltramo et al., 2009). The six modules were placed at identical elevation and positioned according to a 2x3 layout; the solids moved following a chicane path without being hindered by any internal repartition. The multiple spouted bed appeared advantageous in term of heat transfer efficiency (higher gas temperature at the inlet, thanks to a very short contact time between gas and heat sensitive solids) and generated good property polymers. However, the overall operation was troublesome because of difficult control on solids holdup and gas flow rate regulation. From that study sound design hypotheses were drawn to construct the experimental rig appearing in Figure 11A, whose main difference with respect to the previous industrial equipment consists in the possibility of positioning each stage at the desired elevation to facilitate the solids overflow to the downstream stage, as represented in the schematic of Figure 12. The final version of this experimental rig had the possibility of testing all the internals described above.

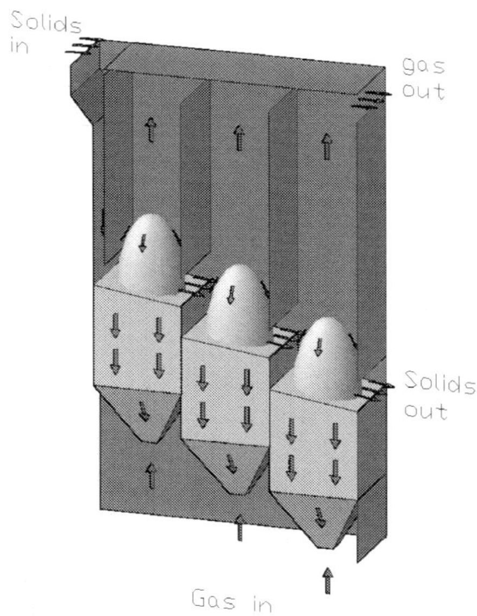

Fig. 12. 3D diagram of a continuously operating unit with a sloped layout of stages

The difference of level between units suitable for continuous and stable operations was evaluated both by running specific tests and by comparing these results against a simple correlation originated by estimating the angle of internal friction of the solids used in the experiments. The tests performed in the triple spouted bed unit aimed at devising the effect of increasing solids feed rates on bulk mass transfer from stage to stage, by measuring the angle of the bed surface with respect to the horizontal level, as well as the effectiveness of the internal baffle positioning. Figure 13 shows three different operating conditions at a mean (snapshot A), corresponding to a mean particle total residence time τ = 13.5 min), high (snapshot B), τ = 4.5 min) solid rates and at a solids throughput far exceeding the nominal system capacity, as required by any foreseen process (snapshot C), τ = 2.5 min). The bed free surface slant increased to a maximum slope (about 15°, as a mean value between inlet and overflow sides) by increasing the flow rate. This angle is in the range of 1/2 to 1/3 of the solids repose angle, which can be measured following literature recommendation (Metcalf, 1965-66). By further enhancing the feed rate, the downcomer flooded unless further raised. The hydrodynamic slope that builds up at the bed surface caused the first stage to work with a solids depth quite higher with respect the last one, this difference increasing with the number of stages. It follows that the fluid dynamic control was much trickier and the overall spouting stability impaired.

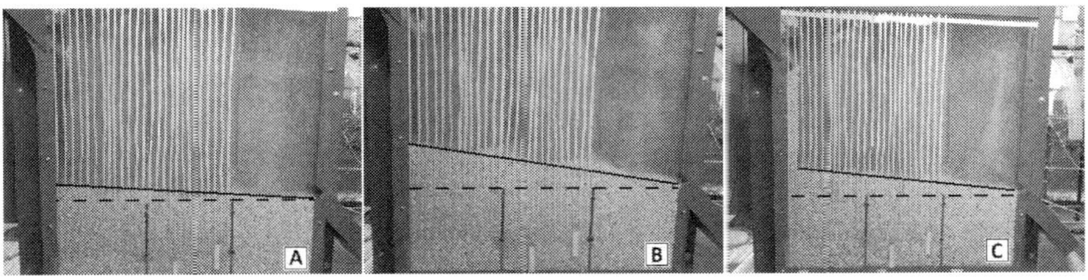

Fig. 13. Continuous operation in the three-module 0.20 m side spouted bed: - - - ideal solids free surface; —— actual solids surface at various solids flow rates: a) 2 kg/min, b) 6 kg/min and c) 10 kg/min, as given in the scheme of Fig. 11 B

This geometrical limitation was overcome by setting each bed at a minimum difference of level equal to:

$$\Delta H = D_c \tan \alpha \tag{3}$$

where α is the angle formed by the actual solids surface with the horizontal level.

A rule of thumb suggests to determine the progressive vertical distance between adjacent spouted beds at:

$$\Delta H = 0.5\, D_c \tag{4}$$

which, compared to the output of Eq. (3), leaves a safe operating margin.

As a conclusion, stable operations in a multiple square-based spouted beds require three types of flat internal baffles: one for properly addressing the solids feed to the bed surface, intermediate baffles in the freeboard to confine each fountain action, submerged baffles, each of them setting the solids overflow level from the upstream to the downstream stage. A non-interfering condition between stages was provided by generating a sloped cascade of independent spouting units.

8.3 Half-sectional spouted bed tests

A rig corresponding to half of the 0.20x0.20 m^2 spouted bed was built to represent a vertical section of the full unit. The axial sectioning included the orifice, the pyramid frustum and the constant cross section sector, thus originating a 0.20x0.10 m^2 column. A flat Perspex wall allowed a direct internal vision of spout, annulus and fountain, as already given in Figure 5.

Specific tests were carried out to demonstrate the close correspondence between data (H_m and U_{ms}) obtained in full and half-sectional columns. Approaching the maximum bed depth an underdeveloped and stable fountain was obtained with $U = 1.01\ U_{ms}$ (Figure 14 a)). Other runs highlighted a relevant bed expansion surmounting a submerged spout when the bed slightly exceeded the maximum bed depth (Figure 14 b)). Identical conditions also revealed occasional instability identified by a submerged wandering spout and some upper slugging, see Figure 14 c).

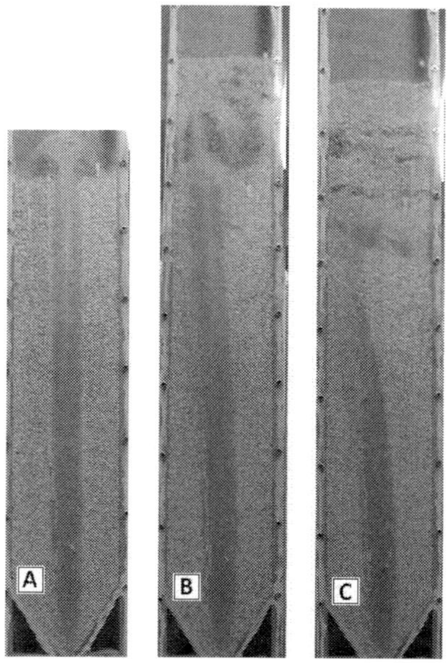

Fig. 14. Half sectional column tests at H: a) external spout and fountain formation at $H \approx 0.95\ H_m$, b) submerged spouting at $H \approx 1.05\ H_m$ c) spouting instability with spout wandering at $H \approx 1.05\ H_m$

Half-sectional units are suitable to measure particle cycle time, define solids streamlines, as well as visualize, at proper frame frequency, zones characterized by a high mixing degree. As far as the downward particle velocities are concerned, the considerations presented in the literature were taken into account, though obtained in semicylindrical vessels (Rovero et al., 1985). Figure 15 shows a sequence of snapshots which make visible the progressive motion of a tracer layer deposited on a fixed bed before starting the spouting process (first image). The second snapshot indicates that the tracer particles have maintained their position ahead of an external spout be formed. The third and fourth images indicate that particles move in the fountain in a piston flow fashion: then, local trajectories, their envelope (i.e. streamlines) and individual particle velocities can be defined. A minor portion of tracer only has been captured by the spout in the travel along the constant cross section of the

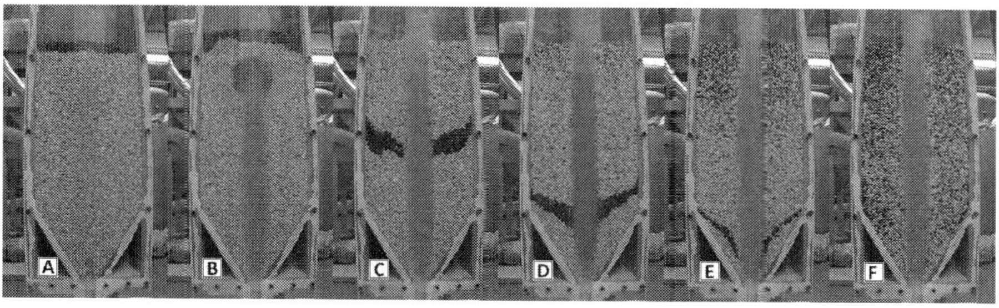

Fig. 15. The snapshot sequence indicates the motion of a tracer layer: A at t = 0 s; B at t = 1 s; C at t = 8 s ; D at t = 16 s; E at t = 22 s and F at t = 28 s

annulus. The fifth image gives evidence of noticeable shear acting on the particles, though no internal mixing occurs as far as particles enter the spout; then, quite an amount of particles is thoroughly mixed in the spout and fountain as they begin recirculating. The last image indicates an overall good mixing condition.

Spout diameter and spout profile were measured at the flat transparent wall. The experimental mean values were compared against the predictions given in the literature (McNab, 1972), with a difference of about 10%, which indicates that squared-based spouted bed findings do match the ones obtained in cylindrical vessels.

9. Fluid dynamics of solids in multiple spouted beds and its modelling

One of the main goals of a spouted bed cascade is to control the mixing degree of the overall system and possibly generate a piston flow of the solid phase to guarantee identical residence time to all particles. These evaluations are carried out by means of stimulus-response techniques after attaining steady state in a continuously operating unit. Since each individual spouted bed appears to have about 90% of its volume in a perfectly mixed state (Epstein & Grace, 2011), the recycle ratio (ratio of internal circulation in the spout referred to feed rate) leaves a small volumetric fraction of the annulus to operate as plug flow. Residence time distribution (RTD) studies in multiple spouted bed were presented in the literature (Saidutta & Murthy, 2000) in small rectangular columns having two or three spout cells. The absence of internals in this system brought to fountain wandering and excessive fountain heights that caused overall mixing higher than the one corresponding to the number of mixed units in series. A detailed RTD study on stable systems and the correlation of the experimental results with respect to the ones predicted by a model can give a relevant contribution in designing these units. Models has gained increasing importance by making use of direct measurements in the half-sectional unit, thus becoming fully predictive.

9.1 Residence time distribution function

The RTD curves represent an effective way to interpret the fluid dynamics of the solid phase in a multiphase continuously operating reactor. These functions describe the elapse of time spent by individual solids fractions in the system and can be modelled by a relatively simple combination of ideal systems, each of them describing a basic element (mixed or plug flow system, dead zone, bypass, recycling).

Two types of curves can be studied. The *E(t)* function describes what a system releases instant after instant, i.e. the volumetric (or mass) fraction of particles whose residence time is between t and t+dt. The *F(t)* function provides the integral of *E(t)*·dt and represents the fraction of elements whose residence time is lower than t.

The most direct way to trace an *E(t)* curve makes use of a physical tracer, whose characteristics are identical or very close to the ones characterizing the bulk of solids travelling the system. Usually, two types of stimulus are adopted, a pulse (given by a definite amount of tracer introduced into the system in the shortest time) or a step (an abrupt change from the normal feedstock to an identical feed made of tracer only). The first one is generally the prompter to use. Right away after the introduction of the pulse, samples are taken at the system exit with a proper scrutiny degree and the tracer concentration is measured and recorded.

The two *E(t)* and *F(t)* functions are defined below:

$$E(t) = \frac{C(t) \cdot \dot{M}}{M_{tr} \cdot \rho_b} \tag{5}$$

$$F(t) = \int_0^t E(t)dt \tag{6}$$

with *C (t)* being the concentration of solid tracer in the discharge, M_{tr} the mass of tracer injected, ρ_b the bulk density of tracer, $\dot{M}$ the mass flow rate of solids travelling the system.

A pulse function (called Dirac function, δ(t)) is given by an instantaneous but finite entity (equal to unity) entering a system at t =0. The Laplace transforms of these functions allow the use of simple algebraic input/output relationships.

In our experiments, the pulse was obtained by quickly introducing in the feed a small amount of tracer (150 to 300 g, equivalent to about 1.5% of the total solids hold-up) made of PET chips doped with some ferromagnetic powder. After sampling the ferromagnetic PET chips were sorted out from the PET bulk by a magnet and their concentration calculated in each sample.

An example of RTD curves is given by Figure 16 which comparatively reports two *E(t)* curves obtained at steady state from tests in the 3-module spouted bed operated with and without submerged baffles, according to the configuration appearing on Figure 11B. The two curves overlap almost perfectly to demonstrate that internal baffles do not alter at all the solids circulation in a multi-unit cascade. The same test also demonstrates that the external solids streamlines do not have any transversal (horizontal) component, as also visible from half column monitoring. The slope of solids at the free surface was modest in these runs, due to a very low throughput (2 kg/min). Thus, it is straightforward assessing that batch operations do not require any repartition between modules. From these considerations, the use of submerged baffles is beneficial to the start-up phase of continuous processes only and becomes fundamental as geometrical boundary to generate the configuration given in Figure 12.

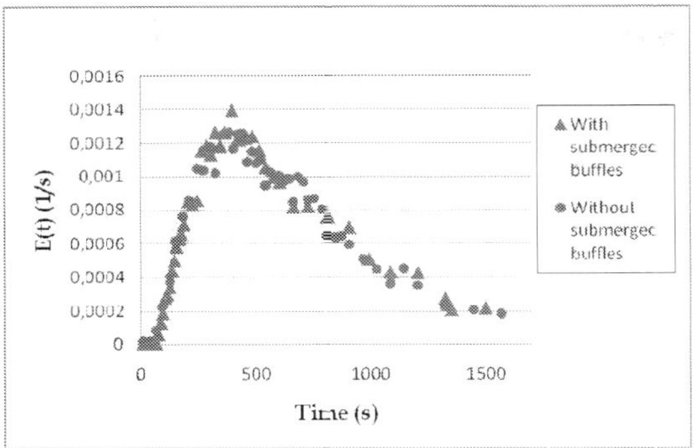

Fig. 16. Comparison between the RTD curves in the 3-module spouted bed with H/D_c=1.72 with and without submerged baffles

9.2 Modelling

The theoretical description of continuous units combines basic elemental models, whose combination gives origin to a system capable of generating an overall response to properly match the actual behaviour of the real system studied. A descriptive model can produce an output without having a strict link with the actual hydrodynamic behaviour and then has to make use of fitting parameters. This approach does not allow sound predictions or extension to more complex reactor structures. A much powerful tool is produced by conceiving a phenomenological model based on experimental observations. These models become fully predictive since all parameters are based on actual measurements. Then, a model interacts with scale-up procedures through the validity of the correlations used rather than its structure.

Dynamic responses were obtained by making use of a Matlab Simulink tool by generating model schemes as given in the below Figures 17 and 18. The fundamental modelling was based on a one-stage spouted bed; a multiple-cell system was then given by a cascade of basic units.

9.2.1 Descriptive model

The initial modelling started by considering an early dynamic description of spouted beds, where the overall behaviour can be portrayed by a well-mixed system with a minor (8 to 10%) portion of plug flow. The corresponding scheme given in Figure 17 A) includes the feed rate F_1, the bypass to fountain F_3 (which become negligible when the inlet diverting baffle is considered, as a consequence $F_2 \equiv F_1$ and $F_5 \equiv F_4$), the total circulation from spout F_4, the net discharge rate $F_7 \equiv F_1$ (for continuity) and $F_6 = F_4 - F_1$. From the experimental conditions F_1 is known and F_4 can be estimated by measuring mean particle velocity at the frontal wall of half-column. As far as the other parameters than appear in Figure 17 B) are concerned, t_d is estimated from particle circulation, $\tau_{annulus}$ follows from holdup in the annulus and $\tau_{spout+fountain}$ is calculated by difference from the known bed holdup. At usual H/D_c ratios adopted in cylindrical columns, the ratio of these two time constants approaches one magnitude order.

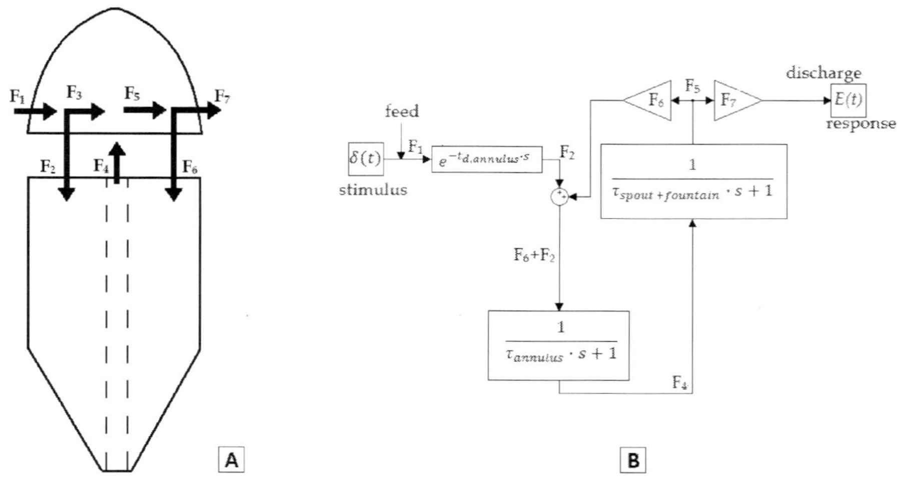

Fig. 17. Descriptive model of one stage two-zone continuously operating spouted bed: A): schematic of flow circulation between annulus and spout/fountain regions; B) Simulink model including pulse stimulus, delay at annulus entrance, two perfectly mixed regions for annulus and spout/fountain volumes and recirculation to bed surface.

Accepting this description also this model does not contain any fitting parameter, once the constitutive elements are assumed. Since the F_4 (internal circulation) to F_7 (net flow) ratio is very large, the overall system approaches a well-mixed unit and in this view the modelling is scarcely sensitive to the hydrodynamic description given to the annulus.

Figure 18 presents the comparison between experimental and modelling results for H/D_c = 1.72. The fitting is excellent, considering the time delay given by the minimum residence time of particles (t_d) and the well-mixed key dynamic component brought by spout recirculation.

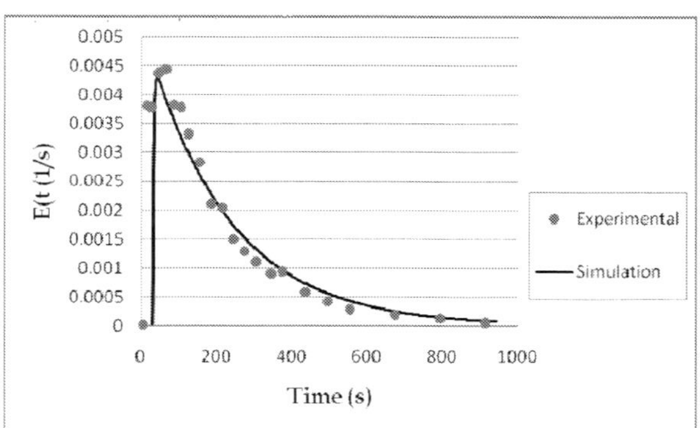

Fig. 18. Comparison between experimental data and the descriptive model

9.2.2 Phenomenological model

The relevant limitation contained by the above model consists in the fact that the annulus, representing the massive part of a spouted bed, has not been given a proper description. By observing it through the flat transparent wall of semicylindrical columns, particles show well-defined trajectories with scarce intermixing, according to consolidated findings (Mathur & Epstein, 1974).

The phenomenological description adopted in the updated model assumes that the squared-section of the annulus (with a cylindrical spout D_s) is divided into three axisymmetric zones, each of them having the same width according to the scheme given in Figure 19 A). Each of these regions receives from the fountain a solids flow rate proportional to its cross sectional area (F_{6A}, F_{6B} and F_{6C}, respectively moving from outside towards the spout). The flow fashion in each region is a piston with particle residence time $t_{d,A}$, $t_{d,B}$ and $t_{d,C}$, from the bed surface down to the cylinder-frustum junction, according to experimental observations. The mixing component acting in the annulus was concentrated in the frustum, which progressively discharges solids into the spout, depending on local streamline length. As a whole, this section was assimilated as far as its dynamics is concerned to a well-mixed volume, accounting to about 20% of the total holdup of the spouted bed. Also in this case, the effect of the F_4/F_7 ratio overcomes the sensitivity of other variables on the model, so that the ratio between frustum to parallelepiped volumes (i.e. plug to well-mixed volume ratio) is not relevant at all.

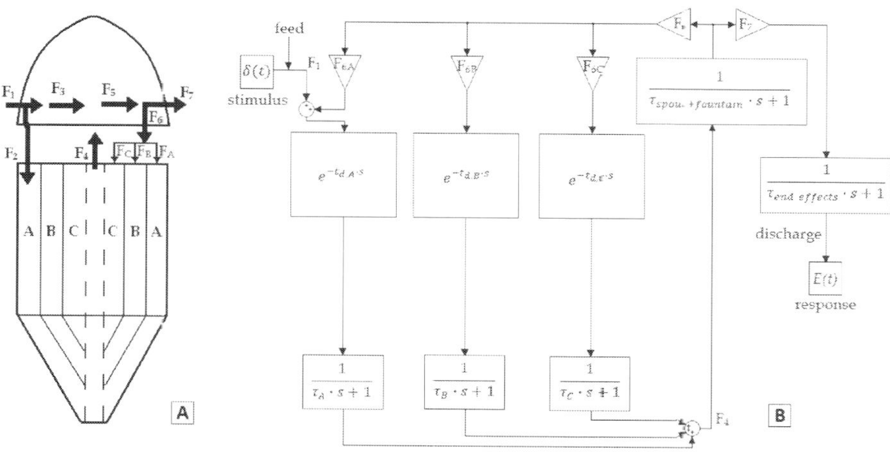

Fig. 19. Phenomenological model of one stage four-zone continuously operating spouted bed: A): schematic of flow circulation between annulus and spout/fountain regions; B) Simulink model including pulse stimulus, three parallel delay times in annulus, followed by three perfectly mixed regions in bottom frustum, one perfectly mixed zone in spout/fountain region and recirculation to bed surface. Small well-mixed volume accounts for sampling end effects.

Figure 19 B) presents the Matlab Simulink scheme, where the RTDF is generated by introducing the pulse into the sector A of the annulus (due to geometrical constrain of the inlet baffle). The overall solids flow rate from fountain travels three parallel annulus pistons, then each portion of solids enters the corresponding well-mixed portion of frustum, respectively characterized by a mean residence time estimated by observations at the flat frontal wall. The spout collects particles from the annulus and mix them in the fountain. A small well-mixed volume characterizes the solids sampling operation to account for end effects.

Figure 20 compares experimental results to the output of the phenomenological model. Any difference can be hardly noted with respect to the previous descriptive model output. Due to plug flow effect, a certain oscillation matching the cycle time frequency is observed. A short sampling time $\tau_{end\ effects}$ was sufficient to damp the greatest part of oscillation.

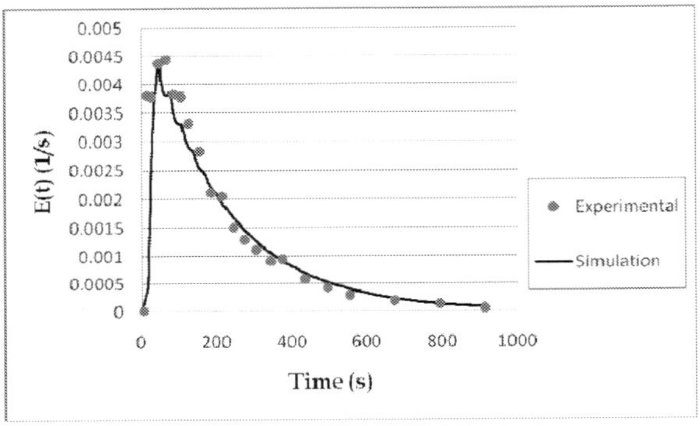

Fig. 20. Comparison between experimental data and the phenomenological model

9.2.3 Model validation for multiple units

The Matlab Simulink description conceived for the phenomenological model of one-stage spouted bed can be replicated a number of times corresponding to the number of cells included in a multistage system. An example is given in Figure 21 for a three-module

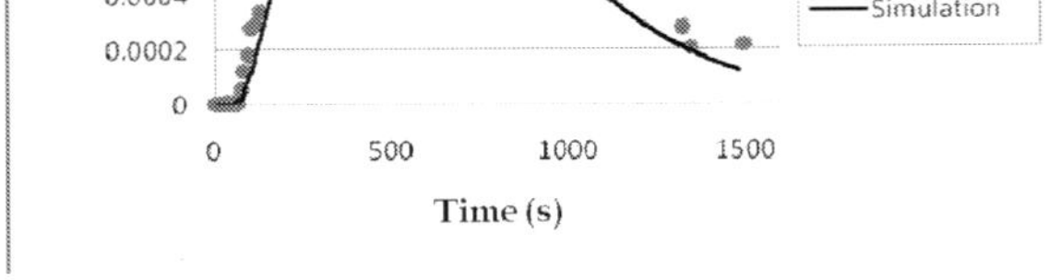

Fig. 21. Comparison between experimental and model RTDs in the 0.20 m side squared-based three-module spouted bed

spouted bed. The agreement is fully satisfactory, even though it may appear that experimental data anticipate the model output moderately and then a tail slightly higher than expected is displayed. This analysis could require to consider some direct bypass from fountain to downstream stage through the gap between submerged and freeboard vertical baffles and a small partially stagnant backwater, possibly existing along edges of frustum walls.

10. Final remarks on multiple bed start-up and shut-down

Both cold tests on the dual and triple module square-based spouted beds and industrial operations on the sextuple module demonstration unit (Beltramo et al., 2009) show that design criteria, operating conditions and step sequence must be defined carefully. In opposite case, start up and transitory from spouting onset in one module to overall stable spouting might be a serious issue. The key design parameters affecting start-up are:

- Cross-sectional area of the gap between vertical baffles: it should be designed to allow a maximum solids flow rate equal to the process solids nominal capacity with an excess of about 50% to avoid solids flooding (if too narrow) or bed emptying (if too large, because of direct discharge from fountain).
- Freeboard baffle height: it has to fully cover the fountain height at spouting onset and, if the structure of an industrial unit is considered, should reach the top deflectors of particles to avoid upper solids bypass.
- Cross-section of particle deflectors on gas phase path in each module (in the upper part of freeboard, above fountain projections): it has to be designed to set a progressive decrease of gas cross-sectional velocity and avoid any upward solids elutriation.
- The design of pyramid frustum base should consider a small gap between the orifice circumference and the slanted wall to help particle circulation. This gap is related to the average size of particulate material with a factor larger than 2. Several experiments have demonstrated that in case of local stagnation at the frustum bottom, this dead zone may spread up to the pyramid/parallelepiped junction. A careful bottom design and/or high gas flow rate counter this undesired phenomenon.
- Spouting modules have to be assembled at sufficient different elevation to guarantee a steady and even solids holdup. A rule of thumb drawn from experimentation suggests that $\Delta H \approx 0.5\ D_c$ surely gives origin to a non-interacting system and prevents bed emptying of downstream modules, or vice versa avoid extra hold up in upstream ones.
- Each module must be provided of independent gas flow regulation.

Given the fact that gas phase pressure drop is maximum at incipient start-up (up to 2 to 3 times of the gas pressure drop at stable spouting for conical or cylindrical shallow beds), to avoid oversize of circuit blower, with consequential dramatic increase of capital and operating costs, one can proceed according to two start up routes:

a. **Start-up with reduced solids holdup**: fill the first module with 1/3 to 1/2 depth with respect to the design bed load; start injecting gas till fountain formation, then begin filling the bed to reach the operating solids bed depth and then continue feeding solids at nominal flowrate to reach overflow discharge to the second module. Continue with the same procedure for all modules.

b. **Start-up with full solids holdup:** fill each module with the design solids bed depth, provide the circuit with a capacitive booster section, suitable to inject high pressure gas to bed orifices for about 10 to 15 seconds and reach external spouting onset; then operate spouting process with the master circuit blower.

In a multiple-bed spouting unit, to have the whole series of spouted beds started up, one has to proceed with a start-up sequence, one by one, from the first bed to the last one, following the solids flow direction. Monitoring the gas phase pressure drop vs. time represent the trigger element for the control system to determine when the start-up of one module is completed and the situation is ready to move and start up next module. The b) procedure is safer since enough head is available to re-start the system in case of failure.

The above defined key design parameters are also suited for shutdown phase. As far as this procedure is concerned, one has to stop solids feeding, allowing holdup of each module to be processed at the steady state operating conditions (as to minimize off-spec), while decreasing each bed depth to a level lower than the overflow weir. At that point, side submerged baffles must be risen, while gas flow continues to be injected to each bed, so to have prompt residual solid emptying.

11. Conclusions

Spouted beds, throughout over half a century studies, have demonstrated to display very interesting features against bubbling conventional fluidization. Thanks their peculiar hydrodynamic structure a relavant gas rate can be saved also operating at the maximum spoutable bed depth. Again, the total frictional pressure drop across a spouting unit can be as lower as one-third of the one in a corresponding particulate material fluidization. The application of spouted bed to relatively coarse solids overcomes undesirable features characteristic of fluidization, namely random gas channelling and solids circulation, slugging and poor contacting between phases.

The scale-up issue of spouting units has remained open since their initial invention and the debate on whether prefer larger or multiple units has struck the opinion of scientists and technologists every time that this problem required a sound solution. This chapter has tackled the scale-up problem by opting for a square-based spouted bed geometry, since constructing a cascade of these vessels is economically advantageous and much more effective for the solids fluid dynamics as well as for insulation problems.

Following the experience gained during an industrial demonstration project, encouraging results came by evaluating the product quality; the potential performance of a multiple spouted bed was thus confirmed. Nevertheless, this unit required an excessive attention to govern its stability over time and the need of several improvements was highlighted.

A new research project required the construction of several apparatuses to pursue a comprehensive strategy which has aimed at:

- comparing the fundamental operating parameters of square-based spouted beds with the corresponding values characteristic of conventional cylindrical columns,
- carrying out an adequate experimental scale-up,

- demonstrating the achievement of fully stable operations by introducing novel concepts (stage segregation with internal baffles, sloped cascade of stages) in designing a multi-cell equipment,
- obtaining a plug flow of solids with a sufficient number of stages, which may implement process scale-up at the same time,
- modelling single unit and multiple square-based spouted beds to predict solids hydrodynamics.

The final structure of a cold model apparatus has demonstrated the achievement of all the listed goals.

12. Acknowledgments

The economical support granted by ENGICO Srl (LT- Italy) allowed construction of the equipment and the research fellowship to one of the authors (M. C.). The authors wish to tank Mr. Alfio Traversino for patiently and carefully constructing the three-module experimental rig.

13. Nomenclature

Ar	Archimedes number
C	concentration of solid tracer in the discharge, kg / m^3
D_c	column diameter, m
d_i	inlet diameter, m
d_p	particle diameter, m
D_s	mean spout diameter, m
$E(t)$	E function
$F(t)$	F function
H	bed depth, m
H_m	maximum spoutable bed depth, m
ΔH	gap between adjacent units, m
$\dot{M}$	mass flow rate of solids, kg / s
M_{tr}	mass of the tracer, kg
ΔP_M	maximum pressure drop across bed. Pa
ΔP_S	spouting pressure drop across bed, Pa
U	fluid velocity, m / s
U_{ms}	minimum spouting velocity, m / s
U_{onset}	fluid onset velocity, m / s

Greek letters:

α	free surface slope, deg
$\delta(t)$	Dirac function
λ	shape factor
ρ_f	fluid density, kg / m^3
ρ_b	bulk solids density , kg / m^3
ρ_s	actual material density, kg / m^3
τ	mean residence time, s

14. References

Beltramo, C.; Rovero, G. & Cavaglià, G. (2009). Hydrodynamics and thermal experimentation on square-based spouted beds for polymer upgrading and unit scale-up. *The Can. J. Chem. Eng.*, Vol.87, 394-402

Cavaglià, G. (2003), "Reactor and process for solid state continuous polymerisation of poly-ethylene terephthalate (PET)" Patent EP 1576028 B1

Chatterjee, A. (1970). Spout-fluid bed technique. *Ind. Eng. Chem. Process Des. Develop*, Vol.9, 340-341

Epstein, N. & Grace, J. (2011). *Spouted and spout-fluid beds*. Cambridge Univ. Press, ISBN 978-0-521-51797-3,New York

Gishler, P.E. & Mathur, K.B. (1957). Method of contacting solid particles with fluids. U.S. Patent No. 2,786,280 to National Research Council of Canada

Grbavčić, Ž.B.; Vuković, D.V.; Hadžismajlovic, D. E.; Garić, R. V. & Littman, H. (1982). Fluid mechanical behaviour of a spouted bed with draft tube and external annular flow. *2nd Int. Symp. on Spouted Beds, 32nd Can. Chem. Eng. Conf.*, Vancouver, Canada

Geldart, D. (1973). Type of gas fluidization. *Powder Tech.*, Vol.7, 285-292

Malek, M.A. & Lu, B.C.Y. (1965). *I&EC Process. Des. Develop.*, Vol.4, 123-127

Mathur, K.B. & Epstein, N. (1974). *Spouted beds*, Academic Press, ISBN 0-12-480050-5, New York

McNab. G.S. (1972). Prediction of spout diameter. *Brit. Chem. Eng & Proc. Techn.*, Vol.17, 532

McNab, G.S. & Bridgwater, J. (1977). Spouted beds - estimation of spouting pressure drop and the particle size for deepest bed. *Proc. of European Council on Particle Technology*, Nuremberg, Germany

Metcalf, J. R. (1965-66). The mechanics of the screw feeder. *Proc. Inst Mech. Eng.*, Vol.180, 131-146

Murthy, D.V.R. & Singh, P.N. (1994). Minimum spouting velocity in multiple spouted beds. *The Can. J. Chem. Eng.*, Vol.72, 235-239

Piccinini, N. (1980). Particle segregation in continuously operating spouted beds, In: *Fluidization III*, J.R Grace & J.M. Matsen,(Eds), 279-285, Plenum Press, ISBN 0-306-40458-3, New York, USA

Rovero, G.; Piccinini, N. & Lupo, A. (1985). Vitesses des particules dans les lits à jet tridimensional et semi-cylindriques. *Entropie*, Vol.124, 43-49

Saidutta, M.B. & Murthy, D.V.R. (2000). Mixing behavior of solids in multiple spouted beds. *The Can. J. Chem. Eng.*, Vol.78, 382-385

17

Techno-Economic Evaluation of Large Scale 2.5-Dimethylfuran Production from Fructose

Fábio de Ávila Rodrigues and Reginaldo Guirardello
State University of Campinas, School of Chemical Engineerging
Brazil

1. Introduction

In an era of increasing oil prices and climate concerns, biofuels have gained more and more attention as potential fuel alternative energy sources. Governments have become active in the target of securing a supply of raw materials and limiting climate change, and many innovative proposals have been made, development work has started and potential candidate fuels have been studied in the energy area (Schaub & Vetter, 2008).

A number of factors must be considered when evaluating biofuels: technical factors (raw materials, supply, conversion and engines), economic (engine modification cost, infrastructure) and ecological/political (greenhouse gases, land use efficiency, oil dependence reduction) (Festel, 2008).

An end-user survey assessed car customer acceptance and attitude toward biofuels and revealed that their main demands are: price (48%), biofuel price should not exceed fossil fuels prices and there should be no cost in engine modification; environment (24%); consumption (19%) and performance (9%) (Festel, 2008).

Since customers consider the final cost as a decisive factor, the economic analysis is an important tool in the assessment of the success of biofuel production process and consequent market success. Achieving economic viability used to be the key to success, but today, other factors are important, such as sustainability.

Leshkov et al. (2007) show a catalytic strategy for the production of 2.5 dimethylfuran (DMF) from fructose (a carbohydrate obtained directly from biomass or by the isomerization of glucose) for use as a liquid transportation fuel. Compared to ethanol, 2.5-dimethylfuran has a higher energy density (by 40 percent), a higher boiling point (by 20K), and is not soluble in water. This catalytic strategy creates a route for transforming abundant renewable biomass resources into a liquid fuel suitable or the transportation sector and it is also a CO_2 free process.

The first step in production is to convert fructose to hydroxymethylfurfural (HMF) using an acid catalyst (HCl) and a solvent with a low boiling point in a biphasic reactor. The reactive aqueous phase in the biphasic reactor contains acid and sugar, and the extractive phase contains a partially miscible organic solvent (eg, 1-butanol) that continuously extracts HMF. The addition of a salt to the aqueous phase improves the partitioning of HMF into the extracting phase, and leads to increased HMF yields without the use of high boiling point solvents. Following, water, HCl and solvent evaporate, leading to precipitation of NaCl. Then, HMF is converted into DMF under a copper-ruthenium based catalyst. The final step involves the separation of DMF from the solvent and the non-reacted intermediates. As described below, the process also involves two separation steps. A schematic diagram of fructose conversion to DMF was reported by Leshkov et al. (2007).

The purpose of this paper is to evaluate economically the process production of DMF from fructose. In the present work the following analysis were carried out: Firstly, thermodynamic process modeling was investigated. Following this, the Process Flow Diagram (PFD) was developed from schematic diagram reported by Leshkov et al. (2007). The simulation stage makes use data from Leshkov et al. (2007). The thermal energy required for each piece of equipment was assessed with material and energy balances for each system using the UNISim™ software. Each piece of equipment is then approximately sized for economic analysis.

2. Thermodynamic modeling

The thermodynamic equilibrium of a system consisted of a multicomponent mixture takes place when temperature, pressure and chemical potential of components are equated between the phases, for each component. Although there are other basic criteria for system equilibrium, the minimization of Gibbs free energy is the condition which ensures equilibrium. Salt can affect the solubility of the system components with the formation of complex associations. In general it can be inferred that the particles (molecules, ions, or both) of dissolved salt tend to attract molecules from one of the system components more strongly than others.

The work of Debye and Huckel (1923) was the first important academic contribution and established a model for long-range interactions between ions based on the concept of ionic strength. A different family of models was developed using another extension of the Debye-Huckel model to represent the different contributions to Gibbs free energy excess. Implementation of the local composition to electrolytes means it is governed by local interactions such as short-range solvent/solvent, short-range ion/solvent and long-range ion/ion interactions that exist around the immediate neighborhood of a central ionic species (Aznar, 1996). For the contribution of short-range the following models of local composition can be used: Non-Random Two Liquid model (NRTL) (Renon and Prausnitz, 1968), UNIQUAC (Abrams and Prausnitz, (1975)) or UNIversal Functional Activity Coefficient (UNIFAC) (Fredenslund et al., 1977). The Debye-Huckel term or one of its variations, such as Fowler and Guggnheim (1949) or the Pitzer (1973) are used for long-range interactions. A series of different combinations have been proposed with these elements.

The hypothesis in this work was that salt should be treated as simple molecule, non-dissociated, rather than as charged ions distributed in the solution. Most works concerning the phase equilibrium in systems containing electrolytes distinguish long-range contributions due to electrostatic forces between ions and between ion and solvent from short-range contributions due to interactions between molecules. Two different models are then used for each contribution type. Considering salt as a simple molecule eliminates both contributions and requires only one appropriate model to describe the interactions between all molecules in solution, solvent or electrolyte (Aznar, 1996) Therefore, no specific model for electrolytes was used in this study. The UNIQUAC model was used to obtain the activity coefficient. According to Mock et al. (1986), although the contribution of long-range interaction of the equation of Pitzer-Debye-Huckel is important to obtain the ionic activity coefficient in the aqueous phase, it has little effect on the behavior of the equilibrium phase of the water-organic solvent system. Thus, the effect of the electrolyte is considered only for non-ideality, represented by the adjustable model parameters.

The binary interaction parameters of UNIQUAC model were estimated from experimental data (Santis et al., 1976a, 1976b), from Catté et al. (1994) and for the UNIFAC method. The tables 2, 3 e 4 show the data experimental used to estimate the binary interactions parameters. A Fortran programming language was used to determine the parameters from experimental data. The UNISIM™ software was used to estimate parameters for the UNIFAC method.

	Water	1-butanol	DMF	Fructose	HCl	HMF	NaCl
Water	-	89.22*	1543′	-153.35**	1160′	1361′	-455*
1-butanol	208*	-	383′	223′	1421′	530′	863*
DMF	249′	-90′	-	73′	-	-146′	-371′
Fructose	324**	91′	892′	-	-197′	1.412′	160′
HCl	-674′	-179′	-	399′	-	702′	-266′
HMF	-121′	-1155′	564′	162′	2,776′	-	2479′
NaCl	-165*	1251*	1793′	354′	2943′	1391′	-

* Binary interaction parameters of UNIQUAC model estimated from experimental data from Santis et al. (1976a, 1976b).
** Catté et al. (1994).
′ Binary interaction parameter estimated from UNIFAC method.

Table 1. Binary interaction parameters of UNIQUAC model

Aqueous Phase			Organic Phase		
Water	1-butanol	NaCl	Water	1-butanol	NaCl
92.60	7.4	-	20.4	79.6	-
92.04	6.8	1.16	18.78	81.2	0.025
91.64	6.1	2.26	17.45	82.5	0.045
90.85	5.8	3.35	16.64	83.3	0.061
90.60	5.0	4.40	15.43	84.5	0.074
89.96	4.6	5.44	14.6	85.3	0.086
89.16	4.4	6.44	14.1	85.8	0.095
87.84	3.7	8.46	13.29	86.6	0.110
86.30	3.3	10.4	12.48	87.4	0.122
85.10	2.7	12.2	11.37	88.5	0.130
83.50	2.5	14.0	10.66	89.2	0.138
82.20	2.0	15.8	8.75	90.1	0.144
80.90	1.7	17.4	9.05	90.8	0.148
79.40	1.5	19.1	8.55	91.3	0.153
78.00	1.3	20.7	7.94	91.9	0.156
76.70	1.1	22.2	7.34	92.5	0.159
75.20	0.9	23.9	7.04	92.8	0.162
74.00	0.8	25.2	6.54	93.3	0.164
73.30	0.8	25.9	6.43	93.4	0.167
73.30	0.8	25.9	6.23	93.6	0.167

Table 2. Liquid-liquid equilibrium in the system water-1-butanol-NaCl (Santis et al., 1976a)

Aqueous Phase			Organic Phase		
water	1-butanol	NaCl	water	1-butanol	NaCl
92.90	7.10	-	20.60	79.4	-
92.43	6.42	1.15	18.77	81.2	0.026
92.04	5.70	2.26	17.75	82.2	0.045
91.22	5.44	3.34	16.74	83.2	0.061
90.59	5.00	4.41	15.92	84.0	0.075
89.96	4.59	5.45	15.31	84.6	0.086
89.28	4.24	6.48	14.70	85.2	0.096
87.9	3.60	8.50	13.69	86.2	0.111
86.46	3.04	10.5	12.68	87.2	0.123
85.01	2.59	12.4	11.87	88.0	0.132
83.35	2.45	14.2	11.26	88.6	0.140
82.19	1.81	16.0	1.15	89.7	0.146
80.79	1.51	17.7	9.35	90.5	0.150
79.67	1.23	19.1	8,74	91.1	0.155
77.97	1.03	21.0	7.84	92.0	0.158
76.51	0.89	22.6	7.44	92.4	0.161
75.15	0.75	24.1	6.94	92.9	0.164
73.81	0.69	25.5	6.63	93.2	0.166
73.34	0.68	26.0	6.43	93.4	0.169

Table 3. Liquid-liquid equilibrium in the system water-1-butanol-NaCl (Santis et al., 1976a)

Aqueous Phase			Organic Phase		
water	1-butanol	NaCl	water	1-butanol	NaCl
93.40	6.60	-	21.40	78.6	-
92.85	5.96	1.19	19.68	80.3	0.021
92.44	5.25	2.31	18.36	81.6	0.041
91.88	4.70	3.42	17.54	82.4	0.060
90.86	4.61	4.53	16.83	83.1	0.072
90.21	4.21	5.58	16.31	83.6	0.085
89.49	3.86	6.65	15.50	84.4	0.095
88.04	3.27	8.69	14.49	85.4	0.111
86.62	2.78	10.6	13.58	86.3	0.122
85.19	2.31	12.5	12.77	87.1	0.131
83.70	1.90	14.4	11.96	87.9	0.139
82.10	1.70	16.2	11.25	88.6	0.146
80.92	1.28	17.8	10.35	89.5	0.151
79.59	1.01	19.4	9.44	90.4	0.156
78.18	0.82	21.0	8.84	91.0	0.160
76.74	0.66	22.6	8.34	91.5	0.163
75.36	0.54	24.1	7.83	92.0	0.166
73.9	0.50	25.6	7.53	92.3	0.170
73.42	0.48	26.1	7.03	92.8	0.173
73.42	0.48	26.1	7.03	92.8	0.173

Table 4. Liquid-liquid equilibrium in the system water-1-butanol-NaCl (Santis et al., 1976b)

3. Simulation

Simulation of DMF plant production was based on the standard conditions by Leshkov et al. (2007) from which it was developed a process flow diagram (PFD). The following unit operations compose the production plant: pumps, heat exchangers, one reactor for conversion of fructose into HMF (CRV-102) and one reactor for conversion of HMF into DMF (CRV-101), two stripping columns (T-100 and T-101), one distillation column (T-102). The volume of feed was of 30% fructose and the ratio between the organic phase and the aqueous phase volume was of 3.1 in the biphasic reactor (CRV-102). The conversion of fructose is 75% and the conversion of HMF to DMF is 100%.

Unreacted fructose was recycled back into the biphasic reactor. 1-Butanol was then separated from the water in the organic biphasic reactor. Cezário et al. (2009) proposed a separation system for water and 1-butanol composed by two stripping columns, one cooler and one settling tank. The formation of heterogeneous azeotrope turns this separation process more difficult and two liquid phases are formed in the decanter. This system can separate 98% of 1-butanol. Literature provides various processes for separating 1-butanol from water but the most traditional recovery process is distillation. Other techniques are adsorption, liquid-liquid extraction, evaporation and reverse osmosis. The energy required to recover 1-butanol by adsorption is of 1948 kcal/kg while the stripping column method requires 5789 kcal/kg. Other techniques such as perevaporation requires 3295 kcal/kg 1-butanol (Qureshi et al., 2005). The last step was to separate DMF from 1-butanol. The proposed separation system was composed by a distillation column (T-102) which separated 92% of DMF. The T-102 operates with reflux rate of 85 kgmol/h and top component (DMF) fraction of 0.9. The 1-butanol recovered in the T-102 was recycled.

Thus, material and energy balances were then solved using UNISim™ software and is showed in Table 5.

	1h	2h	3h	5h	8
Temperature (°C)	25	25	25	25	25
Pressure (kPa)	101	101	101	101	1351
Massic flow (kg/h)	896	1000	864	52	896
Enthalpy (kJ/kgmol)	-275200	-290600	-290500	-326600	-275200
Composition (massic fraction)					
water	0.9	0.7	0.5372	-	0.9
1-butanol	-	-	-	1	-
Fructose	-	0.3	-	-	-
Hmf	-	-	-	-	-
NaCl	-	-	0.4628	-	-
HCl	0.1	-	-	-	0.1
NaOH	-	-	-	-	-
DMF	-	-	-	-	-

	9	10	11	12	13
Temperature (°C)	25	25	180	180	180

Pressure (kPa)	1351	101	1351	1351	1351
Massic flow (kg/h)	1000	864	4142	896	1000
Enthalpy (kJ/kgmol)	-290600	-290500	-326600	-263000	-278700
Composition (massic fraction)					
Água	0.7	0.5372	-	0.9	0.9589
1-butanol	-	-	1	-	-
Frutose	0.3	-	-	-	0.0411
Hmf	-	-	-	-	-
NaCl	-	0.4628	-	-	-
HCl	-	-	-	0.1	-
NaOH	-	-	-	-	-
DMF	-	-	-	-	-

	14	16	15	18	19
Temperature (°C)	180	180	180	180	37
Pressure (kPa)	1351	1351	1351	1351	1351
Massic flow (kg/h)	864	7392	489	7392	7392
Enthalpy (kJ/kgmol)	-260600	-291400	-274800	-277200	-294200
Composition (massic fraction)					
water	0.7901	0.3250	0.8825	0.3250	0.3250
1-butanol	-	0.5603	-	0.5603	0.5603
Fructose	-	0.0484	0.1174	0.0121	0.0121
Hmf	-	-	-	0.0363	0.0363
NaCl	0.2098	0.0541	-	0.0541	0.0541
HCl	-	0.0122	-	0.0218	0.0218
NaOH	-	-	-	0	0
DMF	-	0.00005	-	0	0

	22	29	30	25	24
Temperature (°C)	37	37	37	80	81
Pressure (kPa)	1351	1351	1351	50	50
Massic flow (kg/h)	7592	4876	2716	23575	489
Enthalpy (kJ/kgmol)	-296500	-311400		-240600	-282400
Composition (massic fraction)					
water	0.3353	0.0887	0.7781	0.7995	0.8822
1-butanol	0.5456	0.8444	0.0091	0.1876	-
Fructose	0.0118	0.0118	0.01168	-	0.1178
Hmf	0.0353	0.0549	-	0.01138	-
NaCl	0.0719	-	0.2011	-	-
HCl	-	-	-	-	-
NaOH	-	-	-	-	-
DMF	-	-	-	0.0015	-

	26	27	Organic phase	Aqueous phase	32

Temperature (°C)	74	71	36	36	71
Pressure (kPa)	50	50	50	50	50
Massic flow (kg/h)	78804	55229	59614	19189	55230
Enthalpy (kJ/kgmol)	-224500	-197800	-245800	-284100	-197800
Composition (massic fraction)					
water	0.2585	0.0275	0.0255	0.4822	0.02752
1-butanol	0.1716	0.1647	0.2217	0.01592	0.1647
Fructose	-	-	-	-	-
Hmf	0.00341	-	0.004517	-	-
NaCl	-	-	-	-	-
HCl	-	-	-	-	-
NaOH	-	-	-	-	-
DMF	0.5665	0.8077	0.7483	0.00177	0.8077

	33	34	35	38	39
Temperature (°C)	100	101	219	71	99
Pressure (kPa)	50	1650	1650	50	50
Massic flow (kg/h)	4384	4384	4384	290	4094
Enthalpy (kJ/kgmol)	-390900	-308800	-279700	-215100	-310900
Composition (massic fraction)					
water	-	-	-	-	-
1-butanol	0.9388	0.9388	0.9388	0.07891	0.9999
Fructose	-	-	-	-	-
Hmf	0.0611	0.0611	0.0611	-	-
NaCl	-	-	-	-	-
HCl	-	-	-	-	-
NaOH	-	-	-	-	-
DMF	-	-	-	0.9221	0.000086

	40
Temperature (°C)	-
Pressure (kPa)	0.9388
Massic flow (kg/h)	-
Enthalpy (kJ/kgmol)	0.0611
Composition (massic fraction)	
water	-
1-butanol	-
Fructose	-
Hmf	-
NaCl	-
HCl	0.9388
NaOH	-
DMF	0.0611

Table 5.

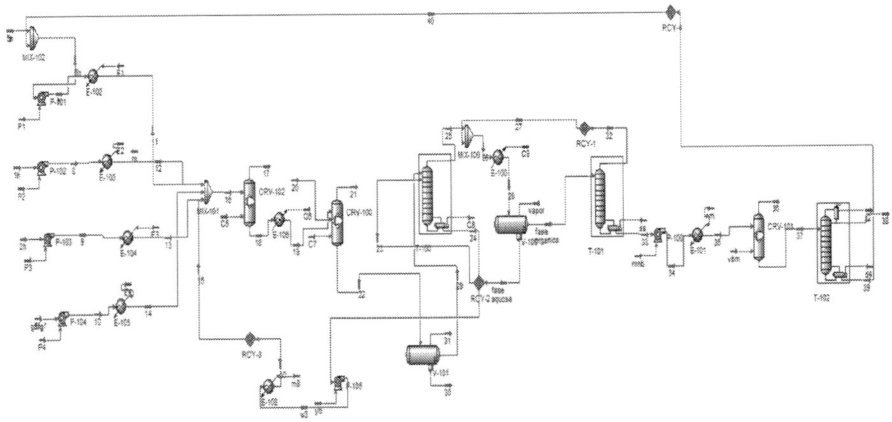

Fig. 1. Material and energy balance for each stream in DMF production plant.

4. Economic evaluation

The economic evaluation was based on the spreadsheets by Peters & Timmethaus (2003). The following steps were used by performed the economic analysis.

i. On the sheet '**Capital Inv**.' The estimated current total purchased cost of the equipament was entered. The results are showed in Table 7.
ii. On the sheet '**Materials & Labor**' the product prices and flowrates, the raw materials prices and flow rates, and the labor requirements were entered. The results are showed in Table 8.
iii. On the sheet '**Utilities**' the quantity of each utility needed annually was entered in appropriate units. The total annual utilities cost is transferred to sheet '**Annual TPC**';
iv. The '**Depreciation**' sheet is used only if the user wishes to change the default (5-year **Modified Accelerated Cost Recovery System (**MACRS) depreciation method);
v. On the '**Annual TPC**' sheet, all values were calculated from information available on other sheets. The Calculated annual TPC was transferred to '**Evaluation**'. The results are showed in Table 9.
vi. The sheet '**Evaluation**' used values from other sheets to calculate the common profitability measures. All calculations in 'Evaluation' are made in current (i.e. inflated) dollars.

Each piece of equipment was roughly sized from material and energy balance and the approximate cost determined. Costs of equipment operating at ambient pressure and using carbon steel, were estimated by Eq. (1) (Turton et al., 2003).

$$\log Cp^{\circ} = K_1 + K_2 \log(A) + K_3(\log(A))^2 \quad (1)$$

Where A is the equipmen t capacity or size parameter and K_1, K_2 and K_3 are the parameters (Turton et al., 2003). The effect of time, operating conditions and material construction on

Equipment number	equipment	Parameter equipment	Equipment cost ($), CEPCI = 499,6
P-101	pump	flow: 4.97 m^3/h	3814
P-102	pump	flow: 0.92 m^3/h	2910
P-103	pump	flow 0.864 m^3/h	2898
P-104	pump	flow: 0.63 m^3/h	2845
P-105	pump	flow: 0.54 m^3/h	2823
P-100	pump	Flow: 5.27 m^3/h	4670
E-102	heater	Heaty duty: 302, 5 KW	64862
E-103	heater	Heaty duty: 160 KW	42107
E-104	heater	Heaty duty: 134 KW	37333
E-105	heater	Heaty duty: 107 KW	320119
E-106	heater	Heaty duty: 8790 KW	637790
E-108	heater	Heaty duty: 100 KW	30610
E-100	cooler	Water flow: 198 m^3/h	11421
E-101	heater	Heaty duty:467 KW	87085
CRV-102	reactor	Heaty duty: 10000 KW	2917722
CRV-100	reactor	Heaty duty: 344,75 KW	22354
CRV-101	reactor	Heaty duty: 10000 KW	2917722
T-100	stripping column	Height: 4 m; diameter: 1m	15003
T-101	stripping column	Height: 4 m; diameter: 1m	15003
T-102	destilation column	Height: 4 m; diameter: 1m	15003
Total			**1609365**

Table 6. Equipment parameter

	Fraction of delivered equipment			% chosed (B)	Calculated values, million ($)
	Solid processing plant (A)	Solid fluid processing plant (B)	Fluid processing plant (C)		
Cost directs					
Purchased equipment, E′					1.609
Delivery, fraction of E′	0.10	0.10	0.10	0.10	0.1604
Purchased equipment instalation	0.45	0.39	0.47	0.39	0.6903
Instrumentation & controls (installed)	0.18	0.26	0.36	0.26	0.460
Piping (installed)	0.16	0.31	0.68	0.31	0.548
Eletrical systems (installed)	0.10	0.10	0.11	0.10	0.177

	Fraction of delivered equipment			% chosed (B)	Calculated values, million ($)
	Solid processing plant (A)	Solid fluid processing plant (B)	Fluid processing plant (C)		
Buildings (incluing services)	0.25	0.29	0.18	0.29	0.513
Yard improvements	0.15	0.12	0.10	0.12	0.212
Service facilities (installed)	0.40	0.55	0.70	0.55	0.973
Total direct costs					5.345
Indirect costs					
Engineering and supervision	0.33	0.32	0.33	0.32	0.566
Construction expenses	0.39	0.34	0.41	0.34	0.602
Legal expenses	0.04	0.04	0.04	0.04	0.071
Contractor's fee	017	0.19	0.22	0.19	0.336
contigency	0.35	0.37	0.44	0.37	0.655
Total indirect costs					2.23
Fixed capital investimento					7.575
Working capital	0.70	0.75	0.89	0.75	1.327
Total capital investment					8.903

Table 7. Estimation of capital investment by percentage of delivered equipment method

Material	classification	Price (US$/kg)	Annual amount (million kg/year)	Annual value (million US$/year)
DMF	Product	variable	2.217	0.88
Fructose	Raw material	0.78 (variable)	2.484	1.94(62%)
NaCl	Raw material	0.015	0.05	0.0007(0.02%)
HCl	Raw material	0.295	0.745	0.22(7.7%)
1-butanol	Raw material	1.72	0.431	0.7906(23%)
water	Raw material	0.08	16.312	1.30(2.6%)
NaOH	Raw material	0.10	25	2.5(2.7%)
H_2	Raw material	10	0.13834	1.3834(1.4%)

Table 8. Annual raw material costs and products values

Item (A)	Factor (B)	Basis (C)	Basis Cost (million) US$/year (D)	Cost (milion) US$/year (E)
1. Raw materials				4.204
2. operating labor (M)				0.885
3. operating supervision	0.15	de (2)	0.885	0.133
4. utilities				0.55
5. maintenance and repairs (MR)	0.06	de FCI	1.407	0.46
6. operating supplies	0.15	de (5)	0.084	0.07
7. laboratory charges	0.15	de (2)	0.885	0.133
8. Royalties	0.01	de c_o	3.090	0.08
9. catalysts and solvents	0			2.5
Variable cost				6.536
10. taxes (property)	0.02	de FCI	1.407	0.156
11. financing (interest)	0	de FCI	1.407	0
12. insurance	0.01	de FCI	1.407	0.014
13. rent	0		1.407	0.078
14. depreciation		Calculated separately		
C				0.233
15. plant overhead, general	0.6		1.102	0.891
16. administration				0.661
17. manufacturing cost				7.660
18. administration	0.2	de (2), (5)	1.102	0.297
19. distribution & selling	0.05	de c_o	3.090	0.437
20. Research & development	0.04	de c_o	3.090	0.350
General expense				0.469
Total product cost without depreciation = c_c				8.744

Table 9. Annual total product cost at 100 % capacity

purchased equipment cost was corrected by time factor (I), material factor (F_M) and conditions factor (F_P). Purchased equipment cost is then expressed by:

$$Cp = Cp^{C} \times F_M \times F_P \times I \qquad (2)$$

In this work, inflation was account ed for by the Chemical Engineering Plant Cost Index (Lozowski, 2010). According the Table 8 the price of raw material and solvent are the more expensive.

With chemical and utility cost were obtained and a discounted cash flow analysis was performed to determine profitability. The quantities of chemical material, utilities and production of DMF were doubled, tripled, etc, from the simulated plant, to achieve sale price and cost DMF similar to gasoline and ethanol. However, the equipment cost increased according to Eq. (3). For all sale price and cost DMF from Tables 5, 6 e 7 the profitability measures were: 15.0 %/year (return on investment) and a 3.6 year payback period.

The conversion and the price fructose were changed too. The tables 10, 11 and 12 show the results.

	Conversion (75%)	Conversion (80%)	Conversion (85%)
Fructose price, 0.78 US$/kg	6.00/3.94	5.90/3.83	5.80/3.75
Fructose price, 0.50 US$/kg	5.70/3.59	5.60/3.49	5.40/3.42
Fructose price, 0.10 US$/kg	5.20/3.10	5.00/3.00	4.90/2.94

Table 10. Sale price of DMF/cost (US $/kg) of DMF to several conversions in reactor CRV-102 and various prices of fructose (standard plan used in the simulation).

	Conversion (75%)	Conversion (80%)	Conversion (85%)
Fructose price, 0.78 US$/kg	3.70/2.68	3.60/2.59	3.52/2.54
Fructose price, 0.50 US$/kg	3.30/2.33	3.20/2.26	3.13/2.22
Fructose price, 0.10 US$/kg	2.70/1.83	2.63/1.78	2.58/1.74

Table 11. Sale price of DMF/cost (US $/kg) of DMF to several conversions in reactor CRV-102 and various prices of fructose (scale factor 12).

	Conversion (75%)	Conversion (80%)	Conversion (85%)
Fructose price, 0.78 US$/kg	2.68/1.95	2.60/1.89	2.55/1.85
Fructose price, 0.50 US$/kg	2.27/1.60	2.20/1.56	2.15/1.52
Fructose price, 0.10 US$/kg	1.68/1.10	1.63/1.07	1.60/1.05

Table 12. Sale price of DMF/cost (US $/kg) of DMF to several conversions in reactor CRV-102 and various prices of fructose (scale factor 30).

In table 12 observes that the sale price of DMF can be compared with the gasoline. The cost of DMF decreases with the increase of the conversion of fructose to HMF and with the price decrease of fructose.

$$\text{Equipament cost} = N^{0.6} \tag{3}$$

Where N is the scale factor with values of N = 2, 5, 10, etc.

Thus, the estimated cost of the equipment was U.S. $ 12 million, fixed capital investment was U.S. $58 million, direct cost were U.S. $41 million, indirect costs were U.S. $17 million, working capital was U.S. $10 million and total capital investment was U.S. $68 million. From economic evaluation the value and cost DMF was 2.68 U.S. $/kg and 1.95 U.S $/kg, respectively. For this analysis, the plant is economically feasible for a scale factor of thirty (N= 30).

5. Conclusions

The following conclusions can be drawn from the facts presented in the above review. In the thermodynamic analysis salt is considered a solute, so it´s possible to use the model UNIQUAC (Mock et al., 1986). The estimation of binary interaction parameters for UNIQUAC in the system water-butanol-salt was carried out with Fortran software from liquid-liquid equilibrium data and UNIFAC (UNIQUAC Functional-group Activity Coefficient) method was used to estimate remain parameters. The separation system (composed by two stripping columns, one cooler and one settling tank) used to separate 1-butanol and water recovery 98% of 1-butanol. The separation system (composed by distillation column) used to separate DMF recovery 92 % of DMF. Economic evaluation showed that a suitable operational plant could work with 12.4 tons/year of fructose. It could produce 11.1 tons/year of DMF. The fixed capital investment in plant and equipment is estimated at U.S. $ 58 million and U.S. $ 12 million, respectively. The DMF value was 2.69 U.$. $/kg. For this analysis, the plant is economically feasible, from comparison with a reference market of 15.0 %/year (return on investment) with a 3.6 year payback period. This analysis suggests that DMF production from fructose deserves serious consideration by investors.

6. References

Abrams, D.S. & Prausnitz, J.M. (1975). Statistical Thermodynamics of Liquid Mixtures: A New Expression for the Excess Gibbs Energy of Partly or Completeley Miscible Systems. *American Institute of Chemical Engineers Journal*, Vol.21, pp. 116-128.

Allen, D.H. (1980). *A Guide to Economic Evaluation of Projects*, The Institutions Chemical Engineers, Rugby, United Kingston.

Aznar, M. (1996). *Equilíbrio Líquido-Vapor de Sistemas com Eletrólitos Via Contribuição de Grupo*, Universidade Federal do Rio de Janeiro, Rio de Janeiro, Brazil.

Catté, M.; Dussap, C.G.; Achard, C. & Gros, J.B. (1994). Excess Properties and Solid-Liquid Equilibria for Aqueous Solutions of Sugars Using a UNIQUAC Model. *Fluid Phase Equilibria*, Vol.96, pp. 33-50.

Cezário, G. L.; M. Filho, R. & Mariano, A. P. (2010). Projeto e Avaliação Energética do Sistema de Destilação de uma Planta de Fermentação Extrativa para a Produção de Biobutanol, *Proceedings of XVII Congresso de Iniciação Científica da Universidade Estadual de Campinas*, Campinas, Brazil.

Debye, P. & Huckel, E. (1923). Zur Theorie der Elektrolyte. *Physics Zeitsch*, Vol.24, pp. 185-206.

Festel, G. W. (2008). Biofuels - Economic Aspects. *Chemical Engineering Technology*, Vol.31, pp. 715-720.

Fowler, R.H. & Guggenheim, E.A. (1949). *Statistical Thermodynamics*, Cambridge University Press, Cambridge.

Fredenslund, A.A.; Gmehling, J. & Rasmussen, P. (1977). *Vapor-Liquid Using UNIFAC*, Elsevier, Amsterdam.

Leshkov, Y.R.; Barrett, C. J.; Liu, Z.Y. & Dumesic, J.A. (2007). Production of Dimethylfuran for Liquid Fuels from Biomass-Derived Carbohydrates. *Nature*, Vol.447, pp. 982-986.

Lozowski, D. (2010). Economic Indicators. *Chemical Engineering*, Vol.117, pp. 55.

Mock, B.; Evans, L.B. & Chen, C.C. (1986). Thermodynamic Representation of Phase Equilibria of Mixed-Solvent Electrolyte Systems. *Association International of Chemical Engineers Journal*, Vol.32, pp. 1655-1664.

Peters, M.S.; Timmerhaus, K.D. & West, R.E.W. (2003), *Plant Design and Economics for Chemical Engineers*, McGraw-Hill, New York.

Pitzer, K.S. (1973). Thermodynamics of Electrolytes I. Theoretical Basis and General Equation. *Journal Physics Chemical*, Vol.77, pp. 268-277.

Qureshi, N.; Hughes, S.; Maddon, I.S. & Cotta, M.A. (2005). Energy-Efficient Recovery of Butanol from Model Solutions and Fermentation Broth by Adsorption. *Bioprocess and Biosystems Engineering*, Vol.27, pp. 215-222.

Renon, H. & Prausnitz, J. M. (1968). Local Compositions in Thermodynamics Excess Functions for Liquid Mixtures. *American Institute of Chemical Engineers Journal*, Vol.14, pp. 135-144.

Turton, R.; Bailie, R.C.; Whiting, W.B. & Shauwitz, J.A. (2003), *Analysis, Synthesis, and Design of Chemical Processes*, Prentice Hall, New Jersey.

Santis, R.; Marrelli, L. & Muscetta, P.N. (1976a). Liquid-Liquid Equilibria in Water-Aliphatic Alcohol Systems in the Presence of Sodium Chloride. *Chemical Engineering Journal*, Vol.11, pp. 207-214.

Santis, R.; Marrelli, L. & Muscetta, P.N. (1976b). Influence of Temperature on the Liquid-Liquid Equilibrium of the Water-n Butyl Alcohol-Sodium Chloride System. *Journal of Chemical and Engineering Data*, Vol.21, pp. 324-327.

Schaub, G. & Vetter, A. (2003). Biofuels for Automobiles - An Overview. *Chemical Engineering Technology*, Vol.31, pp. 721-729.

UNISim Honeywell (2007), http://www.honeywell.com.

18

Inland Desalination: Potentials and Challenges

Khaled Elsaid[1], Nasr Bensalah[1,2,*] and Ahmed Abdel-Wahab[1]
[1]Department of Chemical Engineering,
Texas A&M University at Qatar,
Education City, Doha
[2]Department of Chemistry, Faculty of Sciences of Gabes,
University of Gabes, Gabes
[1]Qatar
[2]Tunisia

1. Introduction

Groundwater is the main source of drinking water in many countries all over the world. In absence of surface water supply, the use of groundwater as the main water source for drinking, industrial, and agricultural use becomes essential especially in the case of rural communities. Underground reservoirs constitute a major source of fresh water, in terms of storage capacity; underground aquifers worldwide contain over 95% of the total fresh water available for human use. Typical groundwater supplies have low coliform counts and total bacterial counts, low turbidity, clear color, pleasant taste, and low odor. Accordingly, groundwater has higher quality than surface water, and the quality is quite uniform throughout the year that makes it easy to treat. A disadvantage of groundwater supplies is that many groundwater aquifers have moderate to high dissolved solids such as calcium, magnesium, iron, sulfate, sodium, chloride, and silica. The high concentration of dissolved solids particularly, sodium chloride, makes the water brackish and thus requires to be desalinated before its use for a certain purpose.

With the growth of membrane science, reverse osmosis RO overtook multi stage distillation MSF as the leading desalination technology. In the last two decades, RO processes have advanced significantly, allowing new brackish groundwater desalination facilities to use RO technology much more economically than distillation. RO treatment plants use semipermeable membranes and pressure to separate salts from water. These systems typically use less energy than thermal distillation, leading to a reduction in overall desalination costs.

The reverse osmosis process enables now the massive production of water with a moderate cost, providing flexible solutions to different necessities within the fields of population supply, industry and agriculture. The great development of reverse osmosis (RO) technology has been a consequence of several factors such as reduction in energy consumption and membrane cost. Nevertheless, the major problem of RO desalination

* Corresponding Author

plants is the generation of a concentrate effluent (brine) that must be properly managed. Disposal of such brines presents significant costs and challenges for the desalination industry due to high cost and environmental impact of brine disposal.

The reject brine from desalination plants not only contains various types of salts at higher concentration but also other types of wastes like pretreatment chemicals (antiscalents, antifouling,...etc). Also, if the feed water includes harmful chemicals such as heavy metals or others, these chemicals are concentrated in the reject brine. Improper disposal of reject brine from inland plants results in several environmental problems.

Although sea disposal of reject brine is a common practice for plants located in coastal areas, it would not be available for inland desalination. Deep well injection is prohibitly expensive and has its own problems such as the possibility of corrosion and subsequent leakage in the well casing, seismic activity which could cause damage to the well and subsequently result in ground water contamination, and uncertainty of the well life. Additionally, when a sewerage system is used for disposal of concentrate brine high in total dissolved solids (TDS) the treated municipal sewage effluent becomes unsuitable for reuse.

While operation and maintenance costs for evaporation ponds are minimal, large land areas are required, and pond construction costs are high. Even in arid climates ideally suited for evaporation, a typical design application rate is only 2 gpm per acre. The construction cost for an evaporation pond with a liner and monitoring system typically ranges between \$100,000 and \$200,000 per acre, exclusive of land cost. Thus a concentrate flow as small as 100 gpm would require a pond area of at least 50 acres and cost \$10 to \$20 million to construct. Consequently, evaporation ponds are often cost prohibitive and impractical for handling any significant concentrate flow. Furthermore, water evaporated from a pond is often a lost resource.

Therefore, the need to protect surface and groundwater resources may in many cases preclude concentrate disposal by the earlier three methods. The alternative is zero liquid discharge (ZLD). In ZLD, concentrate is treated to produce desalinated water and essentially dry salts. Consequently there is no discharge of liquid waste from the process. Most ZLD applications in operation today treat industrial wastewater or power plant cooling water using thermal crystallization, evaporation ponds, or a combination of these technologies. Thermal crystallization is energy-intensive with high capital and operating costs.

Given the need for ZLD and the disadvantages of existing ZLD methods, it is imperative to find alternative ZLD treatment technologies that provide more affordable concentrate management. This article reviews current trends and potential advancements of inland desalination and brine management alternatives.

2. Water and water resources

Water resources present naturally in the environment can be generally divided into freshwater and saline water according to the amount of dissolved solids it contains. Quality and quantity of different water resources are of high importance, many efforts are being made to have good estimates of water resources at both worldwide and country levels. In 1990s "The comprehensive assessment of the freshwater resources of the world" was launched to have estimates on worldwide water resources (United Nations [UN], 1997). The

AQUASTAT program (Food and Agriculture Organization [FAO]) was launched to form global information system on water and agriculture, the main objective of the program is to collect and analyze information on water resources, water uses, and agriculture water management within different countries. Information on the quantity of major water resources is present in table 2.1.

Water Resource	Volume, (1000 km^3)	Percent of total water	Percent of total fresh water
Saline water:			
Oceans/seas	1,338,000	96.54	-
Saline/brackish groundwater	12,870	0.93	-
Saltwater lakes	85	0.006	-
Freshwater:			
Glaciers and permanent snow covers	24,064	1.74	68.70
Fresh groundwater	10,530	0.76	30.06
Fresh lakes	91	0.007	0.26
Wetlands	11.5	0.001	0.03
Rivers	2.12	0.0001	0.006

Table 1. Estimates of major water resources on Earth (Gleick, 2001).

Freshwater is the water naturally found on Earth's surface and in underground aquifers such as surface water, fresh groundwater, and glaciers, and mainly characterized by its low content of dissolved solids. These water sources are considered to be renewable resource, by effect of natural water cycle. The quantity of freshwater present on Earth is around 2.5% only of the total water present on Earth.

Surface water is the water present in rivers, fresh lakes, and wetlands; the main source of surface water is by precipitation in the form of rain, snow....etc. Surface water is characterized by low content of dissolved salts generally below 500 mg/L. Surface water represents only around 0.3 % of the total freshwater present on Earth's surface. Fresh groundwater is the water located under the Earth's surface i.e. subsurface water which is mainly located in pores or spaces of soil and rocks, or in aquifers below the water table. It is mainly characterized by its low suspended solids. In many places groundwater contains high content of dissolved salts compared to that of surface water; with salinity level around 500-2,000 mg/L (Mickley, 2001). Groundwater represents around 0.76 % of the total water present on Earth, and around 30 % of the freshwater available on Earth.

Water or ice present in glaciers, icebergs, and icecaps represents the vast majority freshwater, this huge amount of water is currently unused and locked up in southern and northern poles. Up to date there is no efforts has been made to make use of such water resources due to the high cost associated with its processing as it is mainly present in very distant areas or at very high altitudes.

Brackish groundwater is the water located under the Earth's surface and it is characterized by its higher salinity than that of fresh groundwater with values of 2,000-10,000 mg/L

(Mickley, 2001). It is mainly present in aquifers that are much deeper than that of fresh groundwater. Brackish groundwater represents around 0.93 % of the total water present on Earth.

Saline or salty water is the water that contains considerable amount of salts and it is mainly found in oceans, seas, saline or brackish groundwater, and saltwater lakes. Saline water represents the majority of water resources in terms of quantity with around 97.5 % of the total water present on Earth. the salinity of seawater varying from one location to another, from around 21 g/L in the North Sea to 40 - 45 g/L in the Arabian Gulf and Red Sea, and even up to 300 g/L as in the Dead Sea (Gleick, 2006).

The majority of world population use surface water or groundwater as the main source for domestic, agriculture, and industrial water supplies. The most common surface water sources are rivers, and lakes. However, the most common groundwater sources are pumped wells or flowing artesian wells. In absence of surface water supply, it is clear that the use of groundwater becomes essential especially in the case of rural communities. Underground reservoirs constitute a major source of fresh water, in terms of storage capacity; underground aquifers worldwide contain over 95% of the total fresh water available for human use. In addition when looking to the map of worldwide water stress in figure 2.2, we find that areas that face water stresses are increasing with time with Middle East, North Africa, and Central Asia having the highest water stresses, while when looking to the worldwide groundwater resources map as shown in figure 2.3, we find that most of these areas have access to groundwater resources, which means that groundwater will present the relief to the faced water stress problems, and even can support the different developmental planes of these areas.

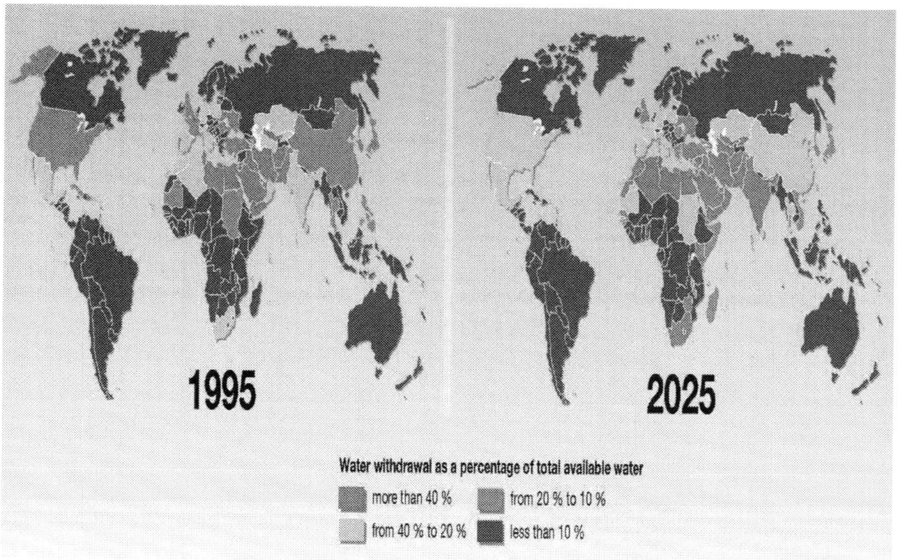

Fig. 1. Worldwide water stresses map (United nations Environmental Programme [UNEP]).

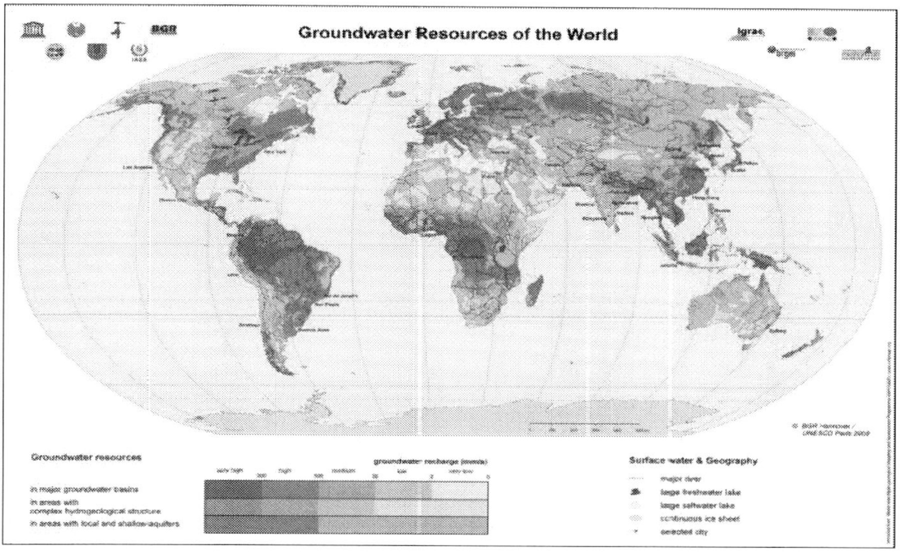

Fig. 2. Worldwide groundwater resources mapping (The Federal Institute for Geosciences and Natural Resources).

3. Groundwater quality

Pure water is a colorless, odorless, and tasteless liquid, water is polar and strong solvent capable of dissolving many natural and synthetic substances, many inorganic and organic compounds, in addition it is able to suspend many solids, and hence it is very hard to find pure water in nature.

The quality of groundwater is mainly determined by its content of dissolved solids and gases, presence of suspended solids, and bacteria present. Usually the nature and concentration of dissolved solids present in groundwater source will depend on characteristics of the aquifer and on travelling time or velocity of groundwater flow through the rock formation (Delleur, 2007).

The physical properties of water are mainly, the total, suspended, and dissolved solids, in addition to turbidity, temperature, color, odor and taste. Typical groundwaters, as they undergo natural filtration while passing through sand formations have very low suspended solids content, low turbidity, clear color, pleasant taste, and low odor. However, while water is travelling through the soil formations, groundwater may carry dissolved solids if the soil formations are relatively soluble. The type and concentration of dissolved solids released from soil to groundwater may vary based on the soil composition, travelling time and flow velocities.

Chemical characteristics are mainly concerned with pH value, cations and anions, alkalinity, acidity, hardness, dissolved gases, and other contaminants such as organic substances and heavy metals that might be present in water. Groundwaters in general have higher hardness when compared to surface waters; this is mainly due to the dissolution of limestone and dolomite formations which in turn increase the content of calcium ions and hence increasing the hardness of water.

Biological characteristics are concerned with the living organisms present in water, mainly bacteria, fungi, algae, and viruses. Microbes are generally absent from groundwater due to natural filtration, and hence groundwater has low coliform and total bacterial counts.

In general groundwater has higher quality than surface waters, and the quality is quite uniform throughout the year making it easy to treat. A disadvantage of groundwater is that many have moderate to high dissolved solids. The high concentration of dissolved solids, particularly sodium chloride, makes the water brackish. The removal of such dissolved solids requires use of desalination process to treat the water to a level of dissolved solids that makes it suitable for a certain use.

Technical and economical evaluations of desalinating groundwater has started early in 1960s. One of the first major brackish groundwater desalination plants was built in Florida, USA with capacity of 2.62 mgd, followed by another one using Reverse Osmosis RO in mid-1970s with capacity of 0.5 mgd (Bart Weiss, 2002). A combination of both field investigations and computer simulation modeling were used to assess the economic suitability of using highly brackish groundwater for large scale abstraction for feeding reverse osmosis desalination plant in many countries around the world (Brich et al., 1985; Hadi, 2002; Sherif et al., 2011; Zubari, 2003). The results from such technical and economical evaluations have shown that brackish groundwater can be considered as high quality source of feedwater for desalination plants, even at higher salinities, which still far below the salinity of seawater. However the evaluations also indicated that the groundwater quality is changing and not constant over long years with continuous increase in dissolved solids content, which should be considered during design stages of such desalination plants.

It was concluded that the major misconception of considering that the groundwater quality is relatively stable over time is very critical for design purposes. The groundwater quality tends to show increase in salinity with time, and hence the initial and future groundwater qualities should be considered (Missimer, 1994). The deteriorating quality of groundwater is becoming of high concern globally, mainly due to human activities such as over abstraction and seawater intrusion. The seawater intrusion is a very common problem worldwide especially, near coastal areas, it is a natural phenomenon in which saline water from sea or ocean moves into the fresh groundwater in coastal aquifers. This behavior is mainly attributed to the density difference and to tidal effects. Seawater intrusion was found to be the main source for the increased salinity of near-coast aquifers in many places worldwide (Amer et al., 2008; Zubari et al., 1994).

4. Inland desalination processes

Desalination is a water treatment process for removing total dissolved solids (TDS) from water. Desalination of seawater and brackish water has become a reliable method for water supply all-over the world and had been practiced successfully for many decades.

The worldwide desalination capacity increased dramatically from around 35 million m^3/d in 2005 (Gleick, 2006) to about 60 million m^3/d by 2009, with the largest desalination plant of 880,000 m³/d at Shoaiba 3 project in Saudi Arabia (International Desalination Association [IDA], 2009). By 2009 there were 14,451 desalination plants online, with further 244 known to be on their way, being under contract or under construction, with an additional capacity of 9.1 million m³/d in 130 countries around the world (IDA, 2009). Desalination in general can be mainly classified according to the feed water source into two main classes as seawater desalination and brackish groundwater desalination. In 2006 around 56 % of the world desalination capacity was for seawater desalination, and 24 % for brackish water, with Saudi Arabia KSA, Unite states US, United Arab Emirates UAE, Spain, Kuwait, and Japan having desalination capacity over 1 million m³/d, with Middle East countries holding over 50 % of the worldwide desalination capacity (Gleick, 2006).

Desalination processes are many and can be generally classified according to the technology used and it mainly classified to thermal desalination, and membrane desalination, ion exchange, and electrodialysis. In addition, other new technologies are still under development. The common processes for desalination have been changed from expensive techniques with extensive energy requirements to a sustainable method for drinking water supply.

The selection of desalination process for a certain purpose depends on many factors such as energy cost, final water quality, fouling propensity, temperature, and overall process cost. Table 2 shows the worldwide, US, and Gulf Corporation Council GCC desalination capacity by processes.

Process Type	Worldwide	United States	GCC Countries
RO	46	69	28
MSF	36	1	54
ED	5	9	-
VC	5	3	-
MED	3	1	9
NF	-	15	-
Others	5	2	9

Table 2. Desalination capacity percentage distribution according to process type (Gleick, 2006).

The quality of desalinated water differs depending on the process used, with thermal desalination producing very high quality product water with salinity about 2-10 mg/L; which usually requires remineralization in the post treatment step (Gabbrielli, 1981). However for membrane process the quality of product water depends on many factors such as the quality of feedwater, design recovery, and membrane properties, but in general the product water will have higher salinity than that produced from thermal desalination.

The quality of the final water product mainly depends on the application in which it is to be used for, ranging from very high quality for process water to specified quality as per the regulation for drinking water, or to a certain quality suitable for agriculture. Most of the desalination plants worldwide are designed to be able to produce high quality water. Drinking water standards vary slightly from country to another. The most widely used is the United States Environmental Protection Agency US-EPA drinking water standards with

500 mg/L for total dissolved solids, and 250 mg/L for each of sulfate and chloride(US EPA, 2009). However the World Health Organization WHO guidelines for drinking water quality suggest that total dissolved solids value of 600-1,000 mg/L is generally acceptable (WHO, 2011). On the other hand the FAO suggests that a total dissolved solids up to 2,000 mg/L is acceptable for irrigation purposes (Ayers & Westcot, 1994).

4.1 Thermal desalination processes

In desalting operations, thermal technologies were the only viable option for long time, and Multi-Stage Flash Distillation (MSF) was established as the baseline technology; however Multiple-Effect Distillation (MED) is now the state-of-the-art thermal technology, but has not been widely implemented yet. Thermal desalination plants have provided the major portion of the world's desalination capacity, as the world's requirements for treated water increase. Thermal desalination is usually used for cases where high salinity feed waters are used i.e. seawater, high recoveries are required, high feedwater temperature, and low energy cost, however the main drawback is the extensive energy consumption (Greenlee et al., 2009).

Thermal desalination or some time called phase change desalination; is a very basic process in concept as it copies the natural process of water cycle, where energy in thermal form through the solar radiation evaporates the water into water vapor, which is condensed later and fall in the form of rains or snow (Gleick, 2001), so in thermal desalination thermal energy or heat is applied to the water present in boiler or evaporator to drive water evaporation, this water vapor is condensed later in condenser by exchanging heat, thus sometimes the process is called phase change desalination as water phase changes from liquid into vapor is encountered. Even though the basic concept is the same however there are many processes which utilize that concept in application today, the main thermal desalination technologies are Multi-Stage Flash Distillation (MSF), Multiple-Effect Distillation (MED), and Vapor Compression (VC).

4.2 Membrane desalination processes

With the growth of membrane science, RO overtook MSF as the leading desalination technology, membrane desalination processes in general and commercial RO processes in particular have been undergoing appreciable development. Important factors in the expansion of commercial RO applications are their favorably low power requirements and the realization of continuous technical improvements in membranes which are used in RO systems, RO was first applied to brackish groundwater with first large scale plants in late 1960s. A decade later RO membrane after further development was suitable for seawater desalination and become a strong competitive to conventional thermal desalination by the 1980s (Vandercaseele & Bruggen, 2002) and hence was able to expand the water sources used for desalination and utilize the brackish groundwater and dominate its market.

In the last two decades, RO processes have undergone major advancements significantly enabling now the massive production of water with a moderate cost, providing flexible solutions to different necessities within the fields of population supply, industry and agriculture. The great development of reverse osmosis (RO) technology has been a consequence of several factors such as energy consumption reduction, improvement of membrane material, and decrease in membrane cost.

Membrane processes are basic in principles, where semi-permeable membrane is used to allow the passage of water but not the salt to under certain driving forces. Different types of membrane processes include reverse osmosis RO, nanofiltration, and forward osmosis FO; the later is still under development mainly at laboratory and pilot scales.

4.2.1 Reverse osmosis RO membrane desalination

Osmosis is a natural physical process, where the solvent (i.e. water) moves through semi-permeable membrane (i.e. permeable to solvent and impermeable to solute) from low solute concentration to higher concentration creating differential pressure called osmotic pressure. The osmotic pressure depends mainly on the concentration difference, temperature, and nature of solute). The process continues till the hydraulic pressure difference due to the liquid column is equal to the osmotic pressure. In reverse osmosis RO, hydraulic pressure in value greater than the osmotic pressure is applied to the concentrated solution, results in reversing the osmosis process and in net solvent or water flow from the concentrate to the dilute solution (Fritzmann et al., 2007).

RO membranes do not have a distinct pore that traverse the membrane; it consists of polymeric material forming a layered, web-like structure characterized by high rejection to most of the dissolved solids present in water, with typical salt rejection above 99 % (Lee et al., 2011). The driving force for RO process is the applied hydraulic pressure which varies considerably from 15-25 bar for brackish water, and 60-80 bar for seawater (Fritzmann et al., 2007).

4.2.2 Nanofiltration NF membrane desalination

The Nanofiltration NF term was first introduced by FilmTech in 1980s to describe RO membranes that allow selectively and purposely some ionic solutes to path through the membrane, the membrane's selectivity was towards solute of about 1 nm cutoff, and hence the term nano comes from (Wang, 2008). Nanofiltration is an intermediate between RO membranes which has a low Molecular Weight Cut Off MWCO of about 100 and ultrafiltration membranes which has MWCO of about 1000 (Eriksson, 1988).

NF membranes have higher permeability for monovalent ions such as Na, K, and Cl and very low permeability to multivalent ions such as Ca, Mg, SO_4 and organics with MWCO of around 300 (Rautenbach & Groschl, 1990), as a result NF membranes were used mainly for removal of hardness and natural organic matter, as pretreatment before RO and MSD for seawater desalination (Al-shammiri et al., 2004; Hassan et al., 1998), and for groundwater quality enhancement (Burggen & Vandercasteele, 2003; Gorenflo et al., 2002; Saitua et al., 2011; Tahaikt et al., 2007). Nanofiltration membranes operating in a similar fashion to that of RO membranes, except of lower driving pressure and hence lower energy requirements, higher flowrates, and lower product water quality (Schaep et al., 1998).

4.2.3 Electrodialysis/ electrodialysis reversal membrane desalination

Electrodialysis/Electrodialysis Reversal membrane desalination (ED/EDR) is one of the oldest tried desalination processes, with ED process since the 1950s. In ED/EDR electrochemical separation is the main phenomena takes place utilizing electrical power as driving force to separate ions through ion-exchange membranes (Gleick, 2001). In typical ED cell, a series of anion- and cation-exchange membranes are a arranged in alternating pattern between the two electrical electrodes, anode and cathode, and hence ion concentrations increase in alternating compartments, and decrease simultaneously in the other compartments (Walha et al., 2007), ED process has been applied successfully but on small scale for brackish water (Adhikary et al., 1991; Brown, 1981; Harkare et al., 1982), and seawater (Sadrzadeh & Mohammadi, 2008; Seto et al., 1978).

In the 1970s, electrodialysis reversal EDR has been introduced as an innovative modification to the conventional electrodialysis, EDR operates on the same principle as ED. However the polarities of the electrodes are reversed for short time at specified time intervals, so that the ions are attracted in the opposite direction, and hence the brine and product channels are switched (Katz, 1979). EDR process has several features that promoted its application such as ability to treat feedwater of different qualities i.e. higher content of dissolved and suspended solids, ability to operate with high salts saturation levels and hence higher scale resistance, chlorine tolerance, cleanability, higher recovery, un affected by non-ionic species such as silica, low chemical pretreatment, and durability (Buros, 1999; Fubao, 1985; Katz, 1979; Katz, 1982; Valcour, 2010). In addition ED/EDR have been integrated successfully with other desalination processes such as RO (Oren et al., 2010).

4.2.4 Forward osmosis FO membrane desalination

Although RO membrane desalination has the major share in membrane desalination plants, the energy and membrane replacement cost are of major concern, and hence there is a search for new low energy and low fouling membrane processes. Forward osmosis (FO) or called direct osmosis (DO), employs the natural physical osmosis process by increasing the osmotic pressure in the permeate side to balance the pressure in the opposite side.

The FO exploits this natural tendency of water to move through the semi-permeable membrane from the saline water to a more concentrated solution called draw solution, the draw solution has significantly higher pressure than the saline water. Draw solutions of different natures have been tested for FO operation. Volatile solutes or gases such as sulfur dioxide which can be stripped out later to have pure water, perceptible salts such as aluminum sulfate which can be treated by lime to precipitate aluminum hydroxide and calcium sulfate, a two-stage system with sulfur dioxide-potassium nitrate used as the draw solutions has been also evaluated (McCutcheon et al., 2005a).

However recently more attention has been given to ammonia-carbon dioxide system due to several advantages such as the high solubility of ammonia and carbon dioxide gasses in water and the solution of formed ammonium bicarbonate have a high osmotic pressure which in turn provide higher water flux and recovery, followed by ease separation of the gases by moderate heating, which will be recycled back to the process (McCutcheon et al., 2005b; McGinnis & Elimelech, 2007).

Unlike other membrane process, FO does not require any hydraulic pressure to be applied and hence less energy requirements, which in turn results in less capital and operating costs. Furthermore the process has much lower fouling propensity (McGinnis & Elimelech, 2007, Phillip et al., 2007). In addition FO has been integrated successfully with other desalination processes such as RO enabling increased recovery, lower energy consumption (Lee et al., 2009; Martinetti et al., 2009; Tang & Ng, 2008; Yangali et al., 2011).

4.3 Other desalination processes

There are many other desalination process that are available today, some of them still at research and development stages, however they did not reach the development level to be commercialized on large scale as the previous processes, although many of them have a very promising features over the now widely used desalination process such as less corrosion and scaling problems, less energy consumption, less need for pre and post treatment. These processes include solar, ion-exchange, freezing, and membrane distillation.

5. Developments in desalination processes

Desalination has been extensively used over the past decades; with the great developments in desalination industry that have led to a higher acceptance and growth worldwide, particularly in arid areas. Although different desalination processes are well established today, further development are needed to resolve its various technical and operational issues which represent the essential key for successful desalination, including feed characterization for fouling and scaling propensity, process development, energy requirements, desalination economics, and finally brine disposal which will be given more attention due to its high importance (Sheikholesami, 2009).

5.1 Scaling and fouling

Fouling is a phenomenon that plagues the operation of desalination units, the deposition of foulants on to the heat (in case of thermal desalination) and mass (in case of membrane desalination) transfer surface results in reducing water productivity and decrease product quality, therefore as the fouling deposit builds up the energy consumption increases to accommodate for the required product flow till the unit is cleaned (Hamrouni & Dhahbi, 2001; Sheikholesami, 2004).

Fouling of membrane surface can be caused by any of the rejected constitutes, and can be generally classified to chemical fouling or scaling caused by sparingly soluble inorganic salts exceeding their saturation level, physical or colloidal fouling caused by particulate matter, biological fouling or biofouling due to the formation of biofilms of microorganisms, and finally organic fouling caused by natural organic matter NOM (Fritzmann et al., 2007). The water recovery is mainly constrained by fouling, and hence it is paramount to mitigate fouling of desalination units (Semiat et al., 2004).

The general approach to avoid scaling and fouling by sparingly soluble salts, is to estimate the saturation level of these salts according to the feed water quality, design recovery, and operation conditions and try to operate below such saturation levels where the solution is stable (Sheikholesami, 2004). Extensive work has been done to study the fouling and to determine the saturation levels of common scale forming sparingly soluble salts at different conditions with focus on calcium, barium, and silica particularly calcium sulfate, barium sulfate, calcium carbonate, and silica.

In seawater desalination main types of fouling is scaling by calcium carbonate, calcium sulfate, and magnesium hydroxide in case of thermal desalination. For membrane desalination, the major fouling of concern is biofouling (Al-Ahmad & Abdil Aleem, 1993). Fouling caused by precipitation of sparingly soluble salts is less likely to occur mainly due to the relative lower recovery, higher ionic strength, and low bicarbonate and sulfate concentration (Reverter et al., 2001). The contribution of each type of fouling in typical seawater RO desalination are 48% for biofouling, 18% for inorganic colloids, 15% for organic matter, 13% for silicates, and only about 6% for mineral deposits (Shon et al., 2009). Hence one of the most important steps in seawater pretreatment for desalination is disinfection with optimized dose of biocide, usually chlorine, in order to reduce biofouling propensity (Fujiwara & Matsuyama, 2008).

Brackish groundwater however has higher quality as it is mainly characterized by low content of suspended solids, low bacterial count, and low content of organic matter; as a result the most found type of fouling is scaling by sparingly soluble inorganic salts such as calcium and barium salts and silica.

The dissolved solids present in groundwater results mainly from chemical weathering or dissolution of geological formations i.e. minerals which can be attributed to the direct contact of groundwater with the calcium carbonate, and calcium sulfate rocks forming the aquifer. In addition sulfate may result from biological oxidation of reduced sulfur species. As a result the different aspects of scaling by calcium sulfate and calcium carbonate has been intensively studied (Sheikholesami, 2003a; Sheikholesami, 2003b; Rahardianato et al., 2008).

Silica originates from the dissolution or chemical weathering of amorphous or crystalline SiO_2 and the major clay minerals (Faust & Aly, 1998). Crystalline silica has a very low solubility in water, however amorphous silica can have solubility up to 120 mg/L at pH 7 and the solubility increases with pH increase reaching around 889 mg/L at pH 10 (Hamrouni & Dhabi, 2001; Sheikholesami & Tan, 1999). Silica in water can be classified into two categories 1) soluble or dissolved silica which contains monomers, dimmers, and polymers of silicic acid, and 2) insoluble or colloidal silica, which results from high polymerization of silicic acid. Due to its severe effect on membrane desalination performance and lifetime of membranes, great attention has been paid to silica fouling (Al-Shammiri et al., 2000; Ning, 2002; Semiat et al., 2003; Sheikholesami et al., 2001,).

In any desalination process, there are three main factors for sustainable operation: 1) proper design, 2) proper pretreatment, and 3) proper operation and maintenance; with the proper pretreatment as the foundation for successful operation (Neofotistou & Demadis, 2004). The primary goal of pretreatment is to lower the fouling propensity during the desalination process, and the required pretreatments depend mainly on the characteristics of the water resource (Greenlee et al., 2009). Scale inhibitors or antiscalents are chemicals that are added to water during pretreatment to prevent scale formation and usually work synergically with dispersant polymers.

However most of traditional antiscalents are successful in scale control for crystalline mineral precipitates but not silica because it is amorphous (Freeman & Majerle, 1995), and hence control of silica scaling requires chemical pretreatment.

5.2 Process development

Desalination is a multi unit process, starting from water intake, pretreatment, desalination, and post-treatment. Desalination plants in the past used to contain one type of desalination processes in the past. However attention has been recently paid to hybridization of different desalination processes together, with main objective of maximizing overall recovery, minimizing energy requirements, and cost reduction. NF has been integrated successfully with RO process mainly as pretreatment step, which resulted in improving the RO performance (Al-Shammiri et al., 2004; Hassan et al., 1998), EDR with RO (Oren et al., 2010), FO with RO (Lee et al., 2009; Martinetti et al., 2009; Tang & Ng, 2008, Yangali et al., 2011) in case of membrane desalination processes, integration of VC with MSF and MED (El-Dessouky et al., 2000; Mabrouk et al., 2007), and even combination of thermal and membrane processes by integrating NF/RO/MSF together (Hamed et al., 2009), and FO and MD to RO (Martinetti et al., 2009).

In membrane processes, development of much better membrane material that can work for wider range of pH, chlorine resistant, high mechanical strength to withstand higher hydraulic pressures, better salt rejection, and scale resistant are considered to be the next breakthrough in membrane desalination development, enabling higher recoveries at lower cost (Sheikholesami, 2009).

5.3 Energy requirements

Desalination processes are known for their intensive energy consumption especially thermal desalination, and hence energy consumption make up the major part of operation cost of any desalination process, the energy consumption differs according to desalination process in use i.e. thermal or membrane, water source and quality i.e. seawater or brackish water, design recovery, system design, plant capacity, and utilization of energy recovery devices.

Energy requirement for desalination processes is generally reported as specific energy consumptions in kWh/m^3 of product water. There are a wide range of reported values for energy consumption in desalination with the most recent values of about 1.8 kWh/m^3 for seawater desalination using MED, 4 kWh/m^3 for MSF with heat recovery mechanism incorporated (Khawaji et al., 2008). However for seawater using RO it went down from 20 kWh/m^3 in early 1970s to 1.6-2 kWh/m^3 recently, and below 1 kWh/m^3 for brackish water with energy recovery devices (Fritzmann et al., 2007; Khawaji et al., 2008).

In RO operation, the main energy consumption is mainly for the high pressure pump to provide hydraulic pressure in excess to the osmotic pressure, most of this pressure is retained by the concentrate stream flowing out of the RO unit. Energy recovery devices ERD have been developed mainly for RO operation to recover some of the energy retained in the concentrate before disposal. There are two main classes of ERD, class I which transfer hydraulic energy from the concentrate stream to the feed stream in one step with net energy transfer of more than 95%. Class II transfer hydraulic energy of the concentrate to centrifugal mechanical energy and then to hydraulic energy in the feed in two steps process (Greenlee et al., 2009).

Integration of desalination plants with power plants or as called cogeneration, which refers to the use of single energy source for multiple needs; mainly encountered with thermal desalination offers better energy utilization. For example, power plants use high pressure steam for power generation by means of turbines, the steam comes out at low pressure which is very suitable for thermal desalination (Gleick, 2001). In addition there are efforts for implementing cogeneration in RO desalination process in order to address the water-electricity demand trade off (Altmann, 1997).

Use of renewable energy sources for driving the desalination process provide another development opportunity, specifically for membrane desalination where less energy is required and for rural communities in which desalination systems are generally small in size with usually non-continuous operation, and hence can be integrated to renewable energy sources. Considerable efforts has been made for integrating desalination processes with different renewable energy sources namely solar (photovoltaic and thermal), wind, and geothermal. The renewable energy can be used in one of two forms thermal or electrical, depending on which one that best match the desalination process (Al-karaghouli et al., 2009; Al-karaghouli et al., 2010; Forstmeier et al., 2007; Mathioulakis et al, 2007).

5.4 Desalination economics

Economics of any process represent the most crucial part for development and application, and hence economical feasibility is a very important factor when considering desalination processes. In desalination processes it is difficult to standardize the economics of process because it is case specific. There are several factors that affect desalination cost such as water source (brackish or seawater), desalination process used (thermal or membrane), energy source (traditional or renewable), and plant size (Dore, 2005; Karagiannis & Soldatos, 2008).

The source, and hence the quality of the feed water plays important rule for determination of both capital and operating cost, and the overall desalination cost. Brackish water has much low salt content and better water quality than seawater and therefore, it incorporates less capital and operating costs. The most recent data for average investment cost for brackish water was around \$200-450/ (m^3/d) with product water cost of \$0.25-0.75/m^3 for RO process which is the process dominating the brackish water desalination market (Vince et al., 2008; Yun et al, 2006).

The product water desalination cost varies significantly according to the salinity of the water sources. For example the product water cost for brackish water with salinity around 3,000 mg/L was found to be \$0.32/m^3. However, for water with salinity around 10,000 mg/L, the desalination cost was \$0.54 /m^3 (Karagiannis & Soldatos, 2008). The same trend was observed also for sea water desalination with desalination cost ranging from \$0.54 /m^3 for Mediterranean seawater to \$0.87/m^3 for Arabic Gulf seawater (Greenlee et al., 2009).

Capacity or size of desalination plant greatly affects the product water cost; table 3 shows the average desalination cost for brackish and seawater desalination plants of different production capacities.

Feed water	Plant size (m^3/ d)	Cost (\$ / m^3)
Brackish	≤ 1,000	0.78 - 1.33
	5,000 – 60,000	0.26 – 0.54
Seawater	< 1,000	2.2 - 11.25
	1,000 – 5,000	0.7 – 3.9
	12,000 – 60,000	0.44 – 1.62
	> 60,000	0.50 – 1.0

Table 3. Size of desalination plant and water production cost (Karagiannis & Soldatos, 2008).

6. Brine disposal from inland desalination

Brine, concentrate, or reject are different names for a stream that is commonly produced from any desalination process. In any desalination process, two streams are produced: 1) product water with high quality, 2) brine or concentrate stream that contains all the salts were originally present in the feed water in addition to the chemicals added in the pretreatment and during desalination such as antiscalents.

To reduce energy consumption, cleaning time and expenditure, loss of production during downtime, it is paramount to mitigate fouling. The general approach is to study the feed water characteristics and couple it with the expected recovery and operating conditions; and to operate at conditions where the solution is stable, hence scaling and fouling is minimized. However working at lower recoveries to avoid the fouling of membrane will increase the flow of brine stream generated, which present the main trade-off for desalination operation.

Brine stream does not contains only 2 or 4-5 folds the salinity of the feed water as in case of seawater, or brackish water respectively, but it contains all the chemicals that has been added to the desalination process during pretreatment. Moreover in case of thermal desalination it will be at high temperature, and hence more attention should be taken when considering the brine discharge method (Ahmed et al., 2002).

The problem of brine discharge is different in sea water and brackish water desalination. In the case of seawater desalination plants the problem is readily solved since these plants are usually placed near the coast, so the discharge method of choose is usually to discharge it back to the sea through brine pipes or submarine emissaries. Encouraging facts to utilize that option are, first the discharged brine is of similar chemistry, even being more concentrated but only by 50-100 %. Second is that the volume of brine stream relative to the water body being discharged to being very small, hence lower drawbacks are expected. However, there are many criteria to be considered such as having the discharge point far enough from intake at good mixing zone so it can be mixed with the main body of seawater.

However, the management of brine from brackish desalination plants i.e. inland desalination can be significant problem in case they are placed far from the coast (inland plants). Some of the conventional options for brine disposal from inland desalination plants are: 1) disposal into surface water bodies, 2) disposal to municipal sewers, 3) evaporation ponds, 4) deep well injection, and 5) irrigation of plants tolerant to high salinities (Ahuja & Howe, 2005). The main factors that influence the selection of a disposal method, among others, are: 1) volume or quantity of concentrate, 2) quality and constitutes present in the concentrate, 3) physical and geographical considerations, 4) capital and operational costs, 5) availability of receiving site, 6) permissibility of the option, facility future expansion plan, and 7) public acceptance. All of these factors together will affect the cost of brine disposal that can ranges from 5 -33 % of the total desalination cost (Ahmed et al, 2001).

Brine disposal method should be considered after the necessary studies and investigations have been performed in order to minimize the brine stream to be disposed off, and hence reduce the cost of subsequent disposal. This is mainly achieved by employing the proper feed water pretreatment, proper desalination process, maximizing the system recovery. However attention to the increased salinity and quality of brine should be considered.

6.1 Disposal to surface water bodies and sewers systems

Disposal of brine to surface water bodies if a available, present the first option to choose as it represent a ready and good solution to the challenge of brine disposal taking into account that the brine stream is diluted by mixing with the water body. However many consideration should be taken into account. The salinity if the receiving body might increase due to the disposal of the high salinity brine, and hence the self-purification capacity of the receiving water should be considered (Ahmed et al., 2000). As a result disposal to surface water should be permitted only if that will avoid dramatic impact on environment.

Another option is to dispose the brine to the local sewage system, which is usually employed by small membrane desalination plants. This option has many advantages such as use of the ready available and installed sewage system, lowering the BOD of the domestic sewage water. However that should be practiced carefully as the salinity of sewage water might increase which might affect the wastewater treatment facility especially biological treatment step. This might also render treated municipal sewage effluent unsuitable for agriculture use when disposing brines with high salinity (Ahmed et al., 2000).

6.2 Disposal to evaporation ponds

In evaporation bonds, the brine is discharged into a large surface area pond, where the water is naturally evaporated. Use of evaporation pond technology is practiced primarily in the arid and semi-arid areas, particularly in Middle East and Australia. Evaporation pond is probably the most widespread method for brine disposal from inland desalination plants.

Simple evaporation ponds have many advantages such as being easy to construct, low maintenance and operation cost, no equipment are needed specifically mechanical. Making it the most appropriate method with lower cost, especially in arid areas with high evaporation rates, low rainfall, and low land cost (Ahmed et al., 2000). Use of evaporation ponds for cultivation of brine shrimps has been studies as well, giving ideal place for brine-shrimp production as it present mono-culture environment under natural conditions with absence of any food competitors or predators (Ahmed et al., 2001).

The basic concern associated with use of evaporation pond for brine disposal is leakage of brine through soil. This may result in subsequent contamination and increasing salinity of the aquifer. Electrical conductivity and concentration of salts in the evaporation ponds can be used as indicators for leakage in the pond, where insignificant increase is a strong indication of brine leakage through the soil (Amed et al., 2001). Deterioration of soil and groundwater quality in areas nearby evaporation ponds used for brine disposal in KSA, UAE, and Oman was investigated and reported as one of the draw backs to use of evaporation bonds (Al-Faifi et al, 2010; Mohamed et al., 2005).

As a result most of the evaporation ponds installed recently are lined with polymeric sheets. Liner installation should be carried out carefully as joints sealing is very important for leakage prevention. Furthermore double lining is strongly recommended with proper monitoring for leakage.

In addition reduction in production from agricultural lands caused by deposition of airborne salts from dried concentrate of evaporation bonds, and formation of eyesores caused by improper disposal of concentrates on nearby land can be another disadvantage of brine disposal to evaporation ponds.

In conclusion, while operation and maintenance costs for evaporation ponds are minimal, large land areas are required which increases as the plant capacity increase, and pond construction costs are high due to lining and monitoring requirements. Consequently, proper evaporation ponds are often cost prohibitive and impractical for handling significant concentrate flow. Furthermore, water evaporated from a pond is often a lost resource.

6.3 Deep well injection

In deep well injection, the brine is injected back underground to depth ranges from few hundreds of meters to thousands of meters, depending on many factors which should be considered while designing, installing, and operating the system. Deep well injection for brine disposal includes permitting considerations, which look for identification of adequate geologic confining unit to prevent upward migration of effluent from the injection area. While design considerations focus generally on the tubing and packing installed inside the final cemented casing of the injection well, compatibility of the concentrate with the tubing material (to avoid corrosion), expected concentrate flow, and leak detection and monitoring systems (Skehan & Kwiatkowski, 2000).

One of the very attractive options with deep well injection is to use depleted oil and gas fields for brine disposal. This encounter many advantages such as making use of the readily available gas and oil wells, long experience encountered with the operation of such wells. However before applying this option the fields should be tested physically and chemically for accepting the brine stream [Mace et al., 2006; Nicot & Cjowdhury, 2005]

Generally site selection for installing of such deep well, is the most important step, and hence hydrological and geological conditions should be considered, as example the wells should never be installed in areas vulnerable to earthquakes (Ahmed et al., 2000). Although of availability of such option to many inland desalination plants, however many factors should be considered with deep well injection for brine disposal which can be summarized as follow (Mickley et al, 2006):

1. Site selection, which is performed through many geological and hydrological studies, to identify the proper area for installing the well,
2. High cost, associated with both capital and operational cost,
3. Possibility of corrosion and subsequent leakage in the well casing,
4. Seismic activity which could cause damage to the well and subsequently result in leakage,
5. Uncertainty of the well life,
6. Pollution of groundwater resources, which may result from high salinity and the presence of other harmful chemicals in the brine.

6.4 Land applications of brine

Land application such as use in irrigation systems that was originally developed for sewage effluents, can be used for brine disposal, and hence helps conserve natural resources. In areas where water conservation is of great importance, spray irrigation is especially attractive option. Concentrate can be applied to cropland or vegetation by sprinkling or surface techniques for water conservation when lawns, parks, or golf courses are irrigated and for preservation and enlargement of greenbelts and open spaces. Crops such as water-tolerant grasses with low potential for economic return but with high salinity tolerance are generally chosen for this type. However soil sanlinization and groundwater contamination should be carefully considered (Mickley et al, 2006).

7. Inland desalination with zero liquid discharge

In many cases of brackish water desalination, brine management is critical and of high concern, and hence the need for affordable inland desalination has become critical in many regions of the world where communities strive to meet rapidly growing water demands with limited freshwater resources.

Where brine disposal and management is a problem, given the disadvantages of existing brine disposal and management methods, it is imperative to find alternative Zero Liquid Discharge ZLD technologies that provide more affordable concentrate management. In ZLD, brine is treated to produce desalinated water and essentially dry salts; therefore there is no discharge of liquid waste from the site. Most ZLD applications in operation today treat industrial wastewater using thermal or membrane separation processes, or a combination of these technologies.

Thermal desalination is a mature technology that has been practiced for long time especially where energy is relatively inexpensive, while it is a proven process that generates high quality product water, thermal desalination is energy-intensive and its capital and operating costs are high. Membrane processes has been proved to provide high quality water, but also has some limitation concerning scaling and maximum hydraulic pressure and cannot alone provide ZLD solution. Advancement of ZLD science and associated reduction of ZLD costs will be of tremendous benefit and will alleviate the water supply challenges faced by many communities worldwide.

ZLD desalination present the perfect solution for the brine disposal and management problem usually encountered with inland desalination plants. In addition applying inland desalination with ZLD provide several advantages, the main advantages can summarized as below:

- Maximize Water Recovery: with ZLD systems approaches 100 % recovery, when compared to the conventional Inland desalination system with regular recovery of about 70-85%. ZLD systems should be able to provide more product water or less plant size by 15-30 %.
- Preserving Natural Resources: for inland desalination with ZLD systems, the natural resources, which are mainly groundwater and land, are preserved both quantitatively and qualitatively, by avoiding the different problems associated with conventional brine disposal methods.
- Byproduct salts: the ZLD system results into two stream, product water, and dry salts, these salts can be treated as added value product rather than solid waste, finding a lot of applications and beneficial uses.
- Integerability and applicability: ZLD system can be integrated to any existing inland desalination plant of any size and location. This is mainly because the system operates

on treat the brine resulted from the existing desalination plants, and hence it can be integrated at any stage from design to operation stages.

There have been many attempts to achieve a successful inland desalination with ZLD, however more attention and further research work and process developments are needed in order develop a full economic-technical feasible ZLD desalination. In the following sections the current efforts for providing a ZLD system, as well as further developments and research needs will be discussed.

7.1 Current zero liquid discharge schemes for inland desalination

Little literature work is available on ZLD systems for inland desalination; however three main schemes can be concluded and summarized as follow:

- Applying thermal processes directly to the brine generated from the primary desalination process, usually RO, followed by thermal processes for brine concentration, then crystallization or drying for final salt production (Mickley et al, 2006).
- Applying chemical treatment to the brine stream, followed by further membrane desalination, brine concentration, and finally crystallization or drying for final salt production (Bond & Veerapaneni, 2008; Mohammadesmaeili et al., 2010).
- Applying ED/EDR process to the brine stream making use of higher recovery encountered with such units, followed by crystallization or drying for final salt production (Greenlee et al., 2009, Oren et al., 2010).

Similarities between schemes are clear, especially for brine concentration and final salt production, with the difference mainly in brine treatment and further desalination. However brine treatment and further desalination results in significant reduction in the volume of brine to undergo the brine concentration and final drying/crystallization. Options for integrating different units in different setups can be investigated with an overall objective of ZLD desalination and production of salts can be generated, and should be evaluated for process optimization (Kim, 2011).

Conventional inland desalination system usually achieve 70-85 % recovery of the feed water, which is the largest recovery increment in single step, the recovery mainly depend on the quality of the feed water. However this recovery is usually limited due to scaling by sparingly soluble salts, typically calcium salts such as calcium sulfate and carbonate, in addition to silica (Freeman & Majerle, 1995: Rhardianato et al., 2008; Sheikholesami, 2003a; Sheikholesami, 2004). With this recovery range, about 15-30 % of the feed stream will be rejected as brine which should be disposed off.

In the first ZLD scheme this 15-30 % is fed to thermal processes using single or multiple effect evaporators or vapor compression evaporators for brine concentration which to be followed by crystallization or drying to obtain final dry salts. In the second scheme the brine is treated chemically to remove most of scale forming constitutes, achieving high removal of such constitutes rendering the treated brine suitable for further membrane separation to recover more water. With the two membrane desalination process with the intermediate brine treatment step recovery up to 95% can be achieved, moreover making use of membrane desalination reduces the cost, and minimizes the energy requirements. Moreover reduces the volume of brine to be handled by final evaporation step which results in lowering the energy requirements and hence the overall process cost. In the last scheme ED/EDR unit are employed, which can operate at high saturation levels of sparingly soluble salts as in the case of brine streams, and where high recovery up to 97% can be achieved.

7.2 Precipitation softening for brine treatment in zero liquid discharge systems

Intermediate brine treatment step as employed in the second ZLD scheme is the one receiving large attention recently. The main objective of this step to remove most of the scale forming constitutes typically calcium, magnesium, carbonate, sulfate, and silica. It is hard to find a chemical treatment process that is able to efficiently remove all of these constitutes. Furthermore most of the tested chemical treatment processes were not able to completely remove such constitutes. However the achieved removal efficiency was good enough to prevent such constitutes from limiting the recovery in the secondary membrane desalination process.

Precipitation softening is one of the widely used processes for reduction of hardness (calcium and magnesium) and alkalinity (mainly bicarbonate) in water treatment plants. The reduction of hardness is mainly achieved by removal of calcium as calcium carbonate $CaCO_3$ and magnesium as magnesium hydroxide $Mg(OH)_2$. This is usually achieved by addition of alkali usually lime, calcium hydroxide $Ca(OH)_2$ in lime softening or sodium hydroxide, NaOH in caustic softening and sodium carbonate Na_2CO_3 depending on the quality of water to be treated (Reynolds & Richards, 1996). Removal of calcium and magnesium present one of the major targets for brine chemical treatment, particularly calcium, as magnesium cause scaling problems only at high pH values forming insoluble magnesium hydroxide $Mg(OH)_2$. However at the normal pH values found in brine streams it will be mainly saturated by calcium sulfate and carbonate (Sheikholesami, 2003a; Sheikholesami, 2004; Rharadianato et al., 2008).

Silica removal during the precipitation softening was extensively studied, and it was found that silica is removal could be by co-precipitation with metal hydroxides, specifically iron, manganese, and magnesium hydroxides, or could be by precipitation as magnesium and calcium silicate (Sheikholesami & Bright, 2002). Furthermore caustic softening using only sodium hydroxide was found to be more effective and more viable in removal of silica over lime-soda softening using lime and soda ash (Al-Rehaili, 2003; Sheikholesami & Bright, 2003). Addition of sodium aluminate and aluminum sulfate was found to enhance the removal of silica during the softening process by co-precipitation with aluminum hydroxide (Cheng et al., 2009; Lindsay & Ryznar, 1939). Conventional softening process is slow, requires extensive space, and generates large volume of sludge which will need dewatering and further treatment later on (Kadem & Zalmon, 1997). As a result a more advanced process designated Compact Accelerated Precipitation Softening CAPS was developed to enhance the performance of the precipitation softening process. In CAPS process the saturated solution is passed through cake of calcium carbonate to enhance crystallization and approach equilibrium rapidly, the process has been found to overcome the different disadvantages encountered in conventional softening process (Gilron et al., 2005; Masarawa et al., 1997; Oren et al., 2001).

Although different precipitation softening processes have been applied basically for surface water treatment and as a pretreatment for membrane processes, specifically NF and RO (Al-

Rehaili, 2003; Cheng et al, 2009; Gilron et al., 2005; Kadem & Zalmon, 1997; Masarawa et al., 1997; Oren et al., 2001; Sheikholeslami & Bright, 2002) showing high efficiency in removal of calcium, magnesium, silica, and heavy metals. However such softening processes were found to be very effective and promising when applied for brine treatment where high calcium, magnesium, and silica removals from brine streams has been achived enabling higher recovery in the subsequent membrane desalination, facilitate reaching zero liquid discharge desalination [Comstock et al., 2011; Gabelich et al., 2007; Ning et al., 2006; Ning & Tryoer, 2009; Rahardianto et al., 2007].

Sulfate usually present in the brine streams in high concentrations, relative to those of calcium, magnesium, carbonate, and silica. However in presence of calcium, saturation and hence scaling due to calcium sulfate is very likely to happen (Rahardianato et al., 2008; Sheikholesami, 2003a; Sheikholesami, 2004). Precipitative softening processes were found to be very effective in removal of calcium, magnesium, carbonate, and silica. However such process had no success for removal of sulfate, even though removal of calcium from brine stream reduces the scaling potential of calcium sulfate. However it will be paramount to remove sulfate completely or partially, converting the brine chemistry typically to monovalent ions i.e. sodium, potassium, and chloride which has no scaling potential at the normal membrane desalination operating conditions.

Several works has been performed on removal of sulfate from industrial wastewater streams such as paper mills, mining, and fertilizers, and several attempts have been performed to reach zero discharge with such streams using membrane and thermal separation processes (Ericsson & Hallmans, 1996). However many attempts has been worked to employ precipitation and crystallization removal of sulfate as calcium sulfate, gypsum, by addition of calcium mainly as calcium hydroxide, lime (Tait et al., 2009), or as calcium chloride (Benatti et al., 2009), which was found to be very effective in removal of sulfate. Removal of sulfate as calcium sulfate below 1300 mg/L was found to be very hard due to solubility limits. However addition of aluminum as aluminum sulfate or alum, aluminum chloride, aluminum nitrate (Christoe, 1976), and sodium aluminate (Batchelor et al, 1985) was found to enhance the removal of sulfate far below this value by formation of more complex solids (Batchelor et al, 1985; Christoe, 1976) which has much lower solubility compared to that of calcium sulfate precipitated by lime addition only.

7.3 Secondary brine concentration and final salt production

The treated brine after being further concentrated in secondary membrane desalination has to be further concentrated reaching zero liquid. Zero liquid and dry salts cannot be produced by membrane desalination such as RO or NF, and hence the concentrated brine has to be subjected to thermal process such as brine concentration followed by crystallization or drying to produce dry salts.

Thermal processes such as single and multi stage evaporators, or vapor compression evaporators are usually employed for further brine concentration. Such units have dual purpose objectives which are further recovery of water with very high quality with salinity about 10 mg/L, and brine concentration up to 250,000 mg/L, with recovery above 90%. Final salt production can be achieved after brine concentration which usually performed in crystallizers or dryers (Mickley, 2006).

In addition to the salts produced from final crystallizer/dryer that can be assumed as byproducts from ZLD desalination, the precipitated solids from brine chemical treatment can be considered as another byproduct or added value product. This precipitate is rich in calcium as calcium carbonate, magnesium as magnesium hydroxide, and silicate of calcium and magnesium, in addition to gypsum or calcium sulfoaluminate in case of sulfate removal, such mixture can find a wide range of applications such as road pavement, cement industry, and any other applications where there is a need for mixture of similar composition.

7.4 Cost associated with zero liquid discharge systems

Reaching inland desalination with zero liquid discharge has to be considered on both scales, technically and economically, while technical ZLD system can be successfully achieved through the different ZLD schemes. However the costs associated with each proposed ZLD system should be carefully considered. It easily noticeable that employing a secondary membrane desalination step is of high importance for reduction of both capital and operating costs over the conventional thermal ZLD systems due to the reduced volume of brine stream to be thermally treated.

A cost comparison for standard bench mark brine treatment by brine concentration and evaporation to advanced brine treatment using secondary RO desalination and final brine concentration and drying for brine of different qualities has been performed. The study showed that a cost reduction ranging from 48-67%, with reduction in energy requirement of 58 - 72% using the advanced ZLD system (Bond & Veerapaneni, 2008) can be achieved. However it worth to mention that the comparison was for the brine management only, not the whole inland desalination system, as the primary RO desalination is a kind of standard step employed for all ZLD schemes.

Inland desalination with zero liquid discharge usually has higher product water cost when compared to conventional inland desalination systems with no brine disposal is employed, but becomes very economically attractive when compared to the different brine disposal methods. The high cost mainly due to the fact that several units such as chemical treatment, secondary membrane desalination, brine concentration, and crystallization/drying are employed to recover only 15-30%. Which increase both capital and operating cost increasing the average product cost compared to single step desalination unit recovering 70-85 % (Greenlee et al., 2009). However due to the different strict regulation on brine disposal using the conventional methods, and the efforts for preserving the groundwater resources, more driving force for advancement of ZLD systems are encouraged (Mickley, 2006).

7.5 Developments and research needs for zero liquid discharge desalination

Desalination with zero liquid discharge is the ultimate achievement for any inland desalination process. This will help to overcome the brine disposal limitations currently faced for applying inland desalination. Although different zero liquid discharge schemes are currently developed or under development, however further development are needed to resolve its various technical, operational, and economical issues. The essential key for successful ZLD inland desalination are brine treatment, process development, energy requirements, and process economics, which should be given more attention and further research and development efforts.

Precipitative softening processes have been widely used for treatment of primary brine stream, however softening process improvements through chemical doses optimization, testing different chemical reagents aiming at high efficiency in removal of scale forming constitutes should help improving the overall process performance. Furthermore other chemical treatment processes should be investigated which can result in better performance.

ZLD system usually employs different units with different nature, such as membrane and thermal process, liquid and solids handling. Process development should look at the different viable and optimum units arrangement and operation conditions with the objective reducing energy requirements and cost.

Thermal processes are usually employed in ZLD systems for further brine concentration up to level that can be handled by crystallizer or dryer. Such processes are known to be energy extensive, and hence reduction in energy requirements and utilization of renewable energy should help in reducing overall energy requirements.

Economics of ZLD process is very important factor in employing the ZLD for inland desalination. Reaching competitive overall cost for inland desalination with ZLD to that of conventional desalination should help in wider application of the process for inland desalination systems.

8. Conclusions

In conclusion, as groundwater presents the main source of potable water to communities that do not have access to surface water, the deterioration of groundwater quality, specifically salinity is of high concern, which leads to the use of desalination techniques to overcome such problem. The use of membrane desalination systems in general and reverse osmosis in particular is very beneficial due to capacity flexibility, lower energy requirements, and in turn lower cost for brackish groundwater desalination. However the generation of brine stream is the main problem facing such systems, and which should be managed properly, there are different ways for brine disposal. However each one has certain advantages and disadvantages that are a matter of question. Approaching inland desalination with zero liquid discharge presents the solution for having a perfect inland desalination system. Given such need it is imperative to find a zero liquid discharge treatment technologies that provide more affordable concentrate management at reasonable cost, and hence a very active area of research is going on to provide such solution.

9. References

Adhikary, S. K.; Narayanan, P. K.; Thampy,S. K.; Dave, N. J.; Chauhan, D. K.; & Indusekhar, V. K. (1991). Desalination of Brackish Water of Higher salinity by Electrodialysis. *Desalination*, Vol. 84, No. 1-3, (October 1991), pp. 189-200, ISSN 0011-9164.

Ahmed, M.; Shayya, W. H.; Hoey, D.; Mahendran, A.; Morris, R.; Al-Handaly, J. (2000). Use of Evaporation Ponds for Brine Disposal in Desalination Plants, *Desalination,* Vol. 130, No. 2, (November 2000), pp. 155-168, ISSN: 0011-9164.

Ahmed, M.; Shayya, W. H.; Hoey, D.; Al-Handaly, J. (2001) Brine Disposal from Reverse Osmosis Desalination Plants in Oman and the United Arab Emirates, *Desalination,* Vol. 133, No. 2, (March 2001),pp. 135-147, ISSN: 0011-9164.

Ahmed, M.; Shayya, W. H.; Hoey, D.; Al-Handaly, J. (2002) Brine Disposal from Inland Desalination Plants, *Water International,* Vol. 27, No. 2, (June 2002), pp. 194 -201.

Ahuja, N.; Howe, K. (2005). Strategies for Concentrate Management from Inland Desalination, *Proceeding of Membrane Technology Conference & Exposition 2005",* pp. 761-776, Phoenix, Arizona, USA, March 6-9, 2005.

Altmann and Water Cogeneration, T. (1997) A New Power Concept with the Application of Reverse osmosis Desalination, *Desalination,* Vol. 114, No. 2 , (December 1997), pp.139-144, ISSN: 0011-9164.

Al-Ahmad, M.; Abdil Aleem, F. (1993) Scale Formation and Fouling Problems Effect on the Performance of MSF and RO Desalination Plants in Saudi Arabia, *Desalination,* Vol. 93, No. 1-3, (August 1993), pp. 287-310, ISSN: 0011-9164.

Al-Faifi, H.; et al. (2010) Soil Deterioration as Influenced by Land Disposal of Reject Brine from Salbukh Water Desalination Plant at Riyadh, Saudi Arabia, *Desalination,* Vol. 250, No. 2, (January 2010), pp. 479-484, ISSN: 0011-9164.

Al-Karaghouli, A.; Renne, D.; Kazmerski, L. L. (2009) Solar and Wind Opportunities for Water Desalination in the Arab Regions, *Renewable and Sustainable Energy Reviews,* Vol. 13, No. 9, (December 2009), pp. 2397-2407, ISSN: 1364-0321.

Al-Karaghouli, A.; Renne, D.; Kazmerski, L. L. (2010) Technical and Economic Assessment of Photovoltaic-Driven Desalination Systems, *Renewable Energy,* Vol. 35, No. 2, (February 2010), pp. 323-328, ISSN: 0960-1481.

Al-Rehaili, A. M. (2003) Comparative Chemical Clarification for Silica Removal from RO Groundwater Feed, *Desalination,* Vol. 159, No. 1, (September 2003), pp. 21-31, ISSN: 0011-9164.

Al-Shammiri, M.; Safar, M.; Al-Dawas, M. (2000) Evaluation of Two Different Antiscalents in Real Operation at the Doha Research Plant, *Desalination,* Vol. 128, No. 1, (March 2000), pp. 1-16, ISSN: 0011-9164.

Al-Shammiri, M.; Ahmed, M.; Al-Rageeb, M. (2004) Nanofilteration and Calcium Sulfate limitation for Top Brine Temperature in Gulf Desalination Plants, *Desalination,* Vol. 167, No. 2, pp. 335-346, ISSN: 0011-9164.

Amer, K. M.; Al-Muraikhi, A.; Rashid, N. (2008) Management of Coastal Aquifers- the Case of a Peninsula- State of Qatar; *20th Salt water Intrusion Meeting,* Florida, USA, June 23-27, 2008.

Ayers, R.S.; Westcot, D.W., (1994), Water Quality for Irrigation 29 Rev.1, accessed 15.08.11, available from:
http://www.fao.org/DOCREP/003/T0234E/T0234E00.htm#TOC

Bart Weiss, P.G. (2002). Brackish Groundwater Desalination Water Supply, *Proceedings of AMTA 2002 Biennial Conference and Exposition "Water Quality Enhancement through Membrane Technology",* pp. 0-44, Tampa, Florida, August 6-9, 2002.

Batchelor, B.; McDevitt, M.; Chan, D. (1985) Removal of Sulfate from Recycled Cooling Water by the Ultra-High Lime Process, *Proceedings water Reuse Symposium III, AWWA Research Foundation,* pp. 798-812, Denver, Colorado, USA, 1985.

Benatti, C. T.; Tavares, C. R. G.; Lenzi, E. (2009) Sulfate Removal from Waste Chemicals by Precipitation, *Journal of Environmental Management,* Vol. 90, No. 1, (January 2009), pp. 504-511, ISSN: 0301-4797.

Bond, R.; Veerapaneni, S. (2008) Zeroing in on ZLD Technologies for Inland Desalination, *American Water Works Association Journal,* Vol. 100, No. 9, (September 2008), pp. 76-89, ISSN: 1551-8833.

Brich, R. P.; Al-Arrayedh M.; Hallmans B. (1985) Bahrain's Fresh Groundwater Situation and the Investigations into Using Bahrain's Brackish water Resources as a Feedwater for the Reverse Osmosis Desalination Programme, *Desalination,* Vol. 55, No. 1, pp. 397-427, ISSN: 0011-9164.

Brown, D. R. (1981) Treating Bahrain Zone C Groundwater using the EDR process, *Desalination,* Vol. 38, (November 1981), pp. 537-547, ISSN: 0011-9164.

Bruggen; B., Vandecasteele, C. (2003) Removal of Pollutants from Surface Water and Groundwater by Nanofiltration: Overview of Possible Applications in the Drinking water Industry, *Environmental Pollution,* Vol. 122, No. 3, (April 2003), pp. 435-445, ISSN: 0269-7491.

Buros, O.K. (1999) The ABCs of Desalting, International Desalination Association, accessed 15.08.11, available from: http://www.idadesal.org/pdf/ABCs1.pdf

Cheng, H.; Chen. S.; Yang, S. (2009) In-Line Coagulation/Ultrafiltration for Silica Removal from Brackish Water as RO pretreatment, *Separation and Purification Technology,* Vol. 70, No. 13, (November 2009), pp. 112-117, ISSN: 1383-5866.

Christoe, J. R. (1976) Removal of Sulfate from Industrial Wastewaters, *Journal of Water Pollution Control Federation,* Vol. 48, No. 12, (December 1976), pp. 2804-2804, ISSN: 0043-1303.

Comstock, S. E. H.; Boyer, T. H.; Graf, K. C. (2011) Treatment of Nanofiltration and Reverse Osmosis Concentrates: Comparison of Precipitative Softening, Coagulation, and anion Exchange, *Water Research,* Vol. 45, No. 16, (October 2011), pp. 4855-4865, ISSN: 0043-1354.

Delleur, J. W. (2nd Ed.). (2007). *The handbook of Groundwater Engineering,* CRC Press, ISBN 0-8493-4316-x, Florida, USA.

Dore, M. H. I. (2005) Forecasting the Economic Costs of Desalination Technology, *Desalination,* Vol. 172, No. 3, (February 2005), pp. 207-214, ISSN: 0011-9164.

El-Dessouky, H. T.; Ettouny, H. M.; Al-Juwayhel, F. (2000) Multiple Effect Evaporation - Vapor Compression Desalination Processes, *Chemical Engineering Research and Design,* Vol. 78, No. 4, (May 2000), pp. 662-676, ISSN: 0263-8762.

Eriksson, P. (1988) Nanofiltration Extends the Range of Membrane Filtration, *Environmental Progress.* Vol. 7, No. 1, (February 1988), pp. 58-62, ISSN: 0278-4491.

Ericsson, B.; Hallmans, B. (1996) Treatment of Saline Wastewater for Zero Discharge at the Debiensko Coal Mines in Poland, *Desalination,* Vol. 105, No. 1-2, (June 1996), pp. 115-123, ISSN: 0011-9164.

Faust, S. D.; Aly, O. M. (1998), *Chemistry of Water Treatment* 2nd Ed., Lewis Publishers CRC Press LLC, ISBN 1-57504-011-5, Florida, USA.

Food and Agriculture Organization of the United Nations FAO, AQUASTAT Program, (n.d.). Accessed 13.08.11, available from: http://www.fao.org/nr/water/aquastat/main/index.stm

Forstmeier, M; et al. (2007) Feasibility Study on Wind-Powered Desalination, *Desalination*, Vol. 203, No. 1-3, (February 2007), pp. 463-470, ISSN: 0011-9164.

Freeman, S. D.; Majerle, R. J. (1995) Silica Fouling Revisited, *Desalination*, Vol. 103, No. 1-2, (November 1995), pp. 113-115, ISSN: 0011-9164.

Fritzmann, C.; Lowenberg, J., Wintgens, T; Melin, T. (2007) State-of-the-Art of Reverse Osmosis Desalination, *Desalination*, Vol. 216, No. 1-3, (October 2007), pp. 1-76, ISSN: 0011-9164.

Fubao , Y. (1985) Study on Electrodialysis reversal (EDR) Process, *Desalination*, Vol. 56, pp. 315-324, ISSN: 0011-9164.

Fujiwara, N.; Matsuyama, H. (2008) Elimination of Biological Fouling in Seawater Reverse Osmosis Desalination Plants, *Desalination*, Vol. 227, No. 1-3, (July 2008), pp. 295-305, ISSN: 0011-9164.

Gabbrielli, E. (1981) A Tailored Process for Remineralization and Potabilization of Salinated Water, *Desalination*, Vol. 39, (December 1981), pp. 503-520, ISSN: 0011-9164.

Gabelich, C. J.; Williams, M. D.; Rahardianto, A.; Franklin, J. C.; Cohen, Y. (2007) High Recovery Reverse Osmosis Desalting Using Intermediate Chemical Demineralization, *Journal of Membrane Science*, Vol. 301, No. 1-2, (September 2007), pp. 131-141, ISSN: 0376-7388.

Gabelich, C. J.; Rahardianto, A.; Northrup, C. R.; Yun, T. I.; Cohen, Y. (2011) Process Evaluation of Intermediate Chemical Demineralization for Water Recovery Enhancement in Production-Scale Brackish Water Desalting, *Desalination*, Vol. 272, No. 1-3, (May 2011), pp. 36-45, ISSN: 0011-9164.

Gilron, J.; Daltrophe, N.; Waissman, M.; Oren, Y. (2005) Comparison Between Compact Accelerated Precipitation Softening (CAPS) and Conventional Pretreatment in Operation of Brackish Water Reverse Osmosis (BWRO), *Industrial Engineering and Chemistry Research*, Vol. 44, No. 15, (June 2005), pp. 5465-5471, ISSN: 0888-5885.

Gleick, P. H. (2001). *The World's water 2000-2001: The Biennial Report on Freshwater Resources*, Island Press, ISBN 1-55963-792-7, Washington, DC, USA.

Gleick, P. H. (2006). *The World's water 2006-2007: The Biennial Report on Freshwater Resources*, Island Press, ISBN 1-59726-108-8, Washington, DC, USA.

Gorenflo, A.; Velazques-Padron, D.; Frimmel, F. H. (2002) Nanofiltration of a German Groundwater of High hardness and NOM content: Performance and Costs, *Desalination*, Vol. 151, No. 3, (January 2003), pp. 253-265, ISSN: 0011-9164.

Greenlee, L. F.; Lawler, D. F.; Freeman, B. D.; Marrot, B.; Moulin, P. (2009) Reverse Osmosis Desalination: Water sources, Technology, and Today's Challenges, *Water Research*, Vol. 43, No. 9, (May 2009), pp. 2317-2348, ISSN: 0043- 1354.

Hadi, Khaled M. B. (2002) Evaluation of the Suitability of Groundwater Quality for Reverse Osmosis Desalination, *Desalination*, Vol. 142, No. 3, (March 2002), pp. 209-219, ISSN: 0011-9164.

Hamed, O. A.; Hassan, A. M.; Al-Shail, K.; Farooque, M. A. (2009) Performance Analysis of a Trihybrid NF/RO/MSF Desalination Plant, *Desalination and Water Treatment*, Vol. 1, No. 1-3, (January 2009), pp. 215-222, ISSN: 1944-3994.

Hamrouni, B.; Dhahbi, M. (2001) Thermodynamics Description of Saline Waters-Prediction of Scaling Limits in Desalination Processes, *Desalination*, Vol. 137, No. 1-3, (May 2001), pp. 275-284, ISSN: 0011-9164.

Hamrouni, B.; Dhabi, M. (2001) Analytical Aspects of Silica in Saline Waters- Application to Desalination of Brackish Waters, *Desalination*, Vol. 136, No. 1-3, (May 2001), pp. 225-232, ISSN: 0011-9164.

Hassan A. M., et al. (1998) A New Approach to Membrane and Thermal Seawater Desalination Processes using Nanofiltration membranes (Part1), *Desalination*, Vol. 118, No. 1-3, (September 1998), pp. 35-51, ISSN: 0011-9164.

Harkare, W. P., et al. (1982) Desalination of Brackish Water by Electrodialysis, *Desalination*, Vol. 42, No. 1, (July 1982), pp. 97-105, ISSN: 0011-9164.

International Desalination Association, (2009). Total World Desalination Capacity close to 60 million m^3/d. In The *international Desalination and water Reuse Quarterly Industry Newsletter*, accessed 15.08.11, available from:
http://www.desalination.biz/news/news_story.asp?id=5121

Karagiannis, I. C.; Soldatos, P. G. (2008) Water Desalination Cost Literature: Review and Assessment, *Desalination*, Vol. 223, No. 1-3, (March 2008), pp. 448-456, ISSN: 0011-9164.

Katz, W. E. (1979) The Electrodialysis Reversal (EDR) Process, *Desalination*, Vol. 28, No. 1, (January 1979), pp. 31-40, ISSN: 0011-9164.

Katz, W. E. (1982) Desalination by ED and EDR - State-of-the-Art in 1981, *Desalination*, Vol. 42, No. 2, (August 1982), pp. 129-139, ISSN: 0011-9164.

Kedem, O.; Zalmon, G. (1997) Compact Accelerated Precipitation Softening (CAPS) as a Pretreatment for Membrane Desalination I. Softening by NaOH, *Desalination*, Vol. 113, No. 1, (November 1997), pp. 65-71, ISSN: 0011-9164.

Khawaji, A. D.; Kutubkhanah, I. K.; Wie, J. (2008) Advances in Seawater Desalination Technologies, *Desalination*, Vol. 221, No. 1-3, (March 2008), pp. 47-69, ISSN: 0011-9164.

Kim, D. H. (2011) A Review of Desalting Process Techniques and Economic Analysis of the Recovery of Salts from Retenates, *Desalination*, Vol. 270, No. 1-3, (April 2011), pp. 1-8, ISSN: 0011-9164.

Lee, K. P.; Arnot, T. C.; Mattia, D. (2011) A Review of Reverse Osmosis Mambrane Material for Desalination - Development to Date and Future Potential, *Journal of Membrane Science*, Vol. 370, No. 1-2, (March 2011), pp. 1-22, ISSN: 0376-7388.

Lee, S.; et al. (2009) Toward a Combined System of Forward Osmosis and Reverse Osmosis for Seawater Desalination, *Desalination*, Vol. 247, No. 1-3, (October 2009), pp. 239-246, ISSN: 0011-9164.

Lindsay, F. K.; Ryznar, J. W. (1939) Removal of Silica from Water by Sodium Aluminate, *Industrial and Engineering Chemistry,* Vol. 31, No. 7, (July 1939), pp. 859-861, ISSN: 0888-5885.

Mabrouk, A. A.; Nafey, A. S.; Fath, H. E. S. (2007) Analysis of a New Design of a Multi-Stage-Flash-Mechanical Vapor Compression Desalination Process, *Desalination*, Vol. 204, No. 1-3, (February 2007), pp. 482-500, ISSN: 0011-9164.

Mace, R. E.; Nicot, J.; Chowdhury, A. H.; Dutton, A. R.; Kalaswad, S., (2006), Please Pass the Salt: Using Oil Fields for the Disposal of Concentrate from Desalination Plants, Texas Water Development Board, Report No. 366, April 2006.

Masarawa, A.; Meyerstein, D.; Daltrophe, N.; Kedem, O. (1997) Compact Accelerated Precipitation Softening (CAPS) as a Pretreatment for Membrane Desalination II. Lime Softening with Contaminant Removal of Silica and Heavy Metals, *Desalination,* Vol. 113, No. 1, (November 1979), pp. 73-84, ISSN: 0011-9164.

Mathioulakis, E.; Belessiotis, V.; Delyannis, E. (2007) Desalination by Using Alternative Energy: Review and State-of-the-Art, *Desalination,* Vol. 203, No. 1-3, (February 2007), pp. 346-365, ISSN: 0011-9164.

Martinetti, C. R.; Childress, A. E.; Cath, T. Y. (2009) High Recovery of Concentrated RO brines using Forward Osmosis and Membrane Distillation, *Journal of Membrane Science*, Vol. 331, No. 1-2, (April 2009), pp. 31-39, ISSN: 0376-7388.

McCutcheon J. R.; McGinnis, R. L.; Elimelech, M. (2005a) A novel Ammonia-Carbon Dioxide Forward (Direct) Osmosis Desalination Process, *Desalination*, Vol. 174, No. 1, (April 2005), pp. 1-11, ISSN: 0011-9164.

McCutcheon J. R.; McGinnis, R. L.; Elimelech, M. (2005b) Desalination by Ammonia-Carbon Dioxide Forward Osmosis: Influence of Draw and Feed Solution Concentrations on Process Performance, *Journal of Membrane Science*, Vol. 278, No. 1-2, (July 2006), pp. 114-123, ISSN: 0376-7388.

McGinnis, R. L.; Elimelech, M. (2007) Energy Requirements of Ammonia-Carbon Dioxide Forward Osmosis Desalination, *Desalination*, Vol. 207, No. 1-3, (March 2007), pp. 370-382, ISSN: 0011-9164.

Mickley, M. C. (2001) Membrane Concentrate Disposal: Practice and Regulations. U.S. Department of the Interior, Bureau of Reclamation, Mickley & Associates.

Mickley, M. C., (2006) Membrane Concentrate Disposal: Practice and Regulation, Desalination and Water Purification Research and Development Program Report No. 123 2nd Ed., April 2006 (Agreement No. 98-FC-81-0054), Prepared for the U.S. Department of the Interior, Bureau of Reclamation, Technical Service Center, Water Treatment Engineering and Research Group.

Missimer T.M. (1994) Groundwater as a Feedwater Source for Membrane Treatment Plants: Hydrogeologic Controls on Water Quality Variations with Time, *Desalination*, Vol. 98, No. 1-3, (September 1994), pp. 451-457, ISSN: 0011-9164.

Mohamed, A. M. O.; Maraqa, M.; Al-Handaly, J (2005) Impact of Land Disposal of Reject Brine from Desalination Plant on Soil and Groundwater, *Desalination,* Vol. 182, No. 1-3, (November 2005), pp. 411-433, ISSN: 0011-9164.

Mohammadesmaeili, F.;Badr, M. K.; Abbaszadegan, M.; Fox, P. (2010) Byproduct Recovery from Reclaimed Water Reverse Osmosis Concentrate Using Lime and Soda-Ash

Treatment, *Water Environment Research,* Vol. 82, No. 4, (April 2010), pp. 342-350, ISSN: 1061-4303.

Neofotistou, E.; Demadis, K. D. (2004) Use of antiscalents for Mitigation of Silica (SiO_2) Fouling and Deposition: Fundamentals and Applications in Desalination Systems, *Desalination,* Vol. 167, (August 2004), pp. 257-272, ISSN: 0011-9164.

Nicot, J.; Chowdhury, A. H. (2005) Disposal of Brackish water Concentrate into Depleted Oil and Gas Fields: a Texas Study, *Desalination,* Vol. 181, No. 1-3, (September 2005), pp. 61-74, ISSN: 0011-9164.

Ning, R. Y. (2002) Discussion of Silica Speciation, Fouling, Control and Maximum Reduction, *Desalination,* Vol. 151, No. 1, (January 2003), pp. 67-73, ISSN: 0011-9164.

Ning, R. Y.; Tarquin, A.; Trzcinski, M. C.; Patwardhan, G. (2006) Recovery Optimization of RO Concentrate from Desert Wells, *Desalination,* Vol. 201, No. 1-3, (November 2006), pp. 315-322, ISSN: 0011-9164.

Ning, R. Y.; Troyer, T. L. (2009) Tandom Reverse Osmosis Process for Zero-Liquid Discharge, *Desalination,* Vol. 237, No. 1-3, (February 2009), pp. 238-242, ISSN: 0011-9164.

Oren, Y.; Katz, V.; Daltrophe, N. (2001) Improved Compact Accelerated Precipitation Softening (CAPS), *Desalination,* Vol. 139, No. 1-3, (September 2001), pp. 155-159, ISSN: 0011-9164.

Oren, Y; et al, (2010) Pilot Studies on High Recovery BWRO-EDR for Near Zero Liquid Discharge Approach, *Desalination,* Vol. 261, No. 3, (October 2010), pp. 321-330, ISSN: 0011-9164.

Phillip, W. A.; Yong, J. S.; Elimelech, M. (2010) Reverse Draw Solute Permeation in Forward Osmosis: Modeling and Experiments, *Environmental Science and Technology,* Vol. 44, No. 13, (July 2010), pp. 5170-5176, ISSN: 0013-936X.

Rahardianto, A.; Gao, J.; Gabelich, C. J.; Williams, M. D.; Cohen, Y. (2007) High Recovery Membrane Desalting of Low-Salinity Brackish Water: Integration of Accelerated Precipitation Softening with Membrane RO, *Journal of Membrane Science,* Vol. 289, No. 1-2. (February 2007), pp. 123-137, ISSN: 0376-7388.

Rahardianato, A; Mccool, B. C.; Cohen, Y. (2008) Reverse Osmosis Desalting of Inland Brackish water of High Gypsum Scaling Propensity: Kinetics and Mitigation of Membrane Mineral Scaling, *Environmental Science and Technology,* Vol. 42, No. 12, (June 2008), pp. 4292-4297, ISSN: 0013-936X.

Rautenbach, R; Groschl, A. (1990) Separation Potential of Nanofiltration membranes, *Desalination,* Vol. 77, (March 1990), pp. 73-84, ISSN: 0011-9164.

Reverter, J. A; Talo, S.; Alday, J. (2001) Las Palmas III – the Success Story of Brine Staging, *Desalination,* Vol. 138, No. 1-3, (September 2001), pp. 207-217, ISSN: 0011-9164.

Reynolds T. D ; Richards P. A., (1996), *Unit Operations and Processes in Environmental Engineering,* PWS Publishing Company, ISBN 0-534-94884-7, Boston, Massachusetts, USA.

Sadrzadeh, M.; Mohammadi, T. (2008) Sea Water Desalination Using Electrodialysis, *Desalination,* Vol. 221, No. 1-3, (March 2008), pp. 440-447, ISSN: 0011-9164.

Saitua, H.; Gil, R.; Padilla, A. P. (2011) Experimental Investigations on Arsenic Removal with a Nanofiltration Pilot Plant from Contaminated Groundwater, *Desalination*, Vol. 274, No. 1-3, (July 2011), pp. 1-6, ISSN: 0011-9164.

Schaep, J; et al. (1998) Removal of Hardness from Groundwater by Nanofiltration, *Desalination*, Vol. 119, No. 1-3, (September 1998), pp. 295-301, ISSN: 0011-9164.

Semiat, R.; Sutzkover, I.; Hasson, D. (2003) Scaling of RO Membranes from Silica Supersaturated Solutions, *Desalination*, Vol. 157, No. 1-3, (August 2003), pp. 169-191, ISSN: 0011-9164.

Semiat, R.; Hasson, D.; Zelmanov, G.; Hemo, I. (2004) Threshold Scaling Limits of RO Concentrates Flowing in a Long Waste Disposal Pipeline, *Water Science and Technology*, Vol. 49, No. 2, pp. 211-219, ISSN: 0273-1223.

Seto, T., et al. (1978) Seawater Desalination by Electrodialysis, *Desalination*, Vol. 25, No. 1, (March 1978), pp. 1-7, ISSN: 0011-9164.

Sheikholesami, R.; Tan, S. (1999) Effects of Water Quality on Silica Fouling of Desalination plants, *Desalination*, Vol. 126, No. 1-3, (November 1999), pp. 267-280, ISSN: 0011-9164.

Sheikholesami, R.; Al-Mutaz, I. S.; Koo, T.; Young, A. (2001) Pretreatment and the Effect of Cations and Anions on Prevention of Silica Fouling, *Desalination*, Vol. 139, No. 1-3, (September 2001), pp. 83-95, ISSN: 0011-9164.

Sheikholesami, R; Bright, J. (2002) Silica and Metals Removal by Pretreatment to Prevent Fouling of Reverse Osmosis Membranes, *Desalination*, Vol. 143, No. 3, (June 2002), pp. 255-267, ISSN: 0011-9164.

Sheikholesami, R. (2003a) Mixed salts - Scaling limits and Propensity, *Desalination*, Vol. 154, No. 2, (April 2003), pp. 117-127, ISSN: 0011-9164.

Sheikholesami, R. (2003b) Nucleation and Kinetics of Mixed Salts in Scaling, *AIChE Journal*, Vol. 49, No. 1, (January 2003), pp. 194-202, ISSN: 001-1541.

Sheikholesami, R. (2004) Assessment of the Scaling Potential for Sparingly Soluble Salts in RO and NF units, *Desalination*, Vol. 167, (August 2004), pp. 247-256, ISSN: 0011-9164.

Sheikholesami, R. (2009) Strategies for Future Research and Development in Desalination-challenges ahead, *Desalination*, Vol. 248, No. 1-3, (November 2009), pp. 218-224, ISSN: 0011-9164.

Sherif, M.; Mohamed, M.; Kacimov, A.; Shetty, A. (2011) Assessment of Groundwater Quality in the Northeastern Coastal area of UAE as Precursor for Desalination, *Desalination*, Vol. 273, No. 2-3, (June 2011), pp. 436-446, ISSN: 0011-9164.

Shon, H. K.; et al. (2009) Physicochemical Pretreatment of Seawater: Fouling Reduction and Membrane Characterization, *Desalination*, Vol. 238, No. 1-3, (November 2009), pp. 10-21, ISSN: 0011-9164.

Skehan, S; Kwiatkowski P. J. (2000) Concentrate Disposal via Injection Wells - Permitting and Design Considerations, *Florida Water Resources Journal*, Vol. 52, No. 5, (May 2000), pp. 19-21, ISSN: 0896-1794.

Tahaikt, M., et al. (2007) Fluoride Removal from Groundwater by Nanofiltration, *Desalination*, Vol. 212, No. 1-3, (June 2007), pp. 46-53, ISSN: 0011-9164.

Tait, S.; Clarke, W. P.; Keller, J.; Batstone, D. J. (2009) Removal of Sulfate from High-Strength Wastewater by Crystallization, *Water Research,* Vol. 43, No. 3, (February 2009), pp. 762-772, ISSN: 0043-1354.

Tang, W.; Ng, H. Y. (2008) Concentration of Brine by Forward Osmosis: Performance and influence of membrane structure, *Desalination,* Vol. 224, No. 1-3, (April 2008), pp. 143-153, ISSN: 0011-9164.

The Federal Institute for Geosciences and Natural Resources, (n.d.). Groundwater Resources of the World, accessed 15.08.11, available from: http://www.bgr.bund.de/EN/Themen/Wasser/Bilder/Was_wasser_startseite_gw_erde_g_en.html?nn=1548136

United Nations UN, (1997). UN Assessment of Freshwater Resources 1997, accessed 13.08.11, available from: http://www.un.org/ecosocdev/geninfo/sustdev/waterrep.htm

United States Environmental Protection Agency, US EPA, (2009) National Primary & Secondary Drinking Water Regulations, accessed 15.08.11, available at: http://water.epa.gov/drink/contaminants/index.cfm

United Nations Environmental Programme UNEP- GRiD, (n.d.). Increased Global Water Stress, accessed 15.08.11, available from: http://www.grida.no/graphic.aspx?f=series/vg-water2/0400-waterstress-EN.jpg

Valcour, H. C. (1985) Recent Applications of EDR, *Desalination,* Vol. 54, pp. 163-183, ISSN: 0011-9164.

Vandercaseele, C.; Van Der Bruggen, B. (2002) Distillation vs. Membrane Filtration: Overview of Process Evolutions in Seawater Desalination, *Desalination,* Vol. 143, No. 3, (June 2002), pp. 207-218, ISSN: 0011-9164.

Vince, F.; Marechal, F.; Aoustin, E.; Bréant, P. (2008) Multi-Objective Optimization of RO Desalination Plants, *Desalination,* Vol. 222, No. 1-3, (March 2008), pp. 96-118, ISSN: 0011-9164.

Walha, K.; Ben Amar, A.; Firdaous, L.; Quemeneur, F.; Jaouen, P. (2007) Brackish Groundwater Treatment by nanofiltration, Reverse Osmosis, and Electrodialysis in Tunisia: Performance and cost comparison, *Desalination,* Vol. 207, No. 1-3, (March 2007), pp. 95-106, ISSN: 0011-9164

Wang, L. K (2008), *Handbook of Environmental Engineering V.13: Membrane and Desalination Technologies,* Humana Press, ISBN 1-58829-940-6, New York, USA.

WHO (2011) Guidelines for Drinking-Water Quality, 4th Ed. Accessed 15.08.11, available from: http://water.epa.gov/drink/contaminants/index.cfm

Yangali-Quintanilla, V.; Li, Z.; Valladares, R.; Li, Q.; Amy G. (2011) Indirect Desalination of Red Sea Water with Forward Osmosis and Low Pressure Reverse Osmosis for water reuse, *Desalination,* doi:10.1016/j.desal.2011.06.066.

Yun, T. I; Gabelich, C. J.; Cox, M. R.; Mofidi, A. A.; Lesan, R. (2006) Reducing Costs for Large-Scale Desalting Plants Using Large-Diameter Reverse Osmosis Membranes, *Desalination,* Vol. 189, No. 1-3, (March 2006), pp. 141-154, ISSN: 0011-9164.

Zubari, W. K. (2003) Assessing the Sustainability of Non-renewable Brackish Groundwater in Feeding an RO Desalination Plant in Bahrain, *Desalination,* Vol. 159, No. 3, (November 2003), pp. 211-244, ISSN: 0011-9164.

Zubari, W. K.; Madany, I. M.; Al-Junaid, S. S.; Al-Manaii, S. (1994) Trends in the quality of Groundwater in Bahrain with Respect to Salinity, 1941-1992., *Environment International*, Vol. 20, No. 6, pp. 739-746, ISSN: 0160-4120.

Part 4

Separation Technology

19

Phase Diagrams in Chemical Engineering: Application to Distillation and Solvent Extraction

Christophe Coquelet[1,2] and Deresh Ramjugernath[1]
[1]*Thermodynamic Research Unit, School of Chemical Engineering, University KwaZulu Natal, Howard College Campus, Durban*
[2]*MINES ParisTech, CEP/TEP - Centre Énergétique et Procédés, Fontainebleau*
[1]*South Africa*
[2]*France*

1. Introduction

By definition, a phase diagram in physical chemistry and chemical engineering is a graphical representation showing distinct phases which are in thermodynamic equilibrium. Since these equilibrium relationships are dependent on the pressure, temperature, and composition of the system, a phase diagram provides a graphical visualization of the effects of these system variables on the equilbrium behavior between the phases. Phase diagrams are essential in the understanding and development of separation processes, especially in the choice and design of separation unit operations, e.g. knowledge about high pressure phase equilibria is essential not just in chemical processes and separation operations, but is also important for the simulation of petroleum reservoirs, the transportation of petroleum fluids, as well as in the refrigeration industry. In order to utilize the knowledge of phase behavior it is important to represent or correlate the phase information via the most accurate thermodynamic models. Thermodynamic models enable a mathematical representation of the phase diagram which ensures comprehensive and reproducible production of phase diagrams. The measurement of phase equilibrium data is necessary to develop and refine thermodynamic models, as well as to adjust them by fitting or correlating their parameters to experimental data. Generally the measurement of phase equilibria is undertaken using two categories of experimental techniques, viz. synthetic and analytic methods. The choice of the technique depends on the type of data to be determined, the range of temperatures and pressures, the precision required, and also the order of magnitude of the phase concentrations expected.

2. Definition of phases, phase transitions, and equilibrium

A phase is a homogeneous space, which can be composed of one or more components or chemical species, in which the thermodynamic properties, e.g. density or composition, are identical. A system can comprise of one or several phases. Depending on whether one or more phases exist, the system can be defined as a monophasic homogenous system in which the composition and thermodynamic properties are identical in the whole space, or as a multiphase heterogenous system for which thermodynamic properties change distinctly at the phase interface. A phase is characterized by its temperature, density, pressure, and other thermodynamics properties, i.e. Gibbs energy, molar enthalpy, entropy, heat capacity, etc.

The concept of a phase can be employed to distinguish between the different states of matter. Matter is generally accepted to exist in three main states, viz. gas, liquid, and solid. Other types of phases do exist, e.g. nematic and smectic phase transitions in liquid crystral, but this will not be relevant to this chapter. Molecular interactions between the components or chemical species which comprise the system are responsible for the different states of matter. Consequently thermodynamics model used to describe phases have to take into account the molecular interactions, or be used at conditions at which they are relevant, e.g. the ideal gas model can only be applied to gas phase at low pressures and sufficiently high temperatures, as at these conditions molecular interactions are negligible.

2.1 Phases transitions

There a number of definitions that can be used to describe a liquid. One of the definitions, which is easy to visualize but may not be entirely thermodynamically correct, is that a liquid is a fluid which takes the form of its containment without necessarily filling the entire containment volume. This characteristic distinguishes a liquid from a solid, as well as a liquid from a gas. As stated, it is not an accurate definition and it is necessary to carefully check the thermophysical microscopic (structure) and macroscopic properties of the state of matter.

A phase change is the transition between different states of matter. It is generally characterized by a sudden change of the internal microscopic structure and the macroscopic properties of the environment. It is probably better to refer to it as a phase transition instead of phase change because a phase transition doesn't imply the change of state of the matter, e.g. liquid-liquid or solid-solid phase transitions. For a solid-solid phase transition there can be a change of the structure of the crystal.

A phase transition can be effected by a change of the composition, temperatures and/or pressures of the system, or by application of an external force on the system. Consequently, composition, density, molar internal energy, enthalpy, entropy, refractive index, and dielectric constant have different values in each phase. However, temperature and pressure are identical for all phases in multiphase systems in compliance with thermodynamic principles. As a result, when two phases (or more) exist, we refer to it as phase equilibria. Table 1 lists all the phase transitions between solid, liquid and vapor phases.

Phase 1	Phase 2	Transition 1-2
liquid	vapor	boiling
liquid	solid	solidification
vapor	liquid	liquefaction
vapor	solid	condensation
solid	vapor	sublimation
solid	liquid	melting

Table 1. Phases transitions.

There are two types of phase transition:

First order: In this type of transition there is a discontinuity in the first derivative of the Gibbs free energy with regard to thermodynamic variables. For this type of transition there is an absorption or release of a fixed amount of energy with the temperature remaining constant.

Second order: For this type of transition, there is continuity in the first derivative of the Gibbs free energy with regard to the thermodynamic variable across the transition, but there is a discontinuity in the second derivative of the Gibbs energy with regard to thermodynamic properties.

2.2 Chemical potentials and equilibrium conditions

2.2.1 Chemical potential

The chemical potential is the one of the most important thermodynamic properties used in the description of phase equilibrium. If one considers a phase with volume V, containing n_c components at temperature T and pressure P, the chemical potential μ_i of component i in the phase is defined by

$$\mu_i(P,T,n_1,n_2,...,n_C) = \left(\frac{\partial G(P,T,n_1,n_2,...,n_C)}{\partial n_i}\right)_{T,P,n_{j\neq i}} \tag{1}$$

Where G=H-TS is the Gibbs free energy of the phase. The expression for the infinitesimal reversible change in the Gibbs free energy is given by

$$dG = VdP - SdT + \sum_i \mu_i dn_i \tag{2}$$

Moreover, from Euler theorem we can write $G = \sum_{i=1}^{N_c} n_i \mu_i$. The Gibbs-Duhem (Eq. 3) equation can be obtained from equations 1 and 2:

$$VdP - SdT + \sum_i n_i d\mu_i = 0 \tag{3}$$

2.2.2 Equilibrium conditions

Considering a multicomponent system in equilibrium between two phases (α and β), at temperature T and pressure P, the Gibbs free energy of this system is $G=G^\alpha+G^\beta$, where G^α and G^β, are the Gibbs free energy of the α and β phases respectively. At equilibrium, the Gibbs free energy must be at the minimum, therefore ($dG=0$):

$$dG = dG^\alpha + dG^\beta = \sum_i \mu_i^\alpha dn_i^\alpha + \sum_i \mu_i^\beta dn_i^\beta = 0 \tag{4}$$

For a closed system, $dn_i^\beta = -dn_i^\alpha$ for each component i between phases α and β. For each chemical species, the following relationship can written:

$$\mu_i^\alpha\left(P,T,n_1^\alpha,n_2^\alpha,...,n_{N_C}^\alpha\right) = \mu_i^\beta\left(P,T,n_1^\beta,n_2^\beta,...,n_{N_C}^\beta\right) \tag{5}$$

Consequently, the equilibrium condition a of multiphase mixture is equality of temperature, pressure, and chemical potential μ_i of each component i in all the phases in equilibrium.

3. Description of phase diagrams

3.1 Pure compound

3.1.1 Description

The phase diagram of a pure compound is characterized by its critical point, triple point, vapor pressure, and melting and sublimation curves (figure 1). At the triple point, the solid, liquid, and vapor phases coexist. The critical point can be defined as the upper limit for the pure component vapor pressure curve. For temperatures and pressures above the critical conditions, there is no possibility to have vapor-liquid equilibrium. The supercritical state can be considered as a "stable state" with no possibility for phase separation.

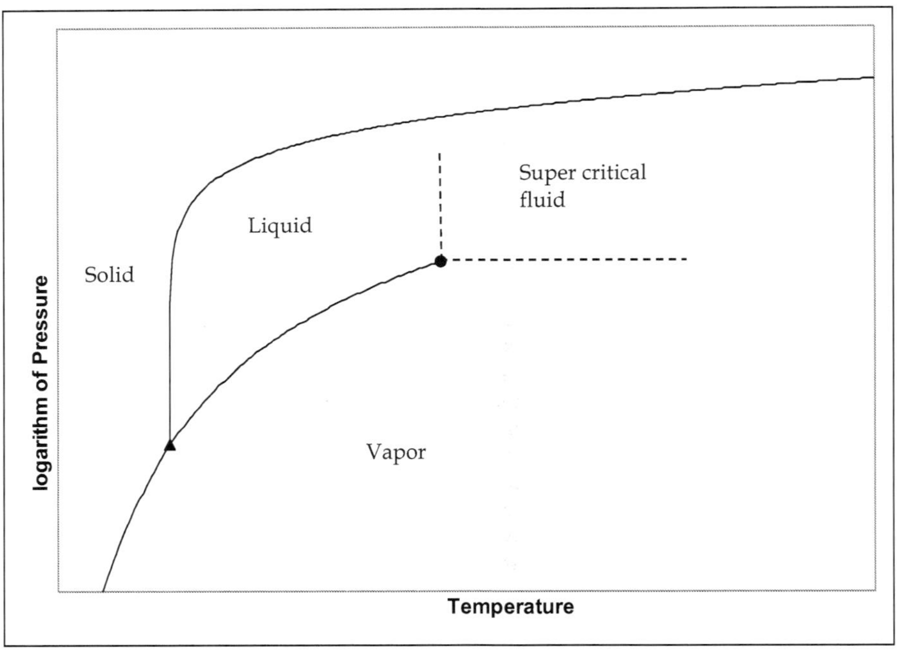

Fig. 1. Pure compound Pressure-Temperature phase diagram. (▲) Triple point, (•) : critical point. Lines : Coexistence curves

The number of intensive variables which have to be specified in order to characterize the system is determined with the Gibbs phases rule (Eq. 6).

$$F = C + 2 - \phi \tag{6}$$

Were, *F*, is the of degrees of freedom, *C*, is the number of components and *Φ*, is the number of phases present. Table 2 illustrates the degrees of freedom for a single component system.

Region of the phase diagram	Number of phases	Degrees of freedom	Variables to be specified
vapor pressure, or melting or sublimation curves	2	1	T or P
liquid, vapor, or solid	1	2	P and T
Triple point	3	0	Everything is fixed
Critical point	2	1	T_C or P_C

Table 2. Degrees of freedom for a pure component (*C=1*)

The phase behavior of a pure component can be represented in a plot of pressure versus temperature. Assuming that there are only two co-exisiting phases, i.e. a liquid and vapor phase, the vapor-liquid equilibrium can be represented as shown in Figure 2.

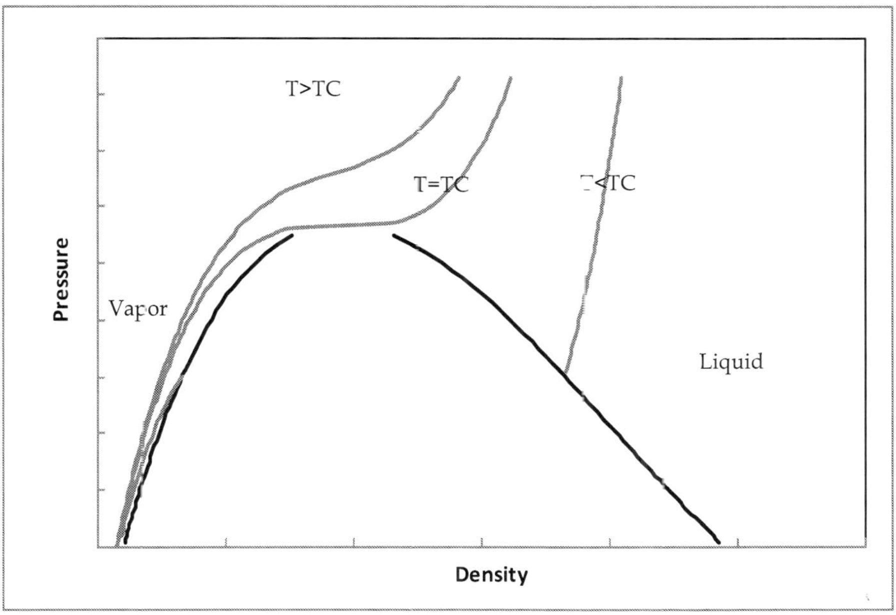

Fig. 2. Pure compound Pressure-Temperature-density phase diagram. Red line: isothermal curve. Black line saturation curve (bubble and dew points lines)

3.1.2 The critical point

The critical point of a pure compound is the upper limit (temperature and pressure) of the pure component vapor pressure curve. For temperatures and pressures below the critical point, phase separation can occur leading to two phases in equilibrium (liquid and vapour). The critical point can be considered as a limit of stability of the supercritical phase.

Critical properties (T_C, P_C, ...) characterize the critical point and the value of the properties are particularly influenced by molecular interactions, i.e. if there are strong attractive interaction between the molecules, the value of the critical temperatures and pressures are greatly increased. Using the example of water [NIST, 2010 #167] (T_C= 647.13 K, P_C= 22.055 MPa) one can clearly observe that the values of the critical properties are far greater when compared to those of methane [NIST, 2010 #167] (T_C= 190.564 K, P_C= 4.599 MPa). Hydrogen bonding between molecules of water is responsible for association effects between water molecules.

At the critical point, due to the fluctuation of the density, there is a disturbance of light waves across the visual spectrum. This phenomenon is called critical opalescence. At the critical point there is a second order type phase transition. In the Pressue-density diagram (figure 2), the isotherm labeled T_C has an inflection point and is characterized by the following relations:

$$P > 0 \, , \; \left(\frac{\partial P}{\partial v}\right)_T = 0 \, , \; \left(\frac{\partial^2 P}{\partial v^2}\right)_T = 0 \, , \; \left(\frac{\partial^3 P}{\partial v^3}\right)_T < 0 \tag{7}$$

Experimentally, it can be observed that along the coexisting curve liquid and vapor densities diverge at the critical point following the relationship $\rho_L - \rho_V \propto (T_C - T)^{\beta}$ where ρ_C and T_C are the critical density and critical temperature respectively and β is an universal critical exponent (value is around 0.326, whatever the pure compound). This mathematical relationship is one of thermodynamics relationships and referred to as scaling laws. For pressure, one can write the following relationship along the critical isotherm curve: $\left|\frac{P - P_C}{P_C}\right| \propto \left|\frac{\rho - \rho_C}{\rho_C}\right|^{\delta}$, where $\delta = 4.800$. With regard to heat capacity, the following equation can be written along the critical isochor: $C_V \propto k_{\alpha 0} + k_{\alpha 1}(\Delta\tau)^{-\alpha}$ for $\rho = \rho_C$ with $\Delta\tau = \left|\frac{T - T_C}{T_C}\right|$.

For isothermal compressibility the relationship along the critical isochors is $\chi \propto k_{\alpha 1}(\Delta\tau)^{-\gamma}$ for $\rho=\rho_C$ with $\Delta\tau = \left|\frac{T - T_C}{T_C}\right|$. The exponents' δ, α and γ are also referred to as critical exponents. Simple relationships can be written between these exponents, e.g. $\alpha + 2\beta + \gamma = 2$. All of these laws are also observed experimentally and can be derived from renormalisation theory. Table 3 shows the optimum values for these critical exponents.

Critical exponent	Values
α	0.110
β	0.3255
γ	1.239
δ	4.800

Table 3. Critical exponents

3.1.3 Triple point

At the triple point, the three states of the matter (solid, liquid, and vapor) are in equilibrium. The temperature and pressure of the triple point is fixed because the degrees of freedom at this point is equal to zero. Triple point values are very useful for the definition of reference points on temperature scales and for the calibration of experimental equipment. The table 4 lists triple point conditions for some for some common chemicals.

Compound	T_{tr}/K	P_{tr}/kPa
Water	273.16	0.6117
Ethanol	150	4.3×10^{-7}
Oxygen	54.36	0.152
Methane	90.67	11.69

Table 4. Triple points of few pure compounds (Ref. : NIST [NIST, 2010 #167]).

3.2 Binary systems

The first equation of state, which could describe the behaviour of both liquid and vapour states of a pure component, was developed by (van der Waals, 1873). Two types of interactions (repulsive and attractive) were considered in this equation.

$$\left(P+\frac{a}{v^2}\right)(v-b)=RT \tag{8}$$

The stability criteria are defined by the following equations:

$$\left(\frac{\partial P}{\partial v}\right)_T<0,\left(\frac{\partial^2 P}{\partial v^2}\right)_T<0 \text{ and } \frac{T}{c_v}>0$$

At the critical point these two previous conditions are equal to zero.

$$\left(\frac{\partial P}{\partial v}\right)_T=0,\left(\frac{\partial^2 P}{\partial v^2}\right)_T=0 \text{ and } \frac{T}{c_v}=0$$

It can be seen that at the critical point, the isochoric heat capacity has infinite value. Phase diagram describing binary mixtures depend of the behaviour of the species. Van Konynenburg & Scott (1980) have classified the phase behaviour of binary mixtures into six types considering van der Waals EoS and quadratic mixing rules. Figure 3 presents the different types of phase diagrams. The transition between each type of phase diagram can be explained by considering the size effects of molecules and the repulsive interactions between them. Figure 4 illustrates the possible transitions between the different types of phase diagrams.

3.2.1 Mixture critical point

The thermodynamic stability of a mixture determines if it would remain as a stable homogenous fluid or split into more stable phases and therefore produce two or more

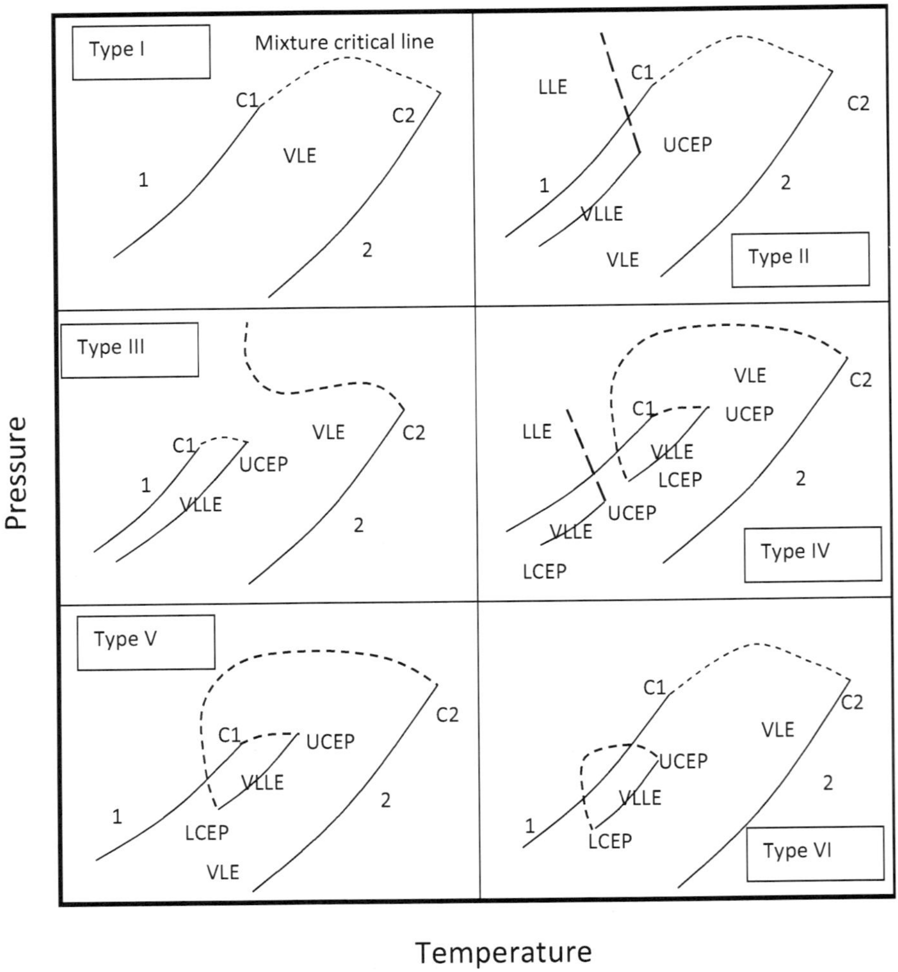

Fig. 3. Six types of phase behaviour in binary fluid systems. C: Critical point, L: Liquid, V: Vapor. UCEP: Upper critical end point, LCEP: Lower critical end point. Dashed curve are critical.

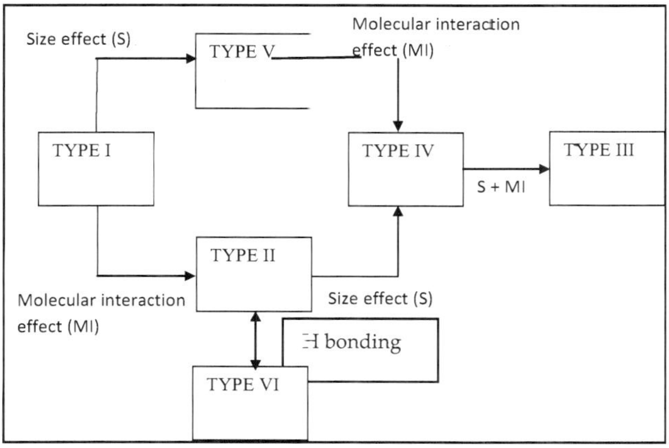

Fig. 4. Evolution of phase diagrams.

phases in equilibrium. By definition, a mixture is considered as stable when the Gibbs or Helmholtz free energy are at their minimum. Figure 5 gives an indication of the change of the Gibbs free energy with composition for a given temperature and pressure from stable condition (a) (T, P) to an unstable condition (b). In figure 5 b, one can observe that there are

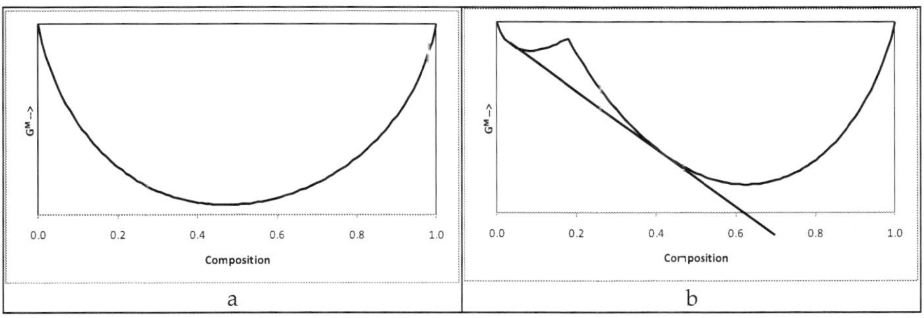

a): stable condition, b): unstable condition with phase equilibria.

Fig. 5. Mixing Gibbs free energy (G) as a function of molar composition (x1) at given T and P.

two minima: this corresponds to phase equilibria. More details on thermodynamic phase stability are given by (Michelsen, 1982a). A Taylor series expansion can be obtained for the Gibbs free energy for a given temperature

$$G=\sum_i n_i^0\mu_i^0+\sum_i \mu_i^0\Delta n+\frac{1}{2}\sum_i\sum_j\left(\frac{\partial^2 G}{\partial n_i\partial n_j}\right)_{T,P}\Delta n_i\Delta n_j+\frac{1}{6}\sum_i\sum_j\sum_k\left(\frac{\partial^3 G}{\partial n_i\partial n_j\partial n_k}\right)_{T,P}\Delta n_i\Delta n_j\Delta n_k+\Theta\left(\Delta n^5\right) \tag{9}$$

With $n_i = n_i^0 + \Delta n_i$ and $V = V^0 + \Delta V$.

The stability condition leads to:

$$G - \sum_i n_i^0 \mu_i^0 - \sum_i \mu_i^0 \Delta n_i \geq 0 \qquad (10)$$

Consequently

$$\frac{1}{2}\sum_i \sum_j \left(\frac{\partial^2 G}{\partial n_i \partial n_j} \right)_{T,P} \Delta n_i \Delta n_j + \frac{1}{6}\sum_i \sum_j \sum_k \left(\frac{\partial^3 G}{\partial n_i \partial n_j \partial n_k} \right)_{T,P} \Delta n_i \Delta n_j \Delta n_k + \Theta\left(\Delta n^5\right) \geq 0 \qquad (11)$$

The critical point can be described as the limit of stability (x is the molar fraction) and coordinates (P, v) can be determined considering the following relations. More details are given by (Michelsen, 1980, 1982b), (Heidmann & Khalil, 1998) and (Stockfleth & Dohrn,1980).

$$G_{2x} = \left(\frac{\partial^2 G}{\partial n_i \partial n_j} \right)_{T,P} = 0 \ ; \ G_{3x} = \left(\frac{\partial^3 G}{\partial n_i \partial n_j \partial n_k} \right)_{T,P} = 0 \ ; \ G_{4x} = \left(\frac{\partial^4 G}{\partial n_i \partial n_j \partial n_k \partial n_l} \right)_{T} > 0 \qquad (11)$$

In their paper, (Baker et al., 1982) present several examples of Gibbs energy analysis. Figure 6 compares phase diagrams (pressure/composition) and the Mixing Gibbs free energy at a given temperature.

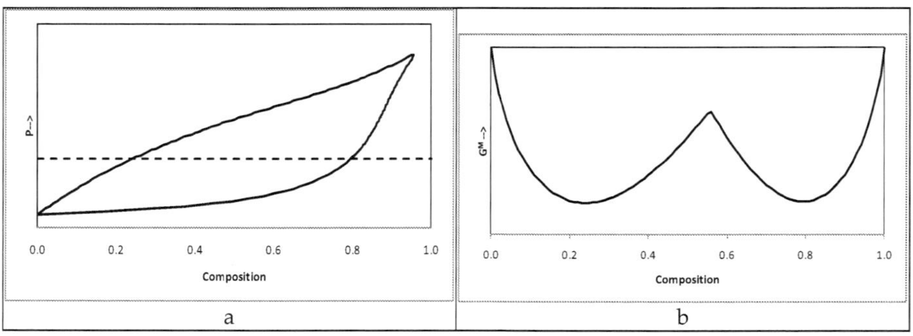

A) Phase diagram at a given temperature. Dashed line: given pressure. B) Gibbs free energy of mixing at the same temperature and given pressure.

Fig. 6. Example of Mixing Gibbs free energy minimum for a binary system.

3.2.2 Phase diagram classification

Type I phase behaviour

It is the simplest type of phase diagram. The mixture critical point line starts at the first pure component critical point and finish at the second pure component critical point. Mixtures which exhibit this behaviour are two components which are chemically similar or have comparable critical properties., e.g. systems with CO_2 and light hydrocarbons, systems with HFC refrigerants, and benzene + toluene binary systems.

Type II phase behaviour

It is similar to type I but at low temperatures, the two components are not miscible in the liquid mixtures. Consequently a liquid - liquid equilibrium appears. The mixture critical point line for liquid - liquid equilibrium starts from the UCEP (upper critical end point). At the UCEP, the two liquid phases merge into one liquid phase. Examples of systems which exhibit this behaviour are those with hydrocarbons and fluorinated fluorocarbons.

Type III phase behaviour

Generally system which have very large immiscibility gaps exhibit this behaviour, e.g. aqueous systems with hydrocarbons. A liquid – liquid - vapour curve appears and a first mixture critical point line starts from the pure component 1 critical point and ends at the UCEP. The second one starts from the infinite pressure ($P \rightarrow \infty$) and ends at the pure component 2 critical point, generally the solvent i.e. water. The slope of this second curve can be positive, negative or positive and negative. Concerning the positive curve, we have two phases at temperatures larger than the critical temperature of pure component two.

Some examples:

Positive slope: helium + water binary system

Negative slope: methane + toluene binary system

Positive and negative slope: nitrogen + ammonia or ethane + methanol binary systems.

Type IV phase behaviour

It is similar to type V behaviour. The vapor-liquid critical point line starts at the critical point of component 2 and ends at the LCEP (lower critical end point). Vapour-liquid-liquid equilibrium (VLLE) exists and is present in two parts. Ethane + n-propanol and CO_2 + nitrobenzene binary systems are examples of systems which show this behaviour.

Type V phase behaviour

It is a modification of type III phase diagram. There are two vapor-liquid critical point lines. One goes from the pure component critical point 1 and ends at the UCEP. The other starts at the pure component critical point 2 and ends at the LCEP. Contrary to type IV systems, below the LCEP the liquids are completely miscible. The ethylene + methanol binary system is an example of a type V system.

Type VI phase behaviour

There are two critical point curves. The first is similar to one presented with the type II diagram: a connection between the two pure component critical points. The second connects the LCEP and the UCEP. Between these two points, there exists VLLE. The system "water + 2-butanol" is a typical example of a type VI system. In fact the main reason is due to the existence of hydrogen bonding for one or both of the pure components (self association) and in the mixture (strong H bonding between the two components). H bonding favours the heat of solution and so miscibility in the liquid state. Above the LCEP hydrogen bonds break and the liquid becomes unstable and a second liquid phase appears.

3.3 Ternary systems

There exists a number of different classifications for ternary systems, but we propose that of Weinstsock (1952). According to this classification the system can exhibit vapor-liquid, liquid-liquid, and also vapor-liquid-liquid equilibrium (VLE, LLE and VLLE). Considering vapor-liquid-liquid equilibria, the phase diagrams can be classified into 3 categories or types. For ternary diagrams, in general, temperature and pressure are fixed.

Type 1 :

It is the most common diagram. In figure 7, one can observe VLE and the monophasic regions. The shape of the phase diagram changes with the pressure. In figure 8, one can see the phase diagram for the ternary system comprising R32 + R227ea + R290 (Coquelet et al., 2004).

Type 2 :

It is an evolution of a type 1 phase diagram: VLLE and LLE appear but the two liquids are partially miscible (see figure 9).

Type 3 :

According to figure 10, there exists strong immiscibility between two species. The size of the LLE region increases with pressure. Consequently, LLE region disappears to form a Liquid Fluid Equilibrium region (if there is a supercritical fluid).

There is also classification for Liquid - liquid equilibrium phase diagram. There are generally three distinguishable categories (Figure 11) :

- One pair of binary components are partially miscible (type 1)
- Two pairs of binary components are partially miscible (type 2)
- All three pairs of binary components are partially miscible (type 3)

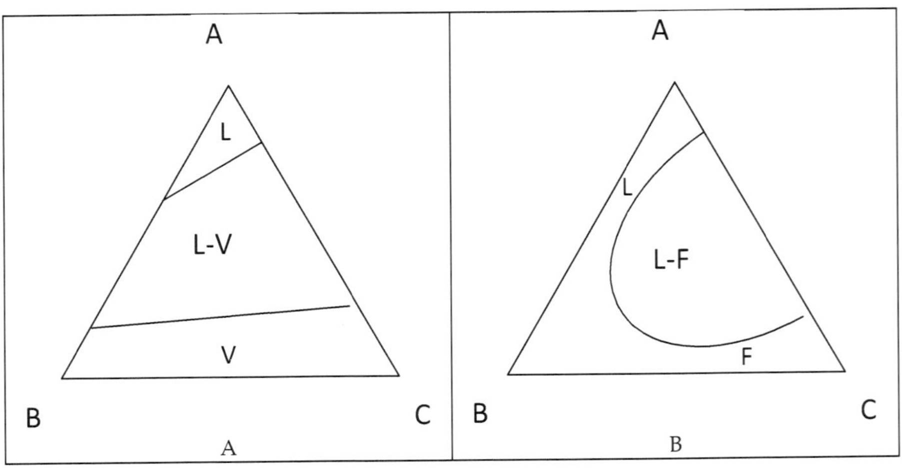

A: Vapor-Liquid Equilibria (VLE), B: if T or P is increased a supercrital fluid (F) appears and so a FLE.

Fig. 7. Ternary diagram: type 1.

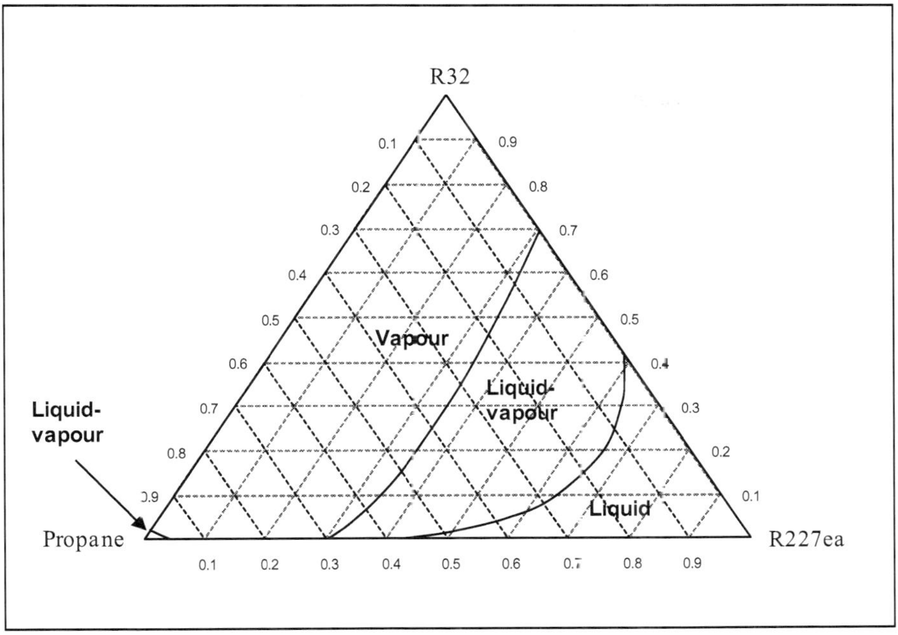

Fig. 8. Phase diagram of the ternary system R32-R290-R227ea at T=293 K and P=8.5 bar (Coquelet et al., 2004).

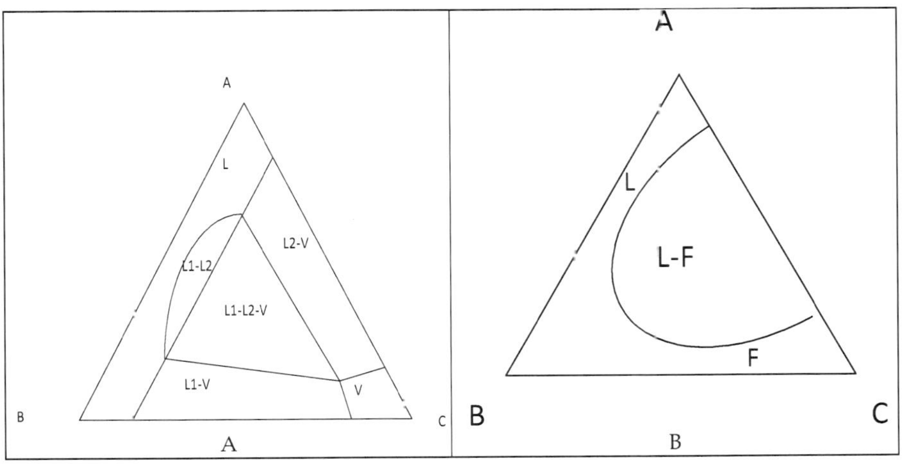

Fig. 9. Ternary diagram: type 2. A: Existence of Vapor Liquid Liquid Equilibrium, B: if we increase pressure or temperature, we have a supercritical fluid in equilibrium with one liquid phase (FLE).

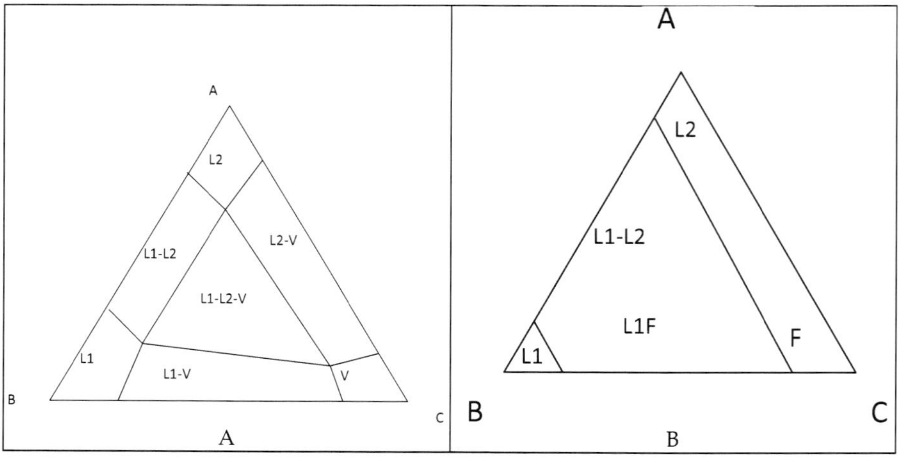

Fig. 10. Ternary diagram: type 3. A: There exists Liquid-Liquid Equilibria (L1L2E) between species A and B: if we increase pressure or temperature, we have a supercritical fluid in equilibrium with one liquid phase (FLE) and L1L2E.

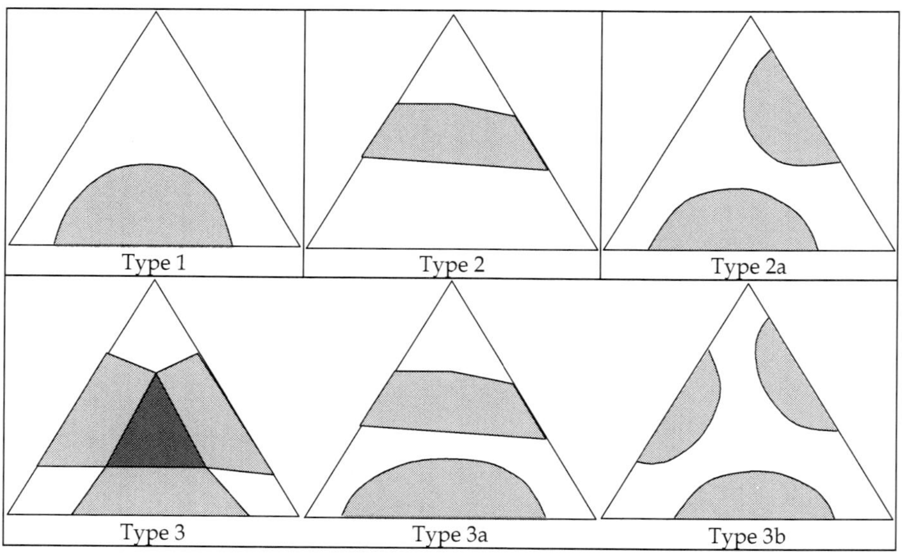

Fig. 11. Liquid - liquid equilibrium : presentation of the 3 configurations of the type 1, 2 and 3 systems.

3.4 Pressure-temperature envelops

The Pressure-Temperature (P-T) envelop is a very interesting way to represent phase diagrams. For a mixture (where composition is known), the P-T envelop represents the limits of the phase equilibrium region. An example of a P-T envelop is illustrated in figure 12. The point which corresponds to the maximum of pressure is called the cricondenbar, and with regard to the maximum of temperature, it is called the cricondentherm. Bubble and dew pressures curves are also presented in such a diagram with the critical point. With such a diagram, phenomenon such as retrograde condensation can be easily explained.

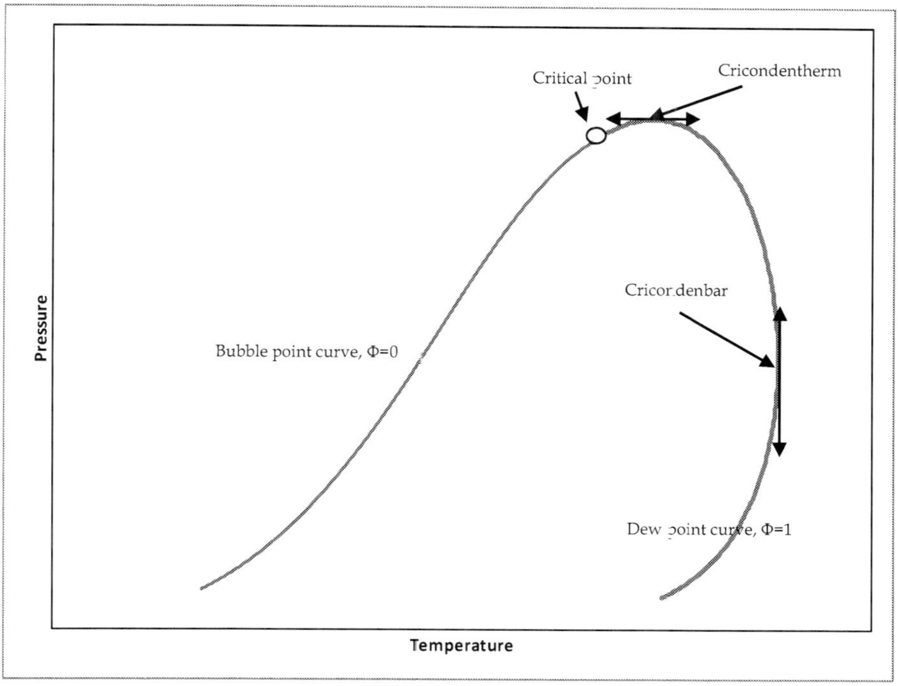

Fig. 12. Example of P-T envelop of hydrocarbons (C1 to C5).

4. Application to distillation

Distillation is the most well known separation unit operation in the world. Utilizing energy, the objective is to create one or more coexisting zones which differ in temperature, pressure, composition and phase state. In order to design a distillation column, the concept of an equilibrium stage is required: at this stage, the vapor and liquid streams which are leaving the stage are in complete equilibrium with each other and thermodynamic relations can be used to determine the temperature and the concentration of the different species for a given pressure. The equilibrium stage can be simulated as a thermodynamic isothermal FLASH (Michelsen, 1982b). Consequently, for a given distillation column, if the number of equilibrium stages is very large it means that the separation is difficult due a relative volatility close to one. The relative volatility of component i with respect to component j is the ratio between the partition coefficient (or equilibrium constant) (Eq. 12).

$$\alpha_{ij} = \frac{K_i}{K_j} = \frac{y_i / x_i}{y_j / x_j} \tag{12}$$

In distillation, three types of binary equilibrium curves are shown in figure 13.

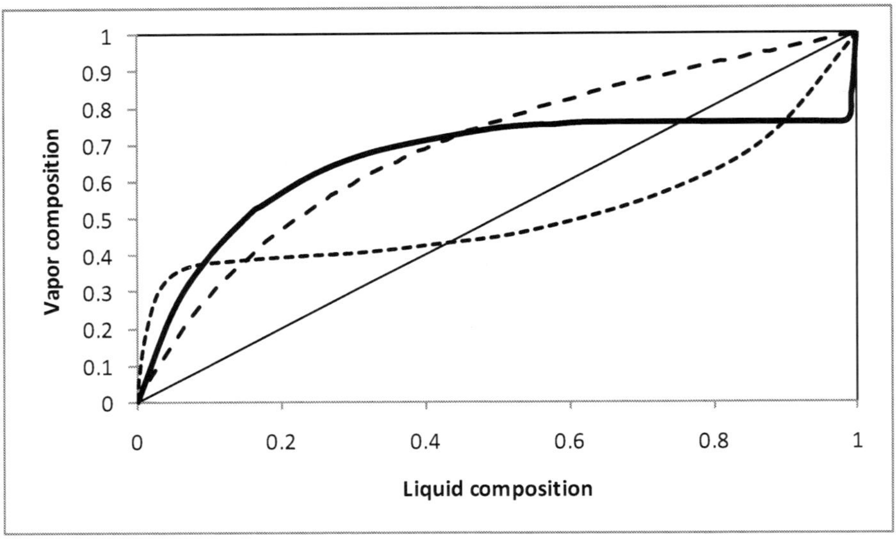

Fig. 13. Typical binary equilibrium curves. Dashed line: system with normal volatility, dotted line: system with homogenous azeotrope and solid line: system with heterogeneous azeotrope

If the relative volatility is equal to one, it is impossible to separate the two components. The phase diagrams are very important for the design of distillation columns: considering the McCabe and Thiele method (Perry & Green, 1997), the knowledge of the y-x curve together with the stream flowrates lead to the determination of the number of theorical equilibrium stage (NTES). The McCabe-Thiele method is based on the assumption of constant molar overflow and the molar heats of vaporization of the feed components being equal; heat effects such as heats of mixing, heat transfer to and from the distillation column are negligible. If the NTES is very large, the separation is difficult. Figure 15 shows phase diagrams for an air separation unit. It can clearly be seen that the separation between O_2 and Ar is more difficult in comparison with O_2 and N_2.

Considering a distillation column with a total reflux, the closer the values of liquid and vapor compositions, the higher is the number of equilibrium stages and more difficult is the

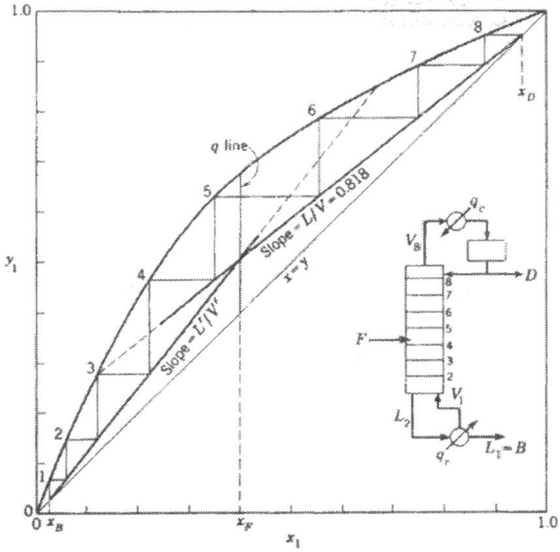

Fig. 14. Example of McCabe and Thiele construction: system Extract from Perry's handbook (Perry & Green, 1997).

phase separation. For a homogeneous azeotrope, the relative volatility is equal to one. The definition of the existence of an azeotrope is as following: $\left(\frac{\partial P}{\partial x_i}\right)_T = 0$ or $\left(\frac{\partial T}{\partial x_i}\right)_P = 0$ or $y_i = x_i$ for the Nth component. Moreover, considering only VLE, the mathematical criterion can determine whether there existences an azeotrope in a system which presents VLLE behavior: in this case there exists a heterogeneous azeotrope. Table 5 describes the different types or categories of azeotropes with example [12].

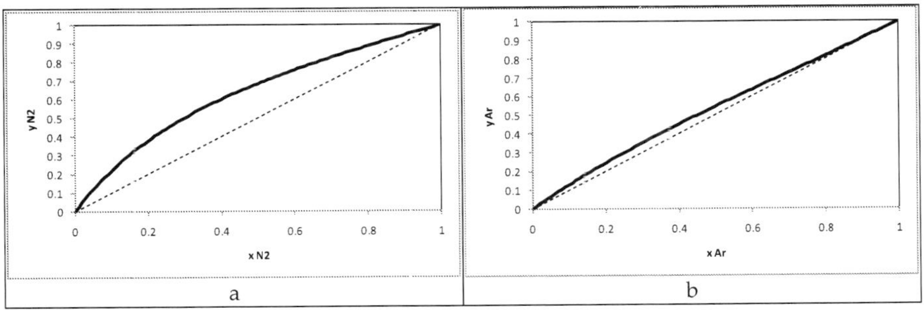

Fig. 15. Comparison between y-x phase diagram of N_2-O_2 (a) and Ar-O_2 (b) binary systems at 110 K.

Category	Type of azeotrope	Examples
I	Homogeneous Maximum of pressure maximale	1-propanol + water ethanol + benzene
II	Heterogeneous azeotrope Maximum of pressure maximale	1-butanol + water water + benzene
III	Homogeneous Minimum of pressure	Trichloromethane + 2-butanone
IV	Homogeneous Maximum of pressure with immiscibility zone	2-butanone + water 2-butanol + water
V	Double azeotrope Local minimum and maximum of pressure	benzene + hexafluorobenzene Diethylamine + methanol
VI	Homogeneous Minimum of pressure with immiscibility zone	Triethylamine + acetic acid

Table 5. Classification of azeotropes (fixed temperature).

Figure 16 presents an example of a heterogeneous azeotropic system (2 butanol + water) at 320 K.

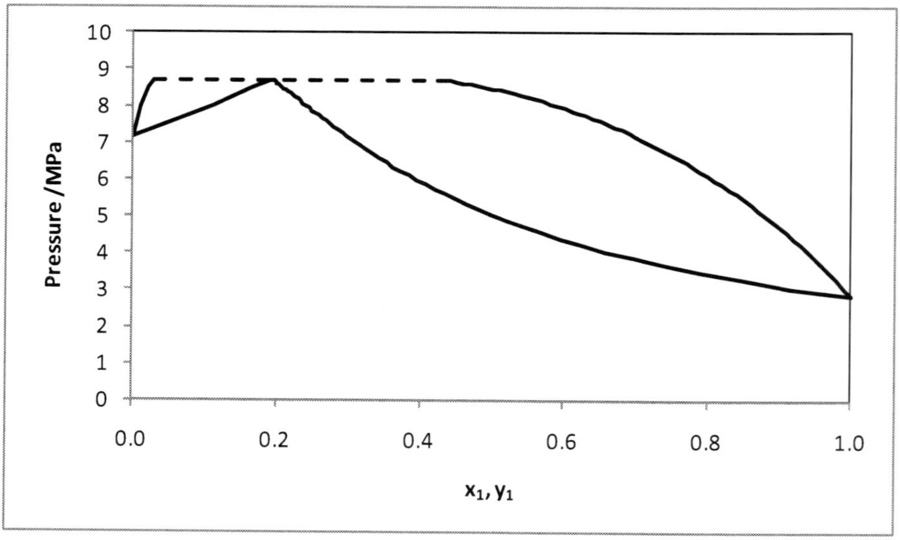

Fig. 16. 2-butanol (1) + water(2) binary system. Prediction using PSRK EoS (Holderbaum & Gmehling, 1991) at 320 K. Dashed line: Vapor Liquid Liquid equilibria (Heterogenous azeotrope)

5. Application to liquid-liquid extraction

Liquid-liquid extraction, which is commonly referred to as solvent extraction, involves the separation of the components that constitute a liquid stream by contacting it with another liquid stream which may be insoluble or partially soluble. Due to some of the components being preferentially more soluble in one of the liquid streams, separation can be effected. The separation effected in a single contacting stage is usually not large and therefore multiple contacting stages are needed to produce a significant separation. In these extraction processes the feed stream (which contains components that are to be separated) is contacted with a solvent stream. Exiting the contacting unit would be solvent-rich stream which is generally referred to as the extract and a residual liquid which is commonly called the raffinate. In general, this contacting removes a solute from the feed stream and concentrates it in the solvent-rich stream, i.e. decreasing the concentration of that particular solute in the raffinate stream.

The typical triangular diagrams which are used to illustrate ternary liquid-liquid equilibria can be converted into more convenient diagrams for visualization and computations in solvent extraction, e.g the distribution diagram, as seen in figure 17. This is undertaken because the phase relationships are generally very difficult to express conveniently algebraically and as a result solvent extraction computations are usually made graphically. The triangular diagram can also be transformed into rectangular coordinates, as seen in Figure 18.

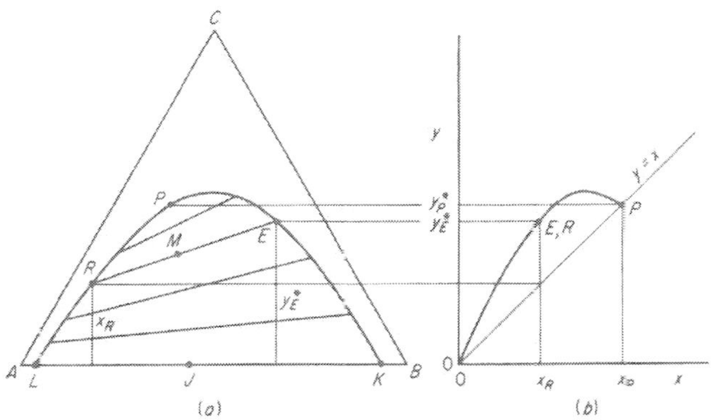

Fig. 17. Triangular (a) and distribution diagram (b) for liquid-liquid equilibria for a system of three components where one pair is partially miscible (extracted from Treybal, 1981).

The principles of separation and utilization of phase diagrams for the sizing of unit operation in solvent extraction is analogous to that which is seen in distillation, e.g. figure 18 illustrates phase diagrams being used to determine the number of theoretical stages for separation. Just as in distillation, the process can be undertaken with reflux. Reflux at the extract end can produce a product which is greater in composition, as is the case in the rectification section of a distillation column. The concept of the operating lines in the diagram, as well as the "stepping-off" in the diagram to determine the number of theoretical stages is similar to the McCabe-Thiele method for distillation.

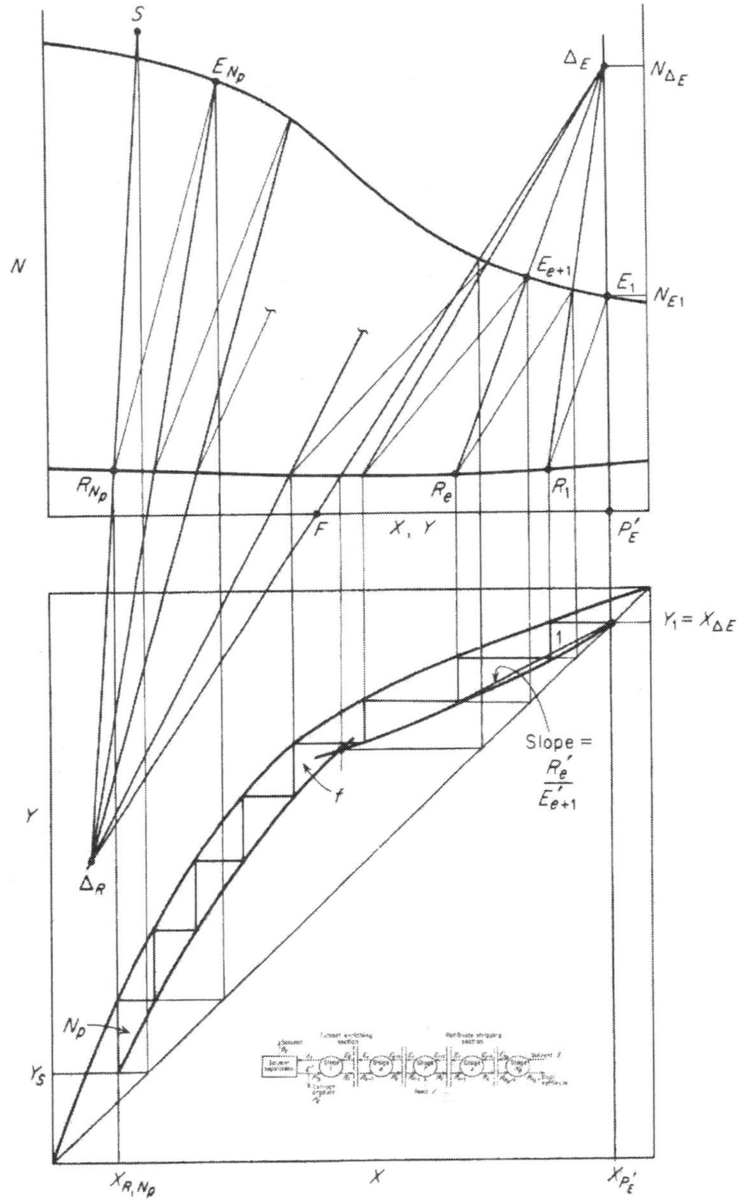

Fig. 18. Illustration of phase diagrams generated for a countercurrent extraction process with reflux (extracted from Treybal, 1981)

As the phase behaviour is affected by the choice of the solvent, and it is highly unlikely that any particular liquid will exhibit all of the properties desirable for solvent extraction, the final choice of a solvent is in most cases a compromise between various properties and parameters, viz. selectivity, distribution coefficient, insolunbility of the solvent, recoverability, density, interfacial tension, chemical reactivity, viscosity, vapor pressure, freezing point, toxicity, flammability, and cost.

Generally, the key parameters (selectivity and distribution coefficient) for determining the best solvent for the separation are calculated from liquid-liquid equilibrium measurements for the system of components concerned (including the solvent).

The selectivity is defined as follows:

$$\beta = \frac{\text{wt fraction of solute in raffinate}}{\text{wt fraction of solute in extract}} \tag{13}$$

Analogous to the relative volatility in distillation, the selectivity must exceed unity (the greater the value away from unity the better) for separation to take place. If the selectivity is unity, no separation is possible. The distribution coefficient (which is effectively the inverse of selectivity) however is not required to be larger than unity, but the larger the distribution coefficient the smaller the amount of solvent which will be required for the extraction.

Excellent summaries of solvent extraction processes are given in Perry and Green (1997), as well as in Treybal (1981) and McCabe et al. (2005).

6. Conclusion

In chemical engineering, the knowledge of the phase behaviour is very important, as the design and the optimization of the separation processes needs a good knowledge of the phase diagrams. Practically, the determination of phase diagrams can be obtained through experimental methods and/or modelling. The readers can refer to the books of Raal and Mühlauber (1998) to have a complete description of experimental methods, and Prausnitz et al. (1999) regarding models and principles. The complexity of phase diagrams is increased drastically if a solid or a polymer phase exists, however the purpose of this chapter was to introduce the reader to phase diagrams encountered by chemical engineers in the most commonly used units operations, viz. distillation and solvent extraction.

7. References

Baker, L.E., Pierce, A.C., Luks, K.D. (1982) Gibbs Energy Analysis of Phase Equilibria, *Society of Petroleum Engineers Journal*, October, pp 731-742, ISSN: 0197-7520

Coquelet, C., Chareton, A., Richon, D. (2004) Vapour–liquid equilibrium measurements and correlation of the difluoromethane (R32) + propane (R290) + 1,1,1,2,3,3,3-heptafluoropropane (R227ea) ternary mixture at temperatures from 269.85 to 328.35 K. *Fluid Phase Equilibria*, Vol. 218, pp 209-214, ISSN 0378-3812

Gmehling, J., Menke, J., Krafczyk, J., Fischer, K. (1994) *Azeotropic Data Part I*, Wiley-VCH, ISBN : 3-527-28671-3, Weinheim, Germany.

Heidemann, R. A. & Khalil, A. M. (1980) The Calculation of Critical Points, *American Institute of Chemical Engineers Journal*, Vol. 26, No.5, pp 769-779, ISSN (printed): 0001-1541. ISSN (electronic): 1547-5905

Holderbaum, T. & Gmehling, J. (1991) PSRK : A group contribution equation of state based on UNIFAC, Fluid Phase Equilibria, Vol. 70, pp 251-265, ISSN 0378-3812

McCabe, W., Smith J., Harriot, P. (2005) Unit Operations of Chemical Engineering 7th edition, McGraw-Hill Book Company, ISBN 0072848235.

Michelsen, M. L. (1980) Calculation of phase envelopes and critical point for multicomponent mixtures, *Fluid Phase Equilibria*, Vol. 4, pp 1-10, ISSN 0378-3812

Michelsen, M. L. (1982a) The isothermal Flash Problem. Part I Stability. *Fluid Phase Equilibria*, Vol. 9, pp 1-19, ISSN 0378-3812

Michelsen, M. L. (1982b) The isothermal Flash Problem. Part II Phase Split Calculation, *Fluid Phase Equilibria*, Vol. 9, pp 21-40, ISSN 0378-3812

Perry, R.H. & Green, D.W. (1997) Perry's chemical Handbook 7th , edition McGraw Hill Companies, ISBN0-07-049841-5, New York, USAPrausnitz, J.M., Lichtenthaler R.N., de Azevedo, E. G. (1999) Molecular thermodynamics of fluid phase equilbria, 3rd edition, Prenticd Hall International Series, Upper Saddle River, USA.

Raal, J. D. & Mühlbauer A. L. (1997) Phase Equilibria, Measurement and Computation, Taylor & Francis, ISBN 1-56032-550-X, London, UK.

Stockfleth, R. & Dohrn, R. (1998) An algorithm for calculating critical points in multicomponent mixtures which can be easily implemented in existing programs to calculate phase equilibria. *Fluid Phase Equilibria*, Vol. 145, pp 43-52, ISSN 0378-3812

Treybal, R.E. (1981) Mass-Transfer Operations 3rd edition, McGraw-Hill Book Company, ISBN-0-07-065176-0, Singapore.

Van der Waals, J.D. (1899) *Over de Continuiteit van den Gas- en Vloestoftoestand. (Über die Kontinuittät des Gas- und Flüssigkeitszustands) 1873*, Dissertation, Universität Leiden, Niederlande, deutsche Übersetzung, Leipzig, Germany.

Van Konynenburg, P. H. & Scott, R.L. (1980) Critical lines and Phase Equilibria in Binary van der Waals mixtures. *Philosophical Transactions of the Royal Society of London*, Vol. 298, pp 495-539, ISSN 0264-3820

Weintsock, J. J. (1952) *Phase equilibrium at elevated pressure in ternary systems of ethylene and water and organic liquids*, Phd dissertation, Princeton University, USA

20

Organic/Inorganic Nanocomposite Membranes Development for Low Temperature Fuel Cell Applications

Touhami Mokrani
University of South Africa
South Africa

1. Introduction

The criteria that are going to influence the evolution of the world energy system in the present century are complex. The most important new factor is the need to preserve the environment, both locally and globally, through the use of new technologies and sustainable use of existing resources. The Kyoto protocol, which put a limit on greenhouse gas emissions (mainly CO_2) from the industrialized countries, is a turning point in the global energy chain. On the other hand, the fuel specifications to control automotive exhaust gas emission obligate fuel producers to look for different ways of making clean fuel. Automakers are also obligated to look for alternative technology to internal combustion engines. The interest in studies on energy sources alternative to fossil fuels is linked both to the reduction of their availability and the increasing environmental impact caused by their use (Goodstein, 1999). In the energy field, an important cause of pollutant emissions is linked to ground transportation. In the last 40 years, some economic, social and cultural changes have encouraged a wide proliferation of vehicles. For example, in Europe, private cars have increased from 232 to 435 per 1000 inhabitants in the period 1971-1995 (Santarelli *et al.*, 2003). Fuel cells are alternative power sources that can meet global emission regulations, and clean production. Although fuel cells have been used since the 1960's for aerospace and military applications, cost was a strong impediment to terrestrial applications.

2. Fuel cell types

Five major types of fuel cells are available and are defined by their electrolyte. These include alkaline (AFC), phosphoric acid (PAFC), molten carbonate (MCFC), solid oxide (SOFC) and proton exchange membrane fuel cells (PEMFC). Table 1 summarizes some characteristics of these fuel cells. Proton exchange membrane fuel cells are the most attractive candidate for alternative automotive and stationary power sources due to their smaller size and much lower operating temperature compared to other fuel cell systems. Low temperature fuel cells are fuel cells operating at temperature less than 100°C. They are H_2-proton exchange membrane fuel cell (H_2-PEMFC), direct methanol fuel cell (DMFC), direct ethanol fuel cell (DEFC) and direct DME fuel cell (DDMEFC).

Type	Electrolyte	Charge carrier in the electrolyte	Temperature (°C)
Alkaline fuel cells (AFC)	aqueous KOH solution	OH^-	<100
Proton exchange membrane fuel cells (PEMFC)	proton exchange membrane	H^+	60-120
Phosphoric acid fuel cells (PAFC)	concentrated phosphoric acid	H^+	160-220
Molten carbonate fuel cells (MCFC)	mixture of molten carbonates (Li_2CO_3/K_2CO_3)	$CO_3{}^{2-}$	600-650
Solid oxide fuel cells (SOFC)	ceramic solid $ZrO_2(Y_2O_3)$	$O_2{}^-$	800-1000

Table 1. Fuel cells systems (Carrette *et al.*, 2001)

3. Low temperature fuel cells

A fuel cell is an electrochemical system which converts chemical energy to electrical energy. A fuel cell differs from a battery in that fuels are continuously supplied and the products are continuously removed. There are two distinct fuels for low temperature fuel cells: hydrogen as used in a H_2-PEMFC, and methanol as used in a DMFC. These fuel cells consist of six major parts: end plates, current collectors, flow channel blocks, gaskets, gas diffusion layers, and a membrane electrode assembly (MEA). The fuel cell principle enables a separation between power and energy. The maximum power required determines the size of the fuel cell; the energy required determines the amount of fuel to be carried. The specific power (W kg^{-1}) of the H_2-PEMFC is roughly twice that of the DMFC (Raadschelders & Jansen, 2001). Because no mobile electrolyte is employed, corrosion problems in low temperature fuel cells are reduced and cell construction is simplified with few moving parts (Bernardi & Verbrugge, 1991). Also, fuel cells operate very quietly, therefore, reducing noise pollution (Kordesch & Simader, 1995). Since the proton exchange membrane used for the electrolyte is a solid phase, it does not penetrate deeply into the electrode as does a liquid one; therefore the reaction area is limited to the contact surface between the electrode and membrane (Shin *et al.*, 2002). The advantage of using solid electrolyte is that no electrolyte leakage will occur (Uchida *et al.*, 1995; Yi & Nguyen, 1999). To meet the requirements of practical application a large number of single cells are assembled together to form a stack. The performance of a stack is different from that of a single cell. The stack has a much higher operating voltage, a greater power and better fuel-energy efficiency (Chu & Jiang, 1999).

4. Fuels for low temperature fuel cells

4.1 Pure hydrogen

H_2-proton exchange membrane fuel cells have existed since the 1960's; in fact they were used in the Gemini aerospace program of the National Aeronautics and Space Administration (NASA) of the United States. The MEA for H_2-PEMFCs consists of five components namely: a porous backing layer, an anode catalyst layer, a proton exchange membrane, a cathode catalyst layer, and a porous backing layer. Hydrogen is oxidized at the anode. The proton formed migrates through the membrane while the electrons flow through the external circuit. In the cathode reaction water is formed from oxygen, protons and electrons.

The two half reactions for the H_2-PEMFC are as follows:

Oxidation half reaction: Anode	$2\,H_2$	$\longrightarrow$	$4H^+ + 4e^-$
Reduction half reaction: cathode	$O_2 + 4H^+ + 4e^-$	$\longrightarrow$	$2\,H_2O$
Cell reaction	$2\,H_2 + O_2$	$\longrightarrow$	$2\,H_2O$

H_2-PEMFCs have attracted the most attention due to their high electrochemical reactivity (Gottesfeld & Zawodzinski, 1997; Parthasarathy *et al.*, 1991; Ralph, 1997) and very low noble catalyst loading since the development of a method at Los Alamos National Laboratory (LANL) to reduce the platinum loading to ca. 0.1 mg/cm^2 (Wilson & Gottesfeld, 1992; Wilson, 1993; Wilson *et al.*, 1995) compared to 35 mg/cm^2 and 4 mg/cm^2 used respectively in the Gemini program and at General Electric in the 1970s (Appleby & Yeager, 1986a,1986b). The efficiency achievable is higher than in power plants and internal combustion engines (Dohle *et al.*, 2002) and there is practically zero pollution. However, the H_2-PEMFC has several disadvantages including hydrogen storage and transportation and the public acceptance of hydrogen as fuel. It is well known that hydrogen and air mixtures are explosive (e.g. the Challenger disaster). Hydrogen safety measures are still one of the major implications when it comes to the commercialization of H_2-PEMFCs. Adequate water content of the membranes is essential to maintain the conductivity of the polymeric proton exchange membrane (Anantaraman & Gardner, 1996; Fontanella *et al.*, 1995; Gavach *et al.*, 1989; Zawodzinski *et al.*, 1991, 1993). During fuel cell operation, water molecules migrate through the membrane under electro-osmotic drag, fluid convection, and molecular diffusion, making it difficult to retain a high water content within the membrane. Generally, humidification is applied to the inlets of the anode and/or cathode in order to supply water to the membrane. However, excessive amounts of liquid water could impede mass transport within the electrode structure (Zawodzinski *et al.*, 1991). A thinner membrane is preferred in H_2-PEMFCs because it can provide an improvement in water management due to the enhanced back-diffusion of production water from the cathode to the anode side (Finsterwalder & Hambitzer, 2001). The oxygen reduction reaction (ORR) is very slow compared to the hydrogen reaction; typically hydrogen electro-oxidation on Pt is shown by an exchange current density of 10^{-3} A cm^{-2} Pt at ambient temperature. This is some 10^7 to 10^9 times more facile than the oxygen reduction at the cathode (Ralph & Hogarth, 2002). Thus, oxygen reduction is a rate limiting factor in H_2-PEMFCs (Gloaguen *et al.*, 1998; Paulus *et al.*, 2001).

4.2 Hydrogen reformate

The question of whether customers will be fuelling their vehicles directly with hydrogen or via the hydrogen-rich carrier (e.g. methanol, ethanol, gasoline, diesel, etc.) still seems to be unanswered. This is a very important issue not just from a refueling infrastructure perspective but also from the public perception and from the gearing up of production, and developing guidelines for dealing with safety issues that will need to put in place for the new fuel (Adamson & Pearson, 2000). In principle, any type of liquid fuel may be employed as a hydrogen source, e.g. gasoline, diesel, methanol, ethanol, etc. Hydrogen is produced by a reforming process. Four distinguish fuels are discussed namely methanol, ethanol, dimethylether (DME) and ammonia.

4.2.1 Methanol reforming

Methanol is produced from steam reformed natural gas and carbon dioxide using copper-based catalyst, and also from renewable biomass sources. Methanol is a leading candidate to provide the hydrogen necessary to power a fuel cell, especially in vehicular applications (Ledjeff-Hey *et al.*, 1998; Mokrani & Scurrell, 2009; Olah *et al.*, 2009). Methanol is currently used as a feed stock for a variety of widely used organic chemicals, including formaldehyde, acetic acid, chloromethane, and methyl tert-butyl ether (MTBE). Methanol is the desired fuel to produce hydrogen on-board. Methanol can be reformed to hydrogen by different processes including steam reforming (Amphlett *et al.*, 1985; Breen & Ross, 1999; Duesterwald *et al.*, 1997; Emonts *et al.*, 1998; C.J. Jiang *et al.*, 1993; Takahashi *et al.*, 1982; Takezawa *et al.*, 1982), partial oxidation (Agrell *et al.*, 2001; Cubiero & Fierro, 1998; Velu *et al.*, 1999) and autothermal reforming (Edwards *et al.*, 1998; Höhlein *et al.*, 1996; L. Ma *et al.*, 1996; Mizsey *et al.*, 2001).

Steam reforming of methanol occurs by two different pathways (Emonts *et al.*, 1998). The first one involves the decomposition of methanol into CO and H_2 through the following reaction:

$$CH_3OH \rightleftharpoons CO + 2H_2$$

followed by a water gas shift reaction:

$$CO + H_2O \rightleftharpoons CO_2 + H_2$$

The second mechanism for methanol steam reforming consists of the reaction of water and methanol to CO_2 and hydrogen:

$$CH_3OH + H_2O \rightleftharpoons CO_2 + 3H_2$$

which can be followed by a reverse shift reaction to establish the thermodynamic equilibrium:

$$CO_2 + H_2 \rightleftharpoons CO + H_2O$$

Methanol steam reforming is endothermic and therefore requires that external heat, typically 300°C, is supplied. Steam reforming of methanol is usually catalyzed over Cu/ZnO type catalyst and can be performed in fixed-bed reactors (Duesterwald *et al.*, 1997).

4.2.2 Ethanol reforming

Among other candidate liquid fuels, ethanol is a particular case, since it can be easily produced in great quantity by the fermentation of sugar-containing raw materials. In addition, in some countries (e.g. Brazil) ethanol is already distributed in gas stations for use in conventional cars with internal combustion engines. Hydrogen is produced from ethanol in a process unit consisting of either a steam reformer (SR) or a partial oxidation (POX) reactor in series with a water-gas shift (WGS) reactor and a reactor for selective oxidation (PROX) of CO (Ioannides & Neophytides, 2000). Product gas from the reformer or the POX reactor, which operates at an exit temperature higher than 677°C, contains a mixture of H_2, CO, CO_2, CH_4 and H_2O. After cooling, this stream enters the WGS reactor, where a large fraction of CO reacts with H_2O towards CO_2 and H_2 at a temperature of 200°C. The product gas of the WGS reactor contains 0.1-1.5% of residual CO and enters the PROX reactor, where CO is totally oxidized - with the addition of a small amount of air - to CO_2 with residual CO being less than 10 ppm. The CO free, hydrogen rich stream is then fed to the H_2-PEMFC.

4.2.3 DME reforming

DME (dimethylether) has become a promising candidate as a hydrogen source for the reforming process, because it has a high hydrogen-to-carbon ratio and a high energy density. DME can be easily handled, stored and transported. Furthermore, the infrastructure of LPG can readily be adapted for DME due to their similar physical properties. Furthermore, DME is not toxic and less explosive. DME can be catalytically reformed at relatively lower temperatures than ethanol and methane. DME can be reformed through three ways, namely steam reforming (SR), partial oxidation (POX) and authothermal reforming (ATR). DME SR proceeds via two moderately endothermic reactions in sequence; hydrolysis of DME to MeOH and steam reforming of MeOH to hydrogen and carbon dioxide. Hydrolysis of DME takes place over acid catalysis, e.g. zeolite and alumina, while MeOH SR proceeds over Cu-, Pt-, or Pd based catalyst. Therefore, bi-functional catalyst containing both acidic and metallic sites are generally needed for DME SR (Faungnawakij *et al.*, 2010; Ledesma & Llorca, 2009; Nishiguchi *et al.*, 2006; Takeishi & Suzuki, 2004). DME POX has been investigated over various metal catalysts such as Pt, Ni, Co and Rh supported on different oxide. Supports such as Al_2O_3, YSZ, $LaGaO_3$-based and MgO were used at a high reaction temperature ranging from 400 to 700°C (S. Wang *et al.*, 2002; Q. Zhang *et al.*, 2005). ATR also can be used to produce hydrogen from DME. ATR is a combination of SR and POX, and catalysis such as $CuFe_2O_4$-Al_2O_3 (Faungnawakij & Viriya-empikul, 2010) and Pd-based (Nilsson *et al.*, 2007, 2009) were investigated.

4.2.4 Ammonia reforming

Anhydrous ammonia is a widely used commodity and is available worldwide in liquid form in low pressure tanks. Procedures for safe handling have been developed in every country. Facilities for storage and transport by barges, trucks and pipelines from producer to ultimate consumer are available throughout the world. Therefore, liquid anhydrous ammonia is an excellent storage medium for hydrogen (Hacker & Kordesch, 2003). Studies demonstrate that hydrogen derived from anhydrous liquid ammonia, via a dissociation and followed by hydrogen purifier, offers an alternative to conventional methods of obtaining pure hydrogen for small scale use (Strickland, 1984). The dissociation rate depends on temperature, pressure and the catalyst being used. An almost complete decomposition of ammonia can take place at approximately 430°C at atmospheric pressure. The influence and kinetic data of materials like porcelain or silica glass, metals like ion, tungsten, molybdenum, nickel, etc. especially noble metals and metal oxides, have been investigated for the dissociation of ammonia. The most used catalysts are nickel oxide and iron oxide (Papapolymerou & Bontozoglou, 1997) and the better cracking efficiencies were obtained with catalysts based on Zr, Mn, Fe and Al/alloys (Boffito, 1999; Rosenblatt & Cohn, 1952; Shikada *et al.*, 1991).

4.3 Direct methanol fuel cell

DMFC technology is relatively new compared to the H_2-PEMFC. However, the direct oxidation of methanol in a DMFC has been investigated over many years and some prototypes were built in the 1960's and early 1970's by the Shell Research Center in England (Glazebrook, 1982; Schatter, 1983) and by Hitachi Research Laboratories in Japan (Tamura *et al.*, 1984; Williams, 1966). These studies were abandoned in the mid-1980's due to the low performance (25 mW cm^{-2} at best) resulting from the use of a liquid acid electrolyte (Glazebrook, 1982; Kordesch & Simader, 1996; Lamy *et al.*, 2001). An alkaline electrolyte was also used, but evolved CO_2 caused carbonation of the electrolyte resulting in decreased efficiency by reducing the electrolyte conductivity and de-polarizing the cathode (Verma, 2000). Currently all the research in DMFCs focuses on using solid proton exchange membranes as electrolyte, largely due to its proliferation in H_2-PEMFCs. The structure of the DMFC is similar to the H_2-PEMFC. At the anode methanol is directly oxidized to carbon dioxide, and the reaction at the cathode is similar to the H_2-PEMFC.

The two main half reactions for the DMFC can be summarized as follows:

Oxidation half reaction: Anode	$CH_3OH + H_2O \longrightarrow CO_2 + 6H^+ + 6e^-$
Reduction half reaction: cathode	$^3/_2\, O_2 + 6H^+ + 6e^- \longrightarrow 3\, H_2O$
Cell reaction	$CH_3OH + {^3/_2}\, O_2 \longrightarrow CO_2 + 2\, H_2O$

The thermodynamic reversible potential for a DMFC is 1.21V at 25°C (Larminie & Dicks, 2000). This value is comparable to that for a H_2-PEMFC, which is 1.23V (Chu & Gilman, 1994; Léger, 2001; Qi & Kaufman, 2002; Scott *et al.*, 1998). In practice, a DMFC has a much lower open circuit voltage (OCV) (Qi & Kaufman, 2002) and electrochemical losses at both electrodes lead to a significant reduction in overall performance from the theoretical thermodynamic maximum (Argyropoulos *et al.*, 1999a). Since methanol is used directly at the anode, and as a consequence, a DMFC requires less auxiliary equipment and is therefore a more simplified system compared to a H_2-PEMFC. Methanol is a liquid made from natural gas or renewable biomass sources, which is relatively cheap. Methanol is also easy to store, transport, and distribute, where advantage can be taken of the existing gasoline infrastructure (Mokrani & Scurrell, 2009; Olah *et al.*, 2009). The anodic reaction is exothermic for both the H_2-PEMFC and the DMFC; heat management is a problem in H_2-PEMFC stacks. In contrast, aqueous methanol acts as a coolant in DMFCs (Hogarth *et al.*, 1997; Hogarth & Ralph, 2002; Lim & C.Y. Wang, 2003; Surampudi *et al.*, 1994).

However, as the DMFC is still in its infancy, many problems need to be overcome to reach the commercialization stage. This includes the very sluggish methanol oxidation reaction, methanol crossover through the polymeric proton exchange membrane, CO_2 evolvement at the anode (Argyropoulos *et al.*, 1999a,1999b; Nordlund *et al.*, 2002; Scott *et al.*, 1998), and cathode flooding (Amphlett *et al.*, 2001; M. Mench *et al.*, 2001; X. Ren & Gottesfeld, 2001; Z.H. Wang *et al.*, 2001). The methanol crossover through the polymer electrolyte leads to a mixed potential at the cathode, which results from the ORR and the methanol oxidation occurring simultaneously. This effect causes a negative potential shift at the cathode and a significant decrease of performance in the DMFC. Methanol crossover also causes fuel loses; it had been found that over 40% of methanol can be wasted in a DMFC across Nafion® membranes (Narayanan *et al.*, 1996). In a DMFC, cathode flooding, which typically occurs unless high cathode stoichiometries are used, can determine to a great extent overall cell performance (Amphlett *et al.*, 2001; M. Mench *et al.*, 2001; X. Ren & Gottesfeld, 2001; Z.H. Wang *et al.*, 2001). Water management in the DMFC is especially critical because anode water activity is near unity due to contact with liquid methanol solution (M.W. Mench & C.Y. Wang, 2003). Thus, unlike a H_2-PEMFC, no back-diffusive flux of water from cathode to anode will occur, and as a result, vapourization into dry cathode flow is the only pathway for removal of excess cathode-side water accumulation from electro-osmotic drag, ORR, and diffusion (M.W. Mench & C.Y. Wang, 2003)

4.4 Direct ethanol fuel cell

Direct fuel utilization will be of interest. Besides methanol, other alcohols, particularly those coming from biomass resources, are being considered as alternative fuels. Ethanol as an attractive fuel for electrical vehicles was investigated in direct ethanol fuel cells (Fujiwara *et al.*, 1999; Gong *et al.*, 2001; Lamy *et al.*, 2001; W.J. Zhou *et al.*, 2004). However, multimetallic catalysts are necessary to orientate the oxidation reaction selectively in the direction of complete combustion to carbon dioxide (Lamy *et al.*, 2001). The reaction mechanisms of anodic oxidation of ethanol are more difficult to elucidate than methanol oxidation, since the number of electrons exchanged greatly increases (12 electrons per ethanol molecule versus 6 electrons for methanol), thus many adsorbed intermediates and products are involved (Lamy *et al.*, 2001). Direct ethanol fuel cell was the second most studied fuel cell after methanol. The proton conductor membranes used are mainly Nafion® membranes. However some investigators used high temperature membrane such as Nafion®/Silica by Aricò *et al.* (1998), and also by J. Wang *et al.* (1995) using a phosphoric acid doped polybenzimidazol (PBI) membrane.

4.5 Direct DME fuel cell

Serov and Kwak (2009) have summarized the recent progress in development of direct DME fuel cell (DDMEFC). DDMEFC could be a valuable direct liquid fuel cell candidate for commercialization. However, compared with PEMFC and DMFC, DDMEFC show poor performances under ambient conditions due to the poor electrooxidation reactivity of DME (Colbow *et al.*, 2000; Kerangueven *et al.*, 2006; Mench *et al.*, 2004; Mizutani *et al.*, 2006; Ueda *et al.*, 2006). On the anode side of DDMEFC, the following oxidation reaction takes place:

$$CH_3OCH_3 + 3H_2O \longrightarrow 2CO_2 + 12\,H^+ - 12\,e^-$$

The number of electron transferred for complete oxidation is 12, this result in a reduced theoretical fuel requirement of DME, when compared to methanol with 6 electrons transferred, and hydrogen with 2 electrons (Mench *et al.*, 2004; Serov & Kwak, 2009; K. Xu *et al.*, 2010). Furthermore, DME has the advantage over methanol in that crossover is much less pronounced (Mench *et al.*, 2004; Serov & Kwak, 2009). DDMEFC performance increase with increasing temperature, since DME electrooxidation is favored at high temperature. Furthermore, increasing the temperature will enhance also oxygen reduction reaction (ORR) (K. Xu *et al.*, 2010). Compared with hydrogen as the fuel for PEMFC, more water is needed for DME electrooxidation reaction and H^+ migration from anode side to cathode side due to the elctro osmotic force of water (Ferrell *et al.*, 2010; K. Xu *et al.*, 2010; Yu *et al.*, 2005).

5. Organic proton conductor membranes

Proton exchange membranes or proton conductor membranes are the most important component of low temperature fuel cells. Since the development of a solid polymer electrolyte, all the research on fuel cells focuses on the use of these types of electrolyte.

5.1 Perfluorinated membranes

The first commercially available perfluorinated membrane material from DuPont was Nafion® 120 (1200 equivalent weight (EW), 250 μm thick) followed by Nafion® 117 (1100 EW, 175 μm thick). These high equivalent weight materials were found to have limited use in fuel cells. In 1988, The Dow Chemical Company developed their own perfluorinated polymer membrane with low equivalent weight, typically in the range of 800-850. Nafion® of DuPont and Dow® membranes have identical backbones and are structurally and morphologically similar, but the side chain is shorter in the Dow polymer. Since the success of Dow Chemical, where it was found that the Dow® membrane performed better than the DuPont membrane in H_2/O_2 fuel cells, DuPont has been active in further developing their membranes with respect to durability and continuous improvement. They increased power densities by further decreasing the equivalent weight from 1100 to 1000 EW and membrane thickness from 175 to 25 μm. Table 2 shows the latest DuPont membranes with some characteristics. Nafion® 117 is the preferred membrane for DMFCs. In the 1990's, Aciplex® perfluorinated ion exchange membranes were introduced by the Asahi Chemical Industry, and the Flemion® series were introduced by Asahi Glass Co. (Yoshida *et al.*, 1998). In general these membranes are in the category of long chain perfluorinated membranes, like Nafion®. Some characteristics of these perfluorinated membranes are summarized in Table 2.

Nafion® membranes are chemically synthesized in four steps according to the DuPont de Nemours process (Grot, 1978): 1) The reaction of tetrafluoroethylene with SO_3 to form the sulfone cycle; 2) The condensation of these products with sodium carbonate followed by co-polymerization with tetrafluoroethylene to form an insoluble resin; 3) The hydrolysis of this resin to form a perfluorosulfonic polymer and 4) The chemical exchange of the counter ion Na^+ with the proton in an appropriate electrolyte. The Dow® membrane is prepared by the co-polymerisation of tetrafluoroethylene with vinylether monomer. The polymer can be described as having a Teflon-like backbone structure with a side chain attached via an ether group. This side chain is characterized by a terminal sulfonate functional group (Savadogo, 1998).

Membrane	Thickness (μm)	Equivalent Weight
Nafion® series (DuPont)		
Nafion® 117	175	1100
Nafion® 115	125	1100
Nafion® 112	50	1100
Nafion® 111	25	1100
Nafion® 1135	87	1100
Nafion® 1035	87	960
Nafion® 105	125	960
Dow Chemicals Co.		
Dow® XUS 13204.10	127	800-850
Flemion® series (Asahi Glass Co.)		
Flemion® R	50	900
Flemion® S	80	900
Flemion® T	120	900
Aciplex® series (Asahi Chemicals Industry)		
Aciplex® 1004	100	1000

Table 2. Perfluorinated membranes

5.2 Partially fluorinated ionomer membranes

5.2.1 Sulfonated copolymer based on the α,β,β-trifluorostyrene monomer membranes

The Canadian Ballard company developed proton conductor membranes based on trifluorostyrene monomer, under the trade name BAM1G and BAM2G (Ballard Advanced Materials first and second generation, respectively). The longevity of these polymers was limited to approximately 500 hours under practical fuel cell operating conditions (Savadogo, 1998). Based on the above work, Ballard developed third generation membranes under the trade name BAM3G (Steck, 1995; Steck & Stone, 1997; Wei *et al.*, 1995a,1995b). The BAM3G membranes consist of sulfonated copolymers incorporating α,β,β-trifluorostyrene and a series of substituted α,β,β-trifluorostyrene co-monomers. These membranes have an equivalent weight ranging between 375 and 920. The water content of the sulfonated BAM3G is much higher than that of Nafion® and Dow membranes. BAM3G membranes demonstrated a lifetime approaching 15,000 hours when tested in a Ballard MK5 single cell and also exhibited performances superior to Nafion® and Dow® membranes in a H_2/O_2 fuel cell. Disadvantages of these membranes include the complicated production process for the monomer α,β,β-trifluorostyrene (Livingston *et al.*, 1956) and the difficult sulfonation procedure (Kerres, 2001; Wei *et al.*, 1995b).

5.2.2 Grafted ionomer membranes

Partially fluorinated membranes can be obtained by using a simultaneous and pre-radiation grafting of monomers onto a base polymer film, and subsequent sulfonation of the grafted component (Brack *et al.*, 2003; Büchi *et al.*, 1995a,1995b; Gode *et al.*, 2003; Gupta *et al.*, 1993; Hatanaka *et al.*, 2002; W. Lee *et al.*, 1996; Lehtinen *et al.*, 1998; Scherer, 1990). These membranes were prepared by pre-irradiation of fluoropolymer films, such as poly(tetrafluoroethylene-*co*-hexafluoropropylene (FEP) or poly(ethylene-*alt*-tetrafluoro-ethylene) (ETFE), using an electron beam or gamma irradiation source. The pre-irradiated films were grafted by exposing them to solutions of styrene and other radically polymerizable monomers. The grafted films are sulfonated using chlorosulfonic acid. The grafting mixture was crosslinked with divinylbenzene (DVB) and tri-allyl cyamirate (TAC) (Gupta *et al.*, 1994; Gupta & Scherer, 1994) or poly(vinylidene fluoride) (Sundholm, 1998). A disadvantage of membranes using styrene and divinylbenzene monomers is that their oxidation stability is limited, due to the tertiary C-H bonds which are sensitive to O_2 and hydrogen peroxide attack (Kerres, 2001).

5.3 Non-perfluorinated membranes

5.3.1 Polybenzimidazole (PBI)

PBI is synthesized from aromatic bis-o-diamines and dicarboxylates (acids, esters, amides), either in the molten state or in solution (Jones & Rozière, 2001). PBI is relatively low cost and is a commercially available polymer known to have excellent oxidation and thermal stability. The commercially available polybenzimidazol is poly-[2,2`-(*m*-phenylene)-5,5`-bibenzimidazole], which is synthesized from diphenyl-iso-phthalate and tetra-aminobiphenyl. Hoel and Grunwald (1977) reported on proton conductivity values of PBI in the range of 2×10^{-4} - 8×10^{-4} S/cm at relative humidities (RH) between 0 and 100%. Other authors observed proton conductivity some two to three orders of magnitude lower (Aharoni & Litt, 1974; Glipa *et al.*, 1997; Powers & Serad, 1986). PBI is a suitable basic polymer which can readily be complexed with strong acids (Jones & Rozière, 2001; Glipa *et al.*, 1997; Y.L. Ma *et al.*, 2004; Powers & Serad, 1986; Samms *et al.*, 1996; Savadogo & B. Xing, 2000; Wainright *et al.*, 1995, 1997; J.T. Wang *et al.*, 1996a,1996b; B. Xing & Savadogo, 1999). The immersion of PBI film in aqueous phosphoric acid leads to an increase in both its conductivity and thermal stability (Powers & Serad, 1986). Savinell and co-workers prepared PBI/H_3PO_4 via two different routes: a) directly casting a film of PBI from a solution containing phosphoric acid; b) preparation by immersion of a preformed PBI membrane in 11M phosphoric acid for several days (Samms *et al.*, 1996; Wainright *et al.*, 1997). The typical thickness for different films was 75 μm. The conductivity depends on the quantity of phosphoric acid in the membrane. Conductivity in the range 5×10^{-3} to 2×10^{-2} S/cm at 130°C and 5×10^{-2} S/cm at 190°C have been reported (Wainright *et al.*, 1995). The conductivity for type "a" membranes is higher than those of type "b" membranes. At a temperature above 150°C, the conductivity of type "a" membranes is similar to that of Nafion® at 80 °C and 100% RH. It was shown that the methanol crossover through doped PBI type "a" membrane, was at least ten times less than that observed with Nafion®. The disadvantage of these membranes is that the H_3PO_4 molecules can diffuse out of the membrane towards basic polymer sites because they are in excess. PBI/H_3PO_4 membranes are suitable for direct methanol fuel cell application at a temperature >100°C. However, they can only be used with a feed of vapourized methanol, because when a liquid contacts the membrane, the phosphoric acid leaches out of the membrane and the proton conductivity drops considerably (Kerres, 2001).

5.3.2 Sulfonated polyimide membranes

The sulfonated polyimide (SPI) membranes were obtained by casting on a glass plate the polymer solution and evaporating the solvent (Cornet *et al.*, 2000; Fan *et al.*, 2002; Faure *et al.*, 1996, 1997; Gebel *et al.*, 1993; Genies *et al.*, 2001; Woo *et al.*, 2003). The polymer solution synthesis was achieved in different ways: The first way was based on the phthalimide-five member imide (4,4′-diamino-biphenyl 2,2′ disulfonic acid (BDSA), 4,4′ oxy-diphthalic dianhydride (ODPA) and 4,4′-oxydianiline (ODA)) at 200°C. The second way was based on the naphthalimide-six member imide ring (BDSA, 1,4, 5,8-naphthalene tetracarboxylic dianhydride (NTDA) and ODA) at 160°C (Faure *et al.*, 1996, 1997; Gebel *et al.*, 1993). The third way was based on the 3,3`,4,4`-benzophenone-tetracarboxylic dianhydride (BTDA), BDSA and ODA (D`Alelio, 1944). The fourth way was based on BDSA/NTDA/mAPI (bis-[3-(Aminophenoxy)-4-phenyl]isopropylidene) (Genies *et al.*, 2001). The water content of membranes at 25°C for the phthalic and naphthalenic sulfonated polyimide membranes is 26% and 30%, respectively. The water content obtained for Nafion® membranes under the same conditions was 20% (Faure *et al.*, 1997; Gebel *et al.*, 1993). It was also claimed that the sulfonated polyimide membranes were 3 times less permeable to hydrogen gas than Nafion® membranes. The lifetime measurements were performed on a 175 μm phthalic polyimide and a 70 μm naphthalenic sulfonated polyimide film at 60°C, 3 bar pressure for H_2 and O_2 and under a constant current density. It was found that the membrane based on the phthalic structure broke after 70 hours whereas the membrane based on the naphthalic polyimide was stable over 3000 hours (D`Alelio, 1944). The proton conductivity of SPI was found to be half of Nafion® 117, typically 4.1×10^{-2} S/cm, and methanol permeability was found to be 7.34×10^{-8} compared to 2.38×10^{-6} cm²/s for Nafion® 117 (Woo *et al.*, 2003).

5.3.3 Phosphazene-based cation-exchange membranes

It was shown that polyphosphazene-based cation-exchange membranes have a low methanol permeability, low water swelling ratios, satisfactory mechanical properties, and a conductivity comparable to that of Nafion® 117 (Allcock *et al.*, 2002a,2002b; Guo *et al.*, 1999; Tang *et al.*, 1999; Wycisk & Pintauro, 1996; X. Zhou *et al.*, 2003). Polyphosphazene-based membranes have been fabricated from poly[bis(3-methylphenoxy)phosphazene] by first sulfonating the base polymer with SO_3 and then solution-casting a thin film (Tang *et al.*, 1999; Wycisk & Pintauro, 1996; X. Zhou *et al.*, 2003). Polymer crosslinking was carried out by dissolving benzophenone photoinitiator in the membrane casting solution and then exposing the resulting films after solvent evaporation to UV light (Guo *et al.*, 1999). The conductivity of the polyphosphazene membranes were either similar to or lower than that of Nafion® 117 membranes (Guo *et al.*, 1999; X. Zhou *et al.*, 2003). However, methanol permeability of a sulfonated membrane was about 8 times lower than that of the Nafion® 117 membrane (X. Zhou *et al.*, 2003). Sulfonated/crosslinked polyphosphazene films showed no signs of mechanical failure (softening) up to 173°C and a pressure of 800 kPa (Guo *et al.*, 1999).

5.3.4 Sulfonated poly(arylethersulfone) membranes

Polysulfone (PSU) is a low cost, commercially available polymer (e.g. PSU Udel™ from Amoco) which has very good chemical stability. The synthesis and characterization of sulfonated polysulfone (SPSU) has been achieved by Johnson *et al.* (1984) and Nolte *et al.* (1993). It was found that membranes cast from SPSU (Udel™ P-1700) solutions were completely water soluble (Nolte *et al.*, 1993) and become very brittle when drying out which can happen in the fuel cell application under intermittent conditions (Kerres *et al.*, 1999).

There are two new but different procedures for the sulfonation of polysulfone. In one procedure, the sodium-sulfonated group was introduced in the base polysulfone via the metalation-sulfination-oxidation process (Kerres *et al.*, 1996, 1998a). In the other procedure, trimethylsilyl chlorosulfonate was used as the sulfonating agent (Baradie *et al.*, 1998). Lufrano *et al.* (2000,2001) prepared SPSU via trimethylsilyl chlorosulfonate with different degrees of sulfonation (DS). Different membranes with sulfonation degree from 23% to 53% (Lufrano *et al.*, 2000) on the one hand and 49%, 61% and 77% (Lufrano *et al.*, 2001) on the other hand were prepared. With a 61% sulfonation degree a proton conductivity for SPSU of 2.7×10^{-2} S/cm at 25°C was reported (Lufrano *et al.*, 2001). This conductivity was 3.5 times lower than Nafion® 117, but was compensated by the lower thickness, 90 μm vs. 210 μm for Nafion® 117. The cell performance obtained by Lufrano *et al.* (2001) was almost the same for SPSU and Nafion® in a H_2/O_2 fuel cell. This is higher than that reported previously by Kerres *et al.* (1998a) and Baradie *et al.* (1998). Y.S. Kim *et al.* (2003) prepared sulfonated poly(arylether sulfone) membranes.

Promising alternatives suggested by Kerres and co-workers, include composite membranes made from blends of acidic and basic polymers (Cui *et al.*, 1998; Jörissen *et al.*, 2002; Kerres *et al.*, 1999, 2000; Walker *et al.*, 1999) or modified PSU via the metalation-sulfochlorination and the metalation-amination routes (W. Zhang *et al.*, 2001) or crosslinked SPSU (Kerres *et al.*, 1997, 1998b,1998c). These alternatives are made by blending acidic polymers such as SPSU with basic polymers such as poly(4-vinylpyridine) (P4VP), polybenzimidazole (PBI) or a basically substituted polysulfone. Crosslinked SPSU blend membranes have been produced via a new crosslinking process. The blends have been obtained by mixing PSU Udel™ Na-sulfonate and PSU Udel™ Li-sulfinate in *N*-methyl pyrrolidone. The membranes have been crosslinked by S-alkylation of PSU sulfinate groups with di-halogenoalkanes. These membranes show very good performance in H_2/O_2 fuel cells and DMFCs (Kerres *et al.*, 2000; Kerres, 2001). These membranes also show a markedly reduced methanol permeability (Kerres, 2001; Walker *et al.*, 1999).

5.3.5 Sulfonated poly(aryletherketone) membranes

The poly(arylether ketones) are a class of non-fluorinated polymers consisting of sequences of ether and carbonyl linkages between phenyl rings, that can either "ether-rich" like PEEK and PEEKK, or "ketone-rich" like PEK and PEKEKK. The most common material is polyetheretherketone (PEEK) which is commercially available under the name Victrex™ PEEK from ICI Advanced Materials. A number of groups are developing proton conducting polymer materials based on this classification of materials including ICI Victrex, Fuma-Tech and Axiva/Aventis/Hoechst. Sulfonated-PEEK (SPEEK) membranes were prepared as proton conductors in PEMFCs by Schneller *et al.* (1993). Sulfonation of polyetherketones can be carried out directly in concentrated sulfonic acid or oleum - the extent of sulfonation being controlled by the reaction time and temperature (Bailly *et al.*, 1987; B. Bauer *et al.*, 1994, 1995). Direct sulfonation of PEEK can give materials with a wide range of equivalent weights to form SPEEK. However, the complete sulfonation of the polymer results in a fully water-soluble product. A sulfonation level of around 60% was found to be a good compromise between the conductivity and mechanical properties of membranes. The backbone of SPEEK is less hydrophobic than the backbone of Nafion®, and the sulfonic acid functional group is less acidic (Kreuer, 2001). Various studies have been made on the conductivity of SPEEK (Alberti *et al.*, 2001; B. Bauer *et al.*, 2000; Kobayashi *et al.*, 1998; Kreuer, 1997, 2001; Linkous *et al.*, 1998; P. Xing *et al.*, 2004; Zaidi *et al.*, 2000). The conductivity increases as a function of the degree of sulfonation, the ambient relative humidity, temperature and thermal history. The conductivity of these materials was found to be high at room temperature (Soczka-Guth *et al.*, 1999). In SPEEK with 65% sites sulfonated, the conductivity was higher than that of Nafion® 117 measured under the same conditions - the conductivity reaching 4×10^{-2} S/cm at 100°C and 100% RH (Linkous *et al.*, 1998). SPEEK membranes exhibit at 160°C and 75% RH, sufficiently high values of protonic conductivity - typically 5×10^{-2} – 6×10^{-2} S/cm - for possible applications in low temperature fuel cells (Alberti *et al.*, 2001). The dependence of the conductivity on RH is more marked for SPEEK than for Nafion® under the same conditions (B. Bauer *et al.*, 2000). Sulfonated polyaryls have been demonstrated to suffer from hydroxyl radical initiated degradation (Hubner & Roduner, 1999). In contrast, SPEEK was found to be durable under fuel cell conditions over several thousand hours by Kreuer (2001). The brittleness of SPEEK makes their handling difficult and may lead to mechanical membrane failure during operation. These types of membranes become very brittle when drying out SPEEK can also be chemically cross-linked to reduce membrane swelling and increase its mechanical strength. Materials prepared by cross-linking are comparable to commercial Nafion® in terms of their mechanical strength and proton conductivity (Yen *et al.*, 1998). Kerres and co-workers prepared novel acid-base polymer blend membranes composed of SPEEK as the acidic compound, and of P4VP or PBI as the basic compounds (Kerres *et al.*, 1999; Jörissen *et al.*, 2002).

6. Organic/inorganic nanocomposite membranes

Membrane electrode assembly (MEA) is the basic component of the single cell of a stack. The proton exchange membrane (PEM) is the key element of this component, which separates the electrode structure to prevent the mixing of reactant gases and the formation of an electrical short. This makes its properties, functionality, cost and reliability very important for real cell operations. Up to now perfluorinated sulfonic acid (PFSA) membranes have been the best choice for commercial low temperature polymer products (<80°C). The advantages of PFSA membranes are:

i. Their strong stability in oxidative and reduction media due to the structure of the polytetrafluoroethylene backbone; and
ii. Their proton conductivity, which can be as high as 0.2 S cm^{-1} in full hydrated polymer electrolyte fuel cell.

When used at elevated temperatures, however, PEMFC performances decrease. This decrease is related to: dehydration of the membrane; reduction of ionic conductivity;

decreases in affinity with water; loss of mechanical strength through a softening of the polymer backbone; and parasitic losses (the high level of gas permeation). There are several reasons for the development of high temperature membranes (Savadogo, 2004):

i. The operation of PEMFC at temperature above 140°C is receiving world-wide attention because fuel selection remains straightforward, and a number of fuels, including reformed hydrogen with high CO content and light hydrocarbons (alcohol, natural gas, propane, etc.) are still being considered for PEMFC application. Accordingly, cell temperature operation at temperatures above 140°C is of great interest because, in this temperature range, anode catalyst poisoning by CO is less important and the kinetics of fuel oxidation will be improved and the efficiency of the cell significantly enhanced. High temperature cell operation will contribute to reducing the complexity of the hydrocarbon fuel cell system. Some other advantages of operating PEMFC at high temperature are: a reduction in the use of expensive catalysts; and minimization of the problems related to electrode flooding. Light hydrocarbons may be potential energy vectors for PEMFC, which may lead to the development of suitable membranes that are stable in high temperature operating conditions and prevent fuel crossover.
ii. Enhancement of gas transport in the electrode layers is also expected because no liquid water will be present in the cell at these temperatures. Membrane proton conductivity should be dependent on water content at these temperatures; consequently, it is not necessary to humidify the gas before it enters the stack. This may help improve the kinetics of mass transport and simplify the fuel cell system. In particular, the kinetics of oxygen reduction reaction could be improved, by at least three orders of magnitude, if we increase the operating temperature from 25 to 130°C. PFSA membranes cannot be used in PEMFC operating above temperatures around 100°C, because at these temperatures they will lose their mechanical properties and their swelling properties will be lowered. They do not perform well above 90°C in a hydrocarbon PEMFC and above 85°C in hydrogen PEMFC. The boiling point of water can be raised by increasing the operating pressure above 3 bar, which may correspond to a boiling point of water of about 135°C. But raising the pressure of PEMFC is undesirable from an efficiency point of view.

One of the main drawbacks of DMFC (direct methanol fuel cell) is the slow methanol oxidation kinetics. An increase in the operating temperature of the DMFC from 90 to about 140°C is highly desirable. Also operation at high temperature will enhanced CO tolerant when a reformate hydrogen is used in H_2-PEMFC. One approach to achieve water retention at high temperature is to fabricate a composite membrane constituted of organic proton conductor and inorganic materials. The organic/inorganic composite proton conductors are developed to overcome the breakdown of the actual state-of-the-art membranes (i.e. PFSA membranes: Nafion® (DuPont), Dow, Flemion® (Asahi Glass Corporation) and Aciplex® (Asahi Chemicals)). Thus, increasing the operating temperature above 100°C, reduced methanol permeability (methanol crossover), increasing the water retention and also increasing the mechanical and thermal stability of the composite membranes.

The method of inclusion of inorganic proton conductor or inorganic particle has involved a bulk powder dispersed in a polymer solution, leading specifically to particles of highly dispersed inorganic fillers of particle size in the sub-micronic range. These methods make use of mild chemistry technique, including intercalation/exfoliation, sol-gel chemistry, and ion-exchange (Bonnet *et al.*, 2000; Jones & Rozière, 2001). Such approach generally avoid any sedimentation of the inorganic component, intimacy of contact between the inorganic and organic components at the molecular level assures the greatest possible interface and, at such small particle size, the mechanical properties can be improved compared with those of a polymer-only membrane (Jones & Rozière, 2001, 2003). In addition, since in many proton conductors of conductivity suitable for electrochemical applications the proton transfer process takes place on the surface of the particles, increase in surface area (small particle size) will increase the conductivity (Jones & Rozière, 2001).

This concept was suggested by Watanabe *et al.* (1995, 1996) and is based on the development of self-humidifying composite membranes. The membranes are fabricated from the dispersion of nano-particles of Pt in a thin Nafion® film (≈ 50 μm). Membranes fabricated based on this concept should not require external humidification and should suppress the crossover of H_2 and O_2. The dispersed particles should catalyze the oxidation and the reduction of the crossover H_2 and O_2 respectively. Water from this reaction is directly used to humidify the membrane. This is supposed to result in a more stable operation of the cell at 80°C without any external humidification of the membrane (Watanabe *et al.*, 1998).

6.1 Organic/silica nanocomposite membranes

Silica as an additive to Nafion® was widely studies. Both recast Nafion® (Adjemian *et al.*, 2002a,2002b; Antonucci *et al.*, 1999; Arimura *et al.*, 1999; Dimitrova *et al.*, 2002a,2002b) and Nafion® film (e.g. Nafion® 117) (Adjemian *et al.*, 2002b; Baradie *et al.*, 2000; Jung *et al.*, 2002) are used in the fabrication of the composite membrane. Organic–silica composite membranes have been prepared according to several methods by casting mixtures such as: using silicon dioxide particles (e.g. Aerosil A380 from Degussa) (Antonucci *et al.*, 1999; Aricò *et al.*, 1998; Arimura *et al.*, 1999; Dimitrova *et al.*, 2002a,2002b), diphenylsilicate (DPS) (Liang *et al.*, 2006), the other one is to introduce silica oxide incorporated via *in situ* sol-gel reaction of tetraethoxysilane (TEOS) (Adjemian *et al.*, 2002b; Baradie *et al.*, 2000; Deng *et al.*, 1998; R.C. R.C. Jiang *et al.*, 2006b; Jung *et al.*, 2002; Mauritz *et al.*, 1995, 1998). Nafion®–silica membranes shows good performance at T > 100°C due to low levels of dehydration. Nafion®–silica membranes were prepared by mixing Nafion® ionomer (5 wt.%) with 3 wt.% SiO_2 followed by a regular membrane casting procedure. In the final stage, the membranes were heat treated at 160°C for 10 min to achieve both a high crystallinity and high mechanical stability (Antonucci *et al.*, 1999). A DMFC utilizing these membranes was tested under galvanostatic conditions at 500 mA cm^{-2}. The voltage initially decreased from 0.42 to 0.36 V but then remained stable for 8 h. The performance decrease is due to adsorption of poisoning species, which appears to be a reversible process at 145°C (removed by short circuit discharging in the presence of water). These membranes demonstrated higher performance with increasing temperature. A nano-particles possessed a core-shell structure consisting of silica core (< 10 nm) and a densely grafted oligometric ionomer layer was incorporated into Nafion® matrix to form a composite membrane. The polyelectrolyte-grafted silica particles [P(SPA)-SiO_2] was dispersed in Nafion® solution and a composite membrane was formed by recasting process. The proton conductivity of Nafion® membrane containing P(SPA)-SiO_2 (4 wt.%) is significantly higher than that of unmodified recast Nafion® in the range 25 to 80°C. The composite membrane offers superior cell performance over unmodified recast Nafion® at both operating temperature of 50 and 80°C. At 50°C, the maximum power output of the composite membrane is about 1.8 times greater than of the Nafion® membrane and at 80°C, the ratio becomes 1.5 (Tay *et al.*, 2008).

Adjemian *et al.* (2002b) prepared composite membranes by either impregnating an extruded film via sol-gel processing of tetraethoxysilane (TEOS), or by preparing a recast film, using solubilized PFSA and silicon oxide polymer/gel. TEOS when reacted with water in an acidic medium undergoes polymerization to form a mixture of silicas (SiO_x) and siloxane polymer with terminal hydroxide and ethoxide groups(SiO_x/-OH/-OEt). When PFSAs are used as the acidic medium, the SiO_x/siloxane polymer forms within the membrane. Composite membranes were tested in fuel cell operating with pre-humidified reactant gases at temperature of 130°C and a pressure of 3 atm. The PFSA/silicon oxide composite membranes shows resistivities 50% lower than their respective unmodified PFSA under the same operating conditions. The observed resistivity trend from best to worst is as follows: Aciplex® 1004/silicon oxide > Nafion® 112/silicon oxide > Nafion® 105/silicon oxide > Aciplex® recast/silicon oxide > Nafion® recast/silicon oxide > Nafion® 115/silicon oxide.

Recently a new approach to make composite membrane was introduced, where a functionalized silica is used as a filler to make the composite membranes (Li *et al.*, 2006; Y.F. Lin *et al.*, 2007; Sambandam & Ramani, 2007 ; Su *et al.*, 2007; Tung & Hwang, 2007). Sol-gel derived sulfonated diphenyldimethoxysilane (SDDS) with hydrophilic $-SO_3H$ functional groups were used as the additive to reduce the methanol permeability of Nafion® (Li *et al.*, 2006). The Nafion®-SDDS nanocomposite membranes were prepared by mixing Nafion® - dimethyl formamide (DMF) solutions with SDDS sol and casting to membranes. The proton conductivity of composite membrane decreased compared with commercial Nafion® membranes. This is partly because (i) the relative low conductivity of organosilica, (ii) the slightly tortuous path through the membrane which is caused by the embedding of the organosilica into the hydrophilic clusters, (iii) and the hydrophobic phenyl groups of the organosilica which change the distribution of the hydrophilic/hydrophobic phases therefore reduce the water content of the membrane. The proton conductivity decreases with the increase of the fillers. On the other hand, the methanol permeability is reduced with the increase of the SDDS content. The methanol permeability drops by a factor of 0.41, 0.61, 0.67 and 0.71 times for nanocomposite with loading of 5, 10, 20 and 25 wt.%, respectively as compared to bare recast Nafion® (Li *et al.*, 2006).

Sulfonic acid functionalized silica was synthesized by condensation of MPTMS (3-mercaptopropyltrimethoxy silane) precursor through a sol-gel approach. Sulfonated poly(ether ether ketone) composite with sulfonic acid functionalized silica were prepared by casting (Sambandam & Ramani, 2007). At 80°C and 75% RH (relative humidity) the measured conductivity was 0.05 S cm^{-1} for SPEEK containing 10% sulfonic acid functionalized silica and 0.02 S cm^{-1} for the plain SPEEK membrane. At 80°C and 50% RH the measured conductivity was 0.018 S cm^{-1} for SPEEK containing 10% sulfonic acid functionalized silica and 0.004 S cm^{-1} for the plain SPEEK membrane.

L64 copolymer-templated mesoporous SiO_2, functionalized with perfluoroalkylsulfonic acid was prepared (Y.F. Lin *et al.*, 2007). A condensation reaction between the surface silanol groups of the mesoporous silicas and 1,2,2-trifluoro-2-hydroxy-1-trifluoromethylethane sulfonic acid Beta-sultone was conducted. Nafion®/functionalized mesoporous silica composite membranes were prepared via homogeneous dispersive mixing and the solvent casting method. The room temperature proton conductivity of full hydration composite membrane was increased from 0.10 to 0.12 S cm^{-1} as the M-SiO_2-SO_3H content increased from 0 to 3 wt.%. Methanol permeability decreases with increasing the content of M-SiO_2-SO_3H, where methanol permeability was 4.5×10^{-6} cm^2 s^{-1}, which was 30% lower than unmodified Nafion®. The current densities measured with composite membranes containing 0, 1, 3 and 5 wt.% M-SiO_2-SO_3H, were 51, 66, 80 and 70 mA cm^{-2}, respectively, at a potential of 0.2 V. Moreover, all composite membranes containing M-SiO_2-SO_3H performed better at high current density than did unmodified Nafion®.

Nanocomposite proton exchange membranes were prepared from sulfonated poly(phtalazinone ether ketone) (SPPEK) and various amounts of sulfonated silica nanoparticles (silica-SO_3H) (Su *et al.*, 2007). The use of silica-SO_3H compensates for the decrease in ion exchange capacity of membranes observed when no-sulfonated nano-fillers are utilized. The strong -SO_3H/-SO_3H interaction between SPPEK chains and silica-SO_3H particles leads to ionic cross-linking in the membrane structure, which increases both the thermal stability and methanol resistance of the membranes. The membrane with 7.5 phr of silica-SO_3H (phr = g of silica-SO_3H / 100 g of SPPEK in membranes) exhibits low methanol crossover, high bound-water content, and a proton conductivity of 3.5 fold increase to that of the unmodified SPPEK membrane. Nafion®/hydrated phosphor-silicate composite membrane was synthesis by Tung and Hwang (2007). The phosphor-silicate glass, with a nominal composition of 30% P_2O_5 and 70% SiO_2 (molar ratio) (called 30P70Si), was prepared by the accelerated sol-gel process, where tetraethylorthosolicate and trimethyl phosphate are used as precursors. It was found that the methanol permeability decreases dramatically with increased SiO_2-P_2O_5 content and the proton conductivity only decreases slightly , as a consequence the selectivity of the hybrid membranes are higher than unmodified Nafion® membrane.

6.2 Organic/heteropolyacid (HPA) nanocomposite membranes

Perfluorosulfonic acid based organic/ inorganic composite membranes with different heteropolyacid (HPA) additives have been investigated as alternate materials for low humidity PEMFC operation (Giordano *et al.*, 1996; Ramani *et al.*, 2004, 2005a,2005b; Tazi & Savadogo, 2000, 2001). Two major factors limiting the performance of Nafion®/HPA composite membranes are (Ramani *et al.*, 2005b): (i) the high solubility of the HPA additive and (ii) the large particle size of the inorganic additive within the membrane matrix (Ramani *et al.*, 2004, 2005a). Stabilization technique have been developed (Ramani *et al.*, 2005a) to limit the solubility of the HPA additive. Recast Nafion® with phosphotungstic acid (PTA) as HPA fillers were prepared by Ramani *et al.* (2005b). Three types of fillers were used PTA with 30-50 nm particle size, PTA with 1-2 μm particle size and TiO_2 with 1-2 μm particle size. The composite membranes had hydrogen crossover currents on the order of 1-5 mA cm^{-2}, with the crossover flux decreasing and approaching the value for recast Nafion® as the particle size was reduced. A 25 μm thick composite membrane with PTA (1-2 μm particle size) had an area-specific resistance of 0.22 Ω cm^{-2} at 120°C and 35% RH, while the corresponding value for a 25 μm thick composite membrane with PTA (30-50 nm) was 0.16 Ω cm^{-2}. The latter membrane compared favorably with recast Nafion®, which had an area-specific resistance of 0.19 Ω cm^{-2} under the same conditions. Savadogo and co-workers (Savadogo, 2004; Tazi & Savadogo, 2000, 2001; Tian & Savadogo, 2005) prepared composite membranes constituted of recast Nafion®and mixed with appropriate concentration of HPA, namely, silicotungstic acid (STA), phosphotungstic acid (PTA) and phosphomolybdic acid (PMA). It was shown that the water uptake of the various membranes increases in this order: Nafion®117 (27%) < Nafion®/STA (60%) < Nafion®/PTA (70%) < Nafion®/PMA (95%). The ionic conductivity increases in the order Nafion®117 (1.3×10^{-2} S cm^{-1})) < Nafion®/PMA (1.5×10^{-2} S cm^{-1}) < Nafion®/PTA (2.5×10^{-2} S cm^{-1}) < Nafion®/STA (9.5×10^{-2} S cm^{-1}). The tensile strength of the membranes decreases in the order: Nafion®117 (15000 Pa) < Nafion®/STA (14000 Pa) < Nafion®/PMA (8000 Pa) < Nafion®/PTA (3000 Pa), while their deformation (ε_{max}) changes in the order : Nafion®/STA (45%) < Nafion®/PMA (70%) < Nafion®/PTA (170%) < Nafion®117 (384%). The current density at 0.600 V of the PEMFCs based on the various membranes varies in the order: Nafion®117 (640 mA cm^{-2}) < Nafion®/STA (695 mA cm^{-2}) < Nafion®/PTA (810 mA cm^{-2}) < Nafion®/PMA (940 mA cm^{-2}).

Tazi and Savadogo (2000) fabricated Nafion® membranes containing silicotungstic acid and thiophene. They reported an increase of up to 60% of water uptake and a considerable improvement in the fuel cell current density, when compared to the plain Nafion® membrane. Dimitrova *et al.* (2002a) prepared a recast Nafion®-based composite membrane containing molybdophosphoric acid. This composite membrane exhibit significantly higher conductivity in comparison to Nafion® 117 and pure recast Nafion®. An enhancement of a factor of 3 in the conductivity at 90°C was observed. Zaidi *et al.* (2000) prepared a series of composite membranes using SPEEK as polymer matrix and tungstophosphoric acid (TPA), its sodium salt (Na-TPA) and molybdophosphoric acid (MoPA) as inorganic fillers. The conductivity of the composite membranes exceeded 10^{-2} S/cm at room temperature and reached values of about 10^{-1} S/cm above 100°C. From the DSC (Differential Scanning Calorimeter) studies, it was indicated that the glass transition temperature of the SPEEK/HPA composite membrane increases due to the incorporation of solid HPA into SPEEK membrane. This increase in the glass transition temperature was attributed to an intermolecular interaction between SPEEK and HPA. Staiti *et al.* (2001) prepared Nafion®(recast)-silica composite membranes doped with phosphotungstic (PWA) and silicotungstic (SiWA) acids for application in direct methanol fuel cell at high temperature (145°C). The phosphotungstic acid-based membrane showed better electrochemical characteristics at high current densities with respect to both silicotungstic acid-modified membrane and silica- Nafion® membrane. The best electrochemical performance is obtained with the PWA-based membrane, which gives a maximum power density of 400 mW cm^{-2} at current density of about 1.4 A cm^{-2} under oxygen feed operation at 145°C. Maximum power density of 340 mW cm^{-2} is obtained from the fuel cell which uses the silica-modified membrane, whereas a lower performance was achieved with the SiWA-based membrane. The maximum power density obtained in air with the PWA-based membrane is 250 mW cm^{-2} at 145°C, and 210 mW cm^{-2} with the Nafion-SiO_2 membrane at the same temperature.

Shao *et al.* (2004) prepared Nafion®/silicon oxide (SiO_2)/phosphotungstic acid (PWA) and Nafion®/silicon oxide composite membranes for H_2/O_2 proton exchange membrane fuel cells operated above 100°C. It was found that the composite membranes showed a higher water uptake compared with the Nafion® recast membrane. The proton conductivity of the composite membranes appeared to be similar to that of the native Nafion® membrane at high temperatures and 100% relative humidity (RH), however, it was much higher at low RH. When the composite membranes viz. Nafion®/SiO_2/PWA and Nafion®/SiO_2 were employed as an electrolyte in H_2/O_2 PEMFC, the higher current density values (540 and 320 mA cm^{-2} at 0.4 V, respectively) were obtained than that of the Nafion®115 membrane (95 mA cm^{-2}), under the operating conditions of 110°C and 70% RH. A similar membrane was prepared by Aricò *et al.* (2003a,2003b). Sulfonic-functionalized heteropolyacid-SiO_2 nanoparticles were synthesized by grafting and oxidizing of a thiol-silane compound onto the heteropolyacid-SiO_2 nanoparticle surface (H.J. Kim *et al.*, 2006). The composite membrane containing the sulfonic-functionalized heteropolyacid-SiO_2 nanoparticles was prepared by blending with Nafion® ionomer. TG-DTA analysis showed that the composite membrane was thermally stable up to 290°C. The DMFC performance of the composite membrane increased the operating temperature from 80 to 200°C. The function of the sulfonic-functionalized heteropolyacid-SiO_2 nanoparticles was to provide a proton carrier and act as a water reservoir in the composite membrane at elevated temperature. The power density was 33 mW cm^{-2} at 80°C, 39 mW cm^{-2} at 160°C, 44 mW cm^{-2} at 200°C, respectively.

SPEEK-silica membranes doped with phosphotungstic acid (PWA) was synthesized by Colicchio *et al.* (2009). The silica is generated *insitu* via the water free sol-gel process of polyethoxysiloxane (PEOS), a liquid hyperbranched inorganic polymer of low viscosity. PEOS was used as silica precursor instead of the corresponding monomeric tetraethoxysilane (TEOS). At 100°C and 90% RH the membrane prepared with PEOS (silica content = 20 wt %) shows two times higher conductivity than the pure SPEEK. The addition of small amount of PWA (2 wt.% of the total solid content) introduce in the early stage of membrane preparation brings to a further increase in conductivity (more than three times the pure SPEEK). Different classes of composite membranes containing HPA and silica were developed, namely, phosphomolybdic acid (PMA)/phosphotungstic acid (PWA)- P_2O_5-SiO_2 glass electrolyte (Uma & Nogami, 2007), poly(vinyl alcohol) (PVA)/sulfosuccinic acid (SSA)/silica hybrid membrane (D.S. Kim *et al.*, 2004), PVA/SiO_2/ SiW (silicotungstic acid) (Shanmugam *et al.*, 2006), PVA/PWA/SiO_2 (W. Xu *et al.*, 2004), polyethylene oxide (PEO)/PWA/SiO_2 (Honma *et al.*, 2002), polyethylene glycol (PEG)/4-dodecylbenzene sulfonic acid (DBSA)/SiO_2 (H.Y. Chang *et al.*, 2003), PWA-doped PEG/SiO_2 (C.W. Lin *et al.*, 2005).

6.3 Organic/TiO_2 nanocomposite membranes

Nanosized titanium oxide was synthesized by sol-gel hydrolyzing an alcoholic solution of $Ti(OiPr)_4$ by Baglio *et al.* (2005). Thermal treatments at different temperature, namely 500, 650 and 800°C, were performed to tailor the oxide powder properties. The crystallite size for the three sample was found to be 12, 22 and 39 nm, respectively. A composite membrane Nafion®/ 5 wt.% TiO_2 was prepared by using the recast procedure. The composite membrane thickness was about 100 μm. A maximum power density of 350 mW cm^{-2} was obtained at 145°C with the cell equipped with the composite membrane containing TiO_2 calcined at 500°C. Sacca *et al.* (2005) synthesized TiO_2 powder by the sol-gel method starting with a $Ti(OiPr)_4$ and calcined a 400°C. This powder was made of spherical particles with a grain size between 5 and 20 nm. A recast Nafion® with 3 wt.% TiO_2 composite membrane was prepared, the thickness of the membrane was 100 μm. The proton conductivity of different membranes were measured at two different values of relative humidity (RH), 100 and 85% RH, respectively, and simulating the cell operating conditions in the temperature range from 80 to 130°C. Nafion® recast (70 μm thickness) has the lower conductivity ranging from 0.12 to 0.14 S cm^{-1}, while the composite Nafion®/TiO_2 showed highest value than Nafion® 115 (125 μm) were value in the range 0.15-0.18 S cm^{-1} . A power density of 0.514 W cm^{-2} for Nafion® / 3 wt.% TiO_2 composite against 0.354 W cm^{-2} for Nafion® 115 at 0.56 V and at T = 110°C was recorded. At 120°C, Nafion® 115 was damaged while the composite Nafion®/TiO_2 membrane continued to work up to 130°C by reaching a power density of about 0.254 W cm^{-2} at 0.5 V.

Hybride membranes based on highly sulfonated poly(ether ether ketone) (SPEEK, DS = 0.9) where titania network was dispersed by *insitu* sol-gel reactions were prepared by Di Vona *et al.* (2007). Titania network was introduced following two routes: route 1 using titanium tetrabutoxide ($Ti(OBu)_4$) and pyridine, and route 2 uses $Ti(OBu)_4$ and 2,4-pentandione. Composite membranes prepared by route 2 showed a good conductivity property that can be attributed to the structural characteristics of the inorganic network generated in the presence of a chelating agent. This membrane shows a stable value ($\sigma = 5.8 \times 10^{-2}$ S cm^{-1}) at 120°C in fully hydrated conditions. Jian-hua *et al.* (2008) prepared a composite Nafion®/TiO_2 membranes by carrying out *insitu* sol-gel reaction of Ti $(OC_4H_9)_4$ followed by hydrolyzation-condensation in Nafion® 112, 1135 and 115. TiO_2 prepared with this method was found to be 4 nm. TiO_2 contents were 1.23, 2.47 and 3.16 wt.% for Nafion® 112/TiO_2, Nafion® 1135/TiO_2 and Nafion® 115/TiO_2, respectively. The polarization characteristics of all three MEAs with the membranes containing TiO_2 were improved significantly comparing with those of pure Nafion® film. A mixture of titanium tetraisopropoxide (TTIP) and PEG 1000 were used to prepare the titania sol by Liu *et al.* (2006). The average particle size of 20 nm was reported. The formed sol was deposited on the surface of Nafion® 112 membranes by spin coating. The TiO_2 film is dense and well attached to the membrane, but some cracks in the membrane coated with diluted titania sol (e.g. 0.002 mg cm^{-2}), while the membrane coated with thick titania sol (e.g. 0.021 mg cm^{-2}) are very dense and cracks free. The proton conductivity of nano-TiO_2-coated Nafion® membranes at 25 and 80°C were recorded with different TiO_2 content. It was found that the maximum conductivity was with uncoated Nafion® 112, with values of 0.027 and 0.041 S cm^{-1} for 25 and 80°C, respectively. The conductivity of coated Nafion® decreases with increasing titania content. On the other hand, methanol permeability of the coated membranes decreases with increasing TiO_2 content, namely from 3.2×10^{-6} to 1.7×10^{-6} cm^2 s^{-1} at 25°C, and from 12.5×10^{-6} to 4.6×10^{-6} cm^2 s^{-1} at 85°C. Thus the rise in temperature leads to a strong increase in permeation by a factor of about 3. The methanol permeability of the unmodified Nafion® 112 membrane was found to be 3.6×10^{-6} and 13×10^{-6} cm^2 s^{-1} at 25 and 85°C, respectively. The cell performance with titania coated membrane with a content of 0.009 mg cm^{-2} exhibits a higher voltage than cells with Nafion® membrane or with the other coated membranes. Nafion® 112 delivered a maximum power density of 37 mV cm^{-2} at a current density of 200 mA cm^{-2}. A maximum power density of 44 mW cm^{-2} is obtained from a fuel cell that employs the titania-coated membrane with 0.009 mg cm^{-2} content.

6.4 Organic/zirconia and sulfated zirconia nanocomposite membranes

Zirconia as an inorganic filler was added to polymeric proton conductor membranes (Aricò *et al.*, 2003b, 2004; Nunes *et al.*, 2002; Sacca *et al.*, 2006; Silva *et al.*, 2005a, 2005b). The incorporation of zirconia should increase the working temperature, water retention and mechanical stability of the composite membrane. Organic / inorganic composite membranes based on SPEK and SPEEK for application in direct methanol fuel cell were synthesized by Nunes and co-workers (Nunes *et al.*, 2002). The inorganic fillers were introduced via *in situ* generation of SiO_2, TiO_2 or ZrO_2. The modification with ZrO_2 led to a 60-fold reduction of the methanol flux. However, a 13-fold reduction of conductivity was also observed.

Recast Nafion® composite membranes containing three different percentages (5%, 10% and 20%, w/w) of commercial ZrO_2 as an inorganic filler were tested in fuel cell in a temperature range of 80-130°C in humidified H_2/air gases at 3.0 bar abs by Sacca *et al.* (2006). The introduction of 5 wt.% ZrO_2 in Nafion® produces no evidence changes in the cell performance, while a better performance with 10 wt.% ZrO_2 in Nafion® was obtained with a power density greater than 600 mW cm-2 at 0.6 V both at 80°C and 110°C. The good performance of 10 wt.% ZrO_2 in Nafion® was maintained at 130°C with gas humidification of 85% RH, with a maximum power density of about 400 mW cm^{-2} was obtained in the potential range of 0.5-0.6 V. Silva *et al.* (2005a,2005b) prepared SPEEK/ZrO_2 composite membranes using *insitu* formation of zirconia with zirconium tetrapropylate as alkoxide and acetyl acetone as chelating agent. The water/alkoxide ratio was always maintained higher than 1 to ensure the formation of a finely dispersed inorganic phase in the polymer solution. The thickness of the prepared membranes with 0.0, 2.5, 5.0, 7.5, 10.0, 12.5 wt.% of zirconium oxide were 188, 175, 133, 146, 128, 106 μm, respectively. The proton conductivity of the composite membranes was measured at 25°C and it was found that it decreases continuously with the ZrO_2 content. Pervaporation experiments at 55°C showed that the membrane permeability towards methanol decreases with the amount of ZrO_2. Composite SPEEK with 5.0, 7.5, 10.0 wt.% of ZrO_2 were tested in fuel cell. The performance of 12.5 wt.% ZrO_2 composite was very low due to the high ohmic resistance of the corresponding MEA. The membrane 5 wt.% ZrO_2 presents the best DMFC performance among all the studied MEAs. Three types of superacidic sulfated zirconia (S-ZrO_2) were prepared by different methods using hydrated zirconia and sulfuric acid by Hara and Miyayama (2004). Their proton conductivities were evaluated at 20-150°C under saturated water vapor pressure. It was found that the concentration on S-ZrO_2 varied largely depending on the method of preparation. The S/Zr atomic ratio changed from 0.046 for the sample prepared through a mixture of hydrated zirconia powder and sulfuric acid to 0.35 for sample prepared through a mixture of hydrated zirconia sol and sulfuric acid. A powder compact of the former S-ZrO_2 showed a proton conductivity of 4×10^{-2} S cm^{-1} at 70°C and 8×10^{-3} S cm^{-1} at 150°C, whereas that of the latter S-ZrO_2 exhibited a high conductivity of 5×10^{-2} S cm^{-1} at 60-150°C.

S. Ren *et al.* (2006) prepared sulfated zirconia/ Nafion® 115 nanocomposite membrane by ion exchange of zirconium ions into the Nafion® followed by precipitation of sulphated ZrO_2 by treatment in H_2SO_4. The incorporation of sulfated zirconia increases water uptake by the Nafion® membrane, and more water is absorbed than an unmodified membrane at high temperatures. The membrane proton conductivity is decreased slightly by ZrO_2 impregnation. The proton conductivity of Nafion® 115 membrane was found to be 1.5×10^{-2} S cm^{-1} at 25°C, while that of S-ZrO_2/ Nafion® 115 membrane is decreased to 5.0×10^{-3} S cm^{-1} at 25°C. At 110°C and above, the proton conductivity of S-ZrO_2/ Nafion® 115 membrane is more than one-half that of the Nafion® 115 membrane. Fine particle superacidic sulfated zirconia (S-ZrO_2) was synthesized by ameliorated method, and composite membranes with different S-ZrO_2 contents were prepared by a recasting procedure from a suspension of S-

ZrO_2 powder and Nafion® solution (Zhai *et al.*, 2006). The results showed that the IEC (Ion Exchange Capacity) of composite membrane increased with the content of S-ZrO_2 and S-ZrO_2 was found to be compatible with the Nafion® matrix. The incorporation of the S-ZrO_2 increased the crystallinity and also improved the initial degradation temperature of the composite membrane. The performance of single cell was the best when the S-ZrO_2 content was 15 wt.% and achieved 1.35 W cm^{-2} at 80°C and 0.99 W cm^{-2} at 120°C based on H_2/O_2 and at a pressure of 2 atm, the performance of the single cell with optimized S-ZrO_2 was far more than that of the Nafion® at the same condition (e.g. 1.28 W cm^{-2} at 80°C and 0.75 W cm^{-2} at 120°C). A self-humidifying composite membrane based on Nafion® hybrid with SiO_2 supported sulfated zirconia particles (SiO_2-SZ) was fabricated and investigated for fuel cell application by Bi *et al.* (2008). The bi-functional SiO_2-SZ particles, possessing hygroscopic property and high proton conductivity, was incorporated in recast Nafion® membrane. The proton conductivity of Nafion®/SiO_2-SZ, Nafion®/SiO_2 and recast Nafion® under dry and wet H_2/O_2 conditions at 60°C were compared. The two composite membranes showed higher proton conductivity in contrast to the recast Nafion® membrane under 0% RH mode with the order Nafion®/SiO_2-SZ > Nafion®/SiO_2 > recast Nafion®. Under 100% RH mode, the Nafion®/SiO_2-SZ composite membrane also exhibited the highest proton conductivity values among the three membranes. The proton conductivity of Nafion®/SiO_2-SZ membrane was 6.95×10^{-2} S cm^{-1} and the value was higher than Nafion®/ SiO_2 membrane of 5.54×10^{-2} S cm^{-1} and recast Nafion® membrane of 6.55×10^{-2} S cm^{-1}. Single cell performance of these composite membranes were tested with wet H_2 and O_2 at 60°C. Nafion®/ SiO_2 composite membrane exhibited the worst output performance (0.864 W cm^{-2}) due to the increased proton conductive resistance caused by incorporated less proton conductivity of SiO_2 particles. In contrast, Nafion®/SiO_2-SZ composite membrane showed similar cell performance to recast Nafion® (1.045 W cm^{-2} vs. 1.014 W cm^{-2}). However, the single cell performance of Nafion®/SiO_2-SZ and Nafion®/SiO_2 membranes with dry H_2 and O_2 at 60°C were 0.980 and 0.742 W cm^{-2}, respectively. These results shows that the composite membrane perform better than unmodified Nafion® (i.e 0.635 W cm^{-2}) under dry condition and the composite membranes manifested a good water retention.

6.5 Organic/zirconium phosphate nanocomposite membranes

Layered zirconium phosphate (ZrP) and phosphonates were used as inorganic fillers of proton conducting polymeric membranes because they are proton conductors with good chemical and thermal stability. Under the most favourable conditions, their conductivity is around 10^{-2} S cm^{-1} for high surface ZrP (Alberti *et al.*, 1978; F. Bauer & Willert-Porada, 2005) and 10^{-1} S cm^{-1} for zirconium phosphate sulfophenylenphosphonates (Alberti *et al.*, 2004,2005a). ZrP can be added to Nafion® (Alberti *et al.*, 2007; F. Bauer & Willert-Porada, 2004,2005,2006a; Casiola *et al.*, 2008; Costamagna *et al.*, 2002; Grot & Rajendran, 1999; Helen *et al.*, 2006,2007; Hou *et al.*, 2008; R. Jiang *et al.*, 2006a; Kuan *et al.*, 2006; H.K. Lee *et al.*, 2004; Mitov *et al.*, 2006; Yang *et al.*, 2001a,2001b,2004), SPEEK (Bonnet *et al.*, 2000; Nunes *et al.*, 2002; Silva *et al.*, 2005b; Tchicaya-Bouckary *et al.*, 2002; Triphathi *et al.*, 2007,2009), SPEK (Nunes *et al.*, 2002; Ruffmann *et al.*, 2003) and difulfonated poly(arulene ether sulfone) (Hill *et al.*, 2006). A similar inorganic material derived from ZrP, named zirconium phosphate sulfophenylen-phophonate was also used as a filler with Nafion® (Y.T. Kim *et al.*, 2004), SPEEK (Krishnan *et al.*, 2006) and PVDF (polyvinyl -lidene fluoride) (Casiola *et al.*, 2005).

F. Bauer & Willert-Porada (2004,2005,2006a) impregnated Nafion® with different ZrP contents. The proton conductivity of unmodified Nafion® 117 and Nafion® 117/ZrP (21 wt.%) was measured at three different temperatures, 80, 100 and 130°C. It was found that the presence of ZrP decreased the proton conductivity in all cases. At high humidity the conductivity first increased from 80 to 100°C and decreased at 130°. The conductivity decrease is more pronounced in case of the unmodified Nafion®. DMFC performance was conducted with Nafion® 117 and Nafion® 117/ZrP composite membrane at 130°C and 4.6 bar at the anode and cathode. It was found that the power output of Nafion® was higher than that of the composite membranes (Nafion® 117/13 wt.% ZrP and Nafion® 117/26 wt.% ZrP). At 0.2 A cm^{-2}, a values of 420, 370 and 370 mV were measured for Nafion® 117, Nafion® 117/13 wt.% ZrP and Nafion® 117/26 wt.% ZrP, respectively. The crossover current was reduced by a factor of two as compared to the unmodified Nafion®. Also the two composite membranes tested exhibited a higher OCV than unmodified Nafion®, which also indicates lower methanol permeability. A values of 725, 768 and 760 mV were reported for Nafion® 117, Nafion® 117/13 wt.% ZrP and Nafion® 117/26 wt.% ZrP, respectively.

Yang and coworkers (Yang *et al.*, 2001a,2001b,2004; Costamagna *et al.*, 2002) introduced ZrP into Nafion® 115 through ion exchange of Zr^{4+} followed by precipitation of ZrP by treatment with phosphoric acid as described by Grot and Rajendran (1999). An MEA employing Nafion® 115/23 wt.% ZrP gave a H_2/O_2 PEMFC performance of about 1000 mA cm^{-2} at 0.45 V at a temperature of 130°C and a pressure of 3 bar, while unmodified Nafion® 115 gave 250 mA cm^{-2} at 0 45 V when operated under the same conditions of temperature and pressure. Similar experiment performed with recast Nafion® and recast Nafion®/36 wt.% ZrP composite confirmed an analogous improvement of performance of the composite membrane over the unmodified ones. The composite recast Nafion®/36 wt.% ZrP gave about 1500 mA cm^{-2} at 0.45 V at a temperature of 130°C and a pressure of 3 bar.

Alberti *et al.* (2005b,2007) prepared a recast Nafion® filled with ZrP according to the procedure described in the patent (Alberti *et al.*, 2005b). Zirconyl propionate was used instead of zirconyl oxychloride and the solutions were dissolved in DMF. The IEC (ion exchange capacity) of the prepared composite membrane was found to be higher than those previously reported for Nafion®/ZrP membranes prepared according to the exchange method (F. Bauer & Willert-Porada, 2006b; Yang *et al.*, 2004). The proton conductivity was found to decrease with increasing the filler loading, which is in agreement with the trend found for Nafion®/ZrP prepared by the exchange method (F. Bauer & Willert-Porada, 2005; Casiola *et al.*, 2008; Yang *et al.*, 2004). At constant RH, the logarithm of conductivity shows approximately the same linear dependence on ZrP loading in the RH range 50-90%. However, at 35% RH, the increase in the ZrP loading results in a larger conductivity decrease than that observed in the above RH range. A similar behavior was also reported for Nafion®/ZrP membranes obtained by the exchange method already at 50% RH (Yang *et al.*, 2004), thus confirming that the same type of filler prepared by using different procedures gives rise to different membrane properties. It was concluded that the main difference between pure Nafion® and composite membranes appear at low RH and high filler loading. It was reported that the Nafion® conductivity undergoes an irreversible decay above certain values of temperature and RH, which was attributed to an anisotropic swelling of the membrane, pressed between the electrodes, in the direction parallel to the membrane surface (Alberti *et al.*, 2001; Casiola *et al.*, 2006). It was also found that, at a given RH value, the decay temperature for composite Nafion®/ZrP membranes was higher than for pure recast Nafion® membranes prepared and thermally treated under the same conditions used for the composite sample. The conductivity of the pure Nafion® starts to decay at temperatures higher than 130°C, while the conductivity of the composite membrane is stable up to 140°C. Nafion® 115/23 wt.% ZrP was prepared by ion exchange and tested for DMFC by Hou *et al.* (2008). It was found that the liquid uptakes of Nafion® 115 and Nafion® 115/23 wt.% ZrP membranes increased linearly with increasing methanol concentration. The slope of the plot for Nafion® 115 was larger than for the composite membrane i.e. the liquid uptake of Nafion® 115 increased from 34.3% in 0M methanol solution to 58.6% in 10M methanol solution, while that of the composite membrane increased from 28.3% to 37.5% in the corresponding methanol solution. When 23 wt.% of ZrP was incorporated into Nafion® 115, the IEC of the resulting membrane increased significantly to 1.93 meq/g from a value of 0.909 meq/g for pure Nafion® 115. The proton conductivity at room temperature of Nafion® 115 and Nafion® 115/23 wt.% ZrP was found to be 0.10 and 0.084 S cm^{-1}, respectively. Also it was found that the methanol crossover through the composite membrane was suppressed. The DMFC test at 75°C and 5M methanol solution shows that the composite membrane performed better that the pure Nafion® 115, with a peak power density of 96.3 and 91.6 mW cm^{-2}, respectively. When the methanol concentration was further increase to 10M, the peak power density of DMFC with composite membrane was 76.19 mW cm^{-2}, which is higher than that for Nafion® (42.4 mW cm^{-2}). However, Bonnet *et al.* (2000) investigated the incorporation of ZrP in SPEEK. It was found that the conductivity of the composite membrane exceeded that of the polymer-only membrane, and increases with the amount of the filler (from 0-30 wt.%) up to 0.08 S cm^{-1} when measured at 100°C and 100% RH. A similar trend was also observed, when the RH varied from 50 to 100%. At all value of RH, the composite membrane SPEEK/20 wt.% ZrP conductivity was higher than that of non-modified SPEEK. A similar membrane was prepared by Tchicaya-Bouckary *et al.* (2002). The conductivity of SPEEK/25 wt.% ZrP was found to be weakly temperature dependence, the conductivity increases from 2×10^{-2} to 5×10^{-2} S cm^{-1} between 20 and 100°C at 100% RH. This composite membrane was tested in H_2/O_2 fuel cell at 100°C at an oxygen pressure of 3.6 bars absolute. A value of 1 A cm^{-2} at 0.6 V was reported. These results are much better than that reported for Nafion® 115/ ZrP (Costamagna *et al.*, 2002) which provided ca. 0.7 A cm^{-2} at 0.6 V, 130°C and 3 bars pressure. Nunes and co-workers studied the incorporation of ZrP in SPEEK and SPEK (Nunes *et al.*, 2002; Ruffmann *et al.*, 2003; Silva *et al.*, 2005c). ZrP was prepared according to the procedure described by Belyakov & Linkov (1999). It was reported that the incorporation of ZrP did not lead to a particular reduction of water and methanol permeability, and the proton conductivity at 25°C was decreased to the same extent (44 mS cm^{-1} for a SPEK/ 20 wt.% ZrP and 50 mS cm^{-1} for a pure SPEK membranes). A good values of proton conductivities were measured for membranes with 70/20/10 and 69/17/14 wt.% SPEK/ZrP/ZrO_2 where a conductivities of 45 and 35 mS cm^{-1} were measured, respectively (Nunes *et al.*, 2002). ZrP pretreated with *n*-propylamine and PBI was incorporated with SPEEK (Silva *et al.*, 2005c), the proton conductivity of the composite membranes decreases with the amount of inorganic incorporation. On the other hand, methanol and water permeability in the pervaporation experiments at 55°C showed that it decrease with the amount of inorganic incorporation. Similar trend was found for the composite membranes permeability towards nitrogen, oxygen and carbon dioxide. The SPEEK composite membranes were tested in a DMFC at 110°C, it was found that the unmodified SPEEK (SD = 42%) membrane presented the maximum power density output. It achieved an output power density value of 10.4 mW cm^{-1} for 51.8 mA cm^{-2}. The unmodified membrane with SD = 68% could not be characterized due to its instability (high swelling or even solubility). However, the SPEEK (SD=68%)/20 wt.% ZrP/11.2 wt.% PBI had even higher power density than the membrane with SD = 42% for current density lower than 25 mA cm^{-2}. When the relative humidity at the cathode feed was increased to 138%, the SPEEK (SD=68%)/20 wt.% ZrP/11.2 wt.% PBI membrane had the best performance, with an output power density value of 14.7 mW cm^{-1} for 58.8 mA cm^{-2} (Silva *et al.*, 2005c). However, the filler addition to SPEEK (SD = 42%) besides reducing the crossover had an excessive (negative) effect on the proton conductivity.

Zirconium phosphate sulfophenylphosphate, a functionalized ZrP, was incorporated in Nafion® (Casiola *et al.*, 2008; Y.T. Kim *et al.*, 2004), SPEEK (Bonnet *et al.*, 2000; Krishnan *et al.*, 2006) and PVA (Casiola *et al.*, 2005). ZrP sulfophenylphosphates (ZrSPP) are a class of layered materials exhibiting proton conductivity comparable with that of Nafion membranes (i.e. 0.07-0.1 S cm^{-1} at 100°C and 100% RH) due to the presence of the $-SO_3H$ groups in the interlayer region (Alberti *et al.*, 2005a). The functionalization of the ZrP nanoparticles with SPP is therefore expected to increase the conductivity of the Nafion®/ZrP membranes. These phosphonates are ideally obtained by partial replacement of SPP groups for the phosphate groups of ZrP (Casiola *et al.*, 2008). Nafion® 117/ZrSPP composite membranes was found to have a higher conductivity than the parent Nafion® 117/ 20 wt.% ZrP and pure Nafion® 117 membrane at 100°C and RH between 30-90%, with highest value approaching 0.1 S cm^{-1} at RH = 90% (Casiola *et al.*, 2008), while appreciable dehydration of Nafion® 117 resulted in drastic reduction of proton conductivity above 100°C (Y.T. Kim *et al.*, 2004). However, the proton conductivity of Nafion®/12.5 wt.% ZrSPP composite membrane slightly increased up to 70°C and remained constant until 140°C, with a conductivity of 0.07 S cm^{-1} (Y.T. Kim *et al.*, 2004).

6.6 Organic/palladium nanocomposite membranes

This approach is to utilize the unique properties of palladium which is permeable to protons, but very resistant to methanol transport. It was suggested first by Pu *et al.* (1995) where they used a palladium foil of 25 μm thick sandwiched between two Nafion® 115 sheets. They proved that with this approach methanol crossover can be reduced, but the cell performance will be lower due to the increase of the membrane thickness.

Choi *et al.* (2001) used the same approach by sputtering metallic palladium on the surface of a Nafion® 117 to plug the pores of Nafion®. The palladium film was found to be 20 nm. Methanol permeability was reduced from 2.392 × 10^{-6} cm^2 s^{-1} in unmodified Nafion® 117 to 1.7 × 10^{-6} cm^2 s^{-1} in Pd-sputtered membrane and the cell performance at 95°C was improved compared to the unmodified Nafion® 117 membrane. The methanol permeability reduction was confirmed by the high OCV obtained with the modified membranes. Similarly, Yoon *et al.* (2002) used sputtering technique to deposit Pd film on the surface of Nafion® 117. It was found that the Pd films thinner than 300 Å were dense and appeared to be well attached to the membrane, but there were many cracks in the 1000 Å films. The 1000 Å films were very unstable and were easily delaminated from the membrane surface. When the composite membrane is immersed in water, the Nafion® membrane swells very much, but the Pd film can not expand as much as the membrane (Yoon *et al.*, 2002). The proton conductivity was found to decrease with increasing Pd thickness. For the Pd-1000 Å film, 30% reduction in conductivity was observed. Methanol permeability at 25°C decreased with increasing Pd thickness and they varied from 2.90×10^{-6} to 2.23×10^{-6} cm^2 s^{-1} by deposition of Pd film of 1000 Å on the Nafion® 117 membrane. For the Pd-1000 Å-Nafion® 115 membrane, permeability decreased as much as 44% from 2.97×10^{-6} to 1.67×10^{-6} cm^2 s^{-1}. The cell performance at 90°C exhibited a slight decrease with the Pd layered Nafion® 115 membranes. The methanol crossover through an MEA is inversely proportional to current density and thus, its effect on the performance is more prominent at low current densities. It was found that at low current densities cell performance increased in Pd film of 1000 Å on the Nafion® 115, where the Pd film act as a barrier to methanol crossover. However, these results are different of that of Choi *et al.* who observed significant increase in DMFC performance in Pd-20 nm film on Nafion® 117.

Z.Q. Ma *et al.* (2003) followed a different approach, where Nafion® membrane was modified by sputtering a thin layer of Pt/Pd-Ag/Pt on its surface. The methanol crossover can be reduced by sputtering Pd-Ag alloys over the polymer electrolyte, furthermore, when hydrogen is absorbed and dissolved in the membrane, the palladium-silver alloy membrane not only reduces the possibility of embrittling due to $\alpha \rightarrow \beta$ phase transition at low temperatures (< 150°C) but also leads to a higher permeability for hydrogen than pure palladium membrane. The composite membranes were prepared as follows: on one side of the Nafion® 117 membrane, a 2 nm Pt film was first deposited. This was followed by a Pd-Ag film with three different thicknesses (0.1, 0.2 and 1 μm). On the top of the Pd-Ag, it was coated with another 2 nm Pt film. Before a MEA was manufactured, a 4-5 μm layer of Nafion® polymer was recast over the surface of the sputtered Pt/Pd-Ag/Pt layer with Nafion® solution. The final membrane prepared was in the form of Nafion®117/Pt/Pd-Ag/Pt/Nafion®, containing 0.0086 mg cm^{-2} Pt, 0.90 mg cm^{-2} Pd, and 0.27 mg cm^{-2} Ag. The cell performance and the OCV increased with increasing the sputtering alloy layer thickness and the best performance and the highest OCV were found with the 1 μm Pd-Ag film. Also it was showed that the performance with the 1 μm Pd-Ag film is higher than that of cell with a Nafion® membrane having catalyst loading twice as high. Palladinized Nafion® composite membrane was prepared via ion-exchange and chemical reduction method (Y.J. Kim *et al.*, 2004). Palladium(II) acetylacetonate and tetraammine-palladium(II) chloride hydride were used as palladium precursors. Nafion® 117 samples were immersed in palladium precursor solutions followed by chemical reduction of palladium precursors by sodium borohydride. The use of tetraamminepalladium(II) chloride hydride formed 40-50 nm of palladium particles while palladium(II) acetylacetonate formed 5-10 nm of particles. For all palladinized samples, water uptake was higher than for unfilled Nafion® whereas above a certain amount of incorporated palladium, methanol uptake was lower than for unfilled Nafion®. Incorporating Pd nanoparticles decreases the proton conductivity and methanol permeability, compared to bare Nafion®, simultaneously. However, the conductivity increased and methanol permeability decreased as the amount of incorporated Pd increased, and above a certain amount the rapid increase of conductivity and permeability appeared. The DMFC performance at 40°C was improved by incorporating Pd.

Different approach for the incorporation of Pd in Nafion® membranes was suggested by Tang *et al.* (2005). Multi-layer self-assembly Nafion® membranes (MLSA Nafion® membranes) were prepared by alternately assembling charged Pd particles and Nafion®

ionomer onto Nafion® 112 membranes. The Pd particles, size of about 1.8 nm in average, are charged by PDDA (polydiallyldimethylammonium chloride) ionomers. The Pd loading of the first-layer MLSA Nafion® membranes was 0.63 μg cm^{-2}, and the surface coverage of the Pd nanoparticles on the Nafion® membrane was estimated as 22%. After 5-double-layer Pd particles/Nafion® ionomers assembling, the Pd loading reached to 2.86 μg cm^{-2}. The methanol crossover current of the original Nafion® membranes and 1-double-layer, 2-double-layer, 3-double-layer, 4-double-layer, 5-double-layer MLSA Nafion® membrane were 0.0495, 3.87×10^{-3}, 1.38×10^{-3}, 7.32×10^{-4}, 5.16×10^{-4} and 4.25×10^{-3} A cm^{-2}, respectively, corresponding conductivities of 0.112, 0.110, 0.105, 0.094, 0.087 and 0.081 S cm^{-2}. No DMFC data were provided, however, it was suggested that the 3-double-layer self-assembly membrane is the best suited for DMFC application, since it has a methanol crossover decreased to 0.86%, and a conductivity remaining at 83.9% comparing to original Nafion® membrane. Electroless plating was also used to deposit Pd layer on Nafion® membranes (Hejze *et al.*, 2005; Sun *et al.*, 2005). Palladium layer can reduce methanol crossover when coated on the surface of Nafion®, i.e. limiting current from methanol permeation through a membrane electrode assembly was reduce from 64 to 57 mA cm^{-2} for 1M methanol, and from 267 to 170 mA cm^{-2} for 5M methanol, for Nafion® 115 and Pd/Nafion® 115 membranes respectively (Sun *et al.*, 2005). Also it was demonstrated that the DMFC performance increase with the incorporation of Pd in Nafion® 115. When 1M methanol was used, the power density increased from 36 to 45 mW cm^{-2}, for Nafion® 115 and Pd/Nafion® 115, respectively. When 5M methanol was used, the maximum power density on Pd/Nafion® 115 was 72 mW cm^{-2}, while the performance of MEA with pure Nafion® 115 membranes was only 32 mW cm^{-2}.

6.7 Organic/montmorillonite nanocomposite membranes

Montmorillonite (MMT) is a type of layered silicate composed of silica tetrahedral and alumina octahedral sheets (J. Chang *et al.*, 2003) and its intercalation into Nafion® membrane can decrease successfully the methanol permeability and improve mechanical property (Jung *et al.*, 2003; Song *et al.*, 2004). Research by J. Chang *et al.* (2003) has showed that layered silicates incorporated into SPEEK membranes helped to reduce swelling significantly in hot water and decrease the methanol crossover without a serious reduction of the proton conductivity. Gaowen and Zhentao (2005) prepared organically modified MMT (OMMT) through ion exchange reaction between alkylammonium cations and metal cations. The nanocomposite membranes (SPEEK/OMMT) were prepared using the solution intercalation technique. The water uptake of SPEEK membrane increased rapidly above 50°C, while the SPEEK/OMMT composite membranes posses of rather constant water uptake up to 80°C. This indicates that MMT layers incorporated into SPEEK matrix prevent extreme swelling of the composite membranes due to the cohesion of the functional groups between SPEEK matrix and MMT layers. The proton conductivity of the membrane was measured at temperature ranging from 22 to 110°C. It was found that the conductivity of SPEEK/OMMT composite is lower than that of the pristine SPEEK and decreases sequentially as the content of OMMT increases, which is due to prolonging the transfer route of proton. The conductivity of SPEEK/OMMT (5 wt.%) approaches the value of Nafion® 115 at 90°C and reaches 1.2×10^{-2} S cm^{-1}. The activation energies of SPEEK/OMMT are higher than that of Nafion® 115, where the value of 32.08 kJ/mol and 10.8 kJ/mol, respectively, were found. Methanol permeability was found to be in the following order: Nafion® 115> SPEEK > SPEEK/OMMT. Therefore, incorporating nano-sized dispersion of OMMT prevents methanol from migrating through the membrane. Song *et al.* (2004) prepared recast Nafion®/MMT nanocomposite membranes at a loading of 3 wt.% clay. It was found that the strength increased more than 35% and the tensile elongation almost doubled. The thermal decomposition behavior of Nafion®/MMT nanocomposite was similar to that of pristine Nafion®, but the major decomposition temperature of polymer main chain shifted to much higher temperature region. The proton conductivity of pristine recast Nafion® with thickness of ca. 100 μm and dry state reached about 0.08 S/cm at room temperature. For Nafion®/MMT nanocomposite membranes, the room temperature conductivity was almost similar to that of neat Nafion® below MMT loading of 2 wt. % and then decreased gradually with the increase in filler content. The methanol permeability of pristine recast Nafion® was 2.3×10^{-6} cm^3 cm/ cm^2 s, while for Nafion®/MMT composite membranes with a thickness of 50 μm significantly decreases to 1.6×10^{-7} cm^3 cm/ cm^2 s by only 1 wt.% organo clay loading, which amounted to more than 90% reduction. Jung *et al.* (2003) prepared dodecylamine-exchanged montmorllonite (m-MMT) by a cation exchange reaction. The thermal resistance of Nafion®/MMT composite was found to be lower than that of the pristine Nafion®. Also the thermal resistance of 5 and 7 wt.% MMT was lower than that of 3 wt.% MMT. On the other hand, Nafion®/MMT displayed higher thermal resistance than that of pristine Nafion®. The thermal resistance of the Nafion®/m-MMT nanocomposite was also increased slightly with increasing the contents of m-MMT in the composite membrane. The methanol permeability of pristine Nafion® was found to be 0.13 mol/l at 1 h. By adding MMT and m-MMT, the methanol permeability decreased to 0.045 and 0.042 mol/l at 1 h, respectively. The proton conductivity of Nafion®/MMT was found to be 8.9×10^{-2}, 7×10^{-2}, 7.2×10^{-2} and 6.7×10^{-2} S/cm at 110°C for a content of MMT of 0, 3, 5 and 7 wt.%, respectively. However, the conductivity of Nafion®/m-MMT was found to be 8.9×10^{-2}, 7.72×10^{-2}, 7.57×10^{-2} and 7.4×10^{-2} S/cm at 110°C for a content of m-MMT of 0, 3, 5 and 7 wt.%, respectively. In general the proton conductivity of the composite membranes decreased slightly with increasing the contents of MMT and m-MMT and lower than pristine Nafion®. Pristine Nafion® performances were 385, 410 and 138.1 mA/cm^2 (at a potential of 0.4 V) at 90, 110 and 125°C, respectively. For Nafion®/3 wt.% MMT performances were 370, 452.6 and 282.86 mA/cm^2 (at a potential of 0.4 V) at 90, 110 and 125°C, respectively. Finally, for Nafion®/3 wt.% m-MMT performances were 367.1, 440 and 290 mA/cm^2 (at a potential of 0.4 V) at 90, 110 and 125°C, respectively. Gosalawit *et al.* (2008) prepared sulfonated montmorillonite (SMMT) with SPEEK. It was found that the inorganic aggregation in SPEEK increased with SMMT loading. The stability in water and in methanol aqueous solution as well as the mechanical stability were enhanced with SMMT loading. Whereas thermal stability improvement did not exist significantly. The methanol permeability was reduced when the SMMT loading increased. The proton conductivity was improved with the incorporation of SMMT. SMMT/SPEEK nanocomposite membranes showed significant cell performance for DMFC as compared to pristine SPEEK and Nafion® 117 membranes.

6.8 Organic/zeolites nanocomposite membranes

Mordenite crystals dispersed in poly(acrylic acid) was prepared by Rao *et al.* (1994) and Libby *et al.* (2001), while mordenite dispersed in Nafion® was prepared by Arimura *et al.* (1999). These membranes displayed proton conductivity about two orders of magnitude lower than that of pristine Nafion®. A study of using ZSM5 as a filler in Nafion® was study by Byun *et al.* (2006). These composite membranes show higher water uptake than Nafion® 115, while methanol permeability has decrease with increasing zeolite contents. The selectivity of nanocomposite membranes was higher than that of Nafion® 115. Zeolite beta was incorporated in Nafion® (Holmberg *et al.*, 2008) and Chitosan (Y. Wang *et al.*, 2008) membranes. Zeolite beta/ Nafion® nanocomposite membranes with loading of 2.5 and 5 wt.% posses proton conductivity/methanol permeability (selectivity) ratios as much as 93% higher than commercial Nafion® 117 at 21°C, and 63% higher at 80°C. These composite membranes outperform Nafion® 117 in DMFC (Holmberg *et al.*, 2008). The incorporation of zeolite beta in Chitosan reduces the methanol permeability. Furthermore, zeolite beta was sulfonated, therefore the methanol permeability was further reduced as a result of the enhanced interfacial interaction between zeolite beta and Chitosan matrix. The cell performance of the composite membrane were comparable to Nafion® 117 at low methanol concentration (2M) and much better at high methanol concentration (12M) (Y. Wang *et al.*, 2008). Other zeolite based nanocomposite membranes investigated are MCM-41 (Bello *et al.*, 2008; Gomes *et al.*, 2008; Karthikeyan *et al.*, 2005; Marschall *et al.*, 2007), zeolite Y (Ahmad *et al.*, 2006), zeolite BEA (Holmberg *et al.*, 2005) and Chabazite and clinoptilolite (Tricoli & Nannetti, 2003). These membranes can maintained the proton conductivity at temperature above 100°C, also methanol permeability is reduced by incorporating these zeolites.

7. Conclusion

High temperature operating fuel cell enhanced the performance, especially for methanol, DME and ethanol fuel cells. Incorporating nano inorganic materials in the organic matrix has several advantages, namely increase membrane thermal and mechanical stabilities, increase the working temperature and water retention. Some inorganic fillers are proton conductor which can increase the conductivity of the composite membrane or at least keep the same conductivity of the pristine organic membrane. Fuel crossover also can be reduced by incorporating these inorganic nano materials.

From all the investigated inorganic fillers, the nano size played a major role to enhance the compatibility and the interaction of the inorganic fillers with the polymeric matrix. Also the optimal inorganic loading was found to be around 5 wt.%, with the majority around 3 wt.%.

However, for commercialization of these nanocomposite membranes, more R&D needed to be done, which include:

- Systematic study of these nanocomposite, especially inorganic loading;
- The interaction between the nano-inorganic materials with the organic matrix need to be understood;
- High temperature fuel cell performance needed to be done with very little or no humidification and no pressure;
- Long term fuel cell performance;
- Long term membrane stability (thermal and mechanical) and also long term membrane leaching; and
- A comprehensive comparative study between all the investigated inorganic materials and the interaction with different organic materials.

8. References

Adamson, K.-A. & Pearson, P. (2000). *J. Power Sources,* 86, pp. 548-555

Adjemian, K.T.; Lee, S.J.; Srinivasan, S.; Benziger, J. & Bocarsly, A.B. (2002a). *J. Electrochem. Soc.,* 149, pp. A256-A261

Adjemian, K.T.; Srinivasan, S.; Benziger, J. & Bocarsly, A.B. (2002b). *J. Power Sources,* 109, pp. 356-364

Agrell, J.; Hsselbo, K.; Jansson, K.; Jaras, S.G. & Boutonnet, M. (2001). *Appl. Catal. A,* 211, pp. 239-250

Aharoni, S.M. & Litt, M.H. (1974). *J. Polym. Sci., Polym. Chem. Ed.,* 12, pp. 639-650

Ahmad, M.I.; Zaidi, S.M.J. & Rahman, S.U. (2006). *Desalination,* 193, pp. 387-397

Alberti, G.; Casiola, M.; Costantino, U.; Levi, G. & Ricciard, G. (1978). *J. Inorg. Nucl. Chem.,* 40, pp. 533-537

Alberti, G.; Casciola, M.; Massinelli, L. & Bauer, B. (2001). *J. Membr. Sci.,* 185, pp. 73-81

Alberti, G.; Casciola, M.; D`Alessandro, E. & Pica, M. (2004). *J. Mater. Chem.,* 14, pp. 1910-1914

Alberti, G.; Casciola, M.; Donnadio, A.; Piaggio, P.; Pica, M. & Sisani, M. (2005a). *Solid State Ionics,* 176, pp. 2893-2898

Alberti, G.; Pica, M. & Tarpanelli, T. (2005b). PCT Patent WO 2005/105667 A1

Alberti, G.; Casiola, M.; Capitani, D.; Donnadio, A.; Narducci, R.; Pica, M. & Sganappa, M. (2007). *Electrochim. Acta,* 52, pp. 8125-8132

Allcock, H.R.; Hofmann, M.A.; Ambler, C.M. & Morford, R.V. (2002a). *Macromolecules,* 35, pp. 3484-3489

Allcock, H.R.; Hofmann, M.A.; Ambler, C.M.; Lvov, S.N.; Zhou, X.Y.; Chalkova, E. & Weston, J. (2002b). *J. Membr. Sci.,* 201, pp. 47-54

Amphlett, J.C.; Evans, M.J.; Mann, R.F. & Weir, R.D. (1985). *Can. J. Chem. Eng.,* 63, pp. 605-611

Amphlett, J.C.; Peppley, B.A.; Halliop, E. & Sadiq, A. (2001). *J. Power Sources,* 96, pp. 204-213

Anantaraman, A.V. & Gardner, C.L. (1996). *J. Electroanal. Chem.,* 414, pp. 115-120

Antonucci, P.L.; Aricò, A.S.; Cretì, P.; Ramunni, E. & Antonucci, V. (1999). *Solid State Ionics,* 125, pp. 431-437

Appleby, A.J. & Yeager, E.B. (1986a). In: *Assessment of Research Needs for Advanced Fuel Cells,* Penner, S.S. (Ed.), Chapter 4, Pergamon, New York

Appleby, A.J. & Yeager, E.B. (1986b). *Energy,* 11, pp. 137-151

Argyropoulos, A.; Scott, K. & Taama, W.M. (1999a). *J. Appl. Electrochem.,* 29, pp. 661-669

Argyropoulos, P.; Scott, K. & Taama, W.M. (1999b). *Electrochim. Acta,* 44, pp. 3575-3584

Aricò, A.S.; Cretì, P.; Antonucci, P.L. & Antonucci, V. (1998). *Electrochem. Solid State Lett.,* 1, pp. 66-68

Aricò, A.S.; Baglio, V.; Di Blasi, A. & Antonucci, V. (2003a). *Electrochem. Communications,* 5, pp. 882-866

Aricò, A.S.; Baglio, V.; Di Blasi, A.; Cretì, P.; Antonucci, P.L. & Antonucci, V. (2003b). *Solid State Ionics,* 161, pp. 251-265

Aricò, A.S.; Baglio, V.; Di Blasi, A.; Modica, E.; Antonucci, P.L. & Antonucci, V. (2004). *J. Power Sources,* 128, pp. 113-118

Arimura, T.; Ostrovskii, D.; Okada, T. & Xie, G. (1999). *Solid State Ionics,* 118, pp. 1-10

Baglio, V.; Arico, A.S.; Di Blasi, A.; Antonucci, V.; Antonucci, P.L.; Licoccia, S.; Traversa, E. & Serraino Fiory, F. (2005). *Electrochim. Acta,* 50, pp. 1241-1246

Bailly, C.; Williams, D.J.; Karasz, F.E. & MacKnight, W.J. (1987). *Polymer,* 28, pp. 1009-1016

Baradie, B.; Dodelet, J.P. & Guay, D. (2000). *J. Electroanal. Chem.,* 489, pp. 101-105

Baradie, B.; Poinsignon, C.; Sanchez, J.Y.; Piffard, Y.; Vitter, G.; Bestaoui, N.; Foscallo, D.; Denoyelle, A.; Delabouglise, D. & Vaujany, M. (1998). *J. Power Sources*, 74, pp. 8-16
Bauer, B.; Menzel, Th. & Kehl, P. (1994). *EP* 0 645 175 A1
Bauer, B.; Rafler, G. & Ulrich, H.-H. (1995). *Ger. Patent* DE 195 38 025
Bauer, B.; Jones, D.J.; Rozière, J.; Tchicaya, L.; Alberti, G.; Casciola, M.; Massinelli, L.; Peraio, A.; Besse, S. & Ramunni, E. (2000). *J. New. Mater. Electrochem Systems*, 3, pp. 93-98
Bauer, F. & Willert-Porada, M. (2004). *J. Membr. Sci.*, 233, pp. 141-149
Bauer, F. & Willert-Porada, M. (2005). *J. Power Sources*, 145, pp. 101-107
Bauer, F. & Willert-Porada, M. (2006a). *Fuel Cells*, 6, pp. 261-269
Bauer, F. & Willert-Porada, M. (2006b). *Solid State Ionics*, 177, pp. 2391-2396
Bello, M.; Javaid Zaidi, S.M. & Rahman, S.U. (2008). *J. Membr. Sci.*, 322, pp. 218-224
Belyakov, V.N. & Linkov, V.M. (1999). *US Patent* 5,932,361
Bernardi, D.M. & Verbrugge, M.W. (1991). *AIChE Journal*, 37, pp. 1151-1163
Bi, C.; Zhang, H.; Zhang, Y.; Zhu, X.; Ma, Y.; Dai, H. & Xiao, S. (2008). *J. Power Sources*, 184, pp. 197-203
Boffito, G. (1999). *US Patent* 5,976,723
Bonnet, B.; Jones, D.J.; Rozière, J.; Tchicaya, L.; Alberti, G.; Casciola, M.; Massinelli, L.; Bauer, D.; Peraio, A. & Ramunni, E. (2000). *J. New Mater. Electrochem. Systems*, 3, pp. 87-92
Brack, H.P.; Wyler, M.; Peter, G. & Scherer, G.G. (2003). *J. Membr. Sci.*, 214, pp. 1-19
Breen, J.P. & Ross, J.R.H. (1999). *Catalysis Today*, 51, pp. 521-533
Büchi, F.N.; Gupta, B.; Haas, O. & Scherer, G.G. (1995a). *Electrochim. Acta*, 40, pp. 345-353
Büchi, F.N.; Gupta, B.; Haas, O. & Scherer, G.G. (1995b). *J. Electrochem. Soc.*, 142, pp. 3044-3048
Byun, S.C.; Jeung, Y.J.; Park, J.W.; Kim, S.D.; Ha, H.Y. & Kim, W.J. (2006). *Solid State Ionics*, 177, pp. 3233-3243
Carrette, L.; Friedrich, K.A. & Stimming, U. (2001). *Fuel Cells*, 1, pp. 5-39
Casiola, M.; Alberti, G.; Ciarletta, A.; Cruccolini, A.; Piaggio, P. & Pica, M. (2005). *Solid State Ionics*, 176, pp. 2985-2989
Casiola, M.; Alberti, G.; Sganappa, M. & Narducci, R. (2006). *J Power Sources*, 162, pp. 141-145
Casiola, M.; Capitani, D.; Comite, A.; Donnadio, A.; Frittella, V.; Pica, M.; Sganappa, M. & Varzi, A. (2008). *Fuel Cells*, 8, pp. 217-224
Chang, H.Y. & Lin, C.W. (2003). *J. Membr. Sci.*, 218, pp. 295-306
Chang, J.; Park, J.H.; Park, G.; Kim, C. & Park, O. (2003). *J. Power Sources*, 124, pp. 18-25
Choi, W.C.; Kim, J.D. & Woo, S.I. (2001). *J. Power Sources*, 96, pp. 411-414
Chu, D. & Gilman, S. (1994). *J. Electrochem. Soc.*, 141, pp. 1770-1773
Chu, D. & Jiang, R. (1999). *J. Power Sources*, 80, pp. 226-234
Colbow, K.M.; Zhang, J. & Wilkinson, D.P. (2000). *J. Electrochem. Soc.*, 147, pp. 4058-4060
Colicchio, I.; Wen, F.; Keul, H.; Simon, U. & Moeller, M. (2009). *J. Membr. Sci.*, 326, pp. 45-57
Cornet, N.; Diat, O.; Gebel, G.; Jousse, F.; Marsaq, D.; Mercier, R. & Pineri, M. (2000). *J. New Mater. Electrochem. Systems*, 3, pp. 33-42
Costamagna, P.; Yang, C.; Bocarsly, A.B. & Srinivasan, S. (2002). *Electrochim. Acta*, 47, pp. 1023-1033
Cubiero, M.L. & Fierro, J.L.G. (1998). *J. Catalysis*, 179, pp. 150-162
Cui, W.; Kerres, J. & Eigenberger, G. (1998). *Sep. Purif. Technol.*, 14, pp. 145-154
D`Alelio, G. (1944). *US Patent* 2,340,111
Deng, Q.; Moore, R.B. & Mauritz, K.A. (1998). *J. Appl. Polym. Sci.*, 68, pp. 747-763
Di Vona, L.M.; Ahmed, Z.; Bellitto, S.; Lenci, A.; Traversa, E. & Licoccia, S. (2007). *J. Membr. Sci.*, 296, pp. 156-161

Dimitrova, P.; Friedrich, K.A.; Stimming, U. & Vogt, B. (2002a). *Solid State Ionics*, 150, pp. 115-122

Dimitrova, P.; Friedrich, K.A.; Vogt, B. & Stimming, U. (2002b). *J. Electroanal. Chem.*, 532, pp. 75-83

Dohle, H.; Divisek, J.; Mergel, J.; Oetjen, H.F.; Zingler, C. & Stolten, D. (2002). *J. Power Sources*, 105, pp. 274-282

Duesterwald, H.G.; Höhlein, B.; Kraut, H.; Meusinger, J.; Peters, R. & Stimming, U. (1997). *Chem. Eng. Technol.*, 20, pp. 617-623

Edwards, N.; Ellis, S.R.; Frost, J.C.; Golunski, S.E.; van Keulen, A.N.J.; Lindewald, N.G. & Reinkingh, J.G. (1998). *J. Power Sources*, 71, pp. 123-128

Emonts, B.; Hansen, J.B.; Jørgensen, S.L.; Höhlein, B. & Peters, R. (1998). *J. Power Sources*, 71, pp. 288-293

Fan, J.; Guo, X.; Harada, S.; Watari, T.; Tanaka, K.; Kita, H. & Okamoto, K. (2002). *Macromolecules*, 35, pp. 9022-9028

Faungnawakij, K. & Viriya-empikul, N. (2010). *Appl. Cat. A*, 382, pp. 21-27

Faungnawakij, K.; Shimoda, N.; Viriya-empikul, N.; Kikuchi, R. & Eguchi, K. (2010). *Appl. Cat. B*, 97, pp. 21-27

Faure, S.; Mercier, R.; Aldebert, P.; Pineri, M. & Sillion, B. (1996). *Fr. Patent* 96 05707

Faure, S.; Cornet, N.; Gebel, G.; Mercier, R.; Pineri, M. & Sillion, B. (1997). In: *Proceeding of the Second International Symposium on New Materials for Fuel Cell and Modern Battery Systems*, Savadogo, O. & Roberge, P.R. (Eds.), pp. 818, Montréal, Canada, July 6-10, 1997

Ferrell, J.R.; Kuo, M.C. & Herring, A.M. (2010). *J. Power Sources*, 195, pp. 39-45

Finsterwalder, F. & Hambitzer, G. (2001). *J. Membr. Sci.*, 185, pp. 105-124

Fontanella, J.J.; Wintersgill, M.C.; Chen, R.S.; Wu, Y. & Greenbaum, S.G. (1995). *Electrochim. Acta*, 40, pp. 2321-2326

Fujiwara, N.; Friedrich, K.A. & Stimming, U. (1999). *J. Electroanal. Chem.*, 472, pp. 120-125

Gaowen, Z. & Zhentao, Z. (2005). *J. Membr. Sci.*, 261, pp. 107-113

Gavach, C.; Pamboutzoglou, G.; Nedyalkov, M. & Pourcelly, G. (1989). *J. Membr. Sci.*, 45, pp. 37-53

Gebel, G.; Aldebert, P. & Pineri, M. (1993). *Polymer*, 34, pp. 333-339

Genies, C.; Mercier, R.; Sillion, B.; Cornet, N.; Gebel, G. & Pineri, M. (2001). *Polymer*, 42, pp. 359-373

Giordano, N.; Staiti, P.; Hocevar, S. & Arico, A.S. (1996). *Electrochim. Acta*, 41, pp. 397-403

Glazebrook, R.W. (1982). *J. Power Sources*, 7, pp. 215-256

Glipa, X.; El Haddad, M.; Jones, D.J. & Rozière, J. (1997). *Solid State Ionics*, 97, pp. 323-331

Gloaguen, F.; Convert, P.; Gamburzev, S.; Velev, O.A. & Srinivasan, S. (1998). *Electrochim. Acta*, 43, pp. 3767-3772

Gode, P.; Ihonen, J.; Strandroth, A.; Ericson, H.; Lindbergh, G.; Paronen, M.; Sundholm, F.; Sundholm, G. & Walsby, N. (2003). *Fuel Cells*, 3, pp. 21-27

Gomes, D.; Marschall, R.; Nunes, S.P. & Wark, M. (2008). *J. Membr. Sci.*, 322, pp. 406-415

Gong, X.; Bandis, A.; Tao, A.; Meresi, G.; Wang, Y.; Inglefield, P.T.; Jones, A.A. & Wen, W.-Y. (2001). *Polymer*, 42, pp. 6485-6492

Goodstein, E.S. (1999). *Economics and Environment*, Upper Saddle River, NJ, Prentice Hall

Gosalawit, R.; Chirachanchai, S.; Shishatskiy, S. & Nunes, S.P. (2008). *J. Membr. Sci.*, 323, pp. 337-346

Gottesfeld, S. & Zawodzinski, T.A. (1997). In: *Advances in Electrochemical Science and Engineering,* Alkire, R.C.; Gerischer, H.; Kolb, D.M. & Tobias, C.W. (Eds.), Vol. 5, Chapter 4, Wiley-VCH, Weinheim

Grot, W. (1978). *Chem. Ing. Tech.,* 50, pp. 299-301; (1975). *Chem. Ing. Tech.,* 47, pp. 617; (1972). *Chem. Ing. Tech.,* 44, pp. 167-169

Grot, W.G. & Rajendran, G. (1999). *US Patent* 5,919,583

Guo, Q.; Pintauro, P.N.; Tang, H. & O'Connor, S. (1999). *J. Membr. Sci.,* 154, pp. 175-181

Gupta, B.; Büchi, F.N.; Scherer, G.G. & Chapiró, A. (1993). *Solid State Ionics,* 61, pp. 213-218

Gupta, B.; Büchi, F.N. & Scherer, G.G. (1994). *J. Polym. Sci. A: Polym. Chem.,* 32, pp. 1931-1938

Gupta, B. & Scherer, G.G. (1994). *Chimia,* 48, pp. 127-137

Hacker, V. & Kordesch, K. (2003). In: *Handbook of Fuel Cells,* Vielstich, W.; Lamm, A. & Gasteiger, H.A. (Eds.), Vol 3, Part 1, Chapter 10, pp. 121-127, Wiley, Chichester, England

Hara, S. & Miyayama, M. (2004). *Solid State Ionics,* 168, pp. 111-116

Hatanaka, T.; Hasegawa, N.; Kamiya, A.; Kawasumi, M.; Morimoto, Y. & Kawahara, K. (2002). *Fuel,* 81, pp. 2173-2176

Hejze, T.; Gollas, B.R.; Sauerbrey, R.K.; Schmied, M.; Hofer. F. & Besenhard, J.O. (2005). *J. Power Sources,* 140, pp. 21-27

Helen, M.; Viswanathan, B. & Srinivasa Murthy, S. (2006). *J. Power Sources,* 163, pp. 433-439

Helen, M.; Viswanathan, B. & Srinivasa Murthy, S. (2007). *J. Membr. Sci.,* 292, pp. 98-105

Hill, M.L.; Kim, Y.S.; Einsla, B.R. & McGrath, J.E. (2006). *J. Membr. Sci.,* 283, pp. 102-108

Hoel, D. & Grunwald, E. (1977). *J. Phys. Chem.,* 81, pp. 2135-2136

Hogarth, M.; Christensen, P.; Hamnett, A. & Shukla, A. (1997). *J. Power Sources,* 69, pp. 113-124

Hogarth, M.P. & Ralph, T.R. (2002). *Platinum Metals Rev.,* 46, pp. 146-164

Höhlein, B.; Boe, M.; Bøgild-Hansen, J.; Bröckerhoff, P.; Colsman, G.; Emonts, B.; Menzer, R. & Riedel, E. (1996). *J. Power Sources,* 61, pp. 143-147

Holmberg, B.A.; Hwang, S.J.; Davis, M.E. & Yan, Y. (2005). *Microp. Mesop. Mat.,* 80, pp. 347-356

Holmberg, B.A.; Wang, X. & Yan, Y. (2008). *J. Membr. Sci.,* 320, pp. 86-92

Honma, I.; Nishikawa, O.; Sugimoto, T.; Nomura, S. & Nakajima, H. (2002). *Fuel Cells,* 2, pp. 52-58

Hou, H.; Sun, G.; Wu, Z.; Jin, W. & Xin, Q. (2008). *Int. J. Hydrogen Energy,* 33, pp. 3402-3409

Hubner, G. & Roduner, E. (1999). *J. Mater. Chem.,* 9, pp. 409-418

Ioannides, T. & Neophytides, S. (2000). *J. Power Sources,* 91, pp. 150-156

Jiang, C.J.; Trimm, D.L.; Wainwright, M.S. & Cant, N.W. (1993). *Appl. Cat. A,* 93, pp. 245-255

Jiang, R.; Kunz, H.R. & Fenton, J.M. (2006a). *Electrochim. Acta.* 51, pp. 5596-5605

Jiang, R.C.; Kunz, H.R. & Fenton, J.M. (2006b). *J. Membr. Sci.,* 272, pp. 116-124

Jian-hua, T.; Peng-fei, G.; Zhi-yuan, Z.; Wen-hui, L. & Zhong-qiang, S. (2008). *Int. J. Hydrogen Energy,* 33, pp. 5686-5690

Johnson, B.C.; Yilgor, I.; Tran, C.; Iqbal, M.; Wightman, J.P.; Llyod, D.R. & McGrath, J.E. (1984). *J. Polym. Sci.,* 22, pp. 721-737

Jones, D.J. & Rozière, J. (2001). *J. Membr. Sci.,* 185, pp. 41-58

Jones, D.J. & Rozière, J. (2001). In: *Handbook of Fuel Cells,* Vielstich, W.; Lamm, A. & Gasteiger, H.A. (Eds.), Vol 3, Part 1, Chapter 35, pp. 447-455, Wiley, Chichester, England

Jörissen, L.; Gogel, V.; Kerres, J. & Garche, J. (2002). *J. Power Sources,* 105, pp. 267-273

Jung, D.H.; Cho, S.Y.; Peck, D.H.; Shin, D.R. & Kim, J.S. (2002). *J. Power Sources,* 106, pp. 173-177

Jung, D.H.; Cho, S.Y.; Peck, D.H.; Shin, D.R. & Kim, J.S. (2003). *J. Power Sources,* 118, pp. 205-211

Karthikeyan, C.S.; Nunes, S.P.; Prado, L.A.S.A.; Ponce, M.L.; Silva, H.; Ruffmann, B & Schulte, K. (2005). *J. Membr. Sci.*, 254, pp. 139-146

Kerangueven, G.; Coutanceau, C.; Sibert, E.; Hahn, F.; Leger, J.M. & Lamy, C. (2006). *J. Appl. Electrochem.*, 36, pp. 441-448

Kerres, J.; Cui, W. & Reichle, S. (1996). *J. Polym. Sci.: Part A: Polym. Chem.*, 34, pp. 2421-2438

Kerres, J.; Cui, W. & Schnurnberger, W. (1997). *D. Patent* 19622337.7; (1997). *Fr. Patent* F. 9706706; (1997). *US Patent* 08/868943

Kerres, J. ; Cui, W.; Disson, R. & Neubrand, W. (1998a). *J. Membr. Sci.*, 139, pp. 211-225

Kerres, J.; Cui, W. & Junginger, M. (1998b). *J. Membr. Sci.*, 139, pp. 227-241

Kerres, J.; Zhang, W. & Cui, W. (1998c). *J. Polym. Sci.: Part A: Polym. Chem.*, 36, pp. 1441-1448

Kerres, J.; Ullrich, A.; Meier, F. & Häring, T. (1999). *Solid State Ionics,* 125, pp. 243-249

Kerres, J.; Ullrich, A.; Häring, T.; Baldauf, M.; Gebhardt, U. & Preidel, W. (2000). *J. New Mater. Electrochem. Systems,* 3, pp. 229-239

Kerres, J.A. (2001). *J. Membr. Sci.,* 185, pp. 3-27

Kim, D.S.; Park, H.B.; Rhim, J.W. & Lee, Y.M. (2004). *J. Membr. Sci.,* 240, pp. 37-48

Kim, H.J.; Shul, Y.G. & Han, H. (2006). *J. Power Sources,* 158, pp. 137-142

Kim, Y.J.; Choi, W.C.; Woo, S.I. & Hong, W.H. (2004). *Electrochim. Acta,* 49, pp. 3227-3234

Kim, Y.S.; Dong, L.; Hickner, M.A.; Pivovar, B.S. & McGrath, J.E. (2003). *Polymer,* 44, pp. 5729-5736

Kim, Y.T.; Song, M.K.; Kim, K.H.; Park, S.B.; Min, S.K. & Rhee, H.W. (2004). *Electrochim. Acta,* 50, pp. 645-648

Kobayashi, T.; Rikukawa, M.; Sanui, K. & Ogata, N. (1998). *Solid State Ionics,* 106, pp. 219-225

Kordesch, K. & Simader, G. (1996). In: *Fuel cells and their Applications*, VCH, Weinheim

Kordesch, K.V. & Simader, G.R. (1995). *Chem. Rev.*, 95, pp. 191-207

Kreuer, K.D. (1997). *Solid State Ionics*, 97, pp. 1-15

Kreuer, K.D. (2001). *J. Membr. Sci.,* 185, pp. 29-39

Krishnan, P.; Park, J.S.; Yang, T.H.; Lee, W.Y. & Kim, C.S. (2006). *J. Power Sources,* 163, pp. 2-8

Kuan, H.C.; Wu, C.S.; Chen, C.Y.; Yu, Z.Z.; Dasari, A. & Mai, Y.W. (2006). *Electrochem. Solid State Lett.*, 9, pp. A76-A79

Lamy, C.; Belgsir, E.M. & Léger, J.-M. (2001). *J. Appl. Electrochem.*, 31, pp. 799-809

Larminie, J. & Dicks, A. (2000). In: *Fuel Cell Systems Explained,* Wiley, Chichester

Ledesma, C. & Llorca, J. (2009). *Chem. Eng. J.*, 154, pp. 281-286

Ledjeff-Hey, K.; Formanski, V.; Kalk, Th. & Roes, J. (1998). *J. Power Sources,* 71, pp. 199-207

Lee, H.K.; Kim, J.I.; Park, J.H. & Lee, T.H. (2004). *Electrochim. Acta,* 50, pp. 761-768

Lee, W.; Shibasaki, A.; Saito, K.; Sugita, K.; Okuyama, K. & Sugo, T. (1996). *J. Electrochem. Soc.*, 143, pp. 2795-2798

Léger, J.-M. (2001). *J. Appl. Electrochem.,* 31, pp. 767-771

Lehtinen, T.; Sundholm, G.; Holmberg, S.; Sundholm, F.; Björnbom, P. & Bursell, M. (1998). *Electrochim. Acta,* 43, pp. 1881-1890

Li, C.; Sun, G.; Ren, S.; Liu, J.; Wang, Q.; Wu, Z.; Sun, H. & Jin, W. (2006). *J. Membr. Sci.,* 272, pp. 50-57

Liang, Z.X.; Zhao, T.S. & Prabhuram, J. (2006). *J. Membr. Sci.,* 283, pp. 219-224

Libby, B.; Smyrl, W.H. & Cussler, E.L. (2001). *Electrochem. Solid State Lett.*, 4, pp. A197-A199

Lim, C. & Wang, C.Y. (2003). *J. Power Sources,* 113, pp. 145-150

Lin, C.W.; Thangamuthu, R. & Chang, P.H. (2005). *J. Membr. Sci.,* 254, pp. 197-205

Lin, Y.F.; Yen, C.Y.; Ma, C.C.M.; Liao, S.H.; Lee, C.H.; Hsiao, Y.H. & Lin, H.P. (2007). *J. Power Sources,* 171, pp. 388-395

Linkous, C.A.; Anderson, H.R.; Kopitzke, R.W. & Nelson, G.L. (1998). *Int. J. Hydrogen Energy*, 23, pp. 525-529

Liu, Z.; Guo, B.; Huang, J.; Hong, L.; Han, M. & Ming Gan, L. (2006). *J. Power Sources*, 157, pp. 207-211

Livingston, D.I.; Kamath, P.M. & Corley, R.S. (1956). *J. Polym Sci.*, 20, pp. 485-490

Lufrano, F.; Squadrito, G.; Patti, A. & Passalacqua, E. (2000). *J. Appl. Polym. Sci.*, 77, pp. 1250-1257

Lufrano, F.; Gatto, I.; Staiti, P.; Antonucci, V. & Passalacqua, E. (2001). *Solid State Ionics*, 145, pp. 47-51

Ma, L.; Jiang, C.; Adesina, A.A.; Trimm, D.L. & Wainwright, M.S. (1996). *Chem. Eng. J.*, 62, pp. 103-111

Ma, Y.-L.; Wainright, J.S.; Litt, M.H. & Savinell, R.F. (2004). *J. Electrochem. Soc.*, 151, pp. A8-A16

Ma, Z.Q.; Cheng, P. & Zhao, T.S. (2003). *J. Membr. Sci.*, 215, pp. 327-336

Marschall, R.; Bannat, I.; Caro, J. & Wark, M. (2007). *Microp. Mesop. Mat.*, 99, pp. 190-196

Mauritz, K.A.; Stefanithis, I.D.; Davis, S.V.; Scheetz, R.W.; Pope, R.K.; Wilkes, G.L. & Huang, H.-H. (1995). *J. Appl. Polym. Sci.*, 55, pp. 181-190

Mauritz, K.A. (1998). *Mater. Sci. Eng. C*, 6, pp. 121-133

Mench, M.; Boslet, S.; Thynell, S.; Scott, J. & Wang, C.Y. (2001). In: *Direct Methanol Fuel Cells*, Narayanan, S.; Zawodzinski, T. & Gottesfeld, S. (Eds.), PV 2001-4, pp. 241, *The Electrochemical Society Proceeding Series*, Pennington, NJ

Mench, M.M.; Chang, H.M. & Wang, C.Y. (2004). *J. Electrochem. Soc.*, 151, pp. A144-A150

Mench, M.W. & Wang, C.Y. (2003). *J. Electrochem. Soc.*, 150, pp. A79-A85

Mokrani, T. & Scurrell, M. (2009). *Cat. Rev., Sci. Eng.*, 51, pp. 1-145

Mitov, S.; Vogel, B.; Roduner, E.; Zhang, H.; Zhu, X.; Gogel. V.; Jorissen, L.; Hein, M.; Xing, D.; Schonberger, F. & Kerres, J. (2006). *Fuel Cells*, 6, pp. 413-424

Mizsey, P.; Newson, E.; Truong, T.-B. & Hottinger, P. (2001). *Appl. Cat. A*, 213, pp. 233-237

Mizutani, I.; Liu, Y.; Mitsushima, S.; Ota, K.I. & Kamiya, N. (2006). *J. Power Sources*, 156, pp. 183-189

Narayanan, S.R.; Kindler, A.; Jeffries-Nakamura, B.; Chun, W.; Frank, H.; Smart, M.; Valdez, T.I.; Surampudi, S.; Halpert, G.; Kosek, J. & Cropley, C. (1996). In: *Proceedings of 11th Annual Battery Conference on Applications and Advances*, pp. 113-122, Long Beach, Calif., Jan. 9-12, 1996

Nilsson, M.; Jozsa, P. & Pettersson, L.J. (2007). *Appl. Cat.* B, 76, pp. 42-50

Nilsson, M.; Jansson, K.; Jozsa, P. & Pettersson, L.J. (2009). *Appl. Cat. B*, 86, pp. 18-26

Nishiguchi, T.: Oka, K.; Matsumoto, T.; Kanai, H.; Utani, K. & Imamura, S. (2006). *Appl. Cat. A*, 301, pp. 66-74

Nolte, R.; Ledjeff, K.; Bauer, M. & Mulhaupt, R. (1993). *J. Membr. Sci.*, 83, pp. 211-220

Nordlund, J.; Roessler, A. & Lindbergh, G. (2002). *J. Appl. Electrochem.*, 32, pp. 259-265

Nunes, S.P.; Ruffmann, B.; Rikowski, E.; Vetter, S. & Richau, K. (2002). *J. Membr. Sci.*, 203, pp. 215-225

Olah, G.A.; Goeppert, A. & Surya Prakash, G.K. (2009). *Beyond oil and Gas: The Methanol Economy*, Wiley-VCH, Weinheim

Papapolymerou, G. & Bontozoglou, V. (1997). *J. Mol. Catal. A*, 120, pp. 165-171

Parthasarathy, A.; Martin, C.R. & Srinivasan, S. (1991). *J. Electrochem Soc.*, 138, pp. 916-920

Paulus, U.A.; Schmidt, T.J.; Gasteiger, H.A. & Behm, R.J. (2001). *J. Electroanal. Chem.*, 495, pp. 134-145

Powers, E.D. & Serad, G.A. (1986). In: *High Performance Polymers: Their Origin and Development,* Seymour, R.B. & Kirschenbaum, G.S. (Eds.), pp. 355, Elsevier, Amsterdam
Pu, C.; Huang, W.; Ley, K.L. & Smotkin, E.S. (1995). *J. Electrochem. Soc.,* 142, pp. L119-L120
Qi, Z. & Kaufman, A. (2002). *J. Power Sources,* 110, pp. 177-185
Raadschelders, J.W. & Jansen, T. (2001). *J. Power Sources,* 96, pp. 160-166
Ralph, T.R. & Hogarth, M.P. (2002). *Platinum Metals Rev.,* 46, pp. 117-135
Ralph, T.R. (1997). *Platinum Metals Rev.,* 41, pp. 102-113
Ramani, V.; Kunz, H.R. & Fenton, J.M. (2004). *J. Membr. Sci.,* 232, pp. 31-44
Ramani, V.; Kunz, H.R. & Fenton, J.M. (2005a). *Electrochim. Acta,* 50, pp. 1181-1187
Ramani, V.; Kunz, H.R. & Fenton, J.M. (2005b). *J. Membr. Sci.,* 266, pp. 110-114
Rao, N.; Anderson, T.P. & Ge, P. (1994). *Solid State Ionics,* 72, pp. 334-337
Ren, S.; Sun, G.; Li, C.; Song, S.; Xin, Q. & Yang, X. (2006). *J. Power Sources,* 157, pp. 724-726
Ren, X. & Gottesfeld, S. (2001). *J. Electrochem. Soc.,* 148, pp. A87-A93
Rosenblatt, E. & Cohn, J. (1952). *US Patent* 2,601,221
Ruffmann, B.; Silva, H.; Schulte, B. & Nunes, S.P. (2003). *Solid State Ionics,* 162-163, pp. 269 -275
Sacca, A.; Carbone, A.; Passalacqua, E.; D'Epifanio, A.; Licoccia, S.; Traversa, E.; Sala, E.; Traini, F. & Ornelas, R. (2005). *J. Power Sources,* 152, pp. 16-21
Sacca, A.; Gatto, I.; Carbone, A.; Pedicini, R. & Passalacqua, E. (2006). *J. Power Sources,* 163, pp. 47-51
Sambandam, S. & Ramani, V. (2007). *J. Power Sources,* 170, pp. 259-267
Samms, S.R.; Wasmus, S. & Savinell, R.F. (1996). *J. Electrochem. Soc.,* 143, pp. 1225-1232
Santarelli, M.G.L.; Calì, M. & Bertonasco, A. (2003). *Energy Conversion and Management,* 44, pp. 2353-2370
Savadogo, O. (1998). *J. New Mater. Electrochem. Systems,* 1, pp. 47-66
Savadogo, O. & Xing, B. (2000). *J. New Mater. Electrochem. Systems,* 3, pp. 343-347
Savadogo, O. (2004). *J. Power Sources,* 127, pp. 135-161
Schatter, M.J. (1983). In: *Fuel Cells,* Young, G.J. (Ed.), Vol. 2, pp. 290, Reinhold, New York
Scherer, G.G. (1990). *Ber. Bunsenges. Phys. Chem.,* 94, pp. 1008-1014
Schneller, A.; Ritter, H.; Ledjeff, K.; Nolte, R. & Thorwirth, R. (1993). *EP* 0574791 A2
Scott, K.; Taama, W.M. & Argyropoulos, P. (1998). *J. Appl. Electrochem.,* 28, pp. 1389-1397
Serov, A. & Kwak, C. (2009). *Appl. Cat.* B, 91, pp. 1-10
Shanmugam, S.; Viswanathan, B. & Varadarajan, T.K. (2006). *J. Membr. Sci.,* 275, pp. 105-109
Shao, Z.-G.; Joghee, P. & Hsing, I.-M. (2004). *J. Membr. Sci.,* 229, pp. 43-51
Shikada, T.; Asanuma, M. & Ikariya, T. (1991). *US Patent* 5,055,282
Shin, S.-J.; Lee, J.-K.; Ha, H.-Y.; Hong, S.-A.; Chun, H.-S. & Oh, I.-H. (2002). *J. Power Sources,* 106, pp. 146-152
Silva, V.S.; Ruffmann, B.; Silva, H.; Gallego, Y.A.; Mendes, A.; Madeira, L.M. & Nunes, S.P. (2005a). *J. Power Sources,* 140, pp. 34-40
Silva, V.S.; Schirmer, J.; Reissner, R.; Ruffmann, B.; Silva, H.; Mendes, A.; Madeira, L.M. & Nunes, S.P. (2005b). *J. Power Sources,* 140, pp. 41-49
Silva, V.S.; Ruffmann, B.; Vetter, S.; Mendes, A.; Madeira, L.M. & Nunes, S.P. (2005c). *Catalysis Today,* 104, pp. 205-212
Soczka-Guth, T.; Baurmeister, J.; Frank, G. & Knauf, R. (1999). *International Patent* WO 99/29763
Song, M.K.; Park, T.S.B.; Kim, Y.T. *et al.* (2004). *Electrochim. Acta,* 50, pp. 639-643
Staiti, P.; Aricò, A.S.; Baglio, V.; Lufrano, F.; Passalacqua, E. & Antonucci, V. (2001). *Solid State Ionics,* 145, pp. 101-107

Steck, A.E. (1995). Membrane materials in fuel cells, In: *Proceedings of the First International Symposium on New Materials for Fuel Cell Systems* 1, Savadogo, O.; Roberge, P.R. & Veziroglu, T.N. (Eds.), pp.74, Montréal, Quebec, Canada, July 9-13, 1995

Steck, A. & Stone, C. (1997). In: *Proceedings of the Second International Symposium on New Materials for Fuel Cell and Modern Battery Systems*, Savadogo, O. & Roberge, P.R. (Eds.), pp. 792-807, Montréal, Canada, July 6-10, 1997

Strickland, G. (1984). *Int. J. Hydrogen Energy*, 9, pp. 759-766

Su, Y.H.; Liu, Y.L.; Sun, Y.M.; Lai, J.Y.; Wang, D.M.; Gao, Y.; Liu, B. & Guiver, M.D. (2007). *J. Membr. Sci.*, 296, pp. 21-28

Sun, H.; Sun, G.; Wang, S.; Liu, J.; Zhao, X.; Wang, G.; Xu, H.; Hou, S. & Xin, Q. (2005). *J. Membr. Sci.*, 259, pp. 27-33

Sundholm, F. (1998). New Polymer Electrolytes for Low Temperature Fuel Cells. In: *Proceedings of the 9th International Conference on Solid State Protonic Conductors SSPC'98*, pp. 155-158, Extended Abstract Book, Bled, Slovenia, 17-21 August, 1998

Surampudi, S.; Narayanan, S.R.; Vamos, E.; Frank, H.; Halpert, G.; LaConti, A.; Kosek, J.; Surya Prakash, G.K. & Olah, G.A. (1994). *J. Power Sources*, 47, pp. 377-385

Takahashi, K.; Takezawa, N. & Kobayashi, H. (1982). *Appl. Cat.*, 2, pp. 363-366

Takeishi, K. & Suzuki, H. (2004). *Appl. Cat. A*, 260, pp. 111-117

Takezawa, N.; Kobayashi, H.; Kamegai, Y. & Shimokawabe, M. (1982). *Appl. Cat.*, 3, pp. 381-388

Tamura, K.; Tsukui, T.; Kamo, T. & Kudo, T. (1984). *Hitachi Hyoron*, 66, pp. 135-138

Tang, H.; Pintauro, P.N.; Guo, Q. & O'Connor, S. (1999). *J. Appl. Polym. Sci.*, 71, pp. 387-399

Tang, H.; Pan, M.; Jiang, S.; Wan, Z. & Yuan, R. (2005). *Colloids and Surfaces A: Physicochem. Eng. Aspects*, 262, pp. 65-70

Tay, S.W.; Zhang, X.; Liu, Z.; Hong, L. & Chan, S.H. (2008). *J. Membr. Sci.*, 321, pp. 139-145

Tazi, B. & Savadogo, O. (2000). *Electrochim. Acta*, 45, pp. 4329-4339

Tazi, B. & Savadogo, O. (2001). *J. New Mater. Electrochem. Systems*, 4, pp. 187-196

Tchicaya-Bouckary, L.; Jones, D.J. & Roziere, J. (2002). *Fuel Cells*, 2, pp. 40-45

Tian, H. & Savadogo, O. (2005). *Fuel Cells*, 5, pp. 375-382

Tricoli, V. & Nannetti, F. (2003). *Electrochim. Acta*, 48, pp. 2625-2633

Triphathi, B.P. & Shahi, V.K. (2007). *J. Colloid and Interface Science*, 316, pp. 612-621

Triphathi, B.P.; Kumar, M. & Shahi, V.K. (2009). *J. Membr. Sci.*, 327, pp. 145-154

Tung, S.P. & Hwang, B.J. (2007). *Fuel Cells*, 7, pp. 32-39

Uchida, M.; Aoyama, Y.; Eda, N. & Ohta, A. (1995). *J. Electrochem. Soc*, 142, pp. 463-468

Ueda, S.; Eguchi, M.; Uno, K.; Tsutsumi, Y. & Ogawa, N. (2006). *Solid State Ionics*, 177, pp. 2175-2178

Uma, T. & Nogami, M. (2007). *Fuel Cells*, 7, pp. 279-284

Velu, S.; Suzuki, K. & Osaki, T. (1999). *Catal. Lett.*, 62, pp. 159-167

Verma, L.K. (2000). *J. Power Sources*, 86, pp. 464-468

Wainright, J.S.; Wang, J.-T.; Weng, D.; Savinell, R.F. & Litt, M. (1995). *J. Electrochem. Soc.*, 142, pp. L121-L123

Wainright, J.S.; Savinell, R.F. & Litt, M.H. (1997). In: *Proceedings of the Second International Symposium on New Materials for Fuel Cell and Modern Battery Systems*, Savadogo, O. & Roberge, P.R. (Eds.), pp. 808-817, Montréal, Canada, July 6-10, 1997

Walker, M.; Baurngdrtner, K.-M.; Kaiser, M. *et al.*, (1999). *J. Appl. Polym. Sci.*, 74, pp. 67-73

Wang, J.; Wasmus, S. & Savinell, R.F. (1995). *J. Electrochem. Soc.*, 142, pp. 4218-4224

Wang, J.-T.; Savinell, R.F.; Wainright, J.; Litt, M. & Yu, H. (1996a). *Electrochim. Acta*, 41, pp. 193-197

Wang, J.-T.; Wainright, J.S.; Savinell, R.F. & Litt, M. (1996b). *J. Appl. Electrochem.*, 26, pp. 751-756

Wang, S.; Ishihara, T. & Takita, Y. (2002). *Appl. Cat. A,* 228, pp. 167-176
Wang, Y.; Yang, D.; Zheng, X.; Jiang, Z. & Li, J. (2008). *J. Power Sources,* 183, pp. 454-463
Wang, Z.H.; Wang, C.Y. & Chen, K.S. (2001). *J. Power Sources,* 94, pp. 40-50
Watanabe, M. (1995). *US Patent* 5,472,799
Watanabe, M.; Uchida, H.; Seki, Y.; Emori, M. & Stonchart, P. (1996). *J. Electrochem. Soc.,* 143, pp. 3847-3852
Watanabe, M.; Uchida, H. & Emori, M. (1998). *J. Electrochem. Soc.,* 145, pp. 1137-1141
Wei, J.; Stone, C. & Steck, A.E. (1995a). *US Patent* 5,422,411
Wei, J.; Stone, C. & Steck, A.E. (1995b). WO 95/08581, 30 March 1995
Williams, K.R. (1966). In: *An Introduction to Fuel Cells,* pp. 152, Elsevier
Wilson, M.S. & Gottesfeld, S. (1992). *J. Electrochem. Soc.,* 139, pp. L28-L30
Wilson, M.S. (1993). *US Patent* 5,211,984
Wilson, M.S.; Valerio, J.A. & Gottesfeld, S. (1995). *Electrochim. Acta,* 40, pp. 355-363
Woo, Y.; Oh, S.Y.; Kang, Y.S. & Jung, B. (2003). *J. Membr. Sci.,* 220, pp. 31-45
Wycisk, R. & Pintauro, P.N. (1996). *J. Membr. Sci.,* 119, pp. 155-160
Xing, B. & Savadogo, O. (1999). *J. New Mater. Electrochem. Systems,* 2, pp. 95-101
Xing, P.; Robertson, G.P.; Guiver, M.D.; Mikhailenko, S.D.; Wang, K. & Kaliaguine, S. (2004). *J. Membr. Sci.,* 229, pp. 95-106
Xu, K.; Lao, S.J.; Qin, H.Y.; Liu, B.H. & Li, Z.P. (2010). *J. Power Sources,* 195, pp. 5606-5609
Xu, W.; Liu, C.; Xue, X.; Su, Y.; Lv, Y.; Xing, W. & Lu, T. (2004). *Solid State Ionics,* 171, pp. 121-127
Yang, C.; Costamagna, P.; Srinivasan, S.; Benziger, J. & Bocarsly, A.B. (2001a). *J. Power Sources,* 103, pp. 1-9
Yang, C.; Srinivasan, S.; Aricò, A.S.; Cretì, P.; Baglio, V. & Antonucci, V. (2001b). *Electrochem. Solid State Lett.,* 4, pp. A31-A34
Yang, C.; Srinivasan, S.; Bocarsly, A.B.; Tulyani, S. & Benziger, J.B. (2004). *J. Membr. Sci.,* 237, pp. 145-161
Yen, S.-P.S.; Narayanan, S.R.; Halpert, G.; Graham, E. & Yavrouian, A. (1998). *US Patent* 5,795,496
Yi, J.S. & Nguyen, T.V. (1999). *J. Electrochem. Soc.,* 146, pp. 38-45
Yoon, S.R.; Hwang, G.H.; Cho, W.I.; Oh, I.H.; Hong, S.A. & Ha, H.Y. (2002). *J. Power Sources,* 106, pp. 215-223
Yoshida, N.; Ishisaki, T.; Watakabe, A. & Yoshitake, M. (1998). *Electrochim. Acta,* 43, pp. 3749-3754
Yu, R.H.; Choi, H.G. & Cho, S.M. (2005). *Electrochem. Commun.,* 7, pp. 1385-1388
Zaidi, S.M.J.; Mikhailenko, S.D.; Robertson, G.P.; Guiver, M.D. & Kaliaguine, S. (2000). *J. Membr. Sci.,* 173, pp. 17-34
Zawodzinski, T.A. Jr.; Neeman, M.; Sillerud, L.O. & Gottesfeld, S. (1991). *J. Phys. Chem.,* 95, pp. 6040-6044
Zawodzinski, T.A. Jr.; Springer, T.E.; Davey, J.; Jestel, R.; Lopez, C.; Valerio, J. & Gottesfeld, S. (1993). *J. Electrochem. Soc.,* 140, pp. 1981-1985
Zhai, Y.; Zhang, H.; Hu, J. & Yi, B. (2006). *J. Membr. Sci.,* 280, pp. 148-155
Zhang, Q.; Li, X.; Fujimoto, K. & Asami, K. (2005). *Appl. Cat. A,* 288, pp. 169-174
Zhang, W.; Tang, C.-M. & Kerres, J. (2001). *Sep. Purif. Technol.,* 22-23, pp. 209-221
Zhou, W.J.; Zhou, B.; Li, W.Z.; Zhou, Z.H.; Song, S.Q.; Sun, G.Q.; Xin, Q.; Douvartzides, S.; Goula, M. & Tsiakaras, P. (2004). *J. Power Sources,* 126, pp. 16-22
Zhou, X.; Weston, J.; Chalkova, E.; Hofmann, M.A.; Ambler, C.M.; Allcock, H.R. & Lvov, S.N. (2003). *Electrochim. Acta,* 48, pp. 2173-2180

21

Membrane Operations for Industrial Applications

Maria Giovanna Buonomenna[1,*], Giovanni Golemme[1,*] and Enrico Perrotta[2]
[1]Department of Chemical Engineering and Materials and INSTM Consortium, University of Calabria, Rende (CS)
[2]Department of Ecology, University of Calabria, Rende (CS)
Italy

1. Introduction

A resource-intensive industrial development, particularly in some Asian countries, characterized the last century. Its main causes can be ascribed to the significant elongation of life expectation, and to the overall increase in the standards characterizing the quality of life. The drawback of these positive aspects is the emergence of problems related to the industrial development: water stress, the environmental pollution, and the increase of CO_2 emissions into the atmosphere.

The need to achieve a knowledge-intensive industrial development is nowadays well recognized. This will permit the transition from an industrial system based on quantity to one based on quality in the framework of a sustainable development.

Sustainable development is a development that meets the needs of the present without compromising the ability of future generations to meet their own needs (United Nations General Assembly, 1987). The "three pillars" of sustainability are the environmental, social and economic demands – (United Nations General Assembly, 1987), which are not mutually exclusive and can be mutually reinforced (Figure 1): both economy and society are constrained by environmental limits. Sustainability is the path of continuous improvement, wherein the products and services required by society are delivered with progressively less negative impacts upon the Earth. Figure 2 is a cartoon depicting the road to sustainability. To benchmark sustainability, the Sustainability Index (SI) accounts for key factors that are fundamental to the industrial process (Cobb *et al*, 2007)

- strategic commitment to sustainability
- safety performance
- environmental performance
- social responsibility
- product stewardship
- innovation
- value-chain management

* Corresponding Authors

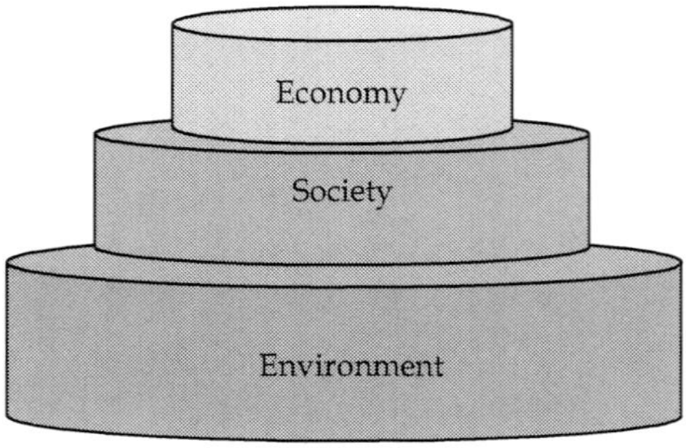

Fig. 1. Relationship between the three pillars of sustainability: environmental limits constrain both society and economy.

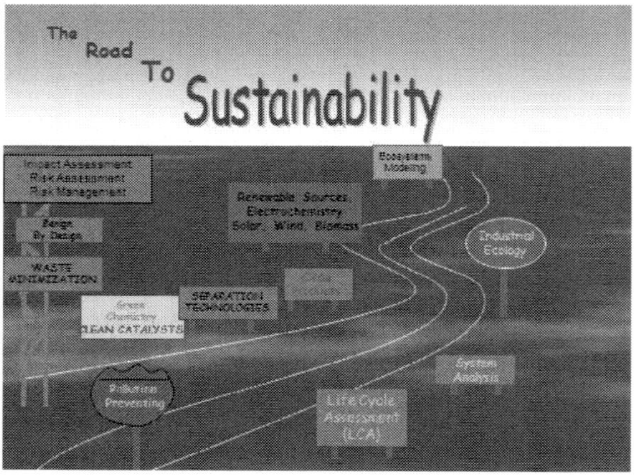

Fig. 2. Cartoon showing the features and the requirements of the road to sustainability

Chemical Engineering faces today the crucial challenge of sustainable growth to find solutions to the increasing demand for raw materials, energy and tailor-made products.

In this context, the rational integration and implementation of innovative technologies, able to increase process performance, save energy, reduce costs, and minimize the environment impact represent interesting answers.

Recently, the logic of process intensification has been suggested as the best process engineering answer to the situation. It consists of innovative equipment, design, and process development methods that are expected to bring substantial improvements in chemical and any other manufacturing and processing, such as decreasing production costs, equipment size, energy consumption, and waste generation, and improving remote control, information fluxes, and process flexibility (Charpentier, 2007).

Membrane operations are, in principle, the most attractive candidates to satisfy the process intensification concepts and requirements.

Their intrinsic characteristics of efficiency and operational simplicity, high selectivity and permeability for the transport of specific components, compatibility between different membrane operations in integrated systems, low energetic requirement, good stability under operating conditions and environmental compatibility, easy control and scale-up, and large operational flexibility, represent an interesting answer for the rationalization of industrial productions (Drioli & Romano, 2001).

The traditional membrane separation operations such as reverse osmosis (RO), microfiltration (MF), ultrafiltration (UF), and nanofiltration (NF), electrodialysis, pervaporation, etc. (Table 1), are largely used in many different applications.

Driving Force	Pressure difference (ΔP)	Concentration difference (ΔC)	Temperature difference (ΔT)	Electrical potential difference (Δf)
Phenomenological Equation	Darcy's law	Fick's law	Fourier's law	Ohm's law
Membrane Operations	Microfiltration (MF)	Gas separation (GS)	Membrane distillation (MD)	Electro-dialysis (ED)
	Ultrafiltration (UF)	Pervaporation (PV)		Electro-osmosis (EO)
	Nanofiltration (NF)	Dialysis (D)		
	Reverse osmosis (RO)	Reverse osmosis (RO)		

Table 1. Classification of membrane processes according to their driving forces

Conventional membrane separation processes have at least two phase interfaces: feed fluid-membrane interface and product/permeate fluid-membrane interface on the two sides of the membrane For example, commercialized membrane separation processes, such as RO, NF, UF, MF, GS, and PV, have two such phase interfaces. Over the last couple of decades, new membranes and membrane-separation techniques have appeared wherein the interface between two bulk phases is allowing the development of a new (improved) membrane, the creation of a novel membrane separation technique, the enhancement of the separation in existing membrane-separation processes, or the enhancement of the separation as such. The impact of such new techniques on conventional equilibrium-based separation processes/techniques is striking. The nature of the phase interface in such techniques is often crucial.

These innovative membrane systems, the separation principle of which is the phase equilibrium and known as membrane contactors, have been studied, realised, and used in integrated membrane processes. In Table 2, a classification of the different types of membrane contactors is given.

	Membrane distillation (MD)	Membrane strippers/ scrubbers	Osmotic distillation (OD)	Membrane emulsifiers	Membrane crystallizers (MCr)	Phase transfer catalysis
Phase 1	LIQUID	GAS/LIQUID	LIQUID	LIQUID	LIQUID	LIQUID
Phase 2	LIQUID	LIQUID	LIQUID	LIQUID	LIQUID	LIQUID
Driving force	PARTIAL PRESSURE GRADIENT	CONCENTR. GRADIENT	PARTIAL PRESSURE GRADIENT	PRESSURE GRADIENT	PARTIAL PRESSURE GRADIENT	CONCENTR. GRADIENT
Limit to transport	TEMPERAT. POLARIZ.	RESISTANCE IN MEMBR. Or LIQUID	CONCENTR. POLARIZ.	RESISTANCE IN MEMBR. Or LIQUID	TEMPERAT. CONCENTR. POLARIZ.	RESISTAN. IN MEMBR. OR LIQUID

Table 2. Classification of membrane contactors

At present, redesigning important industrial production cycles by combining various membrane operations suitable for separation and conversion units, thus realizing highly integrated membrane processes, is an attractive opportunity because of the synergic effects that can be attained.

In various fields, membrane operations are already the dominant technology. Their utilizations as hybrid systems, in combination with other conventional techniques or integrated with different membrane operations, is considered the way forward rationale applications.

In this context, interesting examples are in seawater desalination ; in wastewater treatment and reuse ; and in gas separation.

2. Seawater desalination

Sea and brackish water desalination has been at the origin of the interest for membrane operations, and the research efforts on RO membranes have had an impact on all of the progress in membrane science. Reverse osmosis desalination plants are currently leading the the desalination market, with RO installations representing 60% of the total number of worldwide plants, whereas thermal processes represent just 34.8% (Drioli *et al*, 2011). In Fig. 3 an example of the water desalination processes developed by the Japanese Water Re-Use Promotion Center, in co-operation with Takenaka Corporation and Organo Corporation: this process uses solar energy allowing the installation at location with no external electric energy supply (Drioli *et al*, 2011).

In Table 3 a list of traditional membrane technologies for water treatment is given.

The great flexibility, operational simplicity and mutual compatibility for integration of membrane operations offer the possibility of combining different membrane technologies for minimizing the limits of the single membrane units and for increasing the efficiency of the overall system.

Nowadays the most part of conventional seawater desalination plants use either RO or Multi-Stage Flash (MSF) technology. Thermal desalination is the most frequently applied technology in the Middle East, whilst membrane processes have rapidly developed and now surpass thermal processes in new plant installations due to the lesser energy consumptions (2.2.-6.7 kWh/m^3 for seawater RO, with respect to 17-18 kWh/m^3 for MSF).

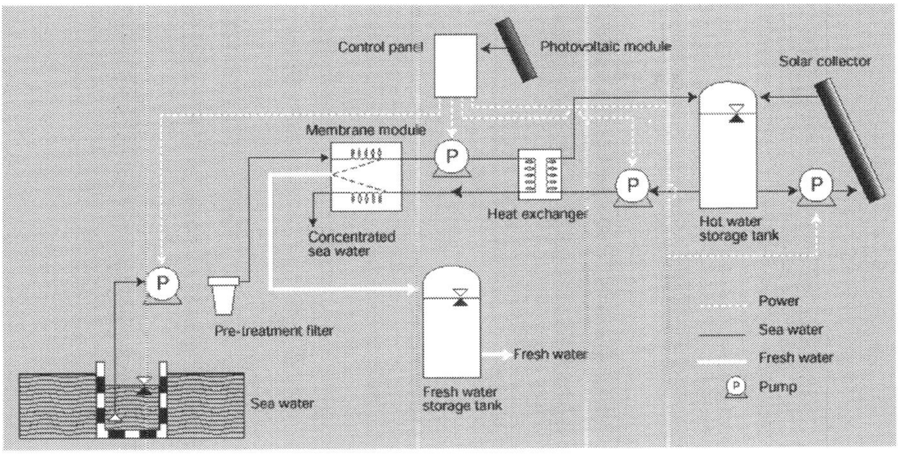

Fig. 3. Water desalination plant based on membrane process using solar energy, developed by the Japanese Water Re-use promotion center, in co-operation with Takenaka Corporation and Organo Corporation (Drioli *et al*, 2011).

However, high quality feedwater is required to ensure stable, long term performance, and an effective pre-treatment is essential for an efficient plant operation. In the past conventional, (i.e. conventional and physical) pre-treatment without the use of membrane technologies has been applied. Nowadays, membrane-based pre-treatments (such as MF, UF, NF) tend to replace conventional pre-treatment systems.

MF is a low energy-consuming technique extensively used to remove suspended solids and to lower chemical oxygen demand (COD)/biochemical oxygen demand (BOD) and silt density index (SDI). UF retains suspended solids, bacteria, macromolecules and colloids and despite of the larger pressure gradient with respect to MF, this membrane separation method is competitive against conventional pre-treatments. In an integrated membrane pre-treatment (Figure 4), the benefits of lower fouling rates of RO membranes compensates the higher cost membrane pre-treatment equipment.

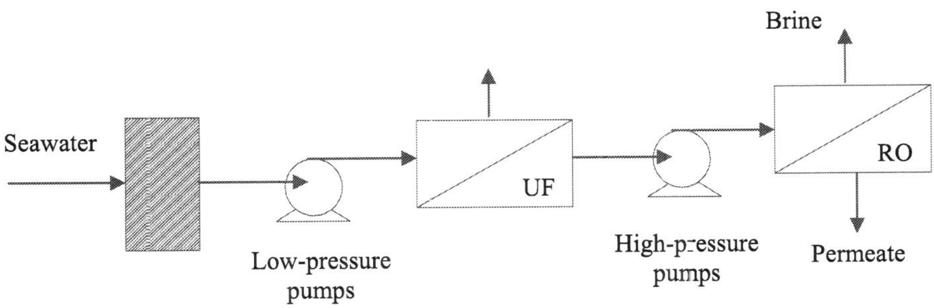

Fig. 4. Integrated (UF/RO) seawater desalination plant: UF as a pre-treatment

Separation process	Membrane type used	Applied driving force	Mode of separation	Applications
Microfiltration (MF)	symmetric macroporous	hydrostatic pressure 0.05-0.2 MPa	size exclusion, convection	water purification, sterilization
Ultrafiltration (UF)	Asymmetric macroporous	hydrostatic pressure 0.1-0.5 MPa	size exclusion, convection	separation of molecular mixtures
Diafiltration (DF)	asymmetric macroporous	hydrostatic pressure 0.1-0.5 MPa	size exclusion and dialysation, diffusion	purification of molecular mixtures, artificial kidney
Nanofiltration (NF)	asymmetric mesoporous	hydrostatic pressure 0.3-3 MPa	size exclusion, diffusion, Donnan-exclusion	separation of molecular mixtures and ions
Reverse osmosis (RO)	asymmetric skin-type, dense or microporous	hydrostatic pressure 1-10 MPa	solution-diffusion mechanism	sea & brackish ,water desalination
Electrodialysis (ED)	symmetric ion-exchange membrane	electrical potential	Donnan-exclusion	water desalination
Donnan Dialysis (DD)	symmetric ion-exchange membrane	concentration of ions	Donnan-exclusion	water softening
Membrane Distillation (MD)	symmetric porous hydrophobic membrane	Vapor pressure	diffusion	water desalination, concentration of solutions

Table 3. Membrane operations used for water treatments

Macedonio and Drioli (2010) analysed an integrated membrane-based desalination plant with membrane crystallization (MCr) as post-treatment for the recovery of salts and water contained in the NF/RO retentate streams of a desalination plant (Fig.5).

The feed water enters into MF membrane modules to be cleaned from suspended solids and large bacteria. After MF, process water is pressurized and then sent to NF membrane modules to be cleaned from turbidity, microorganisms, hardness, multivalent ions and 10-50% of monovalent species. After NF step, the RO step requires that the process water be pressurized to overcome the osmotic pressure. In the RO operation the process water is separated into a permeate and a brine. RO brine enters into the precipitator in which is mixed with Na_2CO_3 for the removal of the Ca^{2+} ions of the RO brine as $CaCO_3$. The process stream enters the MCr modules where it is separated into a permeate, a purge and a salt containing stream.

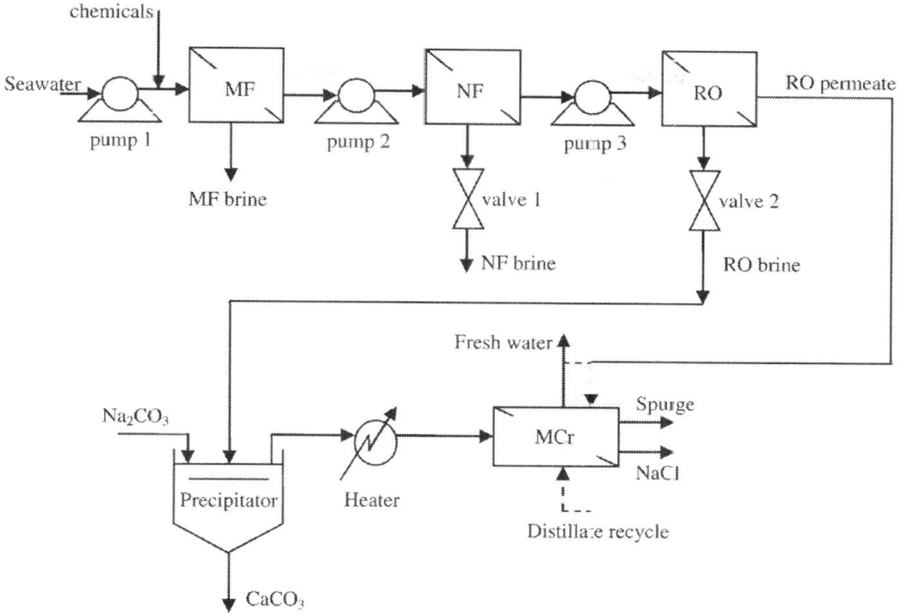

Fig. 5. Flow diagram of integrated desalination membrane-based plant (Macedonio and Drioli, 2010).

Hybrid membrane systems, in combination with conventional separation process, are advantageous in many industrial processes and in particular in desalination systems.

In recent years, the concept of simple hybrid multistage flash-reverse osmosis (MSF/RO) configuration has been applied to a number of existing or new commercial desalination plants. The SWCC Jeddah, Al-Jubail and Yanbu existing Power/Water cogeneration plants are expanded for more water production by combining with new SWRO desalination plants. The simple hybrid desalination arrangement enabled the increase of the water-to-power ratio and utilized effectively the available intake/outfall facilities.

The MSF and RO operate in parallel and are entirely independent. The water product of the single pass RO unit is blended with the MSF product (Hamed, 2006)

The 100 MGD desalination plant in Fujairah (UAE) is one of the largest hybrid MSF (Multistage Flash evaporators) /RO installation in the worlds: it combines a 62.5 MGD MSF and 37.5 MGD SWRO (Fig.6).

This hybrid desalination system is designed to provide significant operational savings by reducing fuel consumption by up to 25 per cent compared with a similar-sized plant based only on MSF technology. Other key criteria influencing the design of the desalination plant were feed water quality, product water requirements and compatibility with the cogeneration of electricity.

Fig. 6. Fujairah desalination plant in United Arab Emirates, from http://fujairahinfocus.blogspot.com/2011/10/fujairah-power-and-desalination-plant.html

The water production system at the Fujairah desalination plant is comprised of five Doosan MSF units producing 57 million l/day (12.5 MGD) each and one RO unit with a design capacity of 171 million l/day (37.5 MGD). The RO unit was supplied by Ondeo Degremont. For drinking water supply, distillate from the MSF units and desalinated water from the RO plant are mixed in a distillate header and treated in a re-mineralization unit before passing into the potable water storage tanks. Prior to export to the water transmission line, potable water is stored in five potable water tanks, each with a capacity of 91 million l (20 million gallons).

A promising approach for pre-treatment of seawater make-up feed to MSF and SWRO desalination processes using NF membranes has been introduced by the R&D Center (RDC) of SWCC.

NF membranes are capable to reduce significantly scale forming ions from seawater, allow high temperature operation of thermal desalination processes, and subsequently increase water productivity.

The developed fully integrated systems NF/MSF and NF/SWRO/MSF result in high water productivity and enhance thermal performance compared to the currently used simple hybrid desalination arrangements (Hamed, 2006).

3. Wastewater treatment and reuse

Considerable advances in MF, UF and NF technologies to recover municipal wastewater have been also achieved. Also in this case, the implementation of integrated membrane systems is growing rapidly with excellent results.

In fig. 7 the main treatment steps to recover municipal wastewater from Kuwait City and the surrounding area are reported: A conventional biological wastewater treatment plant (WWTP) treats the effluent to better than secondary effluent quality. The secondary effluent then flows to the water reclamation plant, which uses UF and RO to further treat the water for reuse. Sludge from the wastewater treatment plant is treated to allow for disposal by

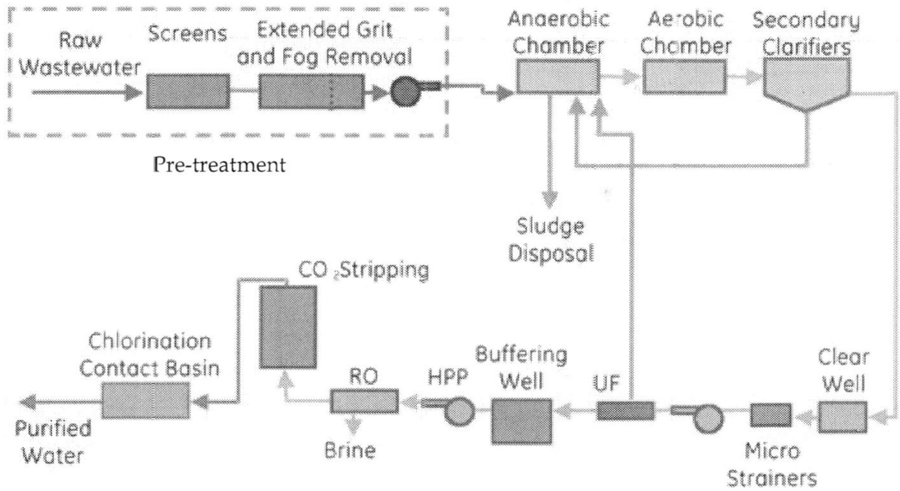

Fig. 7. Wastewater treatment facility at Sulaibiya near Kuwait City from http://www.water-technology.net/projects/sulaibiya/

landfill, incineration, or by composting. Membrane filtration was selected to provide robust pretreatment of the secondary-treated municipal effluent before being fed to the RO. Membrane filtration was chosen over conventional tertiary clarification and filtration because it reduced the plant chemical consumption and could guarantee that low turbidity water is fed to the RO. It is expected that better quality pre-treatment to the RO will lead to longer membrane life, lower operating pressure, and reduced cleaning frequency for the RO system. The UF plant utilizes Norit's X-Flow membranes, which are capillary hydrophilic hollow fibers. The UF units are operated individually. Each unit is backwashed regularly, whereby all suspended matter that is being retained by the membranes is removed from the plant. The backwash water is pumped back upstream of the WWTP to achieve the highest possible overall water recovery for the plant. The salinity of the municipal effluent has an average monthly value of 1,280 mg/l TDS, with a maximum value of 3,014 mg/l. RO is used to desalinate the water to 100 mg/l TDS, as well as provide a second barrier to bacteria and viruses. RO technology is well proven for desalinating municipal effluent. The system consists of 42 identical skids in a 4:2:1 array. Approximately 21,000 membrane modules, provided by Toray of America, were required for this project. The RO product passes through a stripper to remove carbon dioxide to adjust pH with a minimum amount of caustic before distribution, and the product is then chlorinated before leaving the plant. RO brine is disposed of into the Persian Gulf.

Membrane bioreactors (MBR) are a combination of activated sludge treatment and membrane filtration for biomass retention. Low-pressure membrane filtration, either UF or MF, is used to separate effluent from activated sludge. The two main MBR configurations involve either submerged membranes (Fig. 8) or external circulation (side-stream configuration).

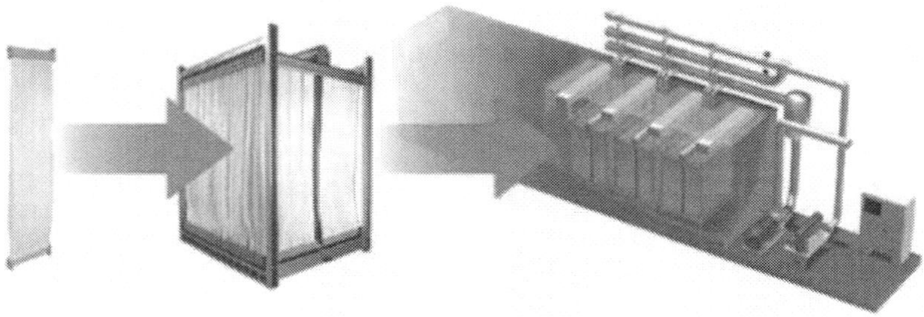

Fig. 8. Submerged membrane module for wastewater treatment. From ZeeWeed® Submerged Membrane System, from http://www.gewater.com.

Since the early MBR installations in the 1990s, the number of MBR systems has grown considerably. One key trend driving this growth is the use of MBR system for decentralized treatment and water reuse. The successful introduction of MBR systems into small scale and decentralized applications has led to the development of packaged treatment solutions from the main technology suppliers. The company Conder Products, UK, designed the package treatment plant Clereflo MBR; Zenon Environmental Inc., now a part of General Electric, produced ZeeMod®.

The Pleiade® Plate & Frame membranes produced by Orelis©, France, which is one of Europe's leading membrane manufacturer, are installed in skids mounted outside the bioreactor and the sludge is circulated through the module in high speed, by pumps (Fig. 9). This configuration and the Orelis membranes have several advantages over other MBR systems.

Fig. 9. Pleiade® MBR membrane bioreactor for treating 1000 m^3/day of effluent from www.vic-ws.it/site/down/PLEENG0411.pdf

The membranes are in the UF range and offer high separation capabilities (0.02 μm, MWCO=40 kD), unlike most MBR membranes which fall under the MF range. As a result, the permeate quality is superb, even in very high sludge concentrations (up to 2%) and in difficult applications. Besides TSS, the membranes effectively reject also bacteria and even viruses.

High flow rate and pressure feed of the reactor content (MLSS) into the Pleiade membrane module, enable high permeate fluxes (60-80 lit/m2*hr). As a result, the membrane area required for treatment of a given WW flow is much lower than in most other MBR membranes.

High circulation speed over the membranes (2 m/sec) reduces fouling accumulation, and membrane cleaning (CIP) demand.

External membrane systems enable full modularity and easy expansion of WW treatment capacity. Gao *et al* (2011) developed a completely green process based on the integration of MBR with UF by treating micro-polluted source water in drinking water treatment. The removal of organic matter is carried by both a biodegradation mechanism in the MBR and by the MF/UF membrane, while the nitrification in MBR removes ammonia (Fig. 10).

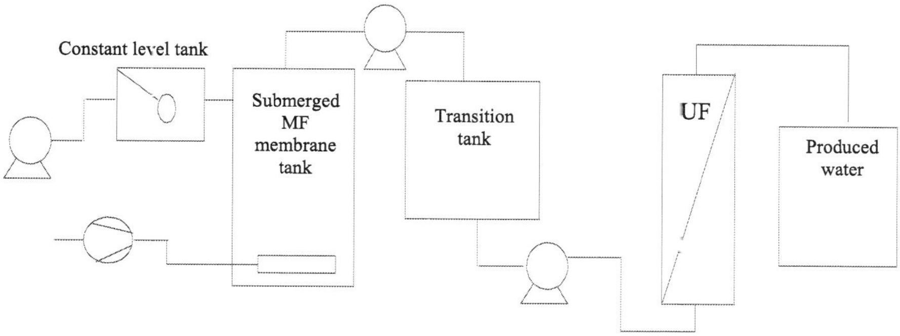

Fig. 10. Schematic diagram of the pilot scale experimental set-up MBR/UF (Gao *et al*, 2011)

The pulp and paper industry is one of the most water-dependent industries (Nurdan and Emre, 2010). The alkaline peroxide mechanical pulping (APMP) process has been widely applied especially in Asia (Liu *et al*, 2011) for the high yield, and the relatively low pollution. To achieve a closed wastewater loop, several APMP plants in the world have attempted to concentrate the total effluent by using a multi-effect evaporation system. Zhang *et al* (2011) studied a hybrid process UF/Multi-effect-evaporation (MEE) to concentrate effluent from APMP plants. With this new membrane concentration process, 88% of the water in the effluent can be removed, 1.4 bilion KWh power could be saved; the capital investment for MEE could also be decreased by 88% correspondingly.

The possibility of redesigning overall industrial production by the integration of various already developed membrane operations is of particular interest: low environmental impacts, low energy consumption, higher quality of final products and new available products are the advantages obtained.

The leather industry might be an interesting case study because of (i) the large environmental problems related to is operation; (ii) the low technological content of its traditional operations; (iii) the tendency to concentrate a large number of small-medium industries in specific districts. In Fig. 11 an ideal process based on integrated membrane operations for the tanning process is showed.

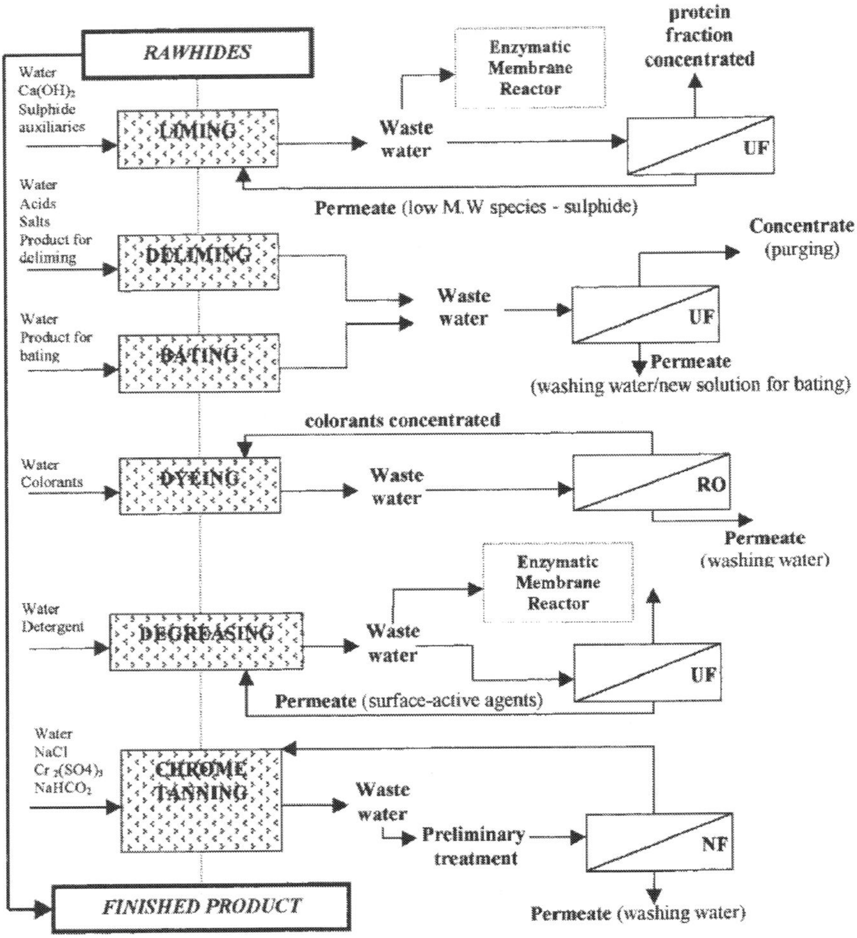

Fig. 11. Integrated membrane process proposed for the tanning process in the leather industry (Drioli & Romano, 2001)

The pollution problems of the leather industry have been minimized one by one at the point where they originate, thereby avoiding the need for huge wastewater treatment plants at the end of the overall production line. In addition, the membrane operations act by physical mechanisms without modification of the chemical procedure at the origin of the final high quality of the leather (Cassano *et al*, 2003)

4. Gas separation

In Table 4 the membrane sales involved for industrial gas separations are shown.

Application	Annual membrane system sales (10^6 U.S. $)	
	Year 2010	Year 2020
Nitrogen from air	100	125
Oxygen from air	10	30
Hydrogen		
H_2/gas (CO, N_2, C_1, C_2)	40	80
H_2/gas (C_{3+}, CO_2)	20	70
Natural gas		
CO_2 removal	60	100
NGL removal and recovery	20	50
N_2 removal, dehydration	10	25
Vapour (C_{2+})/Gas (N_2, Ar)	30	90
Vapour/Vapour (including dehydration)	20	100
Air dehydration/other	30	60
TOTAL	340	730
Annual growth, %	8	8

Table 4. Sales estimates and sales predicted for the principal gas and vapour separation applications

Nowadays, obtaining cheap high purity gases or enriched gas mixtures (the air, in particular) is a very important problem in industry and medicine as well as in everyday life (Bodzek, 2000).

Methods for air separation or oxygen enrichment can be divided into two groups (Freeman *et al*, 2006): cryogenic and non - cryogenic. The gaseous oxygen and nitrogen market is dominated by cryogenic distillation of air, and vacuum swing adsorption (Koros *et al*, 2000). Of the non – cryogenic methods, selective adsorption on zeolites and carbon adsorbents are available (Lin & Guthrie, 2006), and more and more attention is attracted to the membrane separation techniques (Dhingra & Marand, 1998) . The success of polymeric membranes has been largely based on their mechanical and thermal stability, along with good gas separation properties. The process of membrane separation is continuous, has a low capital cost, low power consumption, and the membranes, at least in gas separation, do not require regeneration (Vansant & Dewolfs, 1990).

Nitrogen production today is the largest GS process in use. Nitrogen gas is used in many applications (e.g., to prevent fires and explosions in tanks and piping systems and to prevent equipment degradation, during shutdown periods, in compressors, pipelines and reactors). Single-stage membrane operation is preferred. Air is pressurized and fed into the membrane separators; faster gases (O_2, CO_2, water vapor) permeate through the polymeric fiber walls, are collected and vented to the atmosphere while the slower, non-permeate N_2 gas is available at the other end of the separator.

Oxygen production by membrane systems is still underdeveloped, since most of the industrial O_2 applications require purity higher than 90%, which is easily achieved by adsorption or cryogenic technologies but not by single-stage membranes.

New materials are being developed that could possibly have higher permeabilities than conventional solid electrolytes. Promising oxygen permeation have been obtained in many perovskite systems (Zhu *et al*, 2009; Stiegel, 1999). The dense perovskite type membranes transport oxygen as lattice ions at elevated temperatures with infinite selectivity ratios in O_2 separations.

The oxygen-ion conducting membranes must operate above 700°C: an efficient and cost-effective way to recover the energy contained in the non- permeate, oxygen-depleted stream is that of integrating the membrane system with a gas turbine (Fig. 12) (Drioli & Romano, 2001).

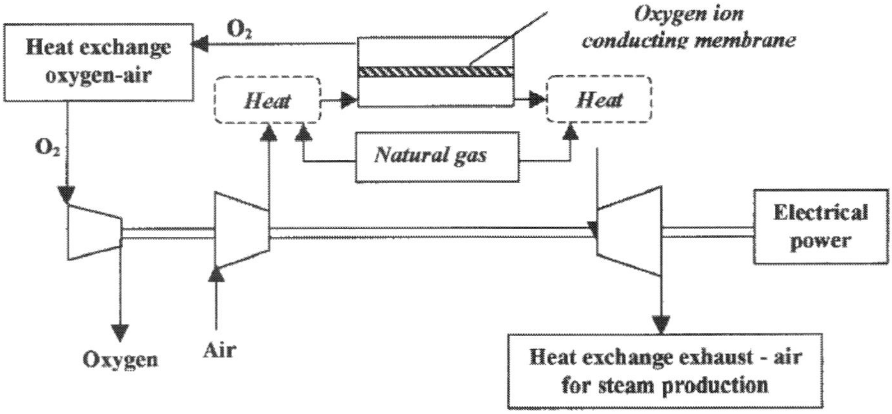

Fig. 12. Integrated system for O_2 and power productions (Drioli & Romano, 2001).

Hydrogen recovery was among the first large-scale commercial applications of membrane GS technology. The commercial success in the mid-1970's in Louisiana of the Permea hollow-fiber Prism system for in-process recycling of hydrogen from ammonia purge gases (Fig. 13) was the starting point of the penetration of membrane technology in large-scale manufacturing. A two-step membrane design was chosen for this ideal application for membrane technology: the ammonia reactor operates at high pressures (ca. 130 bar), thus providing the necessary driving force for separation; the H_2/N_2 membrane selectivity is high and the feed gas is free of contaminants.

This technology has been extended to other situations. In Table 5, some recent H_2 membrane applications are listed. H_2 recovery from refinery streams is an emerging field for membrane GS in the petrochemical industry; it is a key approach to meet the increased demand of hydrogen (for hydrotreating, hydrocracking or hydrodesulfurization processes) owing to new environmental regulations. An example is the H_2 recovery from high pressure purge gas of a hydrotreater.

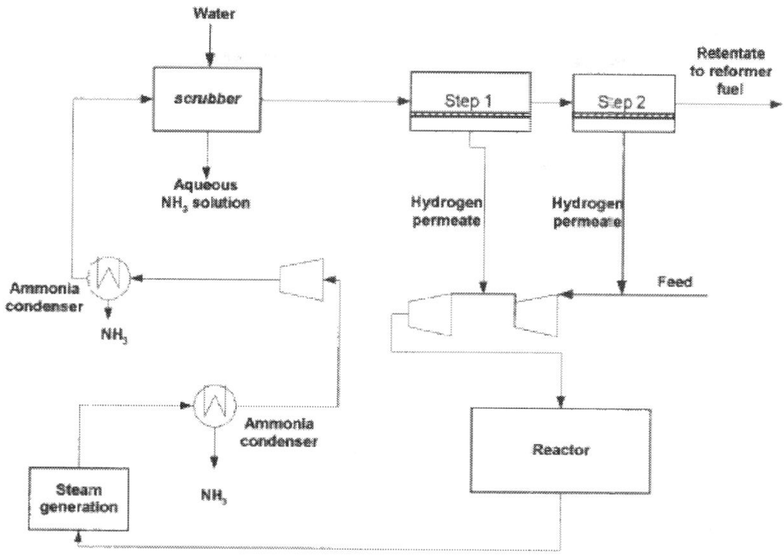

Fig. 13. Plant for H_2 recovery from ammonia synthesis (Drioli & Romano, 2001)

Refining	Chemical/Petrochemical
Catalytic reformer off gases	Synthesis gas composition adjustment (IGCC)
HDT/Recycle purge gases	Methanol purge gases
Refinery fuel gases	Ammonia purge gases
PSA off gases	Polypropylene purge
	Polyethylene purge
	Styrene off gases
	Coke oven gases
	Electrolysis gases
	Cyclohexane plant gases

Table 5. H_2 membrane applications

In a comparison of three separation technologies (membrane, PSA, cryogenic distillation) applied for H_2 recovery from refinery off-gas, Spillman (1989) reported that the use of membranes represent the best choice. An evaluation of these processes is provided in terms of sustainability indexes. The *Energy Intensity* of the membrane system is the lowest; the *Mass Intensity* is less than 50% of that for the conventional separations; the membrane system occupies a tenfold lower area required by PSA and cryogenic distillation of corresponding capacity (highest *Productivity/Footprint ratio*).

Carbon dioxide removal from natural gas (natural gas sweetening) is mandatory to meet pipeline specifications (e.g., down to 2% vol. in U.S.A.), since CO_2 reduces the heating value of natural gas, is corrosive and freezes at a relatively high temperature, forming blocks of dry ice that can clog equipment lines and damage pumps. Membrane technology is attractive for CO_2 and H_2S removal, because many membrane materials are very permeable to these species (enabling a high recovery of the acid gases without significant loss of pressure in the methane pipeline product gases), and because treatment can be accomplished using the high wellhead gas pressure as the driving force for the separation. A high natural gas recovery (>95%) can be achieved in multi-stage systems.

Cynara-NATCO produces hollow fiber modules for CO_2 removal and has provided a membrane system (16 in. modules) for the natural gas sweetening in an offshore platform in the Thailand gulf (830,000 Nm^3/h), which is the biggest membrane system for CO_2 removal. The Natco Group has been awarded in 2008 a $24.9 million contract to provide membrane separation technology and equipment to capture CO_2 for re-injection in Bouri. Eni Oil Limited Libyan Branch operates the Bouri production platforms in the Mediterranean Sea, one of which will house the pre-treatment equipment and membrane systems intended to treat more than 160MM SCFD gas.

The Cynara(R) membrane system has been designed for the selective capture of CO_2 from the gas stream for this retrofit project on the existing platform. Natco will design, engineer, and fabricate four membrane skids and related valving, as well as pretreatment equipment. The contract was awarded in May 2008.

The integration of membranes with other well-established separation processes in the chemical and petrochemical industries, , was considered in different works (Doshi, 1987; Choe *et al*, 1987; Dosh & Dolan, 1995; Bhide *et al*, 1998). Usually, the combination membranes / PSA is considered in H_2 separation, while hybrid membranes + amine absorption are applied to the CO_2 separation.

A comparison of the separation cost for the membrane process with diethanolamine (DEA) absorption showed that the membrane process is more economical for CO_2 concentrations in the feed in the range 5–40 mol% (Bhide *et al*, 1998).

If membrane processes are not economically competitive because of the high H_2S concentration in the feed, the separation cost could be significantly lowered by using hybrid membrane processes. In the block diagram of Fig.14 the membrane unit removes two-thirds of the CO_2 and the amine plant removes the remaining. The combined plant is significantly less expensive than an all-amine or all-membrane plant.

A hybrid system (Cynara membranes + amine absorption) is operating since 1994 in Mallet (Texas, U.S.A.) to perform the bulk removal of CO_2 from associated gas (90% CO_2 and heavy hydrocarbons), before downstream treating. The membrane system offered a 30% reduction in operating cost when compared with a methyl diethanolamine (MDEA) system and significantly reduced the size of the subsequent operations (Blizzard *et al*, 2005).

An integrated system combining membrane permeation and pressure swing adsorption (PSA) has been developed for CO_2/N_2 gas separation (Esteves and Mota, 2007). By using the membrane as a pre-bulk separation unit and coupling it to the PSA, the separation performance of the hybrid scheme is enhanced with respect to that of the two stand-alone

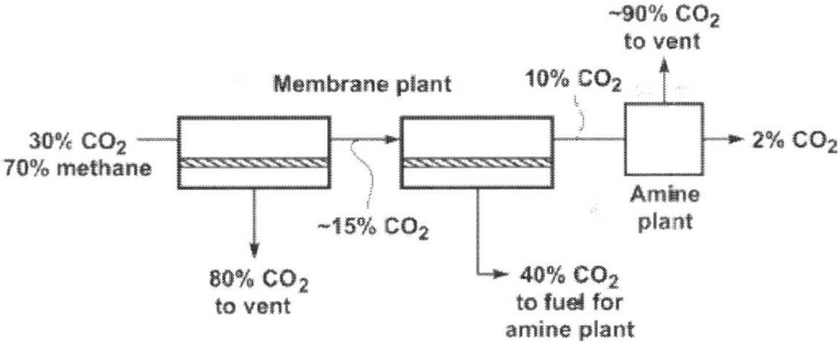

Fig. 14. Membrane/amine hybrid plant for the treatment of natural gas (Baker & Lokhandwala, 2008).

units. Instead of constant-composition regular feed, the PSA is fed with a mixture which is progressively enriched in the more adsorbed component during the pressurization and high-pressure adsorption steps of the cycle. This results in sharper concentration fronts. The hybrid has been applied successfully to the bulk separation of an 30:70 mol% CO_2/N_2 mixture over activated carbon. Process performance is reported in terms of product recovery and purity at cyclic steady state. Numerical simulations were validated by experimental work on a composite membrane and a laboratory-scale PSA unit.

MTR (Membrane Technology and Research, Inc.) has proposed an innovative retro-fit process for the post-combustion carbon capture and sequestration (CCS) of existing coal-fired power plants (Merkel et al., 2010). According to preliminary calculations, a two stage membrane process (Figure 15) should be able to recover 90% of the CO_2 released by the power plant at a competitive cost of 23$/ton CO_2 sequestrated. The selected lower cost configuration uses a blower upstream of the first membrane unit and a vacuum pump downstream in order to boost the driving force for the separation. At the same time, in the second membrane module the air to be fed to the boiler strips part of the CO_2 in the N_2 enriched flue gas from the first membrane module, thereby recycling CO_2 and building a higher CO_2 concentration for the first membrane module: as a consequence, the sequestered CO_2 flux in the first membrane unit is enhanced.

MTR is now involved in a contract with the US Department of Energy for the construction of a membrane skid containing MTR's Polaris membranes capable of 90% CO_2 capture from a 20 tons-of-CO_2/day slipstream of coal fired flue gas. The skid will be operated during a 6 month field test at Arizona Public Service's Cholla power plant. (http://www.mtrinc.com/news.html#DOE14mil).

In the process industry the final choice of a separation process is the result of a balance between the economics, the desired purity and recovery of the product and other conditions and restrictions, such as the desired capacity and the composition of the feed and the possibility of integration with other processes. In this logic, the integration of commercial membrane separation units in the recovery of ethylene or propylene monomer from polyolefin resin degassing vent streams has been proposed by Baker *et al* (1998). A simplified flow scheme of the process that generates such streams is shown in Fig. 16(a).

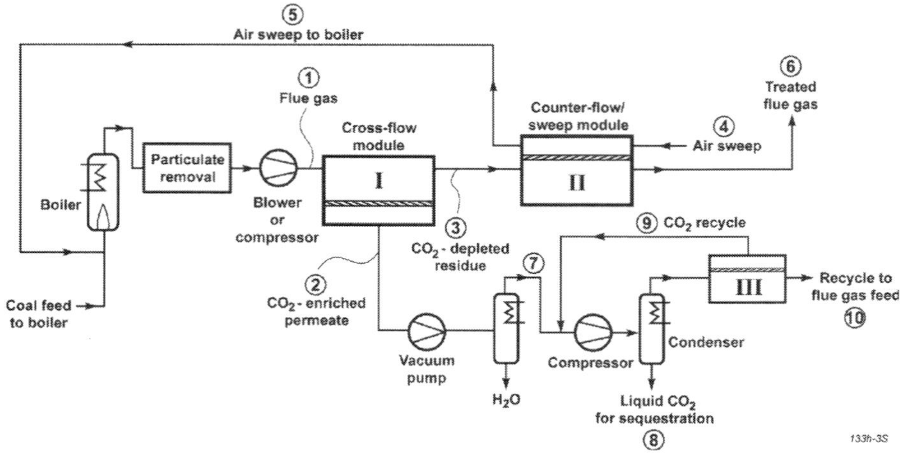

Fig. 15. Simplified flow diagram of a two-step counter-flow/sweep membrane process to capture and sequester CO_2 in flue gas from a coal-fired power plant. The base-case membrane with a CO_2 permeance of 1000 gpu and a CO_2/N_2 selectivity of 50 was used in the calculations (Merkel *et al*, 2010).

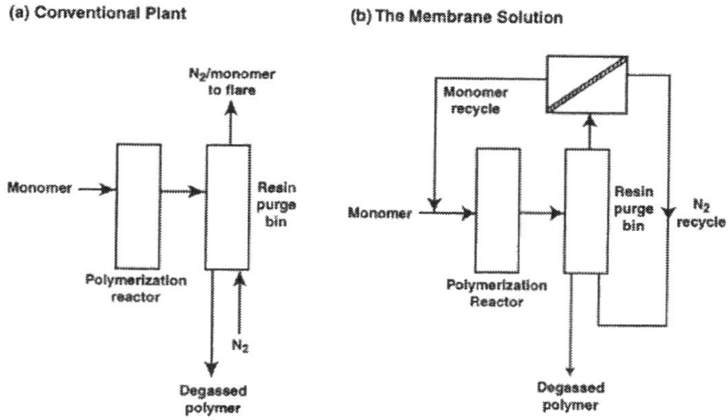

Fig. 16. A schematic diagram of monomer recovery and nitrogen recycle from polyolefin plant resin degassing operations: (a) conventional plant (b) membrane vapor gas separation system (Baker *et al*, 1998)

In a typical polyolefin polymerization plant, after polymerization, the polymer is removed to a low-pressure chamber. amounts of sorbed monomers and processing solvents, which must be removed before the polymer can be used. Therefore, the raw polymer is passed to large resin degassing bins through which nitrogen is circulated.

In the past, no economical method of separating these gases was available, so the stream was used as low grade fuel and the monomer content lost. The value of these potentially recoverable monomers in a typical polymerization plant is very high. The placement of a membrane to recover and recycle the gases is shown schematically in Fig. 16(b).

Membrane vapor gas separation systems have been installed worldwide. The main membrane applications in the petroleum industry are vapor recovery in tank farms and hydrogen recovery in refineries and chemical plants. Other developed applications of the membrane technology in refineries include solvent recovery in lube oil manufacturing (Max-DeWax developed jointly by Grace Division and Exxon Mobil) and aromatics removal from gasoline.

5. Conclusion

The development of innovative processes that follow the process intensification strategy for a sustainable industrial growth is critical to the production by non-polluting, defect-free and safe industrial processes. Membrane operations show a higher efficiency than conventional separation and reaction unit operations. They offer new options for the razionalization of innovative production cycles. Membrane engineering plays a crucial role in water desalination, in municipal water reuse (by MBR), in petrochemicals and in the field of gaseous separations. There are also some interesting opportunities to integrate membrane operations into existing industrial processes to achieve the benefits of process intensification.

6. References

Baker, R.W., Wijmans, J.G. and Kaschemekat, J.H. (1998). The design of membrane Vapor gas separation systems. *J. Membr. Sci.* Vol. 151 pp.55-62 ISSN 0376-7388

Baker, R. & Lokhandwala, K. (2008). Natural Gas Processing with Membranes: An Overview. *Ind. Eng. Chem. Res.* Vol. 47 No 7 pp.2109-2121 ISSN 0888-5885.

Bhide, B.D.; Voskericyan, A.; Stem, S.A. (1998) Hybrid processes for the removal of acid gases from natural gas. *J. Membr. Sci.* Vol. *140* No 1 pp.27-49 ISSN 0376-7388

Blizzard, G.; Parro, G.; Hornback, K. (2005). Mallet gas processing facilities uses membranes to efficiently separate CO_2. *Oil & Gas Journal* Vol. 103 No 14 pp 48-53.

Bodzek, M. (2000) Membrane techniques in air cleaning. *Pol. J. Environm. Stud.* Vol. 9. No.1 , pp. 1-12 ISSN 1230-1485.

Cassano, A; Adzet, J; Molinari, R.; Buonomenna, M.G.; Roig, J; and Drioli, E. (2003) Membrane treatment by nanofiltration of exhausted vegetable tannin liquors from the leather industry. *Water Research*. Vol. 37, pp. 2426-2434 ISSN 0043-1354.

Charpentier, J.C. (2007). In the frame of globalization and sustainability, process intensification, a path to the future of chemical and process engineering (molecules into money). *Chemical Engineering and Processing: Process Intensification,* Vol. 134, No. 1-3, pp.84 92, ISSN 0255-2701

Choe, J.S.; Auvil, S.R.; Agrawal, R. (1987) Process for separating components of a gas stream" US Patent 4,701,187

Cobb, C.; Schuster, D.; Beloff, B.; Tanzil, D. (2007). Benchmarking Sustainability.Chemical Engineering Progress, Vol. 104, No.6, pp. 38-42, ISSN 0360-7275

Dhingra , S.S. & Marand, E. (1998) Mixed gas transport study through polymeric membranes , *J. Membr.Sci.* Vol. 141 No 1 pp 45-63 ISSN 0376-7388

Doshi, K.J. (1987) Enhanced gas separation process, US Patent 4,690,695, Union Carbide.

Doshi, K.J.; Dolan, W.B. (1995) Process for the rejection of CO_2 from natural gas", US Patent 5,411,72, UOP.

Drioli, E; Romano, M. (2001). Progress and New Perspectives on Integrated Membrane Operations for Sustainable Industrial Growth. *Ind. Eng. Chem. Res.* Vol. 40, pp. 1277-1300, ISSN 0888-5885

Drioli, E; Stankiewicz, A; Macedonio, F. (2011). Membrane Engineering in process intensification-An overview. *J. Membr. Sci.* Vol. 380, pp. 1-8, ISSN 0376-7388

Esteves, I.A.A.C., Mota, J.P.B. (2007). Hybrid Membrane/PSA Processes for CO_2/N_2 Separation. *Adsorption Science & Technology* Vol. 25, pp. 693-715 ISSN 0263-6174.

Freeman , B; Yampolskii , Y; Pinnau , I. (2006) Eds., Materials Science of Membranes for Gas and Vapor Separation , John Wiley and Sons, Ltd. , Chichester ISBN 9780470029039.

Gao, W; Liang, H; Wang, L.; Chang, H.-q, , Li, G-b. (2011) Pilot Study of Integrated MF based MBR and UF for Drinking Water Production by Treating Micropolluted Source Water. *IPCBEE* Vol. 14, pp. 11-16, ISSN 2010-4618

Hamed, O.A. (2006) Overview Of Hybrid Desalination Systems - Current Status And Future Prospects. *Desalination* Vol. 186, pp. 207–214, ISSN 0011-9164

Koros , W.J.; Mahajan , R.(2000) Pushing the limits on possibilities for large scale gas separation: which strategies? *J. Membr. Sci.* Vol. 175 No. 1 pp. 181-196 ISSN 0376-7388

Lin, L. & Guthrie, J.T. (2006) Novel oxygen - enhanced membrane assemblies for biosensors. *J. Membr. Sci.* Vol. 278 No 1-2 pp 173-180. ISSN 0376-7388

Macedonio, F.; Drioli, E. (2010). An exergetic analysis of a membrane desalination system. *Desalination* Vol. 261, pp. 293-299, ISSN 0011-9164

Merkel, T.C.; Lin, H.; Wei, X.; Baker, R. (2010). Power plant post-combustion carbon dioxide capture: an opportunity for membranes, *J. Membr. Sci.* Vol. 359 No 1-2 pp 126-139. ISSN 0376-7388

Nurdan, B.; Emre, K. (2010) Economic evaluation of alternative wastewater treatment plant options for pulp and paper industry. *Sci. Total. Environ.* Vol. 408, No.24, pp.6070-6078, ISSN 0048-9697.

Spillman, R. (1989) Economics of gas separation by membranes. *Chem.Eng. Prog.* Vol. 85, pp.41-62, ISSN 0360-7275.

Stiegel, G. J. (1999) Mixed conducting ceramic membranes for gas separation and reaction. *Membr. Technol.* Vol. 1999 No 110 pp 5-7 ISSN 0958-2118

United Nations General Assembly (1987). Report of the World Commission on Environment and Development: Our Common Future

Vansant, E.F. & Dewolfs, R. (1990) *Gas Separation Technology* , Elsevier , 1990 ISBN 0444882308

Zhang, Y; Cao, C-Y; Feng, W-Y; Xue, G-X; Xu, M. (2011) Performance of a pilot scale membrane process for the concentration of effluent from alkaline peroxide mechanical pulping plants. *BioResources*. Vol.6, No.3, pp. 3044-3054, ISSN 19302126.

Zhu, X; Sun, S; Cong, Y; Yang, W. (2009) Operation of perovskite membrane under vacuum and elevated pressures for high-purity oxygen production. *J. Membr. Sci.* Vol.345 pp 47-52 ISSN 0376-7388.

22

Thermal Study on Phase Transitions of Block Copolymers by Mesoscopic Simulation

César Soto-Figueroa[1], Luis Vicente[2]
and María del Rosario Rodríguez-Hidalgo[1,*]
[1]*Departamento de Ciencias Químicas, Facultad de Estudios Superiores Cuautitlán, Universidad Nacional Autónoma de México (UNAM)*
[2]*Departamento de Física y Química Teórica, Facultad de Química Universidad Nacional Autónoma de México (UNAM)*
México

1. Introduction

The block copolymers are an exceptional kind of macromolecules constituted by two or more blocks of different homopolymer chains linked by covalent bonds. These polymeric materials have received much attention over past few years due in large part to their ability to self-assemble in the melted state or in a selective solvent inside a variety of ordered phases or well-defined structures of high regularity in size and shape with characteristic dimensions between 100 and 500 nanometres. These ordered phases and their structural modification are the key to many valuable physical properties which make block copolymers of great industrial and technological interest. The molecular self-assemble and formation of periodic phases in the block copolymers depend of the strength of interblock repulsion and composition, for example, mesoscopic studies of the poly(styrene)-poly(isoprene) (PS-PI) diblock copolymer, have demonstrated that this synthetic material, may generate a series of long-range ordered microdomains when exist a weak repulsion between the unlike monomers isoprene and styrene, as result, the PS-PI diblock copolymer chains tend to segregate below some critical temperature, but, as they are linked by covalent bonds, the phase separation on a macroscopic level is prevented, only a local microphase segregation occurs (Soto-Figueroa et al., 2005, Soto-Figueroa et al., 2007). The phase transition from homogeneous state of polymeric chains to an ordered state with periodic phases is called microphase separation transition (MST) or order-disorder phase transition (ODT) (Leibler, 1980). The phase segregation and generation of ordered structures in the microscopic level of a diblock copolymer via an order-disorder phase transition is illustrated in Fig. 1.

As result of microphase segregation process, the block copolymers can display ordered structures constituted by homopolymer domains that haves only mesoscopic dimensions corresponding to the size of singles blocks. The microphase separation leads to different classes of well-defined periodic structures in dependence on the ratio between the degrees of

* Corresponding Authors

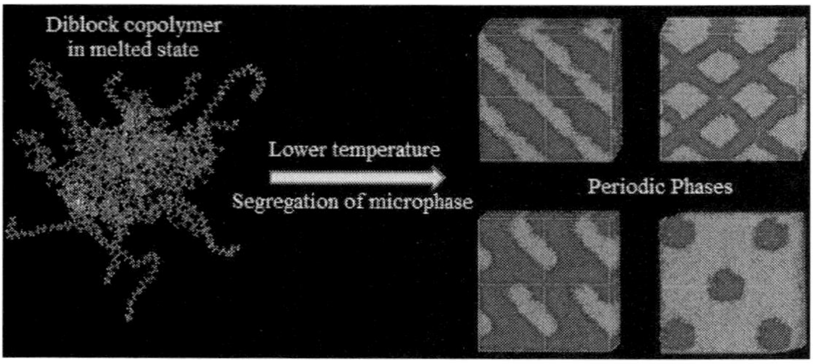

Fig. 1. The microphase segregation process occurs when the PS-PI diblock copolymers in melted state are transformed to a periodic inhomogeneous phase of ordered structures when the temperature diminishes.

polymerization of the component blocks. Periodic phases with specific morphologies such as: spherical, perforated layers, cylindrical, lamellar and Gyroid can be generated manipulating the composition or length of the component blocks (Strobl, 1997; Bates & Fredrickson., 1990).

The order-disorder and order-order phase transitions play an important role in the design and modification of new supramolecular materials and are the key to manipulate the physical and mechanicals properties in these polymeric materials.

In this chapter, attention has been concentrated on the order-disorder and order-order phase transitions that display the PS-PI diblock copolymers and in the mesoscopic simulations methods employed to explore the kinetics transformation pathway of well-defined ordered phases.

1.1 Order-disorder phase transition (ODT)

The order-disorder phase transition is a thermodynamic process controlled by enthalpic and entropic interactions, for example, when a diblock copolymer of type A-B are in a melted state by temperature effect, exhibits a homogeneous phase where all different block segments are completely miscible, in the reverse case when the different block segments are immiscible due to decrease of the temperature displays an heterogeneous state of ordered microphases: this conditions are describes by Gibbs free energy equation of mixing when $\Delta Gm < 0$ and $\Delta Gm > 0$ respectively.

$$\Delta G_m = G_{AB} - (G_A + G_B) \tag{1}$$

where G_A, G_B and G_{AB}, denote the Gibbs free energy of A and B segments in separate states and the mixed state, respectively. Equation (1) in accordance with Flory-Huggins theory can be expressed also as a sum of two thermodynamic contributions (Flory, 1953):

$$\Delta G_m = -T\Delta S_m + \Delta H_m \tag{2}$$

In this equation, ΔH_m and $T\Delta S_m$ exhibits the enthalpic and entropic interactions of mixing at temperature T. The entropic and enthalpic interactions of mixing of two component segments of diblock copolymer are given by:

$$\frac{\Delta S_m}{\bar{R}} = \bar{n}_A \ln\phi_A + \bar{n}_B \ln\phi_B \tag{3}$$

$$\Delta H_m = \bar{R}T\frac{V}{\bar{v}_c}\chi\phi_A\phi_B \tag{4}$$

The equations 3 and 4 can be rewritten as:

$$\Delta S_m = -K[n_A \ln\phi_A - n_B \ln\phi_B] \tag{5}$$

$$\Delta H_m = kT\chi N\phi_A\phi_B \tag{6}$$

where ϕ_A and ϕ_B are volume fraction of A and B components of diblock copolymer, $N=n_A+n_B$ denotes the total number of molecules or degree of polymerization and χ is the Flory-Huggins interaction parameter. The phase behaviour that exhibits the diblock copolymers during an order-disorder phase transition (microphase segregation) is controlled by both entropic and enthalpic interactions. The enthalpic interactions imply the repulsion magnitude between different species via Flory-Huggin's interaction parameter (χ), which represent the chemical incompatibility between different repetitive units and its magnitude is expressed by the type of monomers which integrate the diblock copolymer and has a strong dependence with the temperature:

$$\chi \approx \frac{1}{T} \tag{7}$$

the phase segregation behaviour is controlled by the value of χ, in this way, positives values of interaction parameter lead to incompatibility between different segments and the entropic interactions (ΔS_m) appears to be mostly positives, this generates a positive heat of mixing and therefore a $\Delta G_m > 0$. Negative values of χ lead to homogeneous state and therefore a $\Delta G_m < 0$.

Whereas the entropic interactions involve the configurational and translation displacement of polymeric chains, and are regulated through the degree of polymerization N, architecture constrains and blocks composition (Bates & Fredrickson., 1990, 1999; Hamley, 1998; Balta-Calleja et al., 2000; Thomas et al., 1995). The microphase segregation degree in the diblock copolymers depends of the enthalpic-entropic balance represented by the reduced parameter χN, the ODT occurs at a critical value of χN, the melt phase behaviour is thus governed by composition and a reduced parameter (Leibler et al., 1980; Bates et al., 1990). Three segregation regimes have been identified by Matsen and Bates and have been defined depending on the extent of microphase segregation: the weak ($\chi N \approx 10$), intermediate ($\chi N > 12\text{-}100$) and strong segregation regimes ($\chi N > 100$) (Matsen & Bates, 1996; Bates & Fredrickson, 1990). In the weak segregation regime, the volume fraction of one of the block varies sinusoidally about the average value generating the formation of ordered microphases, this regime is characterized by a diffuse interface between different components and is capable of to be modified by composition effect or for temperature effect (Hamley, 1998; Bates, 1991). In the intermediate segregation regime the composition profile becomes sharper generating ordered microphases with a narrow interface between blocks. The strong segregation regime due to saturation of the blocks composition contains essentially pure components; in this regime, the phase behaviour depends largely on the copolymer composition. Within these segregation regimes it is possible to predict and modify the phase behaviour of block copolymers given χ, N and the segment length (block composition).

A general description of phase behaviour that exhibits the well-defined ordered structures of the PS-PI diblock copolymer via order-disorder phase transitions have been explored by mesoscopic simulation methods. When PS-PI diblock copolymer is in melted state, the polymeric chains assume the lowest free energy configuration, if the diblock copolymer is cooled, the repulsion magnitude expressed by reduced parameter χN increases, when the value of this parameter exceeds a certain value specific ($\chi N \approx 10.5$) for the system under consideration, well-defined periodic structures evolves in the disordered state (Soto-Figueroa et al., 2005).

When the PS-PI diblock copolymer has a symmetric composition (volume fraction of both components are the same) display an ordered phase with lamellar morphology (LAM), Fig. 2(a), however, if the volume fraction of a component increases in relative to other component (asymmetric copolymer), the interface tends to become curve. In this case, the conformational entropy loss of the majority component is too high. Therefore, to gain the conformational entropy, the chains of the majority component tend to expand along the direction parallel to interface. As a result, the PS/PI interface becomes convex towards the minority component. This interface curvature effect is more pronounced when the composition of the diblock copolymer is more asymmetric, Fig. 2(b). The PS-PI diblock copolymers with asymmetric compositions can generate a wide range of ordered structures such as the body-centred-cubic (BCC), hexagonal packed cylinders (HPC), ordered bicontinuous double diamond (OBDD or Gyroid) and lamellar (LAM) arrangement via an order-disorder phase transition process, Fig. 2(c-f).

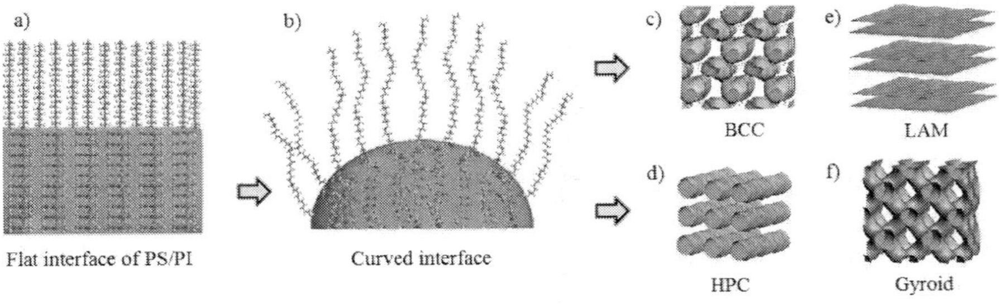

Fig. 2. Schematic representation of composition effect in the phase behaviour of the PS-PI diblock copolymer: a) lamellar phase (symmetric copolymer), b) curvature of PS/PI interface by effect composition (asymmetric copolymer), c) spherical phase, d) cylindrical phase, e) lamellar phase and f) OBDD phase.

The ordered phases with well-defined morphologies that shows the block copolymers are usually describes by the volume fraction of one block (f), the overall degree of polymerization (N), and Flory-Huggins interaction parameter (χ) (Soto- Figueroa et al., 2007). The periodic phases of type BCC, HPC, Gyroid and LAM explored by means of mesoscopic simulations are described in the following sections.

1.1.1 Body centred cubic phase

The body centred cubic (BCC) phase or spherical phase is a classic structure of great thermodynamic stability that exhibits the diblock copolymers of type A-B. This ordered structure shows two specific symmetries: a four-fold symmetry ([100] projection) and the hexagonal symmetry ([111] projection) of cubic array respectively, see Fig. 3.

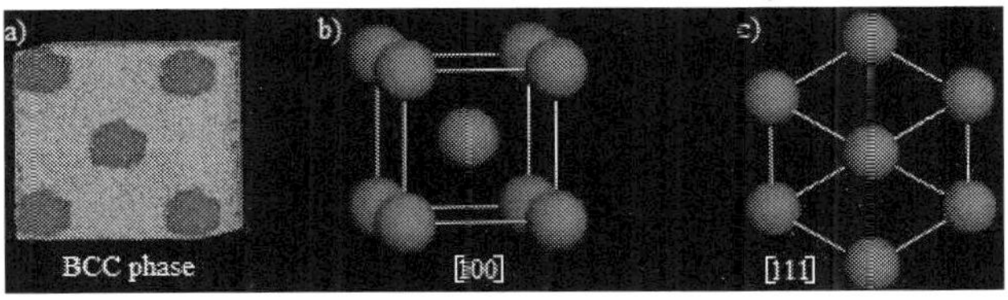

Fig. 3. Snapshots of BCC phase of PS-PI copolymer: a) ordered microdomains of PS and PI chains (red and green regions represent the PS and PI microdomain respectively), b-c) projections [100] and [111] of poly(styrene) microdomains (the poly(isoprene) matrix has been removed for a better visualization).

The BCC phase of diblock copolymers have been studied extensively by both theoretical and experimental methods. For example, the theoretical studies of PS-PI diblock copolymer with linear architecture have demonstrated that this periodic phase can be generated in a composition interval of 0.1 to 0.19 (volume fraction of poly(styrene)) with a stable four-fold symmetry (Soto-Figueroa et al., 2005). The spherical phase of this diblock copolymer is known as a stable phase, however, can be transformed on other ordered structure via an order-order phase transition.

1.1.2 Hexagonal packed cylinders phase

The hexagonally packed cylinders (HPC) phase is also considered as a classic phase of great thermodynamic stability due to its high packing. This ordered structure is characteristic of PS-PI diblock copolymers with linear architecture and can be generated via an order-disorder phase transition in a composition interval of 0.2 to 0.26 (volume fraction of poly(styrene), see Fig. 4.

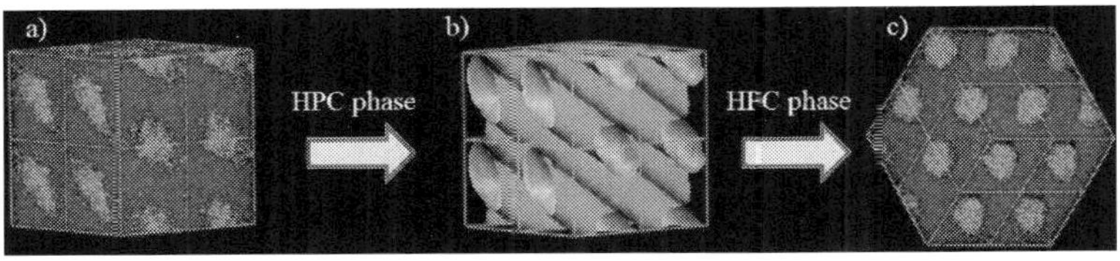

Fig. 4. Snapshots of HPC phase of PS-PI diblock copolymer: a) cylindrical microphase constituted by ordered microdomains of poly(styrene) and poly(isoprene), b) representation of density isosurfaces of microdomains of poly(styrene) and c) poly(styrene) cylinders arranged in a hexagonal lattice.

The HPC phase displays a tetrahedral arrangement of epitaxially cylinders arranged in a hexagonal lattice. In this ordered phase, the cylindrical microdomains of poly(styrene) exhibits an hexagonal arrangement of undulated microdomains immersed in a poly(isoprene) matrix.

1.1.3 Ordered bicontinuous double diamond phase

The ordered bicontinuous double diamond (OBDD) phase (or Gyroid phase) is the ordered structures more complex that exhibits some diblock copolymers of type A-B. For example, The PS-PI diblock copolymer can generate this ordered structure via an order-disorder phase transition, see Fig. 5.

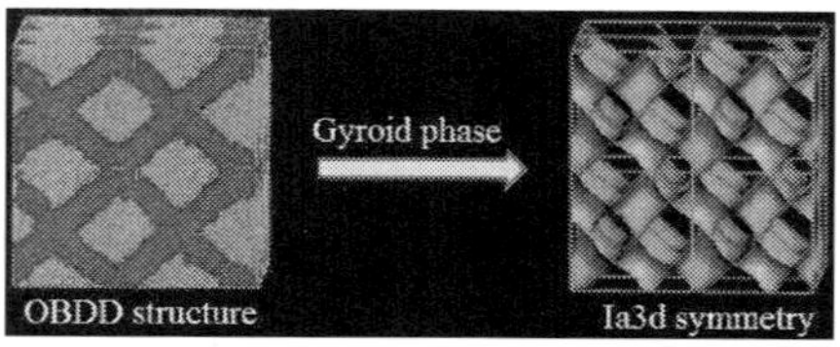

Fig. 5. Snapshots of OBDD arrangement of PS-PI diblock copolymer obtained by mesoscopic simulations: a) ordered microdomains of poly(styrene) and poly(isoprene) chains, and b) representation of density isosurfaces of microdomains of poly(styrene) (Soto-Figueroa et al., 2008).

The OBDD phase has a tetrahedral arrangement of epitaxially cylinders interconnected by channels of Ia3d symmetry (Hajduk et al., 1995). The OBDD phase of PS-PI copolymer exists only in a narrow interval of values (0.37 to 0.4 volume fraction of poly(styrene)) between the regimes of the perforated layer and lamellar phases. The Gyroid arrangement is known as a stable phase, however, can be transformed on other ordered structure via an order-order phase transition.

1.1.4 Lamellar phase

Ordered phase with lamellar morphology are characteristics of block copolymers of symmetric and asymmetric composition. This periodic phase is constituted of alternating layers of different homopolymer microdomains separates by flat interfaces, see Fig 6. For the case of PS-PI diblock copolymer with linear architecture, this displays the lamellar arrangement in a specific predominance interval between 0.47 to 0.68 volume fraction of poly(styrene).

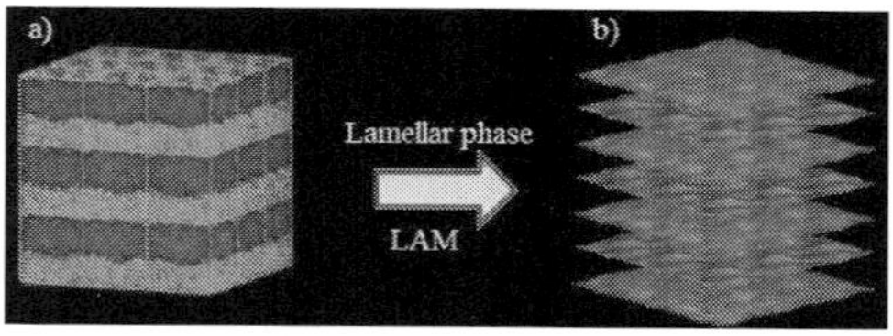

Fig. 6. Snapshots of lamellar phase of PS-PI diblock copolymer, the lamellar structure exhibits undulated microdomains of poly(styrene) and poly(isoprene): a) ordered microdomains of poly(styrene) and poly(isoprene) and b) representation of density isosurfaces of LAM phase.

The lamellar phase can be generated in the three segregation regimes ($\chi N \approx 10$, χN >12-100 and χN >100) via order-disorder phase transitions and can displays flat and undulated interfaces.

1.1.5 Phase diagram of PS-PI diblock copolymer

The ordered structures of PS-PI diblock copolymer represent under specific conditions the states with the lowest Gibbs free energy, these equilibrium phases can be classified through a phase diagram, Fig. 7. The diagram phase depicts the two regions associates constituted by a homogeneous state and an ordered state of five ordered phases different. Each equilibrium phase shows its predominance zone in terms of the volume fraction of poly(styrene) block and reduced parameter χN.

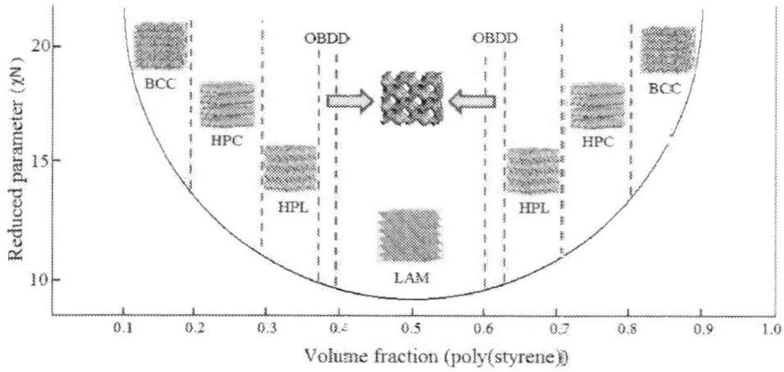

Fig. 7. Phase diagram of PS-PI diblock copolymer, the continuous curve describes the points of phase transition between the homogeneous state and the microphase-separated states. The ordered states split into different classes (BCC, HPC, G, HPL and LAM); the dashed lines show the predominance interval between the different types of ordered phases.

The segregation state in the phases diagram is controlled by the product χN, if χN is mayor than a critical value (typically of the order of 10.5), entropic factors dominate and the diblock copolymers exist in an ordered phases state. On the other hand, an order-to-disorder transition takes place for large values of χN. The phase behaviour of each ordered structures kind on the phase diagram of poly(styrene)-poly(isoprene) diblock copolymer can be modified through order-order phase transitions, this will be tackled in the following section.

2. Order-order phase transition in block copolymers

The diblock copolymers display a wide variety of classic phases with morphologies or defined structures such as: body centred cubic, hexagonal packed cylinders, hexagonal perforated layers, lamellar and the ordered bicontinuous double diamond phase, the morphology of these periodic structures can be controlled and modified by two different routes: 1) order-disorder phase transition and 2) thermotropic order-order phase transition.

In the order-disorder phase transition as was mentioned in the previous section, the phase behaviour in the block copolymers is governed by three experimentally controllable factors during the synthesis process: i) the overall degree of polymerization, ii) architecture of block copolymer, and iii) the interaction parameter between components blocks, of this thermodynamic process can emerge periodic structures with well-defined morphologies (Soto-Figueroa et al., 2005, Soto-Figueroa et al. 2007).

Other way to modify the phase behaviour in the block copolymers is through the temperature; thermally induced phase transitions have the potential to promote the kinetic control in these synthetic materials. Leibler was the first in predicting the phase transition between different ordered structures by temperature effect (Leibler et al., 1980). The thermally induced phase transformations are governed for anisotropic composition fluctuation effects (Ryu et al., 1999), when an ordered phase of specific morphology is subject to a thermal heating process, the homopolymer chains into the ordered microdomains exhibit thermodynamic instability by temperature effect and become less rigid, this entropic process promotes the polymeric chains movement (composition fluctuations) into the homopolymer microdomains and modifies with the time the shape of ordered phase.

Three segregation regimes have been defined to explain the extent of microphase segregation and the thermodynamic stability in the classical phases that exhibits the diblock copolymers, these segregation regimes are : weak regime ($\chi N \approx 10$), intermediate regime (χN >10-100) and strong segregation regime (χN >100) (section 1.1).

Our interest is concentrated in the weak segregation regime, because in this predominance zone, the diblock copolymers are characterized by a widened interface due to enhanced phase mixing. In the vicinity to this regime, thermotropic phase transition between different kinds of ordered phases can be generating (Bates et al., 1990; Matsen et al., 1996).

The transition from one ordered state to another is nowadays denominated as an order-order phase transition (OOT) (Sakurai et al., 1993; Kim et al., 1998; Almadal et al., 1992; Sakurai et al., 1996; Sakamato et al., 1997; Modi et al., 1999). In this way, the ordered phases that exhibit the diblock copolymers inside weak segregation regime can exhibit order-order phase transitions when are subjects to thermal heating cycles.

In the past decade the order-order phase transitions that display the classical phases of diblock copolymers (BCC, HPC, Gyroid and Lamellar) have been investigated by both experimental and theoretical studies. The order-order phase transition is thermoreversible process that develops transient metastable states during phase transformation. The order-order phase transition that exhibits the classical phases with specific morphology such as Gyroid, cylindrical, and lamellar are described next.

2.1 Order-order phase transition of HPC structure

The ordered phase of hexagonal packed cylinders also known as cylindrical structure is characteristic of diblock and triblock copolymers, for example, the poly(styrene)-poly(isoprene) diblock copolymer with linear architecture can generate this ordered arrangement in a specific composition of 0.2/0.8 (volume fraction) of PS/PI via an order-disorder phase transition. When the HPC phase of this diblock copolymer is subject to thermal heating cycles, exhibits an order-order phase transition to body-centred-cubic (BCC) structure (or spherical phase). The order-order transition pathway that exhibits the HPC to BCC phase was recently reported (Modi et al., 1999; Krishnamoorti et al., 2000).

Experimental and theoretical studies have confirmed that HPC phase develops transient metastable phases during thermal heating process (Kimishima et al., 2000; Krishnamoorti et al., 2000; Rodríguez-Hidalgo et al., 2009). The order-order transition pathway from HPC to spherical phase is sketched schematically in Fig. 8.

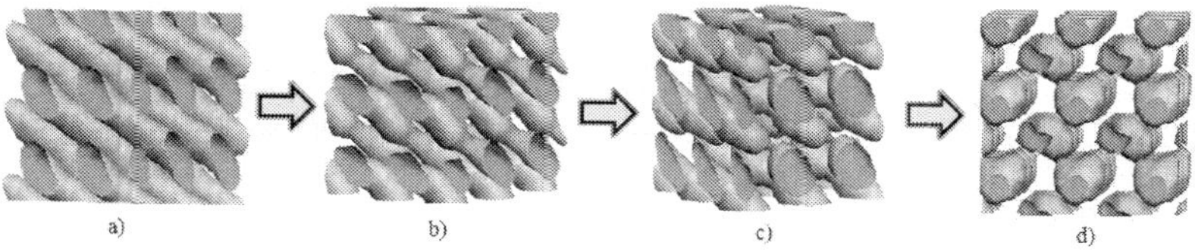

Fig. 8. Snapshots of order-order phase transition process from HPC phase to spherical arrangement monitored by means of mesoscopic simulations: a) cylindrical phase, b) undulation process of poly(styrene) microdomains by temperature effects, c) breakdown of cylindrical microdomains of poly(styrene) and d) formation and stabilization of body-centred-cubic arrangement (BCC phase) (Soto-Figueroa et al., 2008).

Three transitory stages are typical of OOT process from HPC to cylindrical phase: (Stage 1) undulation process of cylindrical microdomains of poly(styrene) in the poly(isoprene) matrix. In this transitory stage the thermal heating induce the thermodynamic instability of HPC phase, where the cylindrical microdomains and the PS/PI interface becomes less rigid due to anisotropic composition fluctuations, this generates the undulation of the poly(styrene) microdomain in the poly(isoprene) matrix. (Stage 2) breakdown of cylindrical microdomains of poly(styrene) by temperature effects. When anisotropic composition fluctuations reach a critical point of thermodynamic instability the cylindrical microdomains become unstable and therefore they are broken in ellipsoids, (Stage 3) formation and stabilization of BCC phase. In this thermally induced stage, the ellipsoids microdomains generated by breakdown of cylindrical microdomains of poly(styrene) evolve to an equilibrium state where the uniform spherical microdomains are stabilized into body-centred-cubic arrangement. The order-order phase transition from HPC to spherical phase is a thermoreversible process. The inverse process from BCC state to HPC arrangement involves deformation and elongation of spheres into ellipsoids and coalescence of ellipsoids into the cylindrical microdomains. This process is driven by thermodynamic instability of the spherical interface caused by decrease of temperature.

2.2 Order-order phase transition of Gyroid structure

The Gyroid phase as was mentioned in the section 1.1.3 exhibits a tetrahedral arrangement of epitaxially cylinders interconnected by channels of Ia3d symmetry and is capable of suffering order-order phase transitions when it is submitted to thermal heating cycles. The poly(styrene)-poly(isoprene) diblock copolymer displays this ordered phase in a narrow interval of specific composition. The thermally-induced phase transition from Gyroid to lamellar phase in PS-PI diblock copolymer was explored in detail by mesoscopic simulation methods (Soto-Figueroa et al., 2008). The phase transformation during thermal annealed proceed in several stages via the generation two metastable phases (HPL and cylindrical). The order-order phase transition process from Gyroid to lamellar phase is sketched schematically in Fig. 9.

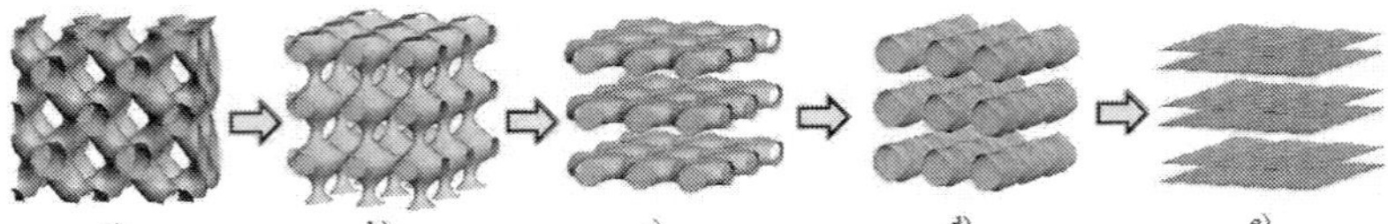

Fig. 9. Transitory stages and metastable phases generated during the order-order phase transition from Gyroid phase to lamellar arrangement: a) undulation of interconnected microdomains of poly(styrene) into the poly(isoprene) matrix, due thermal heating, b) breakdown of interconnected microdomains of Gyroid phase, c) formation of HPL metastable phase, d) formation of cylindrical metastable phase and e) lamellar phase.

When the Gyroid phase of PS-PI copolymer is subject to thermal heating cycles, the interconnected microdomains of poly(styrene) display an undulation process due to anisotropic composition fluctuations, Fig 9(a), with the increase of temperature the anisotropic composition fluctuations induce the breakdown of side interconnections in the Gyroid arrangement, Fig. 9(b), generating the first metastable phase with HPL arrangement, Fig. 9(c). The HPL metastable phase evolves later to cylindrical arrangement by temperature effect, in this transient stage the interconnections of perforated layer microdomains diminish their volume up to the breakdown of interconnections in the HPL arrangement and the formation of cylindrical phase (second metastable phase), Fig 9(d). Finally, the cylindrical metastable phase also changes over time due to anisotropic composition fluctuations. The cylindrical microdomains in this metastable phase are thermodynamically unstable, in order to reach a thermodynamic stability of minimal energy, the cylinders microdomains are joined together to evolve into undulating lamellar phase, Fig. 9(e), the lamellar phase consisting of alternating layers of PS and PI microdomains.

2.3 Order-order phase transition of lamellar structure

The lamellar phase is the ordered structure more simple that exhibit the multiblock copolymers is considered a classical phase of great thermodynamic stability. The lamellar phase consisting of alternating layers of different homopolymer microdomains. The PS-PS diblock copolymer exhibits this ordered arrangement in a specific composition interval (see section 1.1.4). When the lamellar phase of this diblock copolymer with an equivalent composition between their constituent blocks (0.5 volume fraction of poly(styrene) and poly(isoprene)) is subject to thermal heating cycles, it does not generate an order-order phase transition, the lamellar arrangement in this case evolve to a homogeneous state (melted), see Fig. 10.

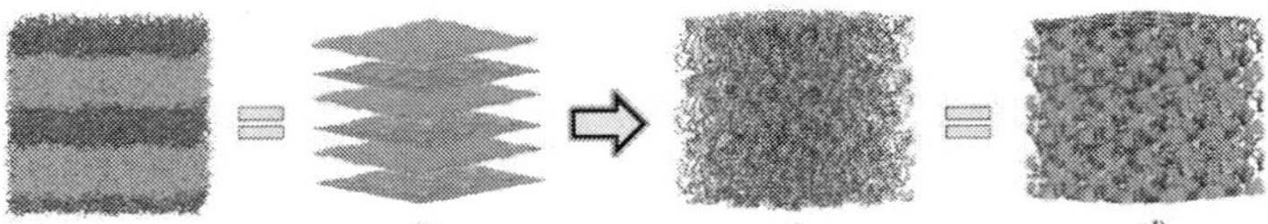

Fig. 10. Snapshots of thermal heating of lamellar phase (diblock copolymer of symmetric composition): a-b) equilibrium phase of lamellar structure (representation of diblock chains and surface isodensities of PS and PI microdomains) and c-d) homogeneous state after of thermal heating (melted phase).

Nevertheless, the lamellar phase generated of block copolymers of asymmetric composition, for example of 0.45/0.55 of PS/PI (volume fraction), it can exhibit a phase transformation to HPL arrangement when is subject to thermal heating via an order-order phase transition process (Mani et al., 2000). The order-order phase transition from lamellar to HPL phase is showed in Fig 11.

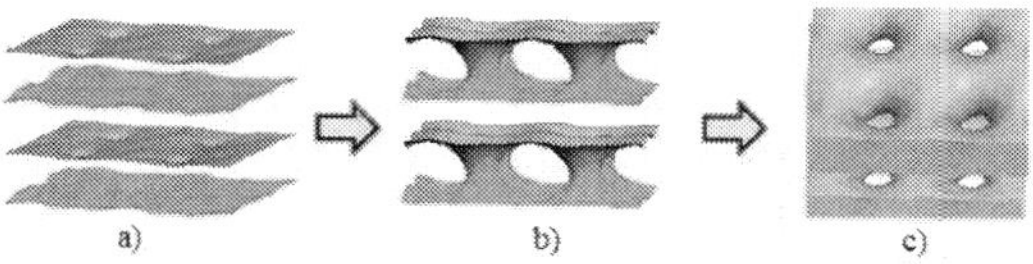

Fig. 11. Order-order phase transition stage from lamellar phase to HPL arrangement: a) undulation process of lamellar microdomains, b) interconnection of parallel microdomains of poly(styrene) and c) formation and stabilization of HPL arrangement.

The order-order phase transition process of this ordered arrangement is governed by the thermodynamic instability between the components microdomains, generated by thermal heating, where the anisotropic composition fluctuations play an important role in the phase transformation. In the initial stage the lamellar microdomains exhibit an undulation process due to thermal heating, Fig. 11(a), with the temperature increase, the lamellar phase becomes thermodynamically unstable, in order to reach a thermodynamic stability of minimal energy, the lamellar alternate microdomains of poly(styrene) are interconnected by means of narrow microdomains (parallels to PS/PI interface), finally the HPL phase evolve in an energetic equilibrium state (Soto-Figueroa et al., 2007), see Fig. 11(b-c).

3. Thermal study of double directionality of order-order phase transitions of hexagonally perforated layers (HPL) phases by mesoscopic simulation

During the past two decades have been reported theoretical and experimental studies of diblock copolymers, where the ODT and OOT transitions play a main role in the structure control and consequently in the physical properties control of these polymeric materials (Kimishima et al., 2000; Kim et al., 2006; Bodycomb et al., 2000; Krishanamoorti et al., 2000; Court et al., 2006; Soto-Figueroa et al., 2008). It is well-known that the poly(styrene)-poly(isoprene) diblock copolymer exhibits a wide variety of classical phases with specific morphologies such as: BCC, HPC, OBDD, LAM and HPL. The order-order phase transitions that exhibit this ordered structures with definite morphologies have been explored by Soto-Figueroa and Rodríguez-Hidalgo, they have confirmed the OOT´s between HPC to BCC, Gyroid to LAM microphase through mesoscopic simulations (Soto-Figueroa et al., 2007, 2008; Rodríguez-Hidalgo et al., 2009), although the OOT´s have been well investigated for the majority of classical phases, the dynamic transformation of HPL phase has not been investigated yet in detail. The HPL phase exhibits a double directionality of order-order phase transition when is subject to thermal process. Experimental evidences suggest that HPL phase of the PS-PI copolymer can evolve to a cylindrical structure and to a Gyroid structure by temperature effect (Park et al., 2005; You et al., 2007).

The double directionality of order-order phase transitions that exhibits the classical phase of PS-PI copolymer is a topic interesting to control the morphology and physical properties of polymeric materials. The order-order phase transitions can be investigated in more detail through mesoscopic simulations than through experimentation. Mesoscopic simulations are efficient methods to investigate the physical processes of soft matter and their interactions with chemical environments (Rodríguez-Hidalgo et al., 2011; Ramos-Rodriguez et al., 2010). Offer a particularly useful way of exploring the matter transfer process and to make predictions that may be of interest for understanding and elucidating complex process such as the double directionality of order-order phase transitions. In the mesoscopic simulations the atoms of each molecule are not directly represented, but they are grouped together into beads (coarse-grained models), where a springs force reproduce the typical nature of them, and therefore can exhibit a real physical behaviour in multicomponent systems. In this section, we explored the double directionality of phase transition of HPL structure of the PS-PI copolymer from a mesoscopic point of view by mesoscopic simulations, where the phase evolution stages and transient ordered states are analysed.

3.1 Model and simulation method

To explore the order-order phase transition of the HPL phase, we employed Dissipative Particle Dynamics (DPD) simulations. The original DPD method was introduced by Hoogerbrugge and Koelman and was later modified by Groot, R.D. (Hoogerbrugge & Koelman, 1992; Koelman & Hoogerbrugge, 1993; Groot & Warren, 1997; Groot & Madden, 1998, 1999). The DPD method allows the study of high-molecular-weight systems as the polymeric materials. The coarse-graine

$$\mathbf{f}_i = m_i \frac{d\mathbf{v}_i}{dt} \tag{8}$$

where $\mathbf{r}_i$, $\mathbf{v}_i$, m_i and $\mathbf{f}_i$ are the position, velocity, mass, and force, respectively, of bead i. Dimensionless units are used in DPD simulations; usually, the mass of each bead is set to 1 DPD mass units, which results in an equation between the force acting on a bead and its acceleration. Each particle is subject to soft interactions with its neighbours via three forces: conservative (F_{ij}^C), dissipative (F_{ij}^D) and random (F_{ij}^R). The total force acting on particle i is:

$$f_i = \sum_{j \neq i} \left(F_{ij}^C + F_{ij}^D + F_{ij}^R \right) \tag{9}$$

The conservative force F_{ij}^C is a soft repulsive force that acts between particles i and j. The dissipative force $F_{ij}{}^D$ corresponds to a frictional force that depends on both the positions and relative velocities of the particles. The random f $\mathbf{v}_i = \frac{\mathbf{r}_i}{dt}$ orce $F_{ij}{}^R$ is a random interaction between a bead i and its neighbour bead j. All forces vanish beyond a cut-off radius r_c, which is usually chosen as the reduced unit of length, $r_c \equiv 1$. The $F_{ij}{}^D$ and $F_{ij}{}^R$ forces act as a thermostat that conserves momentum and gives the correct hydrodynamics at sufficiently large time and length scales. These forces are given by:

$$F_{ij}^{C} = \begin{bmatrix} a_{ij}(1-r_{ij})\hat{r}_{ij} & (r_{ij} \le 1) \\ 0 & (r_{ij} > 1) \end{bmatrix} \tag{10}$$

$$F_{ij}^{D} = \left[-\gamma\omega^{D}(r_{ij})(\mathbf{v}_{ij}\cdot\hat{\mathbf{r}}_{ij})\hat{\mathbf{r}}_{ij}\right] \tag{11}$$

$$F_{ij}^{R} = \left[\sigma\omega^{R}(r_{ij})\xi_{ij}\hat{\mathbf{r}}_{ij}\right] \tag{12}$$

where $r_{ij} = \left|\vec{r}_i - \vec{r}_j\right|$, $\hat{r} = \vec{r}_{ij} / r_{ij}$, γ is the dissipation strength, σ is the noise strength, $\omega^{D}(r_{ij})$ and $\omega^{R}(r_{ij})$ are weight functions of F_{ij}^{D} and F_{ij}^{R} forces, respectively, and a_{ij} is the maximum repulsive force between particle *i* and *j*. For the DPD system to have a well-defined equilibrium state that obeys Boltzmann statistics, the equilibrium temperature is defined as $k_BT = \sigma^2/(2\gamma)$. This condition fixes the temperature of the system and relates with the two DPD parameters γ and σ (k_BT is usually chosen as the reduced unity of energy). The parameter a_{ij} (henceforth referred to as the bead-bead repulsion parameter or simply as the DPD interaction parameter) depends on the underlying atomistic interactions and is related to the parameter χ through:

$$a_{ij} = a_{ii} + \frac{k_BT\chi_{ij}(T)}{0.306} \tag{13}$$

In this way, a connection exists between the molecular character of the coarse-grained model and the DPD parameter. The parameter a_{ij} is given in terms of k_BT (DPD reduced units). Equation (6) implies that, if the species are compatible, $\chi_{ij} \approx 0$ and therefore, $a_{ij} = 25$. Established procedures for mapping between the DPD and physical scales and for choosing the system temperature are not yet available. We therefore use Flory-Huggins theory (through the $\chi(T)$ dependence) to introduce the temperature into the DPD simulations. To calculate the a_{ij} values of Equation (6), we use the Hildebrand relation:

$$\chi = \frac{V_m}{RT}(\delta_1 - \delta_2)^2 \tag{14}$$

where V_m and δ are the mean molar volume and solubility parameter, respectively. In the DPD simulation, the dynamic behaviour of order-order phase transition of the HPL structure is followed by integration of the equations of motion of each species using a modified version of the Verlet algorithm (Verlet, 1967). The integration of the equations of motion for each particle generates a trajectory through the system's phase, from which thermodynamic observables may be constructed by suitable averaging. Based on this information, the order-order phase transition can be observed. In this algorithm, the forces are still updated once per integration, thus there is virtually no increase in computational cost.

3.2 Coarse-grained models and parameterization

The molecular structure of PS-PI copolymer was built by means of the polymer builder module of Accelrys (Accelrys, 2006). The polymeric molecule with linear architecture contains a total of 300 repetitive units in the main chain. The molecular weight of the PS-PI copolymer presented an interval of 20798–25842 g/mol. The diblock copolymer molecule was replaced by a coarse-grained model constituted by 30 beads, where each bead represents a statistical segment (characteristic ratio, ($C_n \approx 10$) (Soto-Figueroa, et al 2005)), see Figure 12. The Equation (15) was used to map the real structure of diblock copolymer to statistical model (mesoscopic model).

$$C_{SGD} = \frac{M_P}{M_m(SSL)} \tag{15}$$

where C_{SGD}, M_P, M_m and SSL are bead number with Gaussian distribution, the molar mass of the block copolymer, molar mass of a repeat unit and means statistical segment level (characteristic ratio (C_n), or persistence length (L_p) or statistical Kuhn segment length (a_k)) respectively (Soto-Figueroa et al., 2007).

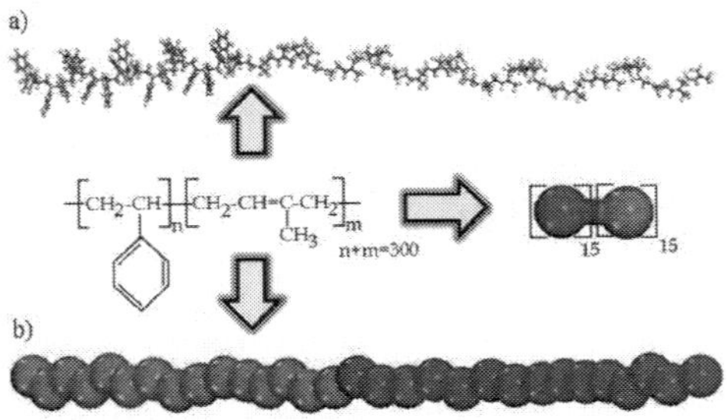

Fig. 12. Schematic representation of PS-PI copolymer: a) chemical structure of a PS-PI chain, n and m represent the polymeric unit number of each block, b) coarse-grained model of PS-PI system with linear architecture.

The chemical and physical nature of the coarse-grained model in mesoscopic simulation is described by interaction parameters (χ_{ij}). In order to study the order-order transition at temperatures different with DPD simulations we take the temperature dependence $\chi_{ij}(T) = \chi(T)$, in this way the real temperature is introduced into the DPD simulation (Rodríguez-Hidalgo et al., 2009). The interaction parameters for Equation (13) were evaluated from bulk atomistic simulations using the Fan, F.C. model (Fan et al., 1992). In this way the parameter interaction was expressed as:

$$\chi(T) = \frac{\Delta G(T)}{RT} = \frac{Z_{12}E_{12}(T) + Z_{21}E_{21}(T) - Z_{11}E_{11}(T) - Z_{22}E_{22}(T)}{2RT} \tag{16}$$

where ΔG denotes the Gibbs free energy, χ is interaction parameter, Z and ΔE_{12} are coordination number and differential energy of interaction of an unlike pair respectively. The Figure 13 show the interaction parameters $\chi(T)$, the temperature interval analysed is from 298 K to 500 K.

The tendencies that exhibit interaction parameters in temperature function are in agreement with the Hildebrand relation and are adequate to explore the formation of HPL phases (via

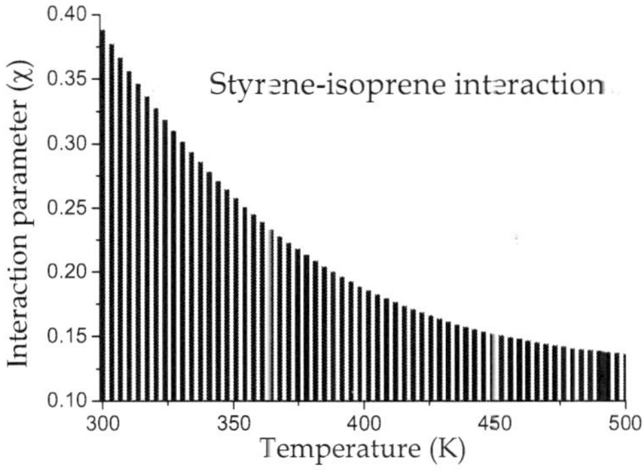

Fig. 13. Interaction parameters between styrene and isoprene molecules at temperatures different obtained by Monte Carlo molecular simulation (Vicente et al. 2006).

order-disorder transition) to room temperature and the double directionality of the HPL phase (via order-order transition) employing the coarse-grained model with predefined architecture proposed in this mesoscopic study.

All DPD simulations were performed in a cubic box that measured 20×20×20 in size, which contained a total of 2.4×10^4 representatives particles, a spring constant of C=4, and a density of ρ=3. The interaction parameters between identical species were then chosen as $a_{ST\text{-}ST} = a_{PI\text{-}PI} = 25$. Each bead was assigned a radius of 1. The coarse-grained number for each chemical species was held constant during the DPD simulations.

3.3 Simulation results and discussion

3.3.1 Hexagonally perforated layers structures

The equilibrium phases formation of PS-PI copolymer is governed by the composition, temperature and immiscibility between their components blocks, these factors were considered in the coarse-grained model via the architecture and interaction parameters.

All simulations start from a disordered state, where the PS-PI chains are in a homogeneous melted phase. First, we set the interactions parameters at T = 298 K. We then let the simulation proceed for 5×10^5 steps and during the temperature relaxation; we observed the microphase segregation process and the generation of equilibrium phases via ODT. Several transient stages were detected in the ODT process; a) melt phase, where the copolymer chains move freely; b) microphase segregation by temperature effect (temperature decrease), c) generation of pure microdomains of poly(styrene) and poly(isoprene), in this stage the ordered phases system attains an equilibrium temperature to room temperature.

To identify the composition region where the HPL structures are formed, we scanned the composition interval from 0.1–0.5 (volume fraction of poly(styrene) with increments of 0.03) at constant temperature of 298 K (i.e. constant χN). All DPD simulations generate a coarse-grained system sufficiently large to observe the classical phases formation in the analysed composition interval. The mesoscopic simulation of PS-PI copolymer exhibits a wide variety of structures or equilibrium phases such as LAM, HPC, BCC, Gyroid and HPL. The equilibrium phases of PS-PI system depend primarily on three factors: (i) the volume fraction of PS and PI blocks (*f*), (ii) the degree of polymerization (*N*), and (iii) the interaction parameter (χ). The results obtained by DPD simulations are in accordance with the mean-field theory (Leibler, 1980).

The HPL phase was obtained into the composition intervals of 0.3 to 0.36 (volume fraction of poly(styrene)), see Figure 14. The tendency of PS-PI chains to self-assemble into HPL structures depends of the previous factors and is governed by thermodynamics interactions (enthalpic and entropic) during the microphase segregation process.

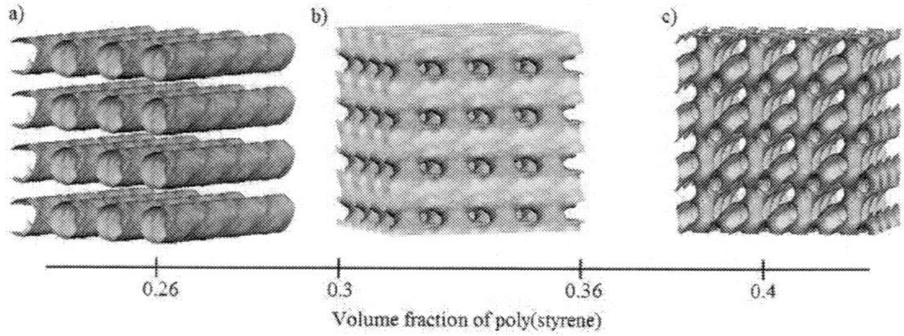

Fig. 14. Specific phases of PS-PI copolymer obtained as the blocks composition (PS/PI) is modified: a) cylindrical, b) HPL and c) Gyroid.

The thermodynamic stability of HPL phase in the predominance region vary with the relative chain length (poly(styrene) composition) of the component blocks. The predominance region where the HPL phase is formed has two composition limits that are contiguous with other equilibrium phases of different morphology. In the low composition limit where the volume fraction of poly(styrene) is close to 0.3, the HPL structure has as neighbour the cylindrical phase, whereas in the height composition limit (close to 0.36), the HPL structure has as neighbour the Gyroid structure. These behaviours between different phases are in accordance with the phase diagram reported by Khandpur (Khandpur et al., 1995).

The order-order phase transitions that exhibits the HPL structures with specific compositions of 0.3 and 0.36 (volume fraction of poly(styrene) are investigated. The HPL structures are modified through temperature increase, keeping constant the PS-PI block composition. The OOT of HPL structure for each specific composition shows a selective directionality (from HPL to cylinders and from HPL to Gyroid) during thermal study.

3.3.2 Order-order phase transition of HPL to cylindrical phase

The HPL phase with a specific composition of 0.3 (volume fraction of poly(styrene) was put at continuous cycles of thermal heating in the temperature interval from 298 to 500 K. A total of 2.0 x 10^5 time steps with step size of Δt = 0.03 were allowed in the mesoscopic simulation to reach the thermodynamic balance of PS-PI system in each temperature increment. The thermally induced phase transition from HPL to cylindrical phase was observed at the temperature of 432 K, this corresponds to the OOT, and this fact is in accordance with theoretical and experimental evidences (You et al., 2007).

The OOT process was visualized during mesoscopic simulation; the results show that the transformation from HPL to cylindrical phase by temperature effect is generated in several stages. When the HPL phase is annealed below the OOT temperature, we observed a slow dynamic motion of the PS and PI microdomains, the thermodynamic stability between microdomains different they are in an energetic barrier that maintain the HPL phase stable.

When the HPL phase is annealed to a higher temperature, T≥ 432 K, the energetic barrier that maintain the HPL phase stable is exceeded, the thermodynamic interactions (enthalpic and entropic) play an important role in the OOT process. The enthalpic interaction is proportional to the Flory-Huggins interaction parameter (Fig. 13), which is found to be inversely proportional to temperature.

During the temperature increase, the interaction parameter between the PS and PI microdomains diminishes, generating the OOT from HPL arrangement to cylindrical phase, the enthalpic interaction in the OOT is accompanied by an increase in entropy. At higher temperatures the entropic interactions dominates and is cause of anisotropic composition fluctuations into polymeric microdomains (Ryu, et al., 1999).

The HPL phase develops short-lived transient states during the OOT process, because of interface instability (PS/PI), combined with fast molecular motion of PS and PI microdomains. In Figure 15(a-d) are shown snapshots of the thermally induced phase transition from HPL to cylindrical structure during the annealed process at temperature, T= 432 K.

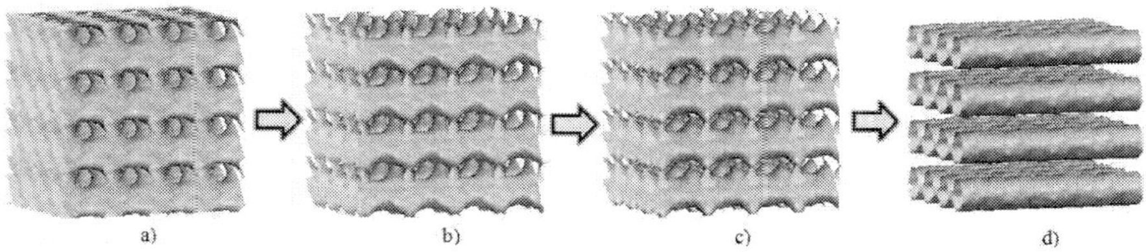

Fig. 15. Snapshots of the OOT pathway from HPL to cylindrical phase: a) undulation process of HPL phase, b-c) instability of HPL phase and d) formation of cylindrical phase.

The snapshot of Figure 15(a) corresponds to the initial equilibrium state of HPL phase, initially, the PS perforated microdomains display an undulation process induced by thermodynamic instability, in this point the PS perforated microdomains and the PS/PI interface become less rigid by temperature effect. The PS microdomains maintain their shape an short-period of time, as time goes, the anisotropic composition fluctuations (thermally-induced) increase quickly, the PI interconnections into PS perforated microdomains enlarge their area, in this stage the PS microdomains are unstable see Figure 15(b-c). When the anisotropic composition fluctuations reach a critical point, the PS unstable microdomains change their structure to a new cylindrical arrangement, see Figure 15(d), in this thermally induced stage the uniform cylindrical microdomains are stabilized.

3.3.3 Order-order phase transition of HPL to Gyroid phase

The HPL phase with a volume fraction of PS, f_{PS}=0.36, (obtained within the first 5x10^5 time steps) was now subject to a thermal heating process for another 5x10^5 time steps. The simulation outcome shows an OOT from HPL to Gyroid phase at the temperature of 438 K. The composition increase of poly(styrene) in HPL microdomains modify the phase transition directionality. The OOT from HPL to Gyroid phase obtained by DPD simulation are consistent with experimental results reported by Insun Park. They have investigated the phase transition behaviour from the hexagonally perforated layer (HPL) to the Gyroid phase in supported thin film of a poly(styrene)-*b*-poly(isoprene) (PS-*b*-PI) diblock copolymer (OOT occur at temperature of 443 K) (Park et al., 2005).

During the thermal heating process, the HPL phase undergoes the microdomains modification of PS and PI, generating transient intermediate stages as are sketched schematically in Figure 16(a-c). Three transient intermediate stages were detected during the OOT: (i) undulation process of pure microdomains and PS/PI interface due to the thermodynamic instability by temperature effect, see Figure 16(a), in this stage, the entropic and enthalpic interactions govern the microphase stability and induce anisotropic composition fluctuations into pure microdomains, (ii) increase of volume of poly(isoprene) interconnections into PS perforates microdomain, Figure 16(b), in this stage the PS perforated microdomains are instable, (iii) formation of parallel interconnections between PS perforated layers, see Figure 16(c), in this stage, the Gyroid phase is formed.

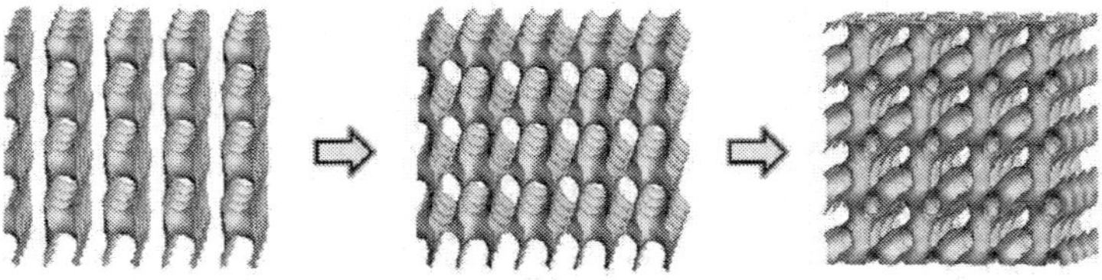

a) b) c)

Fig. 16. Snapshots of evolution process of the OOT from HPL to Gyroid phase: a) undulation process of PS microdomains in the PI matrix by anisotropic composition fluctuations effect, b) volume increase of PI interconnections into PS perforates microdomains, c) formation of Gyroid phase by generation of parallel interconnections between PS perforated microdomains.

The HPL phase of PS-PI diblock copolymer exhibits a double directionality of order-order transition from HPL to cylindrical phase and from HPL to Gyroid phase. The anisotropic composition fluctuations of PS and PI microdomains of HPL phase by thermal heating process induce the order-order phase transitions, however the phase transition directionality towards a specific phase (cylindrical or Gyroid) in the thermal process is governed by small variations of poly(styrene) concentration in the poly(styrene) microdomains of HPL phase. Variations of small composition into structured microdomains according to Leibler´s mean-field theory greatly modify the phase thermodynamic behaviour (separation, segregation and phase transformation).

4. Conclusion

The block copolymers are "smart" materials that have the ability to self-assemble inside a variety of periodic phases of high regularity in size and shape. The phase behaviour (type of structure and morphology) that exhibits these polymeric materials can be controlled by two different processes: order-disorder phase transition and thermotropic order-order phase transition. In the order-disorder phase transition, the phase behaviour is governed by entropic and enthalpic interactions, whereas the order-order phase transition are controlled by anisotropic composition fluctuations of theses thermodynamic processes can emerge supramolecular structures with well-defined morphologies and specific properties.

The ODT and OOT that exhibits of block copolymer can be explored in more detail through mesoscopic simulations than through experimentation. The mesoscopic simulation methods, offer a particularly useful way of exploring the phase behaviour pathway and to make predictions that may be of interest for understanding and elucidating complex process such as order-disorder of order-order phase transitions.

The DPD approach has been successfully applied to the investigation of phase transition processes (ODT and OOT) of HPL structure. The mesoscopic simulation outcomes show that the HPL phase of PS-PI diblock copolymer exhibits a double directionality of order-order phase transition from HPL to cylindrical phase and from HPL to Gyroid phase. This double directionality of OOT is controlled by small variations of poly(styrene) concentration in the poly(styrene) microdomains of HPL phase. Finally, all the simulation outcomes are qualitatively consistent with the experimental results, demonstrating that the DPD method may provide a powerful tool for the investigation and analysis of soft matter transformation process by thermal heating effects.

5. Acknowledgment

This work was supported by the Universidad Nacional Autónoma de México (UNAM), PAPIIT project No. IN104410-2 and IN109712. We also acknowledges the financial support of the Consejo Nacional de Ciencia y Tecnología (CONACYT) Project: 2012.

6. References

Almdal, K.; Koppi, K.A.; Bates, F.S.; Mortensen, K. (1992) Multiple ordered phases in a block copolymer melt. *Macromolecules*, Vol. 25(6), pp. 1743-1751.

Accelrys. (2006), Material studio release, notes, release 5.0; Accelrys Software, Inc.: San Diego, CA.

Bates, F.S. & Fredrickson, G.H. (1990). Block copolymer thermodynamics: Theory and experiment. *Annu. Rev. Phys. Chem.*, Vol. 41, pp. 525-557.

Bates, F.S. & Fredrickson, G.H. (1999), Block copolymers-Designer soft materials. *AIP Phys. Today*, 2, 32-38.

Bates, F. S. (1991). Polymer – Polymer phase behavior. *Science*, Vol. 251, pp. 898.

Bodycomb, J.; Yamaguchi, D.; Hashimoto, T. (2000). A Small-Angle X-ray Scattering Study of the Phase Behavior of Diblock Copolymer/Homopolymer Blends. *Macromolecules*, Vol. 33(14), pp. 5187-5197.

Balta-Calleja, F.J. & Roslaniec, Z. (2000), Block Copolymers; Marcel-Dekker Publishers, (Eds.): New York.

Court, F.; Yamaguchi, D.; Hashimoto, T. (2006). Morphological Studies of Binary Mixtures of Block Copolymers: Temperature Dependence of Cosurfactant Effects. *Macromolecules*, Vol. 39(7), pp. 2596-2605.

Fan, F.C.; Olafson, B.D.; Blanco, M. (1992). Application of molecular simulation to derive phase diagrams of binary mixtures. *Macromolecules*, Vol. 25, pp. 3667-3676.

Flory, P. J. (1953). Principles of Polymer Chemistry, Cornell University Press, Ithaca, New York,

Groot, R. D. & Warren, P. B. (1997). Dissipative particle dynamics: bridging the gap between atomistic and mesoscopic simulation. *J. Chem. Phys.*, Vol. 107, pp. 4423-4435.

Groot, R. D. & Madden, T. J. (1998). Dynamic simulation of diblock copolymer microphase separation. *J. Chem. Phys.*, Vol. 108, pp. 8713-8724.

Groot, R. D. & Madden, T. J., Tildesley, D. J. (1999). On the role of hydrodynamic interactions in block copolymer microphase separation. *J. Chem. Phys.*, Vol. 110, pp.9739-9749.

Hoogerbrugge, P. J. & Koelman, J. M. V. A. (1992). Simulating microscopic hydrodynamic phenomena with Dissipative particle dinamics. *Europhys. Lett.*,Vol. 19, pp.155-160.

Hajduk, D. A.; Harper, P.E.; Gruner, S. M.; Honeker, C. C.; Thomas, E. L.; Fetters, L. J. (1995). A Reevaluation of Bicontinuous Cubic Phases in Starblock Copolymers. *Macromolecules*, Vol. 28, pp. 2570-2573.

Hamley, I.W. (1998). The Physics of Block Copolymers. Oxford University Press, ISBN 019850218 4.

Koelman, J. M. V. A. & Hoogerbrugge, P. J.(1993). Dynamic simulation of hard sphere sispension under steady shear. *Europhys. Lett.*, Vol. 21, 363-368.

Kimishima, K.; Koga, T.; Hashimoto, T. (2000). Order−Order Phase Transition between Spherical and Cylindrical Microdomain Structures of Block Copolymer. I. Mechanism of the Transition. *Macromolecules*, Vol. 33(3), pp. 968-977.

Kim, E.Y.; Lee, D.J.; Kim, J.K. (2006). Phase Behavior of a Binary Mixture of a Block Copolymer with Lower Disorder-to-Order Transition and a Homopolymer. *Macromolecules*, Vol. 39(25), pp. 8747-8757.

Kim, J.K.; Lee, H.H.; Gu, Q.J.; Chang, T.; Jeong, Y.H. (1998). Determination of Order−Order and Order−Disorder Transition Temperatures of SIS Block Copolymers by Differential Scanning Calorimetry and Rheology. *Macromolecules*, Vol. 31(12), pp. 4045-4048.

Krishanamoorti, R.; Modi M.A.; Tse, M.F.; Wang H.-C. (2000). Pathway and Kinetics of Cylinder-to-Sphere Order−Order Transition in Block Copolymers. *Macromolecules*, Vol. 33, pp. 3810-3817.

Krishanamoorti, R.; Silva A. S.; Modi M.A. (2000). Small-Angle Neutron Scattering Study of a Cylinder to Sphere order-order Transition in Block Copolymers. *Macromolecules*, Vol. 33, pp. 3803-3809.

Khandpur, A.K.; Forster, S. ; Bates, F.S.; Hamley, I.W.; Ryan, A.J.; Bras, W.; Almadal, K.; Mortensen, K. (1995). Polyisoprene-Polystyrene Diblock Copolymer Phase Diagram near the Order-Disorder Transition. *Macromolecules*, Vol 28(26), pp. 8796-8806.

Leibler, L. (1980). Theory of Microphase Separation in Block Copolymers. *Macromolecules*, Vol. 13(6), pp. 1602-1607.

Matsen, M. W. & Bates, F. S. (1996). Unifying Weak- and Strong-Segregation Block Copolymer Theories. *Macromolecules*, Vol. 29, pp. 1091-1098.

Modi, M.A.; Krishanamoorti, R.; Tse, M.F.; Wang, H.C. (1999). Viscoelastic Characterization of an Order−Order Transition in a Mixture of Di- and Triblock Copolymers. *Macromolecules*, Vol. 32(12), pp. 4088-4097.

Mani, S.; Weiss, R. A.; Cantino, M. E.; Khairallah, L. H.; Hans, S. F.; Williams, C. E. (2000). Evidence for a Thermally Reversible Order-Order Transition between Lamellar and Perforated Lamellar Microphases in a Triblock Copolymer. *Eur. Polym. J.*, Vol. 36, pp. 215–219.

Park, I.; Lee, B.; Ryu, J.; Im, K.; Yoon, J.; Ree, M.; Chang, T. (2005). Epitaxial Phase Transition of Polystyrene-b-Polyisoprene from Hexagonally Perforated Layer to Gyroid Phase in Thin Film. *Macromolecules*, Vol. 38, pp.10532-10536.

Ramos-Rodriguez, D. A.; Rodriguez-Hidalgo, M. R.; Soto-Figueroa, C.; Vicente, L. (2010). Molecular and mesoscopic study of ionic liquids and their use as solvents of active agents released by polymeric vehicles. *Molecular Physics*, Vol. 108(5), pp. 657–665.

Rodríguez-Hidalgo, M.R.; Soto-Figueroa, C.; Vicente, L. (2011). Mesoscopic simulation of the drug release mechanism on the polymeric vehicle P(ST-DVB) in an acid environment. *Soft Matter*, Vol. 7, pp 8224-8230.

Rodríguez-Hidalgo, M.R.; Soto-Figueroa, C.; Vicente, L. (2009) Mesoscopic study of cylindrical phases of poly(styrene)-poly(isoprene) copolymer: Order-order phase transitions by temperature control. *Polymer*, Vol. 50, pp 4596-4601.

Ryu, C.Y. & Lodge, T.P. (1999). Thermodynamic Stability and Anisotropic Fluctuations in the Cylinder-to-Sphere Transition of a Block Copolymer. *Macromolecules*, Vol. 32, pp. 7190-7201.

Sakurai, S.; Kawada, H.; Hashimoto, T.; Fetters, L.J. (1993).Thermoreversible Morphology Transition between Spherical and Cylindrical microdomains of Block Copolymers. *Macromolecules*, Vol. 26(21), pp. 5796-5802.

Sakurai, S.; Hashimoto, T.; Fetters, L.J. (1996). Thermoreversible Cylinder−Sphere Transition of Polystyrene-block-polyisoprene Diblock Copolymers in Dioctyl Phthalate Solutions. *Macromolecules*, Vol. 29(2), pp. 740-747.

Sakamato, N.; Hashimoto, T.; Han, C.D.; Kim, D.; Vaidya, N.Y. (1997). Order−Order and Order−Disorder Transitions in a Polystyrene-block-Polyisoprene-block-Polystyrene Copolymer. *Macromolecules*, Vol. 30(6), pp. 1621-1632.

Soto-Figueroa, C.; Rodríguez-Hidalgo, M.R.; Martínez-Magadán, J.M. (2005). Molecular simulation of diblock copolymers; morphology and mechanical properties. *Polymer*, Vol. 46, pp. 7485-7493.

Soto-Figueroa, C.; Luis-Vicente.; Martínez-Magadán, J.M.; Rodríguez-Hidalgo, M.R. (2007). Self-Organization Process of Ordered Structures in Linear and Star Poly(styrene)-Poly(isoprene) Block Copolymers: Gaussian Models and Mesoscopic Parameters of Polymeric Systems. *J. Phys. Chem. B*, Vol. 111, pp. 11756-11764.

Soto-Figueroa, C.; Luis-Vicente.; Martínez-Magadán, J.M.; Rodríguez-Hidalgo, M.R. (2007). Mesoscopic simulation of asymmetric-copolymer/homopolymer blends: Microphase morphological modification by homopolymer chains solubilization. *Polymer*, Vol. 48, pp. 3902-3911.

Soto-Figueroa, C.; Rodríguez-Hidalgo, M.R.; Martínez-Magadán, J.M.; Luis-Vicente. (2008). Dissipative Particle Dynamics Study of Order-Order Phase Transition of BCC, HPC, OBDD, and LAM Structures of the Poly(styrene)-Poly(isoprene) Diblock Copolymer. *Macromolecules*, Vol. 41, pp. 3297-3304.

Soto-Figueroa, C.; Rodríguez-Hidalgo, M.R.; Martínez-Magadán, J.M.; Luis-Vicente. (2008). Mesoscopic simulation of metastable microphases in the order–order transition from gyroid-to-lamellar states of PS–PI diblock copolymer. *Chemical Physics Letters*, Vol. 460, pp. 507-511.

Strobl, G. (1997). The Physic of Polymers. Springer, ISBN 3-540-63203-4.

Thomas, E.L. & Lescanec, R.L. (1995). Phase morphology in block copolymer systems, In A. Keller, M. Warner and A.H. Windle (Eds.), Self-order and Form in Polymeric Materials, p.147-164. Chapman & Hall, London.

Verlet, L. (1967). Computer "Experiments" on Classical Fluids. I. Thermodynamical Properties of Lennard- Jones Molecules. *Phys. Rev.*, Vol.159, pp. 98-103.

Vicente, L.; Soto-Figueroa, C.; Pacheco-Sanchez, H.; Hernadez-Trujillo, J.; Martínez Magadán, J. M. (2006). Fluid Phase Equilibria. Vol. 239, pp. 100-106.

You, L. Y.; Chen, L.J.; Qian, H.J.; Lu, Z.Y. Microphase Transitions of Perforated Lamellae of Cyclic Diblock Copolymers under Steady Shear. *Macromolecules*, 2007, Vol. 40(14), pp. 5222-5227.